VOLUME FOUR HUNDRED AND EIGHTY-SIX

METHODS IN ENZYMOLOGY

Research on Nitrification and Related Processes, Part A

METHODS IN ENZYMOLOGY

Editors-in-Chief

JOHN N. ABELSON AND MELVIN I. SIMON

Division of Biology
California Institute of Technology
Pasadena, California

Founding Editors

SIDNEY P. COLOWICK AND NATHAN O. KAPLAN

VOLUME FOUR HUNDRED AND EIGHTY-SIX

Methods in
ENZYMOLOGY

Research on Nitrification and Related Processes, Part A

EDITED BY

MARTIN G. KLOTZ
*Department of Biology, College of Arts and Sciences
Department of Microbiology and Immunology,
School of Medicine, and
Center for Genetics and Molecular Medicine
University of Louisville, Kentucky, USA*

AMSTERDAM • BOSTON • HEIDELBERG • LONDON
NEW YORK • OXFORD • PARIS • SAN DIEGO
SAN FRANCISCO • SINGAPORE • SYDNEY • TOKYO
Academic Press is an imprint of Elsevier

Academic Press is an imprint of Elsevier
525 B Street, Suite 1900, San Diego, CA 92101-4495, USA
30 Corporate Drive, Suite 400, Burlington, MA 01803, USA
32 Jamestown Road, London NW1 7BY, UK

First edition 2011

Copyright © 2011, Elsevier Inc. All Rights Reserved.

No part of this publication may be reproduced, stored in a retrieval system or transmitted in any form or by any means electronic, mechanical, photocopying, recording or otherwise without the prior written permission of the publisher

Permissions may be sought directly from Elsevier's Science & Technology Rights Department in Oxford, UK: phone (+44) (0) 1865 843830; fax (+44) (0) 1865 853333; email: permissions@elsevier.com. Alternatively you can submit your request online by visiting the Elsevier web site at http://elsevier.com/locate/permissions, and selecting *Obtaining permission to use Elsevier material*

Notice
No responsibility is assumed by the publisher for any injury and/or damage to persons or property as a matter of products liability, negligence or otherwise, or from any use or operation of any methods, products, instructions or ideas contained in the material herein. Because of rapid advances in the medical sciences, in particular, independent verification of diagnoses and drug dosages should be made

For information on all Academic Press publications
visit our website at elsevierdirect.com

ISBN: 978-0-12-381294-0
ISSN: 0076-6879

Printed and bound in United States of America
11 12   10 9 8 7 6 5 4 3 2 1

Working together to grow
libraries in developing countries

www.elsevier.com | www.bookaid.org | www.sabre.org

ELSEVIER   BOOK AID International   Sabre Foundation

Contents

Contributors	xiii
Preface	xix
Volumes in Series	xxi

Section 1. Modeling   1

1. Modeling the Role of Nitrification in Open Ocean Productivity and the Nitrogen Cycle   3
Andrew Yool

1. Introduction	4
2. Ocean Carbon Cycle	4
3. Export Production and the f-Ratio	7
4. Plankton Ecosystem Models	12
5. Modeling Nitrification	15
6. Operational Considerations	22
7. Future Modeling	24
8. Summary	27
Acknowledgments	28
References	28

2. Continuous Cultivation and Thermodynamic Aspects of Niche Definition in the Nitrogen Cycle   33
Stefanie Müller and Marc Strous

1. Introduction	33
2. A Minimal Model for Microbial Behavior	35
3. Continuous Cultivation—Background and Experimental Approach	43
4. Conclusion	50
Acknowledgments	50
References	50

Section 2. Cultivation and Cell Physiology 53

3. Isolation, Cultivation, and Characterization of Ammonia-Oxidizing Bacteria and Archaea Adapted to Low Ammonium Concentrations 55
Annette Bollmann, Elizabeth French, and Hendrikus J. Laanbroek

1. Introduction 56
2. General Description of Cultivation Methods for Ammonia-Oxidizing Microorganisms 57
3. Enrichment of Ammonia Oxidizers 63
4. Isolation of Ammonia Oxidizers 68
5. Characterization of Ammonia-Oxidizing Prokaryotes 72
References 84

4. Cultivation, Detection, and Ecophysiology of Anaerobic Ammonium-Oxidizing Bacteria 89
Boran Kartal, Wim Geerts, and Mike S. M. Jetten

1. Introduction 90
2. Enrichment of Anammox Bacteria as Planktonic Cell Suspensions 93
3. Molecular Detection of Anammox Bacteria 97
4. Conclusions 104
Acknowledgments 104
References 105

5. Cultivation, Growth Physiology, and Chemotaxonomy of Nitrite-Oxidizing Bacteria 109
Eva Spieck and André Lipski

1. Introduction 110
2. Recipes for the Culture of NOB 112
3. Cultivation Procedure 114
4. Isolation 116
5. Physiological Investigations 118
6. Chemical Analyses 121
7. Monitoring the Cultures 121
8. Fatty Acid Analyses 124
Acknowledgments 127
References 127

6. Surveying N_2O-Producing Pathways in Bacteria 131
Lisa Y. Stein

1. Introduction 132
2. Bioinformatics 135

3. Surveying Function	143
4. Concluding Remarks	147
References	148

Section 3. Molecular Microbial Ecology and Processes Measurements — 153

7. Stable Isotope Probing with ^{18}O-Water to Investigate Growth and Mortality of Ammonia Oxidizing Bacteria and Archaea in Soil — 155

Karen Adair and Egbert Schwartz

1. Introduction	156
2. Materials and Methods	158
3. Calculation of Growth and Mortality Indices	160
4. Results	161
5. Discussion	165
6. Conclusions	167
Acknowledgments	167
References	168

8. Measuring Nitrification, Denitrification, and Related Biomarkers in Terrestrial Geothermal Ecosystems — 171

Jeremy A. Dodsworth, Bruce Hungate, José R. de la Torre, Hongchen Jiang, and Brian P. Hedlund

1. Introduction	172
2. General Considerations for Measurement of N-Cycle Activities in Terrestrial Geothermal Habitats	177
3. Methods for Measuring Nitrification in Terrestrial Geothermal Habitats	179
4. Methods for Measuring Denitrification in Terrestrial Geothermal Habitats	187
5. Detection and Quantification of Potential Biomarkers for Thermophilic AOA and Denitrifying *Thermus thermophilus*	193
6. Closing Remarks	198
Acknowledgments	198
References	198

9. Determining the Distribution of Marine and Coastal Ammonia-Oxidizing Archaea and Bacteria Using a Quantitative Approach — 205

Annika C. Mosier and Christopher A. Francis

1. Introduction	206
2. Methods for Measuring the Abundance of AOA and AOB Within Marine and Coastal Systems	209

3. Methodological Considerations 213
4. Targeting Specific Ecotypes: Quantifying Shallow and Deep Clades of Marine Water Column AOA 214
Acknowledgments 217
References 218

10. ^{15}N-Labeling Experiments to Dissect the Contributions of Heterotrophic Denitrification and Anammox to Nitrogen Removal in the OMZ Waters of the Ocean 223

Moritz Holtappels, Gaute Lavik, Marlene M. Jensen, and Marcel M. M. Kuypers

1. Introduction 224
2. Theoretical Description of ^{15}N-Incubation Experiments 227
3. In the Field: Supplementary Measurements 233
4. In the Field: Incubation Experiments 237
5. In the Lab: Mass Spectrometry Measurements 240
6. Data Processing and Interpretation 243
7. Common Pitfalls 248
8. Concluding Remarks 249
Acknowledgments 249
References 250

11. Assessment of Nitrogen and Oxygen Isotopic Fractionation During Nitrification and Its Expression in the Marine Environment 253

Karen L. Casciotti, Carolyn Buchwald, Alyson E. Santoro, and Caitlin Frame

1. Introduction 254
2. Methods 259
3. Analytical Techniques 262
4. Data Analysis 264
5. Discussion 272
References 275

12. Identification of Diazotrophic Microorganisms in Marine Sediment via Fluorescence In Situ Hybridization Coupled to Nanoscale Secondary Ion Mass Spectrometry (FISH-NanoSIMS) 281

Anne E. Dekas and Victoria J. Orphan

1. Introduction 282
2. Methods 285
3. Concluding Remarks 300
Acknowledgments 301
References 301

13. Measurement and Distribution of Nitrification Rates in the Oceans — 307
B. B. Ward

1. Introduction — 308
2. Nitrification Rate Measurements in Seawater (Table 13.1) — 310
3. Nitrification Rate Measurements in Sediments — 318
4. Distribution of Nitrification — 319
References — 320

14. Construction of STOX Oxygen Sensors and Their Application for Determination of O_2 Concentrations in Oxygen Minimum Zones — 325
Niels Peter Revsbech, Bo Thamdrup, Tage Dalsgaard, and Donald Eugene Canfield

1. Introduction — 326
2. STOX Sensor Principle — 327
3. Sensor Construction — 328
4. Electronics for STOX Sensors — 332
5. STOX Sensor Calibration and Performance — 333
6. Calculation of Oxygen Concentrations from STOX Data — 335
7. *In Situ* Deployment of STOX Sensors in OMZs — 337
8. Future Fast Response STOX Sensors — 339
9. Using STOX Sensors to Recalibrate Conventional Oxygen Sensors — 339
Acknowledgments — 340
References — 340

15. Regulation and Measurement of Nitrification in Terrestrial Systems — 343
Jeanette M. Norton and John M. Stark

1. Introduction — 344
2. Diversity of the Nitrification Process in Terrestrial Environments — 345
3. Substrates and Products of Nitrification Reactions — 346
4. Controls on Nitrification Rates in Soil Environments — 347
5. Measurements of Nitrification in Terrestrial Environments — 350
6. Measurement of Nitrification Rates in Terrestrial Ecosystems — 351
7. Measurement of Nitrification Kinetics — 357
8. Nitrifier Population Size — 361
9. Modeling Approaches — 361
10. Future Advancements in Nitrification Rate Research in Terrestrial Environments — 362
References — 362

16. Protocol for the Measurement of Nitrous Oxide Fluxes from Biological Wastewater Treatment Plants — 369

Kartik Chandran

1. Introduction — 370
2. Sampling Design for Full-Scale Monitoring — 370
3. Sampling Procedures: Headspace Gas Measurement — 373
4. Sampling Procedures: Measurement of Aqueous N_2O Concentrations — 376
5. Sampling Procedures: Measurement of Advective Gas Flow Rate from Bioreactor Headspace — 376
6. Principles of Real-Time N_2O Measurement — 377
7. Data Analysis: Determination of Fluxes — 377
8. Data Analysis: Determination of Emission Fractions — 378
9. Data Analysis: Calculation of N_2O Emission Factors — 378
10. Standardization of Protocol and Comparison with Established Emissions Flux Measurement Methods — 379
11. N_2O Emission Fluxes from Activated Sludge Processes — 381
12. Triggers for N_2O Emission from Wastewater Treatment Operations — 382
13. Lab-Scale and Field-Scale Adaptation of Protocol N_2O Emission Measurements — 383
14. Concluding Remarks — 383
Acknowledgment — 383
References — 384

Section 4. Genetics, Biochemistry and Biogeochemistry — 387

17. Genetic Transformation of Ammonia-Oxidizing Bacteria — 389

Luis A. Sayavedra-Soto and Lisa Y. Stein

1. Introduction — 390
2. Transformation of AOB — 391
3. Gene Inactivation in AOB — 393
4. Use of Broad-Host Range Plasmids — 398
5. Strain Stability and Maintenance — 399
6. Conclusions — 399
References — 400

18. Dissecting Iron Uptake and Homeostasis in *Nitrosomonas europaea* — 403

Luis A. Sayavedra-Soto, Neeraja Vajrala, and Daniel J. Arp

1. Introduction — 404
2. Use of Bioinformatics to Plan Fe-Related Studies — 406

3.	Methods to Learn About the Physiological Responses to Fe Limitation	407
4.	Methods to Study Siderophore Uptake	418
5.	Genetic Complementation to Study *fur* Homologs	421
6.	Methods for Gene Inactivation of Fe Uptake Systems	422
7.	Lingering Questions that can be Answered by the Above Protocols	423
8.	Conclusion	425
	References	425

19. Production of Recombinant Multiheme Cytochromes *c* in *Wolinella succinogenes* — 429

Melanie Kern and Jörg Simon

1.	Introduction	430
2.	Bacterial Cytochrome *c* Biogenesis Systems	433
3.	Recombinant Cytochrome *c* Production	435
4.	Cytochromes *c* in the Epsilonproteobacterium *W. succinogenes*	435
5.	Three Cytochrome *c* Heme Lyase Isoenzymes in *W. succinogenes*	439
6.	Heterologous Production of Cytochromes *c* in *W. succinogenes*	439
7.	Conclusions and Perspectives	442
	Acknowledgments	443
	References	443

20. Techniques for Investigating Hydroxylamine Disproportionation by Hydroxylamine Oxidoreductases — 447

A. Andrew Pacheco, Jennifer McGarry, Joshua Kostera, and Angel Corona

1.	Introduction	448
2.	General Considerations	450
3.	Ammonia Concentration Determination	454
4.	Nitric Oxide Concentration Determination	457
5.	Nitrite Concentration Determination	458
6.	Nitrous Oxide and Dinitrogen Concentration Determination	460
	References	462

21. Liquid Chromatography—Mass Spectrometry-Based Proteomics of *Nitrosomonas* — 465

Hans J. C. T. Wessels, Jolein Gloerich, Erwin van der Biezen, Mike S. M. Jetten, and Boran Kartal

1.	Introduction	466
2.	LC–MS/MS Instrument Setup	468
3.	Growth of *N. eutropha C91* Pure Culture	469
4.	Sample Preparation	470

5.	C18 Reversed Phase LC–MS/MS Analysis	471
6.	Database Searches and Validation of Results	472
7.	Dataset Description and Protein Identification Example	476
8.	Conclusions	479
	References	479

22. The Geochemical Record of the Ancient Nitrogen Cycle, Nitrogen Isotopes, and Metal Cofactors   483

Linda V. Godfrey and Jennifer B. Glass

1.	Introduction	484
2.	Reconstructing the Nitrogen Cycle and Associated Trace Metal Abundances through Time	489
3.	Determining Changes in the N-Cycle from the Geological Record of N Isotopes and Metal Availability	492
4.	Measurement	497
5.	Concluding Remarks	499
	Acknowledgments	499
	References	499

Author Index   *507*
Subject Index   *533*

Contributors

Karen Adair
Department of Biological Sciences, Northern Arizona University, Flagstaff, Arizona, USA

Daniel J. Arp
Department of Botany and Plant Pathology, Oregon State University, Corvallis, Oregon, USA

Annette Bollmann
Department of Microbiology, Miami University, Oxford, Ohio, USA

Carolyn Buchwald
Woods Hole Oceanographic Institution, Woods Hole, Massachusetts, USA

Donald Eugene Canfield
Nordic Center for Earth Evolution (NordCEE) and Institute of Biology, University of Southern Denmark, Odense, Denmark

Karen L. Casciotti
Woods Hole Oceanographic Institution, Woods Hole, Massachusetts, USA

Kartik Chandran
Department of Earth and Environmental Engineering, Columbia University, New York, USA

Angel Corona
Rufus King High School, Milwaukee, Wisconsin, USA

Tage Dalsgaard
National Environmental Research Institute, Aarhus University, Silkeborg, Denmark

José R. de la Torre
Department of Biology, San Francisco State University, San Francisco, California, USA

Anne E. Dekas
Division of Geological and Planetary Sciences, California Institute of Technology, Pasadena, California, USA

Erwin van der Biezen
Department of Microbiology, Institute of Water and Wetland Research, Radboud University Nijmegen, Heyendaalseweg 135, and Merck Sharp & Dohme BV, 5342 CC Oss, The Netherlands

Jeremy A. Dodsworth
School of Life Sciences, University of Nevada, Las Vegas, Nevada, USA

Caitlin Frame
Woods Hole Oceanographic Institution, Woods Hole, Massachusetts, USA

Christopher A. Francis
Department of Environmental Earth System Science, Stanford University, Stanford, California, USA

Elizabeth French
Department of Microbiology, Miami University, Oxford, Ohio, USA

Wim Geerts
Department of Microbiology, Institute of Water and Wetland Research, Radboud University Nijmegen, Heyendaalseweg, The Netherlands

Jennifer B. Glass
School of Earth and Space Exploration, Arizona State University, Tempe, Arizona, USA

Jolein Gloerich
Department of Laboratory Medicine, Nijmegen Proteomics Facility, Radboud University Nijmegen Medical Centre, Geert Grooteplein-Zuid 10, Nijmegen, The Netherlands

Linda V. Godfrey
Institute of Marine and Coastal Sciences, Rutgers University, New Brunswick, New Jersey, USA

Brian P. Hedlund
School of Life Sciences, University of Nevada, Las Vegas, Nevada, USA

Moritz Holtappels
Max Planck Institute for Marine Microbiology, Bremen, Germany

Bruce Hungate
Department of Biological Sciences, Merriam-Powell Center for Environmental Research, Northern Arizona University, Flagstaff, Arizona, USA

Marlene M. Jensen
Institute of Biology and Nordic Center for Earth Evolution (NordCEE), University of Southern Denmark, Denmark

Mike S. M. Jetten
Department of Microbiology, Institute of Water and Wetland Research, Radboud University Nijmegen, Heyendaalseweg 135, and Department of Biotechnology, Delft University of Technology, Delft, The Netherlands

Hongchen Jiang
Geomicrobiology Laboratory, China University of Geosciences, Beijing, China

Boran Kartal
Department of Microbiology, Institute of Water and Wetland Research, Radboud University Nijmegen, Heyendaalseweg 135, The Netherlands

Melanie Kern
Institute of Microbiology and Genetics, Department of Biology, Technische Universität Darmstadt, Darmstadt, Germany

Joshua Kostera
Department of Chemistry and Biochemistry, University of Wisconsin-Milwaukee, Milwaukee, Wisconsin, USA

Marcel M. M. Kuypers
Max Planck Institute for Marine Microbiology, Bremen, Germany

Hendrikus J. Laanbroek
Department of Microbial Ecology, Netherlands Institute of Ecology, Wageningen, and Utrecht University, Institute of Environmental Sciences, Utrecht, The Netherlands

Gaute Lavik
Max Planck Institute for Marine Microbiology, Bremen, Germany

André Lipski
Rheinische Friedrich-Wilhelms-Universität Bonn, Food Microbiology and Hygiene, Meckenheimer Allee, Bonn, Germany

Stefanie Müller
Max Planck Institute for Marine Microbiology, Bremen, Germany

Jennifer McGarry
Department of Chemistry and Biochemistry, University of Wisconsin-Milwaukee, Milwaukee, Wisconsin, USA

Annika C. Mosier
Department of Environmental Earth System Science, Stanford University, Stanford, California, USA

Jeanette M. Norton
Department of Plants, Soils and Climate, Utah State University, Logan, Utah, USA

Victoria J. Orphan
Division of Geological and Planetary Sciences, California Institute of Technology, Pasadena, California, USA

A. Andrew Pacheco
Department of Chemistry and Biochemistry, University of Wisconsin-Milwaukee, Milwaukee, Wisconsin, USA

Niels Peter Revsbech
Department of Biological Sciences, Microbiology, Aarhus University, Aarhus, Denmark

Alyson E. Santoro
Woods Hole Oceanographic Institution, Woods Hole, Massachusetts, USA

Luis A. Sayavedra-Soto
Department of Botany and Plant Pathology, Oregon State University, Corvallis, Oregon, USA

Egbert Schwartz
Department of Biological Sciences, Northern Arizona University, Flagstaff, Arizona, USA

Jörg Simon
Institute of Microbiology and Genetics, Department of Biology, Technische Universität Darmstadt, Darmstadt, Germany

Eva Spieck
Biocenter Klein Flottbek, University of Hamburg, Department of Microbiology and Biotechnology, Ohnhorststraße, Hamburg, Germany

John M. Stark
Department of Biology and Ecology Center, Utah State University, Logan, Utah, USA

Lisa Y. Stein
Department of Biological Sciences, University of Alberta, Edmonton, Alberta, Canada

Marc Strous
Max Planck Institute for Marine Microbiology, Bremen, and Center for Biotechnology, University of Bielefeld, Bielefeld, Germany

Bo Thamdrup
Nordic Center for Earth Evolution (NordCEE) and Institute of Biology, University of Southern Denmark, Odense, Denmark

Neeraja Vajrala
Department of Botany and Plant Pathology, Oregon State University, Corvallis, Oregon, USA

B. B. Ward
Department of Geosciences, Princeton University, Princeton, New Jersey, USA

Hans J. C. T. Wessels
Nijmegen Centre for Mitochondrial Disorders, Department of Laboratory Medicine, Nijmegen Proteomics Facility, Radboud University Nijmegen Medical Centre, Geert Grooteplein-Zuid 10, Nijmegen, The Netherlands

Andrew Yool
National Oceanography Centre, University of Southampton Waterfront Campus, European Way, Southampton, United Kingdom

Preface

Although not (yet) noticed by most politicians and environmental activists who focus on the carbon footprint, the global nitrogen cycle is the one most impacted by humankind. This has come about as the consequence of a major breakthrough in applied science, namely, the industrial production of ammonium-based fertilizer, synthetically "fixed nitrogen," which has led to almost uncontrolled use worldwide to support the increasing demand for agricultural production. Despite being initiated by the great Sergey Winogradsky more than 100 years ago at the turn of the twentieth century—at the same time when Haber and Bosch invented the industrial synthesis of ammonia—research on the transformation of nitrogen compounds in an environmental context has been stagnant until about 10–15 years ago. In contrast, "reactive nitrogen species" have been a hot topic in biomedical research since quite some time, acknowledged by the Nobel Prize in Physiology and Medicine in 1998 for work on the role of nitric oxide as a signaling molecule in the cardiovascular system of animals. Many of the methods instrumental to the progress of this work were captured in *Methods in Enzymology* Volumes 436 and 437 (2008), entitled "Globins and Other Nitric Oxide-Reactive Proteins," edited by Robert K. Poole, and Volumes 268 and 269 (1996), 301 (199), 359 (2002), 396 (2005), 440 (2008), and 441 (2008) entitled "Nitric Oxide, parts A to G," edited by Enrique Cadenas and Lester Packer. The lack of progress in the environmental arena was mainly attributable to missing instrumentation, tools, and methods at the molecular level rather than underdeveloped theory. For instance, the recent discovery of anerobic ammonia oxidation by bacteria (anammox Process: 1999; molecular verification: 2006) was predicted decades earlier by Engelbert Broda solely based on thermodynamics.

The idea for developing two volumes on "Research on Nitrification and Related Processes" came after a fabulous international meeting on the Nitrogen Cycle organized by the Agouron Institute in Scottsdale, Arizona, in October 2009 (http://agi.org/pdf/nmtg-abstracts/NMtgAbstractsforWeb.pdf), where I presented new insights into N-cycle evolution. The Agouron Nitrogen meeting succeeded another significant event in the N-cycle community: the first International Conference on Nitrification (ICoN1) organized by the Nitrification Network (http://nitrificationnetwork.org/news%20and%20events.php#icon1) under my stewardship in Louisville, Kentucky, in July 2009. ICoN1 was made possible by support from the U.S. National Science Foundation, the Gordon and Betty Moore

Foundation, and the University of Louisville. So, I was well informed about the existing wealth of new approaches and data, which led me to agree when John Abelson and Mel Simon generously invited me to edit a volume of *Methods in Enzymology* to cover research on reactive N in an environmental context.

Acceptance of the three-domain structure of the universal tree of life by the majority of biologists, pioneered by Carl Woese and George Fox, the brilliant concept of primer extension developed by Ray Wu (which is the basis of dideoxy-primer sequencing and PCR) and whole genome sequencing have paved the way for new hot areas in the (micro)biological sciences such as phylogenomics, molecular microbial ecology, and systems microbiology. Newly developed experimental procedures and improvements in instrumentation in context with theoretical work in the last decade have also finally changed our view on many aspects of microbial biogeochemical cycles, including the global nitrogen cycle. Many novel processes and the molecular inventory and organisms facilitating them have been discovered only within the last 5–10 years and this discovery continues today. These new findings suggest that assigning biogeochemical processes exclusively to specific cohorts of (micro)organisms (i.e., "Denitrifiers," "Nitrifiers," "Ammonifiers," etc.) was ill-conceived. Rather, the modular mix and match of metabolic inventory cassettes in diverse genomic backgrounds over long evolutionary time periods ultimately led to an extant diversity of nitrogen cycle organisms, including those with significance to carbon and sulfur cycles. It is thus opportune for *Methods in Enzymology* to present, in two parts, a state-of-the-art update on the methods and protocols dealing with the detection, isolation, and characterization of macromolecules and their hosting organisms that facilitate nitrification and related processes in the nitrogen cycle as well as the challenges of doing so in very diverse environments: Ready-to-use methods for modelers, molecular ecologists, biogeochemists, ecophysiologists, geneticists, environmental engineers, and biochemists are presented.

These two parts would not have been possible without the generous sharing and enthusiasm of so many of my colleagues and friends inside and outside the Nitrification Network. I thank them all, and also Delsy Retchagar and Sujatha Thirugnanasambandam (Elsevier, Chennai, India) and Zoe Kruze (Elsevier, Oxford, UK), for their help and advice in steering both parts to a successful and timely outcome. My particular gratitude and thanks go to my colleague and friend Lisa Yael Stein (University of Alberta-Edmonton, Canada), who has skillfully shared with me the task of coediting part B.

MARTIN GÜNTER KLOTZ

Methods in Enzymology

VOLUME I. Preparation and Assay of Enzymes
Edited by SIDNEY P. COLOWICK AND NATHAN O. KAPLAN

VOLUME II. Preparation and Assay of Enzymes
Edited by SIDNEY P. COLOWICK AND NATHAN O. KAPLAN

VOLUME III. Preparation and Assay of Substrates
Edited by SIDNEY P. COLOWICK AND NATHAN O. KAPLAN

VOLUME IV. Special Techniques for the Enzymologist
Edited by SIDNEY P. COLOWICK AND NATHAN O. KAPLAN

VOLUME V. Preparation and Assay of Enzymes
Edited by SIDNEY P. COLOWICK AND NATHAN O. KAPLAN

VOLUME VI. Preparation and Assay of Enzymes *(Continued)*
Preparation and Assay of Substrates
Special Techniques
Edited by SIDNEY P. COLOWICK AND NATHAN O. KAPLAN

VOLUME VII. Cumulative Subject Index
Edited by SIDNEY P. COLOWICK AND NATHAN O. KAPLAN

VOLUME VIII. Complex Carbohydrates
Edited by ELIZABETH F. NEUFELD AND VICTOR GINSBURG

VOLUME IX. Carbohydrate Metabolism
Edited by WILLIS A. WOOD

VOLUME X. Oxidation and Phosphorylation
Edited by RONALD W. ESTABROOK AND MAYNARD E. PULLMAN

VOLUME XI. Enzyme Structure
Edited by C. H. W. HIRS

VOLUME XII. Nucleic Acids (Parts A and B)
Edited by LAWRENCE GROSSMAN AND KIVIE MOLDAVE

VOLUME XIII. Citric Acid Cycle
Edited by J. M. LOWENSTEIN

VOLUME XIV. Lipids
Edited by J. M. LOWENSTEIN

VOLUME XV. Steroids and Terpenoids
Edited by RAYMOND B. CLAYTON

VOLUME XVI. Fast Reactions
Edited by KENNETH KUSTIN

VOLUME XVII. Metabolism of Amino Acids and Amines (Parts A and B)
Edited by HERBERT TABOR AND CELIA WHITE TABOR

VOLUME XVIII. Vitamins and Coenzymes (Parts A, B, and C)
Edited by DONALD B. MCCORMICK AND LEMUEL D. WRIGHT

VOLUME XIX. Proteolytic Enzymes
Edited by GERTRUDE E. PERLMANN AND LASZLO LORAND

VOLUME XX. Nucleic Acids and Protein Synthesis (Part C)
Edited by KIVIE MOLDAVE AND LAWRENCE GROSSMAN

VOLUME XXI. Nucleic Acids (Part D)
Edited by LAWRENCE GROSSMAN AND KIVIE MOLDAVE

VOLUME XXII. Enzyme Purification and Related Techniques
Edited by WILLIAM B. JAKOBY

VOLUME XXIII. Photosynthesis (Part A)
Edited by ANTHONY SAN PIETRO

VOLUME XXIV. Photosynthesis and Nitrogen Fixation (Part B)
Edited by ANTHONY SAN PIETRO

VOLUME XXV. Enzyme Structure (Part B)
Edited by C. H. W. HIRS AND SERGE N. TIMASHEFF

VOLUME XXVI. Enzyme Structure (Part C)
Edited by C. H. W. HIRS AND SERGE N. TIMASHEFF

VOLUME XXVII. Enzyme Structure (Part D)
Edited by C. H. W. HIRS AND SERGE N. TIMASHEFF

VOLUME XXVIII. Complex Carbohydrates (Part B)
Edited by VICTOR GINSBURG

VOLUME XXIX. Nucleic Acids and Protein Synthesis (Part E)
Edited by LAWRENCE GROSSMAN AND KIVIE MOLDAVE

VOLUME XXX. Nucleic Acids and Protein Synthesis (Part F)
Edited by KIVIE MOLDAVE AND LAWRENCE GROSSMAN

VOLUME XXXI. Biomembranes (Part A)
Edited by SIDNEY FLEISCHER AND LESTER PACKER

VOLUME XXXII. Biomembranes (Part B)
Edited by SIDNEY FLEISCHER AND LESTER PACKER

VOLUME XXXIII. Cumulative Subject Index Volumes I–XXX
Edited by MARTHA G. DENNIS AND EDWARD A. DENNIS

VOLUME XXXIV. Affinity Techniques (Enzyme Purification: Part B)
Edited by WILLIAM B. JAKOBY AND MEIR WILCHEK

VOLUME XXXV. Lipids (Part B)
Edited by JOHN M. LOWENSTEIN

VOLUME XXXVI. Hormone Action (Part A: Steroid Hormones)
Edited by BERT W. O'MALLEY AND JOEL G. HARDMAN

VOLUME XXXVII. Hormone Action (Part B: Peptide Hormones)
Edited by BERT W. O'MALLEY AND JOEL G. HARDMAN

VOLUME XXXVIII. Hormone Action (Part C: Cyclic Nucleotides)
Edited by JOEL G. HARDMAN AND BERT W. O'MALLEY

VOLUME XXXIX. Hormone Action (Part D: Isolated Cells, Tissues, and Organ Systems)
Edited by JOEL G. HARDMAN AND BERT W. O'MALLEY

VOLUME XL. Hormone Action (Part E: Nuclear Structure and Function)
Edited by BERT W. O'MALLEY AND JOEL G. HARDMAN

VOLUME XLI. Carbohydrate Metabolism (Part B)
Edited by W. A. WOOD

VOLUME XLII. Carbohydrate Metabolism (Part C)
Edited by W. A. WOOD

VOLUME XLIII. Antibiotics
Edited by JOHN H. HASH

VOLUME XLIV. Immobilized Enzymes
Edited by KLAUS MOSBACH

VOLUME XLV. Proteolytic Enzymes (Part B)
Edited by LASZLO LORAND

VOLUME XLVI. Affinity Labeling
Edited by WILLIAM B. JAKOBY AND MEIR WILCHEK

VOLUME XLVII. Enzyme Structure (Part E)
Edited by C. H. W. HIRS AND SERGE N. TIMASHEFF

VOLUME XLVIII. Enzyme Structure (Part F)
Edited by C. H. W. HIRS AND SERGE N. TIMASHEFF

VOLUME XLIX. Enzyme Structure (Part G)
Edited by C. H. W. HIRS AND SERGE N. TIMASHEFF

VOLUME L. Complex Carbohydrates (Part C)
Edited by VICTOR GINSBURG

VOLUME LI. Purine and Pyrimidine Nucleotide Metabolism
Edited by PATRICIA A. HOFFEE AND MARY ELLEN JONES

VOLUME LII. Biomembranes (Part C: Biological Oxidations)
Edited by SIDNEY FLEISCHER AND LESTER PACKER

VOLUME LIII. Biomembranes (Part D: Biological Oxidations)
Edited by SIDNEY FLEISCHER AND LESTER PACKER

VOLUME LIV. Biomembranes (Part E: Biological Oxidations)
Edited by SIDNEY FLEISCHER AND LESTER PACKER

VOLUME LV. Biomembranes (Part F: Bioenergetics)
Edited by SIDNEY FLEISCHER AND LESTER PACKER

VOLUME LVI. Biomembranes (Part G: Bioenergetics)
Edited by SIDNEY FLEISCHER AND LESTER PACKER

VOLUME LVII. Bioluminescence and Chemiluminescence
Edited by MARLENE A. DELUCA

VOLUME LVIII. Cell Culture
Edited by WILLIAM B. JAKOBY AND IRA PASTAN

VOLUME LIX. Nucleic Acids and Protein Synthesis (Part G)
Edited by KIVIE MOLDAVE AND LAWRENCE GROSSMAN

VOLUME LX. Nucleic Acids and Protein Synthesis (Part H)
Edited by KIVIE MOLDAVE AND LAWRENCE GROSSMAN

VOLUME 61. Enzyme Structure (Part H)
Edited by C. H. W. HIRS AND SERGE N. TIMASHEFF

VOLUME 62. Vitamins and Coenzymes (Part D)
Edited by DONALD B. MCCORMICK AND LEMUEL D. WRIGHT

VOLUME 63. Enzyme Kinetics and Mechanism (Part A: Initial Rate and Inhibitor Methods)
Edited by DANIEL L. PURICH

VOLUME 64. Enzyme Kinetics and Mechanism
(Part B: Isotopic Probes and Complex Enzyme Systems)
Edited by DANIEL L. PURICH

VOLUME 65. Nucleic Acids (Part I)
Edited by LAWRENCE GROSSMAN AND KIVIE MOLDAVE

VOLUME 66. Vitamins and Coenzymes (Part E)
Edited by DONALD B. MCCORMICK AND LEMUEL D. WRIGHT

VOLUME 67. Vitamins and Coenzymes (Part F)
Edited by DONALD B. MCCORMICK AND LEMUEL D. WRIGHT

VOLUME 68. Recombinant DNA
Edited by RAY WU

VOLUME 69. Photosynthesis and Nitrogen Fixation (Part C)
Edited by ANTHONY SAN PIETRO

VOLUME 70. Immunochemical Techniques (Part A)
Edited by HELEN VAN VUNAKIS AND JOHN J. LANGONE

VOLUME 71. Lipids (Part C)
Edited by JOHN M. LOWENSTEIN

VOLUME 72. Lipids (Part D)
Edited by JOHN M. LOWENSTEIN

VOLUME 73. Immunochemical Techniques (Part B)
Edited by JOHN J. LANGONE AND HELEN VAN VUNAKIS

VOLUME 74. Immunochemical Techniques (Part C)
Edited by JOHN J. LANGONE AND HELEN VAN VUNAKIS

VOLUME 75. Cumulative Subject Index Volumes XXXI, XXXII, XXXIV–LX
Edited by EDWARD A. DENNIS AND MARTHA G. DENNIS

VOLUME 76. Hemoglobins
Edited by ERALDO ANTONINI, LUIGI ROSSI-BERNARDI, AND EMILIA CHIANCONE

VOLUME 77. Detoxication and Drug Metabolism
Edited by WILLIAM B. JAKOBY

VOLUME 78. Interferons (Part A)
Edited by SIDNEY PESTKA

VOLUME 79. Interferons (Part B)
Edited by SIDNEY PESTKA

VOLUME 80. Proteolytic Enzymes (Part C)
Edited by LASZLO LORAND

VOLUME 81. Biomembranes (Part H: Visual Pigments and Purple Membranes, I)
Edited by LESTER PACKER

VOLUME 82. Structural and Contractile Proteins (Part A: Extracellular Matrix)
Edited by LEON W. CUNNINGHAM AND DIXIE W. FREDERIKSEN

VOLUME 83. Complex Carbohydrates (Part D)
Edited by VICTOR GINSBURG

VOLUME 84. Immunochemical Techniques (Part D: Selected Immunoassays)
Edited by JOHN J. LANGONE AND HELEN VAN VUNAKIS

VOLUME 85. Structural and Contractile Proteins (Part B: The Contractile Apparatus and the Cytoskeleton)
Edited by DIXIE W. FREDERIKSEN AND LEON W. CUNNINGHAM

VOLUME 86. Prostaglandins and Arachidonate Metabolites
Edited by WILLIAM E. M. LANDS AND WILLIAM L. SMITH

VOLUME 87. Enzyme Kinetics and Mechanism (Part C: Intermediates, Stereo-chemistry, and Rate Studies)
Edited by DANIEL L. PURICH

VOLUME 88. Biomembranes (Part I: Visual Pigments and Purple Membranes, II)
Edited by LESTER PACKER

VOLUME 89. Carbohydrate Metabolism (Part D)
Edited by WILLIS A. WOOD

VOLUME 90. Carbohydrate Metabolism (Part E)
Edited by WILLIS A. WOOD

VOLUME 91. Enzyme Structure (Part I)
Edited by C. H. W. HIRS AND SERGE N. TIMASHEFF

VOLUME 92. Immunochemical Techniques (Part E: Monoclonal Antibodies and General Immunoassay Methods)
Edited by JOHN J. LANGONE AND HELEN VAN VUNAKIS

VOLUME 93. Immunochemical Techniques (Part F: Conventional Antibodies, Fc Receptors, and Cytotoxicity)
Edited by JOHN J. LANGONE AND HELEN VAN VUNAKIS

VOLUME 94. Polyamines
Edited by HERBERT TABOR AND CELIA WHITE TABOR

VOLUME 95. Cumulative Subject Index Volumes 61–74, 76–80
Edited by EDWARD A. DENNIS AND MARTHA G. DENNIS

VOLUME 96. Biomembranes [Part J: Membrane Biogenesis: Assembly and Targeting (General Methods; Eukaryotes)]
Edited by SIDNEY FLEISCHER AND BECCA FLEISCHER

VOLUME 97. Biomembranes [Part K: Membrane Biogenesis: Assembly and Targeting (Prokaryotes, Mitochondria, and Chloroplasts)]
Edited by SIDNEY FLEISCHER AND BECCA FLEISCHER

VOLUME 98. Biomembranes (Part L: Membrane Biogenesis: Processing and Recycling)
Edited by SIDNEY FLEISCHER AND BECCA FLEISCHER

VOLUME 99. Hormone Action (Part F: Protein Kinases)
Edited by JACKIE D. CORBIN AND JOEL G. HARDMAN

VOLUME 100. Recombinant DNA (Part B)
Edited by RAY WU, LAWRENCE GROSSMAN, AND KIVIE MOLDAVE

VOLUME 101. Recombinant DNA (Part C)
Edited by RAY WU, LAWRENCE GROSSMAN, AND KIVIE MOLDAVE

VOLUME 102. Hormone Action (Part G: Calmodulin and Calcium-Binding Proteins)
Edited by ANTHONY R. MEANS AND BERT W. O'MALLEY

VOLUME 103. Hormone Action (Part H: Neuroendocrine Peptides)
Edited by P. MICHAEL CONN

VOLUME 104. Enzyme Purification and Related Techniques (Part C)
Edited by WILLIAM B. JAKOBY

VOLUME 105. Oxygen Radicals in Biological Systems
Edited by LESTER PACKER

VOLUME 106. Posttranslational Modifications (Part A)
Edited by FINN WOLD AND KIVIE MOLDAVE

VOLUME 107. Posttranslational Modifications (Part B)
Edited by FINN WOLD AND KIVIE MOLDAVE

VOLUME 108. Immunochemical Techniques (Part G: Separation and Characterization of Lymphoid Cells)
Edited by GIOVANNI DI SABATO, JOHN J. LANGONE, AND HELEN VAN VUNAKIS

VOLUME 109. Hormone Action (Part I: Peptide Hormones)
Edited by LUTZ BIRNBAUMER AND BERT W. O'MALLEY

VOLUME 110. Steroids and Isoprenoids (Part A)
Edited by JOHN H. LAW AND HANS C. RILLING

VOLUME 111. Steroids and Isoprenoids (Part B)
Edited by JOHN H. LAW AND HANS C. RILLING

VOLUME 112. Drug and Enzyme Targeting (Part A)
Edited by KENNETH J. WIDDER AND RALPH GREEN

VOLUME 113. Glutamate, Glutamine, Glutathione, and Related Compounds
Edited by ALTON MEISTER

VOLUME 114. Diffraction Methods for Biological Macromolecules (Part A)
Edited by HAROLD W. WYCKOFF, C. H. W. HIRS, AND SERGE N. TIMASHEFF

VOLUME 115. Diffraction Methods for Biological Macromolecules (Part B)
Edited by HAROLD W. WYCKOFF, C. H. W. HIRS, AND SERGE N. TIMASHEFF

VOLUME 116. Immunochemical Techniques (Part H: Effectors and Mediators of Lymphoid Cell Functions)
Edited by GIOVANNI DI SABATO, JOHN J. LANGONE, AND HELEN VAN VUNAKIS

VOLUME 117. Enzyme Structure (Part J)
Edited by C. H. W. HIRS AND SERGE N. TIMASHEFF

VOLUME 118. Plant Molecular Biology
Edited by ARTHUR WEISSBACH AND HERBERT WEISSBACH

VOLUME 119. Interferons (Part C)
Edited by SIDNEY PESTKA

VOLUME 120. Cumulative Subject Index Volumes 81–94, 96–101

VOLUME 121. Immunochemical Techniques (Part I: Hybridoma Technology and Monoclonal Antibodies)
Edited by JOHN J. LANGONE AND HELEN VAN VUNAKIS

VOLUME 122. Vitamins and Coenzymes (Part G)
Edited by FRANK CHYTIL AND DONALD B. MCCORMICK

VOLUME 123. Vitamins and Coenzymes (Part H)
Edited by FRANK CHYTIL AND DONALD B. MCCORMICK

VOLUME 124. Hormone Action (Part J: Neuroendocrine Peptides)
Edited by P. MICHAEL CONN

VOLUME 125. Biomembranes (Part M: Transport in Bacteria, Mitochondria, and Chloroplasts: General Approaches and Transport Systems)
Edited by SIDNEY FLEISCHER AND BECCA FLEISCHER

VOLUME 126. Biomembranes (Part N: Transport in Bacteria, Mitochondria, and Chloroplasts: Protonmotive Force)
Edited by SIDNEY FLEISCHER AND BECCA FLEISCHER

VOLUME 127. Biomembranes (Part O: Protons and Water: Structure and Translocation)
Edited by LESTER PACKER

VOLUME 128. Plasma Lipoproteins (Part A: Preparation, Structure, and Molecular Biology)
Edited by JERE P. SEGREST AND JOHN J. ALBERS

VOLUME 129. Plasma Lipoproteins (Part B: Characterization, Cell Biology, and Metabolism)
Edited by JOHN J. ALBERS AND JERE P. SEGREST

VOLUME 130. Enzyme Structure (Part K)
Edited by C. H. W. HIRS AND SERGE N. TIMASHEFF

VOLUME 131. Enzyme Structure (Part L)
Edited by C. H. W. HIRS AND SERGE N. TIMASHEFF

VOLUME 132. Immunochemical Techniques (Part J: Phagocytosis and Cell-Mediated Cytotoxicity)
Edited by GIOVANNI DI SABATO AND JOHANNES EVERSE

VOLUME 133. Bioluminescence and Chemiluminescence (Part B)
Edited by MARLENE DELUCA AND WILLIAM D. MCELROY

VOLUME 134. Structural and Contractile Proteins (Part C: The Contractile Apparatus and the Cytoskeleton)
Edited by RICHARD B. VALLEE

VOLUME 135. Immobilized Enzymes and Cells (Part B)
Edited by KLAUS MOSBACH

VOLUME 136. Immobilized Enzymes and Cells (Part C)
Edited by KLAUS MOSBACH

VOLUME 137. Immobilized Enzymes and Cells (Part D)
Edited by KLAUS MOSBACH

VOLUME 138. Complex Carbohydrates (Part E)
Edited by VICTOR GINSBURG

VOLUME 139. Cellular Regulators (Part A: Calcium- and Calmodulin-Binding Proteins)
Edited by ANTHONY R. MEANS AND P. MICHAEL CONN

VOLUME 140. Cumulative Subject Index Volumes 102–119, 121–134

VOLUME 141. Cellular Regulators (Part B: Calcium and Lipids)
Edited by P. MICHAEL CONN AND ANTHONY R. MEANS

VOLUME 142. Metabolism of Aromatic Amino Acids and Amines
Edited by SEYMOUR KAUFMAN

VOLUME 143. Sulfur and Sulfur Amino Acids
Edited by WILLIAM B. JAKOBY AND OWEN GRIFFITH

VOLUME 144. Structural and Contractile Proteins (Part D: Extracellular Matrix)
Edited by LEON W. CUNNINGHAM

VOLUME 145. Structural and Contractile Proteins (Part E: Extracellular Matrix)
Edited by LEON W. CUNNINGHAM

VOLUME 146. Peptide Growth Factors (Part A)
Edited by DAVID BARNES AND DAVID A. SIRBASKU

VOLUME 147. Peptide Growth Factors (Part B)
Edited by DAVID BARNES AND DAVID A. SIRBASKU

VOLUME 148. Plant Cell Membranes
Edited by LESTER PACKER AND ROLAND DOUCE

VOLUME 149. Drug and Enzyme Targeting (Part B)
Edited by RALPH GREEN AND KENNETH J. WIDDER

VOLUME 150. Immunochemical Techniques (Part K: *In Vitro* Models of B and T Cell Functions and Lymphoid Cell Receptors)
Edited by GIOVANNI DI SABATO

VOLUME 151. Molecular Genetics of Mammalian Cells
Edited by MICHAEL M. GOTTESMAN

VOLUME 152. Guide to Molecular Cloning Techniques
Edited by SHELBY L. BERGER AND ALAN R. KIMMEL

VOLUME 153. Recombinant DNA (Part D)
Edited by RAY WU AND LAWRENCE GROSSMAN

VOLUME 154. Recombinant DNA (Part E)
Edited by RAY WU AND LAWRENCE GROSSMAN

VOLUME 155. Recombinant DNA (Part F)
Edited by RAY WU

VOLUME 156. Biomembranes (Part P: ATP-Driven Pumps and Related Transport: The Na, K-Pump)
Edited by SIDNEY FLEISCHER AND BECCA FLEISCHER

VOLUME 157. Biomembranes (Part Q: ATP-Driven Pumps and Related Transport: Calcium, Proton, and Potassium Pumps)
Edited by SIDNEY FLEISCHER AND BECCA FLEISCHER

VOLUME 158. Metalloproteins (Part A)
Edited by JAMES F. RIORDAN AND BERT L. VALLEE

VOLUME 159. Initiation and Termination of Cyclic Nucleotide Action
Edited by JACKIE D. CORBIN AND ROGER A. JOHNSON

VOLUME 160. Biomass (Part A: Cellulose and Hemicellulose)
Edited by WILLIS A. WOOD AND SCOTT T. KELLOGG

VOLUME 161. Biomass (Part B: Lignin, Pectin, and Chitin)
Edited by WILLIS A. WOOD AND SCOTT T. KELLOGG

VOLUME 162. Immunochemical Techniques (Part L: Chemotaxis and Inflammation)
Edited by GIOVANNI DI SABATO

VOLUME 163. Immunochemical Techniques (Part M: Chemotaxis and Inflammation)
Edited by GIOVANNI DI SABATO

VOLUME 164. Ribosomes
Edited by HARRY F. NOLLER, JR., AND KIVIE MOLDAVE

VOLUME 165. Microbial Toxins: Tools for Enzymology
Edited by SIDNEY HARSHMAN

VOLUME 166. Branched-Chain Amino Acids
Edited by ROBERT HARRIS AND JOHN R. SOKATCH

VOLUME 167. Cyanobacteria
Edited by LESTER PACKER AND ALEXANDER N. GLAZER

VOLUME 168. Hormone Action (Part K: Neuroendocrine Peptides)
Edited by P. MICHAEL CONN

VOLUME 169. Platelets: Receptors, Adhesion, Secretion (Part A)
Edited by JACEK HAWIGER

VOLUME 170. Nucleosomes
Edited by PAUL M. WASSARMAN AND ROGER D. KORNBERG

VOLUME 171. Biomembranes (Part R: Transport Theory: Cells and Model Membranes)
Edited by SIDNEY FLEISCHER AND BECCA FLEISCHER

VOLUME 172. Biomembranes (Part S: Transport: Membrane Isolation and Characterization)
Edited by SIDNEY FLEISCHER AND BECCA FLEISCHER

VOLUME 173. Biomembranes [Part T: Cellular and Subcellular Transport: Eukaryotic (Nonepithelial) Cells]
Edited by SIDNEY FLEISCHER AND BECCA FLEISCHER

VOLUME 174. Biomembranes [Part U: Cellular and Subcellular Transport: Eukaryotic (Nonepithelial) Cells]
Edited by SIDNEY FLEISCHER AND BECCA FLEISCHER

VOLUME 175. Cumulative Subject Index Volumes 135–139, 141–167

VOLUME 176. Nuclear Magnetic Resonance (Part A: Spectral Techniques and Dynamics)
Edited by NORMAN J. OPPENHEIMER AND THOMAS L. JAMES

VOLUME 177. Nuclear Magnetic Resonance (Part B: Structure and Mechanism)
Edited by NORMAN J. OPPENHEIMER AND THOMAS L. JAMES

VOLUME 178. Antibodies, Antigens, and Molecular Mimicry
Edited by JOHN J. LANGONE

VOLUME 179. Complex Carbohydrates (Part F)
Edited by VICTOR GINSBURG

VOLUME 180. RNA Processing (Part A: General Methods)
Edited by JAMES E. DAHLBERG AND JOHN N. ABELSON

VOLUME 181. RNA Processing (Part B: Specific Methods)
Edited by JAMES E. DAHLBERG AND JOHN N. ABELSON

VOLUME 182. Guide to Protein Purification
Edited by MURRAY P. DEUTSCHER

VOLUME 183. Molecular Evolution: Computer Analysis of Protein and Nucleic Acid Sequences
Edited by RUSSELL F. DOOLITTLE

VOLUME 184. Avidin-Biotin Technology
Edited by MEIR WILCHEK AND EDWARD A. BAYER

VOLUME 185. Gene Expression Technology
Edited by DAVID V. GOEDDEL

VOLUME 186. Oxygen Radicals in Biological Systems (Part B: Oxygen Radicals and Antioxidants)
Edited by LESTER PACKER AND ALEXANDER N. GLAZER

VOLUME 187. Arachidonate Related Lipid Mediators
Edited by ROBERT C. MURPHY AND FRANK A. FITZPATRICK

VOLUME 188. Hydrocarbons and Methylotrophy
Edited by MARY E. LIDSTROM

VOLUME 189. Retinoids (Part A: Molecular and Metabolic Aspects)
Edited by LESTER PACKER

VOLUME 190. Retinoids (Part B: Cell Differentiation and Clinical Applications)
Edited by LESTER PACKER

VOLUME 191. Biomembranes (Part V: Cellular and Subcellular Transport: Epithelial Cells)
Edited by SIDNEY FLEISCHER AND BECCA FLEISCHER

VOLUME 192. Biomembranes (Part W: Cellular and Subcellular Transport: Epithelial Cells)
Edited by SIDNEY FLEISCHER AND BECCA FLEISCHER

VOLUME 193. Mass Spectrometry
Edited by JAMES A. MCCLOSKEY

VOLUME 194. Guide to Yeast Genetics and Molecular Biology
Edited by CHRISTINE GUTHRIE AND GERALD R. FINK

VOLUME 195. Adenylyl Cyclase, G Proteins, and Guanylyl Cyclase
Edited by ROGER A. JOHNSON AND JACKIE D. CORBIN

VOLUME 196. Molecular Motors and the Cytoskeleton
Edited by RICHARD B. VALLEE

VOLUME 197. Phospholipases
Edited by EDWARD A. DENNIS

VOLUME 198. Peptide Growth Factors (Part C)
Edited by DAVID BARNES, J. P. MATHER, AND GORDON H. SATO

VOLUME 199. Cumulative Subject Index Volumes 168–174, 176–194

VOLUME 200. Protein Phosphorylation (Part A: Protein Kinases: Assays, Purification, Antibodies, Functional Analysis, Cloning, and Expression)
Edited by TONY HUNTER AND BARTHOLOMEW M. SEFTON

VOLUME 201. Protein Phosphorylation (Part B: Analysis of Protein Phosphorylation, Protein Kinase Inhibitors, and Protein Phosphatases)
Edited by TONY HUNTER AND BARTHOLOMEW M. SEFTON

VOLUME 202. Molecular Design and Modeling: Concepts and Applications (Part A: Proteins, Peptides, and Enzymes)
Edited by JOHN J. LANGONE

VOLUME 203. Molecular Design and Modeling: Concepts and Applications (Part B: Antibodies and Antigens, Nucleic Acids, Polysaccharides, and Drugs)
Edited by JOHN J. LANGONE

VOLUME 204. Bacterial Genetic Systems
Edited by JEFFREY H. MILLER

VOLUME 205. Metallobiochemistry (Part B: Metallothionein and Related Molecules)
Edited by JAMES F. RIORDAN AND BERT L. VALLEE

VOLUME 206. Cytochrome P450
Edited by MICHAEL R. WATERMAN AND ERIC F. JOHNSON

VOLUME 207. Ion Channels
Edited by BERNARDO RUDY AND LINDA E. IVERSON

VOLUME 208. Protein–DNA Interactions
Edited by ROBERT T. SAUER

VOLUME 209. Phospholipid Biosynthesis
Edited by EDWARD A. DENNIS AND DENNIS E. VANCE

VOLUME 210. Numerical Computer Methods
Edited by LUDWIG BRAND AND MICHAEL L. JOHNSON

VOLUME 211. DNA Structures (Part A: Synthesis and Physical Analysis of DNA)
Edited by DAVID M. J. LILLEY AND JAMES E. DAHLBERG

VOLUME 212. DNA Structures (Part B: Chemical and Electrophoretic Analysis of DNA)
Edited by DAVID M. J. LILLEY AND JAMES E. DAHLBERG

VOLUME 213. Carotenoids (Part A: Chemistry, Separation, Quantitation, and Antioxidation)
Edited by LESTER PACKER

VOLUME 214. Carotenoids (Part B: Metabolism, Genetics, and Biosynthesis)
Edited by LESTER PACKER

VOLUME 215. Platelets: Receptors, Adhesion, Secretion (Part B)
Edited by JACEK J. HAWIGER

VOLUME 216. Recombinant DNA (Part G)
Edited by RAY WU

VOLUME 217. Recombinant DNA (Part H)
Edited by RAY WU

VOLUME 218. Recombinant DNA (Part I)
Edited by RAY WU

VOLUME 219. Reconstitution of Intracellular Transport
Edited by JAMES E. ROTHMAN

VOLUME 220. Membrane Fusion Techniques (Part A)
Edited by NEJAT DÜZGÜNEŞ

VOLUME 221. Membrane Fusion Techniques (Part B)
Edited by NEJAT DÜZGÜNEŞ

VOLUME 222. Proteolytic Enzymes in Coagulation, Fibrinolysis, and Complement Activation (Part A: Mammalian Blood Coagulation Factors and Inhibitors)
Edited by LASZLO LORAND AND KENNETH G. MANN

VOLUME 223. Proteolytic Enzymes in Coagulation, Fibrinolysis, and Complement Activation (Part B: Complement Activation, Fibrinolysis, and Nonmammalian Blood Coagulation Factors)
Edited by LASZLO LORAND AND KENNETH G. MANN

VOLUME 224. Molecular Evolution: Producing the Biochemical Data
Edited by ELIZABETH ANNE ZIMMER, THOMAS J. WHITE, REBECCA L. CANN, AND ALLAN C. WILSON

VOLUME 225. Guide to Techniques in Mouse Development
Edited by PAUL M. WASSARMAN AND MELVIN L. DEPAMPHILIS

VOLUME 226. Metallobiochemistry (Part C: Spectroscopic and Physical Methods for Probing Metal Ion Environments in Metalloenzymes and Metalloproteins)
Edited by JAMES F. RIORDAN AND BERT L. VALLEE

VOLUME 227. Metallobiochemistry (Part D: Physical and Spectroscopic Methods for Probing Metal Ion Environments in Metalloproteins)
Edited by JAMES F. RIORDAN AND BERT L. VALLEE

VOLUME 228. Aqueous Two-Phase Systems
Edited by HARRY WALTER AND GÖTE JOHANSSON

VOLUME 229. Cumulative Subject Index Volumes 195–198, 200–227

VOLUME 230. Guide to Techniques in Glycobiology
Edited by WILLIAM J. LENNARZ AND GERALD W. HART

VOLUME 231. Hemoglobins (Part B: Biochemical and Analytical Methods)
Edited by JOHANNES EVERSE, KIM D. VANDEGRIFF, AND ROBERT M. WINSLOW

VOLUME 232. Hemoglobins (Part C: Biophysical Methods)
Edited by JOHANNES EVERSE, KIM D. VANDEGRIFF, AND ROBERT M. WINSLOW

VOLUME 233. Oxygen Radicals in Biological Systems (Part C)
Edited by LESTER PACKER

VOLUME 234. Oxygen Radicals in Biological Systems (Part D)
Edited by LESTER PACKER

VOLUME 235. Bacterial Pathogenesis (Part A: Identification and Regulation of Virulence Factors)
Edited by VIRGINIA L. CLARK AND PATRIK M. BAVOIL

VOLUME 236. Bacterial Pathogenesis (Part B: Integration of Pathogenic Bacteria with Host Cells)
Edited by VIRGINIA L. CLARK AND PATRIK M. BAVOIL

VOLUME 237. Heterotrimeric G Proteins
Edited by RAVI IYENGAR

VOLUME 238. Heterotrimeric G-Protein Effectors
Edited by RAVI IYENGAR

VOLUME 239. Nuclear Magnetic Resonance (Part C)
Edited by THOMAS L. JAMES AND NORMAN J. OPPENHEIMER

VOLUME 240. Numerical Computer Methods (Part B)
Edited by MICHAEL L. JOHNSON AND LUDWIG BRAND

VOLUME 241. Retroviral Proteases
Edited by LAWRENCE C. KUO AND JULES A. SHAFER

VOLUME 242. Neoglycoconjugates (Part A)
Edited by Y. C. LEE AND REIKO T. LEE

VOLUME 243. Inorganic Microbial Sulfur Metabolism
Edited by HARRY D. PECK, JR., AND JEAN LEGALL

VOLUME 244. Proteolytic Enzymes: Serine and Cysteine Peptidases
Edited by ALAN J. BARRETT

VOLUME 245. Extracellular Matrix Components
Edited by E. RUOSLAHTI AND E. ENGVALL

VOLUME 246. Biochemical Spectroscopy
Edited by KENNETH SAUER

VOLUME 247. Neoglycoconjugates (Part B: Biomedical Applications)
Edited by Y. C. LEE AND REIKO T. LEE

VOLUME 248. Proteolytic Enzymes: Aspartic and Metallo Peptidases
Edited by ALAN J. BARRETT

VOLUME 249. Enzyme Kinetics and Mechanism (Part D: Developments in Enzyme Dynamics)
Edited by DANIEL L. PURICH

VOLUME 250. Lipid Modifications of Proteins
Edited by PATRICK J. CASEY AND JANICE E. BUSS

VOLUME 251. Biothiols (Part A: Monothiols and Dithiols, Protein Thiols, and Thiyl Radicals)
Edited by LESTER PACKER

VOLUME 252. Biothiols (Part B: Glutathione and Thioredoxin; Thiols in Signal Transduction and Gene Regulation)
Edited by LESTER PACKER

VOLUME 253. Adhesion of Microbial Pathogens
Edited by RON J. DOYLE AND ITZHAK OFEK

VOLUME 254. Oncogene Techniques
Edited by PETER K. VOGT AND INDER M. VERMA

VOLUME 255. Small GTPases and Their Regulators (Part A: Ras Family)
Edited by W. E. BALCH, CHANNING J. DER, AND ALAN HALL

VOLUME 256. Small GTPases and Their Regulators (Part B: Rho Family)
Edited by W. E. BALCH, CHANNING J. DER, AND ALAN HALL

VOLUME 257. Small GTPases and Their Regulators (Part C: Proteins Involved in Transport)
Edited by W. E. BALCH, CHANNING J. DER, AND ALAN HALL

VOLUME 258. Redox-Active Amino Acids in Biology
Edited by JUDITH P. KLINMAN

VOLUME 259. Energetics of Biological Macromolecules
Edited by MICHAEL L. JOHNSON AND GARY K. ACKERS

VOLUME 260. Mitochondrial Biogenesis and Genetics (Part A)
Edited by GIUSEPPE M. ATTARDI AND ANNE CHOMYN

VOLUME 261. Nuclear Magnetic Resonance and Nucleic Acids
Edited by THOMAS L. JAMES

VOLUME 262. DNA Replication
Edited by JUDITH L. CAMPBELL

VOLUME 263. Plasma Lipoproteins (Part C: Quantitation)
Edited by WILLIAM A. BRADLEY, SANDRA H. GIANTURCO, AND JERE P. SEGREST

VOLUME 264. Mitochondrial Biogenesis and Genetics (Part B)
Edited by GIUSEPPE M. ATTARDI AND ANNE CHOMYN

VOLUME 265. Cumulative Subject Index Volumes 228, 230–262

VOLUME 266. Computer Methods for Macromolecular Sequence Analysis
Edited by RUSSELL F. DOOLITTLE

VOLUME 267. Combinatorial Chemistry
Edited by JOHN N. ABELSON

VOLUME 268. Nitric Oxide (Part A: Sources and Detection of NO; NO Synthase)
Edited by LESTER PACKER

VOLUME 269. Nitric Oxide (Part B: Physiological and Pathological Processes)
Edited by LESTER PACKER

VOLUME 270. High Resolution Separation and Analysis of Biological Macromolecules (Part A: Fundamentals)
Edited by BARRY L. KARGER AND WILLIAM S. HANCOCK

VOLUME 271. High Resolution Separation and Analysis of Biological Macromolecules (Part B: Applications)
Edited by BARRY L. KARGER AND WILLIAM S. HANCOCK

VOLUME 272. Cytochrome P450 (Part B)
Edited by ERIC F. JOHNSON AND MICHAEL R. WATERMAN

VOLUME 273. RNA Polymerase and Associated Factors (Part A)
Edited by SANKAR ADHYA

VOLUME 274. RNA Polymerase and Associated Factors (Part B)
Edited by SANKAR ADHYA

VOLUME 275. Viral Polymerases and Related Proteins
Edited by LAWRENCE C. KUO, DAVID B. OLSEN, AND STEVEN S. CARROLL

VOLUME 276. Macromolecular Crystallography (Part A)
Edited by CHARLES W. CARTER, JR., AND ROBERT M. SWEET

VOLUME 277. Macromolecular Crystallography (Part B)
Edited by CHARLES W. CARTER, JR., AND ROBERT M. SWEET

VOLUME 278. Fluorescence Spectroscopy
Edited by LUDWIG BRAND AND MICHAEL L. JOHNSON

VOLUME 279. Vitamins and Coenzymes (Part I)
Edited by DONALD B. MCCORMICK, JOHN W. SUTTIE, AND CONRAD WAGNER

VOLUME 280. Vitamins and Coenzymes (Part J)
Edited by DONALD B. MCCORMICK, JOHN W. SUTTIE, AND CONRAD WAGNER

VOLUME 281. Vitamins and Coenzymes (Part K)
Edited by DONALD B. MCCORMICK, JOHN W. SUTTIE, AND CONRAD WAGNER

VOLUME 282. Vitamins and Coenzymes (Part L)
Edited by DONALD B. MCCORMICK, JOHN W. SUTTIE, AND CONRAD WAGNER

VOLUME 283. Cell Cycle Control
Edited by WILLIAM G. DUNPHY

VOLUME 284. Lipases (Part A: Biotechnology)
Edited by BYRON RUBIN AND EDWARD A. DENNIS

VOLUME 285. Cumulative Subject Index Volumes 263, 264, 266–284, 286–289

VOLUME 286. Lipases (Part B: Enzyme Characterization and Utilization)
Edited by BYRON RUBIN AND EDWARD A. DENNIS

VOLUME 287. Chemokines
Edited by RICHARD HORUK

VOLUME 288. Chemokine Receptors
Edited by RICHARD HORUK

VOLUME 289. Solid Phase Peptide Synthesis
Edited by GREGG B. FIELDS

VOLUME 290. Molecular Chaperones
Edited by GEORGE H. LORIMER AND THOMAS BALDWIN

VOLUME 291. Caged Compounds
Edited by GERARD MARRIOTT

VOLUME 292. ABC Transporters: Biochemical, Cellular, and Molecular Aspects
Edited by SURESH V. AMBUDKAR AND MICHAEL M. GOTTESMAN

VOLUME 293. Ion Channels (Part B)
Edited by P. MICHAEL CONN

VOLUME 294. Ion Channels (Part C)
Edited by P. MICHAEL CONN

VOLUME 295. Energetics of Biological Macromolecules (Part B)
Edited by GARY K. ACKERS AND MICHAEL L. JOHNSON

VOLUME 296. Neurotransmitter Transporters
Edited by SUSAN G. AMARA

VOLUME 297. Photosynthesis: Molecular Biology of Energy Capture
Edited by LEE MCINTOSH

VOLUME 298. Molecular Motors and the Cytoskeleton (Part B)
Edited by RICHARD B. VALLEE

VOLUME 299. Oxidants and Antioxidants (Part A)
Edited by LESTER PACKER

VOLUME 300. Oxidants and Antioxidants (Part B)
Edited by LESTER PACKER

VOLUME 301. Nitric Oxide: Biological and Antioxidant Activities (Part C)
Edited by LESTER PACKER

VOLUME 302. Green Fluorescent Protein
Edited by P. MICHAEL CONN

VOLUME 303. cDNA Preparation and Display
Edited by SHERMAN M. WEISSMAN

VOLUME 304. Chromatin
Edited by PAUL M. WASSARMAN AND ALAN P. WOLFFE

VOLUME 305. Bioluminescence and Chemiluminescence (Part C)
Edited by THOMAS O. BALDWIN AND MIRIAM M. ZIEGLER

VOLUME 306. Expression of Recombinant Genes in Eukaryotic Systems
Edited by JOSEPH C. GLORIOSO AND MARTIN C. SCHMIDT

VOLUME 307. Confocal Microscopy
Edited by P. MICHAEL CONN

VOLUME 308. Enzyme Kinetics and Mechanism (Part E: Energetics of Enzyme Catalysis)
Edited by DANIEL L. PURICH AND VERN L. SCHRAMM

VOLUME 309. Amyloid, Prions, and Other Protein Aggregates
Edited by RONALD WETZEL

VOLUME 310. Biofilms
Edited by RON J. DOYLE

VOLUME 311. Sphingolipid Metabolism and Cell Signaling (Part A)
Edited by ALFRED H. MERRILL, JR., AND YUSUF A. HANNUN

VOLUME 312. Sphingolipid Metabolism and Cell Signaling (Part B)
Edited by ALFRED H. MERRILL, JR., AND YUSUF A. HANNUN

VOLUME 313. Antisense Technology
(Part A: General Methods, Methods of Delivery, and RNA Studies)
Edited by M. IAN PHILLIPS

VOLUME 314. Antisense Technology (Part B: Applications)
Edited by M. IAN PHILLIPS

VOLUME 315. Vertebrate Phototransduction and the Visual Cycle (Part A)
Edited by KRZYSZTOF PALCZEWSKI

VOLUME 316. Vertebrate Phototransduction and the Visual Cycle (Part B)
Edited by KRZYSZTOF PALCZEWSKI

VOLUME 317. RNA–Ligand Interactions (Part A: Structural Biology Methods)
Edited by DANIEL W. CELANDER AND JOHN N. ABELSON

VOLUME 318. RNA–Ligand Interactions (Part B: Molecular Biology Methods)
Edited by DANIEL W. CELANDER AND JOHN N. ABELSON

VOLUME 319. Singlet Oxygen, UV-A, and Ozone
Edited by LESTER PACKER AND HELMUT SIES

VOLUME 320. Cumulative Subject Index Volumes 290–319

VOLUME 321. Numerical Computer Methods (Part C)
Edited by MICHAEL L. JOHNSON AND LUDWIG BRAND

VOLUME 322. Apoptosis
Edited by JOHN C. REED

VOLUME 323. Energetics of Biological Macromolecules (Part C)
Edited by MICHAEL L. JOHNSON AND GARY K. ACKERS

VOLUME 324. Branched-Chain Amino Acids (Part B)
Edited by ROBERT A. HARRIS AND JOHN R. SOKATCH

VOLUME 325. Regulators and Effectors of Small GTPases
(Part D: Rho Family)
Edited by W. E. BALCH, CHANNING J. DER, AND ALAN HALL

VOLUME 326. Applications of Chimeric Genes and Hybrid Proteins
(Part A: Gene Expression and Protein Purification)
Edited by JEREMY THORNER, SCOTT D. EMR, AND JOHN N. ABELSON

VOLUME 327. Applications of Chimeric Genes and Hybrid Proteins
(Part B: Cell Biology and Physiology)
Edited by JEREMY THORNER, SCOTT D. EMR, AND JOHN N. ABELSON

VOLUME 328. Applications of Chimeric Genes and Hybrid Proteins (Part C: Protein–Protein Interactions and Genomics)
Edited by JEREMY THORNER, SCOTT D. EMR, AND JOHN N. ABELSON

VOLUME 329. Regulators and Effectors of Small GTPases (Part E: GTPases Involved in Vesicular Traffic)
Edited by W. E. BALCH, CHANNING J. DER, AND ALAN HALL

VOLUME 330. Hyperthermophilic Enzymes (Part A)
Edited by MICHAEL W. W. ADAMS AND ROBERT M. KELLY

VOLUME 331. Hyperthermophilic Enzymes (Part B)
Edited by MICHAEL W. W. ADAMS AND ROBERT M. KELLY

VOLUME 332. Regulators and Effectors of Small GTPases (Part F: Ras Family I)
Edited by W. E. BALCH, CHANNING J. DER, AND ALAN HALL

VOLUME 333. Regulators and Effectors of Small GTPases (Part G: Ras Family II)
Edited by W. E. BALCH, CHANNING J. DER, AND ALAN HALL

VOLUME 334. Hyperthermophilic Enzymes (Part C)
Edited by MICHAEL W. W. ADAMS AND ROBERT M. KELLY

VOLUME 335. Flavonoids and Other Polyphenols
Edited by LESTER PACKER

VOLUME 336. Microbial Growth in Biofilms (Part A: Developmental and Molecular Biological Aspects)
Edited by RON J. DOYLE

VOLUME 337. Microbial Growth in Biofilms (Part B: Special Environments and Physicochemical Aspects)
Edited by RON J. DOYLE

VOLUME 338. Nuclear Magnetic Resonance of Biological Macromolecules (Part A)
Edited by THOMAS L. JAMES, VOLKER DÖTSCH, AND ULI SCHMITZ

VOLUME 339. Nuclear Magnetic Resonance of Biological Macromolecules (Part B)
Edited by THOMAS L. JAMES, VOLKER DÖTSCH, AND ULI SCHMITZ

VOLUME 340. Drug–Nucleic Acid Interactions
Edited by JONATHAN B. CHAIRES AND MICHAEL J. WARING

VOLUME 341. Ribonucleases (Part A)
Edited by ALLEN W. NICHOLSON

VOLUME 342. Ribonucleases (Part B)
Edited by ALLEN W. NICHOLSON

VOLUME 343. G Protein Pathways (Part A: Receptors)
Edited by RAVI IYENGAR AND JOHN D. HILDEBRANDT

VOLUME 344. G Protein Pathways (Part B: G Proteins and Their Regulators)
Edited by RAVI IYENGAR AND JOHN D. HILDEBRANDT

VOLUME 345. G Protein Pathways (Part C: Effector Mechanisms)
Edited by RAVI IYENGAR AND JOHN D. HILDEBRANDT

VOLUME 346. Gene Therapy Methods
Edited by M. IAN PHILLIPS

VOLUME 347. Protein Sensors and Reactive Oxygen Species (Part A: Selenoproteins and Thioredoxin)
Edited by HELMUT SIES AND LESTER PACKER

VOLUME 348. Protein Sensors and Reactive Oxygen Species (Part B: Thiol Enzymes and Proteins)
Edited by HELMUT SIES AND LESTER PACKER

VOLUME 349. Superoxide Dismutase
Edited by LESTER PACKER

VOLUME 350. Guide to Yeast Genetics and Molecular and Cell Biology (Part B)
Edited by CHRISTINE GUTHRIE AND GERALD R. FINK

VOLUME 351. Guide to Yeast Genetics and Molecular and Cell Biology (Part C)
Edited by CHRISTINE GUTHRIE AND GERALD R. FINK

VOLUME 352. Redox Cell Biology and Genetics (Part A)
Edited by CHANDAN K. SEN AND LESTER PACKER

VOLUME 353. Redox Cell Biology and Genetics (Part B)
Edited by CHANDAN K. SEN AND LESTER PACKER

VOLUME 354. Enzyme Kinetics and Mechanisms (Part F: Detection and Characterization of Enzyme Reaction Intermediates)
Edited by DANIEL L. PURICH

VOLUME 355. Cumulative Subject Index Volumes 321–354

VOLUME 356. Laser Capture Microscopy and Microdissection
Edited by P. MICHAEL CONN

VOLUME 357. Cytochrome P450, Part C
Edited by ERIC F. JOHNSON AND MICHAEL R. WATERMAN

VOLUME 358. Bacterial Pathogenesis (Part C: Identification, Regulation, and Function of Virulence Factors)
Edited by VIRGINIA L. CLARK AND PATRIK M. BAVOIL

VOLUME 359. Nitric Oxide (Part D)
Edited by ENRIQUE CADENAS AND LESTER PACKER

VOLUME 360. Biophotonics (Part A)
Edited by GERARD MARRIOTT AND IAN PARKER

VOLUME 361. Biophotonics (Part B)
Edited by GERARD MARRIOTT AND IAN PARKER

VOLUME 362. Recognition of Carbohydrates in Biological Systems (Part A)
Edited by YUAN C. LEE AND REIKO T. LEE

VOLUME 363. Recognition of Carbohydrates in Biological Systems (Part B)
Edited by YUAN C. LEE AND REIKO T. LEE

VOLUME 364. Nuclear Receptors
Edited by DAVID W. RUSSELL AND DAVID J. MANGELSDORF

VOLUME 365. Differentiation of Embryonic Stem Cells
Edited by PAUL M. WASSAUMAN AND GORDON M. KELLER

VOLUME 366. Protein Phosphatases
Edited by SUSANNE KLUMPP AND JOSEF KRIEGLSTEIN

VOLUME 367. Liposomes (Part A)
Edited by NEJAT DÜZGÜNEŞ

VOLUME 368. Macromolecular Crystallography (Part C)
Edited by CHARLES W. CARTER, JR., AND ROBERT M. SWEET

VOLUME 369. Combinational Chemistry (Part B)
Edited by GUILLERMO A. MORALES AND BARRY A. BUNIN

VOLUME 370. RNA Polymerases and Associated Factors (Part C)
Edited by SANKAR L. ADHYA AND SUSAN GARGES

VOLUME 371. RNA Polymerases and Associated Factors (Part D)
Edited by SANKAR L. ADHYA AND SUSAN GARGES

VOLUME 372. Liposomes (Part B)
Edited by NEJAT DÜZGÜNEŞ

VOLUME 373. Liposomes (Part C)
Edited by NEJAT DÜZGÜNEŞ

VOLUME 374. Macromolecular Crystallography (Part D)
Edited by CHARLES W. CARTER, JR., AND ROBERT W. SWEET

VOLUME 375. Chromatin and Chromatin Remodeling Enzymes (Part A)
Edited by C. DAVID ALLIS AND CARL WU

VOLUME 376. Chromatin and Chromatin Remodeling Enzymes (Part B)
Edited by C. DAVID ALLIS AND CARL WU

VOLUME 377. Chromatin and Chromatin Remodeling Enzymes (Part C)
Edited by C. DAVID ALLIS AND CARL WU

VOLUME 378. Quinones and Quinone Enzymes (Part A)
Edited by HELMUT SIES AND LESTER PACKER

VOLUME 379. Energetics of Biological Macromolecules (Part D)
Edited by JO M. HOLT, MICHAEL L. JOHNSON, AND GARY K. ACKERS

VOLUME 380. Energetics of Biological Macromolecules (Part E)
Edited by JO M. HOLT, MICHAEL L. JOHNSON, AND GARY K. ACKERS

VOLUME 381. Oxygen Sensing
Edited by CHANDAN K. SEN AND GREGG L. SEMENZA

VOLUME 382. Quinones and Quinone Enzymes (Part B)
Edited by HELMUT SIES AND LESTER PACKER

VOLUME 383. Numerical Computer Methods (Part D)
Edited by LUDWIG BRAND AND MICHAEL L. JOHNSON

VOLUME 384. Numerical Computer Methods (Part E)
Edited by LUDWIG BRAND AND MICHAEL L. JOHNSON

VOLUME 385. Imaging in Biological Research (Part A)
Edited by P. MICHAEL CONN

VOLUME 386. Imaging in Biological Research (Part B)
Edited by P. MICHAEL CONN

VOLUME 387. Liposomes (Part D)
Edited by NEJAT DÜZGÜNEŞ

VOLUME 388. Protein Engineering
Edited by DAN E. ROBERTSON AND JOSEPH P. NOEL

VOLUME 389. Regulators of G-Protein Signaling (Part A)
Edited by DAVID P. SIDEROVSKI

VOLUME 390. Regulators of G-Protein Signaling (Part B)
Edited by DAVID P. SIDEROVSKI

VOLUME 391. Liposomes (Part E)
Edited by NEJAT DÜZGÜNEŞ

VOLUME 392. RNA Interference
Edited by ENGELKE ROSSI

VOLUME 393. Circadian Rhythms
Edited by MICHAEL W. YOUNG

VOLUME 394. Nuclear Magnetic Resonance of Biological Macromolecules (Part C)
Edited by THOMAS L. JAMES

VOLUME 395. Producing the Biochemical Data (Part B)
Edited by ELIZABETH A. ZIMMER AND ERIC H. ROALSON

VOLUME 396. Nitric Oxide (Part E)
Edited by LESTER PACKER AND ENRIQUE CADENAS

VOLUME 397. Environmental Microbiology
Edited by JARED R. LEADBETTER

VOLUME 398. Ubiquitin and Protein Degradation (Part A)
Edited by RAYMOND J. DESHAIES

VOLUME 399. Ubiquitin and Protein Degradation (Part B)
Edited by RAYMOND J. DESHAIES

VOLUME 400. Phase II Conjugation Enzymes and Transport Systems
Edited by HELMUT SIES AND LESTER PACKER

VOLUME 401. Glutathione Transferases and Gamma Glutamyl Transpeptidases
Edited by HELMUT SIES AND LESTER PACKER

VOLUME 402. Biological Mass Spectrometry
Edited by A. L. BURLINGAME

VOLUME 403. GTPases Regulating Membrane Targeting and Fusion
Edited by WILLIAM E. BALCH, CHANNING J. DER, AND ALAN HALL

VOLUME 404. GTPases Regulating Membrane Dynamics
Edited by WILLIAM E. BALCH, CHANNING J. DER, AND ALAN HALL

VOLUME 405. Mass Spectrometry: Modified Proteins and Glycoconjugates
Edited by A. L. BURLINGAME

VOLUME 406. Regulators and Effectors of Small GTPases: Rho Family
Edited by WILLIAM E. BALCH, CHANNING J. DER, AND ALAN HALL

VOLUME 407. Regulators and Effectors of Small GTPases: Ras Family
Edited by WILLIAM E. BALCH, CHANNING J. DER, AND ALAN HALL

VOLUME 408. DNA Repair (Part A)
Edited by JUDITH L. CAMPBELL AND PAUL MODRICH

VOLUME 409. DNA Repair (Part B)
Edited by JUDITH L. CAMPBELL AND PAUL MODRICH

VOLUME 410. DNA Microarrays (Part A: Array Platforms and Web-Bench Protocols)
Edited by ALAN KIMMEL AND BRIAN OLIVER

VOLUME 411. DNA Microarrays (Part B: Databases and Statistics)
Edited by ALAN KIMMEL AND BRIAN OLIVER

VOLUME 412. Amyloid, Prions, and Other Protein Aggregates (Part B)
Edited by INDU KHETERPAL AND RONALD WETZEL

VOLUME 413. Amyloid, Prions, and Other Protein Aggregates (Part C)
Edited by INDU KHETERPAL AND RONALD WETZEL

VOLUME 414. Measuring Biological Responses with Automated Microscopy
Edited by JAMES INGLESE

VOLUME 415. Glycobiology
Edited by MINORU FUKUDA

VOLUME 416. Glycomics
Edited by MINORU FUKUDA

VOLUME 417. Functional Glycomics
Edited by MINORU FUKUDA

VOLUME 418. Embryonic Stem Cells
Edited by IRINA KLIMANSKAYA AND ROBERT LANZA

VOLUME 419. Adult Stem Cells
Edited by IRINA KLIMANSKAYA AND ROBERT LANZA

VOLUME 420. Stem Cell Tools and Other Experimental Protocols
Edited by IRINA KLIMANSKAYA AND ROBERT LANZA

VOLUME 421. Advanced Bacterial Genetics: Use of Transposons and Phage for Genomic Engineering
Edited by KELLY T. HUGHES

VOLUME 422. Two-Component Signaling Systems, Part A
Edited by MELVIN I. SIMON, BRIAN R. CRANE, AND ALEXANDRINE CRANE

VOLUME 423. Two-Component Signaling Systems, Part B
Edited by MELVIN I. SIMON, BRIAN R. CRANE, AND ALEXANDRINE CRANE

VOLUME 424. RNA Editing
Edited by JONATHA M. GOTT

VOLUME 425. RNA Modification
Edited by JONATHA M. GOTT

VOLUME 426. Integrins
Edited by DAVID CHERESH

VOLUME 427. MicroRNA Methods
Edited by JOHN J. ROSSI

VOLUME 428. Osmosensing and Osmosignaling
Edited by HELMUT SIES AND DIETER HAUSSINGER

VOLUME 429. Translation Initiation: Extract Systems and Molecular Genetics
Edited by JON LORSCH

VOLUME 430. Translation Initiation: Reconstituted Systems and Biophysical Methods
Edited by JON LORSCH

VOLUME 431. Translation Initiation: Cell Biology, High-Throughput and Chemical-Based Approaches
Edited by JON LORSCH

VOLUME 432. Lipidomics and Bioactive Lipids: Mass-Spectrometry–Based Lipid Analysis
Edited by H. ALEX BROWN

VOLUME 433. Lipidomics and Bioactive Lipids: Specialized Analytical Methods and Lipids in Disease
Edited by H. ALEX BROWN

VOLUME 434. Lipidomics and Bioactive Lipids: Lipids and Cell Signaling
Edited by H. ALEX BROWN

VOLUME 435. Oxygen Biology and Hypoxia
Edited by HELMUT SIES AND BERNHARD BRÜNE

VOLUME 436. Globins and Other Nitric Oxide-Reactive Protiens (Part A)
Edited by ROBERT K. POOLE

VOLUME 437. Globins and Other Nitric Oxide-Reactive Protiens (Part B)
Edited by ROBERT K. POOLE

VOLUME 438. Small GTPases in Disease (Part A)
Edited by WILLIAM E. BALCH, CHANNING J. DER, AND ALAN HALL

VOLUME 439. Small GTPases in Disease (Part B)
Edited by WILLIAM E. BALCH, CHANNING J. DER, AND ALAN HALL

VOLUME 440. Nitric Oxide, Part F Oxidative and Nitrosative Stress in Redox Regulation of Cell Signaling
Edited by ENRIQUE CADENAS AND LESTER PACKER

VOLUME 441. Nitric Oxide, Part G Oxidative and Nitrosative Stress in Redox Regulation of Cell Signaling
Edited by ENRIQUE CADENAS AND LESTER PACKER

VOLUME 442. Programmed Cell Death, General Principles for Studying Cell Death (Part A)
Edited by ROYA KHOSRAVI-FAR, ZAHRA ZAKERI, RICHARD A. LOCKSHIN, AND MAURO PIACENTINI

VOLUME 443. Angiogenesis: *In Vitro* Systems
Edited by DAVID A. CHERESH

VOLUME 444. Angiogenesis: *In Vivo* Systems (Part A)
Edited by DAVID A. CHERESH

VOLUME 445. Angiogenesis: *In Vivo* Systems (Part B)
Edited by DAVID A. CHERESH

VOLUME 446. Programmed Cell Death, The Biology and Therapeutic Implications of Cell Death (Part B)
Edited by ROYA KHOSRAVI-FAR, ZAHRA ZAKERI, RICHARD A. LOCKSHIN, AND MAURO PIACENTINI

VOLUME 447. RNA Turnover in Bacteria, Archaea and Organelles
Edited by LYNNE E. MAQUAT AND CECILIA M. ARRAIANO

VOLUME 448. RNA Turnover in Eukaryotes: Nucleases, Pathways and Analysis of mRNA Decay
Edited by LYNNE E. MAQUAT AND MEGERDITCH KILEDJIAN

VOLUME 449. RNA Turnover in Eukaryotes: Analysis of Specialized and Quality Control RNA Decay Pathways
Edited by LYNNE E. MAQUAT AND MEGERDITCH KILEDJIAN

VOLUME 450. Fluorescence Spectroscopy
Edited by LUDWIG BRAND AND MICHAEL L. JOHNSON

VOLUME 451. Autophagy: Lower Eukaryotes and Non-Mammalian Systems (Part A)
Edited by DANIEL J. KLIONSKY

VOLUME 452. Autophagy in Mammalian Systems (Part B)
Edited by DANIEL J. KLIONSKY

VOLUME 453. Autophagy in Disease and Clinical Applications (Part C)
Edited by DANIEL J. KLIONSKY

VOLUME 454. Computer Methods (Part A)
Edited by MICHAEL L. JOHNSON AND LUDWIG BRAND

VOLUME 455. Biothermodynamics (Part A)
Edited by MICHAEL L. JOHNSON, JO M. HOLT, AND GARY K. ACKERS (RETIRED)

VOLUME 456. Mitochondrial Function, Part A: Mitochondrial Electron Transport Complexes and Reactive Oxygen Species
Edited by WILLIAM S. ALLISON AND IMMO E. SCHEFFLER

VOLUME 457. Mitochondrial Function, Part B: Mitochondrial Protein Kinases, Protein Phosphatases and Mitochondrial Diseases
Edited by WILLIAM S. ALLISON AND ANNE N. MURPHY

VOLUME 458. Complex Enzymes in Microbial Natural Product Biosynthesis, Part A: Overview Articles and Peptides
Edited by DAVID A. HOPWOOD

VOLUME 459. Complex Enzymes in Microbial Natural Product Biosynthesis, Part B: Polyketides, Aminocoumarins and Carbohydrates
Edited by DAVID A. HOPWOOD

VOLUME 460. Chemokines, Part A
Edited by TRACY M. HANDEL AND DAMON J. HAMEL

VOLUME 461. Chemokines, Part B
Edited by TRACY M. HANDEL AND DAMON J. HAMEL

VOLUME 462. Non-Natural Amino Acids
Edited by TOM W. MUIR AND JOHN N. ABELSON

VOLUME 463. Guide to Protein Purification, 2nd Edition
Edited by RICHARD R. BURGESS AND MURRAY P. DEUTSCHER

VOLUME 464. Liposomes, Part F
Edited by NEJAT DÜZGÜNEŞ

VOLUME 465. Liposomes, Part G
Edited by NEJAT DÜZGÜNEŞ

VOLUME 466. Biothermodynamics, Part B
Edited by MICHAEL L. JOHNSON, GARY K. ACKERS, AND JO M. HOLT

VOLUME 467. Computer Methods Part B
Edited by MICHAEL L. JOHNSON AND LUDWIG BRAND

VOLUME 468. Biophysical, Chemical, and Functional Probes of RNA Structure, Interactions and Folding: Part A
Edited by DANIEL HERSCHLAG

VOLUME 469. Biophysical, Chemical, and Functional Probes of RNA Structure, Interactions and Folding: Part B
Edited by DANIEL HERSCHLAG

VOLUME 470. Guide to Yeast Genetics: Functional Genomics, Proteomics, and Other Systems Analysis, 2nd Edition
Edited by GERALD FINK, JONATHAN WEISSMAN, AND CHRISTINE GUTHRIE

VOLUME 471. Two-Component Signaling Systems, Part C
Edited by MELVIN I. SIMON, BRIAN R. CRANE, AND ALEXANDRINE CRANE

VOLUME 472. Single Molecule Tools, Part A: Fluorescence Based Approaches
Edited by NILS G. WALTER

VOLUME 473. Thiol Redox Transitions in Cell Signaling, Part A Chemistry and Biochemistry of Low Molecular Weight and Protein Thiols
Edited by ENRIQUE CADENAS AND LESTER PACKER

VOLUME 474. Thiol Redox Transitions in Cell Signaling, Part B Cellular Localization and Signaling
Edited by ENRIQUE CADENAS AND LESTER PACKER

VOLUME 475. Single Molecule Tools, Part B: Super-Resolution, Particle Tracking, Multiparameter, and Force Based Methods
Edited by NILS G. WALTER

VOLUME 476. Guide to Techniques in Mouse Development, Part A Mice, Embryos, and Cells, 2nd Edition
Edited by PAUL M. WASSARMAN AND PHILIPPE M. SORIANO

VOLUME 477. Guide to Techniques in Mouse Development, Part B Mouse Molecular Genetics, 2nd Edition
Edited by PAUL M. WASSARMAN AND PHILIPPE M. SORIANO

VOLUME 478. Glycomics
Edited by MINORU FUKUDA

VOLUME 479. Functional Glycomics
Edited by MINORU FUKUDA

VOLUME 480. Glycobiology
Edited by MINORU FUKUDA

VOLUME 481. Cryo-EM, Part A: Sample Preparation and Data Collection
Edited by GRANT J. JENSEN

VOLUME 482. Cryo-EM, Part B: 3-D Reconstruction
Edited by GRANT J. JENSEN

VOLUME 483. Cryo-EM, Part C: Analyses, Interpretation, and Case Studies
Edited by GRANT J. JENSEN

VOLUME 484. Constitutive Activity in Receptors and Other Proteins, Part A
Edited by P. MICHAEL CONN

VOLUME 485. Constitutive Activity in Receptors and Other Proteins, Part B
Edited by P. MICHAEL CONN

VOLUME 486. Research on Nitrification and Related Processes, Part A
Edited by MARTIN G. KLOTZ

SECTION ONE

MODELING

CHAPTER ONE

MODELING THE ROLE OF NITRIFICATION IN OPEN OCEAN PRODUCTIVITY AND THE NITROGEN CYCLE

Andrew Yool

Contents

1. Introduction	4
2. Ocean Carbon Cycle	4
3. Export Production and the f-Ratio	7
4. Plankton Ecosystem Models	12
5. Modeling Nitrification	15
6. Operational Considerations	22
7. Future Modeling	24
8. Summary	27
Acknowledgments	28
References	28

Abstract

The ocean is an important component of the global carbon cycle, and currently serves as the principal sink for anthropogenic CO_2 from the atmosphere. A key role in the natural oceanic carbon cycle is played by the plankton ecosystem, which acts to elevate the storage capacity of the ocean, but it is believed that this will experience change in the future in response to anthropogenic forcing. One of the approaches used to understand and forecast the oceanic carbon cycle is ecosystem modeling, and this is typically grounded on the nitrogen cycle because of the strong regulatory role this element plays in biological productivity. Nitrification is one of the central processes in the oceanic nitrogen cycle, one whose role may change in the future, but also one with a particular relevance to observational efforts to quantify the biological carbon cycle. Here, we describe and summarize current efforts to model nitrification in pelagic open ocean ecosystems, and look forward to future avenues for progress.

National Oceanography Centre, University of Southampton Waterfront Campus, European Way, Southampton, United Kingdom

Methods in Enzymology, Volume 486 © 2011 Elsevier Inc.
ISSN 0076-6879, DOI: 10.1016/S0076-6879(11)86001-5 All rights reserved.

1. Introduction

This chapter is centered around the representation of nitrification in pelagic open ocean ecosystem models. The chapter frames nitrification in this environment within the context of the carbon cycle and nitrification's relationship with primary production and measuring the biological pump. This is just one framing, and it would be perfectly possible to instead focus on the relationships between open ocean nitrification and other nitrogen cycle processes such as denitrification (e.g., Fernández et al., 2009) or nitrogen fixation (e.g., Casciotti et al., 2008). Nonetheless, much of the methodology discussed would remain similar.

The chapter is structured as follows: First, the ocean carbon cycle and the role of biology in this are briefly introduced. Next, export production, a related metric known as the *f*-ratio, and their relationships with nitrification are outlined. A brief introduction to open ocean plankton ecosystem models is then presented. This is followed by an overview of the approaches by which nitrification is included in such models. Some practical considerations when modeling nitrification and ecosystem processes, in general, are then presented. Finally, potential future avenues for modeling nitrification are discussed.

 ## 2. Ocean Carbon Cycle

In large part because of its role in anthropogenically driven climate change, significant attention is focused on the continuing rise in atmospheric carbon dioxide (CO_2) concentrations. However, the atmosphere is actually one of the smaller reservoirs of carbon in the Earth system; although, being gaseous, it is still the most dynamically active. A much larger reservoir, and the one largely responsible for the partial mitigation of anthropogenic CO_2 emissions to date, is the ocean. Compared to the atmosphere's current CO_2 content of ~ 750 Pg C ($= 750 \times 10^{15}$ g C), the ocean's inventory of 38,100 Pg C of dissolved inorganic carbon (DIC) is vast (Sarmiento and Gruber, 2006). However, the relatively languid turnover of the deep ocean by the thermohaline circulation means that the exchange of CO_2 between the ocean and the atmosphere is insufficiently fast to prevent the ongoing accumulation of CO_2 in the atmosphere (though the ocean is still believed to currently store around one quarter of annual CO_2 emissions; Takahashi et al., 2009).

DIC reaches the deep ocean primarily as a consequence of the activity of two mechanisms: the *solubility pump* and the *biological pump* (Raven and Falkowski, 1999). These so-called "pumps" act to elevate the ocean's interior concentration of DIC beyond that which would naïvely be expected from average surface concentrations. In the case of the solubility pump, its function stems from the coincidence in space of locations of high CO_2 solubility with regions where deep water formation occurs. Deep water formation marks the beginning of the ocean's thermohaline circulation, and is the process whereby cold and salty surface waters convect into the ocean interior. These waters ventilate the deep ocean and carry elevated DIC concentrations there because CO_2 is more soluble at low temperatures. The net effect of the solubility pump is to replenish the deep ocean with water that is laden with atmospheric CO_2 because of these physicochemical processes.

The biological pump is essentially the sum of a series of biological/ecological processes that act to vertically transport carbon, principally as particulate material, into the ocean interior. This carbon occurs in both organic and inorganic forms, with the latter largely composed of the calcium carbonate polymorphs calcite and aragonite.[1] The pump is driven primarily by sinking material, known as "marine snow," that is composed of organisms (or pieces of them), fecal pellets, and aggregates of this material. The varied organic material is originally the product of photosynthesis by phytoplankton in the sunlit surface waters of the ocean. Figure 1.1 shows the geographical distribution of estimated primary production (Behrenfeld and Falkowski, 1997) derived from satellite-observed sea surface chlorophyll (O'Reilly et al., 1998) and averaged for the period 2000–2004 inclusive. Sedimentation of phytoplankton cells, their consumption by grazing zooplankton, and the subsequent predation on zooplankton by higher trophic levels are among the suite of ecological processes that contribute to the "rain" of material into the ocean interior. This sinking material is gradually remineralized down the water column by heterotrophs, including associated and free-living bacteria, leading to elevated DIC concentrations at depth (Buesseler et al., 2007). Similar to the solubility pump, the biological pump is not geographically homogeneous, but instead of occurring in cold, high latitude waters, it is largely focused in regions where nutrients, including nitrogen ones, are abundant in surface waters. These include upwelling zones and areas that experience strong vertical mixing (e.g., seasonal temperate regions).

Since the physical state of the ocean has changed only slightly since the start of the Industrial Revolution (cf. Levitus et al., 2005), the ocean's nutrient cycles have not (yet) significantly changed. As nutrient availability largely governs productivity, the biological pump has not changed in

---

[1] Since calcium carbonate is relatively dense, and does not significantly dissolve above the carbonate compensation depth in the ocean, it is believed that it may act as "ballast" for associated organic material (together with opal, another biomineral) and enhance the flux of organic carbon to the deep ocean (Armstrong et al., 2002).

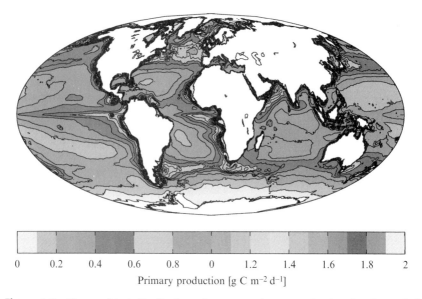

0   0.2   0.4   0.6   0.8   0   1.2   1.4   1.6   1.8   2
Primary production [g C m^{-2} d^{-1}]

Figure 1.1 Geographical distribution of average primary production for the period 2000–2004 inclusive, estimated from SeaWiFS satellite observations of surface chlorophyll using the vertically generalized production model (VGPM; Behrenfeld and Falkowski, 1997). Primary production in g C m^{-2} d^{-1}.

strength and its role in the ocean's uptake of anthropogenic CO_2 is believed to be minor (Gruber et al., 1996; but see also Boyce et al., 2010).[2] Instead, the steadily increasing concentration gradient between the atmosphere and the surface ocean has led to a net flux of CO_2 into the ocean via the solubility pump, with the greatest column inventories occurring in regions of deep water formation, such as the North Atlantic and the Southern Ocean. However, it is believed that future change to the ocean may have consequences for the carbon cycle mediated through the biological pump. For instance, climate change will affect the stratification of the ocean and the vertical supply of nutrients that fuels primary production (van der Waal et al., 2010). Also, the invasion of anthropogenic CO_2 into the ocean has led to the phenomenon of ocean acidification, which is anticipated to have a range of effects as different biological groups are impacted (Raven et al., 2005).

---

[2] This contrasts with the situation on land, where plants are believed to represent an additional sink for carbon (Van Minnen et al., 2009). This occurs because CO_2 acquisition is linked to water loss through evapotranspiration in land plants, and increasing atmospheric CO_2 concentrations have allowed plants to acquire CO_2 for photosynthesis at a decreased cost in water. In the ocean, DIC concentrations are not generally believed to limit phytoplankton growth (but see Beardall and Raven, 2004), and the availability of water is a moot point.

This linkage between the biological pump and the supply of nutrients that fuel it strongly ties the ocean's carbon cycle to those of other major nutrients such as nitrogen and phosphorus. Consequently, the biogeochemical cycles of these other elements can play an important role in how ecosystems regulate the size and distribution of the oceanic carbon reservoir.

3. Export Production and the *F*-Ratio

As far back as 1934, the stoichiometry of coupled elemental cycles in the ocean was recognized to operate along relatively fixed ratios. Redfield (1934) noted an approximate molar ratio of 106 C:16 N:1 P in marine planktonic organic matter, and these elemental relationships have come to be known as the "Redfield ratio." Subsequent work has linked in other elemental cycles, often with less conserved stoichiometry (e.g., the silicon cycle; Yool and Tyrrell, 2003), and has investigated how fixed the Redfield ratio is across a wide range of biogeochemical processes. Consequently, the ratio, and deviations from it, has become widely used in elucidating the spatial (horizontal and vertical) distribution of processes,[3] including estimates of the distribution of anthropogenic CO_2 in the present day ocean (e.g., Gruber *et al.*, 1996).

However, there is an additional connection between the ocean's carbon and nitrogen cycles that potentially provides observational scientists with insight into the strength of the biological pump.

A key flux sought by observational scientists is the so-called "export production" of organic material from the surface ocean to the deep ocean. This quantifies the activity of the biological pump and can be compared with primary production to assess the role of recycling in the plankton community (i.e., if the ratio is high, most production reaches the deep ocean; if low, most is remineralized and nutrients are reused in the surface ocean). Quantifying export production gives insight into the vertical structure of biological activity in the ocean, and has a particular contemporary significance as it is anticipated that this will change in the future. However, accurately measuring export production in the open ocean is a considerable technical challenge. This is normally achieved using instruments such as sediment traps that collect and integrate organic material that sinks into them (Thomas and Ridd, 2004). These instruments are required to operate at demanding depths, sometimes over long periods of time, and they can be difficult to deploy or recover. Furthermore, the measurements they make are complicated by a number of sampling biases, such as contamination by

---

[3] Note that isotopic methods can also be used to infer the spatial distribution of biogeochemical processes in the ocean (e.g., Casciotti *et al.*, 2008).

grazing zooplankton, and they integrate horizontal variability as well as vertical fluxes because of local advection (Buesseler et al., 2007).

Dissolved inorganic nitrogen (DIN) occurs in marine environments as a number of chemical species that are available for uptake and biosynthesis by phytoplankton. These include nitrate, ammonium, nitrite, and organic molecules such as urea and free amino acids (Bradley et al., 2010; Lucas et al., 2007). The majority of DIN occurs as nitrate, with the concentrations of the other species occurring seasonally and at significantly lower values (though they can occur at higher concentrations at times and places where nitrate is depleted). In strongly seasonal locations, biological activity depletes surface nutrient concentrations in the summer when the water column is stratified, and they are replenished with deep nitrate in the winter when the column is mixed to depth. Since ammonium and urea are immediate products of the remineralization of organic material, while nitrate is not, this led to the development of the concept of "new" and "regenerated" primary production (Dugdale and Goering, 1967). The former is production fueled by nitrate, while the latter is that fueled by recently remineralized ammonium.

This concept was used by Eppley and Peterson (1979) who noted that, should the majority of nitrate taken up by phytoplankton in surface waters stem from deeper waters (by upwelling or mixing), then "new" production should closely approximate export production. This is illustrated in Fig. 1.2, where the nitrate and ammonium (which stands in here for all "regenerated" nutrients) consumed by phytoplankton have sources that are vertically separated (deep water in the case of nitrate; shallow regeneration in the case of ammonium). In this picture, the upward flux of nitrate is balanced, at equilibrium, by a corresponding downward flux: export production. Consequently, the use of different nitrogen species by phytoplankton in the surface ocean should allow the transport of organic nitrogen, and carbon, into the ocean interior to be quantified (assuming that other inputs, such as nitrogen fixation, are relatively small). Eppley and Peterson (1979) noted this equivalence, and used it to define a metric, the f-ratio, to quantify the fraction of primary production that was exported from the surface ocean.

$$f - \text{ratio} = \frac{PP_{new}}{PP_{new} + PP_{regen}} \qquad (1.1)$$

where PP_{new} is "new" primary production fueled by nitrate consumption, and PP_{regen} is primary production fueled by "regenerated" nitrogen nutrients. However, in defining the f-ratio, Eppley recognized that surface nitrification, then believed negligible, could distort the f-ratio should it appreciably contribute to surface nitrate supply. Nonetheless, over the following decades, the f-ratio became a standard tool in oceanography for

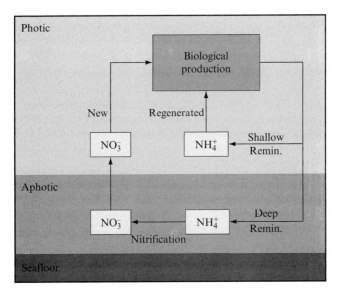

Figure 1.2 A simplified schematic diagram of the nitrogen cycle in the surface and deep ocean. Nitrate and ammonium are consumed by biological activity in surface waters as "new" and "regenerated" production. Ammonium is liberated by remineralization of organic material, but is only nitrified back to nitrate at depth. As a result, the vertical resupply of nitrate matches its surface consumption and export production. Note that the diagram omits sources of fixed nitrogen including rivers, atmospheric deposition and N_2 fixation, and sinks such as denitrification.

estimating export production, normally via the separate measurement of nitrate-fueled primary production (e.g., Lucas *et al.*, 2007).

Over the same period, measurements of nitrification have greatly improved, and there are now a number of different techniques. The most commonly used is a classic pulse-chase experiment with the ammonium (or nitrite) spiked with a ^{15}N isotope which is subsequently tracked. The labeled tracer may be quantified using either isotope ratio mass spectrometry (IRMS; e.g., Fernández, 2003) or, more recently, gas chromatography–mass spectrometry (GCMS; e.g., Clark *et al.*, 2007). As an alternative to spiking, natural isotopic abundances can be tracked instead (e.g., Sutka *et al.*, 2004). Other methods include estimating the carbon assimilated by nitrifying bacteria (e.g., Dore and Karl, 1996) and using parallel incubations one of which is poisoned to inhibit ammonium oxidation (e.g., Bianchi *et al.*, 1997).

One result of these advances in measurement techniques has been a growing pool of evidence concerning the magnitude and distribution of nitrification across the global ocean. Figure 1.3 shows the locations of open ocean studies that were included in Yool *et al.*'s (2007) synthesis study. The studies represented cover the different techniques described above and encompass a diverse range of ocean regions, including upwelling regions,

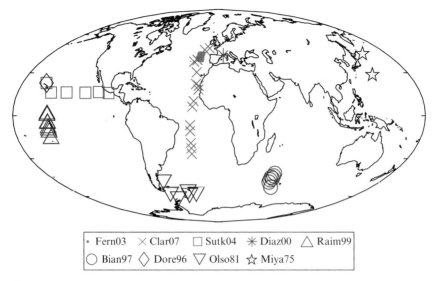

Figure 1.3 The geographical distribution of open ocean nitrification measurements synthesized in Yool et al., 2007. In chronological order, and with the number of observations in [square brackets], the studies shown are Miyazaki et al., 1975 [6]; Olson, 1981 [15]; Dore and Karl, 1996 [5]; Bianchi et al., 1997 [21]; Raimbault et al., 1999 [45]; Diaz and Raimbault, 2000 [17]; Fernández, 2003 [274]; Sutka et al., 2004 [20]; and Clark et al., 2007 [28].

oligotrophic gyres and high-nutrient low-chlorophyll (HNLC) areas such as the Southern Ocean where the biogeochemical cycle of nitrogen is slowed by the availability of the micronutrient iron. As Fig. 1.4 shows, this work found the specific rate of nitrification (d^{-1}; i.e., the rate of nitrification, mmol N m^{-3} d^{-1}, divided by substrate—ammonium or nitrite—concentration, mmol N m^{-3})[4] to vary extremely widely, in fact over several orders of magnitude. But it also shows that the rate is independent of depth, a key finding that undermines Eppley's original f-ratio definition, by suggesting that ammonium is transformed back to nitrate at relatively high rates even in the surface ocean. As per Fig. 1.5, this *qualitatively* breaks the connection between the supply of nitrate from depth and its consumption by phytoplankton in the euphotic zone, with the result that export production need not strongly correlate with the f-ratio.

---

[4] Since studies in marine biogeochemistry routinely involve multiple elemental cycles, concentrations and rates are frequently specified with the specific element identified, that is, mmol N m^{-3} for a nitrogen concentration instead of simply mmol m^{-3}. This avoids any ambiguity whenever an ecological process involving several elemental cycles is quantified. This is less of a concern for observational studies, where the nature of a measurement may dictate the element considered (e.g., ^{14}C and photosynthesis), but it can be an issue in modeling where "processes" are less narrowly defined. For instance, phytoplankton growth, which is driven by photosynthesis, can readily be considered (and is) in terms of carbon, nitrogen, or phosphorus biosynthesis.

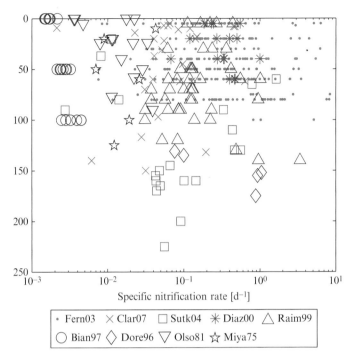

Figure 1.4 The relationship between open ocean measurements of specific nitrification rate (d^{-1}) and depth for the same studies shown in Fig. 1.3. Note that the *x*-axis scale is logarithmic to encompass the large range in nitrification rates measured.

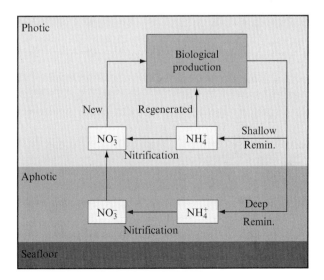

Figure 1.5 Similar to Fig. 1.2, but with surface nitrification of ammonium to nitrate include. Since surface nitrate now has two sources, the equivalence of its consumption by biological activity with export production is no longer qualitatively supported.

However, evaluation of the *quantitative* role of surface nitrification across the whole ocean and over the full annual cycle is more difficult to assess from a observational perspective. Nitrification measurements are not (yet) straightforward to make, and they are still expensive in terms of platform and time costs.[5] One avenue around these restrictions is through the use of appropriately parameterized ecosystem models. These can extrapolate from the limited observations to date and attempt to quantify the magnitude of nitrification, where it occurs both in the water column and geographically, and whether its occurrence poses a threat to the long-established oceanographic tool of the *f*-ratio.

4. Plankton Ecosystem Models

Similar to other biological systems, plankton ecosystems are modeled in a diverse number of ways, with methods largely chosen with regard to the scientific question being addressed. For instance, applications focusing on the long-term behavior of biogeochemical cycles at the global scale typically use simplified models that represent only a few key processes (e.g., Fennel et al., 2005; Tyrrell, 1999). Meanwhile, applications that require detailed (in time and/or space) simulations of a range of ecosystem components use more complex models that separately resolve a large number of processes (e.g., Allen and Clarke, 2007; Fasham et al., 1990). While nitrification is represented across this wide spectrum, attention in this chapter focuses on the latter end of shorter timescales and more detailed ecosystem models.

The origins of currently used plankton ecosystem models lie in pioneering work in the mid-twentieth century by scientists including Riley (1946), Steele (1962), and Dugdale (1967). These and other researchers provided the theoretical foundations for the later blossoming of plankton modeling with the appearance and evolution of increasingly powerful computing platforms. While some early models represented simplified nutrient–phytoplankton–zooplankton systems (NPZ; e.g., Evans and Parslow, 1985; Steele and Henderson, 1981), studies of the ocean's biological pump required more sophisticated models that additionally considered the processing and recycling of organic material. A notable and influential milestone in this regard was the development by Fasham, Ducklow, and McKelvie of a nitrogen-based model that explicitly considered the fates of particulate and dissolved organic material, and included separate pools of nitrate and ammonium for phytoplankton growth (Fasham et al., 1990).

---

[5] Note also that a "standard technique" and measurement protocol for marine nitrification has yet to be determined. So sampling issues are additionally complicated by uncertainty concerning the intercomparison of existing techniques.

This permitted simulation of the export of particulate material from the ocean's surface mixed layer, as well as direct calculation of the f-ratio. Subsequent, and ongoing, developments in plankton ecosystem modeling have included greater ecological complexity (i.e., separation of functionally distinct plankton groups; cf. Hood et al., 2006), a focus on phytoplankton cellular physiology (e.g., Fasham et al., 2006) and so-called "self-assembled" ecosystems (e.g., Follows et al., 2007).

Models typically divide the ecosystem to be modeled into a relatively small number of separately quantified "pools" of model currency. The "currency" chosen is a key biological element (e.g., nitrogen, phosphorus, or carbon) that occurs within, and is transferred between, all ecosystem components of interest. The "pools" that are modeled generally represent a highly simplified version of an ecosystem's trophic structure in which the diversity of the community is aggregated into a series of functionally similar groups such as "primary producers" and "grazers," as well as nonliving reservoirs of currency such as dissolved inorganic nutrients and particulate organic material.[6] These pools are then connected to one another by mathematical functions that dictate how currency moves through the model. The derivation (and parameterization) of these functions is complicated by the large number and diversity of the ecosystem components that they are effectively describing (i.e., a single phytoplankton pool in a model may aim to describe hundreds of functionally similar species, each of which has a large degree of intraspecific diversity). As a result, they typically stem from heuristic or empirical analyses, although there are ongoing efforts to derive them more theoretically (e.g., Baird and Suthers, 2007) and to incorporate a degree of "biodiversity" (e.g., Follows et al., 2007).

The resulting caricature of the plankton ecosystem is described by a series of coupled differential equations that prescribe its evolution in time and, in many cases, space. Each of the ecosystem "pools" is a separate state variable with one of these equations to describe its dynamics. In the simplest case, only temporal evolution of the ecosystem is considered and space is ignored (or, rather, assumed homogeneous). Mixed layer models that represent only average concentrations of state variables within the ocean's surface mixed layer are an example of this class of model (e.g., Fasham et al., 1990). While these models are inexpensive to simulate, and are instructive under certain circumstances, they are also unable to represent any processes that include spatial heterogeneity such as vertical gradients (e.g., in nutrients or phytoplankton).

---

[6] While this simplification of a complex community throws out much important biological detail, one advantage of the resulting model is a system straightforward enough to be comprehensible. The omission of detail prevents a model from capturing all of the subtle behavior of an ecosystem, but in mimicking general trends using a handful of components it becomes possible to elucidate causes and consequences that would be difficult to tease out from a more complete, and more tortuous, description of the ecosystem.

Though there is significant horizontal structure in plankton ecosystems across all scales, from the basin level structure visible in Fig. 1.1 down to submesoscale levels (e.g., Martin et al., 2008), light, nutrients, density stratification, and turbulent mixing mean that vertical space is more significant for plankton ecosystems, and 1D models are a common modeling platform (e.g., Bissett et al., 1999; Denman, 2003; Fujii et al., 2007; Martin and Pondaven, 2003, 2006; Tian et al., 2000). Through physical processes such as light attenuation, differential mixing, and gravitational sinking, 1D models create vertical heterogeneity in biological systems, and allow the simulation of phenomena including nutriclines, deep chlorophyll maxima, and the partitioning of organic matter production and its remineralization. However, by necessity, 1D models omit horizontal processes and the heterogeneity in plankton ecosystems that results from these. At the largest scale, this heterogeneity is represented by features such as oligotrophic gyres[7] and upwelling regions, but variability occurs at all scales and acts to restrict the usefulness of 1D models, particularly where horizontal transports significantly affect the integrity of 1D water columns.

Fully three dimensional general circulation models (GCMs) partially remove these restrictions, since they resolve adjacent water columns and the physical processes that link them. However, the vastly greater computational burden of 3D simulations necessitates choices concerning aspects such as spatial resolution (the scale of the physical features that can be resolved), domain size (global, basin, or regional scale), simulation duration (decadal, century, or millennial scale), and the sophistication of the ecosystem model (NPZD or a phytoplankton functional type model). Nonetheless, 3D models are a powerful tool in ecosystem modeling, and particularly in aquatic ecology where physical processes play a key role in creating and eroding the gradients in light and nutrients that regulate primary production in these systems.

A passing remark to make in the context of spatially resolved models is that it is normal for an embedded ecosystem model to be configured identically (i.e., parameter values, functional forms) throughout the model domain. For instance, specific growth or loss rate parameters would be the same in every grid cell, and differences in the resulting local population dynamics would be the outcome of interactions between different physical forcing (i.e., more/less light; deeper/shallower mixing) and different plankton abundances. Because of a lack of good data to do otherwise, ecosystem models are commonly initialized throughout the geographical domain with identical concentrations of their components, but these arbitrary initial conditions are usually quickly supplanted as model dynamics move plankton

---

[7] Oligotrophic gyres are regions that occur in the subtropics (10–40° latitude) of the Atlantic, Pacific, and Indian basins, which, through large-scale patterns of surface circulation, have very low surface nutrient concentrations and, as a result, low productivity. They can be seen as the pale magenta areas polewards of the equator in Fig. 1.1.

populations toward their equilibria. An exception to the general practice of uniform model configuration can sometimes be found in the specification of plankton models at depth. Since biological productivity is typically confined to the upper 100 m, some ecosystem models feature simplified equations at depth to avoid redundant calculations (e.g., light-driven primary production at 2000 m) that may decrease model performance for no gain (e.g., Popova *et al.*, 2006).

This section has introduced ecosystem modeling, a very large and diverse field, extremely briefly. More thorough (and practical) overviews can be found in the dedicated textbooks of Fennel and Neumann (2004) and Soetaert and Herman (2009), while Sarmiento and Gruber (2006) provide a strong coverage of biogeochemistry to contextualize modeling. Detailed treatments of particular models that serve as good introductions to ecosystem modeling, in general, can be found in Kremer and Nixon (1978; an estuarine system) and Fasham (1993; an ocean system). The reviews of Totterdell *et al.* (1993) and Hood *et al.* (2006), respectively, introduce issues of trophic resolution in ocean systems, and an overview of the current state-of-the-art in plankton ecosystem modeling.

5. MODELING NITRIFICATION

The preceding section described marine ecosystem models in terms of their history, broad structure, and the nature of the physical frameworks within which they are usually simulated. This section introduces a generic NPZD model in which nitrate and ammonium are considered separately, and in which nitrification connects them. Figure 1.6 shows a schematic

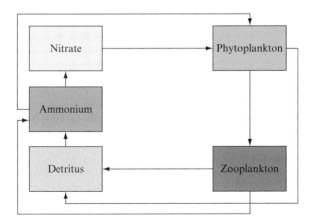

Figure 1.6 A diagrammatic representation of the plankton ecosystem model described by Eqs. (1.2)–(1.6).

representation of this model's components and the processes which connect them. While this example model omits much of the detail that current plankton ecosystem models routinely include, the structure of nutrients (N), primary producers (P), consumers (Z), and nonliving organic material (D) is still largely the core of more sophisticated treatments of plankton ecology. The simplified nature of the example here aims to minimize extraneous detail to permit closer consideration of nitrification.

In passing, it is important to note that not all models of the open ocean ecosystem directly consider nitrification at all. Since most phytoplankton (but potentially not all; cf. Moore *et al.*, 2002) can utilize both nitrate and ammonium for biosynthesis, many modeling applications simply do not require the separate treatment of these molecules, and instead consider only their sum (i.e., DIN; e.g., Moore *et al.*, 2004). And ecosystem models that are ground in a non-nitrogen currency (e.g., Le Quéré *et al.*, 2005) treat nitrification, and wider nitrogen cycle processes, even more implicitly. However, some other models do include both nitrate and ammonium, but nitrification appears only implicitly. For instance, in the 0D model of Fasham *et al.* (1990), mixed layer ammonium is not nitrified, but since mixed layer nitrate is replenished from a fixed reservoir of deep nitrate, nitrification implicitly occurs below the mixed layer. Similarly, in the 1D model of Martin and Pondaven (2003), nitrate and ammonium concentrations are "relaxed" toward observed profiles of the nutrients, with the result that nitrification occurs implicitly.

The partial differential equations below describe the fluxes of nitrogen currency (mmol N m^{-3}) between the five state variables represented: phytoplankton, P; zooplankton, Z; detritus, D; nitrate, N; and ammonium, A. The intermediate species of nitrite is omitted here purely for simplicity; it is assumed that its rate of oxidation to nitrate is at least equal to that of the oxidation of ammonium to nitrite (Yool *et al.*, 2007, found this to be the case for 85% of the relevant measurements they synthesized). Each equation consists of a series of labeled terms that relate to specific ecosystem processes, such as primary production (= phytoplankton growth), grazing, metabolic losses, and remineralization. Table 1.1 presents a summary description of these. These terms may be functions of ambient properties such as temperature or irradiance, and are often nonlinear, but the exact details are unimportant here so are omitted. Exceptions are the separate Michaelis–Menten terms[8] for nitrate and ammonium uptake by phytoplankton, which are shown here because of their connection to the *f*-ratio. Here, an inhibition of nitrate uptake in the presence of ammonium is included to parameterize the preference of the latter for phytoplankton (Dortch, 1990; Fasham *et al.*, 1990; Wroblewski,

---

[8] A Michaelis–Menten term is a rectangular hyperbolic function that originated within the field of enzyme kinetics, but which is widely used within ecology as a heuristic shorthand for processes whose rate saturates with the concentration of some substrate. In plankton ecosystem models it is most commonly used in the context of nutrient limitation of phytoplankton growth (see Eqs. (1.7) and (1.8)).

Table 1.1 Summary table of processes and parameters shown in Eqs. (1.2)–(1.8)

J	Light limitation of phytoplankton growth
G_P	Zooplankton grazing on phytoplankton
μ_P	Phytoplankton nongrazing loss rate
β	Zooplankton assimilation efficiency
μ_{Z1}	Density-independent zooplankton loss rate
μ_{Z2}	Density-dependent zooplankton loss rate
μ_D	Detrital remineralization rate
S	Detrital sinking velocity
Λ_{nitrif}	Nitrification rate
k_N	Nitrate uptake half-saturation concentration
k_A	Ammonium uptake half-saturation concentration
ψ	Nitrate uptake ammonium inhibition factor

Some represent single parameters (e.g., μ_P, S, k_N), while others are nonlinear functions that may contain several parameters (e.g., J, G_P).

1977). Some models favor a multiplicative parameterization in which the affinity of phytoplankton for ammonium (expressed via a lower half-saturation constant; see later) is greater (e.g., Fasham, 1995), while others favor a "most-limiting" relationship where growth is determined solely by the most-limiting nutrient, also known as Liebig's Law (cf. O'Neill et al., 1989).

$$\frac{\partial P}{\partial t} = +\underbrace{[J \cdot Q_N \cdot P]}_{\text{"new" production}} + \underbrace{[J \cdot Q_A \cdot P]}_{\text{"reg" production}} - \underbrace{[G_P]}_{\text{grazing loss}} - \underbrace{[\mu_P \cdot P]}_{\text{metabolic loss}} \quad (1.2)$$

$$\frac{\partial Z}{\partial t} = +\underbrace{[\beta \cdot G_P]}_{\text{grazing}} - \underbrace{[\mu_{Z1} \cdot Z]}_{\text{metabolic loss}} - \underbrace{[\mu_{Z2} \cdot Z^2]}_{\text{predation}} \quad (1.3)$$

$$\frac{\partial D}{\partial t} = +\underbrace{[(1-\beta) \cdot G_P]}_{\text{Z egestion}} + \underbrace{[\mu_P \cdot P]}_{\text{P loss}} + \underbrace{[\mu_{Z2} \cdot Z^2]}_{\text{Z loss}} - \underbrace{[\mu_D \cdot D]}_{\text{remineralization}} - \underbrace{[S \cdot D]}_{\text{sinking}} \quad (1.4)$$

$$\frac{\partial N}{\partial t} = -\underbrace{[J \cdot Q_N \cdot P]}_{\text{"new" production}} + \underbrace{[\Lambda_{nitrif}(A)]}_{\text{nitrification}} \quad (1.5)$$

$$\frac{\partial A}{\partial t} = -\underbrace{[J \cdot Q_A \cdot P]}_{\text{"reg" production}} + \underbrace{[\mu_{Z1} \cdot Z]}_{\text{Z loss}} + \underbrace{[\mu_D \cdot D]}_{\text{D remineralization}} - \underbrace{[\Lambda_{nitrif}(A)]}_{\text{nitrification}} \quad (1.6)$$

where,

$$Q_N = -\underbrace{\left[\frac{N^{-\Psi A}}{k_N + N}\right]}_{NO_3^- \text{ uptake}} \qquad (1.7)$$

$$Q_A = \underbrace{\left[\frac{A}{k_A + A}\right]}_{NH_4^+ \text{ uptake}} \qquad (1.8)$$

In the equations above, nitrification is shown as an unspecified function of ammonium concentration only, $\Lambda_{nitrif}(A)$. In the modeling literature to date, there have been a large number of different formulations of Λ_{nitrif}, some of which include other factors such as ambient light, depth, or oxygen concentration. The following list of functions is not exhaustive but aims to give an overview of these terms. In each case, an example use of the function is indicated below its definition. In all cases, $\Lambda_{nitrif}(A)$ is a flux of ammonium in units of mmol N m^{-3} d^{-1}.

$$\Lambda_{nitrif}(A) = \lambda_i A \qquad (1.9)$$

Anderson and Pondaven (2003)

$$\Lambda_{nitrif}(A) = \begin{cases} 0 & A \leq 0.5 \, \text{mmol N m}^{-3} \\ \lambda_j & \text{otherwise} \end{cases} \qquad (1.10)$$

Walsh and Dieterle (1994)

$$\Lambda_{nitrif}(A) = \lambda_j \frac{A}{k_\lambda + A} \qquad (1.11)$$

Bissett *et al.* (1999)

$$\Lambda_{nitrif}(A) = \lambda_i \exp\{k_T T\} A \qquad (1.12)$$

Fujii *et al.* (2007)

$$\Lambda_{nitrif}(A) = \lambda_i \frac{O_2}{k_{O_2} + O_2} A \qquad (1.13)$$

Fennel *et al.* (2005)

$$\Lambda_{\text{nitrif}}(A) = \lambda_i \exp\{k_T T\} \frac{O_2}{k_{O_2} + O_2} A \quad (1.14)$$

Soetaert *et al.* (2001)

$$\Lambda_{\text{nitrif}}(A) = \begin{cases} \frac{1}{x}\lambda_i A & \text{in euphotic zone} \\ \lambda_i A & \text{below euphotic zone} \end{cases} \quad (1.15)$$

Jiang *et al.* (2003)

$$\Lambda_{\text{nitrif}}(A) = \lambda_i \frac{z^2}{z_{ox}^2 + z^2} A \quad (1.16)$$

Denman (2003)

$$\Lambda_{\text{nitrif}}(A) = \lambda_j \left(1 - \frac{I_z - I_{\min}}{I_z - I_{\min} + K_I}\right) \quad (1.17)$$

McClain *et al.* (1999)

$$\Lambda_{\text{nitrif}}(A) = \begin{cases} \lambda_i 0.01 \frac{I_0}{I_z} A & \text{in euphotic zone} \\ \lambda_i A & \text{below euphotic zone} \end{cases} \quad (1.18)$$

Pätsch and Kühn (2008)

$$\Lambda_{\text{nitrif}}(A) = \begin{cases} 0 & \text{in euphotic zone} \\ \lambda_i \frac{(0.1 I_0) - I_z}{0.1 I_0} A & \text{below euphotic zone} \end{cases} \quad (1.19)$$

Tian *et al.* (2000)

Equation (1.9) shows one of the most straightforward functions in which the transformation of ammonium to nitrate occurs as a fixed specific rate, $\lambda_i(d^{-1})$,[9] multiplied by the concentration of ammonium (mmol N m^{-3}). Equation (1.10) shows an even simpler function in which nitrification occurs at a fixed absolute rate, λ_j (mmol N m^{-3} d^{-1}), until ammonium falls below a threshold concentration (0.5 mmol N m^{-3}) at which point

---

[9] Note that while the parameters λ_i and λ_j are used throughout Eqs. (1.9) to (1.19), their exact values are not necessarily identical, either between the different equations listed above or between different modeling studies using the same equation.

nitrification completely ceases. A similar function is shown in Eq. (1.11), but here the maximum absolute rate of nitrification is scaled by a conventional Michaelis–Menten function to decrease this rate to zero as ammonium is consumed. This function includes a half-saturation constant, k_λ (mmol N m^{-3}), to specify the ammonium concentration at which nitrification occurs at half the maximum value.

Switching attention to other factors, Eq. (1.12) modifies Eq. (1.9) by introducing an exponential relationship with ambient temperature, T (cf. the standard Q_{10} factor in biological systems). In this, the higher the temperature of seawater, the higher the resulting specific rate of nitrification. Equation (1.13) instead focuses on the availability of oxygen, O_2, as a controlling factor in nitrification, and introduces a Michaelis–Menten function (with half-saturation constant k_{O_2}) that decreases the rate of oxidation of ammonium with declining oxygen concentrations. Equation (1.14) combines these factors by linking nitrification to both temperature and oxygen availability. The linkage of nitrification to oxygen is less significant in surface waters, where oxygen is generally at high concentrations, but becomes important hypoxic environments such as oxygen minimum zones (OMZs; e.g., Fernández et al., 2009) and seafloor sediments (e.g., Soetaert et al., 2001).

Equations (1.15)–(1.19) present a range of functions in which the rate of nitrification is related to depth, z, or to the submarine light field, I_z. Although the formulations differ, in each case the effect is to enhance nitrification at depth, but decrease it near the ocean surface. Equation (1.15) is the most straightforward in this regard, and makes euphotic zone nitrification a simple fraction ($1/x$; $x = 4$ in Jiang et al., 2003) of that deeper in the water column. Equation (1.16) uses a Michaelis–Menten function in which the square of depth, z^2, increases nitrification with depth. Equations (1.17)–(1.19) use ambient light, I_z, together with terms such as sea surface light, I_0, to increase nitrification with declining light.

Note, however, that there is little data to justify this diversity in the form by which nitrification is modeled. Some researchers have selected a light- or depth-dependent form because of work that has suggested limited nitrification in surface waters (e.g., Horrigan and Springer, 1990; Lipschultz et al., 2002), while many have found that some formal representation of nitrification is required for modeled nitrate profiles to match observations (e.g., Denman, 2003; Kawamiya et al., 1995; Lévy et al., 1998; Mongin et al., 2003). In synthesizing open ocean measurements, Yool et al. (2007) found little to guide the formulation of a function for nitrification. The commonly assumed inverse relationship with light was not supported by observations, and attempts to find geographical or seasonal correlations were also unsuccessful or inconclusive. For this reason, Yool et al. (2007) made use of one of the simplest functions (Eq. (1.9)), and used sensitivity analyses on the value

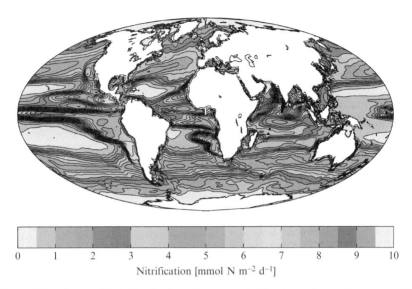

Figure 1.7 Geographical distribution of average nitrification estimated in the modeling study of Yool *et al.* (2007). Nitrification here is vertically integrated and averaged for the period 2000–2004 inclusive. Nitrification in mmol N m^{-2} d^{-1}. Since nitrification is "fueled" by ammonium derived from the remineralization of organic matter, its distribution necessarily correlates with the production of this matter, ultimately via photosynthesis. As such, simulated nitrification broadly follows estimated primary production in Fig. 1.1.

of the specific rate used to assess the significance of nitrification at the global scale (values one order of magnitude smaller and larger than the synthesis dataset median). Figure 1.7 shows the simulated distribution of nitrification that resulted from the use of this function and a default specific nitrification rate of 0.20 d^{-1}. Though highly simplified, this approach allowed Yool *et al.* (2007) to conclude that, even with a specific rate much lower than most observations, surface nitrification breaks the relationship between export production and nitrate use by phytoplankton.

Finally, although data collected to date has not resolved how nitrification is best modeled, that it is a biological process involving the oxidation of ammonium, it may be expected that some temperature- and oxygen-dependent functionality will ultimately be used in ecosystem models. Nonetheless, the huge range in observed nitrification rates, and their lack of obvious relationships with other factors, suggests that all currently used model functions are highly simplified, and that future observational and experimental investigations are crucial.

6. OPERATIONAL CONSIDERATIONS

The preceding sections have provided an overview of open ocean plankton ecosystems, in general, with some specific examples of how the process of nitrification has been implemented within them. In terms of how these models are simulated, a few general points can be made. Note that the following points are not specific to either models that include nitrification, nor to the modeled process itself. Usually, nitrification is just one of many biological processes included within a model, and the following considerations apply to all such processes.

First, the physical framework in which an ecosystem model is embedded is liable to dictate many aspects of a simulation. For instance, a physical model will determine how space is represented and how it is discretized into a grid. In the case of 0D models,[10] since space is not considered explicitly, it does not organize the ecosystem model. However, in 1D and 3D models, the underlying physical model will divide the spatial domain into a finite number of locations with explicit sizes. For example, the study of Yool *et al.* (2007) used a 3D model in which vertical space was divided into 66 levels with thicknesses that increased with depth, and in which horizontal space was divided into $\sim 1°$ latitude by $1°$ longitude regions. Furthermore, since different physical processes come into play at different scales, physical models tend to be parameterized for particular grid dimensions and cannot usually be altered without careful consideration of the resulting associated physical changes. For instance, the resolution used by Yool *et al.* (2007) is fine enough to resolve large-scale oceanographic features such as oligotrophic gyres and western boundary currents, but too coarse to resolve mesoscale features such as fronts and eddies. These so-called subgrid-scale features have sizes close to, or below, that of individual model grid cells, and simply cannot be represented. Since omitting such features can have deleterious effects on the quality of a simulation, some subgrid-scale processes are included via implicit representations, of which the Gent-McWilliams eddy parameterization is probably the most well known (Gent and McWilliams, 1990).[11]

---

[10] In oceanography, 0D models are usually mixed layer models that only consider the concentrations of ecosystem state variables within the upper mixed layer. This layer is assumed to be completely homogenized by turbulent mixing such that no vertical structure occurs. 0D models usually include a seasonal cycle in the depth of this mixed layer to reflect variation in water column stratification and the intensity of wind mixing. Below the mixed layer, concentrations of the modeled state variables are either assumed zero, or at a fixed value that acts to "reset" concentrations during (typically) deep winter mixing. Example 0D ecosystem models include Evans and Parslow (1985) and Fasham *et al.* (1990).

[11] Of course, in reality all ocean models are limited to some degree by subgrid-scale processes, since there are always smaller-scale processes that are omitted by grid resolution. However, since smaller features are generally less quantitatively important, increasing model resolution, and, thus, resolving more features, typically improves the realism of simulations.

Second, and on a related point, the underlying physical model may also dictate the pace of temporal evolution in a simulation. Numerical solution of the water circulation associated with the physical framework generally sets a time-step duration that the biological components follow, and if this period is ill suited to the ecosystem model, there can be theoretical and practical consequences. Since ecosystem models are generally nonlinear, long time-steps can significantly impact the accuracy of the simulation, and in extreme cases can lead to unrealistic results. For example, when running a NPZD ecosystem in a 3D model with a time-step of 1 d (which was not problematic for the underlying physical model), Palmer and Totterdell (2001) found that phytoplankton growth could consume more nutrient than existed within a grid box in a single day. This problem occurred because the instantaneous rate of nutrient consumption calculated at the start of the day could send nutrient concentrations below zero when applied across the full day. Ideally, such issues should be resolved by shorter time-steps (cf. Courant–Friedrichs–Levy conditions), but this may not always be possible. In the case of Palmer and Totterdell (2001), the solution was to invoke an implicit time-stepping function for the biological equations that checked for, and prevented, nutrient concentrations going below zero by adjusting phytoplankton growth.

An exception, again, lies with 0D models, where physics is essentially implicit (i.e., a single, perfectly mixed, homogeneous layer) and the ecosystem dynamics are the sole determinant of simulation time-stepping. In these models, adaptive time-stepping, in which the length of the time-step is chosen to keep the model solution within a certain tolerance, is a desirable option. This allows these models to minimize (an already low) computational burden, while producing a simulation with a known numerical accuracy. However, since this is not an option for other frameworks, where physical processes and scaling may not allow time-step duration to be selected dynamically, comparison of experiments in which different time-step durations are used can determine the magnitude of any errors introduced to the simulated solution of the biological model.

Third, another consideration is the numerical method used to solve the biological equations. As with other factors, this is also normally tied to the physical framework underlying the ecosystem model. In the case of 0D models, their lack of physics means that only the biological dynamics need be considered and relatively sophisticated methods of numerical integration can be used since computational costs are low. As with earlier considerations, the methods used to numerically integrate 1D and 3D models are usually selected on the basis of their solution of the underlying physics. Much as with model time-step, the principal consideration here is that the model solution should be accurate, and not dependent on details of the numerical method used (Butcher, 2008).

Finally, in modeling of all varieties, and much as with experimental or observational science, care should be exercised to ensure that model simulations really are doing what they appear, or claim, to be doing. Setting up and running an ecosystem model usually requires that a long chain of actions are accurately performed, thus providing ample opportunity for errors to creep in. However, there are practical approaches that can be taken to minimize or detect these. For instance, prior to coding a model, checking that the units of its various state variables, functions, and parameters actually multiply through to the rate of change expected in the equations. Another straightforward check concerning parameters is simply to ask the model to report their values during a run in order that errors in initialization or handling within the model can be detected (i.e., are parameters assigned the values expected?). When a model is running, a critical check is that it conserves total currency across its state variables, in order to ensure that there are no "leaks" in the model (i.e., places where currency is gained or lost because model terms are specified incorrectly or even missing). Such inventory checks can be included within the model itself, but they are often best done independently of the model to detect compensating errors that otherwise mask one another. Note, however, that the modeling framework itself can also be responsible for nonconservation of ecosystem currency because of inaccuracies (or even errors) in physical transport calculations. If suspected, this source of nonconservation can be separated from that in biological equations by introducing an additional state variable, but one without any biological fluxes. Should model physics rather than model biology be responsible for currency nonconservation, this fixed inventory variable should uncover it.

As with the earlier section on ecosystem modeling, this has been a brief overview of only a few considerations concerning operational aspects of modeling. More complete treatments can be found in the dedicated textbooks mentioned previously.

7. FUTURE MODELING

Earlier sections have dealt with how nitrification is currently represented in ecosystem models. Here, we consider how this representation may be improved into the future and, based on the cartoon ecosystem model introduced earlier, sketch out how a more mechanistic representation of nitrification may develop. The approach outlined here is analogous to existing models used to describe sediment ecosystems (e.g., Holstein and Wirtz, 2009), but the aim here is to focus on expansion to current open ocean ecosystem models.

First, as the techniques for measuring it continue to develop (and, potentially, to standardize), nitrification is likely to become a more easily, and more widely, measured process. This expansion will allow the buildup of a better picture of its magnitude, where and when it occurs, and which environmental and biological factors it most strongly covaries with. Identifying these factors may help sift the huge variability in nitrification rate shown in Figure 1.4, and point to how it is regulated in the environment. Determining the controlling factors for nitrification will also be key to projecting its future under continuing anthropogenic forcing. For instance, Hutchins *et al.* (2009) have identified a putative "bottleneck" in the nitrogen cycle caused by the influence of ocean acidification on nitrification.

Aside from advances in the quantity, quality, and diversity of field observations, experimentation with cultured ammonia-oxidizing organisms may allow us to more directly characterize nitrification and to develop mechanistic ecosystem models that robustly simulate its distribution and significance. In particular, recent work by Martens-Habbena *et al.* (2009) with the first successfully cultured strain of ammonia-oxidizing archaea (AOA; "*Candidatus* Nitrosopumilus maritimus" strain SCM1), has demonstrated that nitrification can occur both at ammonium concentrations and at rates that potentially allow nitrifiers to successfully compete with marine phytoplankton. Together with earlier work on cultured ammonia-oxidizing bacteria (AOB) strains (e.g., Bollmann *et al.*, 2002; Ward, 1986), and with likely future advances in understanding and characterizing the physiology of a more diverse range of nitrifiers, it is possible to crudely sketch a future structure for nitrification in marine ecosystem models.

In Fig. 1.6 (and in Eqs. (1.5) and (1.6)), nitrification appears as a flux between the ammonium and nitrate state variables. Advances in quantifying nitrification, and in correlating it more completely with environmental variables, will allow us to better specify a function that describes this flux. But advances in understanding the ecophysiology of the organisms that are responsible for nitrification opens up the possibility of an explicit representation. Much as with other ecosystem components responsible for major elemental fluxes, such as phytoplankton, formally representing nitrifiers opens the door to investigations with a population dynamics focus. Since availability of the substrate for nitrification, ammonium, varies seasonally through much of the open ocean, the abundance of nitrifiers will (presumably) also vary, and capturing this behavior may require a different sort of approach to that of Eqs. (1.9)–(1.19). From this viewpoint, Fig. 1.8 instead presents an alternative model structure in which nitrifying microbes are represented by a separate state variable, M, through which ammonium is processed into nitrate (nitrite is again omitted here for simplicity). These nitrifiers compete with model phytoplankton for ammonium, and experience similar pressures from grazing zooplankton. Equations (1.20)–(1.25) describe this new structure, with Eq. (1.21) specifically describing the model nitrifiers.

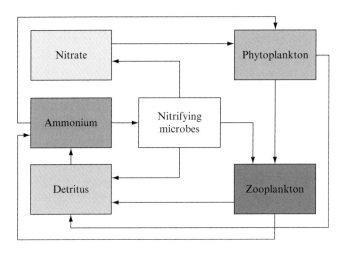

Figure 1.8 A diagrammatic representation of the plankton ecosystem model described by Eqs. (1.20)–(1.25). Whereas nitrification is represented by a flux in Fig. 1.6, here it involves an explicit "nitrifying microbes" state variable (see Eq. (1.21)).

$$\frac{\partial P}{\partial t} = + \underbrace{[J \cdot Q_N \cdot P]}_{\text{"new" production}} + \underbrace{[J \cdot Q_A \cdot P]}_{\text{"reg" production}} - \underbrace{[G_P]}_{\text{grazing loss}} - \underbrace{[\mu_P \cdot P]}_{\text{metabolic loss}} \quad (1.20)$$

$$\frac{\partial M}{\partial t} = + \underbrace{[V_M \cdot Q_M \cdot M]}_{\text{nitrifier growth}} - \underbrace{[G_M]}_{\text{grazing loss}} - \underbrace{[\mu_M \cdot M]}_{\text{metabolic loss}} \quad (1.21)$$

$$\frac{\partial Z}{\partial t} = + \underbrace{[\beta \cdot (G_P + G_M)]}_{\text{grazing}} - \underbrace{[\mu_{Z1} \cdot Z]}_{\text{metabolic loss}} - \underbrace{[\mu_{Z2} \cdot Z^2]}_{\text{predation}} \quad (1.22)$$

$$\frac{\partial D}{\partial t} = + \underbrace{[(1-\beta) \cdot (G_P + G_M)]}_{\text{Z egestion}} + \underbrace{[\mu_P \cdot P]}_{\text{P loss}} + \underbrace{[\mu_M \cdot M]}_{\text{M loss}} + \underbrace{[\mu_{Z2} \cdot Z^2]}_{\text{Z loss}} - \underbrace{[\mu_D \cdot D]}_{\text{remineralization}} - \underbrace{[S \cdot D]}_{\text{sinking}} \quad (1.23)$$

$$\frac{\partial N}{\partial t} = - \underbrace{[J \cdot Q_N \cdot P]}_{\text{"new" production}} + \underbrace{[(\sigma - 1) \cdot V_M \cdot Q_M \cdot M]}_{\text{nitrification}} \quad (1.24)$$

$$\frac{\partial A}{\partial t} = - \underbrace{[J \cdot Q_A \cdot P]}_{\text{"reg" production}} - \underbrace{[\sigma \cdot V_M \cdot Q_M \cdot M]}_{\text{nitrifier consumption}} + \underbrace{[\mu_{Z1} \cdot Z]}_{\text{Z loss}} + \underbrace{[\mu_D \cdot D]}_{\text{D remineralization}} \quad (1.25)$$

where

$$Q_M = \underbrace{\left[\frac{A}{k_M + A}\right]}_{NH_4^+ \text{ uptake}} \quad (1.26)$$

The equations also introduce a number of additional parameters: V_M, nitrifier maximum growth rate (d^{-1}); k_M, Nitrifier ammonium uptake half-saturation concentration (mmol N m^{-3}); σ, Nitrifier ammonium uptake: assimilation ratio (–); μ_M, Nitrifier metabolic loss rate (d^{-1}).

In this model, the nitrifier population grows at a maximum rate of V_M and, like phytoplankton, this growth rate is limited by the availability of ammonium via a Michaelis–Menten term with a half-saturation constant, k_M. Since ammonium is oxidized to provide energy for biosynthesis, each mol of nitrifier growth requires many more mols, σ, of ammonium uptake, of which the majority, ($\sigma - 1$), is returned to the nitrate pool. Finally, also like the phytoplankton, the nitrifiers experience losses to metabolic processes, μ_M.

As well as making the sweeping assumption that nitrifiers can be considered as just another bacterioplankton species that experiences the same grazing and metabolic pressures as phytoplankton, this model outline entirely neglects the separation of nitrification into two stages mediated by different biological communities. It also omits any limitation on nitrifier growth other than ammonium, and disregards any intracellular processes that might act to decouple ammonium uptake and utilization from biosynthesis. Nonetheless, by explicitly representing nitrifiers as a state variable with a standing stock, by placing them in competition with phytoplankton, and by making them a resource that other trophic levels can consume, such a framework may help elucidate the factors that regulate this key nitrogen cycle process.

However, as an ecosystem model—a rarefied, theoretical construct—it is ultimately only as good as the observations, experiments, and biological understanding that underpin it.

8. SUMMARY

- Nitrification plays a central role in governing the use of nitrate ("new") and ammonium ("regenerated") by primary producers in the surface ocean, with an important implication for estimates of carbon export based around the f-ratio.
- Plankton ecosystem models are a key tool used to understand ocean biogeochemical cycles in the present, and to investigate how they may change in the future in response to environmental changes such as global warming and ocean acidification.

- There is a wide diversity in plankton ecosystem models, in the physical frameworks in which they are used, and in their representation of nitrification, but observational data to date has neither strongly constrained these models nor indicated which factors are most important in its regulation.
- Because of their relative simplicity, current models of nitrification present comparable technical challenges to those of other ecosystem processes.
- Advances in understanding of the physiology of ammonia-oxidizing organisms, driven in part by increasing success in culturing, will allow the development and parameterization of ecosystem models in which nitrifiers appear as dynamic variables, through which their wider spatio-temporal role may be explored.

ACKNOWLEDGMENTS

AY would like to thank Camila Fernández I., Darren Clark, and, especially, Adrian Martin for their expertise and assistance with modeling open ocean nitrification. AY is particularly grateful to Martin Klotz and the organizing committee of the First International Conference on Nitrification (ICoN1; University of Louisville, Kentucky, USA; 5–10 July 2009) for their efforts in bringing the diverse nitrification community together and in making ICoN1 an unqualified success.

This work was financially supported by a UK National Environment Research Council (NERC) standard grant (NE/C00387X/1), and by the NERC Oceans 2025 Strategic Marine Science Programme.

REFERENCES

Allen, J. I., and Clarke, K. R. (2007). Effects of demersal trawling on ecosystem functioning in the North Sea: A modelling study. *Mar. Ecol. Prog. Ser.* **336,** 63–75.

Anderson, T. R., and Pondaven, P. (2003). Non-redfield carbon and nitrogen cycling in the Sargasso Sea: Pelagic imbalances and export flux. *Deep Sea Res. I* **50,** 573–591.

Armstrong, R. A., Lee, C., Hedges, J. I., Honjo, S., and Wakeham, S. B. (2002). A new, mechanistic model for organic carbon fluxes in the ocean: Based on the quantitative association of POC with ballast minerals. *Deep Sea Res. II* **49,** 219–236.

Baird, M. E., and Suthers, I. M. (2007). A size-resolved pelagic ecosystem model. *Ecol. Modell.* **203,** 185–203.

Beardall, J., and Raven, J. A. (2004). The potential effects of global climate change in microalgal photosynthesis, growth and ecology. *Phycologia* **43,** 31–45.

Behrenfeld, M. J., and Falkowski, P. G. (1997). Photosynthetic rates derived from satellite-based chlorophyll concentration. *Limnol. Oceanogr.* **42,** 1–20.

Bianchi, M., Feliatra, F., Tréguer, P., Vincedeau, M.-A., and Morvan, J. (1997). Nitrification rates, ammonium and nitrate distribution in upper layers of the water column and in sediments of the Indian sector of the Southern Ocean. *Deep Sea Res. II* **44,** 1017–1032.

Bissett, W. P., Walsh, J. J., Dieterle, D. A., and Carder, K. L. (1999). Carbon cycling in the upper waters of the Sargasso Sea: I. Numerical simulation of differential carbon and nitrogen fluxes. *Deep Sea Res. I* **46,** 205–269.

Bollmann, A., Bär-Gilissen, M.-J., and Laanbroek, H. J. (2002). Growth at low ammonium concentrations and starvation response as potential factors involved in niche differentiation among ammonia-oxidising bacteria. *Appl. Environ. Microbiol.* **68**, 4751–4757.

Boyce, D. G., Lewis, M. R., and Worm, B. (2010). Global phytoplankton decline over the past century. *Nature* **466**, 591–596, 10.1038/nature09268.

Bradley, P. B., Lomas, M. W., and Bronk, D. A. (2010). Inorganic and organic nitrogen use by phytoplankton along Chesapeake Bay, measured using a flow cytometric sorting approach. *Estuar. Coast.* **33**, 971–984.

Buesseler, K. O., Lamborg, C. H., Boyd, P. W., *et al.* (2007). Revisiting carbon flux through the ocean's twilight zone. *Science* **316**, 567–570, 10.1126/science.1137959.

Butcher, J. C. (2008). Numerical Methods for Ordinary Differential Equations. 2nd edn. John Wiley & Sons Ltd, UK, 978-0-470-72335-7.

Casciotti, K. L., Trull, T. W., Glover, D. M., and Davies, D. (2008). Constraints on nitrogen cycling at the subtropical North Pacific Station ALOHA from isotopic measurements of nitrate and particulate nitrogen. *Deep Sea Res. II* **55**, 1661–1672.

Clark, D. R., Rees, A. P., and Joint, I. (2007). A method for the determination of nitrification rates in oligotrophic marine seawater by gas chromatography/mass spectrometry. *Mar. Chem.* **103**, 84–96.

Denman, K. L. (2003). Modelling planktonic ecosystems: Parameterising complexity. *Prog. Oceanogr.* **57**, 429–452.

Diaz, F., and Raimbault, P. (2000). Nitrogen regeneration and dissolved organic nitrogen release during spring in a NW Mediterranean coastal zone (Gulf of Lions): implications for the estimation of new production. *Mar. Ecol. Prog. Ser.* **197**, 51–65.

Dore, J. E., and Karl, D. M. (1996). Nitrification in the euphotic zone as a source for nitrite, nitrate, and nitrous oxide at Station ALOHA. *Limnol. Oceanogr.* **41**, 1619–1628.

Dortch, Q. (1990). The interaction between ammonium and nitrate uptake in phytoplankton. *Mar. Ecol. Prog. Ser.* **61**, 183–201.

Dugdale, R. C. (1967). Nutrient limitation in the sea: Dynamics, identification, and significance. *Limnol. Oceanogr.* **12**, 685–695.

Dugdale, R. C., and Goering, J. J. (1967). Uptake of new and regenerated forms of nitrogen in primary production. *Limnol. Oceanogr.* **12**, 196–206.

Eppley, R. W., and Peterson, B. J. (1979). Particulate organic matter flux and planktonic new production in the deep ocean. *Nature* **282**, 677–680.

Evans, G. T., and Parslow, J. S. (1985). A model of annual plankton cycles. *Biol. Oceanogr.* **3**, 327–347.

Fasham, M. J. R. (1995). Variations in the seasonal cycle of biological production in subarctic oceans: A model sensitivity analysis. *Deep-Sea Res. I* **42**, 1111–1149.

Fasham, M. J. R. (1993). Modelling the marine biota trophic resolution. *In* "The Global Carbon Cycle," (M. Heimann, ed.), pp. 457–504. Springer-Verlag, USA.

Fasham, M. J. R., Ducklow, H. W., and McKelvie, S. M. (1990). A nitrogen-based model of plankton dynamics in the oceanic mixed layer. *J. Mar. Res.* **48**, 591–639.

Fasham, M. J. R., Flynn, K. J., Pondaven, P., Anderson, T. R., and Boyd, P. W. (2006). Development of a robust marine ecosystem model to predict the role of iron in biogeochemical cycles: A comparison of results for iron-replete and iron-limited areas, and the SOIREE iron-enrichment experiment. *Deep Sea Res. I* **53**, 333–366.

Fennel, W., and Neumann, T. (2004). Introduction to the Modelling of Marine Ecosystems. Elsevier, UK, 0-444-51702-2.

Fennel, K., Follows, M., and Falkowski, P. G. (2005). The co-evolution of the nitrogen, carbon and oxygen cycles in the Proterozoic ocean. *Am. J. Sci.* **305**, 526–545.

Fernández, I. C. (2003). Cycle de l'azote et production primaire dans l'Atlantique Nord-Est: Suivi saisonnier et influence de la meso échelle. Ph.D. thesis, Univ. de la Méditerranée, Marseille, France, 331pp.

Fernández, C., Farias, L., and Alcaman, M. E. (2009). Primary production and nitrogen regeneration processes in surface waters of the Peruvian upwelling system. *Prog. Oceanogr.* **83,** 159–168.

Follows, M. J., Dutkiewicz, S., Grant, S., and Chisholm, S. W. (2007). Emergent biogeography of microbial communities in a model ocean. *Science* **315,** 1843–1846.

Fujii, M., Yamanaka, Y., Nojiri, Y., Kishi, M. J., and Chai, F. (2007). Comparison of seasonal characteristics in biogeochemistry among the subarctic North Pacific stations described with a NEMURO-based ecosystem model. *Ecol. Modell.* **202,** 52–67.

Gent, P. R., and McWilliams, J. C. (1990). Isopycnal mixing in ocean circulation models. *J. Phys. Oceanogr.* **20,** 150–155.

Gruber, N., Sarmiento, J. L., and Stocker, T. F. (1996). An improved method for detecting anthropogenic CO_2 in the oceans. *Global Biogeochem. Cycles* **10,** 809–837.

Holstein, J. M., and Wirtz, K. W. (2009). Sensitivity analysis of nitrogen and carbon cycling in marine sediments. *Estuar. Coast. Shelf Sci.* **82,** 632–644.

Hood, R. R., Laws, E. A., Armstrong, R. A., et al. (2006). Pelagic functional group modeling: Progress, challenges and prospects. *Deep Sea Res. II* **53,** 459–512.

Horrigan, S. G., and Springer, A. L. (1990). Oceanic and estuarine ammonium oxidation—Effect of light. *Limnol. Oceanogr.* **35,** 479–482.

Hutchins, D. A., Mulholland, M. R., and Fu, F. (2009). Nutrient cycles and marine microbes in a CO_2-enriched ocean. *Oceanography* **22,** 128–145.

Jiang, M.-S., Chai, F., Dugdale, R. C., Wilkerson, F. P., Peng, T.-H., and Barber, R. T. (2003). A nitrate and silicate budget in the equatorial Pacific Ocean: A coupled physical-biological model study. *Deep Sea Res. II* **50,** 2971–2996.

Kawamiya, M., Kishi, M. J., Yamanaka, Y., and Suginohara, N. (1995). An ecological-physical coupled model applied to Station Papa. *J. Oceanogr.* **51,** 635–664.

Kremer, J. N., and Nixon, S. W. (1978). A Coastal Marine Ecosystem. Springer-Verlag, Germany.

Le Quéré, C., Harrison, S. P., Prentice, C., et al. (2005). Ecosystem dynamics based on plankton functional types for global ocean biogeochemistry models. *Glob. Chang. Biol.* **11,** 2016–2040.

Levitus, S., Antonov, J., and Boyer, T. (2005). Warming of the world ocean, 1955–2003, *Geophys. Res. Lett.* **32,** L02604, doi:10.1029/2004GL021592.

Lévy, M., Mémery, L., and André, J.-M. (1998). Simulation of primary production and export fluxes in the Northwestern Mediterranean Sea. *J. Mar. Res.* **56,** 197–238.

Lipschultz, F., Bates, N. R., Carlson, C. G., and Hansell, D. A. (2002). New production in the Sargasso Sea: History and current status. *Global Biogeochem. Cycles* **16,** 1–17.

Lucas, M., Seeyave, S., Sanders, R., Moore, C. M., Williamson, R., and Stinchcombe, M. (2007). Nitrogen uptake responses to a naturally Fe-fertilised phytoplankton bloom during the 2004/2005 CROZEX study. *Deep Sea Res. II* **54,** 2138–2173.

Martens-Habbena, W., Berube, P. M., Urakawa, H., de la Torre, J. R., and Stahl, D. A. (2009). Ammonia oxidation kinetics determine niche separation of nitrifying Archaea and Bacteria. *Nature* **461,** 976–979.

Martin, A. P., and Pondaven, P. (2003). On estimates for the vertical nitrate flux due to eddy pumping. *J. Geophys. Res.* **108,** 3359, 10.1029/2003JC001841.

Martin, A. P., and Pondaven, P. (2006). New primary production and nitrification in the western subtropical North Atlantic: A modelling study. *Global Biogeochem. Cycles* **20,** 1–12. GB4014, 10.1029/2005GB002608.

Martin, A. P., Zubkov, M. V., Fasham, M. J., Burkill, P. H., and Holland, R. J. (2008). Microbial spatial variability: An example from the Celtic Sea. *Prog. Oceanogr.* **76,** 443–465.

McClain, C. R., Murtugudde, R., and Signorini, S. (1999). A simulation of biological processes in the equatorial Pacific Warm Pool at 165°E. *J. Geophys. Res.* **104,** 18305–18322.

Miyazaki, T., Wada, E., and Hattori, A. (1975). Nitrite production from ammonia and nitrate in the euphotic layer of the western North Pacific Ocean. *Marine Sci. Comm.* **1**, 381–394.

Mongin, M., Nelson, D. M., Pondaven, P., Brzezinski, M., and Tréguer, P. (2003). Simulation of upper-ocean biogeochemistry with a flexible-composition phytoplankton model: C, N and Si cyclingin the western Sargasso Sea. *Deep Sea Res. I* **50**, 1445–1480.

Moore, L. R., Post, A. F., Rocap, G., and Chisholm, S. W. (2002). Utilization of different nitrogen sources by the marine cyanobacteria Prochlorococcus and Synechococcus. *Limnol. Oceangr.* **47**, 989–996.

Moore, J. K., Doney, S. C., and Lindsay, K. (2004). Upper ocean ecosystem dynamics and iron cycling in a global three-dimensional model. *Global Biogeochem. Cycles* **18**, 402810.1029/2004GB002220.

Olson, R. J. (1981). Differential photoinhibition of marine nitrifying bacteria - a possible mechanism for the formation of the primary nitrite maximum. *J. Mar. Res.* **39**, 227–238.

O'Neill, R. V., Angelis, D. L., Pastor, J. J., Jackson, B. J., and Post, W. M. (1989). Multiple nutrient limitation in ecological models. *Ecol. Modell.* **46**, 147–163.

O'Reilly, J. E., Maritorena, S., Mitchell, B. G., Siegal, D. A., Carder, K. L., Garver, S. A., Kahru, M., and McClain, C. (1998). Ocean color chlorophyll algorithms for SeaWiFS. *J. Geophys. Res.* **103**, 24937–24953.

Palmer, J. R., and Totterdell, I. J. (2001). Production and export in a global ocean ecosystem model. *Deep Sea Res.* **48**, 1169–1198.

Pätsch, J., and Kühn, W. (2008). Nitrogen and carbon cycling in the North Sea and exchange with the North Atlantic—A model study. Part I. Nitrogen budget and fluxes. *Cont. Shelf Res.* **28**, 767–787.

Popova, E. E., Coward, A. C., Nurser, G. A., de Cuevas, B., Fasham, M. J. R., and Anderson, T. R. (2006). Mechanisms controlling primary and new production in a global ecosystem model—Part I: Validation of the biological simulation. *Ocean Sci.* **2**, 249–266, 10.5194/os-2-249-2006.

Raimbault, P., Slawyk, G., Boudjellal, B., Coatanoan, C., Conan, P., Coste, B., Garcia, N., Moutin, T., and Pujo-Pay, M. (1999). Carbon and nitrogen uptake and export in the equatorial Pacific at 150°W: Evidence of an efficient regenerated production cycle. *J. Geophys. Res.* **104**, 3341–3356.

Raven, J. A., and Falkowski, P. G. (1999). Oceanic sinks for atmospheric CO_2. *Plant Cell Environ.* **22**, 741–755.

Raven, J. A., Caldeira, K., Elderfield, H., *et al.* (2005). Ocean acidification due to increasing atmospheric carbon dioxide. Royal Society of London, UK.

Redfield, A. C. (1934). On the proportions of organic derivatives in seawater and their relation to the composition of plankton. *In* "James Johnstone Memorial Volume," (R. J. Daniel, ed.)pp. 176–192. University of Liverpool, UK.

Riley, G. A. (1946). Factors controlling phytoplankton populations on Georges Bank. *J. Mar. Res.* **6**, 54–73.

Sarmiento, J., and Gruber, N. (2006). Ocean Biogeochemical Dynamics. Princeton University Press, New Jersey, USA.

Soetaert, K., and Herman, P. M J. (2009). A Practical Guide to Ecological Modelling: Using R as a Simulation Platform. Springer, Germany, 978-1-4020-8623-6.

Soetaert, K., Herman, P. M. J., Middelburg, J. J., Heip, C., Smith, C. L., Tett, P., and Wild-Allen, K. (2001). Numerical modelling of the shelf break ecosystem: Reproducing benthic and pelagic measurements. *Deep Sea Res. II* **48**, 3141–3177.

Steele, J. H. (1962). Environmental control of photosynthesis in the sea. *Limnol. Oceanogr.* **7**, 137–150.

Steele, J. H., and Henderson, E. W. (1981). A simple plankton model. *Am. Nat.* **117**, 676–691.

Sutka, R. L., Ostrom, N. E., Ostrom, P. H., and Phanikumar, M. S. (2004). Stable nitrogen isotope dynamics of dissolved nitrate in a transect from the North Pacific Subtropical Gyre to the Eastern Tropical North Pacific. *Geochim. Cosmochim. Acta* **68,** 517–527.

Takahashi, T., Sutherland, S. C., Wanninkhof, R., et al. (2009). Climatological mean and decadal change in surface ocean pCO_2, and net sea-air CO_2 flux over the global oceans. *Deep Sea Res. II* **56,** 554–577.

Thomas, S., and Ridd, P. V. (2004). Review of methods to measure short time scale sediment accumulation. *Mar. Geol.* **207,** 95–114.

Tian, R. C., Vézina, A. F., Legendre, L., et al. (2000). Effects of pelagic food-web interactions and nutrient remineralization on the biogeochemical cycling of carbon: A modeling approach. *Deep Sea Res. II* **47,** 637–662.

Totterdell, I. J., Armstrong, R. A., Drange, H., Parslow, J. S., Powell, T. M., and Taylor, A. H. (1993). Trophic resolution. *In* "Towards a Model of Ocean Biogeochemical Processes," (G. T. Evans and M. J. R. Fasham, eds.), pp. 71–92. Springer-Verlag, USA.

Tyrrell, T. (1999). The relative influences of nitrogen and phosphorus on oceanic primary production. *Nature* **400,** 525–531.

Van der Waal, D. B., Verschoor, A. M., Verspagen, J. M. H., van Donk, E., and Huisman, J. (2010). Climate-driven changes in the ecological stoichiometry of aquatic ecosystems. *Front. Ecol. Environ.* **8,** 145–152.

Van Minnen, J. G., Goldewijk, K. K., Stehfest, E., Eickhout, B., van Drecht, G., and Leemans, R. (2009). The importance of three centuries of land-use change for the global and regional terrestrial carbon cycle. *Clim. Change* **97,** 123–144.

Walsh, J. J., and Dieterle, D. A. (1994). CO_2 cycling in the coastal ocean. I—A numerical analysis of the southeastern Bering Sea with applications to the Chukchi Sea and the northern Gulf of Mexico. *Prog. Oceanogr.* **34,** 335–392.

Ward, B. B. (1986). Nitrification in marine environments. *In* "Nitrification," (J. I. Prosser, ed.)Vol. 20, pp. 157–184. IRL Press, Oxford, UK.

Wroblewski, J. (1977). A model of plankton plume formation during variable Oregon upwelling. *J. Mar. Res.* **35,** 357–394.

Yool, A., and Tyrrell, T. (2003). The role of diatoms in regulating the ocean's silicon cycle. *Global Biogeochem. Cycles* **17,** 1103.

Yool, A., Martin, A. P., Fernández, C., and Clark, D. R. (2007). The significance of nitrification for oceanic new production. *Nature* **447,** 999–1002.

CHAPTER TWO

Continuous Cultivation and Thermodynamic Aspects of Niche Definition in the Nitrogen Cycle

Stefanie Müller* *and* Marc Strous*,†

Contents

1. Introduction — 33
2. A Minimal Model for Microbial Behavior — 35
3. Continuous Cultivation—Background and Experimental Approach — 43
4. Conclusion — 50
Acknowledgments — 50
References — 50

Abstract

The study of model organisms in pure culture has provided detailed information about the physiology and biochemistry of nitrification and related processes. Metagenomic sequencing of environmental samples is providing information to what extent this understanding also applies to natural microbial communities. Here, we outline a conceptual and experimental strategy that links these two approaches. It consists of the mathematical modeling of nitrification and related processes. The model predictions are subsequently validated experimentally by the study of natural microbial communities in continuous cultures under precisely defined environmental conditions. Combined with calorimetry and metagenomic monitoring this form of "experimental metagenomics" enables the answering of current questions in the ecology of the nitrogen cycle.

1. Introduction

In the past decades, our picture of the nitrogen cycle has progressed from a simple cycle into a complex network. This change was brought about by the discoveries of new microbial processes: anaerobic ammonium

* Max Planck Institute for Marine Microbiology, Bremen, Germany
† Center for Biotechnology, University of Bielefeld, Bielefeld, Germany

oxidation (Jetten *et al.*, 2009), intra-aerobic denitrification (Ettwig *et al.*, 2010), and phototrophic nitrite oxidizers (Griffin *et al.*, 2007). Further, the known diversity of organisms involved in the conventional processes has greatly expanded. For example, a clade of crenarchaea (or thaumarchaea) was identified as nitrifiers (Könneke *et al.*, 2005): nitrification had so far been considered a strictly proteobacterial process (Arp *et al.*, 2007). Finally, progress in the environmental study of conventional processes, such as denitrification and dissimilatory nitrate reduction to ammonia, has been slow. Environmentally relevant denitrifiers or dissimilatory nitrate reducers have hardly been found and their relative importance is generally unknown (but see Dong *et al.*, 2009; Lam *et al.*, 2009).

To understand how the fate of nitrogen is decided in microbial reality, it is important to identify the major remaining unknown microbial actors, particularly for denitrification and dissimilatory nitrate reduction to ammonia. However, it is also essential to understand how the environmental conditions determine the outcome of the competition between the different microbes and the processes they perform. In other words, what factors select for the genes, organisms, and processes that are observed in nature? Is selection important in the first place or is the observed functional redundancy simply ecologically neutral, governed by population-ecological events?

So far, these questions have been addressed by two complementary approaches: the acquisition and study of relevant isolates or enrichments (e.g., Ettwig *et al.*, 2010; Griffin *et al.*, 2007; Könneke *et al.*, 2005; Strohm *et al.*, 2007; Strous *et al.*, 1999) and determination of the *in situ* presence and activity of microbes (e.g., Banfield *et al.*, 2005; Dong *et al.*, 2009; Lam *et al.*, 2009).

In this chapter, we describe a third approach, the study of selection and competition in microbial communities growing in laboratory continuous cultures that mimic the natural environment. This approach overcomes two problems that currently limit the power of the two conventional approaches: First, during isolation of a microbe in pure culture, the information about its ecological niche is lost; for most conventional model organisms we know only the ecological guild, the niche generally remains unknown. Second, *in situ* approaches (such as metagenomics or rate measurements) provide a snapshot of the environment that can only be interpreted by comparison with other environments (e.g., in biogeography; Dinsdale *et al.*, 2008). However, these environments are not under experimental control, differ in many aspects at the same time, and generally have poorly characterized natural histories. Under these circumstances causality necessarily remains elusive. By applying the above-mentioned *in situ* approaches to communities under experimental control (in the laboratory), this problem can be overcome and a single factor can be manipulated under otherwise identical conditions. Further, like in pure cultures, the selected community members can be identified and characterized by metagenomics, transcriptomics, proteomics, biochemical assays, spectroscopy, etc. Although the results will be more ambiguous than for a pure culture, the information about the ecological niche is not lost because now the selective forces are known precisely.

Continuous cultivation is not a high throughput technique. Therefore, generation of hypotheses is essential to make the right choices on what aspect of the environment should be investigated. Such hypotheses can sometimes be derived from existing experimental data such as *in situ* studies (e.g., Béjà *et al.*, 2001) and whole genome sequences. More generally, in the absence of clear leads, they can be derived from first principles and mathematical modeling. In this chapter, we first describe a thermodynamic model that predicts the outcome of microbial competition based on minimal assumptions. This model is used to generate hypotheses predicting the effect of the conditions on the fitness of microbes performing competing processes such as denitrification, dissimilatory nitrate reduction to ammonia, and anaerobic ammonium oxidation. Next the experimental design of the required continuous cultures is described.

2. A Minimal Model for Microbial Behavior

In microbiology, the substrate concentration is often considered a defining aspect of the niche. For example, we discriminate copiotrophs from oligotrophs or K from R strategists by the optimal substrate concentration for growth. However, for the mathematical modeling of microbial behavior, the substrate concentration is not a valid starting point: it is influenced by the microbes themselves, it is not an independent variable. For that reason, we define the influx of substrates into the habitat (e.g., in mol/s) as the independent variable. This immediately leads to an important conclusion: In Darwinian terms, the most successful microbes must be the ones that make best use of this influx of substrates to produce the most offspring (see also Rodriguez *et al.*, 2008). Thermodynamic analyses produce the same conclusion (Fath, 2001): According to Kondeputi and Prigogine (1998) in dissipative systems, far outside thermodynamic equilibrium (e.g., a respiring microbial community), a *steady state* is achieved when the internal entropy production reaches a (local) minimum. Such a *steady state* occurs when the influx of substrates is relatively constant and limits the rate of the microbial processes. It occurs in many well-studied microbial habitats such as sediments and stratified water bodies. This *steady state* is the most simple scenario, without any environmental dynamics. Under that condition, thermodynamics and evolutionary theory agree that the most efficient microbes will outcompete the others and success is determined by the highest biomass yield, not by the highest growth rate (the growth rate is constrained by the substrate influx). A second, more complicated scenario occurs when the conditions are dynamic and substrates periodically accumulate in the habitat. In that case, the outcome of microbial competition is determined by the biological rate. In the following paragraphs, we will explain the assumptions that enable the mathematical modeling of microbial competition for each of these two scenarios.

The first scenario (*steady state*) occurs when the substrate influxes are constant. Microbial populations will grow until substrate consumptions equal the substrate influxes and population sizes become constant, with new growth balanced by decay, maintenance, and export of microbes out of the system (the *steady-state* condition). In such a steady state, microbial competition is decided by "apparent" substrate affinity (apparent denotes the overall affinity of the microbial cell, as opposed to enzymatic affinity). To model substrate affinity, we consider the physical behavior of liquids at micrometer scales and Ficks law of diffusion: the substrate affinity of microbes is a function of cell size (Schulz and Joergensen, 2001) and reaction stoichiometry. Because microbes are smaller than the Kolmogorov scale, they are surrounded by a static body of water. The flux of substrate to a given population is therefore almost completely determined by the size of its cells. The smaller the cells, the better the substrate affinity. However, the stoichiometry is also important: when a cell completely consumes its substrate, the diffusion of substrate toward the cell is linearly dependent on the concentration in the surrounding liquid (Ficks law). For substrates with higher stoichiometric coefficients, a higher concentration in the surrounding liquid is needed to realize the required higher flux toward the cells. For example, in case of aerobic nitrite oxidizers, these bacteria need two molecules of nitrite per molecule of oxygen. To realize a sufficient supply of nitrite, the nitrite concentration in the surrounding liquid has to be twice as high as the oxygen concentration. Therefore, we know that for nitrite oxidizers the "apparent" affinity constant for nitrite is always approximately twice the affinity constant for oxygen. Now the assumption is that for all processes the cell size of the associated microbes is optimal for the specific habitat. That means that in our model, affinity is determined by reaction stoichiometry only. For clarity, we do not consider differences in the affinity constants of enzymes, because this factor only makes a minor difference: because of diffusion limitation an enzyme that has 10 times better affinity would only increase the "apparent" affinity of its host by at most 10%.

In the second scenario, substrate influxes are variable and the system never reaches *steady state*. In this case, periodic high substrate concentrations occur. Under these conditions, we assume that the respiratory speed, the flow of electrons from the electron donor to the electron acceptor via the respiratory complexes, limits the overall microbial conversion. This is a reasonable assumption for the nitrogen cycle, because all its catabolic processes are respiratory processes and the highest material fluxes occur over the respiratory chain, whereas only a small volume (the cell membrane) is available to allocate the necessary catalysts. The intensity of the resulting "traffic jam" is dependent on the number of electrons that need to be transferred. Thus, the maximum rate of each process is assumed to be inversely proportional to the number of electron-pairs transduced. One may argue that the number of respiratory bifurcations is also important, but forked respiratory chains have not yet been implemented in our model.

With these two assumptions, we can describe a system of competing populations of microbes in a simple model. The only knowledge that goes into the model is the reaction stoichiometry, the Gibbs free energy change, and our current understanding of layout and efficiency of the respiratory chain.

In cellular respiration, the available chemical energy is partly dissipated and partly conserved by the generation of a proton motive force or membrane potential. The respiration efficiency can be estimated when the enzyme complexes constituting the respiratory chain are known. These complexes transduce electrons from the electron donor to the terminal electron acceptor and simultaneously remove positive charge from the cytoplasm (Simon *et al.*, 2008). This way, a proton gradient over the membrane is produced (the proton motive force). The amount of energy conserved is the product of the number of protons removed from the cytoplasm and the Gibbs free energy change associated with the translocation of one proton ($\sim$15 kJ/mol at a membrane potential of $\sim$150 mV). The respiratory efficiency is the amount of energy conserved divided by the Gibbs free energy change associated with the overall catabolic reaction. Here, we assume that the conversion of the organic electron donor into NADH and FADH proceeds near thermodynamic equilibrium or that here the dissipation at least burdens all microbes equally. The same assumption is applied to assimilation, although this is arguably not the case when reversed electron transport is required. Apart from the respiratory chain itself, the uncoupled dissipation of the proton motive force is probably the largest additional source of energy dissipation. Also this factor is assumed to affect all microbes equally.

Figure 2.1 illustrates the notion of respiratory efficiency for two processes of the nitrogen cycle: aerobic and anaerobic ammonium oxidation (anammox). These processes are among the most poorly characterized forms of respiration in the nitrogen cycle. Table 2.1 shows the resulting respiratory efficiencies for the known respiratory processes of the nitrogen cycle based on the current understanding of the layout of the responsible respiratory chains. For each enzyme complex, the number of protons removed from the cytoplasm is indicated. Table 2.1 shows that aerobic respiration is the most efficient.

The data of Table 2.1 (reaction stoichiometry, overal Gibbs free energy changes, the respiratory efficiencies, and the number of electrons transferred) are used as input values for a model developed in MATLAB (http://www.mathworks.com). Although MATLAB is frequently used in scientific computing, any other programming environment would be suitable to implement this model. In our MATLAB model, the inputs are encoded in the form of three matrixes: `stoichiometric_coefficients`, `gibbs_free_energy_change`, `respiratory_efficiencies`, and `electrons`. The model further maintains the

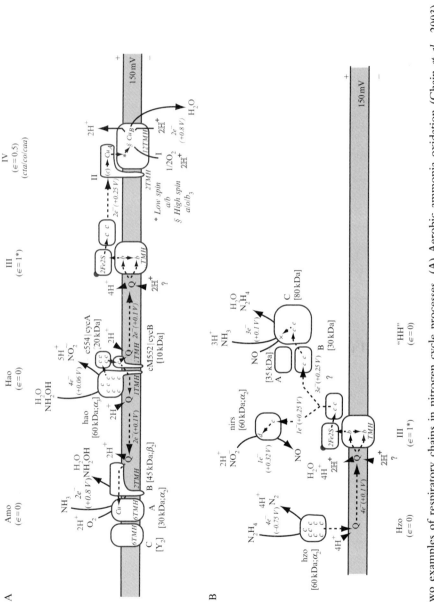

Figure 2.1 Two examples of respiratory chains in nitrogen cycle processes. (A) Aerobic ammonia oxidation (Chain et al., 2003), (B) anaerobic ammonium oxidation (Strous et al., 2006). The respiratory complexes and their thermodynamic efficiencies are shown. Protons that contribute to the proton motive force are drawn outlined. Compared to the other known nitrogen cycle processes the biochemistry of

self explanatory matrixes `substrate_concentrations` and `process_biomass_-concentrations`. Finally, the relative concentrations of substrates entering the system is defined in the matrix `substrate_concentrations_in`. All units are arbritary because we are only interested in the relative success of the processes.

Initially, all biomass and substrate concentrations are initialized to the same value for all processes. Then, fresh substrates start to enter the system as defined by `substrate_concentrations_in` and the competition starts. At each iteration, the model first calculates the rate for each process with Monod kinetics:

```
maximum_rate = 0.001; % a constant indicating the maximum rate
                     % (equal for all processes)
% initialization of a matrix of rates (one for each process):
process_rates(1:length(process_biomass_concentrations)...
,1) = 0.1;
% iterate over the processes:
for i = 1:length(process_rates)
   % iterate over the substrates to implement Monod kinetics:
   % (note: stoichiometric coefficients are used as affinity
   % constants)
   % (note: the rate is determined by the concentration of the
   % limiting substrate)
   for j = 1:length(substrate_concentrations)
      if (stoichiometric_coefficients(i,j) < 0.0)
         new_process_rate = abs(substrate_concentrations(j)...
                          / electrons(i)...
                          / (substrate_concentrations(j)...
                          + stoichiometric_coefficients
                            (i,j)...
                          / electrons(i)));
         if (new_process_rate < process_rates(i))
            process_rates(i) = new_process_rate;
         end
      end
   end
end
process_rates = maximum_rate...
             .* process_rates...
             .* process_biomass_concentrations...
             ./ electrons;
```

Next, the process rates are used in mass balances to update the concentrations for each substrate and for the biomass performing each process:

---

respiration for these two pathways is relatively poorly characterized. Important open questions are indicated by question marks. In nitrification: does complex II facilitate a quinone cycle and are protons produced in the partially enclosed membrane stacks in equilibrium with the external medium? In anammox, the experimentally determined biomass yield cannot be explained by the poor efficiency of the proposed pathway and the midpoint potential of the proposed hydrazine hydrolase reaction is too low to facilitate a quinone cycle by complex III.

Table 2.1 Selected metabolic pathways of the nitrogen cycle with thermodynamic and bioenergetic properties

Pathway	Reaction	e^- [a]	H^+ [b]	ΔH^0 (kJ) reaction	ΔG^0 (kJ) reaction	ΔG^0 (kJ) conserved	ε^c (%)	Enzyme complexes [d]
Anammox	$NO_2^- + NH_4^+ \rightarrow N_2 + 2\,H_2O$	4	1	−334.1	−357.5	−60	17	III
Denitrification NO_2^-	$4\,NO_2^- + 3\,CH_2O + 4\,H^+ \rightarrow 2\,N_2 + 3\,CO_2 + 5\,H_2O$	12	3	−1552.3	−1763.7	−540	31	I, III
Denitrification NO_3^-	$4\,NO_3^- + 5\,CH_2O + 4\,H^+ \rightarrow 2\,N_2 + 5\,CO_2 + 7\,H_2O$	20	3	−2086.2	−2437.8	−900	37	I, III, NAR
DNRA NO_2^-	$2\,NO_2^- + 3\,CH_2O + 4\,H^+ \rightarrow 2\,NH_4^+ + 3\,CO_2 + H_2O$	12	2	−884.2	−1048.7	−360	34	I, NRF
DNRA NO_3^-	$NO_3^- + 2\,CH_2O + 2\,H^+ \rightarrow NH_4^+ + 2\,CO_2 + H_2O$	8	2	−575.5	−692.9	−240	35	I, NAP, NRF
Nitrification (NH_4^+-ox.)	$2\,NH_4^+ + 3\,O_2 \rightarrow 2\,NO_2^- + 4\,H^+ + 2\,H_2O$	4	3	−516.6	−390.4	−180	46	AMO, HAO, IV
Nitrification (NO_2^--ox.)	$2\,NO_2^- + O_2 \rightarrow 2\,NO_3^-$	4	2	−200.0	−142.7	−120	84	IV
Aerobic respiration [e]	$CH_2O + O_2 \rightarrow CO_2 + H_2O$	4	5	−466.9	−479.7	−300	62	I, III, IV
Aerobic respiration [f]	$CH_2O + O_2 \rightarrow CO_2 + H_2O$	4	4	−466.9	−479.7	−240	50	I, III, IV

[a] The number of electrons transduced (passed through the respiratory chain) in the reaction.
[b] H^+ indicates the number of protons removed from the cytoplasm or from the anammoxosome, per electron transferred.
[c] The respiration efficiency (deltaG conserved/deltaG reaction).
[d] Respiratory complexes are given with the following contribution to the proton motive force: Complex I ($4H^+/2e^-$), complex III ($2H^+/2e^-$), complex IV ($4H^+/2e^-$ or $2H^+/2e^-$), NAR ($2H^+/2e^-$), NRF (no H^+), NAP (no H^+), AMO (no H^+, assumed), HAO (no H^+).
[e] Aerobic respiration with cytoplasmic cytochrome c oxidase.
[f] Aerobic respiration with periplasmic cytochrome c oxidase.

Continuous Culture, Thermodynamics and the Niche

```
time = 1.0;                      % a constant indicating the size of
                                 % the numerical time step
mass_transport = 0.01;           % a constant indicating the rate of
                                 % dispersion of substrates
microbe_decay = 0.05;            % a constant indicating the decay or
                                 % maintenance rate
% initial biomass and substrate concentrations:
process_biomass_concentrations(1:length(pathways),1) = 0.1;
substrate_concentrations(1,1:length(...
substrate_concentrations_in)) = 0.01;

% perform the mass balance for the substrates:
for j=1:length(substrate_concentrations)
   % add fresh substrates by influx into the system
   substrate_concentrations(j)= substrate_concentrations(j)
                              + substrate_concentrations_in(j)
                              * mass_transport * time;
   % perform the consumption and production
   for i=1:length(process_rates)
      substrate_concentrations(j) = substrate_concentrations(j)
      + stoichiometric_coefficients(i,j)
      * process_rates(i) * time;
   end
   % remove substrates and products by efflux out of the system:
   substrate_concentrations(j) = substrate_concentrations(j)...
                              - substrate_concentrations(j)...
                              * mass_transport * time;
end

% perform the mass balance for the biomass performing each
% process:
process_biomass_concentrations...
           = process_biomass_concentrations...
           + (process_rates...
           .* respiratory_efficiencies...
           .* gibbs_free_energy_change .* time)...
           - (process_biomass_concentrations...
           .* microbe_decay .* time);
% avoid negative biomass and substrate concentrations:
for i = 1:length(pathways)
   if (process_biomass_concentrations(k,1) < 0.0)
      process_biomass_concentrations(k,1) = 0.0000001;
   end
end
for j = 1:length(substrate_concentrations)
   if (substrate_concentrations(1,1) < 0.0)
      substrate_concentrations(1,1) = 0.0000001;
   end
end
```

We used the model to simulate time courses of substrate and biomass concentrations by iterating the calculations shown above for a time period sufficient to reach *steady state*. In *steady state*, all concentrations approach an equilibrium value. During this process, the model saves the concentration matrixes and relative process activities to file as follows:

```
% save biomass and substrate concentrations in text files
dlmwrite('biomass.txt',process_biomass_concentrations,...
         '-append','delimiter','\t');
dlmwrite('substrate.txt',substrate_concentrations,...
         '-append','delimiter','\t');
dlmwrite('process_rates.txt',process_rates,...
         '-append','delimiter','\t');
```

Figure 2.2 shows an example model prediction: the effect of oxygen influx on nitrogen conversions in a eutrophic freshwater sediment with an

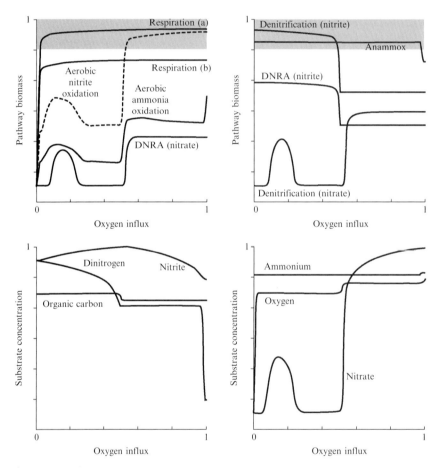

Figure 2.2 An example of the outcome of microbial competition as predicted by the mathematical model. The relative biomass and substrate concentrations are shown (both on a logarithmic scale) as a function of oxygen influx. Other influxes were: nitrite 1, organic carbon 0.5, and ammonia 0.1. For biomass concentrations, the gray area indicates the estimated experimentally detectable range (without using PCR). Production of substrates not provided as influx (e.g., nitrifiers produced nitrate) facilitated the activity of secondary processes (e.g., denitrification of nitrate). For clarity, the biomass concentrations for each process are distributed over the top two panels, and the substrate concentrations over the bottom two panels; all panels relate to a single "experiment."

influx of nitrite. The relative influxes of all the substrates were constant (0.1 for ammonium, 0.5 for organic carbon, 1 for nitrite, 0 for all others) while the oxygen influx was gradually increased from 0 to 1. Thus, the figure shows the predicted activities of all pathways as we move up through the chemocline toward fully oxic conditions. The model predicts that denitrification outcompetes DNRA under oxygen limiting conditions. A regime shift occurs at the point where enough oxygen enters the habitat to oxidize all organic carbon. At this point, the nitrite concentration reaches a maximum and the oxygen concentration increases. From this point onwards, denitrification is fully outcompeted by aerobic respiration and the increased oxygen concentration facilitates aerobic nitrification. The predictions for this simple scenario are generally consistent with experimental observations in nature. This scenario as well as more complicated dynamic scenarios can be used as starting points for experimental studies in continuous culture. In this scenario, oxidized nitrogen was supplied in the form of nitrite rather than nitrate because profiles taken from natural ecosystems (e.g., Lam *et al.*, 2009) often show transient nitrite accumulation in the chemocline, indicating an important role of this compound in the coupling of aerobic and anaerobic nitrogen metabolism.

3. CONTINUOUS CULTIVATION—BACKGROUND AND EXPERIMENTAL APPROACH

Usually, micro-organisms are cultivated in batch. In this approach, a glass flask is filled with medium and inoculated at the beginning of the experiment. It is a closed system. During the experiment, the conditions in the culture change because of metabolic activity of the microbes: substrates are consumed while products accumulate. Therefore, the growth conditions are not constant and not under experimental control.

In contrast to batch culture, a continuous culture is fed continuously with fresh medium after inoculation. The best known example of a continuous culture is the chemostat (Novick and Szilard, 1950). In this system, spent medium and microorganisms are removed from the culture continuously to maintain constant volume of the chemostat vessel; the influent and effluent flow rates are equal and constant. This experimental setup progresses toward a *steady state* (scenario 1 above), with constant substrate and product concentrations inside the culture. In *steady state*, the microbial growth rate is also constant and equal to the "dilution rate," the quotient of influent flow rate and bioreactor volume. The maximum specific growth rate of the microorganism defines an upper limit for the dilution rate. Beyond that limit the cells are washed out more rapidly than they can grow and cell density decreases rapidly. At lower dilution rates, the cell concentration is constant and linearly dependent on the concentration of the "controlling" or limiting substrate in the influent medium (Novick and Szilard, 1950). Gases (e.g., air, argon) are also supplied continuously.

Other examples of continuous culture are the fed batch reactor, the sequencing batch reactor (Strous et al., 1998), and the retentostat (Tappe et al., 1999). In these systems, the dilution rate can be higher than growth rate because the microorganisms are retained by a settler or a filter; such systems are therefore especially suitable to cultivate organisms at very low growth rate (Kartal and Strous, 2008). Low growth rates (weeks to months) are the rule rather than the exception in many natural ecosystems (Jørgensen and Boetius, 2007). In all these forms of continuous cultivation, the concentration of the limiting substrate is very low, as is also the case in our first environmental scenario.

An important difference between the conditions in the chemostat and natural ecosystems is that in the latter the conditions are not always constant (the second scenario described above): In nature, microbes may experience alternating periods of electron donor versus electron acceptor limitation, for example, because of day/night or tidal regimes. In sediments, a steep redox gradient (from fully oxic conditions to sulfate reducing conditions within several millimetres) may shift up and down depending on changes in the hydraulic flow regime. Continuous culture can easily be adapted to create such dynamics by implementing automated changes in the amount of oxygen in the gas supplied or by switching between two media with different substrate concentrations in a defined, cyclic pattern. This way, a *pseudo steady state* will still result: after some time, the cultivated microbes respond the same way each time the dynamic cycle repeats itself. During transitions, the outcome of microbial competition is determined by respiratory speed, not by efficiency as is the case in the chemostat described above.

Such (dynamic) systems can now be used to enforce a defined and realistic selective pressure on a natural microbial community. A single factor, for example, the limiting substrate, the growth rate, or the period of the dynamics can be modulated in several parallel cultures to observe the effect on the nature of the selected community in terms of its conversions, its species, and its functional genes. One can also investigate whether resulting *pseudo steady states* produce stable communities at all or whether they are only "steady" in terms of the observed conversions; at the same time, the identity of selected species and functional genes may still fluctuate and not reach a *steady state* (e.g., Beninca et al., 2008; Graham et al., 2007). The latter situation would point at ecological neutrality or at least weak selective pressure—several populations may be (almost) equally fit under the enforced conditions (Hubbell, 2005). According to our optimality criteria (maximum bioenergetic efficiency and speed), we would expect that during the succession of populations in the experiment, the overall communal efficiency and/or speed would increase, leading to a higher cell concentration, less dissipation of energy, and little accumulation of substrates during transitions.

At this point, it is important to discriminate "selection" from "evolution." The relatively short incubation times (several, at most a hundred generations) will favor selection in the form of the recruitment of well-adapted (the fittest) species already present in the inoculum into the resulting microbial communities. Microevolution may occur and contribute to some extent but the effect will be relatively small. Phenotypical acclimation (e.g., regulatory plasticity) may of course contribute to the fitness of specific populations.

The necessary experimental characterization of the succession of microbial populations during selection consists of the monitoring of (1) the overall microbial conversions, (2) the species and functional genes present, and (3) the overall bioenergetic efficiency. For this reason, the concentrations of relevant compounds in the influent and effluent medium need to be measured (preferably on-line) to make a *mass balance*; the resulting community composition should be monitored by metagenomics. Finally, the heat produced by the culture can be measured by making a heat balance (calorimetry).

Although targeted molecular approaches (e.g., real-time PCR, fluorescence *in situ* hybridization) could also be used instead of metagenomics, such procedures are much more labor intensive and, more importantly, they depend on the availability of specific, comprehensive, and up-to-date primers and probes for each targeted functional gene and taxonomic group. Such oligonucleotides are simply not available in the context of the nitrogen cycle. The bioenergetic efficiency could also be measured by quantifying microbial biomass in the form of cell counts, dry weight, or protein content but these measurements are much less precise (relative errors approximately 10%) than calorimetry (relative errors less than 1%, Schubert *et al.*, 2007). Furthermore, calorimetry is an on-line measurement whereas biomass concentration can only be monitored off-line (measuring periodic snap-shot samples). The experimental characterization of the selected community could ultimately be complemented by metaproteomics and transcriptomics, isotopic labeling, biochemical assays, and off-line activity measurements as necessary.

Calorimetry is generally applied to problems in physics, chemistry, and biology. A number of different approaches exist and it is possible to achieve extreme experimental sensitivity. The principle of isothermal calorimetry is the following (Schubert *et al.*, 2007): The calorimeter consists of a well isolated vessel (in this case the continuous culture) equipped with a cooler and a precision heater. The temperature difference between the vessel and the cooler is constant and the heat transferred to the cooler is compensated by the precision heater. When the heat transfer to the cooler is known, the heat produced inside the vessel can be calculated as the difference between the heat transferred to the cooler and the power applied to the heater. Both these factors can be determined with extremely high experimental accuracy. This principle has been successfully applied to continuous cultures. However, in this case, more factors enter the equation: First, continuous cultures generally have much larger volumes and need to be mixed—and mixing (generally by mechanical stirring) produces heat and noise. Second, the influent medium may have a different temperature than the culture itself. Third, vessels used for continuous cultures often have stainless steel parts (e.g., the lid) with many inserts—influent, effluent, sampling ports, sensors, etc.—and are difficult to isolate. Finally, the supply of gases produces multiple noisy heat effects: expansion, heat transfer, and evaporation. Still, the overall accuracy obtained with standard, commercially available equipment has been impressive ($<1\%$, Schubert *et al.*, 2007).

The calorimetric sensitivity and resolution may be further improved by using customized vessels as follows. First, the stainless steel lid and inserts can be replaced by glass parts. The thermal conductivity of glass is 10 times lower compared to steel. Extra isolation can be added to the outside of the vessel. Further, mixing can be implemented by recycling gas from the top to the bottom of the vessel. Mechanical stirring can then be omitted completely. Without further measures, a gas recycle is probably worse than a mechanical stirrer in terms of its heat effects. However, when the temperature of the recycled gas is controlled these heat effects can be avoided, and that is impossible with a mechanical stirrer. When the recycled gas is cooled slightly (several degrees C) below the temperature of the vessel, the heat effects of gas mixing cancel out: As already mentioned, sparging gas produces three heat effects: (1) heat is generated in the vessel during liquid displacement by the expanding gas; (2) heat is lost from the liquid to the gas because of heat transfer to the expanding gas; (3) heat is lost from the liquid due to evaporation of water into the gas. Because the first positive term is constant and the other two (negative) terms depend on the temperature difference between the gas and the vessel, a certain gas temperature exists where the heat effects cancel out and the noise associated with gas mixing can be effectively removed from the experimental system. The temperature difference between the vessel and the recycled gas at which such a "zero heat" condition occurs depends mainly on the pressure drop during sparging and the temperature of the culture. It is generally several degrees C. See Figure 2.3.

Over time, the monitoring and control procedures necessary to implement mass and heat balancing lead to the compilation of a large amount of data. These data consist of traces of the pH, redox, oxygen, pressure and several temperature sensors, read outs from balances recording the liquid levels in the culture vessel, as well as influent and effluent vessels, data from the mass spectrometer used to monitor the concentrations of relevant compounds in the effluent gas, and the power applied to several heaters.

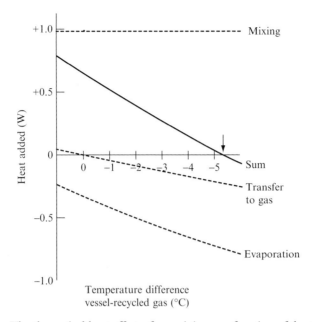

Figure 2.3 The theoretical heat effect of gas mixing as a function of the temperature difference between the recycled gas and the culture liquid. In the example shown, the heat effects cancel out at a temperature difference of 5.3 °C. Vessel temperature 20 °C; pressure drop 225 mbar (mainly caused by the sintered glass plate, see Fig. 2.4). Gas flow 3 l/min. The arrow shows the temperature difference at which no net heat effect occurs.

Together this is even more data than amassed from the metagenomic monitoring. When logging and control would be performed with commercially available equipment, the handling of the experimental data would be nearly impossible because each measurement would need its own controller and associated PC and software for data logging. Therefore, we constructed custom controllers that perform logging for all signals, control all actuators, and generate a single log file per experiment. These custom controllers were built cost-effectively and with relatively little effort with the modular National Instruments Compact RIO system. For mass spectrometry, we used a multiplexed InProcess GAM400 instrument for industrial monitoring with a direct membrane inlet in each cultivation vessel. To control the gas supply to the culture vessels it is necessary to use "mass flow controllers" (not rotameters). These instruments control the gas flow by realizing a constant pressure or temperature drop over a constriction inside the instrument.

The design of the vessels used for continuous cultivation is illustrated in Figure 2.4. The volume of the vessel shown is approximately 3 l and for such a relatively small volume the "inversed erlenmeyer" or "Kluyver flask"

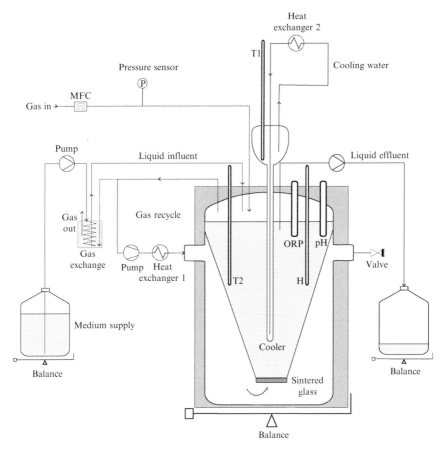

Figure 2.4 Schematic illustration of the experimental setup for continuous cultivation. MFC: mass flow controller (for gas supply). T1 and T2: temperature sensors of the cooler and the bioreactor. H: bioreactor heater ORP: redox sensor.

generates excellent mixing and gas–liquid mass transfer. Two heat exchangers are necessary, one for the calorimetric cooler and one for the gas recycle. Three temperature sensors (thermistors or Pt1000)/heater pairs are necessary, one for the vessel, one for the cooler, and one for the gas recycle.

For the pumping of influent medium into the culture and spent medium out of the culture, it is best to use peristaltic pumps. The advantage of this type of pump is that the tubing can be autoclaved and one can work aseptically. For the circulation of cooling water less expensive aquarium pumps can be used (e.g., Eheim pumps). Liquid volumes (of influent and effluent vessels as well as the culture vessel itself) can be measured with balances, or more cost-effectively with load cells (e.g., LSH load cells).

Pressure sensing is necessary to correct the mass spectrometer data for fluctuations in atmospheric or vessel pressure. A pressure sensor is also very convenient to ascertain gas-tightness of the system at all times. Although the signal of a redox sensor has little value in drawing thermodynamic conclusions, it usually provides direct feedback on the activity of the microbes in the system. Sudden changes in the measured redox potential often reflect a decrease or increase in microbial activity.

While the population adapts to the conditions enforced in the continuous culture, biomass samples are taken and stored for metagenomic sequencing, to document the succession of populations and functional genes. The current technological approaches for massive parallel sequencing, 454 pyrosequencing (0.5 Gb, read length 400 bp for the Titanium model), and Illumina sequencing (currently 18 Gb, read length 70 bp, paired end) are sufficient. The advantage of pyrosequencing is that the individual reads are sufficiently long to enable meaningful phylogenetic characterization and the identification of functional genes without assembly. The advantage of Illumina sequencing is that higher coverage can be achieved at a lower price. However, because the individual reads need to be assembled into longer fragments to enable meaningful interpretation, the sequenced community has to be simple enough to enable assembly. Thus, the complexity of the community determines which sequencing approach is best. A combination of both methods would be the most cost effective. When successful populations comprise in the order of 10% of the cells of the total community selected in the continuous culture, approximately 100 Mb of sequencing would be required per sample. Given the technology currently available, 100 Mb is feasible, with tagging 180 samples could be sequenced on a single Illumina run, for $50 per sample. That means the complexity of the community selected in the continuous cultures needs to be somewhere in between a natural population (dominant community members 1–5% of the total) and a proper enrichment culture (dominant community member up to 80% of the total). With next-next generation sequencing technology, it will become possible to survey more complicated communities metagenomically. The combination of metagenomic sequencing with transcriptomic and proteomic approaches would be even more powerful.

The bioinformatic pipeline used for the analysis of the sequenced samples differs from current general practice in the analysis of metagenomes obtained directly from the natural environment (e.g., Dinsdale *et al.*, 2008). When comparing different continuous cultivations that differ only in a single factor, the metagenome can be interpreted functionally rather than statistically. For example, the pipeline addresses the relative importance of specific respiratory complexes and the presence or absence of specific functional residues that provide clues about the encoded respiratory efficiency. This way the metagenomic data can be integrated with the model predictions and the measured degree of energy dissipation.

4. CONCLUSION

The study of model organisms in pure culture has provided detailed information about the physiology and biochemistry of nitrogen cycle processes such as nitrification. However, it is presently unknown to what extent the studied organisms and respiratory mechanisms are also realized in the environment. That question can currently be addressed by metagenomic sequencing of environmental samples. However, such samples are generally poorly characterized physiologically, have unknown natural history, and differ among each other in many factors simultaneously. Furthermore, the power of current sequencing technology is still too low to reliably detect the relevant functional genes.

Here, we have outlined an approach that consists of the mathematical modeling of nitrogen cycle pathways and the experimental study of natural communities in continuous culture. Combined with calorimetry and metagenomic monitoring this form of "experimental metagenomics" enables the answering of current questions in the ecology of the nitrogen cycle: What environmental factors (dynamics, substrate supply rate) define the fitness of the competing pathways (e.g., anammox vs. nitrification), species (e.g., archaeal nitrifiers vs. proteobacterial nitrifiers, Valentine, 2007), and genes (e.g., "cbb3" vs. "aa3" complex IV in nitrification)? Is the observed functional redundancy ecologically neutral? Is our current understanding of microbial physiology (condensed in the presented mathematical model) sufficient to understand niche differentiation in the nitrogen cycle? And finally, is the respiratory machinery known from model organisms also competitive in nature?

ACKNOWLEDGMENTS

We thank Ingrid van de Leemput for introducing Stefanie Müller to her MATLAB model. Stefanie Müller and Marc Strous are supported by the ERC grant "Masem" (242635) and by the German Federal State of Nordrhein Westfalen.

REFERENCES

Arp, D. J., Chain, P. S., and Klotz, M. G. (2007). The impact of genome analyses on our understanding of ammonia-oxidizing bacteria. *Annu. Rev. Microbiol.* **61,** 503–528.

Banfield, J. F., Verberkmoes, N. C., Hettich, R. L., and Thelen, M. P. (2005). Proteogenomic approaches for the molecular characterization of natural microbial communities. *OMICS* **9,** 301–333.

Béjà, O., Spudich, E. N., Spudich, J. L., Leclerc, M., and DeLong, E. F. (2001). Proteorhodopsin phototrophy in the ocean. *Nature* **411,** 786–789.

Beninca, E., Huisman, J., Heerkloss, R., John, K. D., Branco, P., Van Nes, E. H., Scheffer, M., and Ellner, S. P. (2008). Chaos in a long-term experiment with a plankton community. *Nature* **451**, 822–825.

Chain, P., Lamerdin, J., Larimer, F., Regala, W., Lao, V., Land, M., Hauser, L., Hooper, A., Klotz, M., Norton, J., Sayavedra-Soto, L., Arciero, D., *et al.* (2003). Complete genome sequence of the ammonia-oxidizing bacterium and obligate chemolithoautotroph nitrosomonas europaea. *J. Bacteriol.* **185**, 2759–2773.

Dinsdale, E. A., Edwards, R. A., Hall, D., Angly, F., *et al.* (2008). Functional metagenomic profiling of nine biomes. *Nature* **452**, 629–632.

Dong, L. F., Smith, C. J., Papaspyrou, S., Stott, A., Osborn, A. M., and Nedwell, D. B. (2009). Changes in benthic denitrification, nitrate ammonification, and anammox process rates and nitrate and nitrite reductase gene abundances along an estuarine nutrient gradient (the Colne estuary, United Kingdom). *Appl. Environ. Microbiol.* **75**, 3171–3179.

Ettwig, K. F., Butler, M. K., Le Paslier, D., Pelletier, E., Mangenot, S., Kuypers, M. M. M., Schreiber, F., Dutilh, B. E., Zedelius, J., de Beer, D., Gloerich, J., Wessels, H. J. C. T., *et al.* (2010). Nitrite-driven anaerobic methane oxidation by oxygenic bacteria. *Nature* **464**, 543–548.

Fath, B. (2001). Complementarity of ecological goal functions. *J. Theor. Biol.* **208**, 493–506.

Graham, D. W., Knapp, C. W., Van Vleck, E. S., Bloor, K., Lane, T. B., and Graham, C. E. (2007). Experimental demonstration of chaotic instability in biological nitrification. *ISME J.* **1**, 385–393.

Griffin, B. M., Schott, J., and Schink, B. (2007). Nitrite, an electron donor for anoxygenic photosynthesis. *Science* **316**, 1870.

Hubbell, P. (2005). Neutral theory in community ecology and the hypothesis of functional equivalence. *Funct. Ecol.* **19**, 166–172.

Jetten, M. S., Niftrik, L. V., Strous, M., Kartal, B., Keltjens, J. T., and Op den Camp, H. J. (2009). Biochemistry and molecular biology of anammox bacteria. *Crit. Rev. Biochem. Mol. Biol.* **44**, 65–84.

Jørgensen, B. B., and Boetius, A. (2007). Feast and famine—Microbial life in the deep-sea bed. *Nat. Rev. Microbiol.* **5**, 770–781.

Kartal, B., and Strous, M. (2008). Methods to study consortia and mixed cultures. In "Accessing Uncultivated Microorganisms," (K. Zengler, ed.).978-1-55581-406-9, ASM Press, pp. 205–219.

Kondeputi, D., and Prigogine, I. (1998). Modern thermodynamics—From heat engines to dissipative structures. Wiley, 978-0471-97393-5, pp. 385–404.

Könneke, M., Bernhard, A. E., de la Torre, J. R., Walker, C. B., *et al.* (2005). Isolation of an autotrophic ammonia-oxidizing marine archaeon. *Nature* **437**, 543–546.

Lam, P., Lavik, G., Jensen, M. M., van de Vossenberg, J., Schmid, M., Woebken, D., Gutierrez, D., Amann, R., Jetten, M. S. M., and Kuypers, M. M. M. (2009). Revising the nitrogen cycle in the peruvian oxygen minimum zone. *Proc. Natl. Acad. Sci.* **106**, 4752–4757.

Novick, A., and Szilard, L. (1950). Description of the chemostat. *Science* **112**, 715–716.

Rodriguez, J., Lema, J., and Kleerebezem, R. (2008). Energy-based models for environmental biotechnology. *Trends Biotechnol.* **26**, 366–374.

Schubert, T., Breuer, U., Harms, H., and Maskow, T. (2007). Calorimetric bioprocess monitoring by small modifications to a standard bench-scale bioreactor. *J. Biotechnol.* **130**, 24–31.

Schulz, H. N., and Joergensen, B. B. (2001). Big bacteria. *Annu. Rev. Microbiol.* **55**, 105–137.

Simon, J., Van Spanning, R. J. M., and Richardson, D. J. (2008). The organisation of proton motive and non-proton motive redox loops in prokaryotic respiratory systems. *BBA Bioenergetics* **1777**, 1480–1490.

Strohm, T. O., Griffin, B., Zumft, W. G., and Schink, B. (2007). Growth yields in bacterial denitrification and nitrate ammonification. *Appl. Environ. Microbiol.* **73,** 1420–1424.

Strous, M., Heijnen, J. J., Kuenen, J. G., and Jetten, M. S. M. (1998). The sequencing batch reactor as a powerful tool for the study of slowly growing anaerobic ammonium-oxidizing microorganisms. *Appl. Microbiol. Biotechnol.* **50,** 589–596.

Strous, M., Fuerst, J. A., Kramer, E. H., Logemann, S., *et al.* (1999). Missing lithotroph identified as new planctomycete. *Nature* **400,** 446–449.

Strous, M., Pelletier, E., Mangenot, S., Rattei, T., *et al.* (2006). Deciphering the evolution and metabolism of an anammox bacterium from a community genome. *Nature* **440,** 790–794.

Tappe, W., Laverman, A., Bohland, M., Braster, M., *et al.* (1999). Maintenance energy demand and starvation recovery dynamics of nitrosomonas europaea and nitrobacter winogradskyi cultivated in a retentostat with complete biomass retention. *Appl. Environ. Microbiol.* **65,** 2471–2477.

Valentine, D. L. (2007). Adaptations to energy stress dictate the ecology and evolution of the archaea. *Nat. Rev. Microbiol.* **5,** 316–323.

SECTION TWO

CULTIVATION AND CELL PHYSIOLOGY

CHAPTER THREE

ISOLATION, CULTIVATION, AND CHARACTERIZATION OF AMMONIA-OXIDIZING BACTERIA AND ARCHAEA ADAPTED TO LOW AMMONIUM CONCENTRATIONS

Annette Bollmann,* Elizabeth French,* *and* Hendrikus J. Laanbroek[†,‡]

Contents

1. Introduction 56
2. General Description of Cultivation Methods for Ammonia-Oxidizing Microorganisms 57
   - 2.1. Batch and continuous culture cultivation 57
   - 2.2. General set-up of a batch culture experiment 60
   - 2.3. General set-up of a continuous culture experiment 60
   - 2.4. Media 61
3. Enrichment of Ammonia Oxidizers 63
   - 3.1. Enrichment of ammonia oxidizers in batch culture 63
   - 3.2. Enrichment of ammonia-oxidizing bacteria in continuous culture 65
   - 3.3. Enrichment medium prepared with water from aquatic environments 67
4. Isolation of Ammonia Oxidizers 68
   - 4.1. Isolation of ammonia-oxidizing bacteria using pour plates 68
   - 4.2. Isolation of ammonia-oxidizing bacteria by dilution to extinction in liquid media 69
   - 4.3. Isolation of ammonia-oxidizing bacteria by serial dilution in liquid media 70
   - 4.4. Stock culturing of ammonia-oxidizing bacteria 71

* Department of Microbiology, Miami University, Oxford, Ohio, USA
† Department of Microbial Ecology, Netherlands Institute of Ecology, Wageningen, The Netherlands
‡ Utrecht University, Institute of Environmental Sciences, Utrecht, The Netherlands

4.5. Long-term storage of ammonia-oxidizing bacteria   71
4.6. Isolation of ammonia-oxidizing archaea by serial dilution in
     liquid media   72
5. Characterization of Ammonia-Oxidizing Prokaryotes   72
   5.1. Growth versus activity kinetics   72
   5.2. Determination of growth kinetics of ammonia oxidizers 72
   5.3. Determination of the activity kinetics of ammonia oxidizers 74
References   84

Abstract

Ammonia-oxidizing microorganisms (AOM) generate their energy by the oxidation of ammonia (NH_3) to nitrite (NO_2^-). This process can be carried out by ammonia-oxidizing bacteria (AOB) as well as by the recently discovered ammonia-oxidizing archaea (AOA). In the past, AOB were enriched in batch cultures, often in the presence of rather high concentrations of NH_4^+. Here, we describe methods to enrich, isolate, and investigate the basic physiology of AOB and AOA with emphasis on those that are adapted to low NH_4^+ concentrations. The methods described include enrichment of AOA and AOB in batch cultures and of AOB in continuous cultures, the isolation of AOA by serial dilution and AOB by pour plates or dilution to extinction, and techniques to determine growth and activity of the AOA and AOB. Finally, we incorporated a section with Appendix about the identification of AOA and AOB as well as the measurement of the different inorganic nitrogen species.

1. INTRODUCTION

Ammonia oxidizers are microorganisms that generate energy solely by the oxidation of ammonia to nitrite. This step generates a small amount of energy resulting in low growth rates and yields of these microorganisms, which makes isolation difficult. The first ammonia–oxidizing bacteria (AOB) were discovered at the end of the nineteenth century by Winogradsky (1890, 1892). In the second half of the twentieth century, several groups started cultivating and isolating AOB from a wide variety of environments such as marine waters (Jones et al., 1988; Koops et al., 1976; Watson, 1965), estuarine systems (Stehr et al., 1995), soils (Utaker et al., 1996; DeBoer et al., 1989), and wastewater treatment systems (Suwa et al., 1994). Most of these isolations were conducted at ammonium concentrations that exceeded environmentally relevant concentrations by several orders of magnitude. With the rise of molecular, DNA-based methods, the isolated AOB were phylogenetically characterized. Most of the isolates belong to a distinct phylogenetic group within the *Betaproteobacteria* (Koops and Harms, 1985; Pommerening-Röser et al., 1996; Purkhold et al., 2000, 2003). A few marine AOB, however, belong to the *Gammaproteobacteria*

(Klotz et al., 2006; Woese et al., 1985). Molecular methods targeting the 16S rRNA (Kowalchuk et al., 1997) or the *amoA* gene, the gene that encodes the alpha-subunit of the ammonia monooxygenase, the first enzyme in ammonia oxidation (Rotthauwe et al., 1997) of the AOB were also used to get an insight into the environmental distribution of the different AOB, and revealed several clusters without any cultivated members (Koops et al., 2006; Kowalchuk and Stephen, 2001; Purkhold et al., 2000, 2003). In the beginning of the twenty-first century, metagenomic analysis of soil and marine samples revealed that the presence of *amoA* genes is not simply limited to the domain of Bacteria. *amoA* genes were also found in the domain of Archaea, leading to the discovery of ammonia-oxidizing archaea (AOA; Treusch et al., 2005; Venter et al., 2004). The first AOA *Nitrosopumilus maritimus* was isolated into pure culture from a seawater aquarium in Seattle (Konneke et al., 2005). The whole genome of several AOB—*Nitrosomonas europaea* (Chain et al., 2003), *Nitrosococcus oceani* (Klotz et al., 2006), *Nitrosomonas eutropha* (Stein et al., 2007), *Nitrosospira multiformis* (Norton et al., 2008)—and one AOA—*N. maritimus* (Walker et al., 2010)—have been sequenced and analyzed, revealing many new genes and potential characteristics of these microbes.

Despite cultivation efforts and the knowledge of whole genome sequences, our understanding of the role of AOB and AOA in the environment is still very limited. Right now, it is still impossible to distinguish between AOA and AOB as the microorganisms responsible for ammonia oxidation in the environment. In the future, we will need more efforts to investigate the responses of well-characterized pure and enrichment cultures of AOA and AOB to environmental conditions, to elucidate the function of these microorganisms in their natural environment in detail.

Here, we present a collection of methods to cultivate and investigate AOA and AOB in general, and those adapted to low ammonium concentrations in more detail. We will describe different cultivation methods such as batch and continuous cultures, and give an overview of methods to enrich, isolate, and characterize AOA and AOB.

2. GENERAL DESCRIPTION OF CULTIVATION METHODS FOR AMMONIA-OXIDIZING MICROORGANISMS

2.1. Batch and continuous culture cultivation

Microbes can be grown in the lab in several different ways; two of the most used cultivation methods for AOB are batch and continuous culture cultivation. Batch cultivation methods are relatively easy and convenient to use. In contrast, continuous culture methods are more complex. Fresh growth medium is supplied at a constant rate into the culture vessel, while the microbial cells in spent medium are removed from the vessel at the same rate. The unique properties of continuous culture enable physiological and ecological studies that are not possible in closed systems. For example, continuous culture systems can be used to investigate the influence of low substrate concentrations on the growth and activity of microbes. Conditions in continuous cultures often resemble natural conditions better than conditions in batch cultures. The concept of continuous cultures was first introduced by Monod (1950). Since then, several continuous culture systems have been developed (see Veldkamp, 1976 for an overview).

The single-stage, flow-controlled continuous culture, or chemostat, is the most commonly applied continuous cultivation method used for the study of ammonia oxidizers (Fig. 3.1). During cultivation in a chemostat, the growth vessel is supplied with fresh medium from a reservoir. The medium is prepared such that one of the substrates is growth limiting, that is, it limits the amount of biomass produced in the growth vessel, while the other medium components are present in excess.

The dilution rate of the chemostat (D) equals the flow rate of the fresh medium divided by the volume of the vessel, which is constant. When the dilution rate (D) is set at a fixed value, cell numbers will reach a steady state in which the growth rate and the washout rate of the biomass are in balance. Under these conditions, the specific growth rate (μ) equals the dilution rate. When the dilution rate exceeds the maximum possible growth rate under the prevailing conditions (i.e., μ_{max}), the culture will be washed out of the chemostat.

The specific growth rate μ depends on the concentration of the growth-limiting substrate (s) in the growth vessel according to Monod's equation (Monod, 1950):

$$\mu = \mu_{max}(s/(K_s + s))$$

where K_s is the half saturation constant numerically equal to the concentration of the growth-limiting substrate at 0.5 μ_{max}. Replacing μ by D renders the following equation for s under steady-state conditions:

$$s = K_s D/(\mu_{max} - D)$$

Therefore, the concentration of the growth-limiting substrate remains low over a large range of dilution rates, until D approaches μ_{max}. These low substrate concentrations offer the opportunity to grow bacteria at substrate concentrations that reflect the concentration in their original environment better than batch cultures.

It is important to emphasize that under steady-state conditions, the growth-limiting substrate concentration in the growth vessel (s) is not dependent on the concentration of the growth-limiting substrate in the

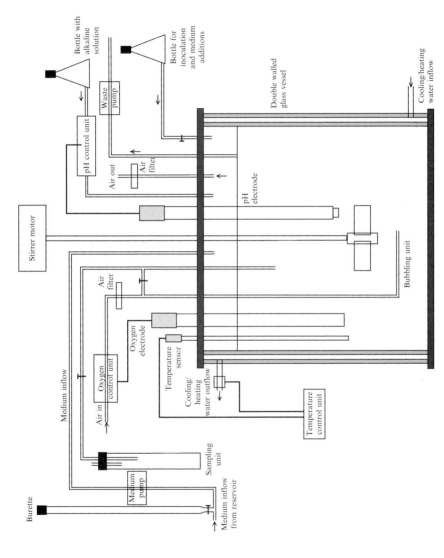

Figure 3.1 Schematic overview of a chemostat used for nutrient-limited cultivation of ammonia-oxidizing bacteria based on Veldkamp (1976).

reservoir. Since the production of biomass is dependent upon the consumption of substrate, the biomass in the growth vessel is determined by the difference between the concentrations of the growth-limiting substrate in the reservoir and in the growth vessel. Consequently, the higher the concentration of substrate in the reservoir, the greater the number of cells in the chemostat. However, one should not increase the amount of ammonium in the reservoir to too high levels, as all of the consumed ammonium is converted to toxic nitrite in the growth vessel, and this will negatively affect the cells.

2.2. General set-up of a batch culture experiment

A test tube or an Erlenmeyer flask is filled one-third with medium and capped with an aluminum cap or a cotton stopper. The use of plastic or rubber caps should be avoided because some ammonia oxidizers don't grow in the presence of plastic or rubber caps. For the best results, liquid batch cultures are inoculated with a late-exponential-phase culture using an amount that equals 10% of the volume of the new inoculated culture. For example, 50 ml of new medium would be inoculated with 5-ml culture. Growth in batch cultures is followed by measurement of the NH_4^+ consumption and NO_2^- or NO_2^- plus NO_3^- production, depending on the absence/presence of nitrite-oxidizing bacteria (NOB).

2.3. General set-up of a continuous culture experiment

Continuous cultures have been used to enrich AOB that are adapted to low ammonium concentration (Bollmann and Laanbroek, 2001), to conduct competition experiments between AOB and heterotrophic bacteria (Verhagen and Laanbroek, 1991), between different AOB (Bollmann et al., 2002), between AOB and NOB (Laanbroek and Gerards, 1993; Laanbroek et al., 1994), and to perform experiments that simulate environmental conditions (Bollmann and Laanbroek, 2002).

Continuous culture studies can be conducted in commercially available chemostats or self-built units (Fig. 3.1). Independent of which approach is used, the following conditions have to be adjusted and kept constant for controlled growth of AOB:

1. *Temperature*: Most continuous culture vessels have a double-wall system so that the temperature can be controlled with an external thermostat. Another option is to place the vessel in a temperature-controlled water bath or room.
2. *Mixing of the culture*: Commercially available chemostats have stirrer systems that mix the culture well. When using a self-built unit, a vessel with a flat bottom should be used. The vessel can be placed on a stirrer

and the liquid is mixed by a stir bar. It is important to observe the chemostat over time to ensure that no wall growth is building up. If the wall growth builds up, the stirrer speed can be increased. However, often a biofilm that has been already developed on the wall cannot be removed by increasing the stirrer speed. Therefore, the experiment should be restarted with a higher stirrer speed to prevent wall growth from the beginning.
3. *O_2 partial pressure*: Depending on the system, the O_2 partial pressure is kept constant by bubbling the culture with air or an O_2-containing gas mixture, or by changing the rate of mixing. Commercially available chemostats often have a unit that can be used to adjust the O_2 partial pressure to a fixed value. Gases for bubbling must be sterilized by filtration through 0.2-μm air filters to ensure that the culture is not contaminated by air-borne bacteria. In a self-built unit, a sparger with an air pump can be used as an alternative; however, these lack the option to adjust the O_2 partial pressure.
4. *pH value*: The pH value is a very important factor while growing ammonia oxidizers. It can be adjusted and kept constant by controlling with a pH electrode and addition of alkaline. If no pH control unit is available, the pH of the culture can be kept constant by using buffer-containing medium. The pH of the culture should be checked regularly with an outside pH electrode and if necessary the pH in the vessel can be readjusted.

2.4. Media

Mineral salt medium with NH_4Cl or $(NH_4)_2SO_4$ as the ammonium source is used as a medium to enrich and grow AOA and AOB. An overview of several different media has been given by Koops *et al.* (2006); most of them with rather high ammonium concentrations, but the general composition is the same. The most important difference among the media is the way by which the pH is maintained, that is, the choice for a certain buffer or alkaline solution. For enrichments and other experiments, we used the mineral salt medium with HEPES as buffering agent (Table 3.1; Verhagen and Laanbroek, 1991).

2.4.1. Preparation of the mineral salt medium

– Dissolve all the components listed in Table 3.1, except the KH_2PO_4.
– Adjust the pH with NaOH or Na_2CO_3 to 7.8.
– Autoclave the medium and add 10 ml/l of a separately autoclaved 40 mM KH_2PO_4 solution aseptically.

Table 3.1 Mineral salt growth medium for AOA and AOB (Verhagen and Laanbroek, 1991)

Ingredients	Concentration
$(NH_4)_2SO_4$	x^a mM
NaCl[b]	0.585 g/l (10 mM)
KH_2PO_4	0.054 g/l (0.4 mM)
KCl	0.075 g/l (1 mM)
$CaCl_2 \cdot 2H_2O$	0.147 g/l (1 mM)
$MgSO_4 \cdot 7H_2O$	0.049 g/l (0.2 mM)
Trace elements solution[c]	1 ml/l
0.04% Bromothymolblue solution (only in batch cultures)	5 ml/l
HEPES buffer[d]	$4x$ mM

[a] Variable amounts depending on strains and experiments.
[b] If the medium is used to enrich AOA or AOB from marine systems 30–40 g/l NaCl or sea salts (Sigma Aldrich, St. Louis, MO, USA), a salt mixture that resembles the composition of ocean water should be added.
[c] See Table 3.2.
[d] Phosphate buffer can be used as alternative buffer to HEPES, but many AOB are sensitive to too high phosphate concentrations and it is not clear how AOA react to phosphate.

The HEPES concentration is dependent on the NH_4^+ concentration. It should be always four times the NH_4^+ concentration on molar basis to provide the medium with a sufficient buffering capacity. If large amounts of media are needed on a regular basis, preparation of stock solutions of the different mineral salts saves a lot of time and effort. Therefore, 100× solutions of NaCl (1 M = 58.5 g/l), KCl (100 mM = 7.5 g/l), $CaCl_2 \cdot 2H_2O$ (100 mM = 14.7 g/l), and $MgSO_4 \cdot 7H_2O$ (20 mM = 4.9 g/l) can be prepared and added to the medium (10 ml/l). The $(NH_4)_2SO_4$ solution should be prepared in two concentrations (1 M NH_4^+ = 66.07 g/l and 10 mM NH_4^+ = 0.6607 g/l) and added to the medium to obtain the final concentration. The stock solutions should be autoclaved before use to prevent any fungal or bacterial growth during storage.

2.4.2. Preparation of the trace elements solution (Table 3.2; Verhagen and Laanbroek, 1991)

— Dissolve the chemicals one after the other.
— Sterilize the solution by filtration through a 0.22-μm filter.
— Store it in the dark at 4 °C.

The trace elements solution can be used for several years.

Table 3.2 Trace elements solution (Verhagen and Laanbroek, 1991)

Ingredient	Concentration
Na_2EDTA	4292 mg/l (11.5 mM)
$FeCl_2 \cdot 4H_2O$	1988 mg/l (10 mM)
$MnCl_2 \cdot 2H_2O$	81 mg/l (0.5 mM)
$NiCl_2 \cdot 6H_2O$	24 mg/l (0.1 mM)
$CoCl_2 \cdot 6H_2O$	24 mg/l (0.1 mM)
$CuCl_2 \cdot 2H_2O$	17 mg/l (0.1 mM)
$ZnCl_2$	68 mg/l (0.5 mM)
$Na_2MoO_4 \cdot 2H_2O$	24 mg/l (0.1 mM)
$Na_2WO_4 \cdot 2H_2O$	33 mg/l (0.1 mM)
H_3BO_3	62 mg/l (1 mM)

2.4.3. Preparation of the contamination test media

Different media with low concentrations of carbon should be used to test the cultures for potential contaminations by heterotrophic bacteria and fungi.

– *Medium 1*: 1 g/l tryptone, 0.5 g/l yeast extract, 0.5 g/l NaCl, pH 7.2.
– *Medium 2*: 3 g/l tryptic soy broth, pH 7.3.
– *Medium 3*: 0.5 g/l yeast extract, 0.5 g/l peptone, 0.5 g/l beef extract, pH 7.3 (Koops *et al.*, 2006).
– Aliquot 5 ml of the media into test tubes.
– Inoculate each tube with 1 ml of the AOA or AOB culture that needs to be tested.
– Incubate the tubes for 2–4 weeks and observe for contaminants visually.

3. Enrichment of Ammonia Oxidizers

In the past, ammonia oxidizers have been enriched mainly in batch cultures. Most of these batch culture enrichments had very high initial NH_4^+ concentrations, up to 20 mM NH_4^+. These are described in detail by Koops *et al.* (2006) and other authors. Here, we will focus on the enrichment of AOA and AOB in batch cultures at low NH_4^+ concentrations as well as enrichment of AOB in chemostats.

3.1. Enrichment of ammonia oxidizers in batch culture

Enrichments in batch cultures are started in the same way for AOA and AOB, but differ in the subsequent steps:

– Prepare mineral salt medium with 0.1–2 mM NH_4^+ and aliquot 50 ml into 125 ml Erlenmeyer flasks with cotton stoppers.

- Inoculate the medium with 1 g sediment or soil or 1–5 ml water or wastewater, depending on the aim of the enrichment.
- Take samples to determine the initial NH_4^+, NO_2^-, and NO_3^- concentrations.
- Incubate the flask in the dark. The incubation temperature depends on the environmental temperature; samples from permanently cold environments should be incubated at low temperatures (i.e., 4 °C), samples from temperate environments at moderate temperatures (i.e., 25 °C), and samples from hot springs at the ambient hot spring temperature.
- Take weekly samples and determine the NH_4^+ concentration. When around 80% of the NH_4^+ has been consumed, split the culture to enrich the AOA and AOB separately.

3.1.1. Enrichment of AOB

- Transfer 5 ml of the culture to 45 ml fresh medium with the same NH_4^+ concentration as the initial enrichment culture.
- Take weekly samples and determine the NH_4^+ concentration. When around 80% of the NH_4^+ is consumed, transfer the culture to fresh medium.
- Repeat this step two to five times until the culture shows reproducible growth.

Repeated transfers can lead to the selection of one dominant AOB competing out all the other AOB. The selection of AOB can be fostered by the use of different NH_4^+ concentrations (Bollmann and Laanbroek, 2001; Suwa et al., 1994) or by the addition of other selecting factors such as salt (Bollmann and Laanbroek, 2002). When the activity of the culture is constant, analyze 50 ml of the culture to determine the identity of the AOB by molecular analysis of the *amoA* and the 16S rRNA gene (see Section 3.1.3 and Appendix A).

3.1.2. Enrichment of AOA

AOA are further enriched by an additional selective step, which can be a filtration of the culture through a 0.45-μm filter, or addition of the antibiotic streptomycin (de la Torre et al., 2008; French et al., unpublished data; Konneke et al., 2005; Mosier, personal communication; Santoro et al., 2008). We tested both selective methods and have had good experience with the use of the filtration step without adding antibiotics; an observation that was confirmed by other studies (Mosier, personal communication; Santoro et al., 2008).

- After splitting the original enrichment culture, filter 5 ml of the original enrichment culture through a 0.45-μm filter into 45 ml of fresh mineral salts medium.

- Monitor the culture for the consumption of NH_4^+, which may need several weeks to months to start again.
- Filter–transfer the culture again when around 80% of the NH_4^+ has been consumed.
- Repeat this filter transfer at least five times to ensure that no AOB are left in the culture.

3.1.3. Molecular analysis of the AOA and AOB enrichment cultures

- Filter 50 ml of the AOA or AOB culture onto a 0.1-μm nitrocellulose filter.
- Store the filters in a -20 or $-80\ °C$ freezer.

Molecular analysis should be done to determine the identity of the enriched AOA or AOB based on the *amoA* gene, and to demonstrate the absence of AOB in the AOA enrichment cultures based on the *amoA* and/or the 16S rRNA gene (see Appendix A). It is very important to verify that the AOA enrichment culture does not contain any AOB. The presence of AOA in AOB enrichment cultures is normally not a problem, because AOB grow faster in the laboratory and outcompete the AOA in the enrichment cultures.

3.2. Enrichment of ammonia-oxidizing bacteria in continuous culture

Chemostats have been used to enrich AOB under controlled conditions and at low substrate availability (Bollmann and Laanbroek, 2001, 2002).

The basic set-up of a chemostat should follow the manufacturer's recommendations. It should include a pH electrode, a bottle with connecting tubing to add alkaline solution, an O_2 electrode (if necessary), a sparger with air filter connected to an air-mixing unit, air outflow, a bottle with connecting tubing to add liquid to the chemostat such as additional medium and inoculum, a sampling unit, media influx tubing with connector including a pipette to determine the flux of the medium, and media efflux tubing.

- Calibrate the pH and O_2 electrodes.
- Prepare mineral salt medium without HEPES buffer and transfer it into the chemostat. HEPES buffer is not added to this medium because the pH is followed using a pH electrode and adjusted by addition of Na_2CO_3 by a pump that is connected to the pH electrode. If no pH electrode and control unit are available, medium with HEPES buffer can be used instead.
- Assemble the chemostat according to the manufacturer's manual.
- Autoclave the complete chemostat. Connecting tubing must be closed except for one tubing. Preferably, this should be the air outflow so that

the vessel does not break due to the overpressure build up during autoclaving. Closing all other tubings is especially important for tubing with in- or outlets below the surface of the medium in the growth vessel. If they are not closed, overpressure during autoclaving will empty the vessel.
- Let the chemostat completely cool down after autoclaving.
- Activate the temperature control, mixing, and bubbling; the latter with the same rate that will be used for the growth of the ammonia oxidizers.
- Activate the pH electrode, add the phosphate solution aseptically to the medium, and adjust the pH with Na_2CO_3 solution.
- Activate the O_2 electrode and adjust the O_2 partial pressure.
- Prerun the chemostat overnight to ensure that everything is working.
- For enrichment experiments, inoculate the chemostat with sediment, water, extracted bacteria, or any other sample from which AOB should be enriched (Bollmann and Laanbroek, 2001, 2002).
- Wrap the chemostat with aluminum foil to exclude light because AOB are sensitive to ultraviolet and blue light (Hooper and Terry, 1974).
- The pump has to be calibrated because a calibration before autoclaving might change during autoclaving. Pump calibration is done with a burette that is inserted into the tubing between the medium reservoir and the pump using a T-piece (Fig. 3.1).
- Connect the burette with the reservoir and fill it with medium.
- Close the tubing between the medium reservoir and the burette so that the pump is pumping medium from the latter.
- Measure the flow rate of the pump per minute.
- Adjust pump rate and repeat the measurement until the pump reaches the desired rate.
- Disconnect the tubing toward the pipette and open the tubing toward the growth vessel so that medium is pumped from the reservoir to the vessel.
- Start the in- and outlet pumps with a dilution rate of 0.01–0.001 h^{-1} to add mineral salt medium from the medium reservoir with an NH_4^+ concentration of 0.25 mM or less.
- Take samples on a daily basis and analyze for NH_4^+, NO_2^-, NO_3^-, and pH.
- Follow the enrichment on the community level using molecular methods once every volume change.

After five volume changes (one volume change equals the addition/removal of one time the complete volume of the liquid in the chemostat), theoretically less than 1% of the original inoculum or of the former steady-state culture is left in the growth vessel. Therefore, a chemostat should always run at least for five volume changes without any disturbances to reach a new steady state. The steady state is determined by sampling the culture on a regular basis to analyze the NH_4^+, NO_2^-, and NO_3^-

concentrations, to measure the pH, and to determine biomass or community composition. For practical reasons, steady state is often assumed after three volume changes when no changes have been occurred in substrate and product concentrations as well as in number of cells. During long-term experiments (up to 2 months), it is important to control all of these factors before and during the growth period on a regular basis. It is also important to monitor the material such as the tubing and the filters to ensure that the material is not wearing out and jeopardizing the experiment.

Here, we described the use of chemostats to enrich AOB. Chemostats can also be used to investigate the competition between AOB and other organisms (Bollmann et al., 2002; Verhagen and Laanbroek, 1991) or to grow pure and enrichment cultures under controlled conditions.

If the chemostat reaches a steady state, the NH_4^+ concentration in the reservoir can be doubled to increase the biomass of the enrichment culture. This step can be done several times; however, after each increase the chemostat should run for three to five volume changes to reach a new steady state and to ensure stability of the community.

If ammonia oxidation and community composition of the AOB in the chemostat are stable, samples can be taken and further enriched in batch culture. Samples (50 ml) from the chemostat are transferred to 125 ml Erlenmeyer flasks with cotton stoppers. NH_4^+ and HEPES (0.5 ml each) will be added from stock solutions (1.65 g/l $(NH_4)_2SO_4$ (25 mM); 23.8 g/l HEPES (100 mM), pH 7.8) to obtain a final concentration of 0.25 mM NH_4^+ and 1 mM HEPES. The initial NH_4^+ concentration should not exceed 0.25 mM because higher NH_4^+ concentrations could favor the growth of AOB that are adapted to higher NH_4^+ concentrations, if they are still present in the enrichment culture.

3.3. Enrichment medium prepared with water from aquatic environments

Even if the conditions of the environment were met as best as possible by adjusting the NH_4^+ concentration in the medium, enrichments often fail. This may be due to the lack of trace nutrients and signaling compounds in the mineral salts medium. An option to initially enrich sensitive AOB from aquatic environments is to use filter sterilized freshwater, marine, and wastewater directly as enrichment medium (Bollmann and Laanbroek, 2002; Konneke et al., 2005).

— Prepare the medium by sterile filtration of the water from the selected environment via a 0.1–0.2-μm filter and add NH_4^+ and HEPES to final concentrations between 0.1–1 mM and 0.4–4 mM, respectively. The water should be sterile filtered rather than autoclaved, because during autoclaving changes in the chemical composition of the water could occur.

- Inoculate 5 ml of the original water in 45 ml medium in a 125-ml Erlenmeyer flask with a cotton plug.
- Follow the NH_4^+ concentration in the culture.
- Transfer to new medium as described in Section 3.1. for AOA and AOB when around 80% of the NH_4^+ is consumed.

When the culture is growing stably, it can be adapted to mineral salt medium by transferring it to a mix of 50% mineral salt medium and 50% of the original water. When the culture grows stably in the mixed medium, it can be transferred to pure mineral salt medium. The cultures can then be further cultivated and analyzed as described in Section 3.3. The same experimental set-up can be used to enrich AOA from freshwater and marine environments by filtering the culture over a 0.45-µm filter during every transfer.

During chemostat enrichments, sterile filtered original water supplemented with NH_4^+ can be used as medium. This was successfully done with water from the Scheldt estuary for the enrichment of AOB adapted to low NH_4^+ concentrations (Bollmann and Laanbroek, 2002).

4. Isolation of Ammonia Oxidizers

AOA and AOB can be isolated into pure culture using different methods such as dilution to extinction, serial dilution, or pour plates (Aakra et al., 1999; Jones et al., 1988; Konneke et al., 2005; Koops et al., 1976, 2006; Stehr et al., 1995; Suwa et al., 1994; Watson, 1965). Isolation is very laborious and time-consuming; however, pure cultures are necessary for specific research questions.

4.1. Isolation of ammonia-oxidizing bacteria using pour plates

Isolation in pour plates is only possible if the AOB have first been adapted to grow at NH_4^+ concentrations of 5 mM or higher; otherwise the colonies are too small to be picked and isolated. To adapt the culture to higher NH_4^+ concentrations, transfer the culture from mineral salt medium with 0.25 mM NH_4^+ to mineral salt medium with 1 mM NH_4^+, and after two transfers to mineral salt medium with 5 mM NH_4^+. The culture needs to grow for two to three transfers and is used at the end of the exponential growth phase.

- Prepare mineral salt medium with 5 mM NH_4^+, no bromothymolblue, and 1.5% Bacto agar in 4 × 100 ml portions for each culture.

- After autoclaving, let the medium cool down to 40–45 °C and add 1 ml of 40 mM KH$_2$PO$_4$ solution.
- Dilute the original culture in mineral salt medium with 5 mM NH$_4^+$ without agar to 10^{-2}–10^{-5} of its original concentration.
- Add 1 ml of the diluted cultures to 100 ml of the agar medium, mix well, and pour into four to five plates with a diameter of 10 cm.
- Close the plates with parafilm to prevent drying and incubate them for 3–6 months at room temperature or at the temperature at which the culture was growing. The best is just to forget about them for a while.
- Screen the plates under a dissecting scope for red-brownish small colonies.
- Prepare mineral salt medium with 5 mM NH$_4^+$ and aliquot it out to test tubes or 12- or 24-well plates.
- Pick red-brownish colonies with a sterile Pasteur pipette and transfer the bacteria to the mineral salt medium.
- Incubate the cultures for 4 weeks, take samples and test them for NH$_4^+$ consumption and NO$_2^-$ production.
- Transfer cultures that consumed around 80% or more of the NH$_4^+$ to fresh medium.
- Test the cultures that did not use 80% of the NH$_4^+$ again every 2 weeks up to 3 months and transfer them too when they consumed 80% or more of the NH$_4^+$.
- Test the transferred cultures again for NH$_4^+$ consumption and/or NO$_2^-$ production after 2–4 weeks of incubation and transfer if necessary.
- During the third transfer, inoculate the culture into contamination test media to look for contamination with heterotrophic bacteria.
- Incubate these contamination tests for at least 3–4 weeks. Cultures that show heterotrophic contamination should be omitted.
- Transfer noncontaminated cultures to 50 ml of mineral salt medium in a 125-ml Erlenmeyer flask in parallel to further cultivation in test tubes.
- When around 80% of the NH$_4^+$ has been used, filter the culture in the Erlenmeyer flask onto a 0.2-μm nitrocellulose filter for molecular analysis to identify the isolated strain. Filters should be stored at -20 or -80 °C.

4.2. Isolation of ammonia-oxidizing bacteria by dilution to extinction in liquid media

This method can be used to isolate AOB at lower NH$_4^+$ concentrations.

- Grow the culture until around 80% of the NH$_4^+$ is consumed.
- Stain the culture with DAPI (Bloem and Vos, 2008) and count the cells under an epifluorescense microscope or in a microscopic counting chamber.
- Dilute the cells in 100 ml mineral salt medium with NH$_4^+$ (NH$_4^+$ concentration is the same as in the enrichment culture) to a final concentration of 2.5 cells/ml.

- Fill out 200 µl of the diluted cells per well into five 96-well plates.
- Cover the plate with parafilm and incubate it for 2–3 months at room temperature or at the temperature at which the culture was growing.
- After the incubation, sample 10 µl per well and pipette it into another 96-well plate.
- Determine NO_2^- in the diluted samples as described in Appendix B.2.
- Transfer positive samples to fresh mineral salt medium and treat as described in Section 4.1 to determine the purity and the identity of the isolate.

Sometimes isolations are not successful. This can have several reasons, such as too few cells or too high NO_2^- concentrations. It is very likely that an enrichment culture contains signaling and other compounds excreted by the microbes that increase the growth rate and reduce the lag phase of AOB as shown for AOB in pure cultures (Batchelor et al., 1997). Therefore, if AOB are not growing due to the low cell number (0.5 cells per well) it could help to replace 50% of the mineral salt medium with spent medium. Spent medium is prepared by sterile filtration of the culture over a 0.22-µm filter. The filtrate is mixed with autoclaved mineral salt medium and the NH_4^+ concentration as well as the pH is adjusted to the original concentration in the medium of the enrichment culture. The mineral salt medium with the spent medium is used in the same way as described in this section for the mineral salt medium alone.

Another problem could be NO_2^- toxicity. Often, freshly isolated AOB are consuming only part of the NH_4^+ in the medium, indicating that these AOB are very sensitive to elevated NO_2^- concentrations. In enrichment cultures, the AOB grow often in coculture with NOB, which consume the NO_2^- produced by the AOB. When isolated into pure culture, the AOB are separated from the NOB and exposed to increasing NO_2^- concentrations in the medium. To overcome this problem, the NH_4^+ concentration in the mineral salt medium used for the isolation and further cultivation can be reduced. Consequently, the NO_2^- concentration will be lower and the AOB might be able to grow.

The use of 96-well plates could be another potential problem because 96-well plates consist of plastic and some AOB are sensitive to plastics. If AOB cannot be isolated in 96-well plates, it might be worthwhile to test the use of glass test tubes with aluminum caps.

4.3. Isolation of ammonia-oxidizing bacteria by serial dilution in liquid media

This method can be used to isolate AOB at lower NH_4^+ concentrations.

- Grow the culture until around 80% of the NH_4^+ is consumed.
- Conduct an MPN dilution series (Alexander, 1982) in a 96-well plate using the medium with the same NH_4^+ concentration as in the enrichment culture. Each step should be a 10 times dilution.

- Cover the plate with parafilm and incubate it for 2–3 months at room temperature or at the temperature at which the culture was growing.
- After the incubation, sample 10 µl per well and pipette it into another 96-well plate.
- Determine NO_2^- in the diluted samples as described in Appendix B.2.
- Use highest positive dilution for another dilution series.
- Repeat the dilution series two to three times.
- Transfer positive samples from the highest positive dilution after at least three consecutive serial dilutions to fresh mineral salt medium and treat as described in Section 4.1 to determine the purity and the identity of the isolate.

4.4. Stock culturing of ammonia-oxidizing bacteria

For stock culturing, AOB should be transferred to fresh mineral salt medium every 4–6 weeks. It is best to use test tubes with aluminum caps filled with 10 ml mineral salt medium and an NH_4^+ concentration at which the culture grows optimally. Before transfer, the consumption of NH_4^+ should be determined even if the pH indicator shows that the culture is growing because sometimes the indicator changes color in the absence of growth. In addition to transferring the cultures to fresh mineral salt medium, they should also be transferred to contamination test media to ensure purity.

4.5. Long-term storage of ammonia-oxidizing bacteria

For long-term storage, the cultures can be stored as glycerol stocks in an $-80\ °C$ freezer. Another possibility is to store them in liquid nitrogen (Koops et al., 2006).

- Grow the culture in 50 ml of mineral salt medium.
- Spin the cells down when they have used around 80% of the NH_4^+.
- Resuspend the pellet in 1 ml mineral salt medium.
- Add 1 ml mineral salt medium with 30% glycerol.
- Transfer 0.5 ml concentrated culture into cryo tubes and freeze them immediately in the $-80\ °C$ freezer. The cultures can be stored for several years.
- To revive the culture, spin-down the cells and remove the mineral salt medium with glycerol because glycerol could inhibit the growth of the AOB.
- Resuspend the pellet in mineral salt medium and transfer it to a test tube with 10 ml of medium.

The culture should start to grow again after 4–8 weeks of incubation. When grown up the cultures should be tested for NH_4^+ consumption and contaminations as described in Section 4.1.

4.6. Isolation of ammonia-oxidizing archaea by serial dilution in liquid media

Only one strain of AOA has been isolated so far: *N. maritimus* (Konneke *et al.*, 2005). The two other strains that have been investigated in more detail, that is, *Candidatii* Nitrososphaera gargensis (Hatzenpichler *et al.*, 2008) and Nitrosocaldus yellowstonii (de la Torre *et al.*, 2008) are still enrichment cultures. Isolation of *N. maritimus* was achieved by a combination of serial dilution (as described in Section 4.3), addition of streptomycin and filtration through a 0.45-μm filter (Konneke *et al.*, 2005). The last two steps help to eliminate bacteria in the enrichment culture. Experience has shown that during enrichment, filtration is more important than the addition of antibiotics to eliminate AOB and other heterotrophs.

5. CHARACTERIZATION OF AMMONIA-OXIDIZING PROKARYOTES

5.1. Growth versus activity kinetics

Growth and activity kinetics are two ways to express the metabolism of microbes. Measuring the growth rate shows how fast microbes are able to divide under specific environmental conditions. Growth involves many different physiological processes such as duplication of DNA and synthesis of new protein (Kovarova-Kovar and Egli, 1998). Kinetic variables that can be determined during growth experiments include specific growth rates, the lag time before exponential growth starts, the half saturation constant of growth (K_s), and the maximum growth rate (μ_{max}). Activity measurements focus on substrate transformation by one enzyme in particular; with whole organisms this usually relates to specific transport processes of substrates. Kinetic parameters that are determined during activity measurements are the transformation rate of the substrate, the half saturation constant of activity (K_m), and the maximum transformation rate of the substrate (V_{max}).

5.2. Determination of growth kinetics of ammonia oxidizers

- Prepare the mineral salt medium with NH_4^+ and the corresponding HEPES concentration in 45 ml portions in 125 ml Erlenmeyer flasks.
- Inoculate the medium with 5 ml of a late-exponential-phase culture.
- Take a sample of 1 ml, spin-down the sample for 10 min at 15,000 rpm, and transfer the supernatant to a new microcentrifuge tube.
- Store the sample at −20 °C.
- Analyze the samples for NH_4^+, NO_2^-, and/or NO_3^-, depending on the sample and the experiment. Depending on the presence of nitrite

oxidizers, the samples will be analyzed for NO_2^- only or for NO_2^- plus NO_3^- (methods in Appendix B).

The sampling pattern has to be adapted to the growth behavior of the culture; slowly growing cultures only need to be sampled every other day, rapidly growing cultures or measurements done at very low substrate concentrations often need to be sampled several times during the day.

5.2.1. Calculation of the growth kinetics

Calculation of the growth kinetics of AOB and AOA is based on the assumption that NO_2^- or NO_2^- plus NO_3^- production is proportional to the growth of the microorganisms determined as cell numbers or as *amoA* copy numbers. This proportionality has been demonstrated for the AOB *N. europaea*, *Nitrosomonas* sp. FH1, *Nitrosospira* sp. AV3, and *Nitrosospira* sp. AV2 (Belser and Schmidt, 1980) and the AOA *N. maritimus* (Konneke et al., 2005). In certain experiments (such as the investigation of the effect of stress factors or growth inhibiting substances on the growth), it is important to determine the biomass yield at the end of the experiment to ensure that the observed growth under different conditions, but in the presence of the same amount of NH_4^+, still results in the same cell numbers. Cell numbers at the end of the experiment can be determined by counting the cells after staining with DAPI in pure cultures (Bloem and Vos, 2008) or FISH probes in enrichment cultures (Schramm, 2003), or determining the cell numbers by quantitative PCR with ammonia oxidizer specific primers (Table 3.3) using an SYBR Green approach (Herrmann et al., 2009; Walker, 2002).

- Log transform the NO_2^- or the sum of the NO_2^- and NO_3^- concentrations and plot against time.
- Determine the specific growth rate $[h^{-1}]$ or $[day^{-1}]$ by calculating the slope of the log transformed data plotted against time during the exponential growth phase.
- The lag phase [h] or [days] is defined as the time before the culture starts to grow exponentially.

5.2.2. Determination of K_s and μ_{max} in batch culture

K_s and μ_{max} of an AOB culture are determined by measuring the growth rates of the culture at different NH_4^+ concentrations. The calculation of K_s and μ_{max} is based on Monod kinetics (Kovarova-Kovar and Egli, 1998).

- Prepare mineral salt medium with NH_4^+ concentrations between 0.1 and 5 mM NH_4^+ and the corresponding HEPES concentrations between 0.4 and 20 mM HEPES.
- Inoculate all media with 1–10% of late-exponential growth phase culture.
- Determine specific growth rates as described in Section 5.2.1 and plot these growth rates against the NH_4^+ concentration.

- Calculate K_s and μ_{max} using Monod kinetics based on the formula.

$$\mu = (\mu_{max} * [S])/(K_s + [S])$$

with μ being the specific growth rate and [S] the substrate concentration. It is not clear whether K_s and μ_{max} can be determined with this method for AOA because the K_m value for NH_3 of *N. maritimus* is around 1000 times lower than the K_m of different AOB (Bollmann et al., 2005; Koops et al., 2006; Martens-Habbena et al., 2009). Therefore, it is very likely that the K_s of AOA is also much lower than the K_s of AOB and might be very difficult to determine. More sensitive methods for detecting growth at low substrate concentrations need to be developed.

5.2.3. Influence of environmental factors on the growth rates of ammonia oxidizers

Changes in the specific growth rate are very good indicators for stress.

- Prepare mineral salt medium with NH_4^+ at which the culture grows optimally. Vary the concentration of the factor that is being investigated, such as for example high or low NH_4^+ concentrations, O_2 concentration, pH value, and addition of carbonate, organic carbon, heavy metals, or nitrite.
- Inoculate 45 ml of mineral salt medium with 5 ml of a late-exponential growth phase culture and determine the specific growth rates as described in Section 5.2.1.

Investigation of the influence of the O_2 partial pressure can be done in gas-tight bottles that are first flushed with nitrogen. Sterile O_2 or air is then added aseptically to obtain a defined O_2 concentration. At low O_2 concentrations, the ratio between gas phase and liquid phase needs to be large enough to ensure that O_2 does not become the limiting factor. During the investigation of the influence of the pH value, it is important to keep in mind that the availability of NH_3 (the substrate of the ammonia monooxygenase) in the medium depends on a combination of NH_4^+ concentration and pH value. When the pH is varied and the NH_4^+ concentration is kept constant, the NH_3 concentration in the medium decreases with decreasing pH value. To keep the NH_3 concentration constant, the NH_4^+ concentration needs to be increased accordingly with decreasing pH.

5.3. Determination of the activity kinetics of ammonia oxidizers

The oxidation of ammonia by AOA and AOB can be determined in two different ways: by measuring the consumption of O_2 (Laanbroek and Gerards, 1993; Laanbroek et al., 1994; Martens-Habbena et al., 2009) or by measuring the production of NO_2^- plus NO_3^- (Bollmann et al., 2005). Both can be done with sensors immersed in concentrated cell suspensions during short-term measurements. The basic difference between these set-ups is that the O_2 consumption should only be measured with pure cultures. When the culture is contaminated, the contaminants will consume O_2 and add to the rates of the overall O_2 consumption. During NH_4^+ uptake measurements in mixed cultures of AOB and NOB, it is assumed that the NOB are not adding to the O_2 uptake during short-term measurements due to their high K_m value for NO_2^- and the low NO_2^- concentration directly after addition of the NH_4^+ (Laanbroek and Gerards, 1993; Laanbroek et al., 1994). Measurement of the production of NO_2^- plus NO_3^- measures only the products of the NH_4^+ oxidation by the AOA or AOB. Because the sensor measures NO_2^- plus NO_3^-, the absence or presence of NOB and heterotrophic bacteria does not have any influence on the ammonia oxidation rates. Here, we will describe two different methods to measure the activity of AOA and AOB: (1) activity of AOB measured by NO_2^- plus NO_3^- biosensor (Bollmann et al., 2005), (2) activity of AOB and AOA measured with an O_2 sensor (Laanbroek and Gerards, 1993; Laanbroek et al., 1994; Martens-Habbena et al., 2009).

5.3.1. Activity of ammonia-oxidizing bacteria using the NO_2^- plus NO_3^- biosensor

The NO_2^- plus NO_3^- biosensor has been developed by Larsen et al. (1996, 1997) and is commercially available from the company Unisense, Aarhus, DK (www.unisence.com).

— Assemble and calibrate the sensor according to the manufacturer's recommendations.
— Grow the AOB cultures (at least 1 l) in mineral salt medium.
— Spin down (20 min at 12,000–15,000 rpm) the culture, wash the cell pellet twice with mineral salt medium without NH_4^+, resuspend in 1% of their original volume in mineral salt medium without NH_4^+, and store on ice.
— Place a flat bottom glass vessel with a small stir bar in a water bath on a stirrer to keep the temperature constant and able to mix the culture well, respectively.
— Add 3.6 ml sterile mineral salt medium without NH_4^+ and 0.4 ml of the concentrated cells into the flat bottom glass vessel.
— Stir and bubble the culture with air.
— Equilibrate the NO_2^- plus NO_3^- biosensor in the culture.
— Add NH_4^+ when the sensor signal is stable and follow the production of NO_2^- plus NO_3^- in time.
— Calculate the NH_4^+ consumption/NO_2^- plus NO_3^- production rate based on the linear increase of the NO_2^- plus NO_3^- concentration over time.
— Take a subsample of the culture to determine the microbial biomass by counting cell numbers after staining with DAPI (Bloem and Vos, 2008)

or by determining the protein content (Bradford, 1976). When working with an enrichment culture, it might be necessary to use FISH probes to count the specific number of AOA or AOB in the sample (Hatzenpichler et al., 2008; Konneke et al., 2005; Schramm, 2003).

5.3.2. Activity of ammonia-oxidizing bacteria using the O_2 sensor

– Assemble the O_2 sensor and the glass vial according to the manufacturer's recommendations and place the glass vial in a water bath on a stirrer to keep the temperature constant and mix the culture.
– Calibrate the O_2 sensor. It is important to ensure that the vial is gastight, because otherwise O_2 could leak into the vial and falsify the results.
– Concentrate the cells as described in Section 5.3.1, add mineral salt solution and cells to the vial, and stir and bubble the culture thoroughly.
– Stop bubbling and place O_2 sensor into the vial so that the vial is airtight.
– Follow the O_2 concentration until stabilized.
– Add NH_4^+ to start the experiment and follow the decrease in the O_2 concentration.
– Calculate the O_2 consumption rate from the linear decrease of the O_2 concentration over time after subtraction of the background O_2 consumption.
– Take a subsample of the culture to determine the microbial biomass as described in Section 5.3.1.

5.3.3. Determination of K_m and V_{max}

Both of these experimental set-ups can be used to determine the K_m and V_{max} of ammonia oxidation and the influence of environmental factors on the activity of the AOB (Laanbroek and Gerards, 1993; Laanbroek et al., 1994). K_m and V_{max} are calculated based on Michaelis–Menten kinetics (Kovarova-Kovar and Egli, 1998). The NO_2^- plus NO_3^- production rates or O_2 consumption rates are plotted against the NH_4^+ concentration and the formula:

$$V = (V_{max}[S])/(K_m + [S])$$

is used to calculate the V_{max} and K_m as defined in Section 5.1.

5.3.4. Influence of environmental factors on the ammonia oxidation rates

Ammonia oxidation rates can be used to determine the influence of environmental factors such as O_2, salt, pH, nitrite, starvation, and heavy metals on the activity of the AOB (Bollmann et al., 2005; Laanbroek and Gerards, 1993; Laanbroek et al., 1994; Verhagen and Laanbroek, 1991).

- The culture will be cultivated in the absence and presence of the stress factor under investigation.
- Take 100 ml sample from the culture.
- Spin down (20 min at 12,000–15,000 rpm) the sample, wash the cell pellet twice with mineral salt medium without NH_4^+, resuspend in 1% of their original volume in mineral salt medium without NH_4^+.
- Treat sample as described in Section 5.3.1 and measure ammonia oxidation rates at medium NH_4^+ concentration (0.1–0.5 mM).

For these kinds of measurements, it is important to ensure that the samples are always handled in the same way to ensure comparability of the results. If the culture that is investigated is still growing or cell numbers could vary for other reasons, samples should be taken for biomass determination as described in Section 5.3.1.

5.3.5. Activity of AOA using the O_2 sensor

Ammonia oxidation kinetics of AOA was determined by a slightly different method based on O_2 consumption with a microrespiration system based on Clark-type O_2 microsensors (Martens-Habbena *et al.*, 2009). *N. maritimus* cells are more sensitive to centrifugation or filtration treatment than cells of AOB. Consequently, the O_2 uptake rates of late-exponential or early-stationary phase cells were measured without concentrating the microbial biomass before the measurement. The NH_4^+ oxidation rates were calculated from O_2 uptake rates according to the ratio of NH_4^+ consumption to O_2 uptake of 1:1.5. Calculation of the K_m value using Michaelis–Menten kinetics resulted in very low K_m values for NH_4^+ and O_2 uptake (Martens-Habbena *et al.*, 2009).

Appendix A. Identification of Ammonia Oxidizers Based on 16S rRNA and Functional Genes

AOA and AOB in enrichment and pure cultures can be identified by the amplification and sequencing of the 16S rRNA gene (DeLong, 1992; Kowalchuk *et al.*, 1997; Lane, 1991) or the *amoA* gene (Francis *et al.*, 2005; Rotthauwe *et al.*, 1997).

A.1. DNA extraction

Before DNA extraction, cultures of AOA and AOB need to be collected on filters, because the cell density in the cultures is very low.

- Filter 20–50 ml of an AOA or AOB culture onto a 0.1- or 0.2-µm nitrocellulose filter, respectively, and store the filters in the −20 or

−80 °C freezer in a screw cap tube that can be used for bead-beating. AOA can be very small and need to be filtered onto 0.1 μm filters to get quantitatively representative samples.
- Add 1 g silica/zirconia beads (diameter 0.1 mm) and 500 μl high salt buffer (1 M NaCl, 5 mM MgCl$_2$, 10 mM Tris, pH 8) to the filter.
- Bead-beat the tubes three times for 30 s at highest speed in the bead-beater. Store samples between runs for at least 15 min on ice.

Further DNA isolation is conducted using the Qiagen DNeasy Blood and Tissue kit (Qiagen, Valencia, CA).

- Add 10 μl of proteinase K and 500 μl buffer AL, mix well and incubate for 20 min at 56 °C.
- Add 500 μl 95% ethanol, mix well, and transfer the liquid in two to three steps to the spin columns provided by the kit.
- Handle the spin columns as described by the manufacturer.
- Elute the DNA using 100 μl of the supplied elution buffer and store the DNA at −20 °C. If the DNA concentration is very low, the elution volume can be decreased.

If the DNeasy Blood and Tissue kit is not available, the DNA can be cleaned up with a phenol/chloroform extraction and precipitation with 70% ethanol (Bollmann and Laanbroek, 2001).

Quantitative PCR demonstrated that the described methods are not as efficient as the modified Power soils DNA isolation method (MoBio, Carlsbad, CA) when isolating DNA from AOA. Therefore the use of the following protocol is recommended for experiments involving quantification of AOA. DNA is isolated using Power soil DNA isolation kit (MoBio, Carlsbad, CA) according to the manufacturers recommendations with the following modifications:

- Filter 20–50 ml of an AOA culture onto 0.1 μm nitrocellulose filters.
- Place the filter in a 2 ml conical tube (cap sealed with o-ring for usage in bead beaters) and decant the beads and buffer provided in the Mo-Bio tubes into the tube with the filter.
- Homogenize the filter in a Bead Beater for 30s at highest speed and place on ice afterwards.
- Add 60 μl of the provided C1 lysis buffer and continue with the recommended protocol provided by Mo-Bio.

A.2. Amplification of the *amoA* and 16S rRNA gene

The gene encoding the structural subunits of the ammonia monooxygenase has been used to identify AOA and AOB. The most widely used primers target the *amoA* genes (Francis *et al.*, 2005; Rotthauwe *et al.*, 1997), but primers targeting the other subunits are also available (Junier et al., 2008a, 2010). The 16S rRNA gene can also be used to detect betaproteobacterial AOB (Kowalchuk et al., 1997), to determine community composition in enrichment cultures (Muyzer et al., 1993), and to identify pure cultures of AOA and AOB (DeLong, 1992; Lane, 1991). Comprehensive overviews of the available primers have been published by Koops et al. (2006) and Junier et al. (2008b, 2010). Primers and PCR conditions to identify AOA and AOB have been summarized in Table 3.3. The variety of kits for PCR on the market is very high and we recommend using the PCR kit that is usually used in your laboratory. The recommended conditions have been selected from papers and experience in the laboratory, but it is absolutely necessary to optimize the PCR with respect to annealing temperature, primer concentration, Mg^{2+} concentration, addition of BSA, and the amount of template, because every PCR mix and machine varies.

A.3. Identification of ammonia oxidizers in enrichment cultures

— Amplify the DNA with archaeal or bacterial *amoA* primers or bacteria AOB 16S rRNA primers (Table 3.3).
— Clone the PCR products with a commercially available cloning kit according to the manufacturer's recommendations.
— Amplify clones with M13 or other primers specific for the amplification of the used vector.
— Clean and sequence PCR products.

If many enrichment cultures need to be screened, the DNA could be amplified with primers with GC clamp and subjected to Denaturing Gradient Gel Electrophoresis (DGGE, Appendix A.5). Based on the DGGE results, a lower number of enrichment cultures can be chosen and further analyzed as described in this section. Additionally, clones can be screened against the original enrichment culture to ensure that all sequences are represented.

A.4. Identification of ammonia oxidizers in pure cultures

A.4.1. Ammonia-oxidizing archaea

— Amplify and sequence the *amoA* gene with the ARCH amoAF and ARCH amoAR primer (Table 3.3).
— Amplify and sequence the 16S rRNA gene with Arch21F and Arch958R primers (Table 3.3).

A.4.2. Ammonia-oxidizing bacteria

— Amplify and sequence the *amoA* gene with the amoA-1F and amoA-2R KS primer (Table 3.3).

Table 3.3 Primers and PCR conditions to identify AOA and AOB in enrichment and pure cultures

Primer	Sequence	Denaturation Temp.	Denaturation Time	Annealing Temp.	Annealing Time	Extention Temp.	Extention Time	Max cycle
Archaeal amoA gene (Francis et al., 2005)								
Arch amoAF	5′-STA ATG GTC TGG CTT AGA CG-3′	95	30	53	60	72	60	35
Arch amoAR	5′-GCG GCC ATC CAT CTG TAT GT-3′							
Bacterial amoA gene (Rotthauwe et al., 1997)[a]								
amoA-1F	5′-GGG GTT TCT ACT GGT GGT-3′	95	30	55	60	72	60	35
amoA-2R KS	5′-CCC CTC KGS AAA GCC TTC TTC-3′							
AOB specific Bacterial 16S rRNA primer (Kowalchuk et al., 1997)[b]								
CTO189F-A	5′-GGA GAA AAG CAG GGG ATC G-3′	95	30	57	60	72	45+1	35
CTO189F-B	5′-GGA GGA AAG CAG GGG ATC G-3′							
CTO189F-C	5′-GGA GGA AAG TAG GGG ATC G-3′							
CTO654R	5′-CTA GCY TTG TAG TTT CAA ACG C-3′							
General bacterial 16S rDNA for identification (Lane, 1991)								
27F	5′-AGA GTT TGA TCC TGG CTC AG-3′	95	30	50	30	72	60	28
1492R	5′-GGT TAC CTT GTT ACG ACT T-3′							
General archael 16S rDNA for identification (DeLong, 1992)								
Arch 21F	5′-TTC CGG TTG ATC CYG CCG GA-3′	95	90	55	90	72	90	30
Arch 958R	5′-YCC GGC GTT GAM TCC AAT T-3′							

[a] The amoA-1F primer can be replaced by the amoA-1F★ primer designed by Stephen et al. (1999). The amoA-1F★ primer amplifies some *Nitrosospira*-like AOB that are not amplified with the amoA-1F primer.

[b] Two other AOB specific bacterial 16S rRNA primer sets have been used frequently: βAMOF/βAMOR (McCaig et al., 1994) and NitA/NitB (Junier et al., 2008a,b; Voytek and Ward, 1995 for review). Both primer sets amplify a larger portion of the 16S rRNA gene than the CTO primers, but are considered to be less specific than the CTO primers. They can be used for screening enrichment cultures with a cloning-sequencing approach instead of CTO primers, but they cannot be used for DGGE.

- Amplify the 16S rRNA gene with the primers 27F and 1492R (Table 3.3) and sequence the PCR products with 357F (5′-CCT ACG GGA GGC AGC AG-3′), 518R (5′-ATT ACC GCG GCT GCT GG), and 518F (5′CCA GCA GCC GCG GTA AT) (Muyzer et al., 1993).

A.4.3. Sequence analysis

- Examine and edit sequences in a sequence-editing program.
- Assemble the sequences if the PCR product has been sequenced with different primers.
- Perform BLAST search to get a first insight into the identity of the strain.
- Align sequences with an alignment program and construct phylogenetic tree.

A.5. Analysis of the ammonia oxidizer community composition

During enrichment experiments, the community composition of the AOA and/or AOB as well as the heterotrophic microorganisms changes. These changes can be followed by DGGE, which separates PCR products based on the sequence and the GC content of the PCR product. DGGE is widely used to follow community shifts in environmental samples and enrichment cultures (e.g., Bollmann and Laanbroek, 2001, 2002). DGGE protocols have been developed for AOA based on the *amoA* gene (Offre et al., 2009), for Archaea based on the 16S rRNA gene (Offre et al., 2009), for AOB based on the *amoA* gene (Nicolaisen and Ramsing, 2002; Oved et al., 2001), also for AOB based on the 16S rRNA gene (Kowalchuk et al., 1997; Speksnijder et al., 1998) and for the total bacterial community based on the 16S rRNA gene (Muyzer et al., 1993; Zwart et al., 1998).

 APPENDIX B. MEASUREMENT OF NH_4^+, NO_2^-, AND NO_3^-

During any kind of experiments with AOA and AOB, it is important to be able to follow the concentrations of the substrate and the products of ammonia oxidation. Here, we present methods that have been adapted to be used in a plate-reader to increase the efficiency.

B.1. NH_4^+ determination (Kandeler and Gerber, 1988)

Reagents

- 12 g/l sodium hydroxide
- Sodium nitroprusside solution: 85 g/l sodium salicylate and 0.6 g/l sodium nitroprusside

- 0.2 g/l sodium dichloroisocyanurate
- 10 mM NH_4^+ calibration solution: 0.6607 g/l $(NH_4)_2SO_4$
- Store all solutions at 4 °C.
- Dilute the calibration solution to 200 µM NH_4^+ in a volumetric flask.
- Dilute the calibration solution further to 10, 20, 50, 100, and 200 µM NH_4^+ and dilute samples to NH_4^+ concentration below 200 µM.

Two different approaches can be used to determine the NH_4^+ concentration: 48-well plates and test tubes. The major difference is that the reagents in the test tubes can be mixed immediately after addition of the different solutions, which makes the NH_4^+ determination more exact and reproducible. However, during enrichment, it is more important to know if NH_4^+ is still available in the culture, rather than knowing the exact concentration. Therefore, it is possible to use the less exact plate-reader method to screen the enrichment cultures for NH_4^+.

Determination of the $NH_4^\pm$ concentration by using the plate-reader/test tubes

- Use of 48-well plates/test tubes to determine the NH_4^+ concentration:
- Add 0.5 ml/2 ml of sample or calibration solution, include blank (water).
- Mix sodium hydroxide and sodium nitroprusside solution in the ratio 2:1 (prepare only the amount required for the determination).
- Add 250 µl/1 ml of the mixture to each sample and shake the plate/vortex the test tube immediately.
- Add 100 µl/0.4 ml of the sodium dichloroisocyanurate solution to each sample and shake the plate/vortex the test tube immediately.
- Incubate in the dark for 30 min.
- Measure the absorbance in the plate-reader at 660 nm/in the spectrophotometer at 660 nm against water.

Calculation of the NH_4^+ concentration [mM]

- Subtract the absorbance of the blank from all absorbance values
- Plot a calibration curve and calculate the slope and Y-intercept
- Calculate ammonium concentration by [S] = ((Absorbance − blank) − Y-intercept)/slope.
- If the sample has been diluted, multiply by the dilution factor.

In samples with very low NH_4^+ concentration, NH_4^+ can be determined with the fluorometric method by Keroul and Aminot (1997) and Holmes *et al.* (1999).

B.2. NO_2^- determination (Keeney and Nelson, 1982)

Reagents

- NO_2^- determination reagent:
    - Add 100 ml H_3PO_4 (85%) to 750 ml H_2O, dissolve 10 g/l sulfanilamide (the solution might need to be heated slightly to dissolve the sulfanilamide), add 0.5 g/l naphthylethylene diamine dichloride (light sensitive, store in foil covered bottle).
    - 10 mM NO_2^- calibration solution: 0.851 g/l KNO_2 or 0.69 g/l $NaNO_2$.
- Store all solutions at 4 °C.
- Dilute the calibration solution to 100 µM NO_2^- in a volumetric flask.
- Dilute the calibration solution further to 10, 25, 50, and 100 µM NO_2^- and dilute samples to NO_2^- concentration below 100 µM.

Determination of the NO_2^- concentration by using the plate-reader (48-well plates)

- Add 0.5 ml of sample or calibration solution, include blank (water).
- Add 125 µl of the reagent to each sample and shake the plate.
- Incubate in the dark for 10 min.
- Measure the absorbance in the plate-reader at 540 nm.

Determination of the NO_2^- concentration by using the plate-reader (96-well plates) to screen isolation experiments

- Use of 96-well plates
- Transfer 10 µl of sample from a 96-well plate using an 8-channel pipette to a new 96-well plate. Use the same plate set-up as in the isolation experiment.
- Add 90 µl water.
- Prepare an extra plate with blank and calibration solutions.
- Add 20 µl of the reagent and shake the plate.
- Incubate in the dark for 10 min.
- Measure the absorbance in the plate-reader at 540 nm.

Calculation of the NO_2^- concentration [mM]: as described above for NH_4^+ concentration (Appendix B.1).

B.3. (NO_2^- plus NO_3^-) determination (Shand *et al.*, 2008)

Reagents

- Catalyst: 35.4 mg/l $CuSO_4 \cdot 5H_2O$ and 900 mg/l $ZnSO_4 \cdot H_2O$
- 40 g/l NaOH

- 1.71 g/l hydrazine sulfate
- Sulfanilamide solution in HCl: prepare 3.5 M HCl and dissolve 10 g/l sulfanilamide (it might be necessary to warm this solution slightly to dissolve the sulfanilamide).
- 1 g/l naphthylethylene diamine dichloride (light sensitive, store in foil covered bottle)
- 10 mM NO_3^- calibration solution: 1.011 g/l KNO_3 or 0.8499 g/l $NaNO_3$
- Store all solutions at 4 °C.
- Dilute the calibration solution to 100 μM NO_3^- in a volumetric flask.
- Dilute the calibration solution further to 25, 50, 75, and 100 μM NO_3^- and dilute samples to (NO_2^- plus NO_3^-) concentration below 100 μM (NO_2^- plus NO_3^-)

Determination of the (NO_2^- plus NO_3^-) concentration by using the plate-reader (48-well plates)

- Add 0.5 ml of sample or calibration solution, include blank (water) per well.
- Add 75 μl catalyst and shake the plate.
- Add 75 μl NaOH and shake the plate.
- Add 75 μl hydrazine sulfate and shake the plate.
- Incubate for 15 min.
- Add 250 μl sulfanilamide solution and shake the plate.
- Add 75 μl NEDD solution and shake the plate.
- Incubate for 10 min.
- Measure the absorbance in the plate-reader at 540 nm.

Calculation of the NO_2^- plus NO_3^- concentration: as described above for NH_4^+ concentration (Appendix B.1).

B.4. NO_3^- determination

The NO_3^- concentration can be determined by subtracting the NO_2^- concentration from the (NO_2^- plus NO_3^-) concentration.

REFERENCES

Aakra, A., Utaker, J. B., Nes, I. F., and Bakken, L. R. (1999). An evaluated improvement of the extinction dilution method for isolation of ammonia-oxidizing bacteria. *J. Microbiol. Methods* **39**, 23–31.

Alexander, M. (1982). Most probable number method for microbial populations. *In* "Methods of Soil Analysis—Part 2," (A. L. Page, ed.), pp. 815–820. American Society of Agronomy, Madison, WI.

Batchelor, S. E., Cooper, M., Chhabra, S. R., Glover, L. A., Stewart, G. S. A. B., Williams, P., and Prosser, J. I. (1997). Cell density-regulated recovery of starved biofilm populations of ammonia-oxidizing bacteria. *Appl. Environ. Microbiol.* **63,** 2281–2286.

Belser, L. W., and Schmidt, E. L. (1980). Growth and oxidation kinetics of 3 genera of ammonia oxidizing nitrifiers. *FEMS Microbiol. Lett.* **7,** 213–216.

Bloem, J., and Vos, A. (2008). Fluorescent staining of microbes for total direct counts. In "Molecular Microbial Ecology Manual," (G. A. Kowalchuk, F. J. de Bruijn, I. M. Head, A. D. L. Akkermans, and J. D. van Elsas, eds.), pp. 861–873. Springer, Dordrecht, NL.

Bollmann, A., and Laanbroek, H. J. (2001). Continuous culture enrichments of ammonia-oxidizing bacteria at low ammonium concentrations. *FEMS Microbiol. Ecol.* **37,** 211–221.

Bollmann, A., and Laanbroek, H. J. (2002). Influence of oxygen partial pressure and salinity on the community composition of ammonia-oxidizing bacteria in the Schelde estuary. *Aquat. Microb. Ecol.* **28,** 239–247.

Bollmann, A., Bär-Gilissen, M.-J., and Laanbroek, H. J. (2002). Growth at low ammonium concentrations and starvation response as potential factors involved in niche differentiation among ammonia-oxidizing bacteria. *Appl. Environ. Microbiol.* **68,** 4751–4757.

Bollmann, A., Schmidt, I., Saunders, A. M., and Nicolaisen, M. H. (2005). Influence of starvation on potential ammonia-oxidizing activity and *amoA* mRNA levels of *Nitrosospira briensis*. *Appl. Environ. Microbiol.* **71,** 1276–1282.

Bradford, M. M. (1976). A rapid and sensitive method for the quantification of microgram quantities of protein utilizing the principle of protein-dye binding. *Anal. Biochem.* **72,** 248–254.

Chain, P., Lamerdin, J., Larimer, F., Regala, W., Lao, V., Land, M., Hauser, L., Hooper, A., Klotz, M., Norton, J., Sayavedra-Soto, L., Arciero, D., et al. (2003). Complete genome sequence of the ammonia-oxidizing bacterium and obligate chemolithoautotroph *Nitrosomonas europaea*. *J. Bacteriol.* **185,** 2759–2773.

De la Torre, J. R., Walker, C. B., Ingalls, A. E., Koenneke, M., and Stahl, D. A. (2008). Cultivation of a thermophilic ammonia oxidizing archaeon synthesizing crenarchaeol. *Environ. Microbiol.* **10,** 810–818.

DeBoer, W., Duyts, H., and Laanbroek, H. J. (1989). Urea stimulated autotrophic nitrification in suspensions of fertilized, acid heath soil. *Soil Biol. Biochem.* **21,** 349–354.

DeLong, E. F. (1992). Archaea in costal marine environments. *Proc. Natl. Acad. Sci. USA* **89,** 5685–5689.

Francis, C. A., Roberts, K. J., Beman, J. M., Santoro, A. E., and Oakley, B. B. (2005). Ubiquity and diversity of ammonia-oxidizing archaea in water columns and sediments of the ocean. *Proc. Natl. Acad. Sci. USA* **102,** 14683–14688.

Hatzenpichler, R., Lebecleva, E. V., Spieck, E., Stoecker, K., Richter, A., Daims, H., and Wagner, M. (2008). A moderately thermophilic ammonia-oxidizing crenarchaeote from a hot spring. *Proc. Natl. Acad. Sci. USA* **105,** 2134–2139.

Herrmann, M., Saunders, A. M., and Schramm, A. (2009). Effect of lake trophic status and rooted macrophytes on community composition and abundance of ammonia-oxidizing prokaryotes in freshwater environments. *Appl. Environ. Microbiol.* **75,** 3127–3136.

Holmes, R. M., Aminot, A., Kerouel, R., Hooker, B. A., and Peterson, B. J. (1999). A simple and precise method for measuring ammonium and marine and freshwater ecosystems. *Can. J. Fish. Aquat. Sci.* **56,** 1801–1809.

Hooper, A. B., and Terry, K. R. (1974). Photo-inactivation of ammonia oxidation in *Nitrosomonas*. *J. Bacteriol.* **119,** 899–906.

Jones, R. D., Morita, R. Y., Koops, H. P., and Watson, S. W. (1988). A new marine ammonium-oxidizing bacterium, *Nitrosomonas cryotolerans* sp.-nov. *Can. J. Microbiol.* **34,** 1122–1128.

Junier, P., Kim, O. S., Molina, V., Limburg, P., Junier, T., Imhoff, J. F., and Witzel, K. P. (2008a). Comparative in silico analysis of PCR primers suited for diagnostics and cloning of ammonia monooxygenase genes from ammonia-oxidizing bacteria. *FEMS Microbiol. Ecol.* **64,** 141–152.

Junier, P., Kim, O. S., Hadas, O., Imhoff, J. F., and Witzel, K. P. (2008b). Evaluation of PCR primer selectivity and phylogenetic specificity by using amplification of 16S rRNA genes from betaproteobacterial ammonia-oxidizing bacteria in environmental samples. *Appl. Envrion. Microbiol.* **74,** 5231–5236.

Junier, P., Molina, V., Dorador, C., Hadas, O., Kim, O. S., Junier, T., Witzel, K. P., and Imhoff, J. F. (2010). Phylogenetic and functional marker genes to study ammonia-oxidizing microorganisms (AOM) in the environment. *Appl. Microbiol. Biotechnol.* **85,** 425–440.

Kandeler, E., and Gerber, H. (1988). Short-term assay of soil urease activity using colorimetric determination of ammonium. *Biol. Fertil. Soils* **6,** 68–72.

Keeney, D. R., and Nelson, D. W. (1982). Nitrogen—Inorganic forms. *In* "Methods of Soil Analysis—Part 2," (A. L. Page, ed.), pp. 643–698. American Society of Agronomy, Madison, WI, USA.

Keroul, R., and Aminot, A. (1997). Fluorometric determination of ammonia in sea and estuarine waters by direct segmented flow analysis. *Mar. Chem.* **57,** 265–275.

Klotz, M. G., Arp, D. J., Chain, P. S., El-Sheikh, A. F., Hauser, L. J., Hommes, N. G., Larimer, F. W., Malfatti, S. A., Norton, J. M., Poret-Peterson, A. T., Vergez, L. M., and Ward, B. B. (2006). Complete genome sequence of the marine, chemolithoautotrophic, ammonia-oxidizing bacterium *Nitrosococcus oceani* ATCC 19707. *Appl. Environ. Microbiol.* **72,** 6299–6315.

Konneke, M., Bernhard, A. E., de la Torre, J. R., Walker, C. B., Waterbury, J. B., and Stahl, D. A. (2005). Isolation of an autotrophic ammonia-oxidizing marine archaeon. *Nature* **437,** 543–546.

Koops, H. P., and Harms, H. (1985). Deoxyribonucleic-acid homologies among 96 strains of ammonia-oxidizing bacteria. *Arch. Microbiol.* **141,** 214–218.

Koops, H. P., Harms, H., and Wehrmann, H. (1976). Isolation of a moderate halophilic ammonia-oxidizing bacterium, *Nitrosococcus mobilis* nov sp. *Arch. Microbiol.* **107,** 277–282.

Koops, H. P., Purkhold, U., Pommerening-Röser, A., Timmermann, G., and Wagner, M. (2006). The lithoautotrophic ammonia-oxidizing bacteria. *In* "The Prokaryotes, Vol. 5," (M. Dworkin, S. Falkow, E. Rosenberg, K.-H. Schleifer, and E. Stackebrandt, eds.), pp. 778–811. Springer Science, New York, NY.

Kovarova-Kovar, K., and Egli, T. (1998). Growth kinetics of suspended microbial cells: From single-substrate-controlled growth to mixed-substrate kinetics. *Microbiol. Mol. Biol. Rev.* **62,** 646–666.

Kowalchuk, G. A., and Stephen, J. R. (2001). Ammonia-oxidizing bacteria: A model for molecular microbial ecology. *Annu. Rev. Microbiol.* **55,** 485–529.

Kowalchuk, G. A., Stephen, J. R., DeBoer, W., Prosser, J. I., Embley, T. M., and Woldendorp, J. W. (1997). Analysis of ammonia-oxidizing bacteria of the beta subdivision of the class Proteobacteria in coastal sand dunes by denaturing gradient gel electrophoresis and sequencing of PCR-amplified 16S ribosomal DNA fragments. *Appl. Environ. Microbiol.* **63,** 1489–1497.

Laanbroek, H. J., and Gerards, S. (1993). Competition for limiting amounts of oxygen between *Nitrosomonas europaea* and *Nitrobacter winogradskyi* grown in mixed continuous cultures. *Arch. Microbiol.* **159,** 453–459.

Laanbroek, H. J., Bodelier, P. L. E., and Gerards, S. (1994). Oxygen-consumption kinetics of *Nitrosomonas europaea* and *Nitrobacter hamburgensis* grown in mixed continuous cultures at different oxygen concentrations. *Arch. Microbiol.* **161,** 156–162.

Lane, D. J. (1991). 16S/23S rRNA sequencing. *In* "Nucleic Acid Techniques in Bacterial Systematics," (E. Stackebrandt and M. Goodfellow, eds.), pp. 115–147. Wiley, New York, NY.

Larsen, L. H., Revsbech, N. P., and Binnerup, S. J. (1996). A microsensor for nitrate based on immobilized denitrifying bacteria. *Appl. Environ. Microbiol.* **62**, 1248–1251.

Larsen, L. H., Kjaer, T., and Revsbech, N. P. (1997). A microscale NO_3^- biosensor for environmental applications. *Anal. Chem.* **69**, 3527–3531.

Martens-Habbena, W., Berube, P. M., Urakawa, H., de la Torre, J. R., and Stahl, D. A. (2009). Ammonia oxidation kinetics determine niche separation of nitrifying Archaea and Bacteria. *Nature* **461**, 976–979.

McCaig, A. E., Embley, T. M., and Prosser, J. I. (1994). Molecular analysis of enrichment cultures of marine ammonia oxidisers. *FEMS Microbiol. Lett.* **120**, 262–268.

Monod, J. (1950). The technique of continuous cultures: Theory and applications. *Ann. Inst. Pasteur Lille* **79**, 390–410, (written in French).

Muyzer, G., de Waal, E. C., and Uitterlinden, A. G. (1993). Profiling of complex microbial populations by denaturing gradient gel electrophoresis analysis of polymerase chain reaction-amplified genes coding for 16S rRNA. *Appl. Environ. Microbiol.* **59**, 695–700.

Nicolaisen, M. H., and Ramsing, N. B. (2002). Denaturing gradient gel electrophoresis (DGGE) approaches to study the diversity of ammonia-oxidizing bacteria. *J. Microbiol. Methods* **50**, 189–203.

Norton, J. M., Klotz, M. G., Stein, L. Y., Arp, D. J., Bottomley, P. J., Chain, P. S., Hauser, L. J., Land, M. L., Larimer, F. W., Shin, M. W., and Starkenburg, S. R. (2008). Complete genome sequence of *Nitrosospira multiformis*, an ammonia-oxidizing bacterium from the soil environment. *Appl. Environ. Microbiol.* **74**, 3559–3572.

Offre, P., Prosser, J. I., and Nicol, G. W. (2009). Growth of ammonia-oxidizing archaea in soil microcosms is inhibited by acetylene. *FEMS Microbiol. Ecol.* **70**, 99–108.

Oved, T., Shaviv, A., Goldrath, T., Mandelbaum, R. T., and Minz, D. (2001). Influence of effluent irrigation on community composition and function of ammonia-oxidizing bacteria in soil. *Appl. Environ. Microbiol.* **67**, 3426–3433.

Pommerening-Röser, A., Rath, G., and Koops, H. P. (1996). Phylogenetic diversity within the genus *Nitrosomonas*. *Syst. Appl. Microbiol.* **19**, 344–351.

Purkhold, U., Pommerening-Röser, A., Juretschko, S., Schmid, M. C., Koops, H. P., and Wagner, M. (2000). Phylogeny of all recognized species of ammonia oxidizers based on comparative 16S rRNA and *amoA* sequence analysis: Implications for molecular diversity surveys. *Appl. Environ. Microbiol.* **66**, 5368–5382.

Purkhold, U., Wagner, M., Timmermann, G., Pommerening-Röser, A., and Koops, H. P. (2003). 16S rRNA and *amoA*-based phylogeny of 12 novel betaproteobacterial ammonia-oxidizing isolates: Extension of the dataset and proposal of a new lineage within the *Nitrosomonads*. *Int. J. Syst. Evol. Microbiol.* **53**, 1485–1494.

Rotthauwe, J. H., Witzel, K. P., and Liesack, W. (1997). The ammonia monooxygenase structural gene *amoA* as a functional marker: Molecular fine-scale analysis of natural ammonia-oxidizing populations. *Appl. Environ. Microbiol.* **63**, 4704–4712.

Santoro, A. E., Francis, C. A., de Sieyes, N. R., and Boehm, A. B. (2008). Shifts in the relative abundance of ammonia-oxidizing bacteria and archaea across physicochemical gradients in a subterranean estuary. *Environ. Microbiol.* **10**, 1068–1079.

Schramm, A. (2003). In situ analysis of structure and activity of the nitrifying community in biofilms, aggregates, and sediments. *Geomicrobiol. J.* **20**, 313–333.

Shand, C., Williams, B., and Coutts, G. (2008). Determination of N-species in soil extracts using microplate techniques. *Talanta* **74**, 648–654.

Stephen, J. R., Chang, Y. J., Macnaughton, S. J., Kowalchuk, G. A., Leung, K. T., Flemming, C. A., and White, D. C. (1999). Effect of toxic metals on indigenous soil beta-subgroup proteobacterium ammonia oxidizier community structure and protection

against toxicity by inoculated metal-resistance bacteria. *Appl. Environ. Microbiol.* **65,** 95–101.
Speksnijder, A. G. C. L., Kowalchuk, G. A., Roest, K., and Laanbroek, H. J. (1998). Recovery of a *Nitrosomonas*-like 16S rDNA sequence group from freshwater habitats. *Syst. Appl. Microbiol.* **21,** 321–330.
Stehr, G., Boettcher, B., Dittberner, P., Rath, G., and Koops, H. P. (1995). The ammonia-oxidizing nitrifying population of the river Elbe estuary. *FEMS Microbiol. Ecol.* **17,** 177–186.
Stein, L. Y., Arp, D. J., Berube, P. M., Chain, P. S. G., Hauser, L., Jetten, M. S. M., Klotz, M. G., Larimer, F. W., Norton, J. M., op den Camp, H. J. M., Shin, M., and Wei, X. (2007). Whole-genome analysis of the ammonia-oxidizing bacterium, *Nitrosomonas eutropha* C91: Implications for niche adaptation. *Environ. Microbiol.* **9,** 2993–3007.
Suwa, Y., Imamura, Y., Suzuki, T., Tashiro, T., and Urushigawa, Y. (1994). Ammonia-oxidizing bacteria with different sensitivities to $(NH_4)_2SO_4$ in activated sludge. *Water Res.* **28,** 1523–1532.
Treusch, A. H., Leininger, S., Kletzin, A., Schuster, S. C., Klenk, H. P., and Schleper, C. (2005). Novel genes for nitrite reductase and Amo-related proteins indicate a role of uncultivated mesophilic crenarchaeota in nitrogen cycling. *Environ. Microbiol.* **7,** 1985–1995.
Utaker, J. B., Bakken, L., Jiang, Q. Q., and Nes, I. F. (1996). Phylogenetic analysis of seven new isolates of ammonia-oxidizing bacteria based on 16S rRNA gene sequences. *Syst. Appl. Microbiol.* **18,** 549–559.
Veldkamp, H. (1976). Continuous culture in microbial physiology and ecology. *In* "Patterns of Progress," (J. G. Cook, ed.). Meadowfield Press Ltd., Durham, England.
Venter, J. C., Remington, K., Heidelberg, J. F., Halpern, A. L., Rusch, D., Eisen, J. A., Wu, D. Y., Paulsen, I., Nelson, K. E., Nelson, W., Fouts, D. E., Levy, S., *et al.* (2004). Environmental genome shotgun sequencing of the Sargasso Sea. *Science* **304,** 66–74.
Verhagen, F. J. M., and Laanbroek, H. J. (1991). Competition for ammonium between nitrifying and heterotrophic bacteria in dual energy limited chemostats. *Appl. Environ. Microbiol.* **57,** 3255–3263.
Voytek, M. A., and Ward, B. B. (1995). Detection of ammonium-oxidizing bacteria of the beta subclass of the class *Proteobacteria* in aquatic samples with the PCR. *Appl. Environ. Microbiol.* **61,** 1444–1450.
Walker, N. J. (2002). A technique whose time has come. *Science* **296,** 557–559.
Walker, C. B., de la Torre, J. R., Klotz, M. G., Urakawa, H., Pinel, N., Arp, D. J., Brochier-Armanet, C., Chain, P. S. G., Chan, P. P., Gollabgir, A., Hemp, J., Huegler, M., *et al.* (2010). *Nitrosopumilus maritimus* genome reveals unique mechanisms for nitrification and autotrophy in globally distributed marine crenarchaea. *Proc. Natl. Acad. Sci. USA* **107,** 8818–8823.
Watson, S. W. (1965). Characteristics of a marine nitrifying bacterium, *Nitrosocystis oceanus* sp. n. *Limnol. Oceanogr.* **10**(Suppl.), R274–R289.
Winogradsky, S. (1890). Recherches sur les organismes de la nitrification. *Ann. Inst. Pasteur* **4,** 213–331.
Winogradsky, S. (1892). Contributions a la morphologie des organnismes de la nitrification. *Arch. Sci. Biol. (St. Petersburg)* **1,** 88–137.
Woese, C. R., Weisburg, W. G., Hahn, C. M., Paster, B. J., Zablein, L. B., Lewis, B. J., Macke, T. J., Ludwig, W., and Stackebrandt, E. (1985). The phylogeny of purple bacteria—The gamma subdivision. *Syst. Appl. Microbiol.* **6,** 25–33.
Zwart, G., Huismans, R., van Agterveld, M. P., Van de Peer, Y., De Rijk, P., Eenhoorn, H., Muyzer, G., van Hannen, E. J., Gons, H. J., and Laanbroek, H. J. (1998). Divergent members of the bacterial division *Verrucomicrobiales* in a temperate freshwater lake. *FEMS Microbiol. Ecol.* **25,** 159–169.

CHAPTER FOUR

Cultivation, Detection, and Ecophysiology of Anaerobic Ammonium-Oxidizing Bacteria

Boran Kartal, Wim Geerts, *and* Mike S. M. Jetten

Contents

1. Introduction	90
2. Enrichment of Anammox Bacteria as Planktonic Cell Suspensions	93
2.1. Substrates and macro and micronutrients	94
2.2. Reactor operation	94
2.3. Membrane maintenance	96
2.4. Mechanism of planktonic cell formation	96
2.5. Physical separation of anammox cells from the other community members of the MBR	97
3. Molecular Detection of Anammox Bacteria	97
4. Conclusions	104
Acknowledgments	104
References	105

Abstract

Anaerobic ammonium-oxidizing (anammox) bacteria oxidize ammonium with nitrite under anoxic conditions. The anammox process is currently used to remove ammonium from wastewater and contributes significantly to the loss of fixed nitrogen from the oceans. In this chapter, we focus on the ecophysiology of anammox bacteria and describe new methodologies to grow these microorganisms. Now, it is possible to enrich anammox bacteria up to 95% with a membrane bioreactor that removes forces of selection for fast settling aggregates and facilitates the growth of planktonic cells. The biomass from this system has a high anaerobic ammonium oxidation rate (50 fmol $NH_4^+ \cdot cell^{-1} \, day^{-1}$) and is suitable for many ecophysiological and molecular experiments. A high throughput Percoll density gradient centrifugation protocol may be applied on this biomass for further enrichment (>99.5%) of anammox bacteria. Furthermore, we provide

Department of Microbiology, Institute of Water and Wetland Research, Radboud University Nijmegen, Heyendaalseweg, The Netherlands an up-to-date list of commonly used primers and introduce protocols for quantification and detection of functional genes of anammox bacteria in their natural environment.

1. INTRODUCTION

Anaerobic oxidation of ammonium (anammox) with nitrite as electron acceptor is associated with a substantial release of Gibbs free energy, -358 kJ mol^{-1} (Broda, 1977). This reaction was overlooked for decades until ammonium disappearance from an anoxic wastewater treatment plant was observed and shown to be mediated by microbes (Mulder *et al.*, 1995; Van de Graaf *et al.*, 1995). The responsible organisms were identified in 1999 (Strous *et al.*, 1999a). So far, the known bacteria with the capacity for anaerobic ammonium oxidation are restricted to a group of microorganisms in the *Planctomycetes* phylum (Jetten *et al.*, 2010). There are five described *candidatus* anammox genera: Brocadia, Kuenenia, Anammoxoglobus, Jettenia, and Scalindua, and these could all be enriched from various wastewater treatment plants (Jetten *et al.*, 2010). These bacteria are not available as pure cultures, but they are cultivated to an enrichment level of 80% in sequencing fed-batch reactors (SBR) or >95% in membrane bioreactors (MBR; Strous *et al.*, 1998; Van der Star *et al.*, 2008).

Anammox bacteria are ubiquitous in man-made and natural oxygen limited ecosystems (Jetten *et al.*, 2009). Many wastewater treatment plants (aerobic or anaerobic) contain an anammox subpopulation albeit usually around 1% of the total population (Kartal *et al.*, 2007b; Schmid *et al.*, 2007). Nitrifying, denitrifying, or even methanogenic sludge may be used as an inoculum to start anammox enrichment cultures (Kartal *et al.*, 2007b; Ni *et al.*, 2010; Tang *et al.*, 2010; Van de Graaf *et al.*, 1995). Depending on the reactor operation regime and the organic content of the influent medium, different anammox species may be enriched from one source of inoculum. For example, a completely inorganic mineral medium sustaining autotrophic growth is the key for the successful enrichment of "*Candidatus* Kuenenia stuttgartiensis" whereas an organic carbon supplement is necessary to enrich "*Candidatus* Anammoxoglobus propionicus" or "*Candidatus* Brocadia fulgida" (propionate or acetate, respectively; Kartal *et al.*, 2007b, 2008).

On the other hand, in most of the natural ecosystems, in particular the oceans, only one genus, "*Candidatus* Scalindua" has been detected (Glud *et al.*, 2009; Kuypers *et al.*, 2003, 2005; Penton *et al.*, 2006; Schmid *et al.*, 2007; Shu and Jiao, 2008; Woebken *et al.*, 2008). When anammox bacteria were first discovered it was assumed that due to their long doubling time (~11 days) these organisms could not be important in the global nitrogen cycling in dynamic ecosystems such as the oceans (Zehr and Ward, 2002).

However, now it is estimated that the anammox bacteria contribute significantly to the loss of fixed nitrogen from the oceans (Dalsgaard *et al.*, 2003; Hamersley *et al.*, 2007; Kuypers *et al.*, 2003, 2005; Lam *et al.*, 2009). Members of the "*Candidatus* Scalindua" genus seem to dominate many clone libraries from the Arctic ice and deep see sediments to tropical mangroves and lakes (Dale *et al.*, 2009; Gihring *et al.*, 2010; Meyer *et al.*, 2005; Rich *et al.*, 2008; Rysgaard *et al.*, 2004; Schmid *et al.*, 2007; Schubert *et al.*, 2006). The representative organisms of this genus, closely related to "*Candidatus* Scalindua," have been enriched in the laboratory from marine sediments after several attempts (Kawagoshi *et al.*, 2010; Nakajima *et al.*, 2008; Van de Vossenberg *et al.*, 2008). Needless to say, not every ecosystem in the oceans has been sampled; therefore, there is still a possibility that other anammox bacteria could be found in selected microniches. Nevertheless, the cell counts fit very well with the activity measurements (via isotope tracing) and leave little room for other microbes to have a significant contribution to the overall nitrogen loss via the anammox pathway. One curious example is the detection of the presence of "*Candidatus* Kuenenia"-like anammox species in the hydrothermal vents in the mid-Atlantic ridge (Byrne *et al.*, 2009), which has completely different geological and chemical properties compared to oxygen minimum zones (OMZ) or other ecosystems that "*Candidatus* Scalindua" species thrive.

Activity measurements, molecular tools (PCR, FISH, etc.), and lipid analysis may be used in tandem to detect anammox bacteria in nature (Kuypers *et al.*, 2006). In recent years, PCR-related methods have been used for quickly screening of samples for anammox cells and the 16S ribosomal rRNA gene has been used as a marker (Schmid *et al.*, 2005). Both fluorescence *in situ* hybridization (FISH) and more recently qPCR may be used to quantify anammox bacteria (Lam *et al.*, 2007; Li *et al.*, 2009; Schmid *et al.*, 2005; Tsushima *et al.*, 2007). Even though FISH has many drawbacks (high detection limit, low signal-to-noise ratio, etc.), it is the only available PCR-independent tool to quantify anammox bacteria with an ever-growing database of available probes (Schmid *et al.*, 2005).

Whatever the molecular tools may suggest, straightforward activity measurements are fundamental to link the presence of an organism to its *in situ* activity. Measurements with the stable nitrogen isotope (^{15}N) have been successfully applied to many ecosystems for determining anammox activity (Kuypers *et al.*, 2006; Risgaard-Petersen *et al.*, 2005). Not only has this technique provided accurate measurements, but it is also assumed to provide a differentiation between anammox and denitrification activity based on the fact that anammox bacteria uses one reduced (NH_4^+) and one oxidized nitrogen molecule (NO_2^-) whereas denitrifiers use two oxidized atoms ($2 \times NO_3^-$ or NO_2^-). In other words, a combination of $^{15}NH_4^+$ and $^{14}NO_2^-$ will yield $^{29}N_2$ through anammox reaction and $^{28}N_2$ through denitrification and the reverse ($^{14}NH_4^+$ and $^{15}NO_2^-$) will yield

$^{29}N_2$ and $^{30}N_2$, respectively, assuming these are the only two nitrogen converting reactions in an anaerobic system (Risgaard-Petersen et al., 2005). However, a mostly overlooked third mechanism, dissimilatory nitrate reduction to ammonium, works also under anaerobic conditions and could completely alter the heavy isotope labeling pattern (Kartal et al., 2007a; Lam et al., 2009). Through such a reaction, $^{15}NO_3^-$ (or $^{15}NO_2^-$) may be converted to $^{15}NH_4^+$ which could then be combined via the anammox reaction with the remaining $^{15}NO_2^-$. This results in the production of $^{30}N_2$ which could lead to an underestimation of anammox activity and an overestimation of denitrification.

In recent years, both the freshwater and marine anammox species were shown to have a more versatile catabolism than previously assumed. Anammox bacteria are able to carry out several redox reactions including the reduction of iron, manganese, nitrate, and nitrite coupled to the oxidation of formate to CO_2 and the oxidation of iron, formate, acetate, propionate, and methylamines coupled to the reduction of nitrate to N_2 (Güven et al., 2005; Kartal et al., 2007b, 2008; Strous et al., 2006). A detailed look into the nitrate reduction pathway of the anammox bacteria revealed that this process proceeds completely different than canonical denitrification: nitrate is reduced to ammonium via nitrite, and then ammonium and nitrite are combined (see above; Kartal et al., 2007a). Surprisingly, the anammox bacteria were able to reduce nitrate, an end product of their metabolism, without the need of an induction and ammonium production did not cease even in the presence of millimolars of ammonium (Kartal et al., 2007a). Thus, anammox bacteria are able to "recycle" a part of their substrates (ammonium and nitrite) via nitrate (Kartal et al., 2007a). A versatile metabolism makes sense especially in the marine environment which is mostly substrate (notably ammonium) limited. The versatile metabolism most likely gives the anammox bacteria a competitive edge over the other chemolithoautotrophs that have a limited substrate spectrum.

As anammox bacteria are not yet available in pure cultures, density gradient centrifugation was applied to obtain highly purified ($>99.5\%$) preparations to prove that reactions such as nitrate reduction or iron oxidation were indeed carried out by the anammox bacteria and not by other community members (Jetten et al., 2005). This process yields an abundance of anammox bacteria up to $\sim 99.5\%$, but it is time consuming (hours of centrifugation) and has a very low yield ($\sim 2–3\%$). Recently, a new type of reactor was developed for growing anammox bacteria as planktonic cell suspensions that allows enrichments up to $>95\%$ anammox bacteria (Van der Star et al., 2008).

In this chapter, we describe the methodology to enrich both freshwater and marine anammox species as planktonic cell suspensions. Furthermore, we describe the amplification of the functional genes of anammox bacteria and provide an updated list of probes and primers used for their detection.

2. Enrichment of Anammox Bacteria as Planktonic Cell Suspensions

Based on the most optimistic and reliable estimation, anammox bacteria need at least 11 days for doubling (Strous *et al.*, 1999b). Therefore, for a successful enrichment of anammox bacteria, biomass retention is vital. Many reactor configurations have been tested in the past to achieve a simple and robust approach for a reproducible strategy to cultivate anammox bacteria (Strous *et al.*, 1997). To this end, a sequencing fed-batch reactor (SBR) has provided the essentials: very efficient biomass retention and stable operation characteristics under substrate limitation (Strous *et al.*, 1998). The SBR works in three basic stages: (1) filling the reactor with a defined volume of medium over a period of time, (2) stopping the feed, letting the biomass settle, and (3) decantation of the supernatant. This scheme allows for a high rate of substrate supply (high influent flow rate), enables the sludge age (i.e., the rate of biomass loss) to be set as long as necessary, and forces the bacteria to form fast settling flocs (or granules). The SBR delivers a large amount of biomass in a relatively short time; however, the produced biomass is not suitable for many fundamental experiments.

First, only about 80% of the enriched microorganisms are anammox bacteria which make it sometimes difficult to draw definite conclusions from physiological experiments. Second, due to the long solids retention time (SRT) the bulk of the biomass in the SBR will not be active at their highest potential. This effect is compounded by the heterogeneity of a biofilm aggregate: the cells on the outside of the aggregate are always more active (proximity to the substrate, less diffusion limitation). Finally, the extracellular polymeric matrix (EPS) amalgamating the aggregates forms a major hurdle for biochemical and other molecular studies. EPS acts as a gelling agent in cell-free extracts, clogs chromatography columns, and makes it difficult to visualize proteins on polyacrylamide gels (Cirpus *et al.*, 2006).

The solution to these problems would be the growth of anammox bacteria as suspended planktonic cells in a bioreactor that does not select for fast settling flocs, but still offers a sufficient retention time. A bioreactor equipped with a membrane unit would remove the necessity for settling because in such a system the effluent is removed via the membrane providing the long SRT. Recently, this technology has been successfully applied for cultivating anammox bacteria (Van der Star *et al.*, 2008).

Normally, the inoculum for starting-up an anammox enrichment culture is activated sludge from a wastewater treatment plant operated with a long sludge age. However, for the start-up of an MBR, the ~80% enriched anammox biomass originating from an SBR culture should be used. In time, as the forces that select for the fast settling aggregates are removed the biomass disintegrates and the anammox bacteria start growing as planktonic cells. The nutrition and growth conditions of the anammox MBR are slightly modified compared to the operation of an SBR and are listed below.

2.1. Substrates and macro and micronutrients

Initially, the influent NO_2^- and NH_4^+ concentrations are set to 15 mM each. In contrast to start-up of an anammox enrichment culture from activated sludge, there is no need for including NO_3^- in the influent because the high activity of the already enriched anammox bacteria is sufficient to sustain a high enough NO_3^- concentration in the reactor. The concentrations of NO_2^- and NH_4^+ are gradually increased up to 60 mM as long as all the NO_2^- is consumed in the MBR. The nutrient supplements for the freshwater and the marine species are different. The freshwater species is fed with a mineral medium containing (in 1 l): 1.25 g $KHCO_3$, 0.05 g NaH_2PO_4, 0.15 g $CaCl_2 \cdot 2H_2O$, 0.1 g $MgSO_4 \cdot 7H_2O$, 0.00625 g $FeSO_4$, 0.00625 g EDTA, and 1.25 ml of trace elements solution containing (in 1 l): 15 g EDTA, 0.43 g $ZnSO_4 \cdot 7H_2O$, 0.24 g $CoCl_2 \cdot 6H_2O$, 0.99 g $MnCl_2 \cdot 4H_2O$, 0.25 g $CuSO_4 \cdot H_2O$, 0.22 g $NaMoO_4 \cdot 2H_2O$, 0.21 g $NaSeO_4 \cdot 10H_2O$, 0.19 g $NiCl_2 \cdot 6H_2O$, 0.05 g $NaWO_4 \cdot 2H_2O$, and 0.014 g H_3BO_4. The Mg^{2+} and Ca^{2+} concentrations are half of the "conventional" anammox medium to prevent nucleus formation for the growth of aggregates due to the insoluble salts of Mg and Ca that could precipitate in the reactor (Trigo et al., 2006). These are also the most abundant bivalent ions in sea salt and to prevent a similar effect in the marine anammox enrichment culture, the salt concentrations were lowered from the previously reported 33 g l^{-1} (Van de Vossenberg et al., 2008). Below 20 g l^{-1} a high anammox activity could not be sustained, so the marine species was fed with 25 g l^{-1} sea salt (Red Sea salt Fish Pharm Ltd., The Seahorse, Arnhem, NL) supplemented with 0.00625 g $FeSO_4$ and 0.05 g NaH_2PO_4. This medium does not contain the $KHCO_3$ that is necessary to buffer reactor. Instead, $KHCO_3$ is supplied from a separate solution (100 g l^{-1}) which is controlled with a pH probe set at 7.3 (see below).

2.2. Reactor operation

The commercially available (Zenon Environmental, Ontario, CA) membrane unit we use is 6 cm × 6 cm × 20 cm and to accommodate it, a 15-l glass vessel with a working volume of 10 l is used for the enrichment of anammox bacteria as planktonic cells (Fig. 4.1A). The MBR is fed continuously with a flow rate 1.4 ml min^{-1} which gradually may be increased to 4.2 ml min^{-1} which corresponds to 6 l day^{-1}. We have not determined the upper limit of the flow rate, but 6 l day^{-1} is the highest amount that is

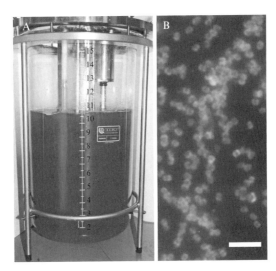

Figure 4.1 (A) The anammox membrane bioreactor. (B) Fluorescence *in situ* hybridization micrograph depicting highly enriched "*Candidatus* Scalindua sp." in pink (combination of BS820 probe counterstained with DNA stain DAPI); scale bar = 5 μm. (See Color Insert.)

feasible to maintain a reasonable frequency for medium preparation. For the homogeneous distribution of substrates, the vessel is stirred at 300 rpm with one 6-bladed Rushton and one marine impeller (74 mm). A heating blanket is used to keep temperature at 33 °C. A gas mixture of Ar/CO_2 (95%/5%) with a flow rate of 10 ml l^{-1} is sparged through the reactor and the medium vessel to maintain anoxic conditions. For the freshwater species, the CO_2 in the gas stream and the $KHCO_3$ in the influent mineral medium stabilize the pH at 7.3–7.4. For the marine species, however, an automated pH control is necessary because of the lack of $KHCO_3$ in the influent medium. To this end, a pH controller unit is used to supply $KHCO_3$ when necessary to keep the pH at 7.3 or 7.4. If desired, dissolved O_2 and redox potential probes may also be used to monitor the dissolved O_2 and oxidation/reduction potential, respectively. A separate pump is used to remove 1 l day^{-1} of biomass to set a doubling time of ~10 days.

The MBR should be operated at a cell concentration of ~10^8 cells ml^{-1} which corresponds to an OD_{600} of 1–1.2. Such a setup will provide enough cells for biochemical and physiological experiments (Fig. 4.1A and B). An average anaerobic ammonium oxidation rate with biomass from the MBR is 50 fmol $NH_4^+ \cdot$cell^{-1} day^{-1} (Fig. 4.2). Higher cell concentrations lead to a faster clogging of the membrane and growth of biomass on the walls and other surfaces in the reactor.

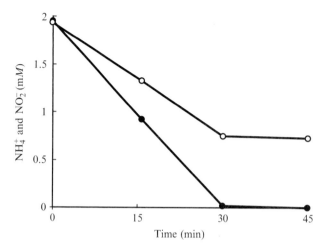

Figure 4.2 Anammox activity test with ammonium (open circles) and nitrite (full circles). The experiment has been conducted in duplicate with $\sim 10^8$ cells ml^{-1}.

2.3. Membrane maintenance

Every 2–4 weeks the membranes should be cleaned thoroughly for an optimal operation life. First the excess biomass on the membrane is removed with pressurized water, and then the membrane is incubated at 30 °C for 2 days in a 1% (w/v) solution of Tergazyme (Alconox, NY, USA) to degrade the biomass within and between the membrane fibers. This is followed by rinsing the unit thoroughly with demineralized water and flushing it with 5 l of 50% ethanol. After this step, the membrane unit may be kept at 4 °C in 50% ethanol until the further use. Before reusing the membrane unit, it should be carefully washed with copious amounts of demineralized water to remove the ethanol.

2.4. Mechanism of planktonic cell formation

In two attempts starting from aggregated cells, 275 days and 400 days were necessary for the enrichment of freshwater and marine anammox bacteria, respectively. A straightforward assumption would be that when the selection for the fast settling is removed, the cells would start growing in suspension. However, most likely, there are several processes working together for the disappearance of aggregates in the MBR reactor. The observations of Van der Star and colleagues identify several factors apart from lack of selection pressure for fast settling biomass (Van der Star et al., 2008). First, the cells are forced to grow at their peak activity through a shorter-than-normal SRT (~ 10 days). Second, the concentrations of calcium and magnesium, which could form salts that do not readily dissolve and lead to nuclei for aggregate formation in the reactor, are half the "normal concentration" (Van der Star *et al.*, 2008). Third, addition of yeast extract and low shear stress (stirring at 160 rpm) are reported to provide an advantage for the growth of planktonic cells (Van der Star *et al.*, 2008). For the enrichment of both types of anammox bacteria, the first two points seem to be vital. Low magnesium and calcium do seem to be important along with a short SRT for the enrichment of both freshwater and marine anammox bacteria as planktonic cell suspensions. However, it is also possible to grow the cells without yeast extract and a higher shear stress (350 rpm). The lengthiness of the enrichment process seems to be mainly due to the slow metabolism of anammox bacteria and the difficult-to-degrade extracellular polymeric substances that form the matrix of the aggregates. At the moment, this mechanism is relatively unknown and based on two sets of observations and there is a need for more experimentation to prove and validate these.

2.5. Physical separation of anammox cells from the other community members of the MBR

For an even higher level of enrichment of anammox bacteria (>99.5), a density gradient centrifugation technique may be applied on the cell suspensions originating from the MBR. This procedure is modified after a previous description (Strous *et al.*, 1999a,b) for a better separation of higher volumes of concentrated planktonic cells.

Percoll (6 ml, Pharmacia) and 3 ml mineral medium (Van de Graaf *et al.*, 1996) or 4-(2-hydroxyethyl)-1-piperazineethanesulfonic acid (HEPES) buffer (20 mM, pH 7.4) are mixed and centrifuged (15 °C, 10.000×g, 30 min) in 15 ml glass tubes (Corex, round bottom) to form a density gradient. About 10–20 ml of anammox cells (10^8 ml^{-1}) are first concentrated to a final volume of 1 ml and then resuspended in mineral medium or in HEPES buffer. This suspension is applied onto the gradient and the top half of the density gradient is gently mixed with the cell suspension (Fig. 4.3A). This is followed by a centrifugation step for another 30 min at 15 °C (6000×g). The target cells are then recovered as a bright red band halfway in the Percoll gradient (Fig. 4.3B). They are washed in mineral medium or HEPES buffer to remove the Percoll. This highly pure final anammox suspension contains one nontarget cell per 500–1000 target cells (Fig. 4.3C).

3. MOLECULAR DETECTION OF ANAMMOX BACTERIA

There are several approaches for the molecular detection of anammox bacteria in the natural environments that may be collected in three groups: molecular tools, lipid analysis, and activity measurements. In the meaning

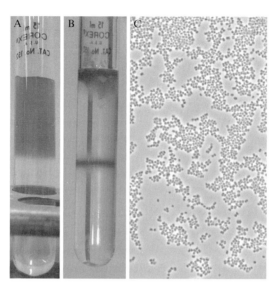

Figure 4.3 Steps of Percoll density centrifugation: (A) Mixture of the cell suspension and top half of the preprepared density gradient. (B) Separation of anammox cells density gradient centrifugation (the bright red band) (C) Phase contrast microscopy of the >99.5% anammox bacteria. (For interpretation of the references to color in this figure legend, the reader is referred to the Web version of this chapter.)

used here, the molecular toolbox contains two basic units: PCR-related procedures and FISH. In the case of anammox bacteria, these methods are almost always used in tandem to get a better understanding of the community structure and minimize the bias arising from either of them. When the quest is to detect possibly "new" (and even known) species, the main limitation is the extent of the database that will be used to design primers or probes. The primer set pla46 630R may be applied to construct a clone library with a more general overview, but these are of limited value for environmental samples (and sometimes highly enriched laboratory cultures) due to the strong primer bias toward *Planctomycetales*. Usually, an anammox-specific amplification needs to be conducted to get a representative picture of the anammox population in the assayed samples which then are analyzed with the existing or newly designed FISH probes. This approach, however, will not extend the span of known anammox bacteria very much. An updated list of most commonly used oligonucleotide probes and primer sets is presented in Tables 4.1 and 4.2. For an extensive description of the application of FISH and PCR targeting the 16S rRNA gene of the anammox bacteria, previous detailed papers are recommended (Jetten *et al.*, 2005; Schmid *et al.*, 2005).

Until recently, due to the lack of available sequence information, PCR on anammox bacteria had to be limited to primers targeting the 16S rRNA

Table 4.1 Commonly used probes to identify known anammox bacteria with probe sequences, target sites, and formamide required for specific *in situ* hybridization

Trivial name	OPD[a] designation	Specificity	Sequence 5'–3'	Target site[b]	% Formamide /mM NaCl[c]
Pla 46 (Neef et al., 1998)	S-P-Planc-0046-a-A-18	Planctomycetes	GACTTGCATGCCTAATCC	46–63	25/159
EUB II (Daims et al., 1999)	S-D-Bact-0338-b-A-18	Specific for Plactomycetales and to be used in combination with EUBI and III	GCAGCCACCCGTAGGTGT	338–355	0/900
Amx 820 (Schmid et al., 2000)	S-*-Amx-0820-a-A-22	"*Cand.* Brocadia anammoxidans" "*Cand.* Kuenenia stuttgartiensis"	AAAACCCCTCTACTTAGTGCCC	820–841	40/56
Kst 1273 (Schmid et al., 2000)	S-*-Kst-1273-a-A-20	"*Cand.* Kuenenia stuttgartiensis"	TCGGCTTTATAGGTTTCGCA	1273–1292	25/159
Amx 368 (Schmid et al., 2003)	S-*-Amx-0368-a-A-18	All anammox organisms	CCTTTCGGGCATTGCGAA	368–385	15/338
Scabr 1114 (Schmid et al., 2003)	S-*-Scabr-1114-a-A-22	"*Cand.* Scalindua brodae"	CCCGCTGGTAACTAAAAACAAG	1114–1135	20/225

(*continued*)

Table 4.1 (continued)

Trivial name	OPD[a] designation	Specificity	Sequence 5′–3′	Target site[b]	% Formamide /mM NaCl[c]
BS 820 (Kuypers et al., 2003)	S-*-BS-820-a-A-22	"Cand. Scalindua wagneri" "Cand. Scalindua sorokinii"	TAATTCCCTCTACTTAGTGCCC	820–841	25/159
Apr 820 (Kartal et al., 2007a,b)	S-*-Apr-0820-a-A-21	"Cand. Anammoxoglobus propionicus" and "Cand. Jettenia asiatica"	AAACCCCTCTACCGAGTGCCC	820–840	40/56
Bfu 613 (Kartal et al., 2008)	S-*-Bfu-0613-a-A-24	Cand. "Brocadia fulgida"	GGATGCCGTTCTTCCGTTAAGCGG	613–637	30/112

[a] Oligonucleotide probe database.
[b] 16S rRNA position E. coli numbering.
[c] % Formamide in the hybridization buffer and mM NaCl in the washing buffer, respectively, required for specific in situ hybridization.

Table 4.2 Primers for constructing planctomycete and anaerobic ammonium oxidizer specific clone libraries and the quantification of anammox bacteria

Primer name	Specificity	Sequence 5′–3′	Target site[a]	T (°C)	References
Pla46F	Planctomycetes	GACTTGCATGCCTAATCC	46–63	58	Neef et al. (1998), Juretschko et al. (1998)
630R	All bacteria	CAKAAAGGAGGTGATCC	1529–1545		
Amx 368F	All anammox bacteria	CCTTTCGGGCATTGCGAA	368–385	56	Schmid et al. (2003)
Amx 820R	"Cand. Brocadia anammoxidans" "Cand. Kuenenia stuttgartiensis"	AAAACCCCTCTACTTAGTGCCC	820–841	56	Schmid et al. (2000)
BS 820R	"Cand. Scalindua wagneri" "Cand. Scalindua sorokinii"	TAATTCCCCTCTACTTAGTGCCC	820–841	56	Kuypers et al. (2003)
Brod541F	"Cand. Scalindua sp."	GAGCACGTAGGTGGGTTTGT	541–560	60	Penton et al. (2006)
Brod1260R		GGATTCGCTTCACCTCTCGG	1260–1279		
AMX809F	Most anammox bacteria	GCCGTAAACGATGGGCACT	809–826	60	Tsushima et al. (2007)
AMX818F		ATGGGCACTMRGTAGAGGGGTTT	818–839		
AMX1066R		AACGTCTCACGACACGAGCTG	1047–1066		
Amx694F	All anammox bacteria	GGGGAGAGTGGAACTTCGG	694–713	60	Ni et al. (2010)
Amx960R		GCTCGCACAAGCGGTGGAGC	960–979		

The last two primer sets are designed for quantitative real-time PCR. Pla46 may be used as a forward primer with Amx368, Amx820, or BS820. Amx368F may be used as a forward primer with Amx820 or BS820 for higher specificity for more defined target groups.
[a] 16S rRNA position E. coli numbering.

gene. However, starting with the genome sequence of the anammox bacterium *Kuenenia stuttgartiensis* and other sequencing efforts more target genes to design specific primers became available (Shimamura et al., 2007; Strous et al., 2006). A unique functional gene target significantly increases the chances of detecting bacteria that constitute only a small part (< 1%) of an ecosystem. Moreover, these primers may be applied on RNA with quantitative reverse transcription PCR to get an impression of transcription levels which could be better correlated to activity rather than presence (Lam et al., 2007, 2009). Nevertheless, it should not be forgotten that the database of available sequence information on functional genes is still lagging behind the available 16S rRNA gene collection.

The anammox metabolism has four core catalytic proteins that appear to be likely options for primer design: nitrite and nitrate reductases, hydrazine hydrolase, and hydrazine dehydrogenase. Currently, the protein with the highest number of available sequences in the database is the hydroxylamine oxidoreductase-like (i.e., hydrazine dehydrogenase, hao, hzo) protein. There are already eight primer sets available for the amplification of the genes that are encoding these proteins (Table 4.3; Li et al., 2010; Quan et al., 2008; Schmid et al., 2008). These primers target three different clusters of hydroxylamine oxidoreductase-like proteins (Klotz et al., 2008; Schmid et al., 2008). Moreover, a primer set designed for the amplification of the cd_1 nitrite reductase (nirS) genes of "*Candidatus* Scalindua sp." is also available now (Table 4.3). This primer set was already used to determine the transcription levels of anammox nitrite reductase genes in the Peruvian upwelling zone (Lam et al., 2009).

For the PCR targeting functional genes, ~50–100 ng of template is necessary. The reaction mixture (50 µl) consists of the template, 20 pmol of reverse and forward primers, 10 pmol of deoxyribonucleotide triphosphates (dNTPs), 1.5 U of *Taq* polymerase with 2 m*M* $MgCl_2$. A typical PCR program for the amplification of the HZO genes would involve and initial denaturation at 94 °C for 5 min, 30 cycles of denaturation at 94 °C for 1 min, annealing at optimal temperatures for different primer sets (Table 4.3) for 1 min and elongation at 72 °C for 1.5 min followed by a final incubation step at 72 °C for 10 min. For the nirS primer set, the following program is recommended: 95 °C for 2 min, 30 cycles of denaturation at 95 °C for 45 s, annealing at 60 °C for 1 min and elongation at 72 °C for 1 min followed by a final incubation step at 72 °C for 15 min (Lam et al., 2009). Negative and positive controls should always be included in all assays.

Recently, real-time PCR protocols for the quantification of anammox bacteria based on 16S rRNA and nirS have also been published (Lam et al., 2009; Ni et al., 2010; Tsushima et al., 2007). The PCR mixture and reaction protocol for the qPCR primers is slightly different. These reactions are performed in 25 µl of final volume and in 96-well plates for higher

Table 4.3 Primers for the amplification of functional genes from anaerobic ammonium oxidizing bacteria

Primer name	Sequence 5'–3'	Specificity	Product size	T (°C)	References
hzocl1F1	TGYAAGACYTGYCAYTGG	hzo	470	50	Schmid et al. (2008)
hzocl1R2	ACTCCAGATRTGCTGACC				
hzocl1F	TGYAAGACYTGYCAYTGGG	hzo	470	53	Schmid et al. (2008)
hzocl1R2	ACTCCAGATRTGCTGACC				
hzocl2aF	GGTTGYCACACAAGGC	hzo	289	50	Schmid et al. (2008)
hzocl2aR1	TYWACCTGGAACATACCC				
hzocl2aF1	GGTTGYCACACAAGGC	hzo	525	48	Schmid et al. (2008)
hzocl2aR2	ATATTCACCATGYTTCCAG				
hzocl2aF2	GTTGTGMTGMWTGTCATGG	hzo	838	48	Schmid et al. (2008)
hzocl2aR1	TYWACCTGGAACATACCC				
Ana-hzo1	ACCTCTTCWGCAGGTGCAT	hzo	1000	53	Quan et al. (2008)
Ana-hzo2R	ACCTCTTCWGCAGGTGCAT				
hzoF1	TGTGCATGGTCAATTGAAAG	hzo	1000	53	Li et al. (2010)
hzoR	CAACCTCTTCWGCAGGTGCATG				
Scnir372F	TGTAGCCAGCATTGTAGCGT	nirS[a]	473	60	Li et al. (2009)
Scnir845R	TCAAGCCAGACCCATTTGCT				

[a] Specific for "*Candidatus* Scalindua sp."

reproducibility. A fluorescent dye should be included for monitoring of the PCR amplification on-line. For the best performance and to reduce contamination from different sources, a premixed SYBR Green Master mix [e.g., the one available from Finnzymes (Finland)] should be used. Both primer sets for qPCR anneal optimally at 60 °C (Table 4.2). For a higher confidence, each sample should always be analyzed three times. In all assays, together with negative controls, a calibration curve that has a concentration of 10^0–10^9 copies per well should be included. A melting curve analysis is a fast, but inconclusive method to determine the purity of the PCR product. Maximum care is extremely important to avoid contamination and reproducible application of qPCR reactions. Glove boxes equipped with UV lights are available for maintaining a sterile working environment and are very well suited to handle such samples.

The integrity of the PCR products may be assayed on an agarose gel. It is recommended to use 0.8% and 1.5% gels for large and small DNA fragments, respectively. The PCR reactions may be followed by several techniques such as denaturant gradient gel electrophoresis (DGGE) or terminal restriction fragment length polymorphism (TRFLP; Dale et al., 2009; Tal et al., 2006). Nevertheless it is strongly recommended that the PCR (normal or quantitative) products are first cloned and sequenced to determine their phylogenetic affiliations. It is extremely important that at least three random samples are cloned and sequenced for each set of qPCR analysis to make sure that the intended target is actually amplified.

The sequences may be aligned with ARB or MEGA programs (Ludwig et al., 2004; Tamura et al., 2007). In our experience, ARB is a very powerful, but difficult to handle tool whereas MEGA is much easier to use but lacks certain features such as displaying the secondary RNA structure. After running an alignment algorithm, the alignments should always be examined by eye.

4. Conclusions

In this overview, we presented an updated list of oligonucleotide probes and described the quantification of anammox bacteria with qPCR and amplification of their functional genes in detail. Moreover, we provided insights for the operation of the MBR that allows the growth of anammox bacteria as highly enriched (>95%) and active cell suspensions.

ACKNOWLEDGMENTS

We thank the many coworkers, colleagues, foreign guests, and funding agencies (ALW, STW, EU, STOWA, DARWIN, Paques BV, Agentschap NL, and ERC 232937) that made the study of many fascinating aspects of anammox bacteria possible.

REFERENCES

Broda, E. (1977). 2 kinds of lithotrophs missing in nature. *Z. Allg. Mikrobiol.* **17,** 491–493.
Byrne, N., Strous, M., Crepeau, V., Kartal, B., Birrien, J. L., Schmid, M., Lesongeur, F., Schouten, S., Jaeschke, A., Jetten, M., Prieur, D., and Godfroy, A. (2009). Presence and activity of anaerobic ammonium-oxidizing bacteria at deep-sea hydrothermal vents. *ISME J.* **3,** 117–123.
Cirpus, I. E. Y., Geerts, W., Hermans, J. H. M., den Camp, H., Strous, M., Kuenen, J. G., and Jetten, M. S. M. (2006). Challenging protein purification from anammox bacteria. *Int. J. Biol. Macromol.* **39,** 88–94.
Dale, O. R., Tobias, C. R., and Song, B. K. (2009). Biogeographical distribution of diverse anaerobic ammonium oxidizing (anammox) bacteria in Cape Fear River Estuary. *Environ. Microbiol.* **11,** 1194–1207.
Daims, H., Bruehl, A., Amann, R., Schleifer, K. H., and Wagner, M. (1999). The domain-specific probe EUB338 is insufficient for the detection of all bacteria: development and evaluation of a more comprehensive probe set. *Syst. Appl. Microbiol.* **22,** 434–444.
Dalsgaard, T., Canfield, D. E., Petersen, J., Thamdrup, B., and Acuna-Gonzalez, J. (2003). N2 production by the anammox reaction in the anoxic water column of Golfo Dulce, Costa Rica. *Nature* **422,** 606–608.
Gihring, T. M., Lavik, G., Kuypers, M. M. M., and Kostka, J. E. (2010). Direct determination of nitrogen cycling rates and pathways in Arctic fjord sediments (Svalbard, Norway). *Limnol. Oceanogr.* **55,** 740–752.
Glud, R. N., Thamdrup, B., Stahl, H., Wenzhoefer, F., Glud, A., Nomaki, H., Oguri, K., Revsbech, N. P., and Kitazato, H. (2009). Nitrogen cycling in a deep ocean margin sediment (Sagami Bay, Japan). *Limnol. Oceanogr.* **54,** 723–734.
Güven, D., Dapena, A., Kartal, B., Schmid, M. C., Maas, B., van de Pas-Schoonen, K., Sozen, S., Mendez, R., Op den Camp, H. J. M., Jetten, M. S. M., Strous, M., and Schmidt, I. (2005). Propionate oxidation by and methanol inhibition of anaerobic ammonium-oxidizing bacteria. *Appl. Environ. Microbiol.* **71,** 1066–1071.
Hamersley, M. R., Lavik, G., Woebken, D., Rattray, J. E., Lam, P., Hopmans, E. C., Damsté, J. S. S., Krüger, S., Graco, M., Gutiérrez, D., and Kuypers, M. M. M. (2007). Anaerobic ammonium oxidation in the Peruvian oxygen minimum zone. *Limnol. Oceanogr.* **52,** 923–933.
Jetten, M. S. M., Op den Camp, H. J. M., Kuenen, G. J., and Strous, M. (2010). Family I. "*Candidatus* Brocadiaceae" fam. nov. *In* "Bergey's Manual of Systematic Bacteriology, Vol. 4," (N. R. Krieg, J. T. Staley, D. R. Brown, B. Hedlund, B. J. Paster, N. Ward, W. Ludwig, and W. B. Whitman, eds.), pp. 596–602. Springer, New York.
Jetten, M., Schmid, M., van de Pas-Schoonen, K., Sinninghe Damsté, J. S., and Strous, M. (2005). Anammox organisms: Enrichment, cultivation, and environmental analysis. *Methods Enzymol.* **397,** 34–57.
Jetten, M. S. M., van Niftrik, L., Strous, M., Kartal, B., Keltjens, J. T., and Op den Camp, H. J. M. (2009). Biochemistry and molecular biology of anammox bacteria. *Crit. Rev. Biochem. Mol. Biol.* **44,** 65–84.
Juretschko, S., Timmermann, G., Schmid, M., Schleifer, K. H., Pommerening-Roser, A., Koops, H. P., and Wagner, M. (1998). Combined molecular and conventional analyses of nitrifying bacterium diversity in activated sludge: *Nitrosococcus mobilis* and Nitrospira-like bacteria as dominant populations. *Appl. Environ. Microbiol.* **64**(8), 3042–3051.
Kartal, B., Kuypers, M. M. M., Lavik, G., Schalk, J., Op den Camp, H. J. M., Jetten, M. S. M., and Strous, M. (2007a). Anammox bacteria disguised as denitrifiers: Nitrate reduction to dinitrogen gas via nitrite and ammonium. *Environ. Microbiol.* **9,** 635–642.

Kartal, B., Rattray, J., van Niftrik, L., van de Vossenberg, J., Schmid, M., Webb, R. I., Schouten, S., Fuerst, J. A., Sinninghe Damsté, J. S., Jetten, M. S. M., and Strous, M. (2007b). *Candidatus* "Anammoxoglobus propionicus" gen. nov., sp. nov., a new propionate oxidizing species of anaerobic ammonium oxidizing bacteria. *Syst. Appl. Microbiol.* **30,** 39–49.

Kartal, B., van Niftrik, L., Rattray, J., de Vossenberg, J., Schmid, M. C., Damste, J. S. S., Jetten, M. S. M., and Strous, M. (2008). Candidatus 'Brocadia fulgida': An autofluorescent anaerobic ammonium oxidizing bacterium. *FEMS Microbiol. Ecol.* **63,** 46–55.

Kawagoshi, Y., Nakamura, Y., Kawashima, H., Fujisaki, K., Furukawa, K., and Fujimoto, A. (2010). Enrichment of marine anammox bacteria from seawater-related samples and bacterial community study. *Water Sci. Technol.* **61,** 119–126.

Klotz, M. G., Schmid, M. C., Strous, M., den Camp, H., Jetten, M. S. M., and Hooper, A. B. (2008). Evolution of an octahaem cytochrome *c* protein family that is key to aerobic and anaerobic ammonia oxidation by bacteria. *Environ. Microbiol.* **10,** 3150–3163.

Kuypers, M. M. M., Sliekers, A. O., Lavik, G., Schmid, M., Jorgensen, B. B., Kuenen, J. G., Sinninghe Damsté, J. S., Strous, M., and Jetten, M. S. M. (2003). Anaerobic ammonium oxidation by anammox bacteria in the Black Sea. *Nature* **422,** 608–611.

Kuypers, M. M. M., Lavik, G., Woebken, D., Schmid, M., Fuchs, B. M., Amann, R., Jorgensen, B. B., and Jetten, M. S. M. (2005). Massive nitrogen loss from the Benguela upwelling system through anaerobic ammonium oxidation. *Proc. Natl. Acad. Sci. USA* **102,** 6478–6483.

Kuypers, M. M. M., Lavik, G., and Thamdrup, B. (2006). Anaerobic ammonium oxidation in the marine environment. *In* "Past and Present Water Column Anoxia," (L. N. Neretin, ed.), pp. 311–335. Springer, The Netherlands.

Lam, P., Jensen, M. M., Lavik, G., McGinnis, D. F., Muller, B., Schubert, C. J., Amann, R., Thamdrup, B., and Kuypers, M. M. M. (2007). Linking crenarchaeal and bacterial nitrification to anammox in the Black Sea. *Proc. Natl. Acad. Sci. USA* **104,** 7104–7109.

Lam, P., Lavik, G., Jensen, M. M., van de Vossenberg, J., Schmid, M., Woebken, D., Dimitri, G., Amann, R., Jetten, M. S. M., and Kuypers, M. M. M. (2009). Revising the nitrogen cycle in the Peruvian oxygen minimum zone. *Proc. Natl. Acad. Sci. USA* **106,** 4752–4757.

Li, X. R., Du, B., Fu, H. X., Wang, R. F., Shi, J. H., Wang, Y., Jetten, M. S. M., and Quan, Z. X. (2009). The bacterial diversity in an anaerobic ammonium-oxidizing (anammox) reactor community. *Syst. Appl. Microbiol.* **32,** 278–289.

Li, M., Hong, Y. G., Klotz, M. G., and Gu, J. D. (2010). A comparison of primer sets for detecting 16S rRNA and hydrazine oxidoreductase genes of anaerobic ammonium-oxidizing bacteria in marine sediments. *Appl. Microbiol. Biotechnol.* **86,** 781–790.

Ludwig, W., Strunk, O., Westram, R., Richter, L., Meier, H., Yadhukumar, H., Buchner, A., Lai, T., Steppi, S., Jobb, G., Forster, W., Brettske, I., *et al.* (2004). ARB: A software environment for sequence data. *Nucleic Acids Res.* **32,** 1363–1371.

Meyer, R. L., Risgaard-Petersen, N., and Allen, D. E. (2005). Correlation between anammox activity and microscale distribution of nitrite in a subtropical mangrove sediment. *Appl. Environ. Microbiol.* **71,** 6142–6149.

Mulder, A., Vandegraaf, A. A., Robertson, L. A., and Kuenen, J. G. (1995). Anaerobic ammonium oxidation discovered in a denitrifying fluidized-bed reactor. *FEMS Microbiol. Ecol.* **16,** 177–183.

Nakajima, J., Sakka, M., Kimura, T., Furukawa, K., and Sakka, K. (2008). Enrichment of anammox bacteria from marine environment for the construction of a bioremediation reactor. *Appl. Microbiol. Biotechnol.* **77,** 1159–1166.

Neef, A., Amann, R., Schlesner, H., and Schleifer, K. H. (1998). Monitoring a widespread bacterial group: in situ detection of planctomycetes with 16S rRNA-targeted probes. *Microbiology UK* **144,** 3257–3266.

Ni, B. J., Hu, B. L., Fang, F., Xie, W. M., Kartal, B., Liu, X. W., Sheng, G. P., Jetten, M., Zheng, P., and Yu, H. Q. (2010). Microbial and physicochemical characteristics of compact anaerobic ammonium-oxidizing granules in an upflow anaerobic sludge blanket reactor. *Appl. Environ. Microbiol.* **76**, 2652–2656.

Penton, C. R., Devol, A. H., and Tiedje, J. M. (2006). Molecular evidence for the broad distribution of anaerobic ammonium-oxidizing bacteria in freshwater and marine sediments. *Appl. Environ. Microbiol.* **72**, 6829–6832.

Quan, Z. X., Rhee, S. K., Zuo, J. E., Yang, Y., Bae, J. W., Park, J. R., Lee, S. T., and Park, Y. H. (2008). Diversity of ammonium-oxidizing bacteria in a granular sludge anaerobic ammonium-oxidizing (anammox) reactor. *Environ. Microbiol.* **10**, 3130–3139.

Rich, J. J., Dale, O. R., Song, B., and Ward, B. B. (2008). Anaerobic ammonium oxidation (Anammox) in Chesapeake Bay sediments. *Microb. Ecol.* **55**, 311–320.

Risgaard-Petersen, N., Meyer, R. L., and Revsbech, N. P. (2005). Denitrification and anaerobic ammonium oxidation in sediments: Effects of microphytobenthos and NO3. *Aquat. Microb. Ecol.* **40**, 67–76.

Rysgaard, S., Glud, R. N., Risgaard-Petersen, N., and Dalsgaard, T. (2004). Denitrification and anammox activity in Arctic marine sediments. *Limnol. Oceanogr.* **49**, 1493–1502.

Schmid, M., Twachtmann, U., Klein, M., Strous, M., Juretschko, S., Jetten, M. S. M., Metzger, J. W., Schleifer, K. H., and Wagner, M. (2000). Molecular evidence for genus level diversity of bacteria capable of catalyzing anaerobic ammonium oxidation. *Syst. Appl. Microbiol.* **23**(1), 93–106.

Schmid, M., Walsh, K., Webb, R., Rijpstra, W.I.C., van de Pas-Schoonen, K., Verbruggen, M.J., Hill, T., Moffett, B., Fuerst, J., Schouten, S., Sinninghe Damsté, J.S., Harris, J., et al. (2003). *Candidatus* "Scalindua brodae", sp nov., *Candidatus* "Scalindua wagneri", sp nov., two new species of anaerobic ammonium oxidizing bacteria. *Syst. Appl. Microbiol.* **26**(4), 529–538.

Schmid, M. C., Maas, B., Dapena, A., de Pas-Schoonen, K. V., de Vossenberg, J. V., Kartal, B., van Niftrik, L., Schmidt, I., Cirpus, I., Kuenen, J. G., Wagner, M., Sinninghe Damsté, J. S., et al. (2005). Biomarkers for in situ detection of anaerobic ammonium-oxidizing (anammox) bacteria. *Appl. Environ. Microbiol.* **71**, 1677–1684.

Schmid, M. C., Risgaard-Petersen, N., van de Vossenberg, J., Kuypers, M. M. M., Lavik, G., Petersen, J., Hulth, S., Thamdrup, B., Canfield, D., Dalsgaard, T., Rysgaard, S., Sejr, M. K., et al. (2007). Anaerobic ammonium-oxidizing bacteria in marine environments: Widespread occurrence but low diversity. *Environ. Microbiol.* **9**, 1476–1484.

Schmid, M. C., Hooper, A. B., Klotz, M. G., Woebken, D., Lam, P., Kuypers, M. M. M., Pommerening-Roeser, A., op den Camp, H. J. M., and Jetten, M. S. M. (2008). Environmental detection of octahaem cytochrome *c* hydroxylamine/hydrazine oxidoreductase genes of aerobic and anaerobic ammonium-oxidizing bacteria. *Environ. Microbiol.* **10**, 3140–3149.

Schubert, C. J., Durisch-Kaiser, E., Wehrli, B., Thamdrup, B., Lam, P., and Kuypers, M. M. M. (2006). Anaerobic ammonium oxidation in a tropical freshwater system (Lake Tanganyika). *Environ. Microbiol.* **8**, 1857–1863.

Shimamura, M., Nishiyama, T., Shigetomo, H., Toyomoto, T., Kawahara, Y., Furukawa, K., and Fujii, T. (2007). Isolation of a multiheme protein from an anaerobic ammonium-oxidizing enrichment culture with features of a hydrazine-oxidizing enzyme. *Appl. Environ. Microbiol.* **73**, 1065–1072.

Shu, Q. L., and Jiao, N. Z. (2008). Profiling Planctomycetales diversity with reference to anammox-related bacteria in a South China Sea, deep-sea sediment. *Mar. Ecol. Evol. Perspect.* **29**, 413–420.

Strous, M., VanGerven, E., Zheng, P., Kuenen, J. G., and Jetten, M. S. M. (1997). Ammonium removal from concentrated waste streams with the anaerobic ammonium oxidation (anammox) process in different reactor configurations. *Water Res.* **31**, 1955–1962.

Strous, M., Heijnen, J. J., Kuenen, J. G., and Jetten, M. S. M. (1998). The sequencing batch reactor as a powerful tool for the study of slowly growing anaerobic ammonium-oxidizing microorganisms. *Appl. Microbiol. Biotechnol.* **50**, 589–596.

Strous, M., Fuerst, J. A., Kramer, E. H. M., Logemann, S., Muyzer, G., van de Pas-Schoonen, K. T., Webb, R., Kuenen, J. G., and Jetten, M. S. M. (1999a). Missing lithotroph identified as new planctomycete. *Nature* **400**, 446–449.

Strous, M., Kuenen, J. G., and Jetten, M. S. M. (1999b). Key physiology of anaerobic ammonium oxidation. *Appl. Environ. Microbiol.* **65**, 3248–3250.

Strous, M., Pelletier, E., Mangenot, S., Rattei, T., Lehner, A., Taylor, M. W., Horn, M., Daims, H., Bartol-Mavel, D., Wincker, P., Barbe, V., Fonknechten, N., et al. (2006). Deciphering the evolution and metabolism of an anammox bacterium from a community genome. *Nature* **440**, 790–794.

Tal, Y., Watts, J. E. M., and Schreier, H. J. (2006). Anaerobic ammonium-oxidizing (anammox) bacteria and associated activity in fixed-film biofilters of a marine recirculating aquaculture system. *Appl. Environ. Microbiol.* **72**, 2896–2904.

Tamura, K., Dudley, J., Nei, M., and Kumar, S. (2007). MEGA4: Molecular Evolutionary Genetics Analysis (MEGA) software version 4.0. *Mol. Biol. Evol.* **24**, 1596–1599.

Tang, C. J., Zheng, P., Zhang, L., Chen, J. W., Mahmood, Q., Chen, X. G., Hu, B. L., Wang, C. H., and Yu, Y. (2010). Enrichment features of anammox consortia from methanogenic granules loaded with high organic and methanol contents. *Chemosphere* **79**, 613–619.

Trigo, C., Campos, J. L., Garrido, J. M., and Mendez, R. (2006). Start-up of the Anammox process in a membrane bioreactor. *J. Biotechnol.* **126**, 475–487.

Tsushima, I., Kindaichi, T., and Okabe, S. (2007). Quantification of anaerobic ammonium-oxidizing bacteria in enrichment cultures by real-time PCR. *Water Res.* **41**, 785–794.

Van de Graaf, A. A., Mulder, A., Debruijn, P., Jetten, M. S. M., Robertson, L. A., and Kuenen, J. G. (1995). Anaerobic oxidation of ammonium is a biologically mediated process. *Appl. Environ. Microbiol.* **61**, 1246–1251.

Van de Graaf, A. A., deBruijn, P., Robertson, L. A., Jetten, M. S. M., and Kuenen, J. G. (1996). Autotrophic growth of anaerobic ammonium-oxidizing micro-organisms in a fluidized bed reactor. *Microbiology-UK* **142**, 2187–2196.

Van de Vossenberg, J., Rattray, J. E., Geerts, W., Kartal, B., van Niftrik, L., van Donselaar, E. G., Sinninghe Damste, J. S., Strous, M., and Jetten, M. S. (2008). Enrichment and characterization of marine anammox bacteria associated with global nitrogen gas production. *Environ. Microbiol.* **10**(11), 3120–3129.

Van der Star, W. R. L., Miclea, A. I., van Dongen, U., Muyzer, G., Picioreanu, C., and Van Loosdrecht, M. C. M. (2008). The membrane bioreactor: A novel tool to grow anammox bacteria as free cells. *Biotechnol. Bioeng.* **101**, 286–294.

Woebken, D., Lam, P., Kuypers, M. M. M., Naqvi, S. W. A., Kartal, B., Strous, M., Jetten, M. S. M., Fuchs, B. M., and Amann, R. (2008). A microdiversity study of anammox bacteria reveals a novel Candidatus Scalindua phylotype in marine oxygen minimum zones. *Environ. Microbiol.* **10**, 3106–3119.

Zehr, J. P., and Ward, B. B. (2002). Nitrogen cycling in the ocean: New perspectives on processes and paradigms. *Appl. Environ. Microbiol.* **68**, 1015–1024.

CHAPTER FIVE

Cultivation, Growth Physiology, and Chemotaxonomy of Nitrite-Oxidizing Bacteria

Eva Spieck* and André Lipski[†]

Contents

1. Introduction	110
1.1. Why are nitrite-oxidizing bacteria not well investigated?	110
1.2. Two groups of NOB deserve different cultivation conditions	111
2. Recipes for the Culture of NOB	112
2.1. Mineral salts medium for lithoautotrophic growth (modified from Bock et al., 1983)	113
2.2. Marine medium (Watson and Waterbury, 1971)	113
2.3. Mixotrophic medium (modified from Bock et al., 1983)	113
2.4. Heterotrophic medium (Nitrobacter; modified from Bock et al., 1983)	113
2.5. Acidic medium	114
2.6. Purity tests (modified from Steinmüller and Bock, 1976)	114
3. Cultivation Procedure	114
3.1. Storage	115
3.2. Growth supported by dissimilatory nitrate reduction	116
4. Isolation	116
4.1. Dilution series	117
4.2. Plating technique	117
4.3. Density gradient centrifugation	117
5. Physiological Investigations	118
5.1. Short-term nitrifying activity	118
5.2. Potential nitrite-oxidizing activity	119
5.3. Short-term activity of a marine biofilter consisting of colonized plastic biocarriers	119
5.4. Potential nitrifying activity of permafrost soil (Sanders, 2006)	119
5.5. Growth parameters	120

* Biocenter Klein Flottbek, University of Hamburg, Department of Microbiology and Biotechnology, Ohnhorststraße, Hamburg, Germany
[†] Rheinische Friedrich-Wilhelms-Universität Bonn, Food Microbiology and Hygiene, Meckenheimer Allee, Bonn, Germany

6. Chemical Analyses 121
7. Monitoring the Cultures 121
   7.1. Electron microscopy 123
8. Fatty Acid Analyses 124
Acknowledgments 127
References 127

Abstract

Lithoautotrophic nitrite-oxidizing bacteria (NOB) are known as fastidious microorganisms, which are hard to maintain and not many groups are trained to keep them in culture. They convert nitrite stoichiometrically to nitrate and growth is slow due to the poor energy balance. NOB are comprised of five genera, which are scattered among the phylogenetic tree. Because NOB proliferate in a broad range of environmental conditions (terrestrial, marine, acidic) and have diverse lifestyles (lithoautotrophic, mixotrophic, and heterotrophic), variation in media composition is necessary to match their individual growth requirements in the laboratory. From *Nitrobacter* and *Nitrococcus* relatively high cell amounts can be achieved by consumption of high nitrite concentrations, whereas accumulation of cells belonging to *Nitrospira*, *Nitrospina*, or the new candidate genus *Nitrotoga* needs prolonged feeding procedures. Isolation is possible for planktonic cells by dilution series or plating techniques, but gets complicated for strains with a tendency to develop microcolonies like *Nitrospira*. Physiological experiments including determination of the temperature or pH-optimum can be conducted with active laboratory cultures of NOB, but the attainment of reference values like cell protein content or cell numbers might be hard to realize due to the formation of flocs and the low cell density. Monitoring of laboratory enrichments is necessary especially if several species or genera coexist within the same culture and due to population shifts over time. Chemotaxonomy is a valuable method to identify and quantify NOB in biofilms and pure cultures alike, since fatty acid profiles reflect their phylogenetic heterogeneity. This chapter focusses on methods to enrich, isolate, and characterize NOB by various cultivation-based techniques.

1. INTRODUCTION

1.1. Why are nitrite-oxidizing bacteria not well investigated?

Nowadays, it is nearly impossible to deposit novel species of NOB in one of the official culture collections. Due to the limited number of isolates (Spieck and Bock, 2005), it was speculated that NOB may remain largely uncultured. The main problem appears to be that culturing of NOB is time-intensive and most of them do not survive standard preservation procedures like freeze-drying. Nevertheless, an increasing diversity of novel strains and even genera became available when the growth parameters became better adapted to natural conditions. For instance, nitrite rarely accumulates to detectable levels in nature, which potentially leads to overexposure to this toxic compound. Most species in the genus *Nitrobacter* were isolated at an initial nitrite concentration of 29 mM whereas cultivation of marine strains (*Nitrospira, Nitrococcus,* and *Nitrospina*) required a lower concentration of 1 mM nitrite (Bock and Koops, 1992; Watson and Waterbury, 1971). Earlier instructions for the quantification of NOB recommended nitrite concentrations as low as 0.05 mM (Woldendorp and Laanbroek, 1989) but it was subsequently observed that different populations of soil NOB have varying nitrite preferences (Both and Laanbroek, 1991). The modification of using a 10-fold lower (2.9 instead of 29 mM) concentration of nitrite resulted in the selective enrichment of *Nitrospira* from activated sludge (Bartosch *et al.*, 1999). Finally, application of yet another 10-fold dilution (0.29 instead of 2.9 mM nitrite) led to the cultivation of the novel "candidate" genus *Nitrotoga* from permafrost-affected soils (Alawi *et al.*, 2007). Simultaneously, it was also found necessary to lower the incubation temperature (10 °C), which is in agreement with the later finding that cold-adapted NOB also occur in moderate climate waste water treatment plants (Alawi *et al.*, 2009). Naturally, higher growth temperatures up to 60 °C are required for cultivation of NOB isolated from hot springs (Lebedeva *et al.*, 2005, in press).

Enrichments of NOB have been obtained from a fascinating variety of habitats (Off *et al.*, 2010), and it is tempting to speculate that the whole diversity of NOB is still not captured. This manual might help to rediscover the often neglected cohort of bacteria that perform the important second step of nitrification.

1.2. Two groups of NOB deserve different cultivation conditions

NOB are Gram-negative bacteria, which gain their energy by converting nitrite to nitrate. Nitrite oxidation is a strictly aerobic process and all C-requirements can be met from assimilation of CO_2. With regard to their physiology, NOB can be separated into two groups: *Nitrobacter* and *Nitrococcus* are relatively easy to handle because they are adapted to high substrate concentrations up to 29 mM of nitrite. Cultures of *Nitrobacter* and *Nitrococcus* reach turbidities visible to the naked eye, which is hard to realize for the second group of NOB, namely *Nitrospira, Nitrospina,* and the new candidate genus *Nitrotoga*. These nitrifiers grow only under nitrite limitation. Additionally, members of *Nitrospira* and *Nitrotoga* possess a high tendency for aggregation and growth is accompanied by the formation of microcolonies, visible as flocs. The grouping correlates with the fact that *Nitrobacter* and *Nitrococcus* contain intracytoplasmic membranes (ICM; Watson *et al.*, 1989) anchoring the key catabolic enzyme, nitrite oxidoreductase (NXR), on the cytoplasmic side of the ICM (Sundermeyer-Klinger et al., 1984). In accordance, a putative nitrite/nitrate antiport system is encoded in all sequenced *Nitrobacter* genomes (Starkenburg et al., 2008a) and in *Nitrococcus*. In contrast, the NXR of the more nitrite sensitive genera of NOB (*Nitrospira, Nitrospina*, and *Nitrotoga*) that lack ICM (Alawi et al., 2007; Watson et al., 1989), is located on the periplasmic side of the cytoplasmic membrane (Spieck et al., 1998); hence nitrite transport systems are not necessary (Lücker et al., 2010). As a consequence, we can speculate that the periplasmic orientation of NXR enables the latter genera to grow effectively at lower concentrations of nitrite (0.29–2.9 mM) than *Nitrobacter* and *Nitrococcus*. The external orientation might also be responsible for the inhibitory effects of high nitrite on growth of these three genera. However, a high tolerance toward nitrite was observed for members of *Nitrospira* retrieved from environments with a high nitrogen load, like a marine biofilter or a wastewater treatment plant (Off et al., 2010). Based on the recently obtained genome sequence, this high nitrite tolerance might be attributed to the presence of two gene copies encoding the α-subunit of the NXR in *Candidatus* Nitrospira defluvii, which are differentially regulated (Daims et al., 2010).

When NOB cultures begin to consume nitrite, it has to be replenished repeatedly in order to get a sufficient quantity of cells for further analyses. For some strains like *Nitrospira marina*, the biomass formation increases further when organic substances are added to the inorganic nutrient media (Watson et al., 1986). Whereas *Nitrobacter* has the ability for heteroorganotrophic growth (Bock et al., 1983), most of the other NOB do not assimilate carbon from organic compounds (Lebedeva et al., 2008).

Another important finding is that NOB can survive starvation from energy and reductant for several years. On the other hand, it can take weeks to months until their metabolism will be reactivated and growth resumes. For this reason, cultures have to be "switched-on" before performing the experiments. During the exponential phase of growth, some NOB grow relatively quickly with minimum generation times of 8 (*Nitrobacter winogradskyi*) or 12 h (*Nitrospira moscoviensis*; Ehrich et al., 1995; Watson et al., 1989).

2. Recipes for the Culture of NOB

In the following section, the composition of various media is described in accordance to the metabolic versatility of NOB. It is most convenient to prepare media by diluting stock solutions. Nitrite serves as sole electron donor and, to avoid toxicity, is added last at respective final concentrations as required by the individual NOB.

2.1. Mineral salts medium for lithoautotrophic growth (modified from Bock et al., 1983)

10-fold concentrated stock solution: 1000 mL distilled or deionized water: 0.07 g $CaCO_3$, 5 g NaCl, 0.5 g $MgSO_4 \cdot 7\ H_2O$, 1.5 g KH_2PO_4. *Trace elements*: 1000 mL 0,01 M HCl, 33.8 mg $MnSO_4 \cdot H_2O$, 49.4 mg H_3BO_3, 43.1 mg $ZnSO_4 \cdot 7H_2O$, 37.1 mg $(NH_4)_6Mo_7O_{24}$, 973 mg $FeSO_4 \cdot 7H_2O$, 25 mg $CuSO_4 \cdot 5H_2O$. *Final medium*: 900 mL distilled water, 100 mL stock solution, 1 mL trace elements, 2 g (29 mM), 0.2 g or 0.02 g $NaNO_2$. The pH should be adjusted to 8.4–8.6 with NaOH or KOH and should drop to 7.4–7.6 after autoclaving at 110 °C for 30 min.

It is helpful to reserve special glassware for NOB cultures and pay careful attention to the quality of water and reagents; use metal or plastic caps for small bottles and cotton stoppers for large culture vessels. Prepare the medium 1–2 days before use so that the pH can regulate and the medium can be saturated with CO_2.

2.2. Marine medium (Watson and Waterbury, 1971)

10-fold concentrated stock solution: 0.05 g $CaCl_2 \cdot 2H_2O$, 1 g $MgSO_4 \cdot 7\ H_2O$, 0.01 g $FeSO_4 \cdot 7H_2O$, 0.017 g KH_2PO_4, 700 mL sea water (purified by filter paper), 300 mL distilled water. *Marine trace elements*: 1000 mL distilled water, 6 mg $CuSO_4 \cdot 5H_2O$, 25 mg $Na_2MoO_4 \cdot 2H_2O$, 50 mg $MnCl_2 \cdot 4H_2O$, 0.5 mg $CoCl_2 \cdot 6H_2O$, 25 mg $ZnSO_4 \cdot 7H_2O$. *Final medium*: 630 mL sea water, 270 mL distilled water, 100 mL stock solution, 1 mL trace elements (marine), 0.2 g or 0.02 g $NaNO_2$. Adjust the pH to 6.5–7.0. It should adapt to 7.4–7.6 within 2 days after autoclaving at 110 °C for 30 min. If you do not have access to natural sea water, you can try artificial sea water, but growth might fail.

2.3. Mixotrophic medium (modified from Bock et al., 1983)

900 mL distilled water, 100 mL mineral stock solution, 1 mL trace elements, 0.02–2 g $NaNO_2$, 0.15 g yeast extract, 0.15 g peptone, 0.055 g Na-pyruvate, pH 8.6. The pH will be 7.4–7.6 after autoclaving. In the case of *Nitrobacter*, the concentrations of the organic matter can be raised by a factor of 10. Prepare plates from mixotrophic medium with 0.2 g $NaNO_2$ per liter and 13 g agarose per liter.

2.4. Heterotrophic medium (*Nitrobacter*; modified from Bock et al., 1983)

900 mL distilled water, 100 mL stock solution for lithoautotrophic growth (see above), 1 mL trace elements for lithoautotrophic growth, 1.5 g yeast extract, 1.5 g peptone, 0.55 g Na-pyruvate. Adjust the pH to 7.4. This pH

value will remain relatively stable during and after autoclaving due to the high amount of organic matter.

2.5. Acidic medium

Although NOB prefer a neutral pH value, Hankinson and Schmidt (1988) succeeded in isolating a strain of *Nitrobacter* growing at pH 5.5.

Solution **1–5** (g/100 mL distilled water): **1**: 2.72 g KH_2PO_4, **2**: 3.48 g K_2HPO_4, **3**: 0.246 g $FeSO_4 \cdot 7H_2O$, 0.331 g Na_2EDTA, **4**: 1.34 g $CaCl_2 \cdot 2H_2O$, **5**: 4 g $MgSO_4 \cdot 7H_2O$. *Trace elements for the acidic medium*: 100 mL distilled water, 10 mg $NaMoO_4 \cdot 2H_2O$, 20 mg $MnCl_2$, 0.2 mg $CoCl_2 \cdot 6H_2O$, 10 mg $ZnSO_4 \cdot 7H_2O$, 2 mg $CuSO_4 \cdot 5H_2O$. *Final medium*: 1 mL S1, 4 mL S2, 1 mL S3, 1 mL S4, 1 mL S5, 0.02 g KNO_2, 1 mL trace elements, make to 1000 mL with distilled water. Adjust the pH to 5.0 with 0.5 M HCL. After sterilization pH should be 5.5.

2.6. Purity tests (modified from Steinmüller and Bock, 1976)

Liquid medium: 1000 mL distilled water, 0.5 g peptone, 0.5 g yeast extract, 0.5 g meat extract, 0.584 g NaCl, pH 7.3–7.4. Inoculate 5 mL of medium with one drop of the culture and check for turbidity for up to 2 weeks. *Agar plates*: 1000 mL distilled water, 2.5 g meat extract, 2.5 g casamino acids, 0.5 g yeast extract, 1 g KH_2PO_4, 0.5 g NaCl, pH 7.3–7.4. Add 15 g agar per liter. *Oligotrophic purity test*: To test for the presence of oligotrophic heterotrophic bacteria, use of a mixotrophic medium (with distilled or 70% sea water) without nitrite is recommended.

Colony formation or turbidity after a few days is typical for heterotrophic contaminants, whereas the absence of colonies and slight turbidity after an incubation period of several weeks indicates heterotrophic growth of *Nitrobacter*.

3. Cultivation Procedure

To start batch cultivation from environmental material use an inoculum of 1 mL or 1 g of material per 150 mL medium in 300 mL Erlenmeyer flasks or reduce the volume to 50 mL in 100 mL flasks for samples with a low cell density. Incubation has to be carried out in the dark without agitation and the temperature depends on the *in situ* conditions (10–46 °C). For thermophilic cultures, restore any moisture losses with sterile distilled/deionized water. Strains of *Nitrospira* grow also in bottles with a screw cap, which were three-quarters full of liquid medium to produce a reduced oxygen content (about 4 mg O_2/L). When using substrate concentrations of 2.9 mM sodium nitrite or lower, you have to replace nitrite every 7–30 days in case of growth. Consumption of nitrite can be checked by the modified Griess–Ilosvay spot test (Schmidt and Belser, 1994) by transmitting an aliquot of the liquid culture on a drop of reagent, provided on a filter paper. The culture is active, if no pink coloring appears and the starting nitrite concentration should be refilled aseptically using a 2.5 M stock solution. Nitrite and nitrate can also be tested semiquantitatively providing a high resolution with Merckoquant test strips (Nitrite test 2–80 mg/L NO_2^-, Nitrate test 10–500 mg/L NO_3^-, Merck, Darmstadt, Germany). Repeated feeding will result in a slight turbidity or the occurrence of dense brownish flocs. Be careful with increasing accumulation of nitrate, because the cultures might be inhibited. For a strain of the marine NOB *Nitrospira* delayed nitrite oxidation was observed in the presence of 10 mM of nitrate. Transfer is recommended every 3–12 month with a 1–10% inoculum. Retain the old cultures from the preceding transfers along with current transfers in the event of contamination or other problems.

Reagent for the modified Griess–Ilosvay Spot test :

Solution 1 (*Diazoting reagent*): 50 mL distilled water, 4 g sulfanilamide, 10 mL orthophosphoric acid. *Solution 2* (*Coupling reagent*): 40 mL distilled water, 0.2 g *N*-(1-Naphthyl)-ethylenediamine-dihydrochloride. Combine both solutions and store in the dark at 4 °C.

Large culture batches (5–10 L) should be stirred when the culture has consumed the substrate for the first time. In the beginning, cultures are sensitive to a high partial pressure of oxygen, but stirring can be raised with increasing cell density. Harvesting of the cells is mostly done by centrifugation (15,000×g). Alternatively, filtration can be used, however, membrane filters can easily be clogged by flocs. The pellet appears brownish when cells possess a high nitrite-oxidizing activity or might be light-colored when flocs were formed. Microcolonies of *Candidatus* Nitrotoga or *Nitrospira* might be lost by centrifugation, when the aggregates do not attach to the walls of the centrifuge bottle. Failure to attachment/sedimentation is attributable to a low specific weight of flocs and to a difference in surface charge between planktonic cells and microcolonies, which are stabilized by divalent cations and hydrophobic interactions (Larsen *et al.*, 2008). As a consequence, larger microcolonies of *Nitrospira* are fairly resistant toward high shear forces and physical/chemical manipulations.

3.1. Storage

Storage of liquid cultures is recommended at 17 °C in the dark and cells of *Nitrobacter* can be reactivated even after several years. This might be difficult for the more sensitive cultures like *Nitrospira*, here a high inoculum of 10% (v/v) should be used and the substrate content can be lowered in the beginning to 0.15 mM sodium nitrite especially for the marine strains. Freezing in liquid nitrogen is suitable for maintenance of stock cultures that are suspended in a cryoprotecting buffer like Hatefi. It is necessary to use cells of about 150 mL medium, concentrated on a filter.

Hatefi: 100 mL distilled water, 0.6 g Tris, 22.6 sucrose, 0.015 g histidine, pH 7.5, freeze in aliquots.

Another possibility for the storage of *Nitrobacter* for several years is the cultivation in 1 L-bottles filled to the top with mixotrophic medium (increase the amount of organic matter by a factor of 10). Glycerol should be used instead of pyruvate to keep the pH stable for a long period. *Nitrobacter* has been shown to grow anaerobically by dissimilatory nitrate reduction, resulting in a high cell amount (Freitag *et al.*, 1987; Starkenburg *et al.*, 2008b). External electrons from organic matter are required as carbon source.

3.2. Growth supported by dissimilatory nitrate reduction

If growing in the absence of dissolved oxygen with nitrate as a terminal electron acceptor, *Nitrobacter* will be inhibited by nitrite at concentrations of more than 0.5 mM. To overcome this problem, a special culture technique using gas permeable membranes in gas tight bottles has been developed (Bock *et al.*, 1988). The components of the medium include 2 g/L sodium nitrite, 1.5 g/L yeast extract, 1.5 g/L peptone, and 0.55 g/L Na-pyruvate in mineralic salts solution (see above). Nitrite and organic substances serve as electron donors, whereas nitrate is the terminal electron acceptor. In the beginning, nitrite is oxidized to nitrate, which in turn is rereduced to nitrite. Other products under these growth conditions are ammonia and nitrogenous gases (NO, N_2O). The reactor is to be operated at 28 °C. Because air is continuously supplied through a silicone tube, an asymmetric biofilm (aerobic/anaerobic) will develop at the surface of the silicone membrane within a few days. This way inhibition of the anaerobically growing cells by nitrite is prevented by aerobically growing nitrite-oxidizing cells directly associated with the membrane surface.

4. Isolation

Successful isolation of NOB can be preceded by extensive and careful enrichment and repeated end-point-dilution.

4.1. Dilution series

NOB can be separated from heterotrophic bacteria using a mineral medium containing nitrite (5–10 mL medium per tube). Triplicate samples and the use of a low nitrite content (0.29 mM) are recommended. Serial dilutions (10^{-1} up to 10^{-8}) of enrichment cultures should be incubated for one to several months and periodically assayed for nitrite consumption. Continue the incubation until there is no more change in the number of positive tubes. To provide an almost homogenous distribution, all tubes were mixed with a Vortex mixer at highest speed before transfer. Especially in the case of *Nitrospira*, microcolonies can be disrupted mechanically by shaking with sterile glass beads (1.7–2 mm diameter) to avoid potential transfer of attached heterotrophic bacteria. Transfer cells from the tube with the highest dilution. Check purity by different complex media (see above).

4.2. Plating technique

Most of the NOB are sensitive to ambient oxygen concentrations and therefore do not grow on agar surfaces. Nevertheless, isolation of two species of *Nitrobacter* was successfully accomplished (Bock *et al.*, 1983; Sorokin *et al.*, 1998). On solid medium, *Nitrobacter* forms small shiny brownish colonies with a round boundary on mixotrophic medium (2.9 mM sodium nitrite) after incubation for several weeks. Plates should be closed by parafilm to avoid drying and can be additionally packed into a plastic bag to reduce the oxygen content. For *Nitrospira*, agarose (13 g/L) should be used instead of agar and irregular brownish colonies appear after 1–3 months. Colonies will usually remain smaller than a millimeter and they have thus to be located under magnification before picking individual colonies with an Eppendorf tip and inoculating them into a small volume of liquid growth medium.

4.3. Density gradient centrifugation

Centrifuge cells from a 1.5-L batch culture or more and resuspend in a small volume of sterile 0.9% NaCl medium (2% NaCl for marine strains). The concentrated cell suspension has to be mixed with 28 mL Percoll and 12 mL NaCl (1.5%) solution and centrifuged for 2 h at $12,000 \times g$ and 4 °C (fixed angle rotor) in 40 mL tubes (10 mL tubes are useful as well). Provided with a sufficient cell density, each cell type will form one brownish band in the gradient. Collect the cells from the tube with sterile Pasteur pipettes, resuspend in 0.9% NaCl, and observe in the light microscope. Dilute serially in the above-mentioned mineral medium with 0.29 mM of nitrite. Incubate for 2–3 months and check for growth. This method is also valuable, if coexisting NOB have to be separated.

5. Physiological Investigations

Pure cultures provide the opportunity to determine the physiological characteristics of environmental relevance. Activity tests can be used to compare the nitrification potentials of natural samples, of different strains growing in varying media or to check inhibition by additives. Nitrite-oxidizing activity has to be measured by chemical analyses and can be performed in different ways: Short-term nitrifying activity or potential nitrite-oxidizing activity. The details for static batch cultures are described by Schmidt and Belser (1994). Instructions for analyses with the aid of chemostats can be found in the chapter by Bollmann et al., this volume (Chapter 3, pp.). An excellent summary of physiological studies with nitrifying bacteria in general is given by Prosser (1989). A most recent review on the metabolism and growth (autotrophic, mixotrophic, and heterotrophic) of NOB in context with an analysis of their genome sequences also provided a discussion of how environmental factors affect their bioenergetics and growth characteristics (Starkenburg et al., 2010).

5.1. Short-term nitrifying activity

The short-term nitrifying activity test reflects the kinetics of aquatic systems, etc. This value does not necessarily express the *in situ* activity, because additional substrate is supplied and aeration is optimized. Here, it is essential to measure the activity over a time-period that does not allow the population to increase significantly. To avoid the latter, activity tests should last ≤ 12 h. It is important to use laboratory cultures in exponential growth phase or active environmental samples. For ammonia-oxidizing bacteria, a shortcut was developed to measure the activity of moderate soils (ISO 15685) using standard conditions (3 mM substrate, 25 °C, 6 h). However, such conditions like shaking will not be beneficial for all NOB and a time-period of only 6 h might be not sufficient for samples with a low cell density and/or activity. Therefore, the test might be conducted over an extended period and nitrite oxidation rates can be measured with varying concentrations of nitrite. Periodically, carry out four to five individual sampling spaced at 1–2 h interval. Under aerobic conditions, the decrease of nitrite and the stoichiometric increase of nitrate reflect the activity of NOB. The maximum activity can be calculated per hour per gram or mL of sample, etc. by determining the slope of decrease in nitrite concentration over time.

5.2. Potential nitrite-oxidizing activity

In order to compare the nitrifying potential or capacity of natural samples, they are amended with substrate and incubated under conditions of favorable temperature, etc. It is necessary to add an unamended sample as control, if mineralization occurs simultaneously. Prepare a duplicate and continue incubation for several weeks. In contrast to the short-term activity test, the potential activity might be measured without agitation and changes in nitrite and nitrate are continuously monitored until all of the substrate is oxidized. When organic matter is present, anoxic microniches might favor denitrification. Here, test vessels should be agitated to enhance oxygen supply. It is essential to analyze both the disappearance of nitrite, and the formation of nitrate. Denitrification will potentially continue until all labile organic matter is consumed.

In the following, there are two examples how nitrite-oxidizing activity can be measured:

5.3. Short-term activity of a marine biofilter consisting of colonized plastic biocarriers

Place 5–10 colonized plastic biocarriers into 25–50 mL of sterile marine medium in individual test vessels with a substrate concentration of 1–2 mM and incubate at an adequate temperature (17–28 °C) on a shaker (170 rpm). Multiple samples should be taken over a period of about 8–12 h, until the decrease of nitrite is clearly measurable.

5.4. Potential nitrifying activity of permafrost soil (Sanders, 2006)

Place 5 g of wet soil material in vessels containing 100 mL sterile terrestrial medium with a substrate concentration of 0.75 mM nitrite. Shake for 1 h at 120 rpm to get a homogenous suspension of soil and incubate afterward without agitation at 17 °C. The lag-phase before growth resumes might take about 2 weeks. It is beneficial to use a parallel soil sample to determine the dry mass in order to express the activity per gram soil.

As the measured activity of liquid cultures is related to the cell amount, numerical estimates of the population are required. The traditionally MPN (most probable number)-approach has been widely used, but it has distinct drawbacks. This indirect enumeration needs prolonged incubation periods of several weeks to months and has a high potential for statistical errors when microcolonies are counted as single cells. For this reason, this method will not be explained in detail here and readers are referred to the instructions given by Schmidt and Belser (1994) and Matulewich *et al.* (1975).

More recently developed molecular methods including real-time fluorescent PCR enable rapid quantification of nitrifier populations in natural and engineered environments such as activated sludge or soils (Attard et al., 2010; Hall et al., 2002).

5.5. Growth parameters

As reviewed by Prosser (1989), product concentration increases exponentially during batch growth of ammonia and nitrite-oxidizing bacteria (NOB). Under the assumption that growth is balanced, the slope of a semilogarithmic plot of product concentration versus time is equivalent to the maximum specific growth rate and provides similar values to those calculated as specific increases in cell concentration. This might be a good approach to estimate doubling times for the nitrite-sensitive strains of NOB, because their cell yield is limited. For example, given a substrate concentration of 8 mM sodium nitrite, the cell concentration increased from 1.2×10^4 cells m/L to 2.2×10^6 cells m/L for a marine strain of *Nitrospira* within 1 month in batch culture. Nevertheless, determination of the protein content was only possible in a culture grown under mixotrophic conditions and after concentrating the samples by a factor of 50–100. Hence the BCA method for protein determination is more suitable than Bradford due to the higher sensitivity (5 µg/mL). Quantification of the cell number might be difficult, if cells are aggregated to microcolonies. Furthermore, cells of *Nitrospira* are relatively small and difficult to detect using a counting chamber with a 40× objective. Improvement can be achieved by using a thin counting chamber (Hawksley, UK) and 100× oil immersions objective.

Growth experiments can be performed with NOB to determine the optimal conditions with regard to temperature, pH, or nitrite tolerance. Figure 5.1 reveals an experiment with *N. defluvii* (Spieck et al., 2006), where different incubation temperatures were compared. To start the experiment, 40 mL volumes of mineral medium containing 0.75 mM nitrite were inoculated with 2 mL aliquots of an active preculture and incubated without agitation. For the determination of the optimal temperature, incubations between 4 and 37 °C were carried out. It is important to correct for evaporation during extended incubations at elevated temperature. Nitrite oxidation was optimal at 32 °C and completed within 7 days, whereas activity was much less at 10 °C. No activity was observed at 4 or 37 °C. To get an optimum curve, it is possible to compare the slope of the individual curves. Alternatively, the amount of consumed substrate (expressed as nitrite oxidation rate per day) can be compared once the first culture has oxidized the whole amount of nitrite.

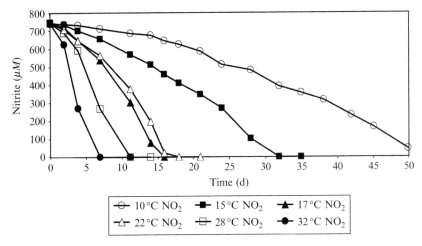

Figure 5.1 Temperature dependence of nitrite oxidation of *Candidatus* Nitrospira defluvii A17. The preculture was incubated at 28 °C.

6. CHEMICAL ANALYSES

Concentrations of nitrite and nitrate can be measured by high-performance liquid chromatography (HPLC) via ion-pair chromatography with a Hypersil ODS C18 column 5 μm (125 mm × 4.6 mm) and ion-pair reagent as appropriate buffer (Meincke et al., 1992). The HPLC system (Hitachi, Maidenhead, GB) consisted of a pump (L-2130), an autosampler (L-2200), a UV-detector (L-2485), and a data system (EZChrom Elite).

To remove cells and particles, samples of 1 mL were taken from the culture and centrifuged for 15 min at 16,000 rpm at room temperature. 600 μL of the supernatant are transferred to a HPLC vial and measured at 255 nm. Peaks of nitrite and nitrate are integrated by their area.

Eluent for measuring nitrite and nitrate by HPLC:

Stock solution: 1000 mL distilled water, 16.98 g tetrabutylammonium hydrogen sulphate (TBAHS), pH 6.4. *Final eluent*: 5 mM TBAHS: 4365 mL double distilled water, 360 mL stock solution, pH 6.4, and 525 mL methanol.

7. MONITORING THE CULTURES

So far, five different genera of NOB are known, which differ in growth rates, biomass, cell yield, and substrate oxidizing activity (Blackburne et al., 2007; Dytczak et al., 2008; Schramm et al., 1999).

Given this diversity, the coexistence of multiple NOB genera or species within the same enrichment culture is feasible. The most abundant cells can be identified by their cell shape using phase contrast microscopy or fluorescence microscopy with DAPI-staining. Characterization is also possible based on the specific ultrastructure (Spieck and Bock, 2005). For instance, electron microscopy (Figs. 5.2 and 5.3) can be used to investigate the presence and arrangement of ICM (Watson *et al.*, 1989) and the extensions of the periplasmic space (Alawi *et al.*, 2007; Watson *et al.*, 1986), in ultrathin sections of *Nitrobacter* and *Nitrococcus*, and of *Nitrospira* and *Nitrotoga*, respectively (see below). Members of *Nitrospina* can be identified by their appearance as long slender rods, but in senescent cultures spherical forms accumulate (Watson and Waterbury, 1971). It is remarkable that planktonic cells alternate with dense microcolonies in a kind of life cycle of *Nitrospira* (Fig. 5.3; Lebedeva *et al.*, 2008). Monoclonal antibodies targeting the NXR are another tool for the differentiation of NOB using immunofluorescence or Western-blotting techniques (Bartosch *et al.*, 1999). Methods for the

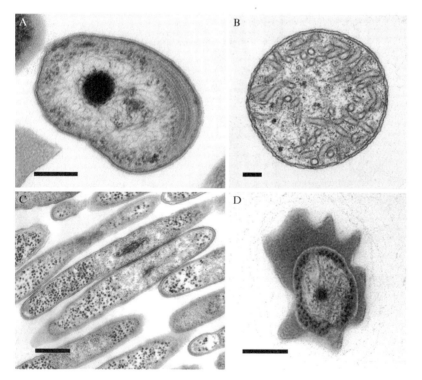

Figure 5.2 Ultrathin sections of four genera of NOB revealing the characteristic cell shape and ultrastructure. (A) *Nitrobacter*, (B) *Nitrococcus*, (C) *Nitrospina*, and (D) *Candidatus* Nitrotoga. Bars: $a, b, d = 200$ nm, $c = 500$ nm.

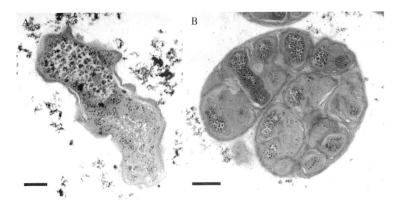

Figure 5.3 Change in morphology of *Nitrospira*. (A) Spiral shaped cell, (B) Condensed microcolony, embedded in an extracellular matrix. Bars: $a = 200$ nm, $b = 500$ nm.

characterization of enrichments of NOB using molecular approaches (e.g., using PCR or FISH) are described by Daims *et al.* (2010). Acid biomass hydrolysis enables analysis of the fatty acid content and leads to chemotaxonomic profiling of NOB congruent with their phylogenetic belonging to different subclasses of the *Proteobacteria* or the deeper branching phylum *Nitrospirae* (Lipski *et al.*, 2001). Further on, these varying chemotaxonomic lipid profiles help to differentiate *Nitrospira* at the species level (see below).

7.1. Electron microscopy

For observation by electron microscopy, cells from at least a 1-L culture are collected at $15,000 \times g$, washed in 0.9% NaCl, and prefixed with 2.5% (v/v) glutaraldehyde for 30 min at room temperature and for 1.5 h on ice. Cells are washed three times with cacodylic acid (Na salt, 75 mM) and embedded in a drop of 2% noble agar. The process will be continued by fixation overnight in 1% (w/v) osmium tetroxide (buffered with cacodylic acid) at 4 °C and washing four times with cacodylic acid. Dehydration has to be performed in an acetone series (15%, 30%, 50%, 70%, 90%, 100%) at 4 °C for 10 min each and finally twice in 100% acetone at room temperature. Embedding is performed in a mixture of Spurr (see below) and acetone at ratios of 1:3 (for 2 h), 1:1 (2 h), 3:1 (17 h), followed by an exposure to pure Spurr for 2.5 h and a final incubation for 6 h in newly provided Spurr (Spurr, 1969). Samples are brought into silicone forms, covered with Spurr and polymerized for 16 h at 70 °C. The dried samples have to be sectioned (50–70 nm) with a diamond knife, collected on copper grids (300 mesh, covered with Mowital polyvinylbutyral in 0.3% chloroform) and double-stained with 2% uranylacetate in distilled water and 2% lead citrate for 10 min, respectively. Examination can be carried out with a transmission electron microscope (Zeiss model Leo 906E with a CCD camera model 794). For whole-cell preparations a few milliliters of the culture have to be centrifuged ($15,000 \times g$), the harvested cells dropped on a copper grid, dried, and stained with uranyl acetate in distilled water (2%).

Spurr resin (Low Viscosity; Serva):

5.0 g Embedding Medium ERL-4221D, 3 g DER 736 (ERL-4206 plasticizer), 13 g NSA (nonenylsuccinic anhydride), mix ERL-4221D, DER 736, and NSA Add 0.2 g DMAE (S-1, dimethylaminoethanol) under stirring.

8. Fatty Acid Analyses

Fatty acid profile analyses emerged to be a useful tool for a rapid and effective differentiation of NOB. These analyses can be used during the course of an enrichment procedure to monitor the differential enrichment of NOB populations, to assess the increasing purity of the enrichment culture, and to predict the systematic position of the finally enriched, almost pure NOB population. The presently known five nitrite-oxidizing genera can be differentiated by characteristic lipid markers (Alawi *et al.*, 2007; Lipski *et al.*, 2001; Spieck *et al.*, 2006). As a member of the *Alphaproteobacteria*, the genus *Nitrobacter* contains the typical dominating fatty acid found in this class, which is *cis*-11-octadecenoic acid (18:1 *cis*-11). This lipid accounts for more than 70% of the whole-cell fatty acid profile, which is characteristic for *Alphaproteobacteria*. Moreover, *cis*-11,12-methyleneoctadecanoic (19:0 cyclo11-12) acid can be present up to 20%. This compound is a derivative of 18:1 *cis*-11; consequently, the sum of both fatty acids is usually between 80% and 90%. The ratio between both lipids depends on the growth phase and growth conditions. Unlike most Gram-negative bacteria but like other closely related genera of the family *Bradyrhizobiaceae* such as the genus *Bradyrhizobium* (Tighe *et al.*, 2000), *Nitrobacter* species lack hydroxy fatty acids and their absence constitutes a negative marker for this NOB genus.

The genus *Nitrococcus* shows a typical profile for *Gammaproteobacteria*. The percentage of 18:1 *cis*-11 is considerably lower than for *Nitrobacter* and hydroxy fatty acids are detected regularly, though percentages for these compounds are low (Table 5.1). The genus *Nitrotoga*, at present taxonomically defined as *Candidatus*, is an example for a fatty acid-guided enrichment strategy. Out of a series of low-temperature cultivated enrichment cultures, one culture showed a clearly distinct fatty acid profile. Following in-depth analysis of this outstanding culture led to the finding of a hitherto uncharacterized NOB and finally to the proposal of the new taxon for this nitrite-oxidizing culture (Alawi *et al.*, 2007).

Table 5.1 Fatty acid profiles of nitrite-oxidizing genera

	Nitrobacter	Nitrococcus	Candidatus Nitrotoga	Nitrospina	Nitrospira
Strains/analyses[a]	7/18[a]	1/3	1/3	2/3	7/11
12:0 3OH	–[b]	tr[c]-1	3–7	–	–
14:1 cis-9	–	–	–	2–9	–
14:0	0–1	tr-1	1–4	29–32	tr-8
16:1 cis-7	0–1	–	0–2	–	3–46
16:1 cis-9	0–4	11–31	24–47	45–54	0–4
16:1 trans-9	–	–	0–1	tr-7	–
16:1 cis-11	–	–	0–1	0–1	0–43
16:0	6–15	23–31	21–27	3–7	15–58
16:0 11methyl	–	–	–	–	0–39
17:0 cyclo	–	tr-2	4–6	1–3	0–7
16:0 3OH	–	–	–	1–3	0–3
18:1 cis-11	72–92	42–52	3–6	tr-1	0–6
18:0	0–2	1–3	1–3	tr-1	tr-9
19:0 cyclo11–12	0–17	tr-3	–	–	0–2

Data from Lipski et al. (2001), Spieck et al. (2006), Alawi et al. (2007). Data are given as percentages.
[a] Number of strains/number of total analyses considered for the range.
[b] Not detected.
[c] tr: $0 < \text{tr} < 1$.

A characteristic property of the genus *Nitrospina* is the high abundance of tetradecanoic acid (14:0), which is in contrast to all other nitrite-oxidizing genera. The genus *Nitrospira* is characterized by the majority presence of cis-7 and cis-11 isomers of hexadecenoic acid and by 11-methyl-hexadecanoic acid (16:0 11methyl; Lipski et al., 2001) and the latter branched chain lipid is unique to strains of this genus. In contrast, cis-9-hexadecenoic acid, which is a constitutive membrane component in almost all bacteria is rarely present in strains of the genus *Nitrospira*. *Nitrospira* species have varying combinations and concentrations of these three marker lipids and therefore this pattern can be used for chemotaxonomic inference at the species level. Considerable amounts of the branched fatty acid 16:0 11methyl are present in three out of five *Nitrospira* species: *N. moscoviensis*, *Candidatus* N. bockiana, and *N. calida*. The latter taxon, so far, identified in two enrichment cultures, appears to be not homogeneous with regard to this lipid. In contrast, the best lipid marker for *N. moscoviensis*, *N. marina*, and *Candidatus* N. defluvii is the fatty acid 16:1 cis-11. Detailed fatty acid profiles for strains of this genus are listed in Table 5.2.

For the analysis of enrichment cultures or cultures of pure isolates whole-cell fatty acid analysis according to the Microbial Identification

Table 5.2 Fatty acid profiles of cultivated strains and enrichment cultures of the genus *Nitrospira*

	Nitrospira moscoviensis	*Nitrospira marina*	*Candidatus Nitrospira defluvii*	*Candidatus Nitrospira bockiana*	*Nitrospira calida*
Strains/analyses[a]	1/3[a]	1/2	1/2	2/2	2/3
14:0	0–1	0–2	7–13	2–3	0–2
16:1 *cis*-7	5–6	33–34	2–3	17–19	9–46
16:1 *cis*-9	0–4	–[b]	1–3	0–2	–
16:1 *cis*-11	28–37	15–23	33–43	–	–
16:0	19–38	30–37	15–17	48–52	42–58
16:0 11methyl	12–34	0–1	–	20–22	0–39
17:0 cyclo	–	–	6–7	–	–
16:0 3OH	0–1	–	2–3	1–3	0–1
18:0	0–2	8–9	0–3	0–3	2–4

Data from Lipski et al. (2001), Spieck et al. (2006), Lebedeva et al. (in press). Data are given as percentages.
[a] Number of strains/number of total analyses considered for the range.
[b] Not detected.

System (MIS; Microbial ID Inc., Newark, NJ, USA) standard procedure is the most favorite method (Sasser, 1990). All fatty acid profiles presented in Tables 5.1 and 5.2 were obtained by this method. Cells are pelleted by centrifugation at $15,000 \times g$ for 30 min when the cultures become turbid or dense brownish flocs develop (see above). The cells from the cultures are washed with 0.9% NaCl (w/v) and stored at $-20\ °C$. The pellets with approximately 20–50 mg wet weight are transferred in Teflon-lined screw-capped tubes and resuspended in 1 mL 15% NaOH in methanol/water (1:1, v/v). For cell lysis and saponification of fatty acids this suspension is incubated at 100 °C for 30 min. For the following methylation of fatty acids, 2 mL 6.0 N HCl/methanol (1.2:1, v/v) are added and the tubes are incubated at 80 °C for 10 min. The resulting fatty acid methyl esters (FAMEs) are extracted with 1.25 mL hexane/methyl *tert*-butyl ether (1:1, v/v) for 10 min by end-over-end mixing with a tube rotator. After phase separation, the lower aqueous phase is discarded. Finally, the remaining organic phase is washed by addition of 3 mL of 1.2% NaOH, and by mixing for 5 min with an end-over-end rotator. Phase separation can be supported by brief centrifugation at $1000 \times g$ for 3 min. About half of the clear upper organic phase is transferred to a GC-vial and can be analyzed directly or stored for several weeks at $-20\ °C$. A reagent control tube without cells should be processed with each analysis to check the purity of the reagents. These preparations include lipids from all cellular membranes, including the taxonomically important amide-bound hydroxy fatty acids.

The FAME extracts are analyzed by gas-chromatography using a 5% phenyl methyl silicone capillary column and helium as carrier gas. The injector temperature is 250 °C and the temperature gradient is from 120 to 240 °C at a rate of 5 °C/min. FAMEs are identified by their equivalent chain lengths calculated from their retention times (Sasser, 1990). As standard a mix containing all straight chain saturated FAMEs between C10 and C20 is used. The data presented in Tables 5.1 and 5.2 were acquired with a Hewlett-Packard model 6890 gas chromatograph with FID and with a 5890 series II gas chromatograph connected to a model 5972 mass selective detector. Use of a GC–MS system is preferred for a reliable identification of FAMEs. Notably, this is important for a reliable localization of the double bond positions in monounsaturated fatty acids. For this, the dimethyl disulfide adducts of the FAMEs are analyzed according to the procedure described by Nichols *et al.* (1986). In particular, this method is recommended if the diagnostic isomers of hexacedenoic acid have to be separated and identified. When using a 5% phenyl methyl silicone capillary column for analysis, which is the prevailing standard, discrimination between *cis*-7, *cis*-9, and *cis*-11 isomers of these FAMEs is low and mixtures of these compounds are difficult to separate especially if one of these compounds dominates the profile.

ACKNOWLEDGMENTS

The authors wish to thank the Deutsche Forschungsgemeinschaft (DFG), the Federal Ministry of Education and Research (BMBF), the Deutsche Bundesstiftung Umwelt (DBU), and the Deutscher Akademischer Austauschdienst (DAAD) for financial support. We express our sincere appreciation to Prof. Eberhard Bock, who established the culture collection of NOB and to Elena V. Lebedeva for her pioneering work in culturing NOB from extreme environments. Elke Woelken is acknowledged for excellent technical assistance in electron microscopy, Peter-Georg Jozsa for HPLC-measurements and Maria Klimova for the determination of the temperature optimum of *N. defluvii*. We thank Sabine Keuter for her methodical input as well as Sandra Off and Yvonne Bedarf for their help in preparing the manuscript.

REFERENCES

Alawi, M., Lipski, A., Sanders, T., Pfeiffer, E. M., and Spieck, E. (2007). Cultivation of a novel cold-adapted nitrite oxidizing betaproteobacterium from the Siberian Arctic. *ISME J.* **1**, 256–264.

Alawi, M., Off, S., Kaya, M., and Spieck, E. (2009). Temperature influences the population structure of nitrite-oxidizing bacteria in activated sludge. *Environ. Microbiol. Rep.* **1**, 184–190.

Attard, E., Poly, F., Commeaux, C., Laurent, F., Terada, A., Smets, B. F., Recous, S., and Le Roux, X. (2010). Shifts between *Nitrospira*- and *Nitrobacter*-like nitrite oxidizers underlie the response of soil potential nitrite oxidation to changes in tillage practice. *Environ. Microbiol.* **12**, 315–326.

Bartosch, S., Wolgast, I., Spieck, E., and Bock, E. (1999). Identification of nitrite-oxidizing bacteria with monoclonal antibodies recognizing the nitrite oxidoreductase. *Appl. Environ. Microbiol.* **65**, 4126–4133.

Blackburne, R., Vadivelu, V. M., Yuan, Z., and Keller, J. (2007). Kinetic characterization of an enriched *Nitrospira* culture with comparison to *Nitrobacter*. *Water Res.* **41**, 3033–3042.

Bock, E., and Koops, H. P. (1992). The genus *Nitrobacter* and related genera. *In* "The Prokaryotes, Vol. 3," (A. Balows, H. G. Trüper, M. Dworkin, W. Harder, and K. H. Schleifer, eds.), 2nd edn. pp. 2302–2309. Springer-Verlag, New York.

Bock, E., Sundermeyer-Klinger, H., and Stackebrandt, E. (1983). New facultative lithoautotrophic nitrite-oxidizing bacteria. *Arch. Microbiol.* **136**, 281–284.

Bock, E., Wilderer, P. A., and Freitag, A. (1988). Growth of *Nitrobacter* in the absence of dissolved oxygen. *Water Res.* **22**, 245–250.

Both, G. J., and Laanbroek, H. J. (1991). The effect of the incubation period on the result of MPN enumerations of nitrite-oxidizing bacteria: Theoretical considerations. *FEMS Microbiol. Ecol.* **85**, 335–344.

Daims, H., Lücker, S., Le Paslier, D., and Wagner, M. (2010). Diversity, environmental genomics and ecophysiology of nitrite-oxidizing bacteria. *In* "Nitrification," (B. B. Ward, D. J. Arp, and M. G. Klotz, eds.), 978-1-55581-481-6. ASM Press, Washington, DC.

Dytczak, M. A., Londry, K. L., and Oleszkiewicz, J. A. (2008). Activated sludge operational regime has significant impact on the type of nitrifying community and its nitrification rates. *Water Res.* **42**, 2320–2328.

Ehrich, S., Behrens, D., Lebedeva, E., Ludwig, W., and Bock, E. (1995). A new obligately chemolithoautotrophic, nitrite-oxidizing bacterium. *Nitrospira moscoviensis* sp. nov. and its phylogenetic relationship. *Arch. Microbiol.* **164**, 16–23.

Freitag, A., Rudert, M., and Bock, E. (1987). Growth of *Nitrobacter* by dissimilatoric nitrate reduction. *FEMS Microbiol. Lett.* **48**, 105–109.

Hall, S. J., Hugenholtz, P., Siyambalapitiya, N., Keller, J., and Blackall, L. L. (2002). The development and use of real-time PCR for the quantification of nitrifiers in activated sludge. *Water Sci. Technol.* **46**(1–2), 267–272.

Hankinson, T. R., and Schmidt, E. L. (1988). An acidophilic and a neutrophilic *Nitrobacter* strain isolated from the numerically predominant nitrite-oxidizing population of an acid forest soil. *Appl. Environ. Microbiol.* **54**, 1536–1540.

Larsen, P., Nielsen, J. L., Svendsen, T. C., and Nielsen, P. H. (2008). Adhesion characteristics of nitrifying bacteria in activated sludge. *Water Res.* **42**, 2814–2826.

Lebedeva, E. V., Alawi, M., Fiencke, C., Namsaraev, B., Bock, E., and Spieck, E. (2005). Moderately thermophilic nitrifying bacteria from a hot spring of the Baikal rift zone. *FEMS Microbiol. Ecol.* **54**, 297–306.

Lebedeva, E. V., Alawi, M., Maixner, F., Jozsa, P. G., Daims, H., and Spieck, E. (2008). Physiological and phylogenetic characterization of a novel lithoautotrophic nitrite-oxidizing bacterium, '*Candidatus* Nitrospira bockiana'. *Int. J. Syst. Evol. Microbiol.* **58**, 242–250.

Lebedeva, E. V., Off, S., Zumbrägel, S., Kruse, M., Shagzhina, A., Lücker, S., Maixner, F., Lipski, A., Daims, H., and Spieck, E. Isolation and characterization of a moderately thermophilic nitrite-oxidizing bacterium from a geothermal spring. *FEMS Microbiol. Ecol.* (in press).

Lipski, A., Spieck, E., Makolla, A., and Altendorf, K. (2001). Fatty acid profiles of nitrite-oxidizing bacteria reflect their phylogenetic heterogeneity. *Syst. Appl. Microbiol.* **24**, 377–384.

Lücker, S., Wagner, M., Maixner, F., Pelletier, E., Koch, H., Vacherie, B., Rattei, T., Sinninghe Damstè, J. S., Spieck, E., Le Paslier, D., and Daims, H. (2010). *Nitrospira* metagenome illuminates the physiology and evolution of globally important nitrite-oxidizing bacteria. *PNAS* **107,** 13479–13484.

Matulewich, V. A., Strom, P. F., and Finstein, M. S. (1975). Length of incubation for enumerating nitrifying bacteria present in various environments. *Appl. Microbiol.* **29,** 265–268.

Meincke, M., Bock, E., Kastrau, D., and Kroneck, P. M. H. (1992). Nitrite oxidoreductase from *Nitrobacter hamburgensis*: Redox centers and their catalytic role. *Arch. Microbiol.* **158,** 127–131.

Nichols, P. D., Guckert, J. B., and White, D. C. (1986). Determination of monounsaturated fatty acid double-bond position and geometry for microbial monocultures and complex consortia by capillary GC–MS of their dimethyl disulphide adducts. *J. Microbiol. Methods* **5,** 49–55.

Off, S., Alawi, M., and Spieck, E. (2010). Enrichment and physiological characterization of a novel *Nitrospira*-like bacterium obtained from a marine sponge. *Appl. Environ. Microbiol.* **76,** 4640–4646.

Prosser, J. I. (1989). Autotrophic nitrification in bacteria. *In* "Advances in Microbial Physiology, Vol. 30," (A. H. Rose and D. W. Tempest, eds.), pp. 125–181. Academic Press, London, New York.

Sanders, T. (2006). Vergleichende Untersuchungen kälteliebender nitrifizierender Bakterien aus Permafrostböden im Lena Delta, Sibirien. Diplomarbeit Universität Hamburg.

Sasser, M. (1990). Identification of bacteria through fatty acid analysis. *In* "Methods in Phytobacteriology," (Z. Klement, K. Rudolph, and D. Sands, eds.), pp. 199–204. Akademiai Kiado, Budapest, Hungary.

Schmidt, E. L., and Belser, L. W. (1994). Autotrophic nitrifying bacteria. *In* "Methods of Soil Analysis. Part 2-Microbiological and Biochemical Properties," (R. W. Weaver, J. S. Angle, and P. J. Bottomley, eds.), pp. 159–177. Soil Science Society of America, Madison, WI.

Schramm, A., de Beer, D., van den Heuvel, J. C., Ottengraf, S., and Amann, R. (1999). Microscale distribution of populations and activities of *Nitrosospira* and *Nitrospira* spp. along a macroscale gradient in a nitrifying bioreactor: Quantification by in situ hybridization and the use of microsensors. *Appl. Environ. Microbiol.* **65,** 3690–3696.

Sorokin, D. Y., Muyzer, G., Brinkhoff, T., Kuenen, J. G., and Jetten, M. S. (1998). Isolation and characterization of a novel facultatively alkaliphilic *Nitrobacter* species *N. alkalicus* sp. nov.. *Arch. Microbiol.* **170,** 345–352.

Spieck, E., and Bock, E. (2005). The lithoautotrophic nitrite-oxidizing bacteria. *In* "Bergey's Manual of Systematic Bacteriology, Vol. 2," (G. M. Garrity, D. J. Brenner, N. R. Krieg, and J. T. Staley, eds.), 2nd edn. pp. 149–153. Springer-Verlag, New York.

Spieck, E., Ehrich, S., Aamand, J., and Bock, E. (1998). Isolation and immunocytochemical location of the nitrite-oxidizing system in *Nitrospira moscoviensis*. *Arch. Microbiol.* **169,** 225–230.

Spieck, E., Hartwig, C., McCormack, I., Maixner, F., Wagner, M., Lipski, A., and Daims, H. (2006). Selective enrichment and molecular characterization of a previously uncultured *Nitrospira*-like bacterium from activated sludge. *Environ. Microbiol.* **8,** 405–415.

Spurr, A. R. (1969). A low viscosity epoxy resin embedding medium of electron microscopy. *J. Ultrastruct. Res.* **26,** 31–43.

Starkenburg, S. R., Larimer, F. W., Stein, L. Y., Klotz, M. G., Chain, P. S., Sayavedra-Soto, L. A., Poret-Peterson, A. T., Gentry, M. E., Arp, D. J., Ward, B., and Bottomley, P. J. (2008a). Complete genome sequence of *Nitrobacter hamburgensis* X14 and comparative genomic analysis of species within the genus *Nitrobacter*. *Appl. Environ. Microbiol.* **74,** 2852–2863.

Starkenburg, S. R., Arp, D. J., and Bottomley, P. J. (2008b). Expression of a putative nitrite reductase and the reversible inhibition of nitrite-dependent respiration by nitric oxide in *Nitrobacter winogradskyi* Nb-255. *Environ. Microbiol.* **10,** 3036–3042.

Starkenburg, S. R., Spieck, E., and Bottomley, P. J. (2010). Metabolism and genomics of nitrite oxidizing bacteria: Emphasis on pure culture studies and *Nitrobacter* species. *In* "Nitrification," (B. Ward, D. J. Arp, and M. G. Klotz, eds.), 978-1-55581-481-6. ASM Press, Washington, DC.

Steinmüller, W., and Bock, E. (1976). Growth of *Nitrobacter* in the presence of organic matter. I. Mixotrophic growth. *Arch. Microbiol.* **108,** 299–304.

Sundermeyer-Klinger, H., Meyer, W., Warninghoff, B., and Bock, E. (1984). Membrane-bound nitrite oxidoreductase of *Nitrobacter*: Evidence for a nitrate reductase system. *Arch. Microbiol.* **140,** 153–158.

Tighe, S. W., de Lajudie, P., Dipietro, K., Lindström, K., Nick, G., and Jarvis, B. D. W. (2000). Analysis of cellular fatty acids and phenotypic relationships of *Agrobacterium*, *Bradyrhizobium*, *Mesorhizobium*, *Rhizobium* and *Sinorhizobium* species using the Sherlock Microbial Identification System. *Int. J. Syst. Evol. Microbiol.* **50,** 787–801.

Watson, S. W., and Waterbury, J. B. (1971). Characteristics of two marine nitrite oxidizing bacteria, *Nitrospina gracilis*, nov. gen. nov. sp. and *Nitrococcus mobilis* nov. gen. nov. sp. *Arch. Microbiol.* **77,** 203–230.

Watson, S. W., Bock, E., Valois, F. W., Waterbury, J. B., and Schlosser, U. (1986). *Nitrospira marina* gen. nov. sp. nov: A chemolithotrophic nitrite-oxidizing bacterium. *Arch. Microbiol.* **144,** 1–7.

Watson, S. W., Bock, E., Harms, H., Koops, H. P., and Hooper, A. B. (1989). Nitrifying bacteria. *In* "Bergey's Manual of Systematic Bacteriology, Vol. 3," (J. T. Staley, M. P. Bryant, N. Pfennig, and J. G. Holt, eds.), pp. 1808–1834. Williams and Wilkins, Baltimore, MD.

Woldendorp, V. W., and Laanbroek, H. J. (1989). Activity of nitrifiers in relation to nitrogen nutrition of plants in natural ecosystems. *Plant Soil* **115,** 217–228.

CHAPTER SIX

Surveying N_2O-Producing Pathways in Bacteria

Lisa Y. Stein

Contents

1. Introduction — 132
2. Bioinformatics — 135
   2.1. Aerobic hydroxylamine oxidation pathway — 138
   2.2. Dissimilatory nitrite reduction pathway — 140
   2.3. Detoxification inventory — 141
   2.4. Identifying motifs and regulatory sequences — 142
3. Surveying Function — 143
   3.1. Following the N — 144
   3.2. Linking activity with gene expression — 146
4. Concluding Remarks — 147
References — 148

Abstract

Nitrous oxide (N_2O) is produced by bacteria as an intermediate of both dissimilatory and detoxification pathways under a range of oxygen levels, although the majority of N_2O is released in suboxic to anoxic environments. N_2O production under physiologically relevant conditions appears to require the reduction of nitric oxide (NO) produced from the oxidation of hydroxylamine (nitrification), reduction of nitrite (denitrification), or by host cells of pathogenic bacteria. In a single bacterial isolate, N_2O-producing pathways can be complex, overlapping, involve multiple enzymes with the same function, and require multiple layers of regulatory machinery. This overview discusses how to identify known N_2O-producing inventory and regulatory sequences within bacterial genome sequences and basic physiological approaches for investigating the function of that inventory. A multitude of review articles have been published on individual enzymes, pathways, regulation, and environmental significance of N_2O-production encompassing a large diversity of bacterial isolates. The combination of next-generation deep sequencing platforms, emerging proteomics technologies, and basic microbial physiology can be used to expand what is known about N_2O-producing pathways in individual bacterial species to

Department of Biological Sciences, University of Alberta, Edmonton, Alberta, Canada discover novel inventory and unifying features of pathways. A combination of approaches is required to understand and generalize the function and control of N₂O production across a range of temporal and spatial scales within natural and host environments.

1. INTRODUCTION

Industrialization has brought many benefits to society including the ability to feed an ever-increasing number of people. Despite this progress, the benefit of abundant food production is now challenged by a massive imbalance of the global nitrogen cycle caused by anthropogenic fixation of dinitrogen for fertilizer production. The anthropogenic contribution of fixed-N to the world's terrestrial ecosystems is now more than double that from natural processes (Röckstrom et al., 2009). Consequences of resulting ecosystem saturation by fixed-N include acceleration of nitrification, increased release of nitric and nitrous oxides, and eutrophication of freshwater, estuaries, and coastal systems from nitrate-rich run-off. Indeed, the nitrogen cycle is so far out of balance between the sources and sinks of N-oxide intermediates that we are rapidly leaving the Holocene epoch altogether for the Anthropocene where human activity is the primary driver of environmental change (Davidson, 2009; Röckstrom et al., 2009).

Within this era of rapid environmental change, more attention has focused on atmospheric accumulation of the potent greenhouse gas, N_2O, which currently accounts for ca. 5–7% of the overall greenhouse effect (IPCC, 2006). The largest source of N_2O is from microbial metabolism in response to modern agricultural practices and ecosystem N-saturation (Wuebbles, 2009). There are several well-characterized pathways of free-living bacteria that create N_2O from the reduction of an NO intermediate including: (1) aerobic hydroxylamine oxidation, usually as part of the ammonia oxidation (nitrification) pathway, (2) dissimilatory nitrite reduction, which may or may not be part of a complete denitrification pathway, and (3) a variety of NOx detoxification pathways that are broadly distributed throughout the bacteria (Fig. 6.1). Overviews on organisms, environmental conditions, and physiological pathways leading to N_2O production can be found elsewhere (Stein, 2010; Stein and Yung, 2003; Zumft, 1997).

Verification of individual processes, diversity of organisms, and range of genetic inventory for N_2O production has largely been accomplished by analyzing a growing number of available genome and metagenome sequences (Arp et al., 2007; Jetten et al., 2009; Klotz and Stein, 2008, 2010). A surprising finding of metagenomic data from the Sargasso Sea (Venter et al., 2004) and soils (Treusch et al., 2005) was that of ammonia-oxidizing Thaumarchaeota,

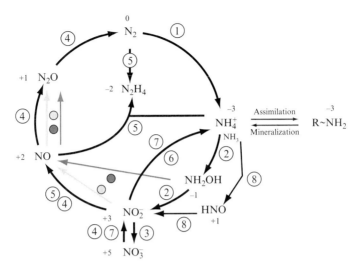

Figure 6.1 Processes in the microbial nitrogen cycle reproduced from Klotz and Stein (2010). The oxidation state of each intermediate is indicated. The pathway for archaeal ammonia oxidation is putative as based on genomic inference (Walker et al., 2010). 1. Dinitrogen fixation. 2. Aerobic oxidation of ammonia to nitrite by bacteria. 3. Aerobic oxidation of nitrite to nitrate by bacteria. 4. Classical denitrification. 5. Denitrifying ammonia oxidation (Anammox). 6. Respiratory ammonification. 7. Assimilatory ammonification. 8. Aerobic oxidation of ammonia to nitrite by archaea. The circles represent: aerobic hydroxylamine oxidation and dissimilatory nitrite reduction pathways of aerobic nitrifier denitrification.

joining the Beta- and Gammaproteobacteria as the only known lineages to include ammonia-oxidizing members. The first obligatorily ammonia-oxidizing mesophilic crenarchaeon isolated in pure culture was *Nitrosopumilus maritimus* (Könneke et al., 2005). Genome-based reconstruction of the ammonia-oxidizing pathway in *N. maritimus* indicated the lack of genes encoding hydroxylamine oxidoreductase (HAO), a key component of both ammonia oxidation and one of the N_2O-producing pathways. Thus, the ammonia-oxidizing archaea utilize either a novel HAO or perhaps produce nitroxyl as the critical intermediate for oxidation by a novel nitroxyl oxidoreductase (Fig. 6.1; Walker et al., 2010). Possibly owing to this difference in inventory, there is currently no evidence for N_2O production by archaeal ammonia oxidizers. There is also no evidence for N_2O production by anaerobic ammonia-oxidizing bacteria (anammox; Kuenen, 2008) or anaerobic denitrifying methanotrophic bacteria (NC10; Ettwig et al., 2010).

Members of the proteobacterial ammonia- and methane-oxidizing bacteria (AOB and MOB, respectively) are capable of aerobic production of N_2O by reducing NO derived from hydroxylamine oxidation and nitrite reduction as a consequence of high N-loading (Cantera and Stein, 2007a; Shaw et al., 2006; Sutka et al., 2003; Yoshinari, 1984; Yu and Chandran, 2010) or, in the case of *Nitrosomonas* spp., during anaerobic growth on ammonia and nitrite (Schmidt *et al.*, 2002, 2004). Nonammonia-oxidizing organisms with denitrification inventory are found in all three kingdoms of life and all are potentially capable of producing N_2O under a range of oxygen levels, although some denitrifiers are restricted to anoxic environments (Stein and Yung, 2003; Zumft, 1997).

The wide-spread distribution and diversity of inventory and environments for N_2O production presents a number of challenges for dissecting, classifying, and assigning N_2O-producing function to individual microbial groups. The following methodologies (outlined in Fig. 6.2) describe the use of cultivated bacterial isolates for *in silico* identification of inventory and regulatory machinery for N_2O-producing pathways through the use of genomic and physiological tools. The ultimate goal for environmental microbiologists is to find unifying features for the function and regulation of common sets of inventory that can be extrapolated to discrete microbial populations in natural ecosystems to elucidate, and perhaps mitigate, further N_2O release to the atmosphere. For pathogenic microbiologists,

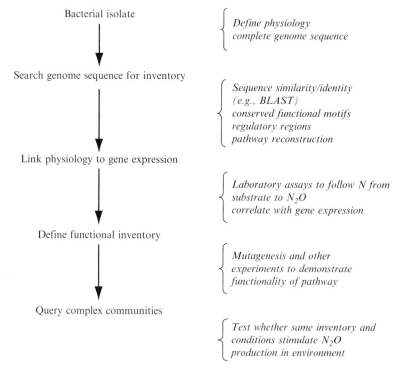

Figure 6.2 Flow diagram for investigating N_2O-producing pathways of bacteria.

understanding of N_2O-producing pathways provides insight into evasion of host defense mechanisms and advancement of disease.

2. BIOINFORMATICS

The following sections describe the characterized inventory with demonstrated involvement in N_2O production from aerobic hydroxylamine oxidation, nitrite reduction, or NOx detoxification using gene and genome sequence databases (Table 6.1). Bioinformatic information can be used to define pathways and their capacity for N_2O production by organisms or populations using physiological and analytical techniques. It must be noted that only a trace amount of ammonia-N is ever converted to N_2O–N from hydroxylamine oxidation during active ammonia oxidation. Isotopic techniques used to discriminate N_2O from either hydroxylamine oxidation or nitrite reduction have been used in both laboratory and environmental studies (Casciotti et al., 2003; Schmidt et al., 2004; Sutka et al., 2006) and some of these methods are described in other chapters of this volume. Using such isotopic techniques, it was shown that N_2O arising from hydroxylamine oxidation is usually eclipsed by the amount of N_2O–N from nitrite reduction (Ostrom et al., 2010), confirming the long-held belief that N_2O is mostly produced under low O_2 tension. However, regulation of N_2O-producing pathways requires intensive study in a broader range of natural environments to gain a better appreciation of their relative contribution to the atmospheric N_2O reservoir. Furthermore, processes that either remove N_2O directly (via N_2O reductase) or indirectly by consuming substrates compete with N_2O production and complicate modeling and budgeting of N_2O emissions. Hence, all inventory for N_2O production and consumption have to be characterized to gain a complete understanding of how and when N_2O is emitted from ecosystems (Fig. 6.3).

Identification of inventory for N_2O production within genome and metagenome sequences is accomplished by utilizing a number of database search tools. The most common starting place is to search for high levels of similarity to characterized genes using the Basic Local Alignment Search Tool (BLAST). Genome and metagenome annotation portals list BLAST results in addition to search results against a number of gene and protein databases. It is highly recommended that these results be closely scrutinized and manually annotated, particularly if only automated annotation is available. Once similar genes are identified, the verification of gene function is furthered by: (1) searching for conserved motifs that have demonstrated functionality from studies using biochemical or genetic methods, (2) examining transcriptional/translational profiles under conditions that stimulate N_2O production, (3) knocking out gene function and showing whether

Table 6.1 Characterized inventory within N_2O-producing pathways of bacteria

Gene ID	Enzyme	Substrate	Pathway[a]	Regulation[b]	References
haoAB	Hydroxylamine oxidoreductase (HAO)	Hydroxylamine[c]	H	NH_3	Bergmann et al. (2005), Poret-Peterson et al. (2008), Sayavedra-Soto et al. (1996)
cycA	Cytochrome c_{554}	Nitric oxide	H	NH_3[d]	Upadhyay et al. (2006)
cytS, cycP	Cytochrome c-beta	Nitric oxide	H, D	NH_3; NO; low O_2	Choi et al. (2005), Cross et al. (2001), Elmore et al. (2007), Poret-Peterson et al. (2008)
nirK	Copper-containing nitrite reductase	Nitrite	H, D	NO_2^-; NO; anoxia	Cantera and Stein (2007a), Rodionov et al. (2005), Suzuki et al. (2000)
nirS	Cytochrome cd(1) nitrite reductase	Nitrite	D	Anoxia; NO	Cutruzzola et al. (2003), Rodionov et al. (2005)
norCB	Cytochrome c nitric oxide reductase (cNOR)		D		Hendriks et al. (2000), Rodionov et al. (2005), Zumft (2005)
norB	Quinol nitric oxide reductase (qNOR)	Nitric oxide	D?, N	Anoxia, nitrosative stress, NO, NH_2OH	Hendriks et al. (2000), Rodionov et al. (2005), Zumft (2005)
norS	Cytochrome c nitric oxide reductase (sNOR)		H?, D?, N?		Cho et al. (2006), Stein et al. (2007)

norVW	Flavorubredoxin nitric oxide reductase (norV) NADH:flavorubredoxin oxidoreductase (norW)	Nitric oxide	N	NO, SNP	Gardner et al. (2003), Vicente et al. (2008)
hmp	NO dioxygenase (flavohemoglobin)	Nitric oxide	N	NO, nitrosative stress, O_2	Poole and Hughes (2000)

[a] "H" denotes genes in the aerobic hydroxylamine oxidation pathway. "D" denotes genes in the dissimilatory nitrite reduction pathway. "N" denotes nitrosative stress pathway.

[b] Molecules are listed that have been shown to regulate gene expression. The general term "nitrosative stress" indicates that the gene products are induced with any variety of NOx compounds (e.g., sodium nitroprusside, NO^+, etc.) that can induce nitrosative stress response. Note that not all homologues are similarly regulated in all bacteria; however, the presence of regulatory protein binding motifs (e.g., Fnr-Crp superfamily of repressor proteins) in promoter regions of the genes is a good indication that expression will be regulated by a specific effector molecule as discussed in the text.

[c] Usually produced from ammonia oxidation by a monooxygenase enzyme. In searching for haoAB in a genome or metagenome sequence, it is also recommended to search for possible monooxygenase sequences or for inventory for hydroxylamine production.

[d] Expression of the cycA gene has not been explicitly studied, although it is assumed to be expressed as a member of the haoAB operon in AOB.

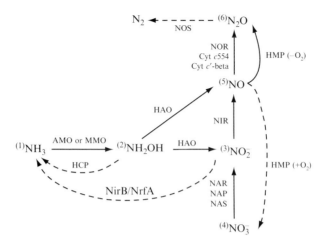

Figure 6.3 Specific pathways for N_2O production and consumption in bacteria. Solid lines indicate pathways that lead to N_2O as an end product and dashed lines indicate pathways that consume N_2O or remove a substrate (NH_2OH, NO, and NO_2^-) for N_2O production. AMO, ammonia monooxygenase; MMO, methane monooxygenase; HAO, hydroxylamine oxidoreductase; HCP, hybrid cluster protein; NirB, siroheme nitrite reductase; NrfA, pentaheme nitrite reductase; NAR, NAP, and NAS, different types of nitrate reductase (González et al., 2006); HMP, flavohemoglobin; NIR, nitrite reductase; NOR, nitric oxide reductase; NOS, nitrous oxide reductase. References for nonisotopic assays for measuring N-pools: (1) NH_3: colorimetric (Clesceri et al., 1998) or ion selective probe (Thermo Scientific Orion Ammonia Electrode), (2) NH_2OH: colorimetric (Dalton, 1977), (3) NO_2^-: colorimetric (Hageman and Hucklesby, 1971), (4) NO_3^- ($+NO_2^-$): cadmium reduction followed by nitrite assay (Clesceri et al., 1998); (5) NO: chemiluminescence detector or NO-specific electrode; (6) N_2O: gas chromatography (thermal conductivity detector or electron capture detector and Hayesep Q 10′ column; helium carrier gas; 120 °C injection/detection temp.; 30 °C column temp.).

N_2O production is compromised, and (4) any additional approaches that connect gene function to N_2O production (Fig. 6.2). Some of these techniques are described below, and other techniques are described in other chapters of this volume.

2.1. Aerobic hydroxylamine oxidation pathway

Both AOB and MOB in the Proteobacteria use a monooxygenase (ammonia monooxygenase or methane monooxygenase, respectively) to generate hydroxylamine from ammonia oxidation, which is then a substrate for N_2O production (Table 6.1 and Fig. 6.3; Cantera and Stein, 2007a; Wrage et al., 2001). Production of N_2O from hydroxylamine by AOB and Gammaproteobacteria MOB involves HAO encoded by the *haoAB* genes

(Arp et al., 2007; Bergmann et al., 2005). AOB (Beta- and Gammaproteobacteria) coexpress *haoAB* with two cytochromes *c* (encoded by *cycAB*) to relay electrons from hydroxylamine oxidation to the electron transport chain for energy production (Arp et al., 2007). This collection of genes is called the Hydroxylamine-Ubiquinone Redox Module (HURM) and its evolutionary history in AOB can be traced back to the Planctomycetes that perform anaerobic ammonia oxidation (anammox; Klotz and Stein, 2008). The c_{554} cytochrome (*cycA*) component of HURM also acts as an NO reductase *in vitro* (Upadhyay et al., 2006), suggesting that it may be the partner with HAO for production of N_2O in the aerobic hydroxylamine oxidation pathway (Fig. 6.3). Another potential candidate for reducing NO produced from hydroxylamine oxidation is NorS (Table 6.1), a homologue of nitric oxide reductases (NOR) of the heme–copper oxidase superfamily (Stein et al., 2007). The gene encoding NorS has been identified in the genome sequences of most AOB and its expression was induced under conditions of nitrosative stress in *Nitrosomonas europaea*, a model AOB strain (Cho et al., 2006).

Many MOB isolates also produce N_2O from hydroxylamine produced during their cometabolism of ammonia (Sutka et al., 2003, 2006). Some Gamma-MOB encode *haoAB* genes but not the adjoining cytochromes *c* as in the AOB (Hou et al., 2008; Klotz et al., 2008; Nyerges and Stein, 2009; Pol et al., 2007; Poret-Peterson et al., 2008), hence ammonia oxidation is not an energy-generating pathway for these bacteria. The *haoAB* genes with uncharacterized function have been identified in a number of other proteobacterial genomes (Bergmann et al., 2005). Interestingly, the HAO enzyme shares an evolutionary origin with pentaheme nitrite reductase, NrfA, that is used in dissimilatory reduction of nitrite to ammonia via hydroxylamine (Einsle et al., 1999).

If the genome sequence of a microbial isolate contains a bona fide *haoAB* that encodes the tyrosine cross-linking residue to form a functional homotrimer (Igarashi et al., 1997), then this organism has potential to produce N_2O from aerobic hydroxylamine oxidation (followed by NO reduction), particularly if it encodes a monooxygenase that can generate hydroxylamine from ammonia. Alternatively, some Alphaproteobacteria MOB do not encode *haoAB* genes yet still produce N_2O from hydroxylamine formed from the oxidation of ammonia by methane monooxygenase (Knowles and Topp, 1988; Sutka et al., 2006). These bacteria have an as-of-yet uncharacterized mechanism for metabolizing hydroxylamine and generating NO and N_2O.

An additional hydroxylamine-oxidizing enzyme is cytochrome P460 that has been identified in some AOB and MOB (Elmore et al., 2007). However, this enzyme alone does not appear to contribute to N_2O production from hydroxylamine, but rather participates in oxidizing NH_2OH and NO to nitrite as a detoxification mechanism. An evolutionarily related enzyme to cytochrome P460, cytochrome c'-beta, is also involved in NO detoxification and metabolism, possibly as an NO reductase (Elmore et al., 2007). Although cytochrome c'-beta is found in many denitrifying and photosynthetic bacteria and is often regulated along with denitrifying inventory, its activity as an NO reductase has been difficult to demonstrate directly in some bacteria (Choi et al., 2005) and more easily in others (Cross et al., 2001). While cytochromes P460 and c'-beta may not be central to N_2O-producing pathways, the presence of *cytL* and *cytS* (*cycP*) genes suggests that the bacterium encoding these enzymes has mechanisms for NOx detoxification, and perhaps metabolism, that can be tested physiologically.

2.2. Dissimilatory nitrite reduction pathway

Dissimilatory reduction of nitrite is more broadly distributed and more significant than aerobic hydroxylamine oxidation for bacterial production of NO and N_2O. This pathway may or may not include nitrate reductase as many bacteria (like *Escherichia coli*) are nitrate respirers that lack dissimilatory enzymes for further reduction of nitrite to NO and N_2O, yet contain enzymes for detoxifying NO to N_2O like flavohemoglobin (Hmp) and flavorubredoxin (NorVW; Gardner et al., 2003). Thus, N_2O-production from bona fide dissimilatory nitrite reduction (rather than detoxification) most commonly involves both nitrite and NOR, and both should be identifiable in a genome sequence for the organism to have this activity. True denitrifying bacteria also have an N_2O reductase that acts as an N_2O sink (Fig. 6.3). It must be noted that some lower eukaryotes can denitrify and some fungi produce substantial amounts of N_2O from a nitrite reduction pathway that can include a copper-containing nitrite reductase with high similarity to the bacterial enzyme (Kim et al., 2009; Ma et al., 2008; Takaya et al., 2003). Two classes of dissimilatory nitrite reductase have been characterized in bacteria and are distinguished by metal content: NirK is the copper-containing enzyme (Suzuki et al., 2000) and NirS contains cytochrome cd(1) (Cutruzzola et al., 2003). Both nitrite reductases are well characterized and can be distinguished bioinformatically from related enzymes by searching translated sequences for conserved motifs and operon structure (Cutruzzola et al., 2003; Yamaguchi et al., 2004).

Identifying homologues to NOR is more complicated owing to the diversity of enzymes capable of performing this process (Table 6.1; Zumft, 2005). The most common bacterial NORs are membrane-bound enzymes in the heme–copper oxidase superfamily that accept electrons from either the cytochrome c (cNOR or short-chain) or quinol pool (qNOR or long-chain; Hendriks et al., 2000; Zumft, 2005). Both cNOR and qNOR are annotated in genome sequences as *norB* genes, although cNOR is a multi-subunit enzyme encoded by at least *norCB*, and is most often expressed with additional operon members (e.g., *norCBQD*). If the entire *norCBQD* operon is present in a genome sequence, there is a very high probability that the bacterium has NO reductase activity. qNOR can be encoded as an orphan, is identifiable by an N-terminal extension (hence the long-chain designation), and is more common to pathogens that use qNOR in defense against host-produced NO (Hendriks et al., 2000). sNOR is a new class of enzymes in the heme–copper oxidase superfamily that is found almost exclusively in the AOB (Stein et al., 2007). Although sNOR (encoded by *norS*, often within a four gene operon) is implicated in aerobic production of N_2O by *N. europaea* (Cho et al., 2006), its physiological function has not been broadly explored. Some of the heme–copper oxidase NOR enzymes have been given alternate gene names, so one must be careful when interpreting the literature and extrapolating information to new genome sequences (Zumft, 2005).

2.3. Detoxification inventory

When exploring genome sequences for N_2O-producing inventory, it is sometimes difficult to ascertain whether enzymes participate in energy-generating or in NOx detoxification pathways. In some instances, enzymes can perform both respiratory and detoxification functions as observed for bacteria that perform aerobic denitrification (Stein, 2010) or aerobic ammonia oxidation (Cantera and Stein, 2007b). Flavohemoglobin (Hmp) is a major player in mediating the nitrosative stress response in bacteria and usually functions as an NO dioxygenase or denitrosylase, forming nitrate under oxic conditions. However, Hmp can also act as an NOR and form N_2O under anoxic conditions (Fig. 6.3; Poole and Hughes, 2000). The regulation of Hmp is of key importance in determining which function is physiologically relevant under oxidative (under SoxR regulation) and/or nitrosative stress (under Fnr regulation; Poole and Hughes, 2000) implying that for each individual bacterium, Hmp expression and activity requires intensive investigation.

Flavorubredoxin proteins encoded by the *norVW* genes are found in anaerobic and facultatively anaerobic microbes where they generally play an NO detoxification role. In *E. coli*, flavorubredoxin reduces NO to N_2O under low O_2, whereas Hmp oxidizes NO to nitrate under high O_2. This way, *E. coli* has protection against nitrosative stress across a spectrum of O_2 levels (Gardner et al., 2003). The contributions of flavorubredoxin and Hmp to global N_2O emissions are likely not as great as that of the heme–copper oxidase NOR enzymes as the latter are central to anaerobic respiration rather than specific to NOx detoxification. However, for pathogenic bacteria, flavohemoglobin and flavorubredoxin provide critical functions for establishment and metabolism in the host environment.

2.4. Identifying motifs and regulatory sequences

Algorithms for identifying the above described inventory have become much more accurate as structural and biochemical details about the enzymes have been discovered and reported. However, it is not uncommon for automated algorithms to improperly assign an activity to a given translated gene based on conserved regions that happen to lack essential residues. To ensure that a gene is actually encoding an enzyme with a specific function, it is important to identify essential, conserved motifs such as heme-binding, copper-binding, or catalytic residues. For instance, while NirK enzymes are within the superfamily of multicopper oxidases, there are both copper-binding and catalytic residues that are absolutely conserved for a multicopper oxidase to perform nitrite reduction (Yamaguchi et al., 2004). To find these motifs, homologues of translated genes for the enzyme in question are aligned to the query sequence using a program like Clustal X (Jeanmougin et al., 1998). Conserved regions will be highlighted. One must then manually identify the ligand-binding, catalytic, and other residues within the query sequence to ensure that all regions are present. If all conserved regions are properly identified, this strongly implies that the query gene product is functional for the assumed activity; however, the actual activity must still be verified within the context of a living system.

A large number of review articles have been written on enzymes in N_2O-producing pathways and several have been cited in this chapter. It is a difficult task to categorize inventory for N_2O production as belonging to a discrete pathway because bacteria can both metabolize and detoxify NOx simultaneously and release N_2O as a byproduct from both pathways. The major controls on N_2O production are O_2 level, with lower levels generally leading to greater expression of N_2O-producing inventory, and levels of nitrosating intermediates like nitrite, hydroxylamine, and NO. As such, several review articles have detailed involvement of the Fnr-Crp superfamily of transcriptional regulators in controlling expression of genes involved in N_2O production (Rodionov et al., 2005; Spiro, 2007; Tucker et al., 2010). Locating conserved repressor protein binding sites involves examination of promoter-proximal regions of the gene or operon in question. These upstream regions can be downloaded from any number of gene viewing platforms. Promoters can be quickly predicted using software like the BPROM function in Softberry (http://linux1.softberry.com/berry.phtml?topic=bprom&group=programs&subgroup=gfindb). Conserved inverted repeats overlapping or near predicted promoters that match the conserved motif for Fnr-Crp or other NOx-responsive regulators can then be quickly located using generic character search engines. There are more advanced methods for locating regulatory binding sites throughout genome sequences that have been described (Rodionov et al., 2005), but in the absence of bioinformatics expertise and for analyzing small numbers of genes, manual searches can be an effective tool.

3. Surveying Function

After predicting inventory, pathways, and regulatory information *in silico* using bioinformatic approaches, the function of the predicted pathways must be demonstrated using pure cultures of the organism from which the genome was sequenced. It may be possible to demonstrate function from a complex community that has an annotated metagenome sequence; however, the more members of a community, the more difficult to ascribe functional changes to individual members. As the primary goal of the described methods is to find unifying features of N_2O-producing pathways, the simplest approach is to work, at least initially, with single cultivated organisms. Figure 6.3 shows much of the inventory that has been characterized in N_2O-producing pathways and lists references for the simplest laboratory-based assays for each N-compound. More advanced techniques, such as the use of isotopic tracers (e.g., feeding a culture $^{15}NH_4^+$ and measuring amounts of ^{15}N ending up in each N-pool), are much more sensitive than the described techniques and require isotope mass spectrometers that can measure ^{15}N in each N-pool. In the absence of such instrumentation, simple assays can be used to approximate the flow of N through a biological system.

Working with pure cultures in a laboratory environment has several advantages over working *in situ*, with field samples, or with microcosm/macrocosm incubations. For instance, all metabolic responses are confined to a single, clonal organism without concern for complex interactions with other organisms, or physicochemical characteristics of the sample. Batch cultures are kept uniform and only one variable is altered at a time. Multiple experiments with replication ensure proper statistical control such that any changes in metabolism can be ascribed to a single effecting molecule. As such, one can determine metabolic pathways and discrete functions by assaying individual processes and changes in gene or protein expression over a growth period or time course. Combined with emerging inexpensive high-throughput deep sequencing platforms, correlating metabolic responses with global transcriptional and translational changes is becoming commonplace. Together, functional data from a multitude of cultivated organisms will allow categorization of unifying features of N_2O producing pathways, which in turn should eventually allow accurate extrapolation to N_2O production by complex communities in natural ecosystems.

Some issues with batch culture studies include: (1) differences in media preparations (e.g., buffering systems, nutrient combinations, trace metals),

(2) physiological changes in cells over a normal growth curve (e.g., lag to exponential to stationary phase), (3) changes in medium as cells metabolize and grow (e.g., pH, nutrient depletion), (4) requirement for much higher concentrations of substrate than would ever be found in a natural system, and (5) accumulation of NOx intermediates (e.g., hydroxylamine, NO, nitrite) that stimulate N_2O production. A few laboratories have mastered functional study of pure cultures in chemostats to control cell growth and media composition by keeping organisms in exponential phase; however, continuous cultures take far more expertise and resources than batch culture experiments. Furthermore, batch cultures are excellent environments for pathway discovery as they are easily replicated and, with careful management, represent the same changes over the growth cycle with each replicate. Thus, for the majority of researchers without expertise or infrastructure for continuous culture, batch cultures are a fine option for the following approaches.

3.1. Following the N

Pathway discovery for N_2O production requires a keen understanding of which substrates potentially lead to N_2O as a product, which transformations of that substrate are likely to occur, the rates of each process, and whether N_2O is a minor side-product or a major end product of the pathway. For aerobic hydroxylamine oxidation, N_2O is a minor side-product that arises from transient hydroxylamine accumulation under conditions where enzyme turnover within the ammonia-oxidizing pathway is unbalanced (Cantera and Stein, 2007b). Thus, when determining whether a bacterium performs this process, N_2O production is maximized by using hydroxylamine as a substrate and assaying for N_2O production, hydroxylamine loss, and nitrite and/or nitrate (and in some cases ammonium) production. The biggest issue with using hydroxylamine as a substrate is its severe toxicity; aside from AOB, most organisms can tolerate no more than ca. 100 μM, and usually much less. It is thus recommended that a range of hydroxylamine concentrations be tested on each organism to gauge its range of tolerance (e.g., ability to grow, respiration rate, other measureable metabolic characteristics in the presence of hydroxylamine) before proceeding. A similar approach is taken to examine the ability of bacteria to reduce nitrite to N_2O. As with hydroxylamine, nitrite and NO are both toxic compounds (particularly NO) that induce nitrosative stress, emphasizing the need to determine substrate tolerance before proceeding.

Experimental assays for determining the ability of bacteria to produce N_2O can be very simple. As mentioned above, the two main factors that generally stimulate expression of N_2O-producing inventory are O_2 level and N-substrate concentration as predicted by the contents of the organism's genome sequence. The bacteria can be inoculated into their normal growth media at low density (e.g., 10^6 cells ml^{-1}) in Wheaton bottles sealed with rubber septated caps to enable gas measurements. O_2 levels are adjusted by sparging with N_2 (needle in/needle out) and adding back pure O_2 to the appropriate level via syringe. Substrates (NH_3, NH_2OH, NO_2^-, NO, etc.) are added from stock solutions via syringe. Note that some N-molecules like NH_2OH and NO are very short lived and must be dispensed fresh before each use. It is recommended to add only one substrate at a time to avoid complications in overlapping N-transformations. The idea is to follow the transformation of a single substrate to N_2O and other possible N-intermediates to demonstrate functionality of the predicted pathways. As the bacteria grow, they will transform the N-substrates, but may also assimilate them. This is one reason why large amounts of substrate well beyond what an organism might experience in its natural environment may be required—to force the bacteria to make N_2O as a detoxification or metabolic "overflow" response. This is particularly true if the inventory is stimulated by nitrosative stress, whereby N-saturation or a critical level of a toxic N-compound is required to induce NOx detoxification. Alternatively, bacteria can be grown to late-log phase, cells harvested, and placed into serum vials at relatively high density (e.g., 10^8–10^9 cells ml^{-1}) with appropriately adjusted O_2 and N-substrate levels. These so-called "resting cell assays" (because the cells do not grow over the brief time course of the assay) are usually short term, that is, over minutes to hours, and can provide a rapid test of pathway function. These assays generally rely on endogenous cellular reductant to complete the process of NO reduction; however, sometimes the addition of an exogenous reductant source (e.g., NADH, cytochrome c, phenazine methosulfate, etc.) can stimulate activity depending on the particular inventory expressed by the bacterium.

The use of appropriate controls for the above assays is critical for proper interpretation of results. Since many NOx compounds are extremely chemically reactive, some N_2O will be released in the absence of biological activity. Controls must include dead cells to demonstrate background reactivity of the chemicals using high enough NOx substrates to simulate the amount experienced by living cells at the height of N_2O production. This is important because living cells will produce NOx intermediates in excess of that added exogenously during the assay. Additional controls should include combinations of NOx compounds in the absence of bacteria and at O_2 levels that simulate the living cell assays. The easiest way to manage both experimental and control assays is to create a matrix whereby each variable is systematically added or altered independently (living or dead cells, O_2, N-substrate) and in each possible combination. This approach results in a large number of assay conditions, but the resting cell approach will easily allow one to set up and control several dozen assays per experiment. The limitations include incubator space, the ability of the researcher to rapidly add materials to each assay without losing too much time, and the ability of the researcher to complete each measurement. While gas measurements must be made in real time, colorimetric or electrode measurements can be performed at a later time on stored supernatants (frozen at $-20\ °C$) once the bacterial cells are removed from the assay for molecular analysis.

The great benefit of using either growth or resting cell assays is that a large sample number can be monitored in a single experiment, offering a larger number of controls and substrate combinations. Some drawbacks include finding appropriate levels of O_2 and substrate to stimulate N_2O production, monitoring activity at the appropriate growth phase, controlling changes to the media as the cells metabolize, and understanding the limits of other parameters of cellular metabolism like pH, media composition, temperature, rate of NOx formation, shaker speed, and other general cultivation factors. Most of these factors are overcome by simple trial and error and extreme patience.

Most of the N-intermediates in N_2O-producing pathways can be measured using standard colorimetric or electrode assays as described and referenced in the legend of Fig. 6.3. The easiest and least expensive assay for measuring N_2O is a gas chromatograph with a thermal conductivity detector (GC-TCD) or electron capture detector (GC-ECD), which measures N_2O in the ppm and ppb ranges, respectively. Packed columns that do not resolve water, such as Hayesep Q (Alltech), are preferred since injections can be made at ca. 2-min intervals. O_2 levels can also be monitored by GC using a molecular sieve column to correlate rates of levels of N_2O production with O_2 availability.

3.2. Linking activity with gene expression

Once parameters for maximal N_2O production are resolved, the bacterial cells can be collected over the course of a physiological assay for extraction of nucleic acids. Examination of mRNA pools allows one to connect activity with the expression of inventory identified in the genome sequence. For examination of a small number of genes, transcriptional assays by reverse transcription quantitative polymerase chain reaction (RT-qPCR) is currently the preferred option. The RT reaction is performed using a kit (e.g., Superscript, Invitrogen) and the resulting cDNA is used for qPCR after ensuring purity and integrity. Specific primers for each gene (collected from the genome sequence) are designed using Primer 3 software (Rozen and Skaletsky, 2000). SYBR green is the preferred fluorescent molecule for single-gene qPCR assays as long as one ensures that primer–dimer pairs do not interfere with detection of gene amplification products late into the reaction. Major benefits of RT-qPCR over other mRNA detection methods like Northern blot are its high sensitivity, specificity, and ability to detect mRNA processing events by designing primers at multiple locations along the predicted mRNA. Several book chapters and review articles have been published for optimizing the qPCR assay and can offer further guidance for optimizing and performing these assays.

Beyond single-gene expression assays, global transcriptome analysis is rapidly approaching cost-effectiveness. Many research groups are abandoning microarray hybridization in favor of next-generation deep sequencing platforms for transcriptome analysis like Roche 454, Illumina GA, Helicos, and ABI SOLiD (Morozova and Marra, 2008). Some benefits of deep sequencing is that most platforms do not require an amplification step and hence avoid PCR bias (better representation of mRNA populations), extremely large amounts of data can be generated in a short period of time, and several samples can be processed in a single run of the instrument. Limitations of deep sequencing commonly arise following data collection as binning and analysis requires intensive bioinformatics expertise and careful data management. Global proteome analysis is also becoming more cost effective, allowing for a comparison between transcriptional and translational profiles to compare with physiological information. Together, this information offers the most complete snap-shot of how a bacterium regulates its genetic machinery to produce N_2O. When transcriptional and translational information is compared between physiological states where N_2O is being produced or not produced, one can ascertain which genes are specific to the process.

Another useful step in linking activity to gene expression is to knockout gene functions that are likely involved in the process and see if N_2O production is altered in the mutated strains. This approach has been broadly applied to demonstrate functionality of individual inventory in a wide range of bacterial strains and is still a cornerstone of the molecular biology tool chest. Specific methodology for creating knockout mutants should be followed for each strain under study. A chapter on mutagenizing AOB is presented in this volume.

4. Concluding Remarks

Although many of the genes and enzymes in N_2O-producing pathways have been well characterized and can be readily identified in bacterial genomes, there are still missing pieces of inventory that have yet to be discovered. For instance, when NirK and NorB functions were disrupted in the AOB, *N. europaea*, the mutant strains still produced N_2O, implying the activity of multiple nitrite- and NO-reducing enzymes (Beaumont et al., 2002, 2004). This is the likely case for other bacterial strains that are evolving under the constant pressure of nitrosative stress. Identification of novel inventory for N_2O production is possible by applying next generation sequencing platforms and expanding research into metagenomics and metatranscriptomics. The major challenge will be to definitively test the function of novel inventory, particularly within complex communities, using discrete physiological tests that can be greatly complicated by multiple overlapping variables.

A potential outcome of surveying and defining all N_2O-producing pathways in bacteria is to extrapolate laboratory experiments to natural environments to solve issues like the increasing atmospheric N_2O reservoir. We are still some intellectual distance from correlating gene or transcript abundance in complex ecosystems with discrete physiological functions of microbial communities. At some point, simultaneous correlation of N_2O emissions with expression of multiple genes involved in coherent pathways may bring us a step closer to understanding how the pathways are regulated over a range of temporal and spatial scales. Our challenge, then, is to eventually move away from single isolate or single-gene approaches and make use of unifying features that distinguish each pathway and its regulation from others.

REFERENCES

Arp, D. J., Chain, P. S. G., and Klotz, M. G. (2007). The impact of genome analyses on our understanding of ammonia-oxidizing bacteria. *Annu. Rev. Microbiol.* **61,** 503–528.

Beaumont, H. J. E., Hommes, N. G., Sayavedra-Soto, L. A., Arp, D. J., Arciero, D. M., Hooper, A. B., Westerhoff, H. V., and van Spanning, R. J. M. (2002). Nitrite reductase of *Nitrosomonas europaea* is not essential for production of gaseous nitrogen oxides and confers tolerance to nitrite. *J. Bacteriol.* **184,** 2557–2560.

Beaumont, H. J. E., van Schooten, B., Lens, S. I., Westerhoff, H. V., and van Spanning, R. J. M. (2004). *Nitrosomonas europaea* expresses a nitric oxide reductase during nitrification. *J. Bacteriol.* **186,** 4417–4421.

Bergmann, D. J., Hooper, A. B., and Klotz, M. G. (2005). Structure and sequence conservation of *hao* cluster genes of autotrophic ammonia-oxidizing bacteria: Evidence for their evolutionary history. *Appl. Environ. Microbiol.* **71,** 5371–5382.

Cantera, J. J. L., and Stein, L. Y. (2007a). Interrelationship between nitrite reductase and ammonia-oxidizing metabolism in *Nitrosomonas europaea*. *Arch. Microbiol.* **188,** 349–354.

Cantera, J. J. L., and Stein, L. Y. (2007b). Role of nitrite reductase in the ammonia-oxidizing pathway of *Nitrosomonas europaea*. *Arch. Microbiol.* **188,** 349–354.

Casciotti, K., Sigman, D., and Ward, B. (2003). Linking diversity and stable isotope fractionation in ammonia-oxidizing bacteria. *Geomicrobiol. J.* **20,** 335–353.

Cho, C. M.-H., Yan, T., Liu, X., Wu, L., Zhou, J., and Stein, L. Y. (2006). Transcriptome of *Nitrosomonas europaea* with a disrupted nitrite reductase (*nirK*) gene. *Appl. Environ. Microbiol.* **72,** 4450–4454.

Choi, P. S., Grigoryants, V. M., Abruna, H. D., Scholes, C. P., and Shapleigh, J. P. (2005). Regulation and function of cytochrome c' in *Rhodobacter sphaeroides* 2.4.3. *J. Bacteriol.* **187,** 4077–4085.

Clesceri, L. S., Greenberg, A. E., and Eaton, A. D. (1998). Standard Methods for the Examination of Water and Wastewater. American Public Health Association, Washington, DC.

Cross, R., Lloyd, D., Poole, R. K., and Moir, J. W. B. (2001). Enzymatic removal of nitric oxide catalyzed by cytochrome c' in *Rhodobacter capsulatus*. *J. Bacteriol.* **183**, 3050–3054.

Cutruzzola, F., Rinaldo, S., Centola, F., and Brunori, M. (2003). NO production by *Pseudomonas aeruginosa* cd(1) nitrite reductase. *IUBMB Life* **55**, 617–621.

Dalton, H. (1977). Ammonia oxidation by the methane oxidising bacterium *Methylococcus capsulatus* strain bath. *Arch. Microbiol.* **114**, 273–279.

Davidson, E. A. (2009). The contribution of manure and fertilizer nitrogen to atmospheric nitrous oxide since 1860. *Nat. Geosci.* **2**, 659–662.

Einsle, O., Messerschmidt, A., Stach, P., Bourenkov, G. P., Bartunik, H. D., Huber, R., and Kroneck, P. M. H. (1999). Structure of cytochrome c nitrite reductase. *Nature* **400**, 476–480.

Elmore, B. O., Bergmann, D. J., Klotz, M. G., and Hooper, A. B. (2007). Cytochromes P460 and c'-beta; a new family of high-spin cytochromes c. *FEBS Lett.* **581**, 911–916.

Ettwig, K. F., Butler, M. K., Le Paslier, D., Pelletier, E., Mangenot, S., Kuypers, M. M. M., Schreiber, F., Dutilh, B. E., Zedelius, J., de Beer, D., Gloerich, J., Wessels, H., *et al.* (2010). Nitrite-driven anaerobic methane oxidation by oxygenic bacteria. *Nature* **464**, 543–548, U94.

Gardner, A. M., Gessner, C. R., and Gardner, P. R. (2003). Regulation of the nitric oxide reduction operon (*norRVW*) in *Escherichia coli*—Role of norR and sigma(54) in the nitric oxide stress response. *J. Biol. Chem.* **278**, 10081–10086.

González, P. J., Correia, C., Moura, I., Brondino, C. D., and Moura, J. J. G. (2006). Bacterial nitrate reductases: Molecular and biological aspects of nitrate reduction. *J. Inorg. Biochem.* **100**, 1015–1023.

Hageman, R. H., and Hucklesby, D. P. (1971). Nitrate reductase in higher plants. *Methods Enzymol.* **23**, 491–503.

Hendriks, J., Oubrie, A., Castresana, J., Urbani, A., Gemeinhardt, S., and Saraste, M. (2000). Nitric oxide reductases in bacteria. *Biochim. Biophys. Acta* **1459**, 266–273.

Hou, S. B., Makarova, K. S., Saw, J. H. W., Senin, P., Ly, B. V., Zhou, Z. M., Ren, Y., Wang, J. M., Galperin, M. Y., Omelchenko, M. V., Wolf, Y. I., Yutin, N., *et al.* (2008). Complete genome sequence of the extremely acidophilic methanotroph isolate V4, *Methylacidiphilum infernorum*, a representative of the bacterial phylum Verrucomicrobia. *Biol. Direct* **3**:26. doi:10.1186/1745-6150-3-26.

Igarashi, N., Moriyama, H., Fijiwara, T., Fukumori, Y., and Tanaka, N. (1997). The 2.8 Å structure of hydroxylamine oxidoreductase from a nitrifying chemoautotrophic bacterium, *Nitrosomonas europaea*. *Nat. Struct. Biol.* **4**, 276–284.

IPCC (2006). IPCC guidelines for National greenhouse gas inventories, prepared by the National Greenhouse Gas Inventories Programme. In "N_2O Emissions from Managed Soils, and CO_2 Emissions from Lime and Urea Application, Vol. 4," H. S. Eggleston, L. Buendia, K. Miwa, T. Ngara, and K. Tanabe (eds.), IGES, Hayama, Japan, Chapter 11.

Jeanmougin, F., Thompson, J. D., Gouy, M., Higgins, D. G., and Gibson, T. J. (1998). Multiple sequence alignment with Clustal X. *Trends Biochem. Sci.* **23**, 403–405.

Jetten, M. S. M., van Niftrik, L., Strous, M., Kartal, B., Keltjens, J. T., and Op den Camp, H. J. M. (2009). Biochemistry and molecular biology of anammox bacteria. *Crit. Rev. Biochem. Mol. Biol.* **44**, 65–84.

Kim, S. W., Fushinobu, S., Zhou, S. M., Wakagi, T., and Shoun, H. (2009). Eukaryotic *nirK* genes encoding copper-containing nitrite reductase: Originating from the protomitochondrion? *Appl. Environ. Microbiol.* **75**, 2652–2658.

Klotz, M. G., and Stein, L. Y. (2008). Nitrifier genomics and evolution of the nitrogen cycle. *FEMS Microbiol. Lett.* **278**, 146–456.

Klotz, M. G., and Stein, L. Y. (2010). Genomics of ammonia-oxidizing bacteria and insights to their evolution. *In* "Nitrification," (B. B. Ward, D. J. Arp, and M. G. Klotz, eds.), ASM Press, Washington, DC, pp. 57–93.

Klotz, M. G., Schmid, M. C., Strous, M., op den Camp, H. J., Jetten, M. S., and Hooper, A. B. (2008). Evolution of an octahaem cytochrome *c* protein family that is key to aerobic and anaerobic ammonia oxidation by bacteria. *Environ. Microbiol.* **10**, 3150–3163.

Knowles, R., and Topp, E. (1988). Some factors affecting nitrification and the production of nitrous oxide by the methanotrophic bacterium *Methylosinus trichosporium* OB3b. *In* "Current Perspectives in Environmental Biogeochemistry," (G. Giovannozzi-Sermanni and P. Nannipieri, eds.), pp. 383–393. Consiglione delle Richerche-I.P.R.A, Rome.

Könneke, M., Bernhard, A. E., de la Torre, J. R., Walker, C. B., Waterbury, J. B., and Stahl, D. A. (2005). Isolation of an autotrophic ammonia-oxidizing marine archaeon. *Nature* **437**, 543–546.

Kuenen, J. G. (2008). Anammox bacteria: From discovery to application. *Nat. Rev. Microbiol.* **6**, 320–326.

Ma, W. K., Farrell, R. E., and Siciliano, S. D. (2008). Soil formate regulates the fungal nitrous oxide emission pathway. *Appl. Environ. Microbiol.* **74**, 6690–6696.

Morozova, O., and Marra, M. A. (2008). Applications of next-generation sequencing technologies in functional genomics. *Genomics* **92**, 255–264.

Nyerges, G., and Stein, L. Y. (2009). Ammonia co-metabolism and product inhibition vary considerably among species of methanotrophic bacteria. *FEMS Microbiol. Lett.* **297**, 131–136.

Ostrom, N. E., Sutka, R., Ostrom, P. H., Grandy, A. S., Huizinga, K. M., Gandhi, H., von Fischer, J. C., and Robertson, G. P. (2010). Isotopologue data reveal bacterial denitrification as the primary source of N_2O during a high flux event following cultivation of a native temperate grassland. *Soil Biol. Biochem.* **42**, 499–506.

Pol, A., Heijmans, K., Harhangi, H. R., Tedesco, D., Jetten, M. S. M., and den Camp, H. (2007). Methanotrophy below pH1 by a new Verrucomicrobia species. *Nature* **450**, 874–878U17.

Poole, R. K., and Hughes, M. N. (2000). New functions for the ancient globin family: Bacterial responses to nitric oxide and nitrosative stress. *Mol. Microbiol.* **36**, 775–783.

Poret-Peterson, A. T., Graham, J. E., Gulledge, J., and Klotz, M. G. (2008). Transcription of nitrification genes by the methane-oxidizing bacterium, *Methylococcus capsulatus* strain bath. *ISME J.* **2**, 1213–1220.

Röckstrom, J., Steffen, W., Noone, K., Persson, A., Chapin, F. S., Lambin, E. F., Lenton, T. M., Scheffer, M., Folke, C., Schellnhuber, H. J., Nykvist, B., de Wit, C. A., *et al.* (2009). A safe operating space for humanity. *Nature* **461**, 472–475.

Rodionov, D. A., Dubchak, I. L., Arkin, A. P., Alm, E. J., and Gelfand, M. S. (2005). Dissimilatory metabolism of nitrogen oxides in bacteria: Comparative reconstruction of transcriptional networks. *PLoS Comp. Biol.* **1**, 0415–0431.

Rozen, S., and Skaletsky, H. J. (2000). Primer3 on the WWW for general users and for biologist programmers. *In* "Bioinformatics Methods and Protocols: Methods in Molecular Biology," (S. Krawetz and S. Misener, eds.), pp. 365–386. Humana Press, Totowa, NJ.

Sayavedra-Soto, L. A., Hommes, N. G., Russell, S. A., and Arp, D. J. (1996). Induction of ammonia monooxygenase and hydroxylamine oxidoreductase mRNAs by ammonium in *Nitrosomonas europaea*. *Mol. Microbiol.* **20**, 541–548.

Schmidt, I., Hermelink, C., van de Pas-Schoonen, K., Strous, M., op den Camp, H. J., Kuenen, G., and Jetten, M. S. M. (2002). Anaerobic ammonia oxidation in the presence of nitrogen oxides (NO_x) by two different lithotrophs. *Appl. Environ. Microbiol.* **68**, 5351–5357.

Schmidt, I., van Spanning, R. J. M., and Jetten, M. S. M. (2004). Denitrification and ammonia oxidation by *Nitrosomonas europaea* wild-type, and NirK- and NorB-deficient mutants. *Microbiology UK* **150**, 4107–4114.

Shaw, L. J., Nicol, G. W., Smith, Z., Fear, J., Prosser, J. I., and Baggs, E. M. (2006). *Nitrosospira* spp. can produce nitrous oxide via a nitrifier denitrification pathway. *Environ. Microbiol.* **8**, 214–222.

Spiro, S. (2007). Regulators of bacterial responses to nitric oxide. *FEMS Microbiol. Rev.* **31**, 193–211.

Stein, L. Y. (2010). Heterotrophic nitrification and nitrifier denitrification. *In* "Nitrification," (B. B. Ward, M. G. Klotz, and D. J. Arp, eds.), ASM Press, Washington, DC, Chapter 5.

Stein, L. Y., and Yung, Y. L. (2003). Production, isotopic composition, and atmospheric fate of biologically produced nitrous oxide. *Annu. Rev. Earth Planet Sci.* **31**, 329–356.

Stein, L. Y., Arp, D. J., Berube, P. M., Chain, P. S. G., Hauser, L., Jetten, M. S. M., Klotz, M. G., Larimer, F. W., Norton, J. M., Op den Camp, H. J., Shin, M., and Wei, X. (2007). Whole-genome analysis of the ammonia-oxidizing bacterium, *Nitrosomonas eutropha* C91: Implications for niche adaptation. *Environ. Microbiol.* **9**, 2993–3007.

Sutka, R. L., Ostrom, N. E., Ostrom, P. H., Gandhi, H., and Breznak, J. A. (2003). Nitrogen isotopomer site preference of N_2O produced by *Nitrosomonas europaea* and *Methylococcus capsulatus* bath. *Rapid Commun. Mass Spectrom.* **17**, 738–745.

Sutka, R. L., Ostrom, N. E., Ostrom, P. H., Breznak, J. A., Gandhi, H., Pitt, A. J., and Li, F. (2006). Distinguishing nitrous oxide production from nitrification and denitrification on the basis of isotopomer abundances. *Appl. Environ. Microbiol.* **72**, 638–644.

Suzuki, S., Kataoka, K., and Yamaguchi, K. (2000). Metal coordination and mechanism of multicopper nitrite reductase. *Acc. Chem. Res.* **33**, 728–735.

Takaya, N., Kuwazaki, S., Adachi, Y., Suzuki, S., Kikuchi, T., Nakamura, H., Shiro, Y., and Shoun, H. (2003). Hybrid respiration in the denitrifying mitochondria of *Fusarium oxysporum*. *J. Biochem.* **133**, 461–465.

Treusch, A. H., Leininger, S., Kletzin, A., Schuster, S. C., Klenk, H. P., and Schleper, C. (2005). Novel genes for nitrite reductase and Amo-related proteins indicate a role of uncultivated mesophilic crenarchaeota in nitrogen cycling. *Environ. Microbiol.* **7**, 1985–1995.

Tucker, N. P., Le Brun, N. E., Dixon, R., and Hutchings, M. I. (2010). There's NO stopping NsrR, a global regulator of the bacterial NO stress response. *Trends Microbiol.* **18**, 149–156.

Upadhyay, A. K., Hooper, A. B., and Hendrich, M. P. (2006). NO reductase activity of the tetraheme cytochrome c_{554} of *Nitrosomonas europaea*. *J. Am. Chem. Soc.* **128**, 4330–4337.

Venter, J. C., Remington, K., Heidelberg, J. F., Halpern, A. L., Rusch, D., Eisen, J. A., Wu, D. Y., Paulsen, I., Nelson, K. E., Nelson, W., Fouts, D. E., Levy, S., *et al.* (2004). Environmental genome shotgun sequencing of the Sargasso Sea. *Science* **304**, 66–74.

Vicente, J. B., Justino, M. C., Goncalves, V. L., Saraiva, L. M., and Teixeira, M. (2008). Biochemical, spectroscopic, and thermodynamic properties of flavodiiron proteins. Globins and Other Nitric Oxide-Reactive Proteins, Part B, Vol. 437, pp. 21–45. Elsevier Academic Press Inc., San Diego.

Walker, C. B., Torre, J. R.d.l., Klotz, M. G., Urakawa, H., Pinel, N., Arp, D. J., Brochier-Armanet, C., Chain, P. S. G., Chan, P., Golabgir, A., Hemp, J., Hügler, M., *et al.* (2010). The genome of *Nitrosopumilus maritimus* reveals a close functional relationship to the globally distributed marine Crenarchaeota. *Proc. Natl. Acad. Sci.* 10.1073/pnas.0913533107.

Wrage, N., Velthof, G. L., van Beusichem, M. L., and Oenema, O. (2001). Role of nitrifier denitrification in the production of nitrous oxide. *Soil Biol. Biochem.* **33**, 1723–1732.

Wuebbles, D. J. (2009). Nitrous oxide: No laughing matter. *Science* **326**, 56–57.

Yamaguchi, K., Kataoka, K., Kobayashi, M., Itoh, K., Fukui, A., and Suzuki, S. (2004). Characterization of two type 1 Cu sites of *Hyphomicrobium denitrificans* nitrite reductase: A new class of copper-containing nitrite reductases. *Biochemistry* **43,** 14180–14188.

Yoshinari, T. (1984). Nitrite and nitrous oxide production by *Methylosinus trichosporium*. *Can. J. Microbiol.* **31,** 139–144.

Yu, R., and Chandran, K. (2010). Strategies of *Nitrosomonas europaea* 19718 to counter low dissolved oxygen and high nitrite concentrations. *BMC Microbiol.* **10,** 70.

Zumft, W. G. (1997). Cell biology and molecular basis of denitrification. *Microbiol. Mol. Biol. Rev.* **61,** 533–616.

Zumft, W. G. (2005). Nitric oxide reductases of prokaryotes with emphasis on the respiratory, heme–copper oxidase type. *J. Inorg. Biochem.* **99,** 194–215.

SECTION THREE

MOLECULAR MICROBIAL ECOLOGY AND PROCESSES MEASUREMENTS

CHAPTER SEVEN

Stable Isotope Probing with ^{18}O-Water to Investigate Growth and Mortality of Ammonia Oxidizing Bacteria and Archaea in Soil

Karen Adair *and* Egbert Schwartz

Contents

1. Introduction	156
2. Materials and Methods	158
2.1. Experimental design	158
2.2. Measuring nitrate and ammonia pools in soil	158
2.3. Incubating soil with ^{18}O-water	159
2.4. DNA extraction from soil	159
2.5. Separation of labeled and nonlabeled DNA	159
2.6. Quantitative real-time PCR analysis of bacterial or archaeal *amoA* genes	160
3. Calculation of Growth and Mortality Indices	160
4. Results	161
5. Discussion	165
6. Conclusions	167
Acknowledgments	167
References	168

Abstract

Ammonia oxidizing bacteria (AOB) and archaea oxidize ammonia to nitrite, the first and rate-limiting step in the important ecosystem process of nitrification. Growth and mortality of ammonia oxidizers in soil are difficult to quantify but accurate measurements would offer important insights into how environmental parameters regulate the population dynamics of these organisms. Stable isotope probing (SIP) is a recently developed technique that can identify microorganisms that assimilate labeled substrates and can be adapted to quantify the growth of organisms in soil. Here, we describe the use of SIP with ^{18}O-water to investigate the growth and mortality of ammonia oxidizers in a soil taken

Department of Biological Sciences, Northern Arizona University, Flagstaff, Arizona, USA from a ponderosa pine forest in northern Arizona, USA. Addition of ammonia to soil stimulated the growth of AOB but not ammonia oxidizing archaea (AOA). The mortality of AOA was increased upon addition of ammonia to soil; however, the variance in these measurements was high. The mortality of AOB, in contrast, was not impacted by addition of ammonia to soil. The results suggest that increased ammonia availability in soil favors AOB over AOA.

1. INTRODUCTION

In the first step of nitrification, ammonia is oxidized to nitrate. This reaction is catalyzed by several groups of organisms, including β or γ Proteobacteria (Kowalchuk and Stephen, 2001) and crenarchaea (Leininger et al., 2006; Nicol and Schleper, 2006). Ammonia oxidizing microorganisms are aerobic and autotrophic, and therefore are thought to thrive in environments where little organic carbon and sufficient oxygen and ammonia are available (Madigan et al., 2005). Within one environment, such as a forest soil, multiple groups of ammonia oxidizers are often found (Adair and Schwartz, 2008; Boyle-Yarwood et al., 2008) and it is unclear if these groups are competing directly with each other or if they occupy distinct niches so that they are active under disparate environmental conditions. One approach to gain insight into the niches of ammonia oxidizing microorganisms is to investigate the impact of changes in specific environmental parameters on their population dynamics. For instance, if the availability of ammonia is increased in soil, do both ammonia oxidizing bacteria (AOB) and ammonia oxidizing archaea (AOA) grow, or does only one group take advantage of the added substrate?

One way to determine how ammonia-oxidizers respond to increased ammonia availability is to compare the growth rates of AOA and AOB between soils with ambient and elevated ammonia levels. It is challenging to accurately quantify the growth rates of ammonia oxidizers in soil. The abundance of microorganisms in soil is dependent upon both the mortality and growth rate, so that an increase in the relative abundance of a gene sequence within the soil metagenome could occur because the organism associated with the gene sequence grew or because other organisms represented in the soil metagenome died. To accurately measure the growth of ammonia oxidizers, it is necessary to label newly formed ammonia oxidizer genomes in soil and consequently to separate labeled DNA from nonlabeled DNA. There are two experimental procedures to label DNA in soil. The first relies on introduction of a brominated nucleotide (Br-Uracil) into newly formed DNA and subsequently separating brominated DNA from nonbrominated DNA through the use of an antibody affixed to magnetic beads (Urbach et al., 1999). The second approach is termed stable isotope probing (SIP) and this technique, to date, is the only labeling approach applied to the study of ammonia oxidizers in soil.

In SIP, a substrate, highly enriched in a stable isotope, such as ^{13}C, ^{15}N, or ^{18}O, is added to soil and, as the microorganisms assimilate the substrate and grow, the newly formed DNA will contain large amounts of the heavier stable isotope (Radajewski et al., 2000). As a result, the buoyant density of the heavy DNA will be higher than that of the light DNA in soil, which contains very few stable isotope atoms. The difference in buoyant density is large enough to separate the newly formed labeled DNA from the older nonlabeled DNA along a cesium chloride gradient setup in an ultracentrifuge (Neufeld et al., 2007). The labeled DNA will hang lower in an ultracentrifuge tube and therefore can be recovered separately from nonlabeled DNA. By analyzing the labeled DNA, it is feasible to identify the organisms that assimilated the substrate added to the soil. Two substrates have been used to investigate ammonia oxidizers in soil with SIP: carbon dioxide (Tourna et al., 2010), the carbon source of autotrophic microorganisms, and water, a substrate used in the formation of nucleic acids by all microorganisms. This chapter discusses the investigation of ammonia oxidizers in soil through SIP with ^{18}O-water.

There are multiple pathways through which oxygen atoms, including the stable isotope ^{18}O, are included in newly formed DNA in growing microorganisms. Oxygen atoms may be incorporated into nucleic acids through food, such as glucose, but oxygen in DNA is also derived from water. Water, inorganic phosphate, pyrophosphate, and ATP exchange oxygen atoms very rapidly and this exchange requires enzymes such as phosphotases and ATPases (Chaney et al., 1972; Cohn and Hu, 1978). Deoxynucleotide triphosphates, once labeled with ^{18}O from ^{18}O-water, transfer the labeled oxygen atom into the phosphodiester linkage of DNA. Thus when Richards and Boyer (1966) grew *Escherichia coli* in ^{18}O-water they found that the branch oxygen atoms in the phosphodiester linkage were highly enriched in ^{18}O, while the C_3 and C_5 bridge oxygen atoms were derived from glucose, the carbon source employed in their studies.

When DNA extracted from soil, incubated with ^{18}O-water for approximately a week, is separated along a cesium chloride gradient, two distinct bands become apparent (Schwartz, 2007, 2009). The top band is DNA that formed prior to the incubation and remained intact during the experiment. The bottom band is newly formed DNA, highly enriched in ^{18}O, and with a higher buoyant density than nonlabeled DNA. The two bands can be extracted from the centrifuge tube separately so that two DNA fractions are obtained after isopycnic centrifugation. The top band represents organisms that were present prior to addition of ^{18}O-water to soil and that survived the incubation. The bottom band is derived from growing organisms that formed their DNA during the incubation. By measuring the abundance of DNA sequences, such as the bacterial or archaeal *amoA* gene (Francis et al.,

2005; Mendum et al., 1999; Rotthauwe et al., 1997), through quantitative real-time PCR (Adair and Schwartz, 2008; Okano et al., 2004) within the labeled DNA fraction it is feasible to calculate a growth index of the organism associated with the target sequence. It is therefore possible with ^{18}O-water SIP to compare the growth rates of AOA and AOB between soils with different levels of ammonia availability.

2. MATERIALS AND METHODS

2.1. Experimental design

To test the impact of ammonia availability on ammonia oxidizers through SIP with ^{18}O-water, two soil treatments were compared to each other. One, referred to as the control samples, contained ambient levels of ammonia while in the second, referred to as ammonia treatment samples, 1 mg of N–NH$_4^+$ per kg soil was added each day for 7 days. DNA of growing organisms was labeled with ^{18}O by adding 200 μL of 99 atom% ^{18}O-water to 1 g of soil in a 15-mL Falcon tube at the beginning of the incubation. The samples were incubated for 1 week at room temperature, which ranged between 20 and 25 °C. After the beginning of the incubation, both control and ammonia treatment received an additional 10 μL of ~9.9 atom% ^{18}O-water daily. The 10-μL heavy water added to the ammonia treatment contained 1 μg N–NH$_4^+$. The ammonia solution was made by dissolving 1 g of N–NH$_4^+$ in 100 mL of ^{16}O-water. Ninety microliters of this solution was mixed with 10 μL of 99 atom% ^{18}O-water to make use of ammonium solution in the experiment. Three replicate incubations of each treatment were included in this study. The control is included in a SIP with ^{18}O-water study to separate the impact on the microbial community of the water addition from that of the NH$_4^+$ treatment.

2.2. Measuring nitrate and ammonia pools in soil

During nitrification, ammonia is converted to nitrate. To measure the impact of ammonia addition on net nitrification rates, the nitrate pool in soil was quantified before and after the incubation of soil with water and NH$_4^+$. Net changes in NH$_4^+$ and NO$_3^-$ pools were measured in separate 10 g soil incubations with ^{16}O-water and proportionately equal amounts of N–NH$_4^+$ as the treatments that received ^{18}O-water. Besides characterizing nitrification kinetics, the ammonia pools were quantified as a control to confirm that increased levels of ammonia were available in the elevated ammonia treatment samples. Ammonia and nitrate concentrations of filtered 2 M KCl soil extractions were quantified by automated spectrophotometric analysis on a QuikChem® 8500 FIA System (Lachat Instruments,

Loveland, CO). To express ammonia and nitrate concentrations in mg N/kg dry soil units, it was necessary to calculate the gravimetric water content of each soil sample. This was determined by calculating mass loss after drying 5 g of soil at 100 °C for 24 h.

2.3. Incubating soil with ^{18}O-water

The top 5 cm of a soil from a ponderosa pine forest located on the San Francisco Peaks in northern Arizona (35°N latitude and 111°W longitude) was collected. Mean annual temperature in the forest, which grows at an elevation of $\sim$2300 m, is 10.2 °C and mean annual precipitation is 660 mm. The soil is classified as a Mollic Eutroboralf, contains 1.11% carbon, 0.07% nitrogen, and has a soil water holding capacity of approximately 37%. When the soil was collected, it had a moisture content of 8.8%.

2.4. DNA extraction from soil

After the incubation the soils were stored in a -45 °C freezer until DNA was isolated from 0.5 g of frozen soil with the PowerSoil^{TM} DNA Isolation Kit (MoBio Laboratories, Carlsbad, CA) and further purified by ethanol precipitation. Not all the soil was used for extraction, leaving 0.5 g of soil which could be used in case SIP analysis failed. Cell lysis was achieved by bead beating for 15 s at setting 5.0 in a FastPrep® Instrument (Q-Bio Gene, Morgan Irvine, CA), replacing step five of the MoBio protocol. Concentration of DNA was determined with a Qubit fluorimeter (Invitrogen, Carlsbad, CA) and all extractions were diluted to 5 ng DNA/μL in TE buffer.

2.5. Separation of labeled and nonlabeled DNA

The separation of ^{18}O-labeled DNA from nonlabeled DNA was performed as described in Schwartz (2009). Briefly, 3.6 mL of a saturated CsCl solution (1.9 g/mL), 0.5 μL of SYBR Green I DNA stain (Invitrogen, Carlsbad, CA), the DNA extracted from soil, and 300 μL of gradient buffer (200 mM Tris pH 8.0, 200 mM KCl, 2 mM EDTA) were added to an ultracentrifuge tube and topped off with H_2O, before the contents were thoroughly mixed. The tubes were loaded into a TLA-110 rotor (Beckman Coulter, Brea, CA) and centrifuged in an Optima MAX ultracentrifuge at 65,000 rpm for approximately 72 h. After centrifugation, two separate bands of DNA were visible in the tubes. The top band was removed by inserting a pipette through the opening of the tube and withdrawing $\sim$400 μL of the CsCl solution containing the unlabeled DNA. The second band was removed by inserting a 21 gauge needle attached to a 1-mL syringe into the side of the tube, immediately below the bottom band, and withdrawing $\sim$400 μL of CsCl solution containing the labeled DNA. Approximately 500 μL of H_2O,

1 μL of glycogen solution (20 mg/mL) and 1 mL of isopropanol were added to each fraction and the DNA was precipitated by centrifuging in a microcentrifuge at 13,000 rpm for 15 min. The DNA was subsequently washed with 500 μL of 70% ethanol.

2.6. Quantitative real-time PCR analysis of bacterial or archaeal *amoA* genes

Real-time PCR analysis was used to quantify the relative abundance of archaeal or bacterial *amoA* genes in the labeled, nonlabeled, and pretreatment ($t = 0$) DNA fractions as described in Adair and Schwartz (2008). The primers used to measure *amoA* abundance were amoA-1F and amoA-2R for bacterial amoA (Rotthauwe *et al.*, 1997) and Arch-amoAF and Arch-amoAR for archaeal amoA (Francis *et al.*, 2005). Standard curves were constructed by cloning fragments of the archaeal or bacterial *amoA* gene into the pCR®4-TOPO® vector (Invitrogen, Carlsbad, CA) and setting up a series of 1:10 dilutions, ranging from 10 to 10^9 copies, which were then analyzed through real-time PCR. Real-time PCR reactions for both bacterial and archaeal *amoA* were run in an DNA Engine Opticon Real-Time PCR System (MJ Research) under the following conditions: 0.2 μM primers, 1 μL of 5 ng/μL soil DNA, 1× PCR buffer, 0.2 mM dNTP's, 4 mM MgCl$_2$, 1.5 units of Platinum *Taq* DNA polymerase, and 0.25× SYBR Green (Invitrogen, Carlsbad, CA) in a final volume of 50 μL. Cycling protocols were 2 min at 94 °C followed by 40 cycles of 30 s at 94 °C, 60 s at 55 °C, 45 s at 72 °C for archaeal *amoA* or 30 s for bacterial *amoA*, and a final extension of 7 min at 72 °C. Results of the real-time PCR reactions were expressed in relative abundance, *amoA* copies per ng of DNA.

3. CALCULATION OF GROWTH AND MORTALITY INDICES

The real-time PCR results, which represent relative abundances of *amoA* genes, were used to calculate growth and mortality indices. The growth index of AOA or AOB was calculated by dividing the number of gene copies per ng DNA in the labeled fraction by the number of gene copies per ng DNA extracted from soil prior to incubation with H$_2$^{18}O:

$$G_{index} = \frac{\text{gene copies per ng DNA}_{(bottom\ band)}}{\text{gene copies per ng DNA}_{(t=0)}}.$$

This index relates the abundance of a gene in labeled DNA to the number of copies present in the soil genome prior to the incubation with ^{18}O-water. Thus an organism represented by a gene that is rare in the $t = 0$

DNA, but abundant in the labeled DNA, will have a higher growth index than an organism associated with a gene that is abundant in both the $t = 0$ DNA and labeled DNA fractions. The mortality indices for AOA and AOB were calculated by dividing the number of gene copies per ng DNA extracted from soil, prior to the incubation, with the number of gene copies per ng DNA of the nonlabeled DNA fraction, obtained after the soil was incubated with heavy water:

$$M_{index} = \frac{\text{gene copies per ng DNA}_{(t=0)}}{\text{gene copies per ng DNA}_{(topband)}}.$$

The growth indices of AOA and AOB were compared to each other by generating a delta growth index (ΔG_{index}). This index was calculated by dividing the growth index of AOB by the growth index of AOA:

$$\Delta G_{index} = \frac{G_{index}AOB}{G_{index}AOA}$$

By comparing the growth indices of two separate genes in the same sample any artifacts such as free DNA in soil, which will influence the relative abundance numbers obtained through real-time PCR analysis, will cancel out so that the ΔG_{index} is a good comparison between the growth rates of AOA and AOB in soil. Similarly, a delta mortality index was calculated by dividing the mortality index of AOB by the mortality index of AOA:

$$\Delta M_{index} = \frac{M_{index}AOB}{M_{index}AOA}.$$

4. RESULTS

The size of the ammonia pool present in the soil prior to incubation was 3.52 (± 0.27) mg $N-NH_4$/kg dry soil (Fig. 7.1). At the end of the 1-week incubation, the ammonia pool in the control samples had increased to 8.12 (± 1.15) mg $N-NH_4$/kg dry soil. The ammonia pools in soil samples that received 1 mg of $N-NH_4$/kg soil per day were lower than the initial pool, averaging 2.71 (± 1.18) mg $N-NH_4$/kg dry soil. The size of the nitrate pool, present in the soil prior to incubation, was 1.92 (± 0.14) mg $N-NO_3$/kg dry soil. The nitrate pool increased slightly to 3.89 (± 2.62) mg $N-NO_3$/kg dry soil in the control samples, while it increased substantially to 17.07 (± 1.51) mg $N-NO_3$/kg dry soil, following incubation, in soils receiving additional ammonia.

After centrifugation of DNA, extracted from soil incubated with ^{18}O-water, two bands were visible in the ultracentrifuge tubes (Fig. 7.2).

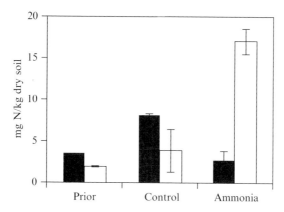

Figure 7.1 The concentration of ammonium (closed bars) and nitrate (open bars) in soil prior to the incubation, in control samples, to which only ^{18}O-water was added, and in ammonia treatment samples, which received both ^{18}O-water and 7 mg/kg soil N–NH_4^+. The error bars represent standard errors calculated from three replicate incubations.

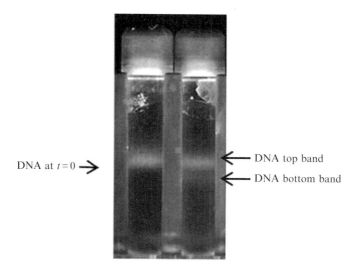

Figure 7.2 Ultracentrifuge tubes containing either DNA extracted from soil prior to the incubation (left) or DNA extracted from soil after incubation with ^{18}O-water (right). The left tube shows only one nonlabeled DNA band, which is referred to as $DNA_{(t=0)}$ in the manuscript. The right tube contains two DNA bands. The top band ($DNA_{(top\ band)}$) is not labeled and represents DNA that was present prior to the incubation and remained intact during the incubation. The bottom band ($DNA_{(bottom\ band)}$) is labeled with ^{18}O and is newly formed DNA of microorganisms that grew during the incubation.

These bands were separated far enough from each other that both bands were retrieved from the ultracentrifuge tube in distinct fractions. In contrast, after isopycnic centrifugation of $t = 0$ DNA, extracted from soil prior to incubation with ^{18}O-water, only one DNA band was observed.

Neither AOA nor AOB were abundant in the soil at the time the incubation was initiated. There were approximately 14 copies per ng DNA of the archaeal *amoA* gene and 6 copies per ng DNA of the bacterial *amoA* gene. Two separate DNA fractions, the nonlabeled DNA that remained intact during the incubation and the labeled DNA that formed during the incubation, were obtained after isopycnic centrifugation of soil samples incubated with ^{18}O-water. In the control samples, to which only ^{18}O-water was added, the concentration of archaeal *amoA* and bacterial *amoA* in the top DNA band was approximately 49 and 24 copies per ng DNA, respectively. The average concentration of archaeal amoA and bacterial amoA in the bottom DNA band was approximately 2 and 3 copies per ng DNA, respectively. In the ammonia treatment samples, which received both ^{18}O-water and ammonia, the concentrations of archaeal *amoA* and bacterial *amoA* in the top DNA band were approximately 32 and 19 copies per ng DNA, respectively, while in the bottom band, on average, there were 2 and 10 copies per ng DNA of archaeal *amoA* and bacterial *amoA*.

The growth indices in the control samples, calculated by dividing the number of gene copies in the labeled DNA fraction by the number of gene copies in the $t = 0$ DNA fraction, were 0.13 and 0.53 for archaeal *amoA* and bacterial *amoA*, respectively (Fig. 7.3). In the ammonia treatment samples, the growth indices were 0.05 and 1.11 for archaeal *amoA* and bacterial *amoA*, respectively. The mortality indices for control samples,

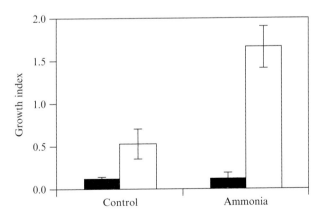

Figure 7.3 The growth indices of ammonia oxidizing archaea (closed bars) and ammonia oxidizing bacteria (open bars) in control and ammonia treatment samples. The error bars represent standard errors calculated from three replicate incubations for ammonia oxidizing archaea and two replicates for ammonia oxidizing bacteria.

calculated by dividing the concentration of gene copies present in soil at $t = 0$ by the concentration of gene copies in the nonlabeled DNA fraction, was 1.05 and 0.30 for archaeal *amoA* and bacterial *amoA*, respectively (Fig. 7.4). The mortality indices for the ammonia treatment samples were 3.65 and 0.35 for archaeal *amoA* and bacterial *amoA*, respectively.

The Δ growth index, which was calculated by dividing the growth index of AOB by that of AOA, was 4.98 and 33.25 for the control and ammonia treatment samples, respectively (Fig. 7.5). The Δ mortality index was 0.92 and 0.65 for the control and ammonia treatment samples, respectively.

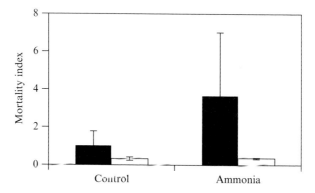

Figure 7.4 The mortality indices of ammonia oxidizing archaea (closed bars) and ammonia oxidizing bacteria (open bars) in control and ammonia treatment samples. The error bars represent standard errors calculated from three replicate incubations.

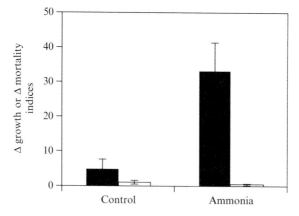

Figure 7.5 The Δ growth (closed bars) and Δ mortality (open bars) indices in control and ammonia treatment samples. The error bars represent standard errors calculated from two replicates for the Δ growth index and three replicates for the Δ mortality index.

5. Discussion

The control samples, to which only ^{18}O-water was added, contained the largest ammonia pools after the incubation. Rewetting a dry soil will release a flush of carbon and nitrogen (Mikha *et al.*, 2005; Miller *et al.*, 2005; Williams and Xia, 2009). This carbon and nitrogen may come directly from the microbial biomass and could consist of osmolytes or biological macromolecules from dead organisms (Halverson *et al.*, 2000; Kieft *et al.*, 1987). The ammonia pools in the ammonia treatment samples were smaller than the control samples, even though the ammonia treatment samples received 7 mg/kg soil ammonia. The large nitrate pool in the ammonia treatment samples indicate that the added ammonia was converted to nitrate through nitrification. The difference between the nitrate pool in the ammonia treatment samples and the ammonia pool in the control samples (8.95 mg N/kg dry soil) is similar to the amount of ammonia added to the ammonia treatment soils.

The abundance of ammonia oxidizer populations was relatively low when the soil was harvested. The soil was harvested during the summer, before the monsoon rains arrived, when soil moisture is low and microbial activity is limited. When the ponderosa pine forest soils are more moist, in the late spring or after the monsoon rains arrive, there may be between 1000 and 10,000 *amoA* copies per ng of DNA (Adair and Schwartz, 2008).

After isopycnic centrifugation of DNA, extracted from soil incubated with ^{18}O-water, two distinct bands of DNA appeared. The bottom band of DNA was labeled with ^{18}O and had formed during the incubation. The abundances of bacterial and archaeal *amoA* genes in the labeled DNA obtained from control samples were similar. However, if ammonia was added to the soil, the abundance of bacterial *amoA* genes in the labeled DNA band was greater than the abundance of archaeal *amoA* genes. Archaeal *amoA* genes are generally more abundant in soils, including the ponderosa pine forest soil, than bacterial *amoA* genes (Adair and Schwartz, 2008; Leininger *et al.*, 2006).

SIP with ^{18}O-water does not determine an absolute growth rate but rather compares the growth rate of one organism to another. To accurately reflect the impact of ammonia on growth of AOA and AOB, it is important to consider the abundance of AOA and AOB in the soil before the incubation was initiated because the real-time PCR results are not absolute but relative. If, for instance, one million copies per ng DNA of a gene were present at $t = 0$, but the concentration of the gene in the labeled DNA band is only 100,000 copies per ng DNA, the organisms associated with the gene grew slower than other organisms in the soil even though, at 100,000 copies per ng DNA, the gene was still abundant in the labeled DNA. The growth index divides the abundance of a gene in the labeled fraction with the abundance of a gene in the soil prior to the initiation of the incubation. As a result, the growth index can be high because a gene was highly abundant in the labeled DNA fraction or because the gene was rare in the $t = 0$ DNA fraction. A growth index equal to one indicates that the targeted organism grew as fast as the average of all the other organisms present during the incubation. In contrast, a growth index higher than one signifies the organism grew faster while a growth index lower than one indicates the organism grew slower relative to the average growth rate of other organisms in the soil.

The growth index for AOB was similar to the growth index for AOA in control samples. However, when ammonia was added to the soil, the growth index of AOB was substantially higher than the growth index of AOA. This observation indicates that AOB compete better than AOA when higher concentrations of ammonia are available. Martens-Habbena et al. (2009) found that AOA were capable of higher ammonia oxidation rates at low ammonia concentrations than AOB. Others have suggested that AOA are more competitive in environments with lower pH (Nicol et al., 2008), where the ratio of ammonium (NH_4^+) to ammonia (NH_3) would be greater. It is thought ammonia-oxidizing organisms use NH_3 as their energy source and ammonium, a cation, would interact with the cation exchange capacity of the soil (Brady, 1990) and therefore may be less available to ammonia oxidizers.

The mortality index was calculated by dividing the concentration of *amoA* genes present at the time the incubation was initiated, by the concentration of genes in the nonlabeled top DNA band. DNA is an excellent source of carbon, nitrogen, and phosphorus for soil microorganisms and during the incubation some of the DNA was likely degraded. This DNA may have originated from organisms that were predated or lysed due to bacteriophage infection. The DNA of an organism that dies during the incubation is present in the soil at the beginning of the incubation but is absent in the nonlabeled top DNA band at the end of the incubation. Therefore, a gene sequence that is abundant in the soil at $t = 0$, but absent in the nonlabeled top DNA band, is associated with an organism that has a high mortality index. However, it is important to note that DNA released from dead organisms can sorb to soil particles and be resistant to degradation (Khanna and Stotzky, 1992). This sorbed DNA may still be extracted from soil and will remain present in the nonlabeled DNA fraction after the incubation even though the organism has died.

The AOA had a higher mortality index than AOB in the control sample and addition of ammonia increased the mortality index further. As water addition to soil releases nutrients, it may be that AOA survive less well in conditions of high nutrient availability. The addition of ammonia to soil further increased the availability of nutrients thereby negatively impacting

AOA further. The AOB, in contrast, did not appear to suffer when ammonia was added to soil. They grew faster upon ammonia addition and their mortality index did not increase substantially. In this experiment, we used 1 g soil incubations, which occupy a relatively small volume. The size of the incubation was constrained by the cost of 99 atom% ^{18}O-water. The small soil volumes could explain the large variances observed in AOA mortality indices.

As SIP with ^{18}O-water relies on the addition of heavy water to soil, the technique is not applicable to the study of very dry soils. Addition of water to soil will retard the diffusion of oxygen into soil, which in turn may impact growth of ammonia-oxidizing organisms. In this study, soil moisture was elevated to 30.6% or 82.7% of the soil's water holding capacity, through addition of ^{18}O-water. After a week of incubation, oxygen availability in the soils of this study may have been limited. Nonetheless, the control samples had identical level of soil moisture as the ammonia treatment samples, so that oxygen availability cannot explain the higher growth rate for AOB in ammonia amended samples.

When comparing the growth and mortality rates of AOA and AOB, it is feasible to eliminate artifacts in the growth and mortality indices due to the presence of free DNA sorbed to soil particles (Khanna and Stotzky, 1992; Ogram et al., 1987) or old plant DNA, which is unable to replicate and grow in a 1-g soil incubation. As this DNA affects the growth indices of AOA and AOB equally, the artifacts cancel each other out when a ratio of the growth or mortality indices is taken. The ratio of the AOB growth index over the AOA index shows that the AOB grew approximately five times more rapidly when only water was added to soil and approximately 33 times faster when water and ammonia was supplied. The results show that high ammonia availability benefits the population dynamics of AOB over AOA.

6. Conclusions

SIP with ^{18}O-water can be used to study the impacts of ammonia addition on growth and death of AOA and AOB in soil. The AOB grew faster and died more slowly than AOA when ammonia availability in soil was increased.

ACKNOWLEDGMENTS

The work described in this manuscript was made possible with support from National Science Foundation, Grant No. EF-0747397.

REFERENCES

Adair, K., and Schwartz, E. (2008). Evidence that ammonia-oxidizing archaea are more abundant than ammonia-oxidizing bacteria in soils along an elevation gradient in northern Arizona, USA. *Microb. Ecol.* **56,** 420–426.

Boyle-Yarwood, S. A., Bottomley, P. J., and Myrold, D. D. (2008). Community composition of ammonia-oxidizing bacteria and archaea in soils under stands of red alder and Douglas fir in Oregon. *Environ. Microbiol.* **10**(11), 2956–2965.

Brady, N. C. (1990). The Nature and Properties of Soils. 10th ed. Macmillan Publishing Company, New York City, NY.

Chaney, S. G., Duffy, J. J., and Boyer, P. D. (1972). Patterns of oxygen interchange between water, substrates and phosphate compounds of *Escherichia coli* and *Bacillus subtilis*. *J. Biol. Chem.* **247**(7), 2145–2150.

Cohn, M., and Hu, A. (1978). Isotopic (^{18}O) shift in ^{31}P nuclear magnetic resonance applied to a study of enzyme-catalyzed phosphate–phosphate exchange and phosphate (oxygen)–water exchange reactions. *Proc. Natl. Acad. Sci. USA* **75,** 200–203.

Francis, C. A., Roberts, K. J., Beman, J. M., Santoro, A. E., and Oakley, B. B. (2005). Ubiquity and diversity of ammonia-oxidizing archaea in water columns and sediments of the ocean. *Proc. Natl. Acad. Sci. USA* **102,** 14683–14688.

Halverson, L. J., Jones, T. M., and Firestone, M. K. (2000). Release of intracellular solutes by four soil bacteria exposed to dilution stress. *Soil Science Society of America Journal* **64,** 1630–1637.

Khanna, M., and Stotzky, G. (1992). Transformation of *Bacillus subtilis* by DNA bound on montmorillonite and effect of DNase on the transforming ability of bound DNA. *Appl. Environ. Microbiol.* **58,** 1930–1939.

Kieft, T. L., Soroker, E., and Firestone, M. K. (1987). Microbial biomass response to a rapid increase in water potential when dry soil is wetted. *Soil Biol. Biochem.* **19**(2), 119–126.

Kowalchuk, G. A., and Stephen, J. R. (2001). Ammonia-oxidizing bacteria: A model for molecular microbial ecology. *Annu. Rev. Microbiol.* **55,** 485–529.

Leininger, S., Urich, T., Schloter, M., Schwark, L., Qu, J., Nicol, G. W., Prosser, J. I., Schuster, S. C., and Schleper, C. (2006). Archaea predominate among ammonia-oxidizing prokaryotes in soils. *Nature* **442,** 806–809.

Madigan, M. T., Martinko, J. M., and Parker, J. (2005). Brock biology of microorganisms. 11th ed. Prentice Hall, Inc., Upper Saddle River, NJ.

Martens-Habbena, W., Berube, P. M., Urakawa, H., de la Torre, J. R., and Stahl, D. A. (2009). Ammonia oxidation kinetics determine niche separation of nitrifying archaea and bacteria. *Nature* **461,** 976–979.

Mendum, T. A., Sockett, R. E., and Hirsch, P. R. (1999). Use of molecular and isotopic techniques to monitor the response of autotrophic ammonia-oxidizing populations of the β subdivision of the class *Proteobacteria* in arable soils to nitrogen fertilizer. *Appl. Environ. Microbiol.* **65,** 4155–4162.

Mikha, M. M., Rice, C. W., and Milliken, G. A. (2005). Carbon and nitrogen mineralization as effected by drying and wetting cycles. *Soil Biol. Biochem.* **37,** 339–347.

Miller, A. E., Schimel, J. P., Meixner, T., Sickman, J. O., and Melack, J. M. (2005). Episodic rewetting enhances carbon and nitrogen release from chaparral soils. *Soil Biol. Biochem.* **37,** 2195–2204.

Neufeld, J. D., Vohra, J., Dumont, M. G., Lueders, T., Manefield, M., Friedrich, M. W., and Murrell, J. C. (2007). DNA stable-isotope probing. *Nat. Protoc.* **2,** 860–866.

Nicol, G. W., and Schleper, C. (2006). Ammonia-oxidising Crenarchaeota: Important players in the nitrogen cycle? *Trends Microbiol.* **14,** 207–212.

Nicol, G. W., Leininger, S., Schleper, C., and Prosser, J. I. (2008). The influence of soil pH on the diversity, abundance and transcriptional activity of ammonia oxidising archaea and bacteria. *Environ. Microbiol.* **10,** 2966–2978.

Ogram, A., Sayler, G. S., and Barkay, T. (1987). The extraction and purification of microbial DNA from sediments. *J. Microbiol. Methods* **7,** 57–66.

Okano, Y., Hristova, K. R., Leutenegger, C. M., Jackson, L. E., Denison, R. F., Gebreyesus, B., Lebauer, D., and Scow, K. M. (2004). Application of real-time PCR to study effects of ammonium on population size of ammonia-oxidizing bacteria in soil. *Appl. Environ. Microbiol.* **70,** 1008–1016.

Radajewski, S., Ineson, P., Parkeh, N. R., and Murrell, J. C. (2000). Stable-isotope probing as a tool in microbial ecology. *Nature* **403,** 646–649.

Richards, O. C., and Boyer, P. D. (1966). ^{18}O labeling of deoxyribonucleic acid during synthesis and stability of label during replication. *J. Mol. Biol.* **19,** 109–119.

Rotthauwe, J.-H., Witzel, K.-P., and Liesack, W. (1997). The ammonia monooxygenase structural gene *amoA* as a functional marker: Molecular fine-scale analysis of natural ammonia-oxidizing populations. *Appl. Environ. Microbiol.* **63,** 4704–4712.

Schwartz, E. (2007). Characterization of growing microorganisms in soil by stable isotope probing with $H_2^{18}O$. *Appl. Environ. Microbiol.* **73,** 2541–2546.

Schwartz, E. (2009). Analyzing microorganisms in environmental samples using stable isotope probing with $H_2^{18}O$. *Cold Spring Harb. Protoc.* 10.1101/pdb.prot5341.

Tourna, M., Freitag, T. E., and Prosser, J. I. (2010). Stable isotope probing analysis of interactions between ammonia oxidizers. *Appl. Environ. Microbiol.* **76,** 2468–2477.

Urbach, E., Vergin, K. L., and Giovannoni, S. J. (1999). Immunochemical detection and isolation of DNA from metabolically active bacteria. *Appl. Environ. Microbiol.* **65,** 1207–1213.

Williams, M. A., and Xia, K. (2009). Characterization of the water soluble soil organic pool following the rewetting of dry soil in a drought-prone tallgrass prairie. *Soil Biol. Biochem.* **41,** 21–28.

CHAPTER EIGHT

Measuring Nitrification, Denitrification, and Related Biomarkers in Terrestrial Geothermal Ecosystems

Jeremy A. Dodsworth,[*] Bruce Hungate,[†] José R. de la Torre,[‡] Hongchen Jiang,[§] *and* Brian P. Hedlund[*]

Contents

1. Introduction	172
2. General Considerations for Measurement of N-Cycle Activities in Terrestrial Geothermal Habitats	177
3. Methods for Measuring Nitrification in Terrestrial Geothermal Habitats	179
3.1. Gross nitrification using $^{15}NO_3^-$ pool dilution approach	180
3.2. Quantification and determination of atom% ^{15}N–NO_x and calculation of gross denitrification rates	183
4. Methods for Measuring Denitrification in Terrestrial Geothermal Habitats	187
4.1. Acetylene block method	187
4.2. $^{15}NO_3^-$ tracer approach	191
5. Detection and Quantification of Potential Biomarkers for Thermophilic AOA and Denitrifying *Thermus thermophilus*	193
5.1. Sample collection and nucleic acid extraction	194
5.2. PCR and qPCR	195
6. Closing Remarks	198
Acknowledgments	198
References	198

---

[*] School of Life Sciences, University of Nevada, Las Vegas, Nevada, USA
[†] Department of Biological Sciences, Merriam-Powell Center for Environmental Research, Northern Arizona University, Flagstaff, Arizona, USA
[‡] Department of Biology, San Francisco State University, San Francisco, California, USA
[§] Geomicrobiology Laboratory, China University of Geosciences, Beijing, China

Abstract

Research on the nitrogen biogeochemical cycle in terrestrial geothermal ecosystems has recently been energized by the discovery of thermophilic ammonia-oxidizing archaea (AOA). This chapter describes methods that have been used for measuring nitrification and denitrification in hot spring environments, including isotope pool dilution and tracer approaches, and the acetylene block approach. The chapter also summarizes qualitative and quantitative methods for measurement of functional and phylogenetic biomarkers of thermophiles potentially involved in these processes.

1. Introduction

Our knowledge of the nitrogen cycle (N-cycle) has changed radically in recent years due to major discoveries, including archaeal ammonia (NH_3) oxidation at low and high temperatures (de la Torre et al., 2008; Hatzenpichler et al., 2008; Könneke et al., 2005), archaeal N_2 fixation at high temperature (Mehta and Baross, 2006), anaerobic ammonium oxidation (anammox) at low and high temperatures (Jaeschke et al., 2009; Strous et al., 1999), and eukaryotic nitrate (NO_3^-) respiration (Risgaard-Petersen et al., 2006). Until a recently, very little was known about the N-cycle at high temperature to the extent that processes such as NH_3 oxidation, nitrite (NO_2^-) oxidation, and anammox had never been addressed. Through a combination of microbial cultivation approaches, process rate measurements, and studies of phylogenetic and functional biomarkers, our knowledge of the N-cycle in terrestrial geothermal habitats is rapidly growing. Table 8.1 summarizes some of the evidence for N-cycle processes and relevant phylogenetic groups at high temperature.

Nitrification and denitrification are two important processes in the N-cycle. Nitrification is the aerobic oxidation of NH_3 to NO_3^- through a NO_2^- intermediate:

$$NH_{3(aq)} \rightarrow NO_{2(aq)}^- \rightarrow NO_{3(aq)}^-$$

No known organism can catalyze both steps of the reaction so it can be valuable to consider these two steps, and the organisms that catalyze them, separately.

The net reaction for NH_3 oxidation to NO_2^- is as follows:

$$NH_{3(aq)} + 1.5 O_{2(aq)} \rightarrow NO_{2(aq)}^- + H_2O + H_{(aq)}^+ (6e^-) \Delta G'_0 = -235 \text{kJ/mol}$$

At moderate temperature, both ammonia-oxidizing bacteria (AOB) and ammonia-oxidizing archaea (AOA) play a role in NH_3 oxidation to NO_2^-. A variety of evidence suggests that AOA are more abundant in many

Table 8.1 Summary of evidence for N-cycle processes at high temperatures, including *in situ* process rate measurements, laboratory cultures, and recovery of possible phylogenetic or functional biomarkers

Process/max. temp. (°C)	Environment	Evidence	References
N₂ fixation			
89/85	Terrestrial, YNP	PCR amplification of putative *nifH* genes; acetylene reduction	Hamilton et al. (submitted)
92	Marine	Pure culture, *Methanocaldococcus* sp. FS406-22	Mehta and Baross (2006)
NH₃ oxidation			
85	Terrestrial, Iceland, Great Basin	$^{15}NO_3^-$ pool dilution	Reigstad et al. (2008), Dodsworth et al. (unpublished data)
74/46	Terrestrial, YNP	Highly enriched cultures, "*Candidatus* Nitrosocaldus yellowstonii" and "*Candidatus* Nitrososphaera gargensis"	de la Torre et al. (2008), Hatzenpichler et al. (2008)
94	Terrestrial, Great Basin, Iceland	Recovery of crenarchaeol (putative biomarker of AOA)	Pearson et al. (2004, 2008), Reigstad et al. (2008)
94	Terrestrial, China, Great Basin	RT-PCR amplification of putative archaeal *amoA* transcripts	Jiang et al. (2010)
97	Terrestrial, "global"	PCR amplification of putative archaeal *amoA* genes	de la Torre et al. (2008), Reigstad et al. (2008), Zhang et al. (2008)
NO₂⁻ oxidation			
85	Terrestrial	$^{15}NO_3^-$ pool dilution	Reigstad et al. (2008)
60	Terrestrial	Activity in enrichment cultures	Lebedeva et al. (2005)
48	Terrestrial	Probable pure culture, "*Candidatus* Nitrospira bockiana"	Lebedeva et al. (2008)
57/69	Terrestrial	PCR amplification of *Nitrospira* 16S rRNA genes and *norB*	Kanokratana et al. (2004), Hirayama et al. (2005)

(*continued*)

Table 8.1 (continued)

Process/max. temp. (°C)	Environment	Evidence	References
Anammox			
85	Marine	Isotope pairing	Byrne et al. (2009)
43	Wastewater	Highly enriched culture	Strous et al. (1999)
65/52	Terrestrial	Ladderane lipids (putative biomarker); 16S rRNA genes	Jaeschke et al. (2009)
Nitrate reduction			
85	Terrestrial	$^{15}NO_3^-$ tracer; acetylene block; qPCR of $narG$	Dodsworth et al. (unpublished data)
113	Marine	Pure culture, *Pyrolobus fumarii* (NH_3 dominant product)	Blöchl et al. (1997)
100	Terrestrial	Pure culture, *Pyrobaculum aerophilum* (N_2O dominant product; capable of N_2 production)	Völkl et al. (1993)

environments (Francis *et al.*, 2005; He *et al.*, 2007; Leininger *et al.*, 2006; Nicol *et al.*, 2008; Shen *et al.*, 2008), except those impacted by high doses of anthropogenic N such as wastewater (Wells *et al.*, 2009). These results may be explained by the competitive advantage of AOA under substrate-limited conditions because AOA have a much higher affinity for NH_3 than AOB (Martens-Habbena *et al.*, 2009). Very little evidence suggests AOB are important at temperatures above 40–50 °C, although this observation needs further verification. Lebedeva *et al.* (2005) reported isolation of AOB from Garga Hot Spring in the Baikal Rift Zone that could grow up to 50 °C, and the two isolates were identified immunochemically as presumptive *Nitrosospira* and *Nitrosomonas*. In addition, *Nitrosomonas amoA* genes have been quantified in a gold mine at temperatures up to 62 °C (Hirayama *et al.*, 2005). Yet, no thermophilic isolates or enrichments capable of ammonia oxidation have been identified definitively as *Bacteria* (e.g., by 16S rRNA gene analysis). In contrast, highly enriched cultures of AOA from Garga Hot Spring and Heart Lake Hot Spring in Yellowstone National Park mediate ammonia oxidation at temperatures up to 46 and 74 °C, respectively (de la Torre *et al.*, 2008; Hatzenpichler *et al.*, 2008). Similarly, homologs of the archaeal ammonia monooxygenase alpha subunit gene, *amoA*, and the biphytanyl lipid crenarchaeol, both tentatively regarded as distinctive biomarkers of AOA, have been described at temperatures up to 97 and 87 °C, respectively (Pearson *et al.*, 2008; Reigstad *et al.*, 2008). However, both of these biomarkers are more reliably found in hot springs below 75 °C, and crenarchaeol was shown to be present at the highest concentrations relative to other archaeal lipids at around 40 °C in hot springs in the US Great Basin (Zhang *et al.*, 2006). Interestingly, crenarchaeol is much more abundant relative to other biphitanyl lipids in Great Basin hot springs, as compared with a variety of Yellowstone National Park hot springs, possibly suggesting a more important role for AOA in Great Basin geothermal ecosystems (Pearson *et al.*, 2008). Recently, two studies recovered *amoA* transcripts from hot spring sediments ranging from 44.5 to 94 °C in several springs in Tengchong, China, and the Great Basin, suggesting that AOA may be active in terrestrial springs at temperatures near boiling (Jiang *et al.*, 2010; Zhang *et al.*, 2008).

In the second step of nitrification, NO_2^- is oxidized to NO_3^-, a process that may also be important in high-temperature environments. The net equation for NO_2^- oxidation to NO_3^- is as follows:

$$NO_{2(aq)}^- + 0.5 O_{2(aq)} \rightarrow NO_{3(aq)}^- (2e^-) \Delta G'_0 = -54\,kJ/mol$$

In moderate-temperature environments, a variety of NO_2^--oxidizing bacteria (NOB) catalyze the oxidation of NO_2^- to NO_3^-, including members of the phylum *Nitrospira* and three classes of *Proteobacteria*, *Alphaproteobacteria*, *Gammaproteobacteria*, and *Deltaproteobacteria*. From an ecological perspective, NO_2^- oxidation has received far less attention than NH_3 oxidation because it is generally contended that the latter is rate limiting in nature. However, this is doubtful at high temperature because some hot springs sourced with NH_3 as the dominant form of inorganic N accumulate high concentrations of NO_2^- (Costa et al., 2009) and AOA grow at much higher temperatures than known NOB. Few investigations have focused on thermophilic nitrite oxidation, though several lines of evidence implicate *Nitrospira* as the dominant nitrifier at moderately elevated temperature. Among cultures of *Nitrospira*, *N. moscovensis* and "*Candidatus* N. bockiana" are the most thermophilic known, with growth temperature optima (T_{opt}) of 39 and 42 °C, respectively, with the latter capable of NO_2^- oxidation and growth up to 48 °C (Ehrich et al., 1995; Lebedeva et al., 2008). However, nitrifying enrichments supporting *Nitrospira* have been reported up to 60 °C (Lebedeva et al., 2005) and *Nitrospira* 16S rRNA genes have been recovered from spring ecosystems from 50 to 57 °C (Kanokratana et al., 2004). In addition, nitrite oxidoreductase genes (*norB*) related to those from *Nitrospira* species have been amplified and quantified in Japanese gold mine samples up to 69 °C (Hirayama et al., 2005).

Very few measurements of rates of oxidative N-cycle processes have been done in geothermal environments. Reigstad et al. (2008) measured gross nitrification at 84 and 85 °C in two acidic Icelandic hot springs with high dissolved clay content using the $^{15}NO_3^-$ pool dilution technique, which yielded rates of 2.8–7.0 nmol NO_3^- N g^{-1} h^{-1} (data converted from volume using reported density). NO_3^- production was stimulated more than twofold by addition of NH_4^+ before incubation, showing that nitrification in these springs was limited by NH_3 supply. The authors have also used the $^{15}NO_3^-$ pool dilution technique to measure gross nitrification at 79–81 °C in two hot springs in the US Great Basin as described in Section 3, where rates of NO_3^- production varied from 0.5 to ~50 nmol NO_3^- N g sediment^{-1} h^{-1} (Dodsworth et al., unpublished data).

Denitrification is the stepwise reduction of NO_3^- to the gaseous products nitric oxide (NO), nitrous oxide (N_2O), and dinitrogen (N_2):

$$NO_{3(aq)}^- \rightarrow NO_{2(aq)}^- \rightarrow NO_{(g)} \rightarrow N_2O_{(g)} \rightarrow N_{2(g)}$$

Denitrification is a respiratory process in which nitrogen oxides serve as electron acceptors and is contrasted with assimilatory NO_3^- reduction in that the former is coupled to energy conservation and growth, whereas the latter serves only to scavenge nitrogen for biosynthesis. A variety of thermophiles and hyperthermophiles from both terrestrial and marine geothermal habitats can respire NO_3^-, although very few studies have focused on this. Among archaea, *Pyrobaculum aerophilum*, *Ferroglobus placidus*, and *Pyrolobus fumarii* have been definitively shown to respire NO_3^- up to temperatures of 80, 95, and 113 °C, respectively, although a number of other thermophiles contain gene homologs for NO_3^- reduction pathways, including the euryarchaeon *Archaeoglobus fulgidus* (Cabello *et al.*, 2004). *P. aerophilum* uses a novel NO_3^- reductase and is capable of complete denitrification to N_2, although N_2O accumulates during growth due to a kinetically inhibited N_2O reductase (Afshar *et al.*, 1998; Cabello *et al.*, 2004; Völkl *et al.*, 1993). *F. placidus* produces NO_2^- and some NO as products during growth (Hafenbradl *et al.*, 1996); however, *in vitro* experiments with cell extracts documented N_2O production (Vorholt *et al.*, 1997). *P. fumarii* stoichiometrically reduces NO_3^- to NH_3 during growth on H_2 (Blöchl *et al.*, 1997). Among thermophilic and hyperthermophilic bacteria, NO_3^- reduction is widespread in the Thermaceae, including several species of the terrestrial genera *Thermus* and *Meiothermus* and the marine genera *Oceanithermus* (Miroshnichenko *et al.*, 2003b) and *Vulcanithermus* (Miroshnichenko *et al.*, 2003c). *Thermus* includes strains described to reduce NO_3^- fully to N_2 (Cava *et al.*, 2008) as well as strains only capable of NO_2^- production (Ramirez-Arcos *et al.*, 1998). *Meiothermus*, *Oceanithermus*, and *Vulcanithermus* are not known to be capable of reduction of NO_3^- past NO_2^-. A variety of *Geobacillus* denitrify completely to N_2 (Mishima *et al.*, 2009). In the Aquificales, *Aquifex pyrophilus* and *Persephonella* spp. denitrify to N_2 (Gotz *et al.*, 2002) and *Thermovibrio ruber* (Huber *et al.*, 2002) reduces NO_3^- to NH_3. The novel bacterium *Caldithrix abysii* also reduces NO_3^- to NH_3 (Miroshnichenko *et al.*, 2003a).

Although many NO_3^--respiring thermophiles exist in culture, very few studies have addressed NO_3^- reduction or denitrification in natural geothermal habitats. Burr *et al.* (2005) measured denitrification by the acetylene block approach, along with N_2 fixation, ammonification, and nitrification, in hot acidic soils at 50, 65, and 80 °C in Yellowstone National Park. N_2O production ranged from 0.34 to 1.1 nmol N_2O N g^{-1} h^{-1} with maximal activity at 65 °C. Activity was dependent on NO_3^- addition and acetylene did not significantly enhance N_2O flux, suggesting that denitrification was NO_3^--limited and that N_2O, rather than N_2, was the major denitrification product. The authors have used the acetylene block method and a $^{15}N-NO_3^-$ tracer method to measure denitrification rates in two hot springs in the US Great Basin, as described in Section 4 (Dodsworth *et al.*, unpublished data).

2. General Considerations for Measurement of N-Cycle Activities in Terrestrial Geothermal Habitats

The following sections describe experimental details of approaches as they have been applied to study N-cycle activities in terrestrial geothermal ecosystems. There are many variations on these general themes and the reader is advised to consult other papers before selecting the strategies that are most appropriate for their experimental system, hypotheses, and available resources (Mosier and Klemedtsson, 1994; Steingruber *et al.*, 2001; Tiedje *et al.*, 1989; Ward and O'Mullan, 2005). Regardless of the approach, the sampling strategy must be carefully conceived. Nitrification and denitrification in soils and sediments are notoriously heterogeneous because of the natural patchiness of resources that affect N-cycling such as soil moisture content, quality of substrates for biofilm formation, organic content, and the availability of N substrates (Tiedje *et al.*, 1989). Denitrification in soils is also incredibly temporally variable, correlating strongly with events that lead soils to be water saturated, such as spring thaws and rain (Tiedje *et al.*, 1989). Spatial heterogeneity is generally addressed by the soil microbiology community by using a highly replicated experimental design to distinguish within- and between-system rates (e.g., 20 replicates per sample; Mosier and Klemedtsson, 1994). However, we find this degree of replication impractical at most geothermal sites due to the relatively small sizes of geothermal features and the strict protection of springs in protected areas such as national parks. Hot spring ecosystems vary widely in terms of their basic hydrology and geochemistry. Although few, if any, springs have been investigated in detail to address spatial heterogeneity, heterogeneity is evident in many springs by the patchiness of conspicuous microbial growth and mineral precipitates. Thus, as in soils research, heterogeneity must be carefully considered. Spring size, shape, and substrate mineralogy are also important practical considerations.

Care should be taken to select the sample site and incubation site. Sample sites should have enough accessible sediment cover within reach of gloved hands to achieve the replication necessary for the experimental design. The incubation site should be of the same temperature as the sampling site and it should be adjacent to the sampling site, but in a place that does not impact the sampling site, for example, immediately downflow from the sampling site. The incubation site should be deep enough so that the liquid phase of the incubation tubes and bottles are completely submerged when sitting in racks. Alternatively, racks can be secured with wire and be suspended in the spring water. For incubation of samples, we commonly use 160 mL serum bottles (#223748; Wheaton, Millvillen, NJ, USA) or Balch tubes (Bellco 2048-00150) sealed with butyl rubber stoppers (Bellco 2048-18150) and aluminum seals (Bellco 2048-11020). Plastic test tube racks (e.g., Bel-Art 187450001) can be cut so that serum bottles fit tightly into the rack.

Regulations, environmental stewardship, and safety are also important concerns for hot spring research. Terrestrial hot springs are rare resources and many are protected in public lands, such as Yellowstone National Park. The relevant regulations must be carefully considered into the experimental design and researchers are urged to follow guidelines of minimum impact research, regardless of the protective status of the research site (Spear, 2005).

Conducting research in active geothermal areas is intrinsically hazardous, due the extreme temperatures, dangerous gasses, and the potential instability and volatility of geothermal features (Whittlesey, 1995). Before any experiments are conducted, the research area should be thoroughly checked for potential hazards, including small or inconspicuous geothermal expressions that should be avoided. It is wise for researchers to minimize time spent standing or walking in close proximity (2–3 m) to springs. In addition, close attention should be paid to any changes in the activity of geothermal features, including changes in flow rate, water level, or outgassing, that could indicate an imminent "eruption" or discharge. Care should also be taken to avoid asphyxiation when working near hot springs, particularly in hot springs or fumaroles in depressions and on calm days when dense gasses (e.g., CO_2) can accumulate (Cantrell and Young, 2009; Whittlesey, 1995). In general, it is advised for the research group to make plans for dealing with potential hazards before reaching the study site, to review those plans at the research site, and to carry appropriate first aid gear in case of an emergency. Field research in geothermal areas should never be done alone.

Working with hot spring sediment and water samples under the safest of circumstances can put researchers at risk for burns. For all of the experiments described below, we find that long cuff PVC-coated gloves (e.g., Grease Monkey™, Big Time Products, Rome, GA) work well for brief manipulations at temperatures to at least 85 °C. We suggest considering sturdy rubber boots to prevent burns if researchers break through shallow crusts at hot spring margins or if the site is wet. Researchers should be aware that hot water trapped against the skin is extremely dangerous and be prepared to remove gloves, boots, socks, and other clothing without hesitation if it gets wet.

3. Methods for Measuring Nitrification in Terrestrial Geothermal Habitats

The $^{15}N-NO_3^-$ pool dilution technique can be used to estimate rates of gross nitrification in water and sediment samples from geothermal environments. This method involves the addition of a small amount (typically 5–10% of the total pool size) of highly enriched $^{15}N-NO_3^-$ to samples and monitoring the atom% ^{15}N in the NO_3^- pool over time. Flux into the NO_3^- pool by transformation of nitrogen species present in the sample at natural isotopic abundance (~0.37 atom% ^{15}N) decreases the relative ^{15}N content of the pool (thus resulting in a "pool dilution"), whereas it is assumed that processes involving consumption of NO_3^- have no effect on the isotope ratio of the NO_3^- pool. While this assumption is not strictly true, it is probably valid for short incubations and where labeled $^{15}N-NO_3^-$

has been added to sufficient excess (Barraclough, 1991; Davidson et al., 1991). Knowledge of the change in atom% ^{15}N–NO$_3^-$ over time and the initial and final size of the NO$_3^-$ pool allow calculation of the rate of gross nitrification, that is, production of NO$_3^-$ (Barraclough, 1991). One advantage of this technique is that the samples are amended with only a small amount of the labeled compound; thus, in situ concentrations are only minimally affected.

Below we describe the preparation and field work necessary to implement the ^{15}N–NO$_3^-$ pool dilution technique. We also describe the analysis of processed field samples for measurement of NO$_3^-$ + NO$_2^-$ concentration (abbreviated as NO$_x$), the determination of atom% ^{15}N–NO$_x$ by GC–MS, and calculation of gross nitrification rates from the resulting data.

3.1. Gross nitrification using ^{15}NO$_3^-$ pool dilution approach

3.1.1. Overview

The following is a protocol for using the ^{15}N–NO$_3^-$ pool dilution technique to estimate gross nitrification rates in hot spring water and sediment slurries. The protocol was designed for environments where the sediment is of a character such that slurries can be made, for example, small particle sizes and relatively little cohesion between sediment particles or aggregates. This protocol estimates gross NH$_3$ oxidation by monitoring the size and isotopic composition of the NO$_x$ pool, rather than distinguishing NO$_2^-$ and NO$_3^-$ individually. Oxidation of NH$_3$ to NO$_2^-$ will dilute the ^{15}N composition of the NO$_x$ pool, whereas oxidation of NO$_2^-$ to NO$_3^-$ will have no effect on either the concentration or the isotopic composition of NO$_x$. Estimating NO$_2^-$ oxidation is possible in this assay, if the concentration and isotopic composition of NO$_3^-$ alone is monitored as well. Briefly, aliquots of spring water or a water–sediment slurry are distributed in bottles containing a predetermined amount of 98+ atom% ^{15}N–NO$_3^-$ such that the resulting NO$_3^-$ pool is 5–10 atom% ^{15}N. Sealed bottles (with ~5:1 headspace: sample volume) are incubated in the field at ambient temperatures by submersing them in spring water and the bottles are sampled over time. At each time point, samples are cooled and shaken in 1M KCl to extract NO$_x$. After extraction, samples are centrifuged, the supernatant passed through a 0.2 μm filter, and the filtrate is frozen or stored at 4 °C until analysis within 28 days (US-EPA, 1993a). Sediment samples are saved for determination of sediment dry weight. In the lab, filtrate is assayed for [NO$_x$] using automated colorimetry. ^{15}N–NO$_x$ is determined by isotope ratio mass spectrometry (IRMS) of the ammonified NO$_x$ pool or by coupled gas chromatography–isotope ratio mass spectrometry (GC–IRMS) of N$_2$O generated from the NO$_x$ pool by *Pseudomonas aureofaciens* (Sigman et al., 2001).

3.1.2. Preparation

As with all field work, it is recommended that as much preparation as possible be done prior to travel to the field site. For incubation of samples, we have used 160 mL serum bottles sealed with butyl rubber stoppers to prevent loss of liquid by evaporation, which can be excessive at high temperatures even over a short incubation time. Individual researchers may choose to use different incubation vessels depending on available resources and convenience. Incubation bottles and stoppers, as well as a large vessel for obtaining water and sediment slurry samples, should be acid washed, rinsed thoroughly, and sterilized by autoclaving. While only two time points are necessary for rate calculations, we typically prepare enough bottles for four time points (0, 3, 6, and 12–24 h incubations) with suitable replication at each time point appropriate for the experimental design. An alternative incubation approach is to prepare a single, larger flask, per replicate, and to remove subsamples from this flask over time. This has the advantage of reducing variability driven by differences between flasks, but also involves disturbing the entire sample when subsamples are removed. In our work to date, the two approaches have yielded quantitatively similar results. Using either approach, multiple time points are recommended, as they allow for more accurate determination of rates and are useful for determining whether rates are constant throughout the incubation or only for a subset of time points. A concentrated stock of 98+ atom% $^{15}N–KNO_3^-$ should be prepared, sterilized by passage through a 0.2 μm filter, and diluted to 1 mM or some other appropriate concentration such that attaining 5–10 atom% $^{15}N–NO_3^-$ requires addition of a small volume of the stock relative to the sample volume (e.g., 0.2% or less). If desired, a similarly concentrated solution of NH_4Cl (e.g., 500 mM) can be prepared for amendment of some samples with NH_4^+. For extraction of NO_3^- from samples after incubation, a 3 M solution of KCl should be made in a quantity of at least half the total volume of all samples to be processed. Additional equipment to bring to the field includes reagents and equipment for determination of NH_4^+, NO_3^-, and NO_2^- levels (see below); pipetmen and sterile pipet tips appropriate for adding microliter volumes; 25 mL pipets and a pipet bulb; racks suitable for suspending the sample bottles in the spring during sample preparation and incubation; aluminum crimps and a crimping tool for sealing stoppered bottles; 60 mL syringes and 23 G needles for periodically exchanging sample bottle headspace during incubation; sterile 50 mL polypropylene tubes for mixing sacrificed samples with KCl and a centrifuge compatible with these tubes to clarify sediment slurry samples; an orbital or rotary shaker for extraction of NO_3^- with KCl with racks compatible with the 50 mL tubes; syringes and 0.2 μm filters for sterilizing extracted samples; additional 50 mL polypropylene tubes for collecting the filtered, extracted samples; and wet and dry ice for cooling samples before extraction and freezing filtered samples, respectively.

3.1.3. Field work

The amount of ^{15}N–KNO$_3^-$ to be added to attain the proper atom% ^{15}N–NO$_3^-$ during incubation should be determined before sample preparation by quantifying the [NO$_x$] in the environment to be sampled, or by knowledge of concentrations typically found for the study system. Bulk spring water and water/sediment slurries should be considered separately, as concentrations of NO$_3^-$ and NO$_2^-$ are likely to differ between these habitats. It is also useful to determine the [NH$_4^+$], as this potentially serves as the primary "diluent" to the NO$_x$ pool. Knowledge of NH$_4^+$ levels will help inform the experimenter as to whether to amend some samples with NH$_4^+$, and how much should be added. In the field, we routinely use a Smart2 handheld spectrophotometer and colorimetric kits for determination of NH$_4^+$ (#3642-SC), NO$_2^-$ (#3650-SC), and NO$_3^-$ (#3649-SC; all products of LaMotte, Chestertown, MD, USA). Care should be taken to correct the calculated [NO$_3^-$] for interference by NO$_2^-$ as suggested by the manufacturer and this [NO$_3^-$] protocol may be problematic in environments in which [NO$_2^-$] > [NO$_3^-$] (Hedlund and Dodsworth, unpublished observation). Some springs may require that water is filtered prior to analysis (e.g., 0.2 µm Supor polysulfone filters (Pall)); samples for these three analyses should be assayed immediately after cooling to $\sim$25 °C. In environments where [NO$_x$] is at or below the reliable detection limit (<1 µM), we suggest either adding the ^{15}N–KNO$_3^-$ to 0.5 µM or diluting the ^{15}N–KNO$_3^-$ with unlabeled KNO$_3^-$ to 10 atom% ^{15}N and adding this solution to 10 µM total KNO$_3^-$. It is convenient to add the appropriate amounts of ^{15}N–KNO$_3^-$ and NH$_4$Cl to bottles prior to addition of samples.

Once sample bottles have been prepared with amendments, place them in racks in the spring water to bring the incubation bottles to ambient spring temperature before the incubation begins. Collect enough spring water for all samples in a large, sterile vessel, such as a 2 L glass flask. Add 25 mL spring water or water/sediment slurry to sample bottles by pipet. In cases where both spring water and sediment slurry will be used, collect the spring water and add it to bottles prior to preparation of the sediment slurry. This avoids unnecessary disturbance of the sediment and the unwanted collection of suspended sediment in samples intended for spring-water-only incubations. After addition of spring water to appropriate bottles, collect sediment (e.g., from the top $\sim$1 cm of the sediment–water interface) and add to the vessel containing spring water until a desired amount of sediment is obtained. If later calculation to surface area is desired, several shallow sediment cores with known diameter can be pooled or attention can be paid to the surface area sampled. When using this protocol, we typically make the slurry at a 4:1 volume ratio of water:sediment. Seal and incubate the bottles in the racks suspended in spring water. Immediately remove bottles corresponding to the initial time point and cool on wet ice to ambient temperature. For bottles that are incubated longer than 2 h, potential O$_2$ limitation can be avoided by exchanging the headspace at regular intervals by flushing with 120 mL atmosphere using a 60 mL syringe and needles, or by briefly unsealing and then resealing the bottles or incubation vessels.

At each time point, remove a replicate set of bottles (or sample from the single, large incubation) and cool them on wet ice to ambient temperature (~25 °C). Decant the contents into 50 mL polypropylene tubes containing 12.5 mL of 3 M KCl. Seal the tubes, transfer them to a rack on a rotary or orbital shaker, and extract for 1 h with shaking at 120 rpm. If an electrical outlet is not available, the tubes may be shaken intermittently by hand for 15 min–1 h. After extraction, centrifuge the tubes containing sediment slurry samples for 10 min at $1500 \times g$, and use the supernatant fraction for filtration. Alternatively, pass the sediment slurry samples through filter paper (e.g., Whatman no. 42) in a funnel if it is not practical to bring or power a centrifuge at the field site. Pass all samples through a 0.2 μm filter and collect the filtrate in 50 mL polypropylene tubes. Samples may either be frozen on dry ice or acidified to pH < 2 with H_2SO_4 and stored at 4 °C for up to 28 days before analysis (US-EPA, 1993a). Tubes containing sediment pellets should be saved and the sediment dry weight determined. Wash the sediment with water to remove excess KCl (two repetitions of resuspension in 45 mL water, followed by centrifugation as above) and dry to constant weight to determine the mass of sediment in the incubation.

3.2. Quantification and determination of atom% ^{15}N–NO_x and calculation of gross denitrification rates

3.2.1. Quantification of NO_x and NH_4^+

Samples can be analyzed for NO_x using cadmium reduction and automated colorimetry (APHA, 1992; US-EPA, 1993a). In this method, NO_3^- is reduced to nitrite by cadmium reduction; nitrite is then determined by diazotizing with sulfanilamide and coupling with N-(1-naphthyl)-ethylenediamine dihydrochloride, which forms a dye that can be measured colorimetrically. Nitrite and NO_3^- can be determined separately by first conducting the procedure with and subsequently without the cadmium reduction step.

The Berthelot reaction (Searle, 1984), or modifications thereof (e.g., Rhine *et al.*, 1998), can be used for analysis of NH_4^+, if required. In the classical approach, NH_4^+ in the sample reacts with phenol and hypochlorite, producing indophenol blue in proportion to the NH_4^+ concentration, which is measured colorimetrically (Searle, 1984). NH_4^+ can also be measured using semiautomated colorimetry (US-EPA, 1993b).

For analysis of NO_3^- plus NO_2^-, and for NH_4^+, samples can be preserved for up to 28 days by acidification to pH < 2 with sulfuric acid and storage at 4 °C. For separate analysis of NO_2^- and NO_3^-, store at 4 °C, do not acidify, and analyze within 24 h. Alternatively, samples can be frozen for long-term storage (Avanzino and Kenedy, 1993; Bremner and Keeney, 1966). To avoid interference with the cadmium column, samples should not be preserved with mercuric chloride.

3.2.2. Determination of atom% $^{15}N-NO_x$

There are several procedures used for preparing aqueous NO_x samples for isotopic analysis. These procedures either convert dissolved NO_x into a solid or gaseous phase, in a form suitable for analysis by IRMS. The diffusion technique concentrates NO_x-N (as NH_4^+) onto a glass fiber filter disk, which can be analyzed by Dumas' combustion and subsequent GC–IRMS (Brooks et al., 1989; Stark and Hart, 1996). The anion-exchange method traps NO_x on anion-exchange resin column followed by precipitation as silver nitrate, also suitable for analysis by combustion GC–IRMS (Chang et al., 1999; Silva et al., 2000). Two other approaches convert aqueous NO_x to N_2O, either chemically (McIlvin and Altabet, 2005) or biologically (Sigman et al., 2001), and the N_2O is analyzed by GC–IRMS. Here, we briefly describe the diffusion and denitrifier techniques for analyzing the ^{15}N composition of NO_x.

3.2.3. Diffusion technique for $^{15}N-NO_x$ analysis

A volume of aqueous sample ideally containing 20 µg NO_x-N is placed in a clean (acid-washed) plastic container with an air-tight removable top. The first step of the procedure removes NH_4^+ from the sample, so the top of the container is removed to allow gas exchange (the NH_4^+ is removed as NH_3. Note: if ^{15}N determination of NH_4^+-N is also desired, this step can be modified to include a Teflon sandwich enclosed acid trap and incubation with sealed top, trapping the NH_4^+-N as described below for NO_x-N). Enough powdered magnesium oxide is added to saturate the solution (approximately 10 g L^{-1}), which increases the pH to around 9 (check), favoring phase change of NH_4^+ to NH_3, with subsequent volatilization and loss of NH_3 from the solution. The samples are incubated with moderate rotary shaking for 5–7 days, although modified procedures can reduce the required time period (Chen and Dittert, 2008). A few samples may be checked at this point to ensure NH_4^+ concentrations in the samples are below detection limits. During the incubation, prepare acid traps. First, cut Whatman glass–fiber filters into small disks, about 8 mm diameter, the size of a standard hole punch. Then, add 20 µL of 2.5 M $KHSO_4$ to the filter disk. Enclose the acidified disk between two layers of polytetrafluoroethylene (PTFE) tape, creating a seal around the disk by pressing the two layers of PTFE tape together using a glass test tube. To avoid accumulating ambient NH_4^+, the PTFE-enclosed acid traps should be stored in a sealed desiccator containing an open vial of H_2SO_4 or some other suitable acid. Avoid long-term storage of the acid traps.

Once the PTFE-enclosed traps have been prepared, and the preincubation of the samples is complete such that NH_4^+ has been removed, place one PTFE-enclosed acid trap into each sample container, to each add a scoop of finely ground Devarda's alloy (about 10 g L^{-1}; this metal alloy reduces both NO_2^- and NO_3^- to NH_3), and tighten the lid to ensure an air-tight seal. The sealed samples can be placed on a shaker–incubator at elevated temperature to promote rapid reduction to NH_3, which readily diffuses through the PTFE tape to the acidified filter disk, where it is trapped as NH_4^+. After 5–7 days, remove the PTFE-trap packets from the samples and place them in a sealed desiccator (with liquid acid trap and desiccant) for at least 24 h, or until analysis. On the day the samples are to be analyzed on the mass spectrometer, open the PTFE packets and retrieve the acidified filter disk using clean forceps, placing the disk into tin capsules suitable for Dumas' combustion coupled to GC–IRMS as described (Brooks et al., 1989; Stark and Hart, 1996). The disks are acidic and will corrode the tin if left too long, so analyze the samples immediately after they have been sealed in the tin capsules (the same day is optimal). If storage is necessary, store them in the PTFE packets in the desiccator.

3.2.4. Denitrifier method for ^{15}N–NO_x analysis

P. aureofaciens is a facultative denitrifying organism, but lacks the enzyme nitrous oxide reductase, the enzyme that converts N_2O to N_2 during denitrification, so the reaction stops at N_2O. Because of the lower background of N_2O in the atmosphere compared to N_2, N_2O is a more convenient analyte for IRMS. In this method, *P. aureofaciens* is used to convert NO_x in the sample to N_2O, which is then analyzed by IRMS (Casciotti et al., 2002; Révész and Casciotti, 2007; Sigman et al., 2001). The major advantages of this method are: (1) it is fairly rapid, and, compared to the diffusion method, involves fewer steps, and (2) it can be used to measure simultaneously the isotopic composition of both N and O. The disadvantage is that the method depends on a biological enzyme system, and thus involves keeping a pure culture of *P. aureofaciens*.

P. aureofaciens cultures are grown in tryptic soy broth amended with 10 m*M* NO_3^- and 15 m*M* NH_4^+. After 4–7 days of growth, cultures are centrifuged and resuspended in NO_3^--free medium to achieve a 10-fold concentration. Eight milliliter of the concentrated suspension is added to each 20 mL vial, sealed with gas-tight septa. Each vial is purged with He for an hour. This flushing procedure is designed to promote anaerobic conditions, and to remove any residual NO_x (which will be converted to N_2O) from the broth. The flushing procedure is conveniently done using an autosampler such as the Thermofinnigan CombiPAL (Thermo Fisher Scientific, Wlatham, MA, USA), so that the purging process can be automated.

NO_x concentrations in the samples must be known, so that the appropriate volume of sample can be added to each 20 mL vial. The sample is introduced to the vial through the septum using syringe and needle.

Sufficient sample is added to obtain enough N for analysis, which for some laboratories is as little as 10 nmol (Révész and Casciotti, 2007). In our experience, higher amounts are optimal (50 nmol).

Several drops of antifoaming agent (e.g., Antifoam B Emulsion, Dow Corning, Midland, MI, USA) are added to each vial to reduce bubble formation during the reaction. The vials are allowed to incubate for 8 h, during which time NO_3^- is converted completely to N_2O. After the 8-h period, 0.1 mL of 10 N NaOH is added to each vial to stop the reaction, and to absorb CO_2, which can interfere with N_2O analysis (since CO_2 has the same masses as N_2O: 44, 45, and 46). The samples are then placed on an autosampler tray for preconcentration prior to isotope analysis. In this step, headspace is withdrawn from each vial, passed through water and CO_2 traps (Nafion drier and Ascarite or equivalent, respectively), a cryogenic purification trap (liquid N_2), a GC column, and into an open split, which interfaces with the IRMS. Typical preconcentration systems are the Thermo Scientific GasBench, SerCon Cryo-Prep, and the Isoprime Trace Gas preconcentrator. The mass IRMS is equipped with a universal triple collector suitable for masses 44, 45, and 46 (e.g., Thermo Scientific DeltaV, SerCon GEO 20-20, or IsoPrime100 IRMS). Standards of known $\delta^{15}N$ and $\delta^{18}O$ must be included in the autosampler tray, such as USGS32, USGS 34, USGS 35, and IAEA NO3. Mass ratios of 45:44 and 46:44 distinguish $\delta^{15}N$ and $\delta^{18}O$ signatures, respectively.

3.2.5. Calculation of gross nitrification rates

Once the size of the NO_x pool and the atom% ^{15}N in this pool for each time point are known, the following equations, modified from Barraclough (1991), can be used to calculate gross nitrification rates for either a constant or changing NO_x pool size. In cases where the NO_x pool size is changing over time, the gross nitrification rate, n, in units of moles per hour, can be calculated as follows:

$$n = -\theta \left(\ln\%^{15}N_t - \ln\%^{15}N_0 \right) / \left(\ln(1 + \theta \times t/N_0) \right)$$

where N_0 is the initial NO_x pool size in moles, θ is the rate of change in the NO_x pool in moles per hour, $\%^{15}N_0$ and $\%^{15}N_t$ are the atom% ^{15}N–NO_x at the initial time point and at time t (in hours), respectively, and ln is the natural logarithm. If data from multiple time points are available, the slope calculated from a plot of ln (atom% ^{15}N in excess of natural abundance) versus time (in hours) can be used to replace (ln $\%^{15}N_t$ − ln $\%^{15}N_0$) in the above equation, where $t = 1$. In cases where the NO_x pool size is constant, the above equation is invalid and the following one can be used:

$$n = -N_0 \left(\ln\%^{15}N_t - \ln\%^{15}N_0 \right) / t$$

where the symbols have the same meaning and the slope, if multiple time points are used, can be inserted as described above. Calculations can further be normalized per gram dry weight of sediment or per milliliter of spring water, depending on whether the incubations included a sediment slurry or spring water only. In cases where [NO$_x$] cannot be reliably determined, the pool size can be estimated at the initial time point using the atom% ^{15}N–(NO$_x$) determined for this time point and the known amount of labeled ^{15}N–NO$_3^-$ added.

4. Methods for Measuring Denitrification in Terrestrial Geothermal Habitats

Many approaches for measurement of denitrification *in situ* are available to microbiologists studying geothermal habitats (Mosier and Klemedtsson, 1994; Steingruber *et al.*, 2001; Tiedje, 1994; Tiedje *et al.*, 1989). Here, we describe two of the most common, the acetylene block method and the ^{15}NO$_3^-$-tracer approach. Together, these two approaches should allow the researcher to bound rates of *in situ* denitrification and quantitatively distinguish the fates of ^{15}NO$_3^-$ respired in geothermal ecosystems.

4.1. Acetylene block method

The acetylene block is a simple, inexpensive, and effective approach to measure denitrification in the field that is based on the observation that acetylene inhibits the reduction of N$_2$O to N$_2$, causing an accumulation of N$_2$O that can be measured by GC (Balderson *et al.*, 1976; Federova *et al.*, 1967; Yoshinari and Knowles, 1976). This method is extremely sensitive at low denitrification rates since N$_2$O is present at a low atmospheric concentration, about 310 ppb, as compared with N$_2$ at about 78%. Limitations of the approach have been described in detail elsewhere (Tiedje *et al.*, 1989) but the most significant may be the inhibition of nitrification, which can lead to underestimation of denitrification rates in habitats with low NO$_x$ concentration where denitrification is tightly coupled to nitrification. Therefore, we recommend making measurements with and without NO$_3^-$ amendments. The latter measurement addresses the denitrification rate based on the *in situ* NO$_x$ pool in the sediment porewater in addition to substrate diffusing into sediments from the overlying water column but will miss any contribution of NO$_x$ from nitrification. If sufficient substrate is added, the former measurement with excess NO$_3^-$ will provide an upper bound on denitrification, taking into account all sources of NO$_x$.

Coordinated experiments with either NO_3^- or NO_2^- can distinguish the relative importance of the two electron acceptors in denitrification.

A number of specific protocols for application of the acetylene block to soils or sediments have been used including the static core protocol, the gas-phase recirculation core protocol, the closed chamber protocol, sediment slurry protocols, and aqueous flow-through protocols (Mosier and Klemedtsson, 1994; Tiedje, 1994; Tiedje et al., 1989). We have only used the sediment slurry approach, which is the simplest approach commonly used for aquatic sediments (Miller et al., 1986; Oremland et al., 1984; Sorensen, 1978). This approach may either overestimate or underestimate denitrification rates in hot spring sediments, depending on the possible stimulatory effects of sediment disruption by relieving diffusion limitations in stratified sediments or on possible inhibitory effects due to oxygen exposure during sample processing. However, the strong linear N_2O production rates without lag we have observed suggest the latter is not a major problem (Dodsworth and Hedlund, unpublished). Thus, care should be taken to minimize manipulations in the field in order to maximize efficient use of field time, mistakes due to time pressures, and potential chemical and biological contamination.

4.1.1. Overview

Samples from the sediment/water interface are collected with a coring device and quickly extruded under a stream of N_2 into 28 mL Balch tubes containing 5 mL of a preheated anaerobic liquid phase (e.g., filtered spring water). The headspace is flushed with N_2 for 5 min, and tubes are sealed with butyl rubber stoppers secured with aluminum crimp caps. Freshly prepared acetylene is added to a volume of 10% and the tubes are shaken for 20 s. Tubes are incubated in the hot spring under aluminum foil and sampled destructively by cooling to ~ 25 °C, shaking vigorously, and removing a sample into an evacuated bottle. In the lab, N_2O is quantified by comparison with a standard curve by gas chromatography using a ^{63}Ni electron capture detector (GC-ECD).

As with nitrification measurements, it is best to establish linearity of N_2O production to determine the appropriate incubation time frame before doing complex experiments. Since NO_3^-/NO_2^- concentrations in geothermal springs are typically low (<200 μM; Holloway et al., 2004; Shock et al., 2010), incubations of a few hours are generally recommended over long-term experiments, particularly for experiments without NO_x amendment (e.g., 0, 2, 4, 8, and 16 h incubations).

4.1.2. Preparation

Twenty eight milliliters Balch tubes containing an appropriate liquid phase should be set up prior to the trip. We have used spring water collected from a previous trip by ultrafiltration (30 kDa molecular weight cutoff) as a medium. The ultrafiltrate is dosed with amendments, if desired (e.g., 30–1000 μM NO_3^-), sparged for 30 min with N_2 in a bottle with minimal headspace, and transferred to an anaerobic chamber (e.g., Coy Type B, Grass Lake, MI). In the chamber, ultrafiltrate is added to Balch tubes (BellCo Glass, Vineland, NJ) at 5 mL per tube. Following removal from the chamber, hydrogen and other gasses are removed by three cycles of vacuum and gassing (to 1 atm of overpressure) with N_2 and pressurized to 0.5 atm of overpressure. The overpressure will enable detection of major leaks due to punctured stoppers and will ease stopper removal in the field; however, it should be recognized that pressurized tubes are a hazard and safety glasses should be used during transportation and manipulation. Finally, the tubes are autoclaved for 30 min at 121 °C. The pH of the fully prepared spring water should be checked to ensure it is close to *in situ* pH. The concentration of NO_3^- and NO_2^- should also be checked by ion chromatography or colorimetric assays, as described above. In our experience, all NO_3^-, NO_2^-, and NH_4^+ disappears in the ultrafiltrate within months of collection, presumably due to microbial consumption. Ideally, NO_3^- and NO_2^- in "unamended" ultrafiltrate should be amended to match the substrate concentrations in the hot spring water column during the time of the field experiments by adding aliquots of concentrated stock solutions following NO_3^- and NO_2^- field measurements.

As an alternative to the use of ultrafiltrate, a synthetic mineral salts medium can be prepared to simulate the hot spring water or freshly collected hot spring water can be used for the aqueous phase. However, if the latter is used, researchers should consider either sparging the spring water (e.g., with N_2) or replacing the sediment slurry approach with a static core approach because contact of the sediment microbial community with dissolved oxygen, if present, may inhibit denitrification.

Preparations should also be made for work with gasses in the field. Bottles for fresh acetylene can be prepared in the lab by adding $\sim$1 g of calcium carbide (CaC_2, Sigma 270296) to serum bottles (e.g., Wheaton 160 mL bottles, 223748), which are subsequently stoppered, sealed, and evacuated. Acetylene is later prepared in the field by adding 1–5 mL of distilled water. Care should be taken not to add water too fast because this could explode the serum bottle. Safety glasses are required. Acetylene produced from CaC_2 contains traces of H_2, CH_4, C_2H_4, and PH_3 (Hyman and Arp, 1987), though these contaminants do not appear to influence denitrification (Tiedje *et al.*, 1989). For application of Hungate technique in the field, we use a portable field gassing manifold with 25 G needles and bent 18 G needles for gassing probes.

Other important equipment includes aluminum seals, a crimper and decrimper, cutoff and autoclaved syringes for sediment sampling, sterile spatulas for manipulating cored sediments, needles and syringes (5 or 10 mL), vacutainer needles for sampling gases, distilled water, evacuated sample collection receptacles (e.g., Wheaton 10 mL bottles, 223739), extra racks for *in situ* incubations, aluminum foil, and sturdy gloves for working in the hot spring.

4.1.3. Field work

Measure NO_3^- and NO_2^- concentrations immediately after cooling a small spring water sample to $\sim$25 °C and make amendments if needed. Decrimp a tube and release overpressure while minimizing oxygen exposure by constant flushing with N_2 using a cannula, following the techniques of Hungate (Hungate, 1950, 1969). Remove a sediment core from the shallows of the spring and extrude the top $\sim$0.5 cm (or other strata if desired) directly into Balch tubes under a stream of N_2 using the techniques of Hungate. Replace the stopper, crimp, and flush for 5 min with N_2, using a needle in the stopper to allow gas to flush out. Release overpressure. Dose the tube with a 1/10 volume of acetylene (2.5 mL), shake the tube 20 s to disperse sediment and encourage acetylene solubilization, and return to the spring for incubation. Alternatively, a single anaerobic sediment slurry can be prepared for addition to all tubes.

Following the incubation period, Balch tubes are removed from the spring and cooled for $\sim$5 min in ambient air and then in a $\sim$25 °C water bath. The actual temperature should be noted and used to select the appropriate Bunsen coefficient, as described below. The tubes are then shaken vigorously for 30 s once every 3 min for three cycles ($\sim$10 min) to equilibrate N_2O between the aqueous phase and the headspace. A vacutainer is used to allow gas to flow to the evacuated collection vial, and 10 mL of distilled water is added to the Balch tube to alleviate the vacuum. Pressurize the collection vials to 1 atm overpressure with N_2. Collection vials receiving any liquid should be marked or discarded because microbial activity may influence N_2O concentrations during transport and storage. Transport the Balch tubes intact to the lab for determination of sediment dry weight (e.g., after filtration onto Whatman filter paper) after drying to constant weight.

4.1.4. Analysis and rate calculations

N_2O is measured in the lab as soon as possible by GC-ECD in comparison with purified standards. Here, we describe a protocol using a GC-2014 Nitrous Oxide Analyzer (Shimadzu, Moorpark, CA). If using a different GC-ECD system, we refer the reader to other descriptions of systems that have been used for N_2O measurements (Loftfield *et al.*, 1997; Mosier and Klemedtsson, 1994; Mosier and Mack, 1980). The volume needed to flush the sample loop ($\sim$5 mL with Shimadzu GC-2014) can be decreased with a short length of stainless steel tubing ($\sim$0.25 mL) to reduce the volume of sample needed to 2 mL, allowing multiple injections per sample.

The standard factory-ready protocol uses high-quality (99.999% purity) N_2 at 25 mL min^{-1} as the carrier gas. P5 (argon/methane, 95/5 v/v) at 2.5 mL min^{-1} can be used as the make-up gas. One milliliter of sample from the sample loop (injector temperature 250 °C) is injected serially onto four columns, all at 80 °C: Haysep T (80/100 mesh; 1 m), Haysep D (80/100 mesh; 4 m), Haysep N (80/100 mesh; 1.5 m), and Shimalite Q (80/100 mesh; 0.4 m). Purges are pneumatically controlled to prevent CO_2 or water from interfering with the ECD. The ECD is programmed at 325 °C, 2 nA, and a 200 ms time constant. This protocol yields a linear standard curve for N_2O concentrations up to 100 ppm (Mosier and Klemedtsson, 1994). At concentrations > 100 ppm, we have found that a nonlinear equation can be applied. Alternatively, dilutions can be made prior to analysis or N_2O can be quantified by GC using a thermal conductivity detector (TCD; Ryden et al., 1987). The amount of N_2O produced in any sample, X, can be calculated by the following equation:

$$X = ([N_2O])(2)\left[(V_g + (V_{aq})(\alpha)\right]/(24.5)(W)$$

where [N_2O] is the concentration of N_2O in the 10 mL collection vials (in parts per million), obtained by comparison with a standard curve; 2 accounts for the 1:1 dilution of the sample by pressurizing to 2 atm with N_2; V_g and V_{aq} are volumes (in liters) of the gas and aqueous phase in the Balch tubes, respectively; 24.5 is the liters of gas per mole using the ideal gas law at 25 °C (the constant should be adjusted if tubes were equilibrated at a different temperature); W is the dry weight of the sediment sample (in grams); and α is the Bunsen absorption coefficient of N_2O at the temperature at which the headspace sample was removed from the Balch tube. The value of α is as follows for the following temperatures: 5 °C, 1.06; 10 °C, 0.88; 15 °C, 0.74; 20 °C, 0.63; 25 °C, 0.54; 30 °C, 0.47; and 35 °C, 0.41 (Mosier and Klemedtsson, 1994; Tiedje, 1982). The resulting units are μmol N_2O/g dry weight of sediment. For experiments with spring water, replace W with the volume of the water in the Balch tube in milliliter, resulting in units of μmol N_2O/mL. To assess linearity, the amount of N_2O–N can be plotted versus time and used to calculate a slope and associated statistics.

4.2. $^{15}NO_3^-$ tracer approach

4.2.1. Overview

Hot spring water or anaerobic sediment slurry is amended with 98+ atom% ^{15}N–NO_3^- so that the resulting NO_3^- pool is 5–10 atom% ^{15}N and incubated in stoppered bottles *in situ* with suitable replication. Replicate bottles are sacrificed by cooling to ambient temperature (~25 °C) and shaking to equilibrate N_2 and N_2O. Gas samples are collected for

GC–IRMS to determine the amount of N_2 and N_2O and ^{15}N enrichment, and sediment slurries are extracted with KCl and used for IRMS to determine the amount of NO_x and NH_4^+ pools and ^{15}N enrichment. Changes in the pool flux and the flow of ^{15}N into different N pools can be plotted versus time to calculate rates of different NO_x reduction processes.

4.2.2. Preparation
Preparation is similar to the approaches described above. Necessary items include: incubation bottles (e.g., 160 mL serum bottles), incubation baskets, stoppers, aluminum seals, crimper and decrimper, portable gassing station and N_2 tank, 60 mL syringes and needles, evacuated gas-tight gas collection vials, concentrated stocks of $N-NO_3^-$ and $^{15}N-NO_3^-$, a 3 M solution of KCl of at least half the total volume of samples to be processed, 50 mL polypropylene tubes and a compatible centrifuge, an orbital or rotary shaker with racks compatible with the 50 mL tubes, syringes and 0.2 μm filters, and wet and dry ice for cooling samples before extraction and freezing filtered samples, respectively.

4.2.3. Field work
As described for the acetylene block above, we suggest using sterile, anaerobic ultrafiltrate from the spring of interest for a medium, in which case serum bottles can be amended to the *in situ* $[NO_3^-]$ measured at the time of the experiment at 5–10 atom% $^{15}N-NO_3^-$. Alternatively, an artificial medium or freshly collected spring water can be used and amended to 5–10 atom% $^{15}N-NO_3^-$, with the caveats discussed above. We suggest a total aqueous volume of at least 60 mL. The prepared serum bottles are preincubated in a wire basket *in situ*. Either freshly collected sediment cores or a small amount of sediment in an anaerobic sediment slurry is added to the serum bottles. If the former is used, the bottle should be flushed for 5 min with N_2 to achieve a headspace of 1 atm N_2. The bottles are incubated for the desired time and removed from the spring to cool to ambient temperature ($\sim$25 °C). Bottles are pressurized to 2 atm N_2 and shaken vigorously for 30 s once every 3 min for three cycles ($\sim$10 min) to equilibrate N_2O and N_2 between the aqueous phase and the headspace. Safety glasses are required. The overpressure in the headspace is removed using a 60 mL syringe to fill evacuated gas sample vials. The sediment slurry is poured into three 50 mL polypropylene tubes already containing 20 mL of 3 M KCl and stored on ice until extraction (a few hours). KCl extraction is carried out and extracted samples are filtered and stored on dry ice, as described in Section 3.1. Sediments should be washed, dried, and weighed as described in Section 3.1 for normalization to sediment dry weight.

4.2.4. Analysis and calculations

N_2O concentrations are measured with a gas chromatograph equipped with an electron capture detector (Section 4.1). The ^{15}N composition of N_2O, if desired, is measured using the trace-gas preconcentration and IRMS procedure described in Section 3.2. NO_3^- concentrations are determined colorimetrically, and ^{15}N content of NO_3^- can be determined using the diffusion or denitrifier procedures described in Section 3.2. N_2 concentrations and ^{15}N composition can be measured by injecting headspace samples into a GC coupled to an IRMS equipped with a universal triple collector (identical to those described above, Section 3.2). Masses 30, 29, and 28 are used to determine ^{15}N–N_2.

The rate of denitrification is estimated as the accumulation of N_2 and N_2O that reflects the ^{15}N composition of the NO_3^- pool during the incubation:

$$\text{Denitrification} = ((\text{atom\%}^{15}N - N_2) \times [N_2] \times V)/(S \times \text{atom\%}^{15}N - NO_3^-)$$

where atom% ^{15}N–N_2 is the ^{15}N composition of N_2 gas, $[N_2]$ is the concentration of N_2 gas, atom% ^{15}N–NO_3 is the ^{15}N composition of the NO_3^- pool, expressed as atom% (either measured or estimated based on the amount of ^{15}N–NO_3^- added and the initial ambient concentrations), V is the volume of the headspace, and S is the mass of sediment (or surface area of sediment, or volume of water used in the incubation, depending on how results are to be expressed).

5. DETECTION AND QUANTIFICATION OF POTENTIAL BIOMARKERS FOR THERMOPHILIC AOA AND DENITRIFYING *THERMUS THERMOPHILUS*

Analysis of genes or transcripts involved in microbial processes of interest, so-called functional genes, can serve as valuable biomarkers to link community activity measurements and microorganisms possibly carrying out the activities. Various studies involving the authors have used conserved or degenerate primers to amplify and sequence *amoA* genes or transcripts from large numbers of hot springs using PCR or quantitative real-time PCR (qPCR), providing insights into relationships between diversity or quantity of functional groups and geochemistry or geographical location (de la Torre et al., 2008; Jiang et al., 2010; Reigstad et al., 2008; Zhang et al., 2008). More recently, we have used knowledge of the microbial community composition in well-studied springs to combine measurements of nitrification and denitrification with enumeration of specific populations using specific qPCR primers (Dodsworth et al., unpublished data). Below, we describe protocols for sample collection and nucleic acid extraction and primers and procedures for diversity studies of *amoA* genes as well as specific primers and protocols for qPCR for *amoA* and 16S rRNA gene of "*Candidatus* Nitrosocaldus yellowstonii" and the nitrate reductase large subunit, *narG*, of denitrifying *T. thermophilus*.

5.1. Sample collection and nucleic acid extraction

Sediment is collected using a sterile coring device, polypropylene tube, or other sterile sampling device. For collection of planktonic cells, bulk water can be filtered through 0.1 or 0.2 µm Supor polysulfone, 25 mm diameter filters (Pall) contained in presterilized filter cartridges (Pall). Prior to freezing, individual filters are transferred to 1.5 mL polypropylene tubes containing 0.1 mL of TE (10 mM Tris pH 8, 1 mM EDTA). To maximize both the quantity and quality of extracted nucleic acids and other biomolecules, it is recommended that samples be frozen as quickly as possible after collection in the field. Flash freezing by immersion in liquid N_2 is ideal. Alternatively, samples may be frozen on crushed dry ice or by immersion in an ethanol bath cooled by dry ice. Frozen samples should be stored and transported on dry ice and transferred to an ultracold freezer (-80 °C) for storage in the laboratory until processing. If transport or maintenance of liquid nitrogen or dry ice is not practical, addition of RNA later (Applied Biosystems/Ambion, Austin, TX, USA) or a sucrose lysis buffer (SLB; 20 mM EDTA, 200 mM NaCl, 0.75 M sucrose, 50 mM Tris–HCl, pH 9) may help prevent nuclease activity for subsequent extraction of RNA or DNA, respectively (Grant et al., 2006; Hall et al., 2008). As a general rule, it can be useful to freeze multiple identical samples (e.g., 1.5 mL polypropylene tubes each with 1 cc of sediment or multiple filters) as this allows rapid freezing and will allow for more efficient use of the samples, either for the extraction of different analytes or for their dissemination to collaborators. In cases where biomarkers are to be analyzed alongside community activity measurements, the same fraction should be collected to enable direct comparisons.

Nucleic acid extraction is performed on samples immediately after their removal from storage at -80 °C. There are a variety of protocols and commercially available kits that can be used to extract nucleic acids from environmental samples (Purdy, 2005), including geothermal environments (Herrera and Cockell, 2007; Mitchell and Takacs-Vesbach, 2008). For extractions of DNA and RNA, we and others (Costa et al., 2009; de la Torre et al., 2008; Jiang et al., 2009, 2010; Reigstad et al., 2008; Vick et al., 2010) have had success with the FastDNA® SPIN Kit for Soil and FastRNA® Pro Soil-Direct Kit, respectively (MP Biomedicals, Solon, OH,

USA; formerly made by Bio101 and Q-biogene). Both of these commercially available kits utilize a benchtop FastPrep Instrument (MP Biomedicals; formerly made by Bio101/ThermoSavant and Q-biogene) for cell disruption and sample homogenization. These kits allow rapid and reproducible DNA or RNA extractions from a variety of sample types (e.g., sediments, microbial mats, biomass collected on filters). Prepared DNA and RNA should be stored at -20 and -70 °C, respectively, until use. It is best to freeze samples in multiple aliquots to avoid potential loss of nucleic acids due to repeated freeze–thaw cycles. To prepare crude RNA samples for use in RT-PCR, DNA is digested by treatment of the sample with RNase-free DNase I (Takara, Japan). The DNase-treated samples are then checked for potential genomic DNA contamination by PCR amplification with primer sets specific for archaeal and bacterial 16S rRNA and archaeal *amoA* genes according to Jiang *et al.* (2010). DNA-free RNA samples are reverse transcribed into cDNA using the Promega AMV reverse transcription system (Promega Corporation, Madison, WI) as previously described (Jiang *et al.*, 2009).

5.2. PCR and qPCR

PCR and qPCR are useful techniques for detection and quantification, respectively, of AOA in geothermal environments, especially in samples that contain relatively low biomass, which might not yield sufficient amounts of lipids or other biomarkers. Many studies have used primers specific to the archaeal *amoA* to detect and/or quantify AOA in a variety of environments (de la Torre *et al.*, 2008; Francis *et al.*, 2005; Jiang *et al.*, 2009; Leininger *et al.*, 2006; Zhang *et al.*, 2008). Primers designed by Francis *et al.* (2005) are most commonly used and have led to amplification of extremely diverse putative *amoA* genes and transcripts from a wide variety of geothermal habitats (Jiang *et al.*, 2010; Zhang *et al.*, 2008). However, these primers are minimally degenerate and the forward primer has two mismatches near the 3′ end with the *amoA* sequence of "*Ca.* N. yellowstonii." Thus, de la Torre *et al.*, (2008) modified the forward primer to be more degenerate. However, we have recently found that neither the original primer pair of Francis nor the modified primer pair of de la Torre were successful in amplifying *amoA* genes from certain high temperature (>73 °C) spring sources in the US Great Basin, despite the high abundance of 16S rRNA genes $>98\%$ identical to the AOA "*Ca.* N. yellowstonii" in clone libraries in some of these hot springs (Costa *et al.*, 2009). We therefore designed a more degenerate set of primers (DegAamoA-F and DegAamoA-R; Table 8.2), which targets a region conserved between "*Ca.* N. yellowstonii" and all *amoA* cluster IV sequences, as described by de la

Torre et al. (2008). PCR using these primers yielded product of the predicted size, ~450 bp. Sequences obtained from this product were aligned with the putative *amoA* gene from "*Ca.* N. yellowstonii," and nondegenerate primers specific for this set of sequences for use in qPCR were designed (CNY amoA-F and CNY amoA-R, Table 8.2). We recommend using an approach similar to that described above to attempt to obtain *amoA* sequences for primer design from geothermal environments in cases where other less degenerate primer sets fail to yield product. In general, this type of result also mandates that some caution be used in interpreting results of studies dependent on functional gene PCR, particularly when the study site has not been investigated using a metagenomics approach.

The following is a protocol optimized for primer sets specific for the 16S rRNA gene (CNY 16S-F and -R; Table 8.2) and putative *amoA* (CNY amoA-F and -R) of "*Ca.* N. yellowstonii" and close relatives detected in US Great Basin hot springs (Costa et al., 2009). This is designed for use with an iCycler iQ Multicolor Real-Time PCR Detection System (BioRad, Hercules, CA, USA), using SYBR Green to detect PCR product. For each template, reactions should be prepared in triplicate and coupled with negative controls containing no template. Standard curves are obtained by using linearized plasmid containing the target sequence as template in 10-fold dilutions ranging from $\sim 10^2$ to 10^7 copies/reaction. Prepare individual reactions (25 µL) in individual wells of a iQ 96-well PCR plates (223-9441, BioRad): PerfeCta SYBR Green SuperMix for iQ (Quanta Biosciences, Gaithersburg, MD, USA), 12.5 µL; 2.5 µL each of forward (F) and reverse (R) primer (4 µM); 2.5 µL template DNA; and 5 µL sterile, nuclease-free water. The following cycling conditions are used: an initial melt cycle (95 °C for 3 min) followed by 45 cycles of melting (94 °C for 15 s), annealing (58 °C for 15 s), and extension (72 °C for 35 s), with data collection using a SYBR-490 filter enabled during the 72 °C step, followed by a melt curve 55–95 °C by 0.5 °C increments (10 s each step). Threshold cycles are calculated and data analyses are performed using version 3.1 of the iCycler iQ Optical System Software (BioRad). This same protocol can be used for quantification of *amoA* transcripts by using cDNA generated from RNA, as described in Section 5.1, as template and including RNA samples not treated with reverse transcriptase as a negative control.

We have used a similar qPCR protocol for *narG* using primers TnarG-F and TnarG-R (Table 8.2), which target the plasmid-borne *T. thermophilus* HB8 *narG* sequence, as well as *T. thermophilus* isolates from Great Basin hot springs (Hedlund, unpublished data). These primers have several mismatches and are not expected to amplify *narG* from *Meiothermus* or other genera. Currently, no other *Thermus narG* sequences are available for primer design. We urge reevaluation of the primers as additional *Thermus narG* sequences become available.

Table 8.2 Primers targeting putative *amoA* from AOA, 16S rRNA genes of "*Ca.* N. yellowstonii," and putative *narG* from denitrifying *Thermus thermophilus* strains

Primer name	Sequence (target position in "*Ca.* N. yellowstonii" *amoA* or *T. thermophilus* HB8 *narG*)	Application	Reference
Arch-amoAF	5′ STA ATG GTC TGG CTT AGA CG (−3-17)	PCR	Francis et al. (2005)
Arch_amoA_F	5′ AAT GGT CTG GST TAG AMG (−1-17)	PCR	de la Torre et al. (2008)
Arch-amoAR	5′ GCG GCC ATC CAT CTG TAT GT (616-635)	PCR	Francis et al. (2005)
DegAamoA-F	5′ ATH AAY GCN GGN GAY TA (73-89)	PCR	This study
DegAamoA-R	5′ ACY TGN GGY TCD ATN GG (502-518)	PCR	This study
CNY amoA-F	5′ ATA TTC TAC TCY GAC TGG ATG (91-111)	qPCR	This study
CNY amoA-R	5′ TAT GGG TAK CCT AAG CCT CC (265-284)	qPCR	This study
CNY 16S-F	5′ TAG CTG AAA TCT ATA TGG CCC	qPCR	This study
CNY 16S-R	5′ ATT CTC CAG CCT TTT TAC AGC	qPCR	This study
TnarG-F	5′ GGG TCT GGT TCA TCT GGC (2024-2041)	qPCR	This study
TnarG-R	5′ TTC CTG TAG ACC ACC TCC (2151-2168)	qPCR	This study

6. Closing Remarks

Recent advances in our understanding of nitrogen cycling processes at high temperatures have sparked new interest in the field and highlight how little is known regarding these processes in geothermal environments. Some terrestrial hot springs may harbor large populations of AOA (Costa *et al.*, 2009), but their relative contribution to primary production and the energy budget in these systems are not understood. Although nitrite oxidation apparently occurs at temperatures up to 85 °C (Reigstad *et al.*, 2008), the organisms responsible for this process at these temperatures are not known. Furthermore, although a diverse array of thermophilic and hyperthermophilic *Bacteria* and *Archaea* are capable of nitrate reduction, their relative contributions to this process *in situ* are unknown. The authors hope that the application of existing techniques for quantifying nitrification and denitrification to terrestrial geothermal ecosystems, as described in this chapter, will help further the understanding of the N-cycle in these systems and the microorganisms responsible for these processes.

ACKNOWLEDGMENTS

J. A. D. and B. P. H. were supported by NSF Grant Numbers EPS-9977809 and MCB-0546865. B. P. H., B. H., H. J., and J. R. T. are funded by NSF OISE-0968421. J. R. T. is also supported by NSF award MCB-0949807. A grant from the UNLV Research Infrastructure Award Program (RIAP) to B. P. H. provided the GC-ECD described in this chapter.

REFERENCES

Afshar, S., Kim, C., Monbouquette, H. G., and Schroder, I. I. (1998). Effect of tungstate on nitrate reduction by the hyperthermophilic archaeon *Pyrobaculum aerophilum*. *Appl. Environ. Microbiol.* **64**, 3004–3008.

APHA (1992). Method 4500-NO$_3$ F. Standard Methods for the Examination of Water and Wastewater 18th edn. pp. 4–91. American Public Health Association, Washington, DC.

Avanzino, R. J., and Kenedy, V. C. (1993). Long-term frozen storage of stream water samples for dissolved orthophosphate, nitrate plus nitrite, and ammonia analysis. *Water Resour. Res.* **29**, 3357–3362.

Balderson, W. L., Sherr, B., and Payne, W. J. (1976). Blockage by acetylene of nitrous oxide reduction in *Pseudomonas perfectomarinus*. *Appl. Environ. Microbiol.* **31**, 504–508.

Barraclough, D. (1991). The use of mean pool abundances to interpret ^{15}N tracer experiments. *Plant Soil* **131**, 89–96.

Blöchl, E., Rachel, R., Burggraf, S., Hafenbradl, D., Jannasch, H. W., and Stetter, K. O. (1997). *Pyrolobus fumarii*, gen. and sp. nov., represents a novel group of archaea, extending the upper temperature limit for life to 113 degrees C. *Extremophiles* **1**, 14–21.

Bremner, J. M., and Keeney, D. R. (1966). Determination and isotope-ratio analysis of different forms of nitrogen in soils: 3. Exchangeable ammonium, nitrate and nitrite by extraction-distillation methods. *Soil Sci. Soc. Am. Proc.* **30**, 577–582.

Brooks, P. D., Stark, J. M., McInteer, B. B., and Preston, T. (1989). Diffusion method to prepare soil extracts for automated nitrogen-15 analysis. *Soil Sci. Soc. Am. J.* **53**, 1707–1711.

Burr, M. D., Botero, L. M., Young, M. J., Inskeep, W. P., and McDermott, T. R. (2005). Observations concerning nitrogen cycling in a Yellowstone thermal soil environment. *In* "Geothermal Biology and Geochemistry in Yellowstone National Park," (W. P. Inskeep and T. R. McDermott, eds.), pp. 171–182. Montana State University Publications, Bozeman, MT.

Byrne, N., Strous, M., Crepeau, V., Kartal, B., Birrien, J. L., Schmid, M., *et al.* (2009). Presence and activity of anaerobic ammonium-oxidizing bacteria at deep-sea hydrothermal vents. *ISME J.* **3**, 117–123.

Cabello, P., Roldan, M. D., and Moreno-Vivian, C. (2004). Nitrate reduction and the nitrogen cycle in archaea. *Microbiology* **150**, 3527–3546.

Cantrell, L., and Young, M. (2009). Fatal fall into a volcanic fumarole. *Wilderness Environ. Med.* **20**, 77–79.

Casciotti, K. L., Sigman, D. M., Galanter Hastings, M., Bohlke, J. K., and Hilkert, A. (2002). Measurement of the oxygen isotopic composition of nitrate in seawater and freshwater using the denitrifier method. *Anal. Chem.* **74**, 4905–4912.

Cava, F., Zafra, O., da Costa, M. S., and Berenguer, J. (2008). The role of the nitrate respiration element of *Thermus thermophilus* in the control and activity of the denitrification apparatus. *Environ. Microbiol.* **10**, 522–533.

Chang, C. C. Y., Langston, J., Riggs, M., Campbell, D. H., Silva, S. R., and Kendall, C. (1999). A method of nitrate collection for $\delta^{15}N$ and $\delta^{18}O$ analysis from waters with low nitrate concentrations. *Can. J. Fish. Aquat. Sci.* **56**, 1856–1864.

Chen, R. R., and Dittert, K. (2008). Diffusion technique for ^{15}N and inorganic N analysis of low-N aqueous solutions and Kjeldahl digests. *Rapid Commun. Mass Spectrom.* **22**, 1727–1734.

Costa, K. C., Navarro, J. B., Shock, E. L., Zhang, C. L., Soukup, D., and Hedlund, B. P. (2009). Microbiology and geochemistry of Great Boiling and Mud Hot Springs in the United States Great Basin. *Extremophiles* **13**, 447–459.

Davidson, S. K., Hart, S. C., Shanks, C. A., and Firestone, M. K. (1991). Measuring gross nitrogen mineralization, immobilization, and nitrification by ^{15}N isotopic pool dilution in intact soil cores. *J. Soil Sci.* **42**, 335–349.

de la Torre, J. R., Walker, C. B., Ingalls, A. E., Konneke, M., and Stahl, D. A. (2008). Cultivation of a thermophilic ammonia oxidizing archaeon synthesizing crenarchaeol. *Environ. Microbiol.* **10**, 810–818.

Ehrich, S., Behrens, D., Lebedeva, E., Ludwig, W., and Bock, E. (1995). A new obligately chemolithoautotrophic, nitrite-oxidizing bacterium, *Nitrospira moscoviensis* sp. nov. phylogenetic relationship. *Arch. Microbiol.* **164**, 16–23.

Federova, R. I., Melekhina, E. I., and Ilyuchina, N. I. (1967). Evaluation of the method of "gas metabolism" for detecting extra terrestrial life. Identification of nitrogen-fixing organisms. *Izv. Akad. Nauk SSSR Ser. Biol.* **6**, 791.

Francis, C. A., Roberts, K. J., Beman, J. M., Santoro, A. E., and Oakley, B. B. (2005). Ubiquity and diversity of ammonia-oxidizing archaea in water columns and sediments of the ocean. *Proc. Natl. Acad. Sci. USA* **102**, 14683–14688.

Gotz, D., Banta, A., Beveridge, T. J., Rushdi, A. I., Simoneit, B. R., and Reysenbach, A. L. (2002). *Persephonella marina* gen. nov., sp. nov. and *Persephonella guaymasensis* sp. nov., two novel, thermophilic, hydrogen-oxidizing microaerophiles from deep-sea hydrothermal vents. *Int. J. Syst. Evol. Microbiol.* **52**, 1349–1359.

Grant, S., Grant, W. D., Cowan, D. A., Jones, B. E., Ma, Y., Ventosa, A., and Heaphy, S. (2006). Identification of eukaryotic open reading frames in metagenomic cDNA libraries made from environmental samples. *Appl. Environ. Microbiol.* **72,** 135–143.

Hafenbradl, D., Keller, M., Dirmeier, R., Rachel, R., Rossnagel, P., Burggraf, S., *et al.* (1996). *Ferroglobus placidus* gen. nov., sp. nov., a novel hyperthermophilic archaeum that oxidizes Fe^{2+} at neutral pH under anoxic conditions. *Arch. Microbiol.* **166,** 308–314.

Hall, J. R., Mitchell, K. R., Jackson-Weaver, O., Kooser, A. S., Cron, B. R., Crossey, L. J., and Takacs-Vesbach, C. D. (2008). Molecular characterization of the diversity and distribution of a thermal spring microbial community using rRNA and metabolic genes. *Appl. Environ. Microbiol.* **74,** 4910–4922.

Hamilton, T. L., Boyd, E. S., and Peters, J. W. (submitted). Physicochemical distribution of *nifH* and nitrogen fixation activity in Yellowstone National Park.

Hatzenpichler, R., Lebedeva, E. V., Spieck, E., Stoecker, K., Richter, A., Daims, H., and Wagner, M. (2008). A moderately thermophilic ammonia-oxidizing crenarchaeote from a hot spring. *Proc. Natl. Acad. Sci. USA* **105,** 2134–2139.

He, J. Z., Shen, J. P., Zhang, L. M., Zhu, Y. G., Zheng, Y. M., Xu, M. G., and Di, H. (2007). Quantitative analyses of the abundance and composition of ammonia-oxidizing bacteria and ammonia-oxidizing archaea of a Chinese upland red soil under long-term fertilization practices. *Environ. Microbiol.* **9,** 2364–2374.

Herrera, A., and Cockell, C. S. (2007). Exploring microbial diversity in volcanic environments: A review of methods in DNA extraction. *J. Microbiol. Methods* **70,** 1–12.

Hirayama, H., Takai, K., Inagaki, F., Yamamoto, Y., Suzuki, M., Nealson, K. H., and Hoidoshi, K. (2005). Bacterial community shift along a subsurface geothermal water stream in a Japanese gold mine. *Extremophiles* **9,** 169–184.

Holloway, J. M., Smith, R. L., and Nordstrom, D. K. (2004). Nitrogen transformations in hot spring runoff, Yellowstone National Park, USA. *In* "Water-Rock Interaction," (R. R. Seal and R. B. Wanty, eds.), pp. 145–148. Taylor and Francis Group, London.

Huber, H., Diller, S., Horn, C., and Rachel, R. (2002). *Thermovibrio ruber* gen. nov., sp. nov., an extremely thermophilic, chemolithoautotrophic, nitrate-reducing bacterium that forms a deep branch within the phylum Aquificae. *Int. J. Syst. Evol. Microbiol.* **52,** 1859–1865.

Hungate, R. E. (1950). The anaerobic mesophilic cellulytic bacteria. *Bacteriol. Rev.* **14,** 1–49.

Hungate, R. E. (1969). A roll tube method for cultivation of strict anaerobes. *In* "Methods in Microbiology," (J. R. Norris and W. Ribbons, eds.), pp. 117–132. Academic Press, New York.

Hyman, M. R., and Arp, D. J. (1987). Quantification and removal of some contaminating gases from acetylene used to study gas-utilizing enzymes and microorganisms. *Appl. Environ. Microbiol.* **53,** 298–303.

Jaeschke, A., Op den Camp, H. J., Harhangi, H., Klimiuk, A., Hopmans, E. C., Jetten, M. S., *et al.* (2009). 16S rRNA gene and lipid biomarker evidence for anaerobic ammonium-oxidizing bacteria (anammox) in California and Nevada hot springs. *FEMS Microbiol. Ecol.* **67,** 343–350.

Jiang, H. C., Dong, H. L., Yu, B. S., Lv, G., Deng, S. C., Berzins, N., and Dai, M. H. (2009). Diversity and abundance of ammonia-oxidizing archaea and bacteria in Qinghai Lake, Northwestern China. *Geomicrobiol. J.* **26,** 199–211.

Jiang, H., Huang, Q., Dong, H., Wang, P., Wang, F., Li, W., and Zhang, C. (2010). RNA-based investigation of ammonia-oxidizing archaea in hot springs of Yunnan Province, China. *Appl. Environ. Microbiol.* **76,** 4538–4541.

Kanokratana, P., Chanapan, S., Pootanakit, K., and Eurwilaichitr, L. (2004). Diversity and abundance of *Bacteria* and *Archaea* in the Bor Khleung Hot Spring in Thailand. *J. Basic Microbiol.* **44,** 430–444.

Könneke, M., Bernhard, A. E., de la Torre, J. R., Walker, C. B., Waterbury, J. B., and Stahl, D. A. (2005). Isolation of an autotrophic ammonia-oxidizing marine archaeon. *Nature* **437,** 543–546.

Lebedeva, E. V., Alawi, M., Fiencke, C., Namsaraev, B., Bock, E., and Spieck, E. (2005). Moderately thermophilic nitrifying bacteria from a hot spring of the Baikal rift zone. *FEMS Microbiol. Ecol.* **54,** 297–306.

Lebedeva, E. V., Alawi, M., Maixner, F., Jozsa, P. G., Daims, H., and Spieck, E. (2008). Physiological and phylogenetic characterization of a novel lithoautotrophic nitrite-oxidizing bacterium, 'Candidatus Nitrospira bockiana'. *Int. J. Syst. Evol. Microbiol.* **58,** 242–250.

Leininger, S., Urich, T., Schloter, M., Schwark, L., Qi, J., Nicol, G. W., et al. (2006). Archaea predominate among ammonia-oxidizing prokaryotes in soils. *Nature* **442,** 806–809.

Loftfield, N., Flessa, H., Augustin, J., and Beese, F. (1997). Automated gas chromatographic system for rapid analysis of the atmospheric trace gases methane, carbon dioxide, and nitrous oxide. *J. Environ. Qual.* **26,** 560–564.

Martens-Habbena, W., Berube, P. M., Urakawa, H., de la Torre, J. R., and Stahl, D. A. (2009). Ammonia oxidation kinetics determine niche separation of nitrifying Archaea and Bacteria. *Nature* **461,** 976–979.

McIlvin, M. R., and Altabet, M. A. (2005). Chemical conversion of nitrate and nitrite to nitrous oxide for nitrogen and oxygen isotopic analysis in freshwater and seawater. *Anal. Chem.* **77,** 5589–5595.

Mehta, M. P., and Baross, J. A. (2006). Nitrogen fixation at 92 degrees C by a hydrothermal vent archaeon. *Science* **314,** 1783–1786.

Miller, L. G., Oremland, R. S., and Paulsen, S. (1986). Measurement of nitrous oxide reductase activity in aquatic sediments. *Appl. Environ. Microbiol.* **51,** 18–24.

Miroshnichenko, M. L., Kostrikina, N. A., Chernyh, N. A., Pimenov, N. V., Tourova, T. P., Antipov, A. N., et al. (2003a). Caldithrix abyssi gen. nov., sp. nov., a nitrate-reducing, thermophilic, anaerobic bacterium isolated from a Mid-Atlantic Ridge hydrothermal vent, represents a novel bacterial lineage. *Int. J. Syst. Evol. Microbiol.* **53,** 323–329.

Miroshnichenko, M. L., L'Haridon, S., Jeanthon, C., Antipov, A. N., Kostrikina, N. A., Tindall, B. J., et al. (2003b). Oceanithermus profundus gen. nov., sp. nov., a thermophilic, microaerophilic, facultatively chemolithoheterotrophic bacterium from a deep-sea hydrothermal vent. *Int. J. Syst. Evol. Microbiol.* **53,** 747–752.

Miroshnichenko, M. L., L'Haridon, S., Nercessian, O., Antipov, A. N., Kostrikina, N. A., Tindall, B. J., et al. (2003c). Vulcanithermus mediatlanticus gen. nov., sp. nov., a novel member of the family Thermaceae from a deep-sea hot vent. *Int. J. Syst. Evol. Microbiol.* **53,** 1143–1148.

Mishima, M., Iwata, K., Nara, K., Matsui, T., Shigeno, T., and Omori, T. (2009). Cultivation characteristics of denitrification by thermophilic Geobacillus sp. strain TDN01. *J. Gen. Appl. Microbiol.* **55,** 81–86.

Mitchell, K. R., and Takacs-Vesbach, C. D. (2008). A comparison of methods for total community DNA preservation and extraction from various thermal environments. *J. Ind. Microbiol. Biotechnol.* **35,** 1139–1147.

Mosier, A. R., and Klemedtsson, L. (1994). Measuring denitrification in the field. In "Methods of Soil Analysis. Part 2," (R. W. Weaver, ed.), p. 1047. SSSA, Madison, WI.

Mosier, A. R., and Mack, L. (1980). Gas chromatographic system for precise, rapid analysis of nitrous oxide. *Soil Sci. Soc. Am. J.* **44,** 1121–1123.

Nicol, G. W., Leininger, S., Schleper, C., and Prosser, J. I. (2008). The influence of soil pH on the diversity, abundance and transcriptional activity of ammonia oxidizing archaea and bacteria. *Environ. Microbiol.* **10,** 2966–2978.

Oremland, R. S., Umberger, C., Culbertson, C. W., and Smith, R. L. (1984). Denitrification in San Francisco Bay intertidal sediments. *Appl. Environ. Microbiol.* **47,** 1106–1112.

Pearson, A., Huang, Z., Ingalls, A. E., Romanek, C. S., Wiegel, J., Freeman, K. H., *et al.* (2004). Nonmarine crenarchaeol in Nevada hot springs. *Appl. Environ. Microbiol.* **70,** 5229–5237.

Pearson, A., Pi, Y., Zhao, W., Li, W., Li, Y., Inskeep, W., *et al.* (2008). Factors controlling the distribution of archaeal tetraethers in terrestrial hot springs. *Appl. Environ. Microbiol.* **74,** 3523–3532.

Purdy, K. J. (2005). Nucleic acid recovery from complex environmental samples. *Methods Enzymol.* **397,** 271–292.

Ramirez-Arcos, S., Fernandez-Herrero, L. A., and Berenguer, J. (1998). A thermophilic nitrate reductase is responsible for the strain specific anaerobic growth of *Thermus thermophilus* HB8. *Biochim. Biophys. Acta* **1396,** 215–227.

Reigstad, L. J., Richter, A., Daims, H., Urich, T., Schwark, L., and Schleper, C. (2008). Nitrification in terrestrial hot springs of Iceland and Kamchatka. *FEMS Microbiol. Ecol.* **64,** 167–174.

Révész, K., and Casciotti, K. (2007). Determination of the $\delta(^{15}N/^{14}N)$ and $\delta(^{18}O/^{16}O)$ of Nitrate in Water: RSIL Lab Code 2900. *In* "Methods of the Reston Stable Isotope Laboratory: Reston, Virginia, U.S. Geological Survey, Techniques and Methods," (K. Révész and T. B. Coplen, eds.), pp. 34. U.S. Geological Survey, Reston, VA.

Rhine, E. D., Sims, G. K., Mulvaney, R. L., and Pratt, E. J. (1998). Improving the Berthelot reaction for determining ammonium in soil extracts and water. *Soil Sci. Soc. Am. J.* **62,** 473–480.

Risgaard-Petersen, N., Langezaal, A. M., Ingvardsen, S., Schmid, M. C., Jetten, M. S., Op den Camp, H. J., *et al.* (2006). Evidence for complete denitrification in a benthic foraminifer. *Nature* **443,** 93–96.

Ryden, J. C., Skinner, J. H., and Nixon, D. J. (1987). Soil core incubation system for the field measurement of denitrification using acetylene-inhibition. *Soil Biol. Biochem.* **19,** 753–757.

Searle, P. L. (1984). The Berthelot or Indophenol reaction and its use in the analytical-chemistry of nitrogen—A review. *Analyst* **109,** 549–568.

Shen, J. P., Zhang, L. M., Zhu, Y. G., Zhang, J. B., and He, J. Z. (2008). Abundance and composition of ammonia-oxidizing bacteria and ammonia-oxidizing archaea communities of an alkaline sandy loam. *Environ. Microbiol.* **10,** 1601–1611.

Shock, E. L., Holland, M., Meyer-Dombard, D. R., Amend, J. P., Osburn, G. R., and Fischer, T. P. (2010). Quantifying inorganic sources of geochemical energy in hydrothermal ecosystems, Yellowstone National Park. *Geochim. Cosmochim. Acta* **74,** 4005–4043.

Sigman, D. M., Casciotti, K. L., Andreani, M., Barford, C., Galanter, M., and Bohlke, J. K. (2001). A bacterial method for the nitrogen isotopic analysis of nitrate in seawater and freshwater. *Anal. Chem.* **73,** 4145–4153.

Silva, S. R., Kendall, C., Wilkison, D. H., Ziegler, A. C., Chang, C. C. Y., and Avanzino, R. J. (2000). A new method for collection of nitrate from fresh water and analysis of the nitrogen and oxygen isotope ratios. *J. Hydrol.* **228,** 22–36.

Sorensen, J. (1978). Capacity for denitrification and reduction of nitrate to ammonia in a Coastal Marine sediment. *Appl. Environ. Microbiol.* **35,** 301–305.

Spear, J. R. (2005). What is minimum impact research? *In* "Geothermal Biology and Geochemistry in Yellowstone National Park," (W. P. Inskeep and T. R. McDemott, eds.), pp. 343–352. Montana State University Publications, Bozeman, MT.

Stark, J. M., and Hart, S. C. (1996). Diffusion technique for preparing salt solutions, Kjeldahl digests, and persulfate digests for nitrogen-15 analysis. *Soil Sci. Soc. Am. J.* **60,** 1846–1855.

Steingruber, S. M., Friedrich, J., Gachter, R., and Wehrli, B. (2001). Measurement of denitrification in sediments with the ^{15}N isotope pairing technique. *Appl. Environ. Microbiol.* **67,** 3771–3778.

Strous, M., Fuerst, J. A., Kramer, E. H., Logemann, S., Muyzer, G., van de Pas-Schoonen, K. T., *et al.* (1999). Missing lithotroph identified as new planctomycete. *Nature* **400,** 446–449.

Tiedje, J. M. (1982). Denitrification. *In* "Methods of Soil Analysis. Part 2," (A. L. Page, ed.) 2nd edn. pp. 1011–1026. ASA and SSSA, Madison, WI.

Tiedje, J. M. (1994). Denitrifiers. Methods of Soil Analysis, Part 2. Microbiological and Biochemical Properties SSSABook Series, no. 5, pp. 245–267. Soil Society of America, Madison, WI.

Tiedje, J. M., Simkins, S., and Groffman, P. M. (1989). Perspectives on measurement of denitrification in the field including recommended protocols for acetylene based methods. *Plant Soil* **115,** 261–284.

US-EPA (1993a). Method 353.2, Rev. 2. Determination of nitrate-nitrite nitrogen by automated colorimetry. J. O'Dell (ed.). *In* "Methods for the Determination of Inorganic Substances in Environmental Samples". Inorganic Chemistry Branch, Chemistry Research Division, United States Environmental Protection Agency, Cincinnati, OH. http://www.epa.gov/waterscience/methods/method/files/353_2.pdf,14pp.

US-EPA (1993b). Method 350.1, Rev. 2. Determination of ammonium nitrogen by semi-automated colorimetry. J. O'Dell (ed.). *In* "Methods for the Determination of Inorganic Substances in Environmental Samples". Inorganic Chemistry Branch, Chemistry Research Division, United States Environmental Protection Agency, Cincinnati, OH. http://www.epa.gov/waterscience/methods/method/files/350_1.pdf,15pp.

Vick, T. J., Dodsworth, J. A., Costa, K. C., Shock, E. L., and Hedlund, B. P. (2010). Microbiology and geochemistry of Little Hot Creek, a hot spring environment in the Long Valley Caldera. *Geobiology* **8,** 140–154.

Völkl, P., Huber, R., Drobner, E., Rachel, R., Burggraf, S., Trincone, A., and Stetter, K. O. (1993). *Pyrobaculum aerophilum* sp. nov., a novel nitrate-reducing hyperthermophilic archaeum. *Appl. Environ. Microbiol.* **59,** 2918–2926.

Vorholt, J. A., Hafenbradl, D., Stetter, K. O., and Thauer, R. K. (1997). Pathways of autotrophic CO_2 fixation and of dissimilatory nitrate reduction to N_2O in *Ferroglobus placidus*. *Arch. Microbiol.* **167,** 19–23.

Ward, B. B., and O'Mullan, G. D. (2005). Community level analysis: Genetic and biogeochemical approaches to investigate community composition and function in aerobic ammonia oxidation. *Methods Enzymol.* **397,** 395–413.

Wells, G. F., Park, H. D., Yeung, C. H., Eggleston, B., Francis, C. A., and Criddle, C. S. (2009). Ammonia-oxidizing communities in a highly aerated full-scale activated sludge bioreactor: Betaproteobacterial dynamics and low relative abundance of Crenarchaea. *Environ. Microbiol.* **11,** 2310–2328.

Whittlesey, L. H. (1995). Death in Yellowstone: Accidents and Foolhardiness in the First National Park. The Court Wayne Press, Boulder, CO.

Yoshinari, T., and Knowles, R. (1976). Acetylene inhibition of nitrous oxide reduction by denitrifying bacteria. *Biochem. Biophys. Res. Commun.* **69,** 705–710.

Zhang, C. L., Pearson, A., Li, Y. L., Mills, G., and Wiegel, J. (2006). Thermophilic temperature optimum for crenarchaeol synthesis and its implication for archaeal evolution. *Appl. Environ. Microbiol.* **72,** 4419–4422.

Zhang, C. L., Ye, Q., Huang, Z., Li, W., Chen, J., Song, Z., *et al.* (2008). Global occurrence of archaeal amoA genes in terrestrial hot springs. *Appl. Environ. Microbiol.* **74,** 6417–6426.

CHAPTER NINE

Determining the Distribution of Marine and Coastal Ammonia-Oxidizing Archaea and Bacteria Using a Quantitative Approach

Annika C. Mosier *and* Christopher A. Francis

Contents

1. Introduction	206
2. Methods for Measuring the Abundance of AOA and AOB Within Marine and Coastal Systems	209
2.1. Sample collection	209
2.2. DNA extraction	209
2.3. RNA extraction and cDNA synthesis	209
2.4. PCR screening and gene sequencing	210
2.5. Quantifying archaeal and bacterial amoA genes and transcripts	210
2.6. Other potential target genes for ammonia oxidizers	212
3. Methodological Considerations	213
4. Targeting Specific Ecotypes: Quantifying Shallow and Deep Clades of Marine Water Column AOA	214
Acknowledgments	217
References	218

Abstract

The oxidation of ammonia to nitrite is the first and often rate-limiting step in nitrification and plays an important role in both nitrogen and carbon cycling. This process is carried out by two distinct groups of microorganisms: ammonia-oxidizing archaea (AOA) and ammonia-oxidizing bacteria (AOB). This chapter describes methods for measuring the abundance of AOA and AOB using ammonia monooxygenase subunit A (*amoA*) genes, with a particular emphasis on marine and coastal systems. We also describe quantitative measures designed

Department of Environmental Earth System Science, Stanford University, Stanford, California, USA

Methods in Enzymology, Volume 486
ISSN 0076-6879, DOI: 10.1016/S0076-6879(11)86009-X

© 2011 Elsevier Inc.
All rights reserved.

to target two specific clades of marine AOA: the "shallow" (group A) and "deep" (group B) water column AOA.

1. INTRODUCTION

Nitrification—the microbial conversion of ammonia to nitrate—links the fixation of atmospheric nitrogen (N_2 gas to ammonium) to nitrogen removal processes [denitrification and anaerobic ammonium oxidation (anammox)], which ultimately convert nitrate or ammonium to N_2 gas. Through tight coupling with nitrogen removal processes, nitrification plays a critical role in reducing the associated risks of elevated dissolved inorganic nitrogen (DIN: ammonia, nitrate, and nitrite) often found in coastal systems, including loss of biodiversity, nuisance/toxic algal blooms, and depleted dissolved oxygen (e.g., Bricker et al., 1999). In fact, it is estimated that over 50% of external DIN inputs to estuaries are removed by these microbial processes (e.g., Seitzinger, 1988). Besides representing a key branch in the global nitrogen cycle, nitrification also significantly impacts the biogeochemical cycling of carbon; this chemoautotrophic process not only directly fixes CO_2 into biomass in the absence of light, but also contributes to oceanic primary production by supplying nitrate to phytoplankton in the euphotic zone (Yool et al., 2007).

Nitrification is carried out by two functional groups (or guilds) of chemoautotrophic organisms: ammonia oxidizers convert ammonia to nitrite and then nitrite oxidizers convert nitrite to nitrate. Ammonia oxidation is thought to be the rate-limiting step of the overall reaction, in part because the free energy of the reaction is significantly greater than that of nitrite oxidation and because nitrite rarely accumulates in the environment. Two distinct groups of microbes are capable of ammonia oxidation: (1) the recently discovered ammonia-oxidizing archaea (AOA); and (2) the ammonia-oxidizing bacteria (AOB), including betaproteobacteria (β-AOB) from the genera *Nitrosomonas* and *Nitrosospira*, as well as gammaproteobacteria (γ-AOB) from the genus *Nitrosococcus*.

Despite the biogeochemical importance of ammonia oxidation, only recently have quantitative molecular approaches been employed to determine the relative abundance of the key "players" responsible for this process—the AOA and AOB. AOB *amoA* genes appear to be more abundant than AOA *amoA* genes in many coastal and estuarine sediments based on quantitative PCR (qPCR) estimates (Caffrey et al., 2007; Magalhaes et al., 2009; Mosier and Francis, 2008; Santoro et al., 2008; Wankel et al., in press). However, AOA *amoA* genes are more abundant than AOB *amoA* genes in other estuaries and salt marshes (Abell et al., 2010; Bernhard et al., 2010; Caffrey et al., 2007; Moin et al., 2009). The relative distribution of these two ammonia-oxidizing groups in estuarine environments has been shown to depend on salinity (Mosier and Francis, 2008; Santoro et al., 2008), sediment C:N ratio (Mosier and Francis, 2008), oxygen concentration (Santoro et al., 2008), and chlorophyll-a (Abell et al., 2010). Other physical and geochemical factors shown to specifically affect AOA, AOB, or overall nitrification rates will also likely impact the ratio of AOA:AOB, such as sulfide concentrations (Joye and Hollibaugh, 1995), Fe(III) content (Dollhopf et al., 2005), light (Horrigan and Springer, 1990), temperature (Berounsky and Nixon, 1993), and exogenous inputs from rivers and wastewater treatment plants (Dang et al., 2010). Based on a review of the current literature, Erguder et al. (2009) proposed specific AOA and AOB niches corresponding to varying dissolved oxygen, ammonia, pH, phosphate, and sulfide levels. Substrate availability likely plays a major role in the relative distribution of AOA versus AOB, particularly since it was recently demonstrated that the cultivated ammonia-oxidizing crenarchaeote, *Nitrosopumilus maritimus* SCM1, has a far greater affinity (lower K_m) for ammonium than many AOB (Martens-Habbena et al., 2009). The community composition of AOA and AOB should be also considered when evaluating the relative abundance of these two groups across different sites, because ammonia oxidizers often exhibit distinct spatial structure in coastal/estuarine systems with different phylotypes dominating at different sites (e.g., Beman and Francis, 2006; Bernhard et al., 2005; Francis et al., 2003; Freitag et al., 2006; Mosier and Francis, 2008).

In the open ocean, fixed nitrogen typically enters the deep ocean as ammonia (NH_3/NH_4^+) but it accumulates as nitrate (Karl, 2007). Nitrifiers rapidly convert the ammonia into nitrate, and thus are responsible for producing the large nitrate reservoir (20–40 μM) in the deep waters. Nitrification also plays a critical role in the upper ocean, both as a source of NO_3^- to fuel primary production (Wankel et al., 2007; Yool et al., 2007) and as a source of the greenhouse gas N_2O to the atmosphere (Dore et al., 1998). Interestingly, a number of studies have now shown that AOA *amoA* genes are often several 100- to 1000-fold more abundant than AOB *amoA* genes in the open ocean (Beman et al., 2008, 2010; Mincer et al., 2007; Santoro et al., 2010; Wuchter et al., 2006).

It appears that the vast majority of crenarchaea in the open ocean (particularly within the epi- and mesopelagic zones) possess ammonia monooxygenase genes and thus may be capable of oxidizing ammonia to nitrite. In fact, several oceanic water column studies have found that archaeal *amoA* gene copy numbers are equal to or even greater than archaeal 16S rRNA gene copies, based on qPCR estimates. For instance, archaeal *amoA* and marine Group I (MGI) crenarchaeal 16S rRNA genes have a ratio of ≥ 1 at the San Pedro Ocean Time-series (SPOT) site in the Southern California Bight (Beman et al., 2010), the Guaymas and Carmen Basins in the Gulf of California (Beman et al., 2008), the central California Current

(Santoro et al., 2010), Monterey Bay (Mincer et al., 2007), Station ALOHA near Hawaii (Mincer et al., 2007), and across a large (~5200 km) region of the central Pacific Ocean (Church et al., 2010). Wuchter et al. (2006) also found similar results when comparing *amoA* gene copy numbers to crenarchaeal cell counts generated by catalyzed reporter deposition–fluorescence *in situ* hybridization (CARD–FISH). In a whole-genome shotgun library from 4000-m depth at Station ALOHA, Konstantinidis et al. (2009) showed that the ratio of individual crenarchaeal ammonia monooxygenase sequence reads to crenarchaeal 16S rRNA sequence reads was consistent with a 1:1 *amoA*:16S rRNA ratio. Taken together, all these results imply that most MGI crenarchaea contain *amoA* genes and thus are likely capable of oxidizing ammonia. Indeed, MGI crenarchaeal 16S rRNA gene copies and "group A" (see below) archaeal *amoA* gene copies were significantly correlated with measured $^{15}NH_4^+$ oxidation rates in the Gulf of California (Beman et al., 2008), providing strong evidence for marine crenarchaeal nitrification. Considering that previous studies have shown that MGI crenarchaea constitute ~10–40% of the total microbial community in the ocean below the euphotic zone (Herndl et al., 2005; Karner et al., 2001; Kirchman et al., 2007; Teira et al., 2006), crenarchaea are likely the dominant ammonia oxidizers in the open ocean and ultimately responsible for the deep nitrate reservoir.

In contrast, studies in the North Atlantic (Agogué et al., 2008), Eastern Mediterranean Sea (de Corte et al., 2009), and Antarctic (Kalanetra et al., 2009) have suggested that some crenarchaea lack the *amoA* gene, based on low 16S rRNA:*amoA* gene ratios, and thus may be incapable of ammonia oxidation. However, all three of these studies used *amoA* qPCR primers developed by Wuchter et al. (2006), which have several mismatches to marine AOA and therefore may lead to spurious quantitative results (primer mismatches analyzed and discussed by Church et al., 2010; Konstantinidis et al., 2009; Santoro et al., 2010). Nevertheless, some marine crenarchaea may utilize alternate metabolic pathways. For instance, isotopic (Herndl et al., 2005; Ouverney and Fuhrman, 2000; Teira et al., 2006; Varela et al., 2008) and genomic (Blainey/Mosier et al., in revision; Hallam et al., 2006; Walker et al., 2010) studies have suggested that AOA may be capable of mixotrophic growth. Additionally, urease genes have been reported in the *Cenarchaeum symbiosum* A genome (Hallam et al., 2006) and within a crenarchaeal genomic scaffold from the Pacific Ocean (Konstantinidis et al., 2009), suggesting that at least some marine AOA may have the potential to use urea as an energy source. It is also possible that the archaeal AMO enzymes act on substrates other than, or in addition to, ammonia. Further studies, including cultivation-based approaches, will be necessary to confirm whether or not AOA derive energy from sources other than ammonia.

2. Methods for Measuring the Abundance of AOA and AOB Within Marine and Coastal Systems

2.1. Sample collection

Sediment cores for molecular analyses are generally collected using sterile, cutoff 5-cc syringes. Larger cores can be collected using larger syringes (e.g., 60-cc), PVC piping, or acrylic tubing. Water samples are collected at discrete depths with a Niskin or multi-bottle rosette sampler. Water samples (1–4 l) are vacuum filtered onto a 25 mm, 0.2 μm filter and filters are placed in bead-beating tubes. The water samples can also be size fractionated using filters with different pore sizes (e.g., 10 μm pore size filter to remove larger plankton). For filters specified for RNA extraction, 600 μl RLT Buffer (Qiagen) with 1% β-mercaptoethanol is added to the tube containing the filter (Santoro et al., 2010). Water filters and sediment cores for DNA and RNA analyses are immediately frozen on liquid nitrogen or dry ice until permanent storage at $-80\,^\circ$C.

2.2. DNA extraction

We routinely use the FastDNA SPIN Kit for Soil (MP Biomedicals) to recover total community DNA from sediment samples (Beman and Francis, 2006; Francis et al., 2005; Mosier and Francis, 2008; Santoro et al., 2008). The kit produces "PCR-ready" genomic DNA after bead-beating lysis, sample homogenization and protein solubilization, and DNA purification. DNA is extracted from ∼0.5 g of sediment from the upper 5 mm of the syringe cores. All steps in the manufacturer's protocol are followed, except that the purified DNA is eluted in 100 μl DNase/RNase-free water.

For water samples, we use the DNA extraction method described by Santoro et al. (2010). Briefly, sucrose–EDTA lysis buffer and 10% sodium dodecyl sulfate are added to the bead-beating tubes containing the frozen filters. Following bead beating, proteinase K is added and filters are incubated at 55 °C for ∼4 h. The supernatant is then purified using DNeasy columns (Qiagen) following the manufacturer's protocol with the incorporation of an additional wash step with wash buffer AW2 (Qiagen). The purified DNA is eluted in 200 μl of DNase/RNase-free water.

2.3. RNA extraction and cDNA synthesis

Total RNA extracts are analyzed to determine which *amoA* genes are actively expressed in the environment. RNA is extracted from sediment samples using the RNA PowerSoil Total RNA Isolation Kit (MO BIO).

RNA from filtered water samples is extracted according to the methods described by Santoro et al. (2010), which are based on modifications of the RNeasy Kit (Qiagen). RNA extracts are immediately treated with DNase I to remove any residual DNA. Complimentary DNA (cDNA) is generated from mRNA using the SuperScript III first strand cDNA synthesis kit (Invitrogen).

2.4. PCR screening and gene sequencing

Mixed template DNA and cDNA is PCR amplified with primers targeting betaproteobacterial and archaeal *amoA*: amoA-1F★ (Stephen *et al.*, 1999) or amoA-1F and amoA-2R (Rotthauwe *et al.*, 1997) for β-AOB; Arch-amoAF and Arch-amoAR (Francis *et al.*, 2005) for AOA. Clone libraries of *amoA* gene fragments are constructed using a TOPO TA cloning kit (Invitrogen) and individual clones are sequenced. Both nucleic acid and amino acid sequences are subjected to phylogenetic analysis. Sequences are manually aligned with GenBank sequences using MacClade (http://macclade.org) and phylogenetic trees are constructed in ARB (Ludwig *et al.*, 2004).

2.5. Quantifying archaeal and bacterial amoA genes and transcripts

qPCR is used to estimate the number of archaeal and bacterial *amoA* gene or transcript copies in sediment or water samples. It is advisable to analyze replicate DNA or RNA extractions for each sediment or water sample to minimize biases associated with sample heterogeneity and extraction efficiency. Reactions are carried out in a qPCR thermal cycler such as the StepOnePlus™ Real-Time PCR System (Applied Biosystems).

A standard curve is prepared from a dilution series of control template of known concentration, such as a plasmid containing a cloned gene of interest, genomic DNA, or cDNA. We use plasmids containing cloned *amoA* PCR amplicons that have been extracted with the Qiagen Miniprep Spin Kit. Plasmids are linearized with the *Not*I restriction enzyme, purified, and quantified using the Quant-iT™ Broad Range DNA assay with the Qubit Fluorometer (Invitrogen). The standard curve should include at least four data points spanning the range of template concentrations in a given set of samples (often several orders of magnitude). Following amplification, the threshold cycle (C_t) value for each standard dilution is plotted against the logarithm of the copy number of each dilution, generating the standard curve. The efficiency of the qPCR reaction is calculated using the equation below:

$$\text{Efficiency} = \left[\left[10^{-1/\text{slope}}\right] - 1\right]100$$

When the efficiency is perfect (100%), the slope of the standard curve is −3.32, indicating that there is a perfect doubling of target amplicon every cycle. Correlation coefficients (R^2) for the standard curve should be very close to 1 (ideally > 0.985). Melt curves (for SYBR assays) and gels should be analyzed to check the specificity of amplification.

Sample nucleic acids are simultaneously amplified alongside the standards. Sample C_t values are compared to the standard curve to estimate the copy numbers within each reaction. All sample reactions are performed in triplicate and an average value is calculated. An outlier may be removed from some triplicate measurements to maintain a standard deviation of less than 10–15% for each sample. The number of genes or transcripts in the original sample can then be calculated using the initial volume of water filtered or weight of sediment extracted, the total amount of nucleic acids recovered from the extraction, and the amount of nucleic acids added to each qPCR reaction.

SYBR Green qPCR assays are commonly used for amplification of archaeal and bacterial *amoA* genes or transcripts. Although TaqMan assays have advantages over SYBR assays, divergence of the archaeal and bacterial *amoA* genes precludes development of universal TaqMan probes that capture the full phylogenetic diversity of each gene. Primers Arch-amoAF and Arch-amoAR (Francis *et al.*, 2005) are frequently used for AOA *amoA* quantification and amoA-1F and amoA-2R (Rotthauwe *et al.*, 1997) for β-AOB *amoA* quantification. Other archaeal *amoA* primers used for qPCR in marine studies (e.g., Bernhard *et al.*, 2010; Church *et al.*, 2010; Mincer *et al.*, 2007; Moin *et al.*, 2009) include CrenAmoAQ-F (Mincer *et al.*, 2007) and CrenAmoAQModF (Moin *et al.*, 2009). Variations on the amoA-1F and amoA-2R primers have been used in some marine studies (e.g., Dang *et al.*, 2010; Magalhaes *et al.*, 2009) for β-AOB *amoA* quantification: amoA-1F★ (Stephen *et al.*, 1999) and amoA-2R′ (Okano *et al.*, 2004). Primer amoA-3F and amoB-4R (Purkhold *et al.*, 2000) have been used to quantify γ-AOB (Lam *et al.*, 2009).

Reaction conditions for SYBR assays using primers Arch-amoAF and Arch-amoAR for AOA *amoA* quantification and amoA-1F and amoA-2R for β-AOB *amoA* quantification follow. qPCR reactions are prepared on a cold block (∼4 °C) to avoid nonspecific priming. AOA *amoA* qPCR is performed in a 25 μl reaction mixture with a known amount of template DNA (typically 0.2–10 ng), 0.4 μM of each primer, 2 mM MgCl$_2$, 1.25 Units AmpliTaq DNA polymerase (Applied Biosystems), 1 μl passive reference dye, 40 ng μl^{-1} BSA, and 12.5 μl Failsafe Green Premix E (Epicenter). The qPCR protocol is as follows: 3 min at 95 °C, then 32 cycles consisting of 30 s at 95 °C, 45 s at 56 °C, and 1 min at 72 °C with a detection step at the end of each cycle. β-AOB *amoA* qPCR is performed in the same manner but with 0.3 μM of each primer, Failsafe Green Premix F (Epicenter), and no addition of BSA or $MgCl_2$. The PCR protocol is as follows: 5 min at 95 °C, then 34 cycles consisting of 45 s at 94 °C, 30 s at 56 °C, 1 min at 72 °C, and a detection step for 10 s at 80.5 °C.

2.6. Other potential target genes for ammonia oxidizers

The sheer number of *amoA* sequences in GenBank (e.g., currently over 8000 archaeal *amoA* sequences alone) makes these genes excellent candidates for phylogenetic and quantitative studies. However, there are also a number of other genes that can potentially be used to target ammonia oxidizers. Primers have been designed to amplify the *amoB* and *amoC* subunits of the ammonia monooxygenase enzyme for both AOA and AOB (Junier et al., 2008a,b; Könneke et al., 2005; Norton et al., 2002; Purkhold et al., 2000). Genes encoding other enzymes involved in key nitrogen transformations have been amplified from AOB, including the copper-containing dissimilatory nitrite reductase gene (*nirK*; Casciotti and Ward, 2001) and the nitric oxide reductase gene (*norB*; Casciotti and Ward, 2005). However, primers for these genes must be optimized for qPCR assays to ensure that they are only targeting AOB, while still capturing the full range of sequence diversity within AOB. Notably, putative *nirK* homologues have recently been identified in marine (Walker et al., 2010), low-salinity (Blainey/Mosier et al., in revision), and soil AOA (Bartossek et al., 2010; Treusch et al., 2005), which may prove to be powerful molecular markers for characterizing and quantifying AOA in marine and estuarine systems (Lund and Francis, unpublished).

Beyond nitrogen cycling genes, carbon fixation genes can also be targeted for phylogenetic and quantitative studies of ammonia oxidizers. While RuBisCO genes can be used to examine AOB (Sinigalliano et al., 2001; Utåker et al., 2002), AOA apparently lack RuBisCO genes (Blainey/Mosier et al., in revision; Hallam et al., 2006; Walker et al., 2010). Instead, genes encoding the acetyl-CoA carboxylase enzyme (e.g., *accC*) involved in the 3-hydroxypropionate/4-hydroxybutyrate autotrophic carbon fixation cycle have been used as molecular markers for CO_2 assimilation linked to archaeal autotrophy (Auguet et al., 2008). While this enzyme is found in all sequenced AOA genomes (Blainey/Mosier et al., in revision; Hallam et al., 2006; Walker et al., 2010), it is also found in other crenarchaea (that do not oxidize ammonia), so the primer sets should be further refined for a qPCR assay specific to the AOA. In general, we prefer using C-cycling functional genes as a complement to (rather than a replacement for) *amoA*-based quantitative analyses of AOA and AOB in marine and coastal systems, as *amoA* genes are directly linked to the process of ammonia oxidation.

3. Methodological Considerations

It is important to note that different DNA extraction protocols may yield different gene copy numbers. For instance, Leininger et al. (2006) demonstrated that the use of different bead-beating times yielded different abundances of AOA and AOB amoA gene copies. In four different soil samples, bead-beating times of 30 s for quantification of AOA amoA and between 120 and 150 s for quantification of bacterial amoA yielded the highest abundances. Thus, care should be taken when comparing absolute copy numbers across different samples and/or studies; however, general patterns and trends can still yield valuable insights into the dynamics of microbial populations.

PCR inhibition should also be considered in quantitative assays. The effects of inhibition can be assessed by comparing the amplification of a positive control with the amplification of the positive control mixed with environmental DNA template: inhibitors in the environmental template would decrease the amplification efficiency of the positive control. Additionally, PCR inhibition can be detected by screening amplification of varying concentrations of template DNA (e.g., 0.5, 1, 5, 10 ng); the presence of inhibitors at higher DNA concentrations can result in lower quantification values.

When setting up a new qPCR experiment, careful consideration should be given to primer and probe choice and design. The specificity of the primers/probes is critical to ensure that the assay only quantifies the target gene of interest. If the primers are too specific, the qPCR results will underestimate the abundance of the target gene; however, if the primers are too broad, various nontarget sequences will be amplified (e.g., amplify both amoA and pmoA sequences) and abundance will be overestimated. We design new primers/probes (generally between 15 and 30 bp in length) by manually visualizing a multisequence nucleotide alignment and determining candidate oligonucleotide regions. After taking G + C content, potential secondary structure, and annealing temperature into consideration, we run a BLAST search on the primer/probe sequences to determine which targets they might anneal to. Candidate primers/probes are then scanned against the entire database of sequences using ARB to determine specificity in silico.

The qPCR assay and thermal cycle program are then optimized. Particular attention is given to the choice of qPCR premix, primer/probe concentrations, and $MgCl_2$ concentrations. A variety of qPCR premixes are commercially available (e.g., Epicenter's Failsafe Green Premix A–L, Invitrogen's AccuPrime SuperMix I–II, Applied Biosystem's Power SYBR Green PCR Master Mix, etc.) and each is made with a proprietary recipe with varying concentrations of $MgCl_2$, PCR enhancers (e.g., betaine),

BSA, dNTPs, etc. Therefore, it is necessary to test several different premixes when optimizing a new qPCR assay because some will undoubtedly work better than others. Primer concentrations influence the amplification efficiency and specificity of the reaction. Low concentrations can decrease PCR efficiency, whereas high concentrations can cause nonspecific priming and primer-dimer formation. $MgCl_2$ concentrations influence DNA polymerase activity and fidelity, DNA denaturation temperature, primer annealing, PCR specificity, and primer-dimer formation. Generally speaking, excess magnesium can cause nonspecific amplification and insufficient magnesium reduces the overall yield. Finally, it is imperative to sequence the amplification products from the qPCR assay to ensure that only the product of interest is amplified under the optimized conditions.

FISH is an alternative approach for quantifying AOA and AOB in marine and coastal systems. FISH requires 16S rRNA gene probes that have high specificity for AOA or AOB and yet capture the full phylogenetic diversity within each group. Functional gene- and mRNA-based FISH approaches are in development, but are still not widely accessible. We prefer to use qPCR to quantify AOA and AOB, because of the high-throughput nature of the method, the ability to rapidly examine multiple genes/organisms in the same samples, and the ease of use with multiple sample types (e.g., sediment, water column, soil).

4. TARGETING SPECIFIC ECOTYPES: QUANTIFYING SHALLOW AND DEEP CLADES OF MARINE WATER COLUMN AOA

The majority of archaeal *amoA* genes from marine water columns fall into two phylogenetically distinct clades: group A and group B (first identified by Francis *et al.*, 2005). These clades are distinct from *N. maritimus*, which falls in the marine/coastal sediment clade ($\sim$82% nucleotide identity between group A and *N. maritimus amoA* sequences and 75% identity between group B and *N. maritimus*). Group A *amoA* sequences appear to represent a primarily "shallow" water ecotype (found at depths <200 m) and group B *amoA* sequences represent a deep water ecotype (found predominantly at depths >200 m; Beman *et al.*, 2008; Hallam *et al.*, 2006; Mincer *et al.*, 2007; Santoro *et al.*, 2010). Notably, crenarchaeal depth-specific groups have also been identified from phylogenetic analyses of 16S rRNA gene sequences (García-Martínez and Rodríguez-Valera, 2000; Massana *et al.*, 2000). Although deep group B AOA are found primarily in waters below $\sim$200 m, they may periodically be transported to the surface waters during upwelling/mixing events. Along a transect spanning the central California Current (off the coast of Monterey Bay), the deep group B *amoA* sequences were abundant in clone libraries from the upper water column (25–100 m) at a site experiencing upwelling, whereas group B sequences comprised only a small percentage of the sequences (down to 150 m) at a more offshore site without upwelling (Santoro et al., 2010). Interestingly, deep group B *amoA* gene transcripts were not detected at 25 m depth at the upwelling site, suggesting that these AOA may indeed be specifically adapted to deep waters (Santoro et al., 2010).

In a study of the water column of the Gulf of California—a biologically productive subtropical sea bordered by the Baja California peninsula and mainland Mexico—Beman et al. (2008) designed *amoA* PCR primers/assays to specifically quantify shallow (group A) versus deep (group B) clades of planktonic AOA, using SYBR Green reaction chemistry (Table 9.1). Clade-specific qPCR analysis using these assays revealed that AOA communities in the upper water column (0–100 m) were composed exclusively of group A. In contrast, group B *amoA* genes comprised over 50% of the archaeal *amoA* copies within most of the deeper sampling depths within the oxygen minimum zone (OMZ; 300–650 m). Notably, the sum of the group A and B assays correlated with the general AOA *amoA* qPCR assay ($r^2 = 0.89$, slope = 1.15). Thus, although the extensively used ArchamoAF/R primer set was designed based on a limited number of sequences (Francis et al., 2005) and undoubtedly miss some archaeal *amoA* sequences (de la Torre et al., 2008), these primers still work remarkably well as a general proxy for AOA.

For certain applications (e.g., high-throughput and/or automated qPCR screening), it is particularly useful, if not essential, to have TaqMan assays available for quantifying specific genes of interest. Toward this end, we have designed TaqMan-based qPCR assays to target the shallow group A and deep group B archaeal *amoA* clades. The assays are specific for *amoA* sequence types from each group with no cross-amplification. For both assays, the qPCR protocol is as follows: 94 °C for 75 s, then 42 cycles consisting of 94 °C for 15 s and 55 °C for 30 s. Each 25 μl reaction contains 12.5 μl AccuPrime™ SuperMix I (Invitrogen), 1 μl template DNA, 2.5 mM MgCl$_2$, forward primer, reverse primer, and probe. Primers and probes for the individual assays are shown in Table 9.1.

In addition to standard qPCR analyses, these AOA qPCR assays can be deployed on the Environmental Sample Processor (ESP; Scholin et al., 2001, 2009; http://www.mbari.org/ESP/). The ESP is a field-deployable system that can quantify specific genes of interest (Scholin et al., 2009; http://www.mbari.org/esp/mfb/mfb.htm; http://www.mbari.org/ESP/PCR/pcr.htm), detect 16S rRNAs without amplification (recently demonstrated in Monterey Bay, CA; Preston et al., 2009), and measure relevant environmental parameters using bundled chemical and physical sensors. In collaboration with Christopher Scholin's group (MBARI), our qPCR assays are being used on the ESP to assess fine scale variations in the abundance of the shallow group A

Table 9.1 Primers and probes for archaeal *amoA* assays specific to water column groups A and B marine sequences

Primer/probe	Sequence (5′–3′)	Rxn. Conc. (µM)	Reference
Taqman assay: shallow water column group A archaeal amoA			
WCA-amoA-F	ACACCAGTTTGGCTWCCDTCAGC	0.5	Modified from Beman et al. (2008)
WCA-amoA-R	TCAGCCACHGTGATCAAATTG	0.333	This study
WCA-amoA-P	FAM-ACTCCGCCGAACAGTATCA-BHQ1	0.4	This study
Taqman assay: deep water column group B archaeal amoA			
Arch-amoAFB	CATCCRATGTGGATTCCATCDTG	1.5	Beman et al. (2008)
WCB-amoA-R	AAYGCAGTTTCTAGYGGATC	1.0	This study
WCB-amoA-P	FAM-CCAAAGAATATYAGCGARTG-BHQ1	0.4	This study
SYBR assay: shallow water column group A archaeal amoA			
Arch-amoAFA	ACACCAGTTTGGYTACCWTCDGC	0.4	Beman et al. (2008)
Arch-amoAR	GCGGCCATCCATCTGTATGT	0.4	Francis et al. (2005)
SYBR assay: deep water column group B archaeal amoA			
Arch-amoAFB	CATCCRATGTGGATTCCATCDTG	0.4	Beman et al. (2008)
Arch-amoAR	GCGGCCATCCATCTGTATGT	0.4	Francis et al. (2005)

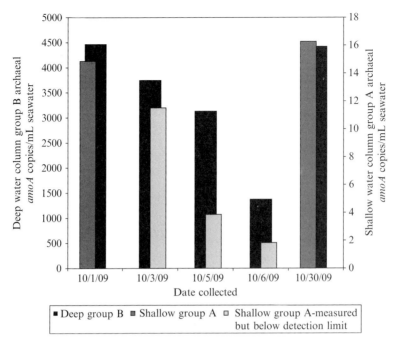

Figure 9.1 Abundance of the shallow (group A) and deep (group B) archaeal *amoA* clades within seawater collected at a depth of 891 m in Monterey Bay, CA. Note difference in scales (y-axis) for group A versus B qPCR data. (See Color Insert.)

and deep group B archaeal *amoA* clades in surface and deep waters in Monterey Bay, CA. Preliminary data shows that, at 891 m, deep group B archaeal *amoA* gene copies are several orders of magnitude more abundant than the shallow group A clade (Fig. 9.1). The abundance of both the shallow and deep clades fluctuates significantly over time (Fig. 9.1). Persistent measurements of these *amoA* genes in Monterey Bay and elsewhere will illuminate marine AOA dynamics on a time continuum that cannot be captured by traditional discrete shipboard sampling. Quantitative data generated by the ESP has the potential to refine, if not dramatically change, our understanding of the molecular microbial ecology of the oceanic nitrogen cycle.

ACKNOWLEDGMENTS

We thank two anonymous reviewers for constructive feedback on this chapter and Christopher Scholin for useful comments on the ESP text. We also thank both Christopher Scholin and Christina Preston for providing access to samples collected in Monterey Bay. This work was supported in part by National Science Foundation grants MCB-0604270, OCE-0825363, and OCE-0847266 (to C. A. F.), as well as an EPA STAR Graduate Fellowship (to A. C. M.).

REFERENCES

Abell, G. C. J., Revill, A. T., Smith, C., Bissett, A. P., Volkman, J. K., and Robert, S. S. (2010). Archaeal ammonia oxidizers and *nirS*-type denitrifiers dominate sediment nitrifying and denitrifying populations in a subtropical macrotidal estuary. *ISME J.* **4**, 286–300.

Agogué, H., Brink, M., Dinasquet, J., and Herndl, G. J. (2008). Major gradients in putatively nitrifying and non-nitrifying Archaea in the deep North Atlantic. *Nature* **456**, 788–791.

Auguet, J.-C., Borrego, C. M., Bañeras, L., and Casamayor, E. O. (2008). Fingerprinting the genetic diversity of the biotin carboxylase gene (*accC*) in aquatic ecosystems as a potential marker for studies of carbon dioxide assimilation in the dark. *Environ. Microbiol.* **10**, 2527–2536.

Bartossek, R., Nicol, G. W., Lanzen, A., Klenk, H.-P., and Schleper, C. (2010). Homologues of nitrite reductases in ammonia-oxidizing archaea: Diversity and genomic context. *Environ. Microbiol.* **12**, 1075–1088.

Beman, J. M., and Francis, C. A. (2006). Diversity of ammonia-oxidizing archaea and bacteria in the sediments of a hypernutrified subtropical estuary: Bahia del Tobari, Mexico. *Appl. Environ. Microbiol.* **72**, 7767–7777.

Beman, J. M., Popp, B. N., and Francis, C. A. (2008). Molecular and biogeochemical evidence for ammonia oxidation by marine Crenarchaeota in the Gulf of California. *ISME J.* **2**, 429–441.

Beman, J. M., Sachdeva, R., and Fuhrman, J. A. (2010). Population ecology of nitrifying Archaea and Bacteria in the Southern California Bight. *Environ. Microbiol.* **12**, 1282–1292.

Bernhard, A. E., Donn, T., Giblin, A. E., and Stahl, D. A. (2005). Loss of diversity of ammonia-oxidizing bacteria correlates with increasing salinity in an estuary system. *Environ. Microbiol.* **7**, 1289–1297.

Bernhard, A. E., Landry, Z. C., Blevins, A., De La Torre, J. R., Giblin, A. E., and Stahl, D. A. (2010). Abundance of ammonia-oxidizing archaea and bacteria along an estuarine salinity gradient in relation to potential nitrification rates. *Appl. Environ. Microbiol.* **76**, 1285–1289.

Berounsky, V. M., and Nixon, S. W. (1993). Rates of nitrification along an estuarine gradient in Narragansett Bay. *Estuaries* **16**, 718–730.

Blainey, P. C., Mosier, A. C., Potanina, A., Francis, C. A., and Quake, S. R. Genome of low-salinity ammonia-oxidizing archaeon determined by single-cell and metagenomic analysis. *PLoS ONE* (in revision).

Bricker, S. B., Clement, C., Pirhalla, D., Orlando, S., and Farrow, D. (1999). National Estuarine Eutrophication Assessment: Effects of Nutrient Enrichment in the Nation's Estuaries NOAA, National Ocean Service, Special Projects Office and the National Centers for Coastal Ocean Science, Silver Spring, MD71pp.

Caffrey, J. M., Bano, N., Kalanetra, K., and Hollibaugh, J. T. (2007). Ammonia oxidation and ammonia-oxidizing bacteria and archaea from estuaries with differing histories of hypoxia. *ISME J.* **1**, 660–662.

Casciotti, K. L., and Ward, B. B. (2001). Dissimilatory nitrite reductase genes from autotrophic ammonia-oxidizing bacteria. *Appl. Environ. Microbiol.* **67**, 2213–2221.

Casciotti, K. L., and Ward, B. B. (2005). Phylogenetic analysis of nitric oxide reductase gene homologues from aerobic ammonia-oxidizing bacteria. *FEMS Microbiol. Ecol.* **52**, 197–205.

Church, M. J., Wai, B., Karl, D. M., and Delong, E. F. (2010). Abundances of crenarchaeal *amoA* genes and transcripts in the Pacific Ocean. *Environ. Microbiol.* **12**, 679–688.

Dang, H., Li, J., Chen, R., Wang, L., Guo, L., Zhang, Z., and Klotz, M. G. (2010). Diversity, abundance, and spatial distribution of sediment ammonia-oxidizing

Betaproteobacteria in response to environmental gradients and coastal eutrophication in Jiaozhou Bay, China. *Appl. Environ. Microbiol.* **76,** 4691–4702.

De Corte, D., Yokokawa, T., Varela, M. M., Agogué, H., and Herndl, G. J. (2009). Spatial distribution of Bacteria and Archaea and *amoA* gene copy numbers throughout the water column of the Eastern Mediterranean Sea. *ISME J.* **3,** 147–158.

de la Torre, J. R., Walker, C. B., Ingalls, A. E., Konneke, M., and Stahl, D. A. (2008). Cultivation of a thermophilic ammonia oxidizing archaeon synthesizing crenarchaeol. *Environ. Microbiol.* **10,** 810–818.

Dollhopf, S., Hyun, J., Smith, A., Adams, H., O'Brien, S., and Kostka, J. (2005). Quantification of ammonia-oxidizing bacteria and factors controlling nitrification in salt marsh sediments. *Appl. Environ. Microbiol.* **71,** 240–246.

Dore, J. E., Popp, B. N., Karl, D. M., and Sansone, F. J. (1998). A large source of atmospheric nitrous oxide from subtropical North Pacific surface waters. *Nature* **396,** 63–66.

Erguder, T. H., Boon, N., Wittebolle, L., Marzorati, M., and Verstraete, W. (2009). Environmental factors shaping the ecological niches of ammonia-oxidizing archaea. *FEMS Microbiol. Rev.* **33,** 855–869.

Francis, C. A., O'Mullan, G. D., and Ward, B. B. (2003). Diversity of ammonia monooxygenase (*amoA*) genes across environmental gradients in Chesapeake Bay sediments. *Geobiology* **1,** 129–140.

Francis, C. A., Roberts, K. J., Beman, J. M., Santoro, A. E., and Oakley, B. B. (2005). Ubiquity and diversity of ammonia-oxidizing archaea in water columns and sediments of the ocean. *Proc. Natl. Acad. Sci.* **102,** 14683–14688.

Freitag, T. E., Chang, L., and Prosser, J. I. (2006). Changes in the community structure and activity of betaproteobacterial ammonia-oxidizing sediment bacteria along a freshwater–marine gradient. *Environ. Microbiol.* **8,** 684–696.

García-Martínez, J., and Rodríguez-Valera, F. (2000). Microdiversity of uncultured marine prokaryotes: The SAR11 cluster and the marine Archaea of Group I. *Mol. Ecol.* **9,** 935–948.

Hallam, S. J., Mincer, T. J., Schleper, C., Preston, C. M., Roberts, K., Richardson, P. M., and DeLong, E. F. (2006). Pathways of carbon assimilation and ammonia oxidation suggested by environmental genomic analyses of marine Crenarchaeota. *PLoS Biol.* **4,** e95.

Herndl, G. J., Reinthaler, T., Teira, E., van Aken, H., Veth, C., Pernthaler, A., and Pernthaler, J. (2005). Contribution of Archaea to total prokaryotic production in the deep Atlantic Ocean. *Appl. Environ. Microbiol.* **71,** 2303–2309.

Horrigan, S., and Springer, A. (1990). Oceanic and estuarine ammonium oxidation – effects of light. *Limnol. Oceanogr.* **35,** 479–482.

Joye, S., and Hollibaugh, J. (1995). Influence of sulfide inhibition of nitrification on nitrogen regeneration in sediments. *Science* **270,** 623–625.

Junier, P., Kim, O.-S., Hadas, O., Imhoff, J. F., and Witzel, K.-P. (2008a). Evaluation of PCR primer selectivity and phylogenetic specificity by using amplification of 16S rRNA genes from betaproteobacterial ammonia-oxidizing bacteria in environmental samples. *Appl. Environ. Microbiol.* **74,** 5231–5236.

Junier, P., Kim, O. S., Molina, V., Limburg, P., Junier, T., Imhoff, J. F., and Witzel, K. P. (2008b). Comparative *in silico* analysis of PCR primers suited for diagnostics and cloning of ammonia monooxygenase genes from ammonia-oxidizing bacteria. *FEMS Microbiol. Ecol.* **64,** 141–152.

Kalanetra, K. M., Bano, N., and Hollibaugh, J. T. (2009). Ammonia-oxidizing Archaea in the Arctic Ocean and Antarctic coastal waters. *Environ. Microbiol.* **11,** 2434–2445.

Karl, D. M. (2007). Microbial oceanography: Paradigms, processes and promise. *Nat. Rev. Microbiol.* **5,** 759–769.

Karner, M. B., DeLong, E. F., and Karl, D. M. (2001). Archaeal dominance in the mesopelagic zone of the Pacific Ocean. *Nature* **409,** 507–510.

Kirchman, D. L., Elifantz, H., Dittel, A. I., Malmstrom, R. R., and Cottrell, M. T. (2007). Standing stocks and activity of Archaea and Bacteria in the western Arctic Ocean. *Limnol. Oceanogr.* **52**, 495–507.

Könneke, M., Bernhard, A. E., de la Torre, J. R., Walker, C. B., Waterbury, J. B., and Stahl, D. A. (2005). Isolation of an autotrophic ammonia-oxidizing marine archaeon. *Nature* **437**, 543–546.

Konstantinidis, K. T., Braff, J., Karl, D. M., and Delong, E. F. (2009). Comparative metagenomic analysis of a microbial community residing at a depth of 4,000 meters at station ALOHA in the North Pacific subtropical gyre. *Appl. Environ. Microbiol.* **75**, 5345–5355.

Lam, P., Lavik, G., Jensen, M. M., van de Vossenberg, J., Schmid, M., Woebken, D., *et al.* (2009). Revising the nitrogen cycle in the Peruvian oxygen minimum zone. *Proc. Natl. Acad. Sci.* **106**, 4752–4757.

Leininger, S., Urich, T., Schloter, M., Schwark, L., Qi, J., Nicol, G. W., *et al.* (2006). Archaea predominate among ammonia-oxidizing prokaryotes in soils. *Nature* **442**, 806–809.

Ludwig, W., Strunk, O., Westram, R., Richter, L., Meier, H., Yadhukumar, H., *et al.* (2004). ARB: A software environment for sequence data. *Nucleic Acids Res.* **32**, 1363–1371.

Magalhaes, C. M., Machado, A., and Bordalo, A. A. (2009). Temporal variability in the abundance of ammonia-oxidizing bacteria vs. archaea in sandy sediments of the Douro River estuary, Portugal. *Aquat. Microb. Ecol.* **56**, 13–23.

Martens-Habbena, W., Berube, P. M., Urakawa, H., de La Torre, J. R., and Stahl, D. A. (2009). Ammonia oxidation kinetics determine niche separation of nitrifying Archaea and Bacteria. *Nature* **461**, 976–979.

Massana, R., DeLong, E. F., and Pedrós-Alió, C. (2000). A few cosmopolitan phylotypes dominate planktonic archaeal assemblages in widely different oceanic provinces. *Appl. Environ. Microbiol.* **66**, 1777–1787.

Mincer, T. J., Church, M. J., Taylor, L. T., Preston, C. M., Karl, D. M., and DeLong, E. F. (2007). Quantitative distribution of presumptive archaeal and bacterial nitrifiers in Monterey Bay and the North Pacific Subtropical Gyre. *Environ. Microbiol.* **9**, 1162–1175.

Moin, N. S., Nelson, K. A., Bush, A., and Bernhard, A. E. (2009). Distribution and diversity of archaeal and bacterial ammonia oxidizers in salt marsh sediments. *Appl. Environ. Microbiol.* **75**, 7461–7468.

Mosier, A. C., and Francis, C. A. (2008). Relative abundance and diversity of ammonia-oxidizing archaea and bacteria in the San Francisco Bay estuary. *Environ. Microbiol.* **10**, 3002–3016.

Norton, J. M., Alzerreca, J. J., Suwa, Y., and Klotz, M. G. (2002). Diversity of ammonia monooxygenase operon in autotrophic ammonia-oxidizing bacteria. *Arch. Microbiol.* **177**, 139–149.

Okano, Y., Hristova, K. R., Leutenegger, C. M., Jackson, L. E., Denison, R. F., Gebreyesus, B., *et al.* (2004). Application of real-time PCR to study effects of ammonium on population size of ammonia-oxidizing bacteria in soil. *Appl. Environ. Microbiol.* **70**, 1008–1016.

Ouverney, C. C., and Fuhrman, J. A. (2000). Marine planktonic archaea take up amino acids. *Appl. Environ. Microbiol.* **66**, 4829–4833.

Preston, C. M., Marin, R., Jensen, S. D., Feldman, J., Birch, J. M., Massion, E. I., *et al.* (2009). Near real-time, autonomous detection of marine bacterioplankton on a coastal mooring in Monterey Bay, California, using rRNA-targeted DNA probes. *Environ. Microbiol.* **11**, 1168–1180.

Purkhold, U., Pommerening-Roser, A., Juretschko, S., Schmid, M. C., Koops, H.-.P., and Wagner, M. (2000). Phylogeny of all recognized species of ammonia oxidizers based on comparative 16S rRNA and *amoA* sequence analysis: Implications for molecular diversity surveys. *Appl. Environ. Microbiol.* **66**, 5368–5382.

Rotthauwe, J. H., Witzel, K. P., and Liesack, W. (1997). The ammonia monooxygenase structural gene *amoA* as a functional marker: Molecular fine-scale analysis of natural ammonia-oxidizing populations. *Appl. Environ. Microbiol.* **63**, 4704–4712.

Santoro, A. E., Francis, C. A., de Sieyes, N. R., and Boehm, A. B. (2008). Shifts in the relative abundance of ammonia-oxidizing bacteria and archaea across physicochemical gradients in a subterranean estuary. *Environ. Microbiol.* **10**, 1068–1079.

Santoro, A. E., Casciotti, K. L., and Francis, C. A. (2010). Activity, abundance and diversity of nitrifying archaea and bacteria in the central California Current. *Environ. Microbiol.* **12**, 1989–2006.

Scholin, C. A., Massion, E. I., Wright, D. K., Cline, D. E., Mellinger, E., and Brown, M. (2001). Aquatic Autosampler Device. US patent 6187530.

Scholin, C. A., Doucette, G. J., Jensen, S., Roman, B., Pargett, D., Marin, R., et al. (2009). Remote detection of marine microbes, small invertebrates, harmful algae, and biotoxins using the Environmental Sample Processor (ESP). *Oceanography* **22**, 158–167.

Seitzinger, S. P. (1988). Denitrification in freshwater and coastal marine ecosystems: Ecological and geochemical significance. *Limnol. Oceanogr.* **33**, 702–724.

Sinigalliano, C. D., Kuhn, D. N., Jones, R. D., and Guerrero, M. A. (2001). *In situ* reverse transcription to detect the *cbbL* gene and visualize RuBisCO in chemoautotrophic nitrifying bacteria. *Lett. Appl. Microbiol.* **32**, 388–393.

Stephen, J. R., Chang, Y. J., Macnaughton, S. J., Kowalchuk, G. A., Leung, K. T., Flemming, C. A., and White, D. C. (1999). Effect of toxic metals on indigenous soil beta-subgroup proteobacterium ammonia oxidizer community structure and protection against toxicity by inoculated metal-resistant bacteria. *Appl. Environ. Microbiol.* **65**, 95–101.

Teira, E., Lebaron, P., van Aken, H., and Herndl, G. J. (2006). Distribution and activity of Bacteria and Archaea in the deep water masses of the North Atlantic. *Limnol. Oceanogr.* **51**, 2131–2144.

Treusch, A. H., Leininger, S., Kletzin, A., Schuster, S. C., Klenk, H.-.P., and Schleper, C. (2005). Novel genes for nitrite reductase and Amo-related proteins indicate a role of uncultivated mesophilic crenarchaeota in nitrogen cycling. *Environ. Microbiol.* **7**, 1985–1995.

Utåker, J. B., Andersen, K., Aakra, A., Moen, B., and Nes, I. F. (2002). Phylogeny and functional expression of ribulose 1, 5-bisphosphate carboxylase/oxygenase from the autotrophic ammonia-oxidizing bacterium Nitrosospira sp. isolate 40KI. *J. Bacteriol.* **184**, 468–478.

Varela, M. M., van Aken, H. M., Sintes, E., and Herndl, G. J. (2008). Latitudinal trends of Crenarchaeota and Bacteria in the meso- and bathypelagic water masses of the Eastern North Atlantic. *Environ. Microbiol.* **10**, 110–124.

Walker, C. B., La Torre, J. R. D., Klotz, M. G., Urakawa, H., Pinel, N., Arp, D. J., et al. (2010). *Nitrosopumilus maritimus* genome reveals unique mechanisms for nitrification and autotrophy in globally distributed marine crenarchaea. *Proc. Natl. Acad. Sci.* **107**, 8818–8823.

Wankel, S. D., Kendall, C., Pennington, J. T., Chavez, F. P., and Paytan, A. (2007). Nitrification in the euphotic zone as evidenced by nitrate dual isotopic composition: Observations from Monterey Bay, California. *Glob. Biogeochem. Cycles* **21**, RB2009.

Wankel, S. D., Mosier, A. C., Hansel, C. M., Paytan, A., and Francis, C. A. Spatial variability in nitrification rates and ammonia-oxidizing microbial communities in the agriculturally-impacted Elkhorn Slough estuary. *Appl. Environ. Microbiol.* doi:10.1128/AEM.01318-10.

Wuchter, C., Abbas, B., Coolen, M. J. L., Herfort, L., van Bleijswijk, J., Timmers, P., et al. (2006). Archaeal nitrification in the ocean. *Proc. Natl. Acad. Sci.* **103**, 12317–12322.

Yool, A., Martin, A. P., Fernandez, C., and Clark, D. R. (2007). The significance of nitrification for oceanic new production. *Nature* **447**, 999–1002.

CHAPTER TEN

^{15}N-Labeling Experiments to Dissect the Contributions of Heterotrophic Denitrification and Anammox to Nitrogen Removal in the OMZ Waters of the Ocean

Moritz Holtappels,* Gaute Lavik,* Marlene M. Jensen,[†] *and* Marcel M. M. Kuypers*

Contents

1. Introduction	224
2. Theoretical Description of ^{15}N-Incubation Experiments	227
2.1. Experiment 1 (adding ^{15}NH$_4^+$)	229
2.2. Experiment 2 (adding ^{15}NO$_2^-$)	229
2.3. Experiment 3 (adding ^{15}NO$_2^-$ + ^{14}NH$_4^+$)	231
2.4. Experiment 4 (adding ^{15}NO$_3^-$)	232
2.5. Additional experiments to detect nitrification	232
3. In the Field: Supplementary Measurements	233
3.1. Detecting the boundaries of the OMZ	233
3.2. Sampling of NH$_4^+$, NO$_2^-$, and NO$_3^-$	235
4. In the Field: Incubation Experiments	237
5. In the Lab: Mass Spectrometry Measurements	240
5.1. Mass spectrometers	241
5.2. Inlet systems	241
5.3. The preparation line	242
5.4. Handling of the sample	243
6. Data Processing and Interpretation	243
6.1. From N$_2$ peak areas to N$_2$ concentrations	244
6.2. Rate calculations and interpretations	246

* Max Planck Institute for Marine Microbiology, Bremen, Germany
[†] Institute of Biology and Nordic Center for Earth Evolution (NordCEE), University of Southern Denmark, Denmark

7. Common Pitfalls 248
8. Concluding Remarks 249
Acknowledgments 249
References 250

Abstract

In recent years, ^{15}N-labeling experiments have become a powerful tool investigating rates and regulations of microbially mediated nitrogen loss processes in the ocean. This chapter introduces the theoretical and practical aspects of ^{15}N-labeling experiments to dissect the contribution of denitrification and anammox to nitrogen removal in oxygen minimum zones (OMZs). We provide a detailed description of the preparation and realization of the experiments on board. Subsequent measurements of N_2 isotopes using gas chromatography mass spectrometry as well as processing of data and calculation of anammox and denitrification rates are explained. Important supplementary measurements are specified, such as the measurement of nanomolar concentrations of ammonium, nitrite, and nitrate. Nutrient profiles and ^{15}N-experiments from the Peruvian OMZ are presented and discussed as an example.

1. INTRODUCTION

Nitrogen (N) limits primary production in most of the oceans and potentially controls the biological CO_2 sequestration and thus the biological carbon pump (Falkowski et al., 1998; Gruber, 2004). The marine N-cycle is mostly mediated by microorganisms. More than 50% of the bioavailable nitrogen ("fixed" nitrogen, NO_3^-, NO_2^-, NH_4^+) in the oceans derives from the microbial assimilation of gaseous dinitrogen (N_2), whereas most of the fixed nitrogen (~90%) is lost from the ocean by the microbial production of N_2 (Codispoti, 2001; Gruber and Sarmiento, 1997; Middelburg et al., 1996). Roughly, 50% of the global N-loss from the ocean is believed to occur in so-called oxygen minimum zones (OMZs) that, altogether, constitute only about 0.1% of the ocean volume (Codispoti et al., 2001). The most prominent OMZs are found in the equatorial Eastern Pacific Ocean, the Arabian Sea, and the Benguela upwelling system off Namibia. For decades, microbial N_2 production was attributed to heterotrophic denitrification only (Fig. 10.1). Canonical denitrification is the respiratory reduction of nitrite through a well-defined cascade of reduction steps to nitrous oxide (N_2O) or N_2 (Codispoti et al., 2005):

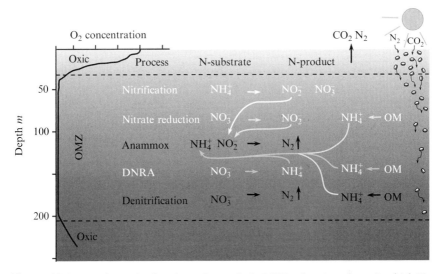

Figure 10.1 A schematic drawing of a typical OMZ, showing the microbial N_2 production pathways (anammox and denitrification) and important N-cycling processes providing the N-substrate for anammox (nitrification, nitrate reduction, and DNRA). In addition, NH_4^+ is released during heterotrophic nitrate reduction, DNRA and denitrification (OM->NH_4^+).

$$NO_3^- \to NO_2^- \to NO \to N_2O \to N_2$$

where nitrate reduction to nitrite often precedes the sequential reduction of nitrite to N_2O and N_2 in the same cell (Zumft, 1997). The accumulation of intermediates and products (NO_2^-, N_2O, and N_2) in OMZs as well as decreased concentrations of "fixed" nitrogen with respect to phosphorus (i.e., N*, see Gruber and Sarmiento, 1997) has often been used as evidence for denitrification. However, this interpretation is challenged by the recent findings of a second microbial N-loss pathway, the anaerobic ammonium oxidation (Fig. 10.1; Mulder et al., 1995). Anammox is a chemolithoautotrophic pathway that yields energy from the anaerobic oxidation of NH_4^+ with NO_2^- to N_2:

$$NO_2^- + NH_4^+ \to N_2 + 2H_2O$$

Anammox has been found to be the predominant N-loss pathway in the OMZs off Peru, Chile, and Namibia (Hamersley et al., 2007; Kuypers et al., 2005; Thamdrup et al., 2006). These studies used ^{15}N-labeling experiments to dissect heterotrophic denitrification from anammox and to investigate the interactions between different N-cycling pathways in OMZs (Lam et al., 2009). The studies detected only minor to insignificant N_2 production attributable to denitrification. In contrast, recent studies report substantial N-loss from the Arabian Sea OMZ due to heterotrophic denitrification (Ward et al., 2009).

As nitrogen is metabolized through the food web, it returns to the dissolved pool as ammonium. Heterotrophic denitrification was previously considered as the key remineralization pathway in the oxygen-free OMZ, providing anammox with ammonium (Dalsgaard et al., 2003). However, the low contribution or even absence of detectable heterotrophic denitrification in OMZ's raises the question how the anammox bacteria obtain their nitrogen sources (NH_4^+ and NO_2^-). Recent studies of a denitrification-free N-cycle in the ETSP OMZ off Peru indicate that anammox is fueled by nitrite ammonification (Kartal et al., 2007; Lam et al., 2009):

$$NO_3^- \rightarrow NO_2^- \rightarrow NH_4^+$$

This stand-alone process is also known as dissimilatory nitrate/nitrite reduction to ammonium (DNRA; see Fig. 10.1). Like denitrification, it is generally coupled to the reduction of nitrate to nitrite ($NO_3^- \rightarrow NO_2^-$), which is also a stand-alone process given that it is performed by various microorganisms that do not have enzymes for the further reduction of nitrite (Zumft, 1997). Lam and others detected increased rates of nitrate reduction, providing anammox with 67% of its nitrite requirement in the Peruvian OMZ (Lam et al., 2009). In addition, anammox need for nitrite was satisfied through aerobic ammonium oxidation, the first step in nitrification, in or at the upper and lower boundaries of the OMZ (Lam et al., 2007). Nitrification is a two-step pathway in the oxidative part of the N-cycle:

$$NH_4^+ \rightarrow NO_2^- \rightarrow NO_3^-$$

where ammonium in the presence of oxygen is converted to nitrite and further to nitrate by chemolithoautotrophic ammonium oxidizers and nitrite oxidizers, respectively (Fig. 10.1).

Research on the marine N-cycle in OMZs has increased during the past decades as a result of recent discoveries on novel processes and players (Lam and Kuypers, 2011). Furthermore, an assumed consequence of global warming is the increase of water masses that are completely devoid of oxygen (Stramma et al., 2008). Thus, the need for appropriate methods to quantify N-loss pathways and to fully understand the complex nature of the N-cycle has grown. This chapter describes the use of ^{15}N-labeled nitrogen compounds to study microbial N-cycling with emphasis on the pelagic N-loss pathways. We provide a protocol to distinguish between anammox and denitrification and to measure the activity of these two processes.

2. THEORETICAL DESCRIPTION OF ^{15}N-INCUBATION EXPERIMENTS

The abundance of the ^{15}N stable isotope is low in natural environments (0.366%) and, therefore, ^{15}N provides a powerful tool to trace different pathways in the N-cycle. Direct tracer-based measurements have played a crucial role in the discovery of anammox in natural settings (Dalsgaard et al., 2003; Kuypers et al., 2003; Thamdrup and Dalsgaard, 2002; Trimmer et al., 2003) and are now widely used to quantify anammox and denitrification rates (see references in Dalsgaard et al., 2005; Francis et al., 2007). The power of using ^{15}N lies in the unique labeling patterns which result from the distinct biochemistry of the two processes. In the stepwise nitrite reduction by denitrifiers, N_2O and N_2 are produced through random pairing of N atoms during the reduction of NO, while anammox generates N_2 by one-to-one pairing of N from nitrite and ammonium. Appropriate ^{15}N amendments for the detection and quantification of anammox and denitrification activity in OMZ waters are as follows (see also Table 10.1):

Experiment 1	$^{15}NH_4^+$
Experiment 2	$^{15}NO_2^-$
Experiment 3	$^{15}NO_2^- + ^{14}NH_4^+$
Experiment 4	$^{15}NO_3$

The N-substrates are added until the labeling percentage is above 50% (usually to a final concentration of 3–5 μM). The production of ^{15}N-labeled N_2 (i.e., $^{29}N_2$ and $^{30}N_2$) measured in experiment 1–4 can be assigned to specific N-loss pathways. The production of N_2 through anammox and denitrification is calculated from the production of $^{29}N_2$ and $^{30}N_2$ and the mole fraction of the ^{15}N label in the respective NH_4^+, NO_2^-, or NO_3^- pools ($F_{\text{N-Substrate}}$) (Thamdrup and Dalsgaard, 2002). The latter is calculated according to

$$F_{\text{N-Substrate}} = \frac{^{15}\text{N} - \text{Substrate}}{^{15}\text{N} - \text{Substrate} + ^{14}\text{N} - \text{Substrate}} \quad (10.1)$$

using the measured concentrations before and after addition of the tracer.

Table 10.1 The table relates N-loss pathways to added ^{15}N-substrates (experiments 1–4) and measured ^{15}N products (^{29}N$_2$ and ^{30}N$_2$)

Experiment	Added N-substrate	Indication from the sole production of ^{29}N$_2$	Indication from the additional production of ^{30}N$_2$	Additional measurements
1	^{15}NH$_4^+$	Anammox (Eq. (10.2))	Coupled Nitrification–Anammox (Eq. (10.3)) (only if coupled nitrification–denitrification can be excluded)	Nitrification measurable from increase of ^{15}NO$_2^-$ and ^{15}NO$_3^-$ (only if NO$_x^-$ pool is big)
2	^{15}NO$_2^-$	Anammox (Eq. (10.2))	Denitrification (Eqs. (10.4) and (10.5)) Intracellular or intercellular coupled DNRA–Anammox (Eq. (10.3)) (intercellular only if NH$_4^+$ pool is small)	DNRA measurable from increase of ^{15}NH$_4^+$ (only if NH$_4^+$ pool is big)
3	^{15}NO$_2^-$ + ^{14}NH$_4^+$	Anammox (Eq. (10.2))	Denitrification (Eqs. (10.4) and (10.5)) Intracellular coupled DNRA–Anammox (Eq. (10.3))	DNRA measurable from increase of ^{15}NH$_4^+$
4	^{15}NO$_3^-$	Anammox (Eq. (10.2)) (only if nitrite pool is very small)	Denitrification (Eqs. (10.4) and (10.5)) (only if NO$_2^-$ pool is very small) Coupled DNRA–Anammox (Eq. (10.3))	Nitrate reduction measurable from increase of ^{15}NO$_2^-$ (only if NO$_2^-$ pool is big)

Substrate pool sizes are considered as "big" or "small" when the ratio of the incubation time over the average residence time of a substrate molecule in the pool is either much below 1 or above 1, respectively.

2.1. Experiment 1 (adding $^{15}NH_4^+$)

Anoxic incubations amended with $^{15}NH_4^+$ provide an experimental evidence for anammox, as bacteria with anammox capacity will produce two possible isotopic N_2 species, which is $^{28}N_2$ and $^{29}N_2$ through random pairing of one molecule from each of the NO_2^- and NH_4^+ pools (Fig. 10.2). It should be noted that NO_2^- in this experiment derives from a natural pool of $^{14}NO_2^-$ and/or the reduction of $^{14}NO_3^-$ to $^{14}NO_2^-$ during the incubation. Because of the high background concentrations of $^{28}N_2$, only the labeled N_2 species can be measured (see Sections 5 and 6). Thus, the production of total N_2 (i.e., $^{28}N_2$ and $^{29}N_2$) through anammox (A) is estimated from the $^{29}N_2$ production using the following expression:

$$A = \frac{^{29}N_2}{F_{NH_4}} \qquad (10.2)$$

where F_{NH_4} denotes the mole fraction of ^{15}N ammonium calculated from Eq. (10.1).

Jensen et al. (2008) showed that anammox bacteria tolerate oxygen concentrations of up to $\sim 12\ \mu M$. The possible co-occurrence of microaerobic nitrification and microaerotolerant anammox (Lam et al., 2007) would lead to the production of $^{30}N_2$. In this case, the oxidation of $^{15}NH_4^+$ to $^{15}NO_2^-$ by nitrification is followed by a one-to-one pairing of $^{15}NO_2^-$ and the added $^{15}NH_4^+$ by anammox. Similarly, $^{29}N_2$ may be produced by pairing of native $^{14}NH_4^+$ and $^{15}NO_2^-$ generated through nitrification. In the case of the co-occurrence of nitrification and anammox, Eq. (10.2) is not valid to calculate anammox rates. If denitrifying bacteria are not active based on the results from experiments 2 and 3 (see below), it is possible to derive total anammox rates from $^{15}NH_4^+$ incubations through the accumulation of both $^{29}N_2$ and $^{30}N_2$, according to (Jensen et al., 2008):

$$A = \left(^{29}N_2 + 2 \times {}^{30}N_2 - \frac{^{30}N_2}{F_{NH_4}}\right) \times \frac{1}{F_{NH_4}} \qquad (10.3)$$

2.2. Experiment 2 (adding $^{15}NO_2^-$)

The contribution of anammox and denitrification to the total N_2 production can be estimated from anoxic incubations with added $^{15}NO_2^-$. Anammox is assumed to produce $^{28}N_2$ and $^{29}N_2$ following addition of $^{15}NO_2^-$ (Fig. 10.2). Denitrification will produce all three possible isotopic N_2 species through random isotope pairing, with $^{28}N_2$, $^{29}N_2$, and $^{30}N_2$ forming at ratios of $(1 - F_{NO_2})^2 : 2 \times (1 - F_{NO_2}) \times (F_{NO_2}) : (F_{NO_2})^2$. The production of $^{29}N_2$ in the absence of $^{30}N_2$ would indicate that anammox is the

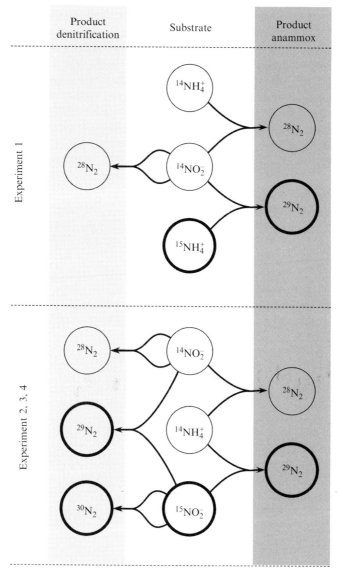

Figure 10.2 N_2 isotopes produced via denitrification (left) and anammox (right) from labeled and nonlabeled N-substrates amended in experiments 1–4. Please note that coupled N-cycling–N-loss processes are not included.

predominant pathway for N loss. In this case, Eq. (10.2) (using F_{NO_2}) is used to calculate the anammox rates. The anammox rates are expected to be similar to those measured in experiment 1. Any significant differences indicate a limitation by either ammonium or nitrite (see also experiment 3).

The production of $^{30}N_2$ is attributed to denitrification only. In the presence of denitrification, the measured $^{29}N_2$ production includes the $^{29}N_2$ production of both anammox and denitrification (Fig. 10.2). Thus, concentrations of N_2 produced by anammox (A) and denitrification (D) are calculated as follows:

$$D = \frac{^{30}N_2}{(F_{NO_2})^2} \qquad (10.4)$$

$$A = \frac{^{29}N_2}{(F_{NO_2})^2} - D \times 2 \times \left(1 - F_{NO_2}\right) \qquad (10.5)$$

For full rationale, the reader is referred to (Thamdrup and Dalsgaard, 2000, 2002; Thamdrup et al., 2006). Measuring denitrification from the reduction of $^{15}NO_2^-$ assumes that denitrifiers preferably consume the pool of free nitrite rather than performing the full nitrate reduction to N_2 within the same cell. However, nitrate reduction and the complete denitrification within a cell can be investigated using $^{15}NO_3^-$ as labeled N-substrate (experiment 4).

In the presence of significant DNRA rates and a close coupling between DNRA and anammox, anammox bacteria could also produce $^{30}N_2$ by combining DNRA substrate ($^{15}NO_2^-$) with DNRA product ($^{15}NH_4^+$). This $^{30}N_2$ production might be significant not only if DNRA and anammox are performed within the same cell (intracellular coupled DNRA–anammox; see Kartal et al., 2007) but also if DNRA and anammox are performed from distinct organisms in the presence of a small ambient ammonium pool (intercellular coupled DNRA–anammox). At a given DNRA rate ($^{15}NO_2^- \rightarrow {}^{15}NH_4^+$), the $^{15}NH_4^+$ enrichment of a small ambient ammonium pool is much faster than the $^{15}NH_4^+$ enrichment of a large ambient ammonium pool. The production of $^{30}N_2$ via intercellular coupled DNRA–anammox is, therefore, more likely in the presence of a small rather than a large ammonium pool (see also experiment 3). If denitrification can be excluded, the coupled DNRA–anammox rates can be calculated from Eq. (10.3) using F_{NO_2} instead of F_{NH_4}.

2.3. Experiment 3 (adding $^{15}NO_2^- + {}^{14}NH_4^+$)

In this experiment, the addition of ammonium would considerably dilute the $^{15}NH_4^+$ generated from DRNA so that the $^{30}N_2$ production via coupled DNRA–anammox is unlikely. Thus, any $^{30}N_2$ production from $^{15}NO_2^-$ is a strong indication for denitrification. Similar to experiment 2, the Eq. (10.2) (sole $^{29}N_2$ production) or the Eqs. (10.4) and (10.5) ($^{29}N_2$ and $^{30}N_2$ production) are used to calculate anammox and denitrification rates, whereas the Eq. (10.3) is used to calculate coupled DNRA–anammox rates in the absence of denitrification.

As the ammonium pool in OMZs is often at nanomolar levels, the question arises if anammox rates are stimulated by the addition of NH_4^+. Any significant increase in anammox rates obtained from incubations experiments using $^{15}NO_2^- + {}^{14}NH_4^+$ (experiment 3) compared to those using only $^{15}NO_2^-$ (experiment 2) would indicate that anammox is NH_4^+ limited (Dalsgaard et al., 2003).

2.4. Experiment 4 (adding $^{15}NO_3^-$)

Nitrate is the most abundant species of fixed N in oceanic OMZs (up to $\sim 30~\mu M$) and represents the highest oxidation state. The fate of nitrate in OMZs is investigated using experiments with labeled $^{15}NO_3^-$. The determination of N-loss rates from $^{15}NO_3^-$ labeling experiments assumes that NO_3^- reduction and the subsequent reduction steps of NO_2^- to N_2 via denitrification or anammox are tightly coupled. In the case of denitrification, full reduction of NO_3^- to N_2 can occur within the same cell, as found for chemolithotrophic denitrification driven by sulfide oxidation (Lavik et al., 2009). However, if any of the produced $^{15}NO_2^-$ is mixed with a large pool of ambient $^{14}NO_2^-$ before it is further reduced to N_2, the ^{15}N-signal is diluted in the subsequent reduction steps and, thus, experiment 4 is not adequate to measure denitrification and anammox rates. For this reason, a parallel incubation with $^{15}NO_2^-$ (experiment 2) is necessary. Nevertheless, experiments with labeled $^{15}NO_3^-$ are important to investigate the coupling of nitrate reduction to the subsequent reduction steps. Furthermore, nitrate reduction to nitrite as a stand-alone process can be determined from the accumulation of ^{15}N in the nitrite pool (e.g. Lam et al., 2009).

2.5. Additional experiments to detect nitrification

Both anammox and denitrification depend on nitrite as a substrate. To elucidate the importance of microaerobic nitrification as a source of nitrite in the upper and lower OMZ, additional incubations with $^{15}NH_4^+ + {}^{14}NO_2^-$ are necessary. Rates of aerobic ammonium oxidation can be determined from the ^{15}N isotopic composition of NO_2^-. The incubation should be done with and without allylthiourea (ATU), a selective inhibitor of bacterial aerobic ammonium oxidation to nitrite at a concentration of 86 μM (Bedard and Knowles, 1989; Hall, 1984). Because ATU has no marked effect on the rates of anammox and denitrification (Jensen et al., 2007) it can also be used to avoid the coupling of nitrification and anammox in experiment 1 ($^{15}NH_4^+$). However, the effect of ATU on archaeal nitrifiers is not known.

In several publications (Hamersley *et al.*, 2007; Kuypers *et al.*, 2003, 2005; Ward *et al.*, 2009), evidence for the presence and activity of denitrifying and anammox bacteria in OMZs was provided by a combination of different approaches including ^{15}N rate measurements, oxygen and nutrient concentration profiles (i.e., NH_4^+, NO_2^-, NO_3^- and N★), molecular techniques (DNA sequencing, fluorescent *in situ* hybridization), and membrane lipid analysis. The main focus of this article is on the ^{15}N-labeling experiments. In addition, we will describe the measurement of oxygen and nutrient concentration profiles because (i) they are needed to initially detect the OMZ and to decide for the appropriate sampling depths and (ii) they provide the backbone for a reliable interpretation of the rate measurements. Molecular techniques are very powerful especially for linking the turnover rates with the abundance and gene expression of specific microorganisms. However, a method description of the various molecular techniques would go beyond the scope of this article. We refer the interested reader to the work of Woebken *et al.* (2007, 2008), Lam *et al.* (2007, 2009), and Ward *et al.* (2009).

3. IN THE FIELD: SUPPLEMENTARY MEASUREMENTS

3.1. Detecting the boundaries of the OMZ

It is important to have precise and reliable measurements of oxygen concentrations to detect the vertical boundaries of the OMZ and select appropriate sampling depths. Since the boundaries of OMZs usually coincide with density gradients, the physical properties of seawater, that is, temperature, salinity, and pressure are important variables, too. The vertical extent of the OMZ is determined using a Conductivity, Temperature, and Depth probe (CTD) that has an oxygen sensor attached. Although oxygen sensing has become a routine measurement nowadays it is still a challenge to resolve oxygen concentrations in the low micromolar and nanomolar range (<2 μM). With the increasing numbers of studies focusing on suboxic processes in OMZ waters the need for methods of high-accuracy oxygen sensing has become inevitable. The issue of oxygen sensing is, therefore, discussed at length by Revsbech *et al.* in Chapter 14. For now we want to stress that the most widely used methods for oxygen measurements, that is, Winkler titration and polarographic membrane sensors (such as provided by Sea-Bird Electronics) should be carefully calibrated because their accuracy (percent deviation of the measured value from the true value) is rapidly decreasing at low concentrations (Revsbech *et al.*, 2009).

Often, a first estimate of the extent of the oxygen-deficient zone can be found in the literature. Figure 10.3 (left plot) shows the density and oxygen profiles from the Peruvian OMZ (Station 4, April 17, 2005, Hamersley

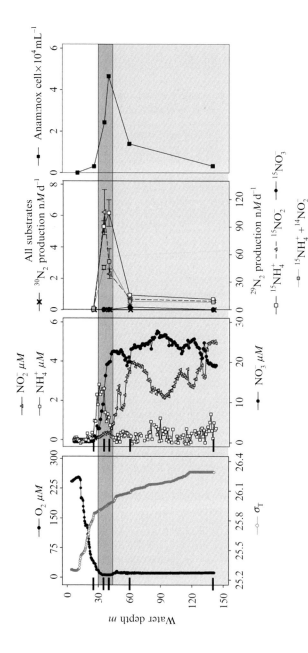

Figure 10.3 Density, oxygen concentration (1st panel) and NH_4^+, NO_2^-, NO_3^- concentrations (2nd panel) in the Peruvian OMZ (Station 4, see Hamersley et al., 2007). $^{29}N_2$ and $^{30}N_2$ production rates from different experiments (3rd panel) and anammox cell counts (4th panel) at five different depths, (also marked by black bars on the vertical axis of panel 1+2). The dark gray area denotes the upper OMZ, while the light gray area denotes the core and lower OMZ.

et al., 2007). A strong density gradient was found in the water column between 10 and 30 m depth, which was also the layer of the strongest oxygen concentration gradient. Constant low values of oxygen ($< 10 \,\mu M$) were found below 30 m depth, which marks the upper boundary of the OMZ. The lower boundary was set by the seafloor at 146 m depth. Anammox rates can be expected at oxygen concentrations of $\sim 12 \,\mu M$ and below (Jensen *et al.*, 2008). Although anammox is an autotrophic process, its supply of NH_4^+ and NO_2^- depends on heterotrophic processes such as nitrate reduction (release of NO_2^-, NH_4^+), DNRA (release of NH_4^+), and, if oxygen is still available, microaerobic respiration (Lam *et al.*, 2009). Therefore, high N-loss rates, either from heterotrophic denitrification or from anammox, are expected just below the oxycline, where oxygen concentrations are below the inhibition concentrations and the supply of fresh organic material from sinking particles is high. In the Peruvian upwelling (Fig. 10.3), a total of five sampling depths were chosen for rate measurements. Two depths were sampled just below the oxycline at 35 and 40 m depth. Another two depths at 60 and 140 m were chosen to cover the range of the OMZ and to investigate the effect of the sediments. At the uppermost sampling depth (25 m), oxygen concentrations (40 μM) were expected to inhibit all N-loss processes. This depth was chosen as reference where anaerobic processes are unlikely.

More information is provided by combining the oxygen profile with profiles of NH_4^+, NO_2^-, and NO_3^- (Fig. 10.3, middle plot). However, these data are generally not available when sampling depths for rate measurements have to be chosen. Nevertheless, nutrient profiles are extremely helpful in the interpretation of rate measurements and to link N-cycling processes to respective nitrogen sinks and sources (see Section 6). In the following, we present sampling gears and methods to measure nitrate and submicromolar concentrations of nitrite and ammonium.

3.2. Sampling of NH_4^+, NO_2^-, and NO_3^-

To date, there exist no *in situ* sensors for measuring submicromolar concentrations of nitrate, nitrite, and ammonium. Therefore, water samples are taken from specific depths and are subsequently analyzed for NH_4^+, NO_2^-, and NO_3^- using fluorometric and photometric methods. The depth resolution of concentration profiles should be as high as possible to resolve the sinks and sources in the OMZ and the adjacent layers. The CTD-Rosette is the most commonly used equipment for water column sampling and can be used for this purpose, too. However, the use of the CTD-Rosette limits the resolution of the profile and, therefore, if possible, a so-called Pump-CTD should be used for continuous nutrient concentration profiling (see Strady *et al.*, 2008 for any details). Depending on the pump rate and lowering speed of the Pump-CTD, continuous profiles with a resolution of less than 1 m can be obtained. By knowing the exact traveling time of the water sample through the hose and by recording the sampling time of the respective nutrient sample in the laboratory, the measured nutrient concentrations and the depth recording of the CTD can be merged into a nutrient depth profile. Additional sampling (e.g., N_2O, H_2S) and sensor measurements (e.g., O_2, and salinity) in the laboratory can be used to get more information on the *in situ* conditions and help to ensure the correct depth interpolation of the nutrient samples.

Ammonium and nitrite are often found in the nanomolar to micromolar range, which makes sampling of these nutrients delicate. Especially ammonium samples can easily be contaminated via air and material contact. Therefore, sample vials should always be kept capped prior to the sampling. For each depth, we use one 50 ml centrifuge tube (Sarstedt) for all nutrients and another one for ammonium only. The latter should be washed beforehand with a 10% HCl solution, rinsed with deionized water and, just prior to the sampling, rinsed again with the working reagent (see below). Centrifuge tubes are filled with at least 40 ml of water sample. A 5 ml subsample is transferred from the nutrient sample into a 10 ml Sarstedt tube for nitrite measurement. The remaining 45 ml of the nutrient sample can be stored at $-20\ °C$ for further measurements of nitrate and, for example, phosphate and silicate in the lab.

For ammonium measurements, we use a fluorometric method described in detail by Holmes *et al.* (1999). To 40 ml of sample volume, 10 ml of working reagent is added and mixed with the sample (Holmes and coworkers suggest 80 ml sample volume and 20 ml working reagent). The active ingredient in the working reagent is orthophthaldialdehyde (OPA) that forms a fluorescent complex with dissolved ammonium. After adding the working reagent, the samples have to be incubated for 3–8 h at room temperatures before measuring the fluorescence at excitation and emission wave lengths of 350 and 400 nm, respectively. Standards should be treated as samples for each individual depth profile. The detection of low concentrations is often limited by the contamination of standards with trace concentrations of ammonium. We prepare the standards from deionized water, which is purified a second time on board using a portable water purification system (Simplicity Millipore). By using the method by Holmes *et al.* (1999), we successfully measured ammonium concentrations down to 10 nM.

For the nitrite measurements, we use a photometric method described in detail by Grasshoff *et al.* (1999). Just after sampling, sulfanilamide solution and NED solution (*N*-(1-naphthyl)-ethylenediamine dihydrochloride) are added to the 5 ml subsample. Subsequently samples are incubated for 1–12 h in the dark at room temperature before they are measured photometrically at 542 nm. We use a flow-through cuvette of 5 cm length to increase the sensitivity of the photometric reading. Standards should be treated as samples and prepared for each profile. By using the method by Grasshoff et al. (1999), we successfully measured nitrite concentrations down to 10 nM.

4. IN THE FIELD: INCUBATION EXPERIMENTS

To collect samples for rate measurements, the Pump-CTD (or CTD-Rosette) is lowered to the depths of interest. At each depth, the hose of the Pump-CTD has to be flushed before sampling for rate measurements. Using a CTD-Rosette, the bottles should remain at constant depth for a few minutes to have them flushed before closing the lid. At each depth, sub-samples to measure the ambient concentrations of NH_4^+, NO_2^-, NO_3^-, and O_2 have to be taken. The main steps of the incubation experiments are summarized in a workflow diagram (Fig. 10.4). We strongly suggest practicing all steps in the lab before going into the field. The steps are described in the following:

Step 1: From the outlet of the Pump-CTD or from the bottles of the CTD-Rosette, the samples are transferred into 250 ml serum bottles avoiding any contamination with the atmosphere. We use the following sampling technique. To transfer the sample, we use gas tight tubes (Ø 5–6 mm), which are first flushed with the sample until all air bubbles are completely removed. Then the flow is reduced either by using a valve or by simply squeezing the tube. The tube is introduced into the serum vial, down to the bottom of the vial. The flow is gently increased leaving the tube opening at the bottom until ∼30% of the vial is filled. The tube is pulled upward slowly so that the tube opening stays just below the water surface. When the serum vial overflows with water, the tube is pushed down again to the bottom of the vial and pulled up with the same speed as before. This is repeated at least two times while checking that no bubbles are trapped in the vial. The tube is slowly pulled out of the serum vial, which is then immediately capped using gray butyl stoppers and crimps. At each depth, 1–2 extra serum vials should be filled as a backup. The serum vials should be stored in the dark at *in situ* temperatures until further processing.

Step 2: During the second step, the samples are purged with helium for 15 min to lower the N_2 background concentration and decrease the concentration of any oxygen that accidentally contaminated the sample during the sampling procedure. This treatment will reduce the N_2 and O_2 concentrations to approximately 20% of their initial value. Subsequently, the ^{15}N-labeled substrate and eventually the unlabeled substrates and inhibitors are added to the serum bottle where they are properly mixed with the water sample.

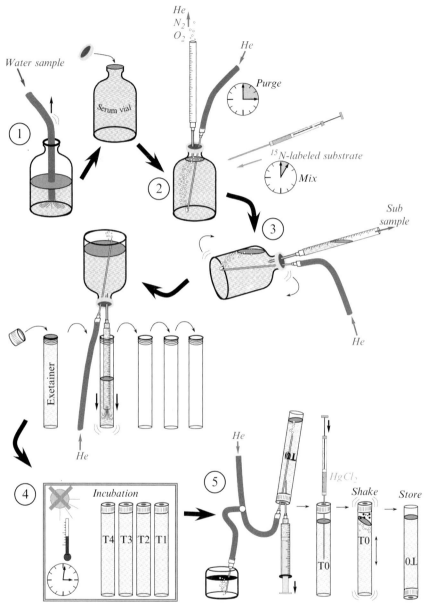

Figure 10.4 A workflow diagram of the incubation experiment. (1) The serum vial is filled with the sample, capped, and stored in the dark at *in situ* temperatures. (2) The sample is purged with helium for 15 min to remove N_2 and O_2. The substrate is added and the sample is purged for 5 min to ensure proper mixing. (3) The sample is subdivided into five Exetainers. (4) The Exetainers are incubated in the dark at *in situ* temperatures. (5) Microbial activity is stopped after preassigned time intervals by introducing a helium headspace and adding $HgCl_2$. Vigorous shaking ensures that most N_2 is trapped in the gas phase.

In detail, helium (He, ultra high purity, 5.0) is introduced into the vial (Fig. 10.3) using a long needle (length 120 mm, Ø 0.8 mm), whereas the gas exhaust is built using a short thick needle (length 38 mm, Ø 1.6 mm) and a 1 ml syringe (Henke Sass Wolf) from which the plunger is removed and the top handle is cut off. The exhaust needle is pushed through the stopper extending not more than 1 cm into the vial. A gas tight tube is connected to the helium source and properly flushed with helium before the flow is reduced and the long needle is attached. The long needle is pushed through the stopper down to the bottom of the vial. The helium flow will push out some sample while a helium headspace is created in the vial. Subsequently, the sample is purged with helium for 10 min at a flow rate of ~ 0.4 l min^{-1}. If the sample is purged too long, the removal of CO_2 from the sample would affect the pH.

After degassing, the helium flow is decreased. Using 50 µl and 100 µl gas tight syringes (SGE Analytical Science), all substrates and eventually the inhibitor (here we use ATU) are taken from the respective stock solution and injected into the serum vial (usually to a final concentration of 5 µM and 86 µM, respectively). Each syringe should be used only for one specific substrate to avoid cross contamination. The sample and the substrates are mixed by gently increasing the helium flow again to ~ 0.4 l min^{-1} and bubbling for another 5 min.

We suggest performing step 2 simultaneously for all samples from various depths that undergo the same incubation experiment and that are treated with the same substrate. To purge several serum vials at the same time, we use a tube with several three-way valves in series that split the helium stream. Step 2 (and also step 3) should be carried out in a sink that is large enough to handle several serum vials at once and, most important, the sink should be far away from any laboratory measuring natural ^{15}N-isotope abundance.

Step 3: During step 3, the sample in the serum bottle is subdivided into five 12 ml gas-tight glass vials (Exetainers, LabCo) avoiding any contamination with the atmosphere and leaving no gasphase in the Exetainers. Step 3 requires the most care because two persons have to work hand in hand. Beforehand, the Exetainers should be properly labeled and decapped. The helium flow rate should be at 0.15–0.25 l min^{-1}. The first serum vial (e.g., experiment 1, depth 1) is tilted so that the helium headspace is moving away from the bottle neck (Fig. 10.4) and the sample is pressed out of the exhaust needle and out of the attached syringe. The serum vial is turned upside down and the first Exetainer is quickly pushed over the exhaust syringe. The Exetainer is filled in the same manner as the serum vial was filled (see step 1): The opening of the exhaust syringe is first at the bottom of the Exetainer. While the serum vial stays fixed, the Exetainer is slowly pulled downward leaving the opening of the exhaust syringe below the water surface. While the samples overflow, the Exetainer is pushed a second time over the syringe and is slowly pulled downward. Finally, the filled Exetainer is handed to a second person in exchange for the next empty Exetainer. The filled Exetainer is immediately capped. When all five Exetainers are filled, the helium supply for the empty serum vial is cut off using the 3-way valve. This procedure is repeated with the next serum vial (e.g., experiment 1, depth 2). Most difficult during step 3 is the timing. The Exetainers should be properly flushed with the sample but the amount of sample in the serum vial is limited. At the end of step 3, five Exetainers are filled for each experiment and depth, which have the same initial substrate and background concentrations.

Step 4: The Exetainers are incubated in the dark and at *in situ* temperatures.

Step 5: Microbial activity in the Exetainers is stopped after preassigned time intervals (usually 0, 6, 12, 24, 48 h after step 3) by introducing a helium headspace and adding 100 µl of saturated $HgCl_2$ solution. The first Exetainer of each time series (T0 at 0 h) is treated immediately after step 3.

The following technique is used to introduce a helium headspace and draw a 2 ml subsample from the Exetainer. To avoid overpressure of the helium headspace, the helium stream is split into two lines (Fig. 10.4). One tube ends in a control vessel that is filled with water so that the helium flow can be observed from the bubble formation. The other tube is connected to a long needle which is introduced into the Exetainer. The helium flow is regulated so that a gentle bubble stream can be observed in the control vessel. The Exetainer is turned upside down and, using a 2 ml syringe with a short needle, a subsample of 2 ml volume is drawn from the Exetainer and replaced by a helium headspace. The subsample is transferred into labeled cryo vials (Sarstedt) and frozen at $-20\ °C$ for later measurements of NO_2^- and NO_3^- using the method described in Braman and Hendrix (1989). Subsequently, 100 µl of saturated $HgCl_2$ solution is injected into the Exetainer. The Exetainer is vigorously shaken to mix the $HgCl_2$ and to transfer the dissolved N_2 into the headspace. Finally, the Exetainer is stored upside down so that the headspace is not in contact with the cap and its perforated septum. The rest of the Exetainers (i.e., T1–T4) are treated the same way after preassigned time intervals. The Exetainers are stored in the dark at stable temperatures and shipped back to the laboratory for N_2 gas analysis.

5. In the Lab: Mass Spectrometry Measurements

Returning from the field work, the N_2 isotope composition of the headspace should be analyzed within a few weeks. In the following, we briefly discuss different kinds of mass spectrometers (MS) and inlet systems and describe the setup of a gas chromatography mass spectrometer (GCMS), which is optimized for the detection of the stable isotopes $^{28}N_2$, $^{29}N_2$, and $^{30}N_2$. We do not explain the principle of mass spectrometry. We assume that the reader is familiar with the method.

5.1. Mass spectrometers

In general, different kinds of MS such as sector field MS and quadrupole MS can be used, and combined with various gas-inlet systems such as membrane inlets, dual inlets, gas bench/gas boxes with purge and trap systems, or gas chromatography continuous flow inlets.

Although quadrupole MS are able to measure the absolute quantity of gases with high accuracy, they are still not as accurate as sector field MS in measuring the relative isotope abundance. However, quadrupole MS have some advantages. The range, in which the MS shows a linear response, is several orders of magnitude larger for a quadrupole MS than for a sector field MS. This can be of advantage if high N_2 production rates are measured in experiments with, for example, sediments and bacterial cultures. Furthermore, low nitrous oxide background, easier maintenance and operation as well as lower investment costs are advantages of a quadrupole MS. In combination with GC inlet and peak integration software, a quadrupole MS system could probably be used to determine N_2 production rates as low as a few nmol N l^{-1} d^{-1}. However, to cover the range of N_2 production rates reported from the water column (0.1 nmol N l^{-1} d^{-1} (Hamersley *et al.*, 2007 and Jensen, unpublished results) to 1000 nmol N l^{-1} d^{-1} (Hannig *et al.*, 2007; Lavik *et al.*, 2009)) the use of a sector field MS system in combination with one of the following inlet systems and preparation lines is necessary.

5.2. Inlet systems

In general, one can choose between dual inlet MS and continuous flow MS. In dual inlet MS the entire gas volume of the sample is extracted and trapped, which is favorable when the amount of sample volume is very limited. However, extracting and purifying the gas for dual inlet MS is labor- and time-intensive. Applying the N_2 incubation experiments described above, the amount of N_2 in the headspace is usually not limiting. Considering the background concentrations of N_2, rather too much than too little N_2 is available. Therefore, we suggest the easiest and most direct approach, which is the continuous flow MS coupling gas chromatography with mass spectrometry. Figure 10.5 shows a sketch of the setup. The sample is injected into a stream of carrier gas, passes a water trap and an O_2 reduction oven (see below) before the remaining gases (N_2 and NO) are separated using a packed GC column (Porapak Q, 6 ft, 1/8 in, 80/100 mesh) and enter the MS via an

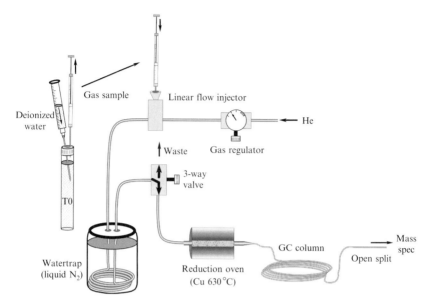

Figure 10.5 A sketch showing the sampling of the headspace and the GCMS-preparation line. The subsample is injected via the linear flow injector into the helium stream (25–30 ml min^{-1}). The liquid N_2 trap removes H_2O and the copper column (630 °C) reduces the O_2 before the gas sample is separated by the GC column (Porapak Q, 6 ft, 1/8 in) and enters the MS via an open split.

open split. The helium flow is regulated to 25–30 ml min^{-1}. Using this setup, the time to analyze a single sample is approximately 3 min.

In an open split system, the high vacuum at the inlet of the MS extracts the sample from the continuous gas flow in the GC column. The amount of sample entering the MS is defined by the flow rate and the length and diameter of the GC column. The exact dimensions of the GC column can differ from one system to another and have to be tested empirically. The guiding principles are: The higher the carrier gas flow the higher the fraction of the sample that is lost at the open split and the lower the fraction that enters the MS. If the carrier gas flow is decreased, the time it takes for the sample to pass the GC column and enter the MS is increased. If the carrier gas flow is too low, the MS will start to extract atmospheric air from the open split.

5.3. The preparation line

To increase the accuracy and stability of the measurement it is most important to separate and trap all unwanted gases and elements before the sample enters the MS. For example, O_2 and water molecules that remain in the sample gas and enter the MS are potentially ionized and split into reactive oxygen species, which in turn recombine with nitrogen to form nitrous monoxide (NO). NO has a mass of 30 and, therefore, greatly increases the detection limit of the $^{30}N_2$ measurement. Similar negative implications come from the contamination with CO (mass 28) from the atmosphere.

Using helium as a carrier gas for the N_2 sample, a liquid nitrogen trap is ideal to trap water, CO_2, CO, and many other gases except N_2, O_2, He, and Ar. The liquid nitrogen trap should have a purge valve to clean the trap after a series of measurements (Fig. 10.5). The O_2, which passes the liquid nitrogen trap, has to be removed by a reduction oven, where O_2 is reduced at the surface of copper granules at temperatures of 600–650 °C.

Efficient trapping of water and oxygen is most important to reduce the NO background. For many MS, the mass 30 background is considerably higher than the background of mass 29. This makes any detection of trace amounts of $^{30}N_2$ extremely difficult. Besides the trapping of O_2 and water, the NO background can be lowered by reducing the ionization/acceleration current (and the magnet current) of the MS. Because mass 30 is not considered in natural abundance $\delta^{15}N$ measurements, the tuning recommended by the MS manufacturer is often not ideal.

5.4. Handling of the sample

When the GCMS is set up and tuned, the sample has to be introduced into the stream of carrier gas. A syringe of deionized water (Fig. 10.5, left side) is first introduced into the Exetainer allowing to equilibrate the vacuum that sometimes builds up due to variable temperatures during the storage and that progressively builds up during the sampling of the headspace. A 1 ml subsample is drawn from the headspace of the Exetainers using a 1 ml gas tight syringe, which was previously flushed with helium to remove the air from dead volumes (e.g., from the needle). When the gas syringe is pulled out of the Exetainer, it is important to avoid a vacuum in the syringe because atmospheric gas would enter the syringe immediately and contaminate the sample. The sample is quickly injected into the linear flow injector. As an N_2 standard, we inject 25 µl of air using a 25 µl gas tight syringe. After every five samples, one air standard is measured. All gas tight syringes can easily clog and should, therefore, be tested after every injection by filling the syringes with air (or helium) and slowly pushing the gas into some water. If a bubble stream appears immediately, the needle is not clogged.

6. Data Processing and Interpretation

From the MS measurement, we derive the integrated peak areas of $^{28}N_2$, $^{29}N_2$, and $^{30}N_2$. In the following, we describe how the peak areas are processed to derived $^{29}N_2$ and $^{30}N_2$ concentrations in the Exetainers and how N_2 production rates are calculated and interpreted.

6.1. From N_2 peak areas to N_2 concentrations

The measured peak area of a specific isotope (e.g., $^{29}N_2$) is the sum of its background abundance in the sample and its amount produced in the course of the experiment. We can state the following equations:

$$^{28}A_M = {}^{28}A_N + {}^{28}A_B \qquad (10.6)$$

$$^{29}A_M = {}^{29}A_N + {}^{29}A_B \qquad (10.7)$$

$$^{30}A_M = {}^{30}A_N + {}^{30}A_B \qquad (10.8)$$

where A denotes the peak area and the superscript number marks the specific isotope. The index M denotes the measured peak area, which is the sum of the background peak area (index B) and the peak area due to N_2 production in the experiment (index N). In order to calculate the variables of interest ($^{29}A_N$ and $^{30}A_N$) from Eqs. (10.7) and (10.8), the background isotope abundance in the sample ($^{29}A_B$ and $^{30}A_B$) has to be determined.

As a first approximation, the relative isotopic composition of the background N_2 is assumed to be identical with the relative isotopic composition of the air standard. Any difference between the isotopic compositions of background N_2 and air standard is caused by the isotopic fractionation during the microbial production of N_2. However, the change of isotopic composition during the labeling experiment is several orders of magnitudes greater than any possible fractionation effects, which are, therefore, neglected. Hence, the ratios $^{29}A_B / {}^{28}A_B$ and $^{30}A_B / {}^{28}A_B$ can be determined from the respective ratios of the air standards ($^{29}A_{Air} / {}^{28}A_{Air}$ and $^{30}A_{Air} / {}^{28}A_{Air}$).

Furthermore, we assume that the measured $^{28}N_2$ peak area ($^{28}A_M$) is approximately equal to the background $^{28}N_2$ peak area ($^{28}A_B$). This approximation can be made, because the N_2 background concentration in the sample is high ($\sim 100\ \mu M$) compared to the expected range of N_2 produced from unlabeled N-substrate (100 nM to 1 μM during 48 h of incubation). The potential error of this approximation is discussed below. The following equations exemplify how $^{29}A_N$ is calculated. In the same way $^{30}A_N$ is calculated (not shown here). First, Eq. (10.7) is rearranged to:

$$^{29}A_N = {}^{29}A_M - \frac{{}^{29}A_B}{{}^{28}A_B}{}^{28}A_B \qquad (10.9)$$

The ratio $^{29}A_B / {}^{28}A_B$ in Eq. (10.9) is known from the ratio $^{29}A_{Air} / {}^{28}A_{Air}$ of the air standard, which results in

$$^{29}A_N = {}^{29}A_M - \frac{{}^{29}A_{Air}}{{}^{28}A_{Air}}{}^{28}A_B \qquad (10.10)$$

By assuming $^{28}A_B \approx {}^{28}A_M$ we derive

$$^{29}A_N \approx {}^{29}A_M - \frac{^{29}A_{\text{Air}}}{^{28}A_{\text{Air}}} {}^{28}A_M \qquad (10.11)$$

Equation (10.11) can be rearranged to

$$^{29}A_N \approx \left(\frac{^{29}A_M}{^{28}A_M} - \frac{^{29}A_{\text{Air}}}{^{28}A_{\text{Air}}}\right) {}^{28}A_M \qquad (10.12)$$

The term inside the brackets of Eq. (10.12) is also known as excess ratio (Nielsen, 1992). This way of calculating $^{29}A_N$ takes advantage of the high accuracy of sector field MS in measuring the relative abundance of stable isotopes. The systematic error, which is introduced by assuming $^{28}A_B \approx {}^{28}A_M$, can be calculated from Eqs. (10.10) and (10.11):

$$\text{Error} = \frac{^{29}A_{\text{Air}}}{^{28}A_{\text{Air}}} \frac{^{28}A_N}{^{29}A_N} \qquad (10.13)$$

The first term in Eq. (10.13) is known from the air standards ($^{29}A_{\text{Air}} / {}^{28}A_{\text{Air}} \approx 0.0077$). The second term is the expected ratio of $^{28}N_2$ over $^{29}N_2$ production, which depends on the ratio of unlabeled over labeled N-substrate in the experiment (i.e., $F_{\text{N-Substrate}}$, see Eq. (10.2)). Using the suggested ^{15}N concentrations of 5 µM, the ratio $^{28}A_N / {}^{29}A_N$ is generally below 1. Therefore, the systematic error is generally below 0.7% and can be neglected. In case of a high ratio of unlabeled over labeled N-substrate, leading to $^{28}A_N / {}^{29}A_N \gg 1$, the results can be corrected using Eq. (10.13).

So far, the calculated $^{29}A_N$ is expressed in unit peak area. From this value, the moles of $^{29}N_2$ injected into the MS is calculated using the air standard, of which we know the ratio of mol N_2 to peak area:

$$^{29}M_N = {}^{29}A_N \frac{^{tot}M_{\text{Air}}}{^{tot}A_{\text{Air}}} \qquad (10.14)$$

Here, M denotes the moles of N_2 injected into the MS. In Eq. (10.14), $^{tot}A_{\text{Air}}$ is calculated according to

$$^{tot}A_{\text{Air}} = {}^{28}A_{\text{Air}} + {}^{29}A_{\text{Air}} + {}^{30}A_{\text{Air}} \qquad (10.15)$$

while $^{tot}M_{\text{Air}}$ is calculated using the molar volume of air (at 20 °C and 1013 hPa) and the fraction of N_2 in the atmosphere:

$$^{tot}M_{\text{Air}} = 0.79 \frac{V_{\text{Air}}}{24.055} \qquad (10.16)$$

The injected volume of the air standard (V_{Air}) is usually 25 μl so that $^{tot}M_{Air} = 0.821$ μmol. From the injected moles of $^{29}N_2$ ($^{29}M_N$), the initial $^{29}N_2$ concentration in the Exetainer is calculated according to

$$^{29}C_N = {}^{29}M_N \frac{V_H + V_W/F_{g/w}}{V_I V_W} \quad (10.17)$$

using the injection volume V_I (1 ml), the headspace volume V_H (2 ml), the water volume V_W (10 ml), and the fractionation factor of N_2 between gas and water phase $F_{g/w}$ (~76 for a sample with salinity 35 at 20 °C). Finally, we can combine Eqs. (10.12), (10.14), and (10.17) to calculate concentrations of $^{29}N_2$ and $^{30}N_2$ according to

$$^{29}C_N = \left(\frac{^{29}A_M}{^{28}A_M} - \frac{^{29}A_{Air}}{^{28}A_{Air}}\right){}^{28}A_M \frac{^{tot}M_{Air}}{^{tot}A_{Air}} \frac{V_H + V_W/F_{g/w}}{V_I V_W} \quad (10.18)$$

$$^{30}C_N = \left(\frac{^{30}A_M}{^{28}A_M} - \frac{^{30}A_{Air}}{^{28}A_{Air}}\right){}^{28}A_M \frac{^{tot}M_{Air}}{^{tot}A_{Air}} \frac{V_H + V_W/F_{g/w}}{V_I V_W} \quad (10.19)$$

It is possible that the $^{29}N_2$ and $^{30}N_2$ concentrations calculated from Eqs. (10.18) and (10.19) are negative, especially at time point T0. The reason for this is the calculation of the excess ratio (see Eq. (10.12)), which can result in negative numbers when the air ratios ($^{29}A_{Air}/{}^{28}A_{Air}$) are greater than the measured ratios ($^{29}A_M/{}^{28}A_M$) in the beginning of the experiment. However, the change of concentrations with time (i.e., the rates) remains unaffected.

6.2. Rate calculations and interpretations

The production rates of $^{29}N_2$ and $^{30}N_2$ are calculated from the increase of $^{29}N_2$ and $^{30}N_2$ concentrations over time. The slope of the linear regression, the standard error, and the coefficient of determination (R^2) are calculated. The hypothesis that the slope is significantly different from zero is tested using the "Student's t-test." To improve the statistics, replicate incubations or replicate time points from the same incubation can be performed. However, as the number of incubation experiments that are carried out at each station is often limited, replicate incubations can be at the expense of the number of depths (or experiments) that are measured. The measurement of replicate time points from the same incubation requires more Exetainers that have to be filled from a single 250 ml serum vial. In this case, the use of larger serum vials is suggested. The additional Exetainers can be used either to measure replicates at each time point or to increase the number of time points. However, using five time points and no replicates as

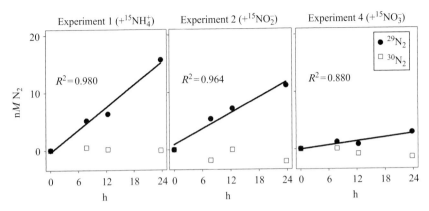

Figure 10.6 The plots show the $^{29}N_2$ and $^{30}N_2$ concentration over time derived from experiments 1, 2, and 4. The samples were taken in 60 m depth in the Peruvian OMZ (Station 4, Hamersley et al., 2007; see above Fig. 10.3).

described here was sufficient to calculate a significant linear increase over time in most studies (Hamersley et al., 2007; Jensen et al., 2008; Kuypers et al., 2005).

Figure 10.6 shows an example of $^{29}N_2$ and $^{30}N_2$ concentrations over time using samples from 60 m depth at station 4 (see Fig. 10.3). In the experiments 1, 2, and 4, the concentrations of $^{29}N_2$ show a significant increase over time ($p = 0.010$, 0.018, and 0.062, respectively), whereas the increase of $^{30}N_2$ is not significant ($p > 0.38$). The production of only $^{29}N_2$ indicates anammox as the predominant N-loss pathway. Therefore, anammox rates in experiments 1, 2, and 4 were calculated according to Eq. (10.2) resulting in 17, 19, and 5 nM N$_2$ d^{-1}, respectively.

It is possible that $^{29}N_2$ or $^{30}N_2$ concentrations do not increase linearly but exhibit an exponential increase. This behavior indicates that the substrate pool of the N-loss process is enriched with ^{15}N-substrate in the course of the experiment. For example, the addition of $^{15}NO_2^-$ can produce $^{30}N_2$ via intercellular coupled DNRA–anammox (see Table 10.1). In this case, the gradual reduction of $^{15}NO_2^-$ via DNRA leads to the gradual enrichment of $^{15}NH_4^+$ in the ammonium pool. The random one to one pairing of nitrite and ammonium via anammox leads to an exponential increase of $^{30}N_2$ concentration over time. In contrast, if the addition of $^{15}NO_2^-$ produces $^{30}N_2$ via denitrification, the increase over time should be linear from the start of the experiment.

At station 4, the production of $^{30}N_2$ could not be detected in any of the experiments and depths (Fig. 10.3). As indicated by the sole production of $^{29}N_2$, anammox was the predominant N-loss pathway throughout the OMZ. Anammox rates were significantly increased at the upper boundary of the OMZ (i.e., at 35 and 40 m depth). In these depths, the ammonium profile exhibits a distinct peak that overlaps with increasing nitrite and nitrate concentrations. The supply with sufficient N-substrate in the upper OMZ provides a favorable environment for the growth of anammox bacteria. The increased rates were supported by the increased number of anammox cells (Fig. 10.3, right panel) stained by fluorescent *in situ* hybridization (FISH). In the core and at the lower boundary of the OMZ, anammox rates and cell abundance were decreased probably due to the lower ammonium supply.

The ammonium profile at lower depths shows decreased and very variable concentrations. The scattering can be attributed to the high detection limit and the low accuracy of the former method (Grasshoff et al., 1999), which was used to measure ammonium concentrations in course of the study (Hamersley et al., 2007). The high scattering complicates a thorough interpretation of the ammonium profile and stresses the need for a more accurate method such as presented by Holmes et al. (1999) (see above). In contrast, the concentration profiles of nitrite and nitrate demonstrate the value of accurate high-resolution nutrient profiling. Nitrate and nitrite concentrations show a significant negative correlation in the OMZ suggesting that nitrate reduction is the main process connecting the pools of nitrate and nitrite. In fact, high rates of nitrate reduction were detected at this station (Lam et al., 2009) by measuring the increase of $^{15}NO_2^-$ in incubation experiment with $^{15}NO_3^-$ (i.e., experiment 4).

7. Common Pitfalls

Anaerobic incubations with ^{15}N-labeled substrates are delicate, comprising quite a few pitfalls, of which we will mention the most common ones. Most important is to avoid atmospheric contamination of the sample during the sample transfers. Enzymes in facultative anaerobic bacteria might be inhibited or destroyed and interfering oxic processes appear when O_2 concentrations are elevated. Once the sample is contaminated with oxygen, it is difficult to remove the oxygen efficiently, even during the degassing step. The purpose of degassing the sample at the start of the incubation is to reduce the N_2 background and not primarily to remove O_2. Degassing with He for 15 min will reduce the background concentration of N_2 and O_2 to about 20% of the original concentration. Consequently, considerable amount of O_2 can be found even after degassing a contaminated sample.

The samples should not be exposed to direct sunlight or large temperature changes, which can also affect microbial processes. The incubations should be started within a couple of hours after sampling, because the microbial community will change as soon as the microbial community is cut off from their natural environment (bottle effect).

It is recommended to keep control of the proper labeling of the samples, syringes, and stock solutions. Mixing the samples from various depths and mixing or contaminating the N-substrates will be fatal for the evaluation of the experiment even though turnover rates can be measured afterwards. We recommend using a color code to identify the syringes and the N-substrate. In case of any uncertainties, discard the syringe and use a new one.

Once the headspace is set, great care is required to keep the gas sample uncontaminated. Make sure that no air is introduced when the headspace is set. Store the samples upside down in the dark at stabile temperatures and measure the isotope composition within a few weeks. N-loss rates can still be determined from samples that are stored more than 1 year but the accuracy will be reduced and the detection limit increased.

Because the rate measurements are based on volume calculations (see Eq. (10.17)), the exact headspace volume and the volume injected into the MS should be known. If the MS injections are done manually, clogged syringes cause the most trouble. Therefore, the syringes should be checked after every injection.

8. CONCLUDING REMARKS

The method described here takes advantage of the low abundance of ^{15}N in nature and the high-accuracy measuring isotope ratios with mass spectrometry. Using this experimental approach, it is possible to detect N-turnover rates down to the nanomolar range. In contrast to common N-budget calculations (derived from N★ or N_2/Ar profiles), the experimental approach is capable to trace and quantify the various processes involved in the N-cycle. This "inside view" is essential if we want to understand the rates and regulations of the complex N-cycle in OMZs and their response on climate change. Of course, the combination of several parallel experiments using different ^{15}N-labeled substrates (here experiment 1–4) can be changed to address the various N-turnover processes. Certainly, the experimental approach will be modified as new questions arise from the growing knowledge of the N-cycle.

ACKNOWLEDGMENTS

We are grateful to Gabriele Klockgether and Daniela Franzke for analytical assistance. The study was funded by the Max Planck Society and the Danish National Research Foundation.

REFERENCES

Bedard, C., and Knowles, R. (1989). Physiology, biochemistry, and specific inhibitors of CH_4, NH_4^+, and CO oxidation by methanotrophs and nitrifiers. *Microbiol. Rev.* **53**, 68–84.

Braman, R. S., and Hendrix, S. A. (1989). Nanogram nitrite and nitrate determination in environmental and biological-materials by vanadium(III) reduction with chemi-luminescence detection. *Anal. Chem.* **61**, 2715–2718.

Codispoti, L. A. (2001). The oceanic nitrogen cycle: A double-edged agent of environmental change? *In* "Waters in Peril," (L. Bendell-Young and P. Gallaugher, eds.), pp. 73–101. Kluwer Academic, Norwell.

Codispoti, L. A., *et al.* (2001). The oceanic fixed nitrogen and nitrous oxide budgets: Moving targets as we enter the anthropocene? *Sci. Mar.* **65**, 85–105.

Codispoti, L. A., *et al.* (2005). Suboxic respiration in the oceanic water column. *In* "Respiration in Aquatic Ecosystems," (P. J. Le.B. Williams and P. A. del Giorgio, eds.), pp. 225–247. Oxford University Press, Oxford.

Dalsgaard, T., *et al.* (2003). N_2 production by the anammox reaction in the anoxic water column of Golfo Dulce, Costa Rica. *Nature* **422**, 606–608.

Dalsgaard, T., *et al.* (2005). Anaerobic ammonium oxidation (anammox) in the marine environment. *Res. Microbiol.* **156**, 457–464.

Falkowski, P. G., *et al.* (1998). Biogeochemical controls and feedbacks on ocean primary production. *Science* **281**, 200–206.

Francis, C. A., *et al.* (2007). New processes and players in the nitrogen cycle: The microbial ecology of anaerobic and archaeal ammonia oxidation. *ISME J.* **1**, 19–27.

Grasshoff, K., *et al.* (1999). Methods of seawater analysis. Wiley-VCH, Weinheim, New York.

Gruber, N. (2004). The dynamics of the marine nitrogen cycle and its influence on atmospheric CO_2 variations. *In* "The ocean carbon cycle and climate," (M. Follows and T. Oguz, eds.), pp. 97–148. Kluwer Academic, Dordrecht.

Gruber, N., and Sarmiento, J. L. (1997). Global patterns of marine nitrogen fixation and denitrification. *Glob. Biogeochem. Cycles* **11**, 235–266.

Hall, G. H. (1984). Measurement of nitrification rates in Lake sediments—Comparison of the nitrification inhibitors Nitrapyrin and Allylthiourea. *Microb. Ecol.* **10**, 25–36.

Hamersley, M. R., *et al.* (2007). Anaerobic ammonium oxidation in the Peruvian oxygen minimum zone. *Limnol. Oceanogr.* **52**, 923–933.

Hannig, M., *et al.* (2007). Shift from denitrification to anammox after inflow events in the central Baltic Sea. *Limnol. Oceanogr.* **52**, 1336–1345.

Holmes, R. M., *et al.* (1999). A simple and precise method for measuring ammonium in marine and freshwater ecosystems. *Can. J. Fish. Aquat. Sci.* **56**, 1801–1808.

Jensen, M. M., *et al.* (2007). Effects of specific inhibitors on anammox and denitrification in marine sediments. *Appl. Environ. Microbiol.* **73**, 3151–3158.

Jensen, M. M., *et al.* (2008). Rates and regulation of anaerobic ammonium oxidation and denitrification in the Black Sea. *Limnol. Oceanogr.* **53**, 23–36.

Kartal, B., *et al.* (2007). Anammox bacteria disguised as denitrifiers: Nitrate reduction to dinitrogen gas via nitrite and ammonium. *Environ. Microbiol.* **9**, 635–642.

Kuypers, M. M. M., *et al.* (2003). Anaerobic ammonium oxidation by anammox bacteria in the Black Sea. *Nature* **422**, 608–611.

Kuypers, M. M. M., *et al.* (2005). Massive nitrogen loss from the Benguela upwelling system through anaerobic ammonium oxidation. *PNAS* **102**, 6478–6483.

Lam, P., *et al.* (2007). Linking crenarchaeal and bacterial nitrification to anammox in the Black Sea. *PNAS* **104**, 7104–7109.

Lam, P., et al. (2009). Revising the nitrogen cycle in the Peruvian oxygen minimum zone. *PNAS* **106**, 4752–4757.

Lam, P., and Kuypers, M. M. M. (2011). Microbial Nitrogen Cycling Processes in Oxygen Minimum Zones. *Annu. Rev. Mar. Sci.* Vol 3. (see http://www.annualreviews.org/catalog/pubdates.aspx).

Lavik, G., et al. (2009). Detoxification of sulphidic African shelf waters by blooming chemolithotrophs. *Nature* **457**, 581, U86.

Middelburg, J. J., et al. (1996). Denitrification in marine sediments: A model study. *Glob. Biogeochem. Cycles* **10**, 661–673.

Mulder, A., et al. (1995). Anaerobic Ammonium Oxidation Discovered in a Denitrifying Fluidized-Bed Reactor. *FEMS Microbiol. Ecol.* **16**, 177–183.

Nielsen, L. P. (1992). Denitrification in sediment determined from nitrogen isotope pairing. *FEMS Microbiol. Ecol.* **86**, 357–362.

Revsbech, N. P., et al. (2009). Determination of ultra-low oxygen concentrations in the oxygen minimum zone by the STOX sensor. *Limnol. Oceanogr. Methods* **7**, 371–381.

Strady, E., et al. (2008). PUMP-CTD-System for trace metal sampling with a high vertical resolution. A test in the Gotland Basin, Baltic Sea. *Chemosphere* **70**, 1309–1319.

Stramma, L., et al. (2008). Expanding oxygen-minimum zones in the tropical oceans. *Science* **320**, 655–658.

Thamdrup, B., and Dalsgaard, T. (2000). The fate of ammonium in anoxic manganese oxide-rich marine sediment. *Geochim. Cosmochim. Acta* **64**, 4157–4164.

Thamdrup, B., and Dalsgaard, T. (2002). Production of N_2 through anaerobic ammonium oxidation coupled to nitrate reduction in marine sediments. *Appl. Environ. Microbiol.* **68**, 1312–1318.

Thamdrup, B., et al. (2006). Anaerobic ammonium oxidation in the oxygen-deficient waters off northern Chile. *Limnol. Oceanogr.* **51**, 2145–2156.

Trimmer, M., et al. (2003). Anaerobic ammonium oxidation measured in sediments along the Thames estuary, United Kingdom. *Appl. Environ. Microbiol.* **69**, 6447–6454.

Ward, B. B., et al. (2009). Denitrification as the dominant nitrogen loss process in the Arabian Sea. *Nature* **461**, 78, U77.

Woebken, D., et al. (2007). Potential interactions of particle-associated anammox bacteria with bacterial and archaeal partners in the Namibian upwelling system. *Appl. Environ. Microbiol.* **73**, 4648–4657.

Woebken, D., et al. (2008). A microdiversity study of anammox bacteria reveals a novel *Candidatus scalindua* phylotype in marine oxygen minimum zones. *Environ. Microbiol.* **10**, 3106–3119.

Zumft, W. G. (1997). Cell biology and molecular basis of denitrification. *Microbiol. Mol. Biol. Rev.* **61**, 533–616.

CHAPTER ELEVEN

Assessment of Nitrogen and Oxygen Isotopic Fractionation During Nitrification and Its Expression in the Marine Environment

Karen L. Casciotti, Carolyn Buchwald, Alyson E. Santoro, *and* Caitlin Frame

Contents

1. Introduction	254
1.1. Nitrification and the marine N cycle	254
1.2. Stable isotopes and the marine N cycle	255
1.3. Isotopic fractionation	255
1.4. Nitrification and $\delta^{18}O_{NO_3}$	257
2. Methods	259
2.1. N and O isotopic fractionation and exchange during ammonia oxidation	259
2.2. N and O isotopic fractionation and exchange during nitrite oxidation	260
2.3. O isotopic exchange in mixed cultures and field populations	261
3. Analytical Techniques	262
3.1. Concentration analyses	262
3.2. $\delta^{15}N_{NH_3+NH4}$ analysis	262
3.3. $\delta^{15}N_{NO_2}$ and $\delta^{18}O_{NO_2}$ analyses	263
3.4. $\delta^{15}N_{NO_3}$ and $\delta^{18}O_{NO_3}$ analyses	263
3.5. $\delta^{18}O_{H_2O}$ analysis	264
4. Data Analysis	264
4.1. N isotopic fractionation during ammonia oxidation and nitrite oxidation	264
4.2. O isotope fractionation and exchange during ammonia oxidation	265
4.3. O_2 enrichment experiments	267
4.4. O isotopic fractionation and exchange during nitrite oxidation	268

Woods Hole Oceanographic Institution, Woods Hole, Massachusetts, USA

Methods in Enzymology, Volume 486   © 2011 Elsevier Inc.
ISSN 0076-6879, DOI: 10.1016/S0076-6879(11)86011-8   All rights reserved.

4.5. O isotopic exchange in mixed cultures and field populations	269
4.6. Tracking nitrification in the environment using natural N and O isotopic distributions	270
5. Discussion	272
5.1. Conditions for interpretation using the Rayleigh model	272
5.2. Fractionation during H_2O and O_2 incorporation	273
5.3. N_2O production fractionation factors, branching effects	274
5.4. Isotope effects for archaea	275
References	275

Abstract

Nitrification is a microbially-catalyzed process whereby ammonia (NH_3) is oxidized to nitrite (NO_2^-) and subsequently to nitrate (NO_3^-). It is also responsible for production of nitrous oxide (N_2O), a climatically important greenhouse gas. Because the microbes responsible for nitrification are primarily autotrophic, nitrification provides a unique link between the carbon and nitrogen cycles. Nitrogen and oxygen stable isotope ratios have provided insights into where nitrification contributes to the availability of NO_2^- and NO_3^-, and where it constitutes a significant source of N_2O. This chapter describes methods for determining kinetic isotope effects involved with ammonia oxidation and nitrite oxidation, the two independent steps in the nitrification process, and their expression in the marine environment. It also outlines some remaining questions and issues related to isotopic fractionation during nitrification.

1. INTRODUCTION

1.1. Nitrification and the marine N cycle

Nitrification occurs as a two-step process, whereby ammonia-oxidizing bacteria (AOB) and ammonia-oxidizing archaea (AOA) convert ammonia (NH_3) to nitrite (NO_2^-) and nitrite-oxidizing bacteria (NOB) convert NO_2^- to nitrate (NO_3^-). The process of nitrification forms an important link in the nitrogen cycle, converting the most reduced form of nitrogen to the most oxidized form, as well as producing substrates for denitrification and anaerobic ammonia oxidation (anammox), which remove fixed (bioavailable) forms of nitrogen (N) from the environment.

Ammonia oxidation also is linked to production of nitrous oxide (N_2O), a climatically important trace gas that is involved in the "greenhouse effect" and in stratospheric ozone chemistry (Ravishankara et al., 2009). The amount of N_2O emitted from the ocean comprises approximately one-third of the natural flux, or 20% of the present-day (anthropogenically enhanced) flux to the atmosphere (IPCC, 2007). Understanding the sources of N_2O in the ocean is an important step toward understanding how this flux may increase with climate-induced changes in productivity and anthropogenic perturbation of the N cycle. The relative contributions of nitrification and denitrification to the oceanic source of N_2O are not well understood, particularly in low oxygen environments where both may play a role. Evidence for nitrification as a major source of oceanic N_2O comes from relationships of O_2 with N_2O (Butler et al., 1989; Cohen and Gordon, 1978; Yoshinari, 1976), association of N_2O accumulation with high nitrification rates (Dore and Karl, 1996), and from isotopic measurements of N_2O (Dore et al., 1998; Kim and Craig, 1990; Ostrom et al., 2000; Popp et al., 2002).

1.2. Stable isotopes and the marine N cycle

N and oxygen (O) stable isotope ratios in dissolved inorganic N molecules (NH_4^+, NO_2^-, NO_3^-, and N_2O) have provided many insights into the biogeochemical processes comprising the marine N cycle, including nitrification. Because enzymatic processes tend to alter the distribution of isotopes between substrate and product pools, variations in the stable isotope ratios of these molecules can be used to constrain rates and pathways of N transformation. For example, the balance of N inputs and losses from the ocean is revealed in the isotope ratios of deep ocean NO_3^- (Altabet, 2007; Brandes and Devol, 2002; Deutsch et al., 2004), and the balance of uptake and regeneration is expressed in surface ocean NO_3^- isotopes (Wankel et al., 2007). Large signals are also observed in the isotope ratios of NO_2^- (Casciotti and McIlvin, 2007) and N_2O (McIlvin and Casciotti, 2010; Popp et al., 2002; Yamagishi et al., 2007; Yoshida et al., 1984; Yoshinari et al., 1997), which reflect the unique isotopic fractionation expressed during processes such as nitrification and denitrification in the subsurface. In order to correctly interpret the distributions of N and O isotope ratios in nature, knowledge about the isotopic fractionation associated with key microbial processes is needed.

1.3. Isotopic fractionation

Kinetic fractionation factors (α_k) arise from small differences in the rates at which isotopically substituted molecules react, and they have provided important insights into the mechanisms of enzyme activity (Cleland, 1987; Northrop, 1981). Kinetic fractionation factors are defined by the instantaneous change in the isotope ratio of the reaction product (R_{PI}) for a given substrate isotope ratio (R_S): $\alpha_k = R_S/R_{PI}$ (Mariotti et al., 1981), where $^{15}R = {}^{15}N/^{14}N$ for N isotopes and $^{18}R = {}^{18}O/^{16}O$ for O isotopes. The kinetic fractionation factor may also be represented by the ratio of rate constants for reaction of heavy and light isotopes. Assuming first order reaction kinetics, $\alpha_k = k_1/k_2$, where "1" and "2" refer to the light and heavy isotopes, respectively (Bigeleisen and Wolfsberg, 1958).

Kinetic isotopic fractionation during enzymatic reaction is understood in the framework of transition state theory, where bonds are made and/or broken to the isotopically substituted site (Cleland, 1987; Northrop, 1981; Thornton and Thornton, 1978). If the isotopically substituted species is more stably bonded in the substrate than in the transition state, then molecules containing the heavy isotope will react more slowly ($k_2 < k_1$). This behavior is termed "normal" kinetic isotope fractionation and typically occurs in bond-breaking reactions. However, if the isotopically substituted species is more stably bonded in the transition state, which may be the case in some bond-forming reactions, then molecules containing the heavy isotope may react more quickly ($k_2 > k_1$). This phenomenon is referred to as "inverse" kinetic isotope fractionation and has only been unequivocally observed during the process of nitrite oxidation, which is an elegantly simple bond-forming reaction (Buchwald and Casciotti, 2010; Casciotti, 2009).

Kinetic isotope fractionation defines the isotopic imprint of microbial activity in the environment and may be an important geochemical expression of microbial diversity (Buchwald and Casciotti, 2010; Casciotti et al., 2003, 2010; Detmers et al., 2001; Scott et al., 2004). In order to use natural isotopic distributions to infer the pathways of microbial N transformation, it is important to know the kinetic fractionation factors for each of the relevant processes. The fractionation factors have been determined for many of the N cycle transformations (Sigman et al., 2009b), with more recent progress on the isotopic fractionation involved with nitrification (Buchwald and Casciotti, 2010; Casciotti, 2009, 2010; Frame and Casciotti, 2010), which will be discussed in more detail below.

Kinetic fractionation factors are most simply determined from laboratory experiments with pure cultures of appropriate organisms. These experiments have the benefit of providing a tightly controlled, closed system with a single dominant process that can be (relatively) easily interpreted. However, they carry the risk of producing biased estimates of the isotope effects, due to the often nonrepresentative strains grown under nonrepresentative culture conditions. Field experiments may provide more relevant estimates of the fractionation factor, but careful accounting must be made of the sources and alternate fates of the compounds of interest in a mixed community. Some examples of this approach will be given below.

The Rayleigh model describes the evolution of the isotope ratio of the substrate ($\delta^{15}N_S$ or $\delta^{18}O_S$, where $\delta^{15}N_S = (^{15}R_S/^{15}R_{standard} - 1) \times 1000$ and $\delta^{18}O_S = (^{18}R_S/^{18}R_{standard} - 1) \times 1000$), instantaneous product ($\delta^{15}N_{PI}$ or $\delta^{18}O_{PI}$), and accumulated product ($\delta^{15}N_{PA}$ or $\delta^{18}O_{PA}$) in a

Assessment of Nitrogen and Oxygen Isotopic Fractionation

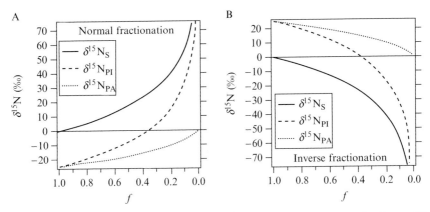

Figure 11.1 Expression of N isotopic fractionation in a closed system for "normal" kinetic isotopic fractionation (A) and "inverse" kinetic isotopic fractionation (B) as a function of f, the fraction of initial substrate remaining. In both panels, the $\delta^{15}N$ of the substrate ($\delta^{15}N_S$) is shown in solid lines, while the $\delta^{15}N$ of the instantaneous product ($\delta^{15}N_{PI}$) is shown in dashed lines, and the $\delta^{15}N$ of the accumulated product ($\delta^{15}N_{PA}$) is shown in dotted lines. In (A), $^{15}\varepsilon = +25‰$, and in (b) $^{15}\varepsilon = -25‰$.

closed system as the substrate is consumed (Fig. 11.1). If a transformation proceeds with a constant isotope effect ($\varepsilon(‰) = (\alpha - 1) \times 1000$) and if the substrate pool is not replenished during the transformation process, then the N isotopic composition of a substrate should evolve according to Eq. (11.1), the instantaneously generated product Eq. (11.2), and the accumulated product pool Eq. (11.3):

$$\delta^{15}N_S = \delta^{15}N_{S,0} - {}^{15}\varepsilon \times \ln(f) \quad (11.1)$$

$$\delta^{15}N_{PI} = \delta^{15}N_S - {}^{15}\varepsilon \quad (11.2)$$

$$\delta^{15}N_{PA} = \delta^{15}N_{S,0} + {}^{15}\varepsilon \times f \times \ln(f)/(1-f) \quad (11.3)$$

where f is the fraction of the initial substrate remaining at the point of measurement and, $^{15}\varepsilon$ is the N isotope effect. This simplified form of the Rayleigh model is often used to describe isotopic fractionation in batch cultures as well as some naturally occurring events, such as the uptake of NO_3^- in ocean surface waters (Sigman et al., 1999).

1.4. Nitrification and $\delta^{18}O_{NO_3}$

In addition to N isotope effects, there are also O isotope effects involved with the nitrification process (Buchwald and Casciotti, 2010; Casciotti et al., 2010). During bacterial nitrification, the biochemical sources of O atoms

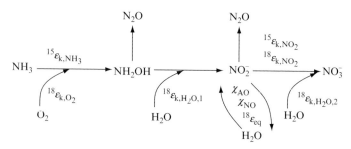

Figure 11.2 Schematic of isotopic fractionation and exchange during nitrification. During ammonia oxidation, N isotopic fractionation occurs at ammonia monooxygenase ($^{15}\varepsilon_{k,NH_3}$), and O isotopic fractionation occur during O_2 and H_2O incorporation ($^{18}\varepsilon_{k,O_2}$ and $^{18}\varepsilon_{k,H_2O,1}$, respectively) and exchange (x_{AO}, $^{18}\varepsilon_{eq}$). During nitrite oxidation, N and O isotopic fractionation occur at nitrite oxidoreductase ($^{15}\varepsilon_{k,NO_2}$ and $^{18}\varepsilon_{k,NO_2}$, respectively), and O isotopic fractionation occurs during H_2O incorporation ($^{18}\varepsilon_{k,H_2O,2}$) and exchange ($x_{NO}$, $^{18}\varepsilon_{eq}$). N and O isotopic fractionation may also occur during N_2O production from NH_2OH and NO_2^- by ammonia oxidizers (not shown).

are dissolved oxygen (O_2) and water (H_2O) (Andersson and Hooper, 1983; Kumar et al., 1983). O_2 is incorporated during the oxidation of NH_3 to hydroxylamine (NH_2OH), while H_2O is incorporated during the oxidation of NH_2OH to NO_2^- and NO_3^- (Fig. 11.2). While the ratio of 1:2 O atoms from O_2 and H_2O implied by these observations is commonly used to interpret the O isotopic content of NO_3^-, the utilization of this ratio involves the assumptions that exchange and fractionation of O isotopes during nitrification are minimal. In some cases, observed $\delta^{18}O$ values of NO_3^- ($\delta^{18}O_{NO3}$) do appear to be explained by the biochemical stoichiometry (Mayer et al., 2001); however, in the ocean NO_3^- does not appear to retain much of an isotopic imprint from O_2 (Casciotti et al., 2002; Sigman et al., 2009a). It is important to know if and when this stoichiometry can be used to interpret the $\delta^{18}O$ values of NO_2^-, NO_3^-, and N_2O.

In addition to variations in the $\delta^{18}O$ of the O atom donors (O_2 and H_2O), the $\delta^{18}O$ of newly produced NO_3^- is affected by O isotopic exchange and fractionation (Fig. 11.2). Kinetic isotope effects occur during (1) selection of NH_2OH ($^{18}\varepsilon_{k,NH_2OH}$), (2) selection of NO_2^- ($^{18}\varepsilon_{k,NO_2}$), (3) incorporation of O from O_2 ($^{18}\varepsilon_{k,O_2}$), and incorporation of O from H_2O during (4) hydroxylamine oxidation and (5) nitrite oxidation ($^{18}\varepsilon_{k,H_2O,1}$ and $^{18}\varepsilon_{k,H_2O,2}$, respectively) (Buchwald and Casciotti, 2010; Casciotti et al., 2010). In addition to these isotope effects, equilibrium fractionation ($^{18}\varepsilon_{eq}$) may occur during O isotope exchange between NO_2^- and H_2O during ammonia oxidation (x_{AO}) and nitrite oxidation (x_{NO}). Below we describe the set of methods that has been used to determine the N and O isotope systematics for nitrification in cultures of AOB and NOB, and present guidelines on extending these methods to mixed communities, including enrichment cultures of AOA, as well as field populations.

2. METHODS

2.1. N and O isotopic fractionation and exchange during ammonia oxidation

Determination of N ($^{15}\varepsilon_{k,NH_3}$; Fig. 11.2) and O ($^{18}\varepsilon_{k,O_2}$, $^{18}\varepsilon_{k,H_2O,1}$, and x_{AO}; Fig. 11.2) isotope systematics during ammonia oxidation can be done using simple batch culture experiments (Casciotti et al., 2003, 2010; Mariotti et al., 1981). Cells for ammonia oxidation experiments are harvested from appropriate maintenance cultures, then washed and resuspended in sterile DIW or seawater, depending on the culture's salinity requirements. Experiments are initiated by the injection of washed cell suspension into replicate serum bottles containing the appropriate medium: Walker medium (Soriano and Walker, 1968) for terrestrial species, and Watson medium (Watson, 1965) for marine species.

For tests of O isotope fractionation and exchange, the $\delta^{18}O$ of the H_2O in the medium ($\delta^{18}O_{H_2O}$) is adjusted to different values between -6‰ and $+88$‰ in replicate bottles. The $\delta^{18}O_{H_2O}$ adjustment is achieved by addition of ^{18}O-enriched H_2O to the base medium before dividing the medium into aliquots. To prepare the ^{18}O–H_2O spike, 1 g of 95 atom% ^{18}O–labeled H_2O (Cambridge Isotope Laboratories) is diluted in 99 mL distilled deionized water to produce a stock solution with $\delta^{18}O_{H_2O}$ of approximately $+5000$‰. This stock solution is added in 2.5, 5.0, or 10.0 mL volumes to 500 mL prepared medium to achieve final media $\delta^{18}O_{H_2O}$ values between -6‰ and $+88$‰. Each batch of medium is then divided among replicate 500 mL bottles, sealed with gray butyl septa, and autoclaved. The medium should be adjusted to pH 8.2 after autoclaving by addition of sterile K_2CO_3 to minimize abiotic exchange between NO_2^- and H_2O during the experiment (Casciotti et al., 2007). If other variables are to be tested, the appropriate conditions should be prepared prior to inoculation. For example, tests of fractionation and exchange at differing dissolved O_2 levels (5–200 µM) are conducted in serum bottles that have been autoclaved and sparged with the appropriate N_2:O_2 mixture.

To track abiotic production of NO_2^- in N isotope experiments and abiotic equilibration of NO_2^- in O isotope experiments, controls consisting of uninoculated medium are incubated with NH_4^+ (N isotope experiments) or NH_4^+ and NO_2^- (O isotope experiments), both adjusted to match the initial [NH_4^+] in the experimental bottles. In all, an experimental test of N fractionation for a single species, [NH_4^+], and pO_2 contains four bottles: two replicate inoculated bottles and two replicate uninoculated bottles. Tests of O isotopic fractionation and exchange involve 16 bottles: two replicate inoculated bottles and two replicate uninoculated bottles at each of four $\delta^{18}O_{H_2O}$ values.

Time course experiments are subsampled periodically until all the NH_4^+ has been oxidized to NO_2^-. The overall duration of the time course depends on the species of microorganism, cell density, and the relative concentrations of O_2 and NH_4^+ in the experiment. Samples (15 mL) are collected at the beginning of the experiment for $[NH_4^+]$, $[NO_2^-]$, $\delta^{15}N_{NO_2}$, $\delta^{18}O_{NO_2}$, and $\delta^{18}O_{H_2O}$ analyses. Subsequent time points are subsampled for $[NH_4^+]$, $[NO_2^-]$, $\delta^{15}N_{NO_2}$, and $\delta^{18}O_{NO_2}$ analyses (15 mL total). Liquid volumes removed from the bottles via syringe are replaced with gas injections of the same $N_2:O_2$ ratio as the original headspace to maintain a constant pressure and pO_2 in the bottle. All samples are filtered through a 0.22 μm pore size filter immediately upon collection. For O isotope experiments, concentrations and isotopic compositions of NO_2^- are analyzed within 1 h of sampling to alleviate uncertainty from preservation of $\delta^{18}O_{NO_2}$ samples (Casciotti et al., 2007). Otherwise, samples for $[NH_4^+]$, $[NO_2^-]$, $\delta^{15}N_{NO_2}$, $\delta^{18}O_{H_2O}$ can be frozen until analysis.

2.2. N and O isotopic fractionation and exchange during nitrite oxidation

In order to measure the N and O isotope effects during nitrite oxidation, experiments are conducted with batch cultures of NOB as previously described (Buchwald and Casciotti, 2010). For O incorporation fractionation and exchange experiments, H_2O with a $\delta^{18}O$ value of approximately +5000‰ versus VSMOW (prepared as described above) is added in amounts of 3–33 mL per liter of medium to achieve $\delta^{18}O_{H_2O}$ values for the media of −5‰ to +150‰ versus VSMOW. The ^{18}O-labeled H_2O is added prior to autoclaving to ensure adequate mixing of the labeled H_2O throughout the media. After autoclaving, filter-sterilized $NaNO_2$ is added and the medium is neutralized to pH 8.2 with sterile K_2CO_3.

The maintenance cultures (300–500 mL) are harvested either by centrifugation (4000 rpm for 30 min) or filtration (0.22 μm pore size filter), depending on the density of the culture (denser cultures are centrifuged, rather than filtered). The harvested bacteria are then washed and resuspended with 0.22 μm-filtered artificial seawater. The experiments are initiated with inoculation of the bacteria into the prepared media. All experiments utilize replicate experimental (inoculated) and sterile control bottles to check for abiotic oxidation of NO_2^- to NO_3^- and abiotic equilibration of NO_2^- with water at each $\delta^{18}O_{H_2O}$ value.

After inoculation, the flasks are subsampled (10–15 mL) immediately and then periodically throughout conversion of NO_2^- to NO_3^- and again after all NO_2^- has been consumed in the experimental bottles. Each subsample is 0.22 μm filtered immediately after collection. NO_2^- concentrations are measured immediately after sampling, and then used to calculate the sample volume needed to obtain 5–20 nmol of NO_2^- for isotopic analysis. Azide reactions for NO_2^- isotope analyses (see below for details) are also conducted within 1 h of sampling to avoid abiotic exchange of O atoms between NO_2^- and H_2O during storage of samples (Casciotti et al., 2007). The remainder of each subsample is stored frozen until analyses for NO_3^- concentration and isotopic composition.

2.3. O isotopic exchange in mixed cultures and field populations

Experiments to determine the O isotopic exchange during nitrification in mixed communities are similar in design to ammonia oxidation experiments in pure culture. The mixed communities may be as simple as two cultured organisms combined in a coculture or as complex as a laboratory enrichment culture or natural microbial community. If the mixed community contains ammonia oxidizers and nitrite oxidizers and is capable of transforming NH_3 to NO_3^-, with or without the accumulation of NO_2^-, the quantity of O isotope exchange during nitrification can be readily determined. Since x_{NO} is negligible (Buchwald and Casciotti, 2010), these experiments may essentially be used to determine x_{AO} for uncultured ammonia oxidizers, as well as for cultured ammonia oxidizers under more natural conditions.

In a defined coculture experiment, medium is prepared as described above for ammonia oxidation experiments, and ^{18}O-labeled H_2O is added prior to inoculation to achieve $\delta^{18}O_{H_2O}$ values between $-5‰$ and $+100‰$. In a field experiment, water samples would simply be collected and spiked with ^{18}O labeled H_2O, and NH_4^+, if necessary. Controls would include uninoculated medium or filter-sterilized water samples with added NO_2^- to track abiotic equilibration between NO_2^- and H_2O at each $\delta^{18}O_{H_2O}$ value.

Samples (~ 30 mL total) should be collected immediately and 0.22-μm filtered for analyses of $[NH_4^+]$, $[NO_2^-]$, $[NO_3^-]$, $\delta^{18}O_{H_2O}$, as well as $\delta^{18}O_{NO_2}$ and $\delta^{18}O_{NO_3}$ if sufficient quantities (>0.5 μM) are present. Thereafter, samples (~ 10 mL) are collected and filtered to track the course of the experiment in $[NH_4^+]$, $[NO_2^-]$, and $[NO_3^-]$. $[NO_2^-]$ analyses are made immediately, and the remainder of the sample is frozen for $[NH_4^+]$ and $[NO_3^-]$ analyses. When sufficient NO_2^- or NO_3^- has been produced, samples (~ 30 mL) are also collected and filtered for $\delta^{18}O_{NO_2}$ and $\delta^{18}O_{NO_3}$ analyses. Analyses of $\delta^{18}O_{NO_2}$ are conducted immediately using the azide method (see below), and the remainder of the sample is frozen for later $\delta^{18}O_{NO_3}$ analysis. To determine x_{AO}, it is not necessary for NH_4^+ to be completely oxidized to NO_3^-, only that sufficient NO_2^- and/or NO_3^- be produced for precise isotopic analysis (>0.5 μM). If there is preexisting NO_2^- or NO_3^-, its contribution to the measured $\delta^{18}O_{NO_2}$ and $\delta^{18}O_{NO}$ values will need to be subtracted in order to examine only the material produced in the experiment (in the presence of ^{18}O-labeled H_2O).

As both ammonia-oxidizing and nitrite-oxidizing microbes require O_2 for growth, the ambient O_2 concentration can affect the growth rate of nitrifying organisms (Carlucci and McNally, 1969) and the efficiency of coupling between ammonia oxidation and nitrite oxidation (Abeliovich, 2006). Decoupling ammonia oxidation from nitrite oxidation may subsequently lead to accumulation of NO_2^- and higher levels of exchange. Therefore, O_2 concentration may be an important variable in determining the amount of O exchange occurring during nitrification, and it should be considered when comparing ^{18}O labeled H_2O experiments in field populations.

3. Analytical Techniques

3.1. Concentration analyses

Concentrations of NH_4^+ and NO_2^- are analyzed spectrophotometrically by the indophenol blue assay (Solorzano, 1969) and the Greiss–Ilosvay assay (Strickland and Parsons, 1972), respectively. NO_3^- plus NO_2^- concentration is analyzed using a chemiluminescent NOx detector, following reduction to nitric oxide (NO) in a hot (95 °C) vanadium chloride bath (Cox, 1980; Garside, 1982). NO_3^- concentration is calculated by difference between $NO_3^- + NO_2^-$ and NO_2^- concentrations. Bracketing standards (0–50 μM) are produced gravimetrically and analyzed in parallel for NH_4^+, NO_2^-, and NO_3^-.

3.2. $\delta^{15}N_{NH_3+NH4}$ analysis

In determining the isotope effects for ammonia oxidation, it is necessary to follow the concentration and $\delta^{15}N$ value of the substrate (NH_3), product (NO_2^-), or both if possible. Given that the substrate is more sensitive to isotope effects than the accumulated product (Mariotti et al., 1981; Fig. 11.1), it would be beneficial to measure $\delta^{15}N_{NH_3}$. However, methods for determining $\delta^{15}N_{NH_3}$ actually group NH_3 and NH_4^+ in the analysis ($\delta^{15}N_{NH_3+NH4}$). The two most commonly applied methods for determination of $\delta^{15}N_{NH_3+NH4}$ are ammonia distillation (Velinsky et al., 1989) and diffusion (Holmes et al., 1998). These techniques rely on separation of gaseous NH_3 from a liquid sample and collection of NH_4^+ in acidified media. Because these standard methods rely on N_2 as the analyte, accurate $\delta^{15}N_{NH_3+NH4}$ analyses require 1–10 μmol of N, which sets limitations on the applicability of these techniques in systems where NH_4^+ concentrations are less than 1 μmol L^{-1}. A more recently developed technique based on oxidation of NH_3 and NH_4^+ to NO_2^-, followed by conversion of NO_2^- to N_2O by reaction with sodium azide (Zhang et al., 2007), has lowered the sample requirement and removed the need for distillation or diffusion. However, many steps are still involved with the conversion of NH_4^+ to N_2O and blanks need to be carefully evaluated at every step.

3.3. $\delta^{15}N_{NO_2}$ and $\delta^{18}O_{NO_2}$ analyses

Traditional methods for analysis of $\delta^{15}N_{NO_2}$ rely on complexation of NO_2^- as an analine dye, followed by solvent extraction (Olson, 1981) or solid phase extraction (Kator et al., 1992; Ward and O'Mullan, 2005). Using these methods, the N isotope ratio is measured after combustion of complexed NO_2^- to N_2. Newer techniques involving the conversion of NO_2^- to N_2O using the "azide method" are rapid and require only 10 nmol of nitrite per analysis (McIlvin and Altabet, 2005). After it is produced, the N_2O analyte is purged and trapped on a cryogenic preconcentration system (Casciotti et al., 2002), then released to an isotope ratio mass spectrometer. Each sample is analyzed in duplicate and is reported in delta notation relative to air ($\delta^{15}N_{NO_2}$) or VSMOW ($\delta^{18}O_{NO_2}$) by normalizing to NO_2^- isotopic reference materials N7373, N23, and N10219 (Casciotti et al., 2007), which are analyzed in parallel. Precisions for $\delta^{15}N_{NO_2}$ and $\delta^{18}O_{NO_2}$ measurements are approximately 0.5‰ for replicate analyses (McIlvin and Altabet, 2005). NO_2^- can also be selectively analyzed with similar sensitivity and precision using a modified denitrifier method with a bacterium that lacks nitrate reductase and nitrous oxide reductase (Böhlke et al., 2007).

3.4. $\delta^{15}N_{NO_3}$ and $\delta^{18}O_{NO_3}$ analyses

$\delta^{15}N$ analyses of NO_3^- can be made using an adaptation of the ammonia diffusion method with reduction of NO_3^- to NH_4^+ using Devarda's alloy (Sigman et al., 1997), high temperature on-line combustion of precipitated $AgNO_3$ (Revesz and Böhlke, 2002), the denitrifier method (Casciotti et al., 2002; Sigman et al., 2001), or the Cd/azide method (McIlvin and Altabet, 2005). $\delta^{18}O$ analyses of NO_3^- can be made using all of these except the ammonia diffusion method. In addition, although the $AgNO_3$ combustion technique works well for fresh water samples, separation of NO_3^- from a seawater matrix by anion exchange may prove difficult (Silva et al., 2000). Sample requirements for the combustion-based methods are 1–10 μmol NO_3^-. The denitrifier method (Casciotti et al., 2002; Sigman et al., 2001) and the azide method with prior Cd reduction (McIlvin and Altabet, 2005) provide high precision, high sensitivity analyses of $\delta^{15}N_{NO_3}$ and $\delta^{18}O_{NO_3}$ with N_2O as the analyte. These N_2O-based methods have lower sample requirements (5–20 nmol NO_3^-) and can be readily applied to seawater samples with simple purge and trap separation technologies.

In samples containing both NO_2^- and NO_3^-, these methods will all yield isotope ratios that reflect the combination of both species. Therefore, in order to obtain analyses of NO_3^- alone in the presence of significant amounts of NO_2^- (>1%), the NO_2^- should be removed using sulfamic acid (Granger and Sigman, 2009; Olson, 1981) prior to analysis of the

NO_3^- sample. It is also possible to determine the isotopic composition of NO_3^- by subtracting the NO_2^- contribution from the $NO_3^- + NO_2^-$ analysis (Casciotti and McIlvin, 2007), but this approach has larger uncertainties than direct measurement of NO_3^- isotopes after sulfamic treatment.

For NO_3^- isotopic analyses using the denitrifier method, the sample volumes are calculated using NO_3^- concentrations to obtain a constant amount (fixed at 5–20 nmol) of NO_3^-–N in each vial. Six replicate sets of NO_3^- standards USGS32, USGS34, and USGS35 (Böhlke et al., 2003) are analyzed at the same level of N with each denitrifier run to allow correction for blank and exchange during N and O isotopic analyses (Casciotti and McIlvin, 2007; Casciotti et al., 2002, 2007; Sigman et al., 2001). Precision on replicate $\delta^{15}N_{NO_3}$ and $\delta^{18}O_{NO_3}$ analyses is approximately 0.3‰ and 0.5‰, respectively (Casciotti et al., 2002; Sigman et al., 2001).

3.5. $\delta^{18}O_{H_2O}$ analysis

Methods for analysis of $\delta^{18}O_{H_2O}$ based on equilibration with CO_2 are well established (Epstein and Mayeda, 1953). They require approximately 1 mL of H_2O for analysis and have precision for $\delta^{18}O_{H_2O}$ analysis of 0.1‰. Newer techniques include laser spectroscopic techniques (Gianfrani et al., 2003) and H_2O equilibration with NO_2^-, followed by conversion of NO_2^- to N_2O using a modified azide method (McIlvin and Casciotti, 2006). For studies of oxygen isotope systematics of nitrification, the nitrite equilibration technique can easily be incorporated using the same equipment used for NO_2^- and NO_3^- analyses. In this technique, each sample is analyzed in duplicate using samples as small as 0.5 mL, and data are reported in delta notation relative to VSMOW by normalizing to three H_2O standards that are analyzed in triplicate (McIlvin and Casciotti, 2006). Precision for $\delta^{18}O_{H_2O}$ measurements by this technique is approximately 0.5‰ for replicate analyses.

4. DATA ANALYSIS

4.1. N isotopic fractionation during ammonia oxidation and nitrite oxidation

The general approach for estimating kinetic fractionation factors in batch culture using a Rayleigh model was discussed above. Here we discuss the specific application of this model to determine N fractionation factors during ammonia oxidation ($^{15}\varepsilon_{k,NH_3}$) and nitrite oxidation ($^{15}\varepsilon_{k,NO_2}$).

During ammonia oxidation, the substrate (NH_3) becomes enriched in ^{15}N as it is consumed according to the Rayleigh equation (Eq. (11.1)). The main product (NO_2^-) is depleted in ^{15}N relative to the substrate pool but also becomes enriched in ^{15}N over time (Eq. (11.3); Fig. 11.1A).

$^{15}\varepsilon_{k,NH_3}$ could then theoretically be estimated from $\delta^{15}N_{NH_3}$ according to the Rayleigh equation (Eq. (11.1)), or from $\delta^{15}N_{NO_2}$ according to the accumulated product equation (Eq. (11.3)). However, concentration and isotopic analyses generally group NH_3 and NH_4^+ which are connected by a pH-dependent equilibrium. At circumneutral pH, NH_4^+ (rather than NH_3) is the dominant form. This has implications for the estimation of $^{15}\varepsilon_{k,NH_3}$, estimated from the Rayleigh equation (Eq. (11.1)), which will be discussed in more detail below. Nevertheless, $^{15}\varepsilon_{k,NH_3}$ is rarely estimated from the Rayleigh distillation equation, but is most often calculated from the accumulated product equation (Eq. (11.3)) using $\delta^{15}N_{NO_2}$ measured at multiple points during the course of ammonia oxidation in batch culture. In this application, f is the fraction of initial $NH_3 + NH_4^+$ remaining ($[NH_3 + NH_4^+]/[NH_3 + NH_4^+]_0$).

During nitrite oxidation, both the N and O isotopes of the substrate pool (NO_2^-) are fractionated as it is consumed, and in this case the substrate pool becomes depleted in the heavy isotopes (^{15}N and ^{18}O) during the course of nitrite oxidation (Buchwald and Casciotti, 2010; Casciotti, 2009). This pattern is the result of an inverse kinetic isotope effect. As a result, the main product (NO_3^-) is enriched in ^{15}N relative to the substrate pool (Fig. 11.1B). Measurement of $\delta^{15}N$ of NO_2^- or NO_3^- at multiple points during the course of nitrite oxidation in batch culture yields the N isotope effect for nitrite oxidation ($^{15}\varepsilon_{k,NO_2}$) according to the Rayleigh equation (Eq. (11.1)), or the accumulated product equation (Eq. (11.3)), respectively. Similar equations can also be written for O isotope fractionation in a closed system, although $\delta^{18}O_{NO_3}$ produced via nitrite oxidation is also affected by incorporation of O from H_2O (see below). In both cases, f is the fraction of initial NO_2^- remaining ($[NO_2^-]/[NO_2^-]_0$).

4.2. O isotope fractionation and exchange during ammonia oxidation

The $\delta^{18}O_{NO_2}$ values produced during ammonia oxidation depend on the $\delta^{18}O$ of substrates that are incorporated biochemically (O_2 and H_2O), as well as isotopic fractionation during O atom incorporation (Fig. 11.2). In addition, exchange between NO_2^- and H_2O during ammonia oxidation (x_{AO}) has the effect of increasing the dependence of $\delta^{18}O_{NO_2}$ on $\delta^{18}O_{H_2O}$ and lowering its dependence on $\delta^{18}O_{O_2}$. If there is isotopic fractionation during this equilibration ($^{18}\varepsilon_{eq}$), then the oxygen atom exchange may lead to a systematic offset between $\delta^{18}O_{H_2O}$ and $\delta^{18}O_{NO_2}$ as occurs abiotically (Casciotti et al., 2007).

The fractionation factor for H_2O incorporation during ammonia oxidation ($^{18}\varepsilon_{k,H_2O,1}$) is not easily determined from changes in $\delta^{18}O_{H_2O}$ using a standard Rayleigh model since the fractional consumption of H_2O during these reactions is so small, and changes in the $\delta^{18}O$ of the substrate ($\delta^{18}O_{H_2O}$) are negligible. Determining the isotope effects for O_2

incorporation ($^{18}\varepsilon_{k,O_2}$) based on isotopic enrichment of O_2 is also challenging given that the reaction catalyzed by ammonia monooxygenase (AMO) is not the only sink for O_2 during ammonia oxidation. Respiration of O_2 by cytochrome oxidase serves as a simultaneous sink of O_2 in ammonia-oxidizing microbes. It is theoretically possible to disentangle the isotope effects for cytochrome oxidase and AMO by conducting parallel experiments of ammonia oxidizers grown on NH_4^+ (AMO and cytochrome oxidase) and NH_2OH (only cytochrome oxidase), if the stoichiometry is well known (Bock et al., 1989). However, growth of ammonia oxidizers on NH_2OH is not always robust. The approach we have taken to determine the combined isotope effect for H_2O and O_2 incorporation during ammonia oxidation is to compare the produced $\delta^{18}O_{NO_2}$ with the substrates $\delta^{18}O_{O_2}$ and $\delta^{18}O_{H_2O}$, as shown in Eq. (11.4):

$$\delta^{18}O_{NO_2} = \left[\frac{1}{2}\left(\delta^{18}O_{O_2} - {}^{18}\varepsilon_{k,O_2}\right) + \frac{1}{2}\left(\delta^{18}O_{H_2O} - {}^{18}\varepsilon_{k,H_2O,1}\right)\right](1 - x_{AO})$$
$$+ \left[\delta^{18}O_{H_2O} + {}^{18}\varepsilon_{eq}\right](x_{AO}) \quad (11.4)$$

where $^{18}\varepsilon_{eq}$ is the equilibrium isotope effect for NO_2^- equilibration with H_2O. This equation can be rearranged to group terms containing $\delta^{18}O_{H_2O}$, resulting in Eq. (11.5):

$$\delta^{18}O_{NO_2} = \left[\frac{1}{2}(1 + x_{AO})\right]\left(\delta^{18}O_{H_2O}\right)$$
$$+ \frac{1}{2}\left(\delta^{18}O_{O_2} - {}^{18}\varepsilon_{k,O_2} - {}^{18}\varepsilon_{k,H_2O,1}\right)(1 - x_{AO}) + \left({}^{18}\varepsilon_{eq}\right)(x_{AO}) \quad (11.5)$$

Based on Eq. (11.5), $\delta^{18}O_{NO_2}$ is expected to vary linearly with $\delta^{18}O_{H_2O}$ in parallel batch cultures that differ only in their $\delta^{18}O_{H_2O}$ values. The slope of the linear regression yields an expression for x_{AO} (Fig. 11.3A), while the intercept relies on $\delta^{18}O_{O_2}$, $^{18}\varepsilon_{k,O_2}$, $^{18}\varepsilon_{k,H_2O,1}$, $^{18}\varepsilon_{eq}$, and x_{AO}. Once x_{AO} is determined from the slope, it can be used to solve for $^{18}\varepsilon_{k,O_2} + {}^{18}\varepsilon_{k,H_2O,1}$ from the intercept term, with measurements or assumptions of $\delta^{18}O_{O_2}$ and $^{18}\varepsilon_{eq}$.

We typically assume that the value of $^{18}\varepsilon_{eq}$ that has been determined for abiotic oxygen atom exchange between nitrite and water (14.4‰; Casciotti et al., 2007) applies to the enzymatically catalyzed O atom exchange during ammonia oxidation and nitrite oxidation. This is partially justified since enzymes are understood to accelerate the approach to equilibrium but do not change the equilibrium point of a system.

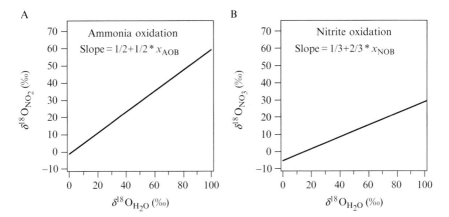

Figure 11.3 $\delta^{18}O_{H_2O}$ dependence of $\delta^{18}O_{NO_2}$ produced during ammonia oxidation (A) and $\delta^{18}O_{NO_3}$ produced during nitrite oxidation (B). The slopes depend on the fraction of O incorporated from H_2O and the exchange between NO_2^- and H_2O during ammonia oxidation (x_{AO}) and nitrite oxidation (x_{NO}). The intercepts depend on the O isotopic fractionation during incorporation of O_2 and H_2O (see text).

The term $^{18}\varepsilon_{k,O_2} + {}^{18}\varepsilon_{k,H_2O,1}$ represents the overall fractionation factor for O atom incorporation during ammonia oxidation. Although we cannot currently separate the isotope effects of O_2 and H_2O incorporation, the overall kinetic isotopic fractionation for O atom incorporation during ammonia oxidation is useful for predicting $\delta^{18}O_{NO_2}$ and $\delta^{18}O_{NO_3}$ values from known values of $\delta^{18}O_{O_2}$ and $\delta^{18}O_{H_2O}$.

These calculations assume that O atom exchange occurs between NO_2^- and H_2O, rather than NH_2OH and H_2O. Exchange of O isotopes between NH_2OH and H_2O probably cannot be completely ruled out. However, studies that determined the O atom source for NH_2OH did not observe exchange between NH_2OH and H_2O in *N. europaea* cultures (Dua *et al.*, 1979; Hollocher *et al.*, 1981). If exchange between NH_2OH and H_2O does occur to some extent in these experiments, it would not change estimates for the amount of O atom exchange, but it could change the interpretation of the $\delta^{18}O_{NO_2}$ versus $\delta^{18}O_{H_2O}$ intercept.

4.3. O_2 enrichment experiments

$^{18}O_2$ enrichment experiments could be conducted with AOB to examine the dependence of $\delta^{18}O_{NO_2}$ on the $\delta^{18}O_{O_2}$ value. Rearranging Eq. (11.4) so that $\delta^{18}O_{O_2}$ is the independent variable gives Eq. (11.6):

$$\delta^{18}O_{NO_2} = \frac{1}{2}(1 - x_{AO})(\delta^{18}O_{O_2}) + \frac{1}{2}(1 - x_{AO}) \\ (\delta^{18}O_{H_2O} - {}^{18}\varepsilon_{k,H_2O,1} - {}^{18}\varepsilon_{k,O_2}) + (\delta^{18}O_{H_2O} + {}^{18}\varepsilon_{eq})(x_{AO}) \quad (11.6)$$

where terms are as defined in Eqs. (11.4) and (11.5). Based on Eq. (11.6), it can be seen that the slope between $\delta^{18}O_{NO_2}$ and $\delta^{18}O_{O_2}$ would be dependent on x_{AO}, but in this case exchange would have the effect of lowering the slope or dependence on $\delta^{18}O_{O_2}$. The intercept would have similar dependencies as for $\delta^{18}O_{H_2O}$ experiments, and separation of $^{18}\varepsilon_{k,O_2}$ and $^{18}\varepsilon_{k,H_2O,1}$ would still be unachievable. Only if the $\delta^{18}O$ value of NH_2OH were measured could $^{18}\varepsilon_{k,O_2}$ and $^{18}\varepsilon_{k,H_2O,1}$ be separated in their effects on nitrification end products.

4.4. O isotopic fractionation and exchange during nitrite oxidation

The O isotope systematics for nitrite oxidation can be described by the following equation:

$$\delta^{18}O_{NO_3,final} = \frac{2}{3}((1-x_{NO})\delta^{18}O_{NO_2,initial}$$
$$+x_{NO}(\delta^{18}O_{H_2O} + {}^{18}\varepsilon_{eq})) + \frac{1}{3}(\delta^{18}O_{H_2O} - {}^{18}\varepsilon_{k,H_2O,2}) \quad (11.7)$$

In this equation, x_{NO} is the fraction of O atoms in NO_2^- that have been enzymatically exchanged with water prior to conversion to NO_3^-, $\delta^{18}O_{NO_3,final}$ is the final $\delta^{18}O_{NO_3}$ value after all NO_2^- has been oxidized to NO_3^-, $\delta^{18}O_{NO_2,initial}$ is the $\delta^{18}O$ value of the initial substrate, and $^{18}\varepsilon_{k,H_2O,2}$ is the kinetic isotope effect for H_2O incorporation during nitrite oxidation, and other terms are as defined above. A nonzero value for x_{NO} causes the $\delta^{18}O$ of the reacting NO_2^- to change over time, which introduces an additional dependence on $\delta^{18}O_{H_2O}$ in Eq. (11.7). As above, we assume that an equilibrium isotope effect ($^{18}\varepsilon_{eq}$) of 14.4‰ applies.

We can determine values for x_{NO} and $^{18}\varepsilon_{k,H_2O,2}$ by analyzing data from parallel incubations of NOB conducted at a variety of $\delta^{18}O_{H_2O}$ values. This becomes apparent after rearranging Eq. (11.7) with $\delta^{18}O_{H_2O}$ as the independent variable:

$$\delta^{18}O_{NO_3,final} = \left[\frac{2}{3}(x_{NO}) + \frac{1}{3}\right]\delta^{18}O_{H_2O}$$
$$+ \left[\frac{2}{3}[(1-x_{NO})\delta^{18}O_{NO_2,initial} + x_{NO}({}^{18}\varepsilon_{eq})] - \frac{1}{3}{}^{18}\varepsilon_{k,H_2O,2}\right] \quad (11.8)$$

If x_{NO} and $^{18}\varepsilon_{k,H_2O,2}$ are constant for a given experiment (do not change with different $\delta^{18}O_{H_2O}$ values), then the slope (m) and intercept (b) of this linear equation are constants. Therefore, if $\delta^{18}O_{NO_2,initial}$, $^{18}\varepsilon_{eq}$, and

$\delta^{18}O_{NO_3,final}$ are known then x_{NO} can be calculated from the slope and $^{18}\varepsilon_{k,H_2O,2}$ from the intercept of Eq. (11.8) (as shown in Fig. 11.3B).

The kinetic isotope effect for H_2O incorporation during nitrite oxidation ($^{18}\varepsilon_{k,H_2O,2}$) can be calculated from the intercept of Eq. (11.8) after incorporating measured values of b, x_{NO}, and $\delta^{18}O_{NO_2,initial}$ and an estimate of $^{18}\varepsilon_{eq}$ (+14.4‰, see above). Low amounts of exchange are typically observed during nitrite oxidation (Buchwald and Casciotti, 2010; Dispirito and Hooper, 1986; Friedman et al., 1986) so that estimates of $^{18}\varepsilon_{k,H_2O,2}$ are not very sensitive to the assumed value of $^{18}\varepsilon_{eq}$, which affects estimates of $^{18}\varepsilon_{k,H_2O,2}$ by only 0.05‰ for every 1‰ change in $^{18}\varepsilon_{eq}$.

4.5. O isotopic exchange in mixed cultures and field populations

O isotopic exchange during ammonia oxidation can also be determined in mixed culture or field populations incubated with ^{18}O-labeled H_2O because H_2O is only incorporated into these pools during nitrification. Consumption of NO_2^- and NO_3^- by co-occurring processes may lead to fractionation of NO_2^- and NO_3^-, but *variations* of $\delta^{18}O_{NO_2}$ ($\Delta\delta^{18}O_{NO_2}$) and $\delta^{18}O_{NO_3}$ ($\Delta\delta^{18}O_{NO_3}$) in parallel incubations differing only in $\delta^{18}O_{H_2O}$ should depend only on $\delta^{18}O_{H_2O}$ and the ratios of O_2 and H_2O incorporated during nitrification. These ratios, in turn, depend on the biochemical incorporation ratio and the exchange of O atoms during ammonia oxidation, according to Eqs. (11.9) and (11.10):

$$\Delta\delta^{18}O_{NO_2} = \left(\frac{1}{2} + \frac{1}{2}x_{AO}\right) \times \Delta\delta^{18}O_{H_2O} \quad (11.9)$$

$$\Delta\delta^{18}O_{NO_3} = \left(\frac{2}{3} + \frac{1}{3}x_{AO}\right) \times \Delta\delta^{18}O_{H_2O} \quad (11.10)$$

where $\Delta\delta^{18}O_{H_2O}$ is the $\delta^{18}O_{H_2O}$ difference between parallel incubations and x_{AO} is the fraction of enzymatically exchanged O atoms by the ammonia-oxidizing population. These equations assume that x_{NO} is negligible, as observed for several species of NOB (Buchwald and Casciotti, 2010). It should be noted that these variations in $\delta^{18}O_{NO_2}$ and $\delta^{18}O_{NO_3}$ would only be observed in the newly produced NO_2^- and NO_3^-, and therefore the $\delta^{18}O$ contributions from preexisting NO_2^- and NO_3^- should be subtracted from the measured values prior to interpretation via Eqs. (11.9) and (11.10). In some cases, the signals might be too small to resolve due to low rates of production of NO_2^- and NO_3^- relative to preexisting pools.

4.6. Tracking nitrification in the environment using natural N and O isotopic distributions

Natural distributions of N isotopes in NH_4^+ and N and O isotopes in NO_2^-, NO_3^- and N_2O can provide important constraints on the biogeochemical cycling of nitrogen. Two primary applications of natural abundance isotopes have been used to track nitrification in the environment: (1) estimation of fractionation factors for NH_4^+ consumption, consistent with nitrification (Cifuentes et al., 1989; Horrigan et al., 1990a; Mariotti et al., 1984), and (2) application of previously estimated fractionation factors to constrain nitrification fluxes (Casciotti and McIlvin, 2007; Sigman et al., 2005; Wankel et al., 2007).

The large N isotope effect associated with ammonia oxidation is expected to lead to production of ^{15}N-depleted nitrification products (NO_2^-, NO_3^-, N_2O) where nitrification plays a dominant role in their biogeochemistry. In addition, the O isotopic systematics of nitrification sets the source signatures of NO_2^-, NO_3^-, and in some cases, N_2O. Expression of N isotopic fractionation of nitrification in the environment depends on the extent of NH_4^+ consumption and involves processes occurring at the cellular-level, as well as physical mixing, input of new NH_4^+ sources, and alternate fates for NH_4^+ and NO_2^- that need to be considered in the interpretation of their isotopic distributions. Because of this, a good understanding of the physical, chemical, and biological setting is critical for proper interpretation of stable isotope patterns in any system.

Horrigan et al. (1990b) used ^{15}N tracer incubations to establish the instantaneous rates of DIN transformation on two cruises (Spring and Fall 1984) in the Chesapeake Bay. In fall, they observed rapid storm-induced mineralization of NH_4^+ followed by ammonia and nitrite oxidation. In many cases, the rates of the exchanges between DIN pools exceeded uptake by phytoplankton. After determining that ammonia oxidation was a dominant pathway for NH_4^+ consumption, they then used natural abundance $\delta^{15}N_{NH_3+NH_4}$ and $\delta^{15}N_{NO_2}$ measurements to calculate an N isotope effect for these processes (Horrigan et al., 1990a). Focusing on one station where $\delta^{15}N_{NH_3+NH_4}$ showed an inverse relationship to $[NH_4^+]$, they were able to derive isotope effects for nitrification of 12–17‰, based on the offset and temporal changes in $\delta^{15}N_{NH_3+NH_4}$ and $\delta^{15}N_{NO_2}$. Remarkably, these estimated fractionation factors are within the range of estimates for AOB (Casciotti et al., 2003; Mariotti et al., 1981; Miyake and Wada, 1971).

This study was feasible because they had independently determined that ammonia oxidation was a dominant process in the system. In general, if fates for NH_3 other than oxidation to NO_2^- and NO_3^- are also important, they need to be considered in interpreting $\delta^{15}N_{NH_3+NH_4}$, $\delta^{15}N_{NO_2}$, and $\delta^{15}N_{NO_3}$ variations. For example, concurrent assimilation of NH_4^+ by phytoplankton will lead to a lower apparent fractionation factor for ammonia oxidation when estimated from the $\delta^{15}N_{NH_3+NH_4}$ and $\delta^{15}N_{NO_3}$, because this fraction of the NH_4^+ pool will be consumed with a smaller isotope effect. In addition, alternative fates for nitrification products NO_2^- and NO_3^- may also affect the expression of the isotope effects for nitrification based on the $\delta^{15}N_{NO_2}$ and $\delta^{15}N_{NO_3}$ values. For example, if NO_2^- produced by ammonia oxidation is subsequently oxidized or assimilated, $\delta^{15}N_{NO_2}$ may be raised or lowered, depending on the magnitude and sign of the kinetic fractionation for NO_2^- consumption. If subsequent processes like nitrite oxidation lower $\delta^{15}N_{NO_2}$, then the fractionation for ammonia oxidation would be *overestimated*. If the processes such as assimilation raise $\delta^{15}N_{NO_2}$, then the fractionation for ammonia oxidation would be *underestimated*.

Another powerful application of stable isotope ratio measurements is to estimate the ratio of N fluxes through different pathways without prior knowledge of those fluxes. This requires building a biogeochemical model of the system accounting for possible N transformations based on a parallel system of equations describing the isotopic dynamics. Such models incorporate estimates of the isotope effects based on culture or those derived from other systems (Casciotti and McIlvin, 2007; Sigman et al., 2005; Wankel et al., 2007). Since co-occurring processes may produce overlapping or counteracting signals in any one isotope system, the use of multiple isotope systems ($\delta^{15}N$ and $\delta^{18}O$ in NH_4^+, NO_2^-, NO_3^-, or N_2O) is necessary to constrain fluxes where multiple processes are occurring.

A good example of the application of this approach to track nitrification is using coupled N and O isotope variations of nitrate in the euphotic zone, which enables the tracking of nitrification co-occurring with nitrate uptake (Wankel et al., 2007). Nitrate assimilation produces ^{15}N and ^{18}O enrichment in residual NO_3^- in a 1:1 ratio (Granger et al., 2004, 2010), whereas nitrification tends to decouple variations in $\delta^{15}N_{NO_3}$ and $\delta^{18}O_{NO_3}$. This decoupling occurs because the N isotope signature of nitrification is constrained by the $\delta^{15}N$ of organic matter being remineralized, and thus carries with it some "memory" of the NO_3^- uptake. In contrast, the new O atoms are added to regenerated N from O_2 and H_2O pools, independently of other N cycle processes. Careful modeling of the of the N and O isotope deviations of NO_3^- can then be used to determine the relative rates of ammonia oxidation and uptake, as well as the fraction of NO_3^- uptake that is supplied by nitrification in the euphotic zone (Wankel et al., 2007). This approach is highly dependent on knowledge of the N and O isotope effects for assimilation and nitrification and is best coupled with ancillary data such as instantaneous rate measurements of productivity and export, nutrient ratios (N:P stoichiometry or N deficits), redox conditions (degree of oxygenation), and assessment of the microbial communities. In this case, a considerable amount of ancillary data is generally needed to constrain the interpretation of isotopic data in any field setting. It is also important that the time scales being recorded by the ancillary measurements are consistent with those for changes in the isotope ratios. For example, the isotopes of inorganic N pools with long residence times (hundreds of years) may not be easily related to instantaneous measurements of microbial activity, which may change on time scales of hours, days, or weeks. However, some pools of inorganic N (such as NO_3^- or NH_4^+ in oligotrophic surface waters) do turn over on short time scales and are more amenable to pairing with analyses of microbial community structure and activity.

5. Discussion

Despite many of the advances described above for determining isotope effects involved with nitrification in culture and in the environment, there are still many aspects of kinetic isotope fractionation during nitrification that we do not understand. Here we discuss some of the biggest remaining uncertainties in the isotopic systematics of nitrification.

5.1. Conditions for interpretation using the Rayleigh model

The Rayleigh model only strictly holds for a single-step unidirectional reaction occurring in a closed system. Although the kinetic isotope effect for ammonia oxidation is often measured in pure batch cultures, it involves multiple steps and has several opportunities to deviate from Rayleigh behavior, including: (1) pH-dependent isotopic equilibration between NH_3 and NH_4^+, (2) variable rates of oxidation relative to rates of transport (NH_4^+) or diffusion (NH_3), and (3) transient accumulation of intermediates.

Although isotopic analyses do not separate NH_3 from NH_4^+, these pools are isotopically distinct. Equilibration between NH_3 and NH_4^+ leads to a large isotopic offset between NH_3 and NH_4^+, whereby NH_3 (the substrate for ammonia oxidation) is depleted in ^{15}N by 20–35‰ relative to NH_4^+ (Bigeleisen, 1965; Hermes et al., 1985). Because NH_4^+ ($pK_a = 9.3$) is the dominant species at circumneutral pH, its $\delta^{15}N$ value ($\delta^{15}N_{NH_4}$) is similar to measured $\delta^{15}N_{NH_3+NH4}$, while $\delta^{15}N_{NH_3}$ would be 20–35‰ lower. Consumption of the minor, ^{15}N-depleted pool (NH_3) should then lead to an apparent isotope effect of 20–35‰ for ammonia oxidation, even with no enzyme-level effect. This would apply to $^{15}\varepsilon_{k,NH_3}$ estimated from either $\delta^{15}N_{NH_3+NH4}$ (Rayleigh equation) or $\delta^{15}N_{NO_2}$ (accumulated product equation), relative to initial $\delta^{15}N_{NH_3+NH4}$. This effect most likely contributes to the large isotope effects observed for ammonia oxidation when diffusion is not limiting.

In some systems, however, expressed fractionation has been shown to be dictated by a balance of transport and enzyme-level effects (Goericke et al., 1994; Laws et al., 1997; Needoba et al., 2004). The AMO enzyme is thought to be positioned in the cell membrane with its active site facing the periplasm

(Whittaker et al., 2000). Therefore, an ammonia-oxidizing cell must rely on diffusion of NH_3 across the outer membrane or transport of NH_4^+ into the periplasm to supply the substrate for reaction. In order for fractionation associated with NH_3/NH_4^+ equilibration or the enzyme itself to be expressed, there must be leakage of partially consumed substrate from the periplasm back into the growth medium. If enzymatic activity occurs slowly relative to transport or diffusion, then the isotope effect associated with equilibration and the enzyme will be maximally expressed. However, if enzyme activity occurs rapidly relative to transport or diffusion, the isotopic fractionation associated with transport/diffusion will be expressed instead. Given that fractionation associated with transport is generally smaller than fractionation caused by enzymatic activity (O'Leary, 1981), and would certainly be smaller than the equilibrium isotope effect between NH_3 and NH_4^+, the rate of ammonia oxidation should strongly affect the expressed kinetic isotope effect. These cellular-level effects may lead to different apparent isotope effects for ammonia oxidation among nitrifying organisms (also among different experiments with the same organism, or in a single experiment as NH_4^+ is drawn down) if they express high- or low-affinity transport systems or if the rates of ammonia oxidation differ. Putative ammonia/ammonium transporte genes have been identified in the genomes of cultivated AOB (Chain et al., 2003; Klotz et al., 2006; Norton et al., 2008) and two putative Amt-type transporters have been identified in the genome of the AOA *Nitrosopumilus maritimus* (Walker et al., 2010).

Finally, the oxidation of NH_3 by AMO produces NH_2OH, which is later converted to NO_2^- by HAO. NH_2OH is not generally thought to accumulate, although few studies have attempted to firmly establish this. If NH_2OH does accumulate, even temporarily, then the expressed fractionation will reflect the activity of both AMO and HAO, as well as equilibrium effects (discussed above). This could lead to an overestimation of the N isotope effect for AMO based on $\delta^{15}N_{NO_2}$ because any isotope effect involved in the second step (NH_2OH oxidation to NO_2^-) would increase the $\delta^{15}N$ difference between the substrate (NH_3) and ultimate product (NO_2^-). Measurement of $\delta^{15}N_{NH_3+NH4}$ in this case would lead to better estimates of $^{15}\varepsilon_{k,NH_3}$. If AMO catalyzes the rate-limiting step and fractionation associated with HAO is not expressed, calculations of $^{15}\varepsilon_{k,NH_3}$ based on $\delta^{15}N_{NO_2}$ would reflect only equilibrium and enzyme-level effects, as expected.

5.2. Fractionation during H_2O and O_2 incorporation

As discussed above, it is now possible to make a first-order prediction of the $\delta^{18}O_{NO_2}$ and $\delta^{15}N_{NO_3}$ values produced wherever bacteria are responsible for ammonia oxidation and the O isotope ratios of H_2O and O_2 are known. Such predictions are possible using the overall kinetic isotope effect for O atom incorporation by AOB. Defining the O isotope systematics for N_2O

production, however, will require separation of the isotope effects for O_2 and H_2O incorporation because N_2O generated from NH_2OH is only affected by the isotope effect of O_2 incorporation. Furthermore, isotope effects for O atom incorporation for archaeal ammonia oxidation are still needed to predict NO_2^- and NO_3^- isotope ratios where archaea dominate the nitrification process. Therefore, additional work is needed to fully characterize the O isotopic systematics for nitrification. The experiments described above may be used to address part of this, but new approaches will be needed to separate the fractionation factors for O incorporation from O_2 and H_2O.

5.3. N_2O production fractionation factors, branching effects

N_2O is produced by AOB through two pathways: (1) NH_2OH decomposition and (2) NO_2^- reduction, or "nitrifier–denitrification" (Stein and Yung, 2003). The isotopic composition of N_2O produced by AOB therefore depends on the proportion from each pathway and the isotope effects between the substrates, NH_2OH and NO_2^-, and N_2O (Frame and Casciotti, 2010). Fractionation may occur during substrate binding by the enzyme, as well as during bond breaking to produce the asymmetrical N_2O molecule. Initial estimates have been made for the net fractionation between O_2 and N_2O for the NH_2OH-derived pathway and between NO_2^- and N_2O for the nitrifier–denitrification pathway in the marine AOB *Nitrosomonas* C-113a (Frame and Casciotti, 2010). In that study, it was estimated that C-113a expressed a net fractionation of oxygen isotopes between O_2 and N_2O of 2.9‰, and -8.4‰ between NO_2^- and N_2O. In other words, N_2O produced from NH_2OH was *depleted* in ^{18}O by 2.9‰ relative to the O_2 substrate, and N_2O produced from NO_2^- was *enriched* in ^{18}O by 8.4‰ relative to NO_2^-. The enrichment can be understood as a "branching fractionation" between NO_2^- and N_2O where ^{16}O is preferentially lost from the nitrogen oxide pools during breakage of N–O bonds (Casciotti *et al.*, 2007; Frame and Casciotti, 2010). This may also occur in NH_2OH decomposition to N_2O where it could be offset by a large kinetic fractionation for O_2 incorporation into NH_2OH.

Several aspects of the O isotope systematics of N_2O production are still poorly constrained. The isotope effects determined for N_2O production by C-113a are combined isotope effects that encapsulate fractionation during O_2 incorporation into NH_2OH, as well as fractionation during NH_2OH oxidation to N_2O. Measurement of $\delta^{18}O$ in NH_2OH is necessary to determine the isotope effect for O incorporation by AMO and subsequent fractionation leading to N_2O. Beyond that, we still need to know whether AOA produce N_2O, and if so, by what mechanism(s).

5.4. Isotope effects for archaea

AOA are increasingly recognized as being important players in the marine nitrogen cycle, and there are still many open questions about their biochemistry and physiology, including the mechanism of ammonia oxidation to NO_2^- (De La Torre *et al.*, 2008; Hatzenpichler *et al.*, 2008; Martens-Habbena *et al.*, 2009; Walker *et al.*, 2010). The ammonia monooxygenase subunit a genes (*amoA*) found in AOA are quite different from the bacterial *amoA*, and AOA appear to lack a recognizable hydroxylamine oxidoreductase (HAO; Klotz *et al.*, 2008; Walker *et al.*, 2010), suggesting that even if they share a common pathway for ammonia oxidation with AOB, the fractionation factors may be quite different. Variation among organisms in active site characteristics may be reflected in their isotopic fractionation (Casciotti *et al.*, 2003). If the pathway for ammonia oxidation in AOA is different from AOB, then the differences in their N and O isotope systematics, or the relationships between $\delta^{18}O_{H_2O}$, $\delta^{18}O_{O_2}$, and $\delta^{18}O_{NO_2}$, could be even more dramatic. The O atom sources for AOA are currently unknown, but most likely involve O_2 and H_2O (Walker *et al.*, 2010).

Experiments with ^{18}O-labeled H_2O (or ^{18}O-labeled O_2) could be used to determine the sources of O, as well as the amount of exchange and fractionation during ammonia oxidation by AOA in much the same way as described above for AOB. Such experiments would provide insight into the mechanism of archaeal ammonia oxidation. This would ideally be done with pure cultures, but similar experiments may also be successful in enrichment culture. In this case, care must be taken to evaluate whether mass balance is maintained between NH_4^+, NO_2^-, and NO_3^-. If there are alternate fates for NH_4^+, NO_2^-, or NO_3^- in the system, then the effect of these additional pathways would need to be considered in any interpretation of fractionation factors. However, as described above for mixed culture experiments, incorporation and exchange of O atoms during nitrification would still be the only factors affecting the dependence of $\delta^{18}O_{NO_2}$ and $\delta^{18}O_{NO_3}$ on $\delta^{18}O_{H_2O}$.

REFERENCES

Abeliovich, A. (2006). The nitrite-oxidizing bacteria. *The Prokaryotes* **5**, 861–872.

Altabet, M. A. (2007). Constraints on oceanic N balance/imbalance from sedimentary ^{15}N records. *Biogeosciences* **4**, 75–86.

Andersson, K. K., and Hooper, A. B. (1983). O_2 and H_2O are each the source of one O in NO_2^- produced from NH_3 by *Nitrosomonas*: ^{15}N-NMR evidence. *FEBS Lett.* **164**, 236–240.

Bigeleisen, J. (1965). Chemistry of Isotopes. *Science* **147**, 463–471.

Bigeleisen, J., and Wolfsberg, M. (1958). Theoretical and experimental aspects of isotope effects in chemical kinetics. *Adv. Chem. Phys.* **1**, 15–76.

Bock, E., Koops, H.-P., and Harms, H. (1989). Nitrifying bacteria. *In* "Autotrophic bacteria," (H. G. Schelgel and B. Bowien, eds.), pp. 81–96. Springer-Verlag, Berlin.

Böhlke, J. K., Mroczkowski, S. J., and Coplen, T. B. (2003). Oxygen isotopes in nitrate: new reference materials for O-18: O-17: O-16 measurements and observations on nitrate-water equilibration. *Rapid Commun. Mass Spectrom.* **17,** 1835–1846.

Böhlke, J. K., Smith, R. L., and Hannon, J. E. (2007). Isotopic analysis of N and O in nitrite and nitrate by sequential selective bacterial reduction to N_2O. *Anal. Chem.* **79,** 5888–5895.

Brandes, J. A., and Devol, A. H. (2002). A global marine fixed nitrogen isotopic budget: Implications for Holocene nitrogen cycling. *Glob. Biogeochem. Cycles* **16,** 1120.

Buchwald, C., and Casciotti, K. L. (2010). Oxygen isotopic fractionation and exchange during bacterial nitrite oxidation. *Limnol. Oceanogr.* **55,** 1064–1074.

Butler, J. H., Elkins, J. W., and Thompson, T. M. (1989). Tropospheric and dissolved N_2O of the West Pacific and East Indian Oceans during the El-Nino Southern Oscillation event of 1987. *J. Geophys. Res. Atmos.* **94,** 14865–14877.

Carlucci, A. F., and McNally, P. M. (1969). Nitrification by marine bacteria at low concentrations of substrate and oxygen. *Limnol. Oceanogr.* **14,** 736–739.

Casciotti, K. L. (2009). Inverse kinetic isotope fractionation during bacterial nitrite oxidation. *Geochim. Cosmochim. Acta* **73,** 2061–2076.

Casciotti, K. L., and McIlvin, M. R. (2007). Isotopic analyses of nitrate and nitrite from reference mixtures and application to Eastern Tropical North Pacific waters. *Mar. Chem.* **107,** 184–201.

Casciotti, K. L., Sigman, D. M., Galanter Hastings, M., Böhlke, J. K., and Hilkert, A. (2002). Measurement of the oxygen isotopic composition of nitrate in seawater and freshwater using the denitrifier method. *Anal. Chem.* **74,** 4905–4912.

Casciotti, K. L., Sigman, D. M., and Ward, B. B. (2003). Linking diversity and stable isotope fractionation in ammonia-oxidizing bacteria. *Geomicrobiol. J.* **20,** 335–353.

Casciotti, K. L., Böhlke, J. K., McIlvin, M. R., Mroczkowski, S. J., and Hannon, J. E. (2007). Oxygen isotopes in nitrite: Analysis, calibration, and equilibration. *Anal. Chem.* **79,** 2427–2436.

Casciotti, K. L., McIlvin, M. R., and Buchwald, C. (2010). Oxygen isotopic exchange and fractionation during bacterial ammonia oxidation. *Limnol. Oceanogr.* **55,** 753–762.

Chain, P. *et al.* (2003). Complete genome sequence of the ammonia-oxidizing bacterium and obligate chemolithoautotroph Nitrosomonas europaea. *J. Bact.* **185,** 2759–2773.

Cifuentes, L. A., Fogel, M. L., Pennock, J. R., and Sharp, J. H. (1989). Biogeochemical factors that influence the stable nitrogen isotope ratio of dissolved ammonium in the Delaware Estuary. *Geochim. Cosmochim. Acta* **53,** 2713–2721.

Cleland, W. W. (1987). The use of isotope effects in the detailed analysis of catalytic mechanisms of enzymes. *Bioinorg. Chem.* **15,** 283–302.

Cohen, Y., and Gordon, L. I. (1978). Nitrous oxide in the oxygen minimum of the eastern tropical North Pacific: Evidence for its consumption during denitrification and possible mechanisms for its production. *Deep Sea Res.* **6,** 509–525.

Cox, R. D. (1980). Determination of nitrate and nitrite at the parts per billion level by chemiluminescence. *Anal. Chem.* **52,** 332–335.

De La Torre, J. R., Walker, C. B., Ingalls, A. E., Konneke, M., and Stahl, D. A. (2008). Cultivation of a thermophilic ammonia oxidizing archaeon synthesizing crenarchaeol. *Environ. Microbiol.* **10,** 810–818.

Detmers, J., Bruchert, V., Habicht, K. S., and Kuever, J. (2001). Diversity of sulfur isotope fractionations by sulfate-reducing prokaryotes. *Appl. Environ. Microbiol.* **67,** 888–894.

Deutsch, C., Sigman, D. M., Thunell, R. C., Meckler, A. N., and Haug, G. H. (2004). Isotopic constraints on glacial/interglacial changes in the oceanic nitrogen budget. *Global Biogeochem. Cycles* **18,** ARTN GB4012.

Dispirito, A. A., and Hooper, A. B. (1986). Oxygen exchange between nitrate molecules during nitrite oxidation by *Nitrobacter*. *J. Biol. Chem.* **261**, 10534–10537.

Dore, J. E., and Karl, D. M. (1996). Nitrification in the euphotic zone as a source for nitrite, nitrate, and nitrous oxide at Station ALOHA. *Limnol. Oceanogr.* **41**, 1619–1628.

Dore, J. E., Popp, B. N., Karl, D. M., and Sansone, F. J. (1998). A large source of atmospheric nitrous oxide from subtropical North Pacific surface waters. *Nature* **396**, 63–66.

Dua, R. D., Bhandari, B., and Nicholas, D. J. D. (1979). Stable isotope studies on the oxidation of ammonia to hydroxylamine by *Nitrosomonas europaea*. *FEBS Lett.* **106**, 401–404.

Epstein, S., and Mayeda, T. (1953). Variation of O^{18} content of waters from natural sources. *Geochim. Cosmochim. Acta* **4**, 213–224.

Frame, C. H., and Casciotti, K. L. (2010). Biogeochemical controls and isotopic signatures of nitrous oxide production by a marine ammonia-oxidizing bacterium. *Biogeosci. Discuss.* **7**, 3019–3059.

Friedman, S. H., Massefski, W., and Hollocher, T. C. (1986). Catalysis of intermolecular oxygen atom transfer by nitrite dehydrogenase of *Nitrobacter agilis*. *J. Biol. Chem.* **261**, 10538–10543.

Garside, C. (1982). A chemiluminescent technique for the determination of nanomolar concentrations of nitrate, nitrite, or nitrite alone in seawater. *Mar. Chem.* **11**, 159–167.

Gianfrani, L., Gagliardi, G., Van Burgel, M., and Kerstel, E. R. (2003). Isotope analysis of water by means of near-infrared dual-wavelength diode laser spectroscopy. *Opt. Express* **11**, 1566–1576.

Goericke, R., Montoya, J. P., and Fry, B. (1994). Physiology of isotope fractionation in algae and cyanobacteria. *In* "Stable Isotopes in Ecology and Environmental Science," (K. Lakitha and B. Michener, eds.), Blackwell Scientific, Oxford.

Granger, J., and Sigman, D. M. (2009). Removal of nitrite with sulfamic acid for nitrate N and O isotope analysis with the denitrifier method. *Rapid Commun. Mass Spectrom.* **23**, 3753–3762.

Granger, J., Sigman, D. M., Needoba, J. A., and Harrison, P. J. (2004). Coupled nitrogen and oxygen isotope fractionation of nitrate during assimilation by cultures of marine phytoplankton. *Limnol. Oceanogr.* **49**, 1763–1773.

Granger, J., Sigman, D. M., Rohde, M. M., Maldonado, M. T., and Tortell, P. D. (2010). N and O isotope effects during nitrate assimilation by unicellular prokaryotic and eukaryotic plankton cultures. *Geochim. Cosmochim. Acta* **74**, 1030–1040.

Hatzenpichler, R., et al. (2008). A moderately thermophilic ammonia-oxidizing crenarchaeote from a hot spring. *Proc. Natl. Acad. Sci. USA* **105**, 2134–2139.

Hermes, J. D., Weiss, P. M., and Cleland, W. W. (1985). Use of nitrogen-15 and deuterium isotope effects to determine the chemical mechanism of phenylalanine ammonia-lyase. *Biochemistry* **24**, 2959–2967.

Hollocher, T. C., Tate, M. E., and Nicholas, D. J. D. (1981). Oxidation of ammonia by *Nitrosomonas europaea*. *J. Biol. Chem.* **256**, 10834–10836.

Holmes, R. M., Mcclelland, J., and Sigman, D. M. (1998). Measuring $^{15}N-NH_4^+$ in marine, estuarine, and fresh waters: an adaptation of the ammonium diffusion method for samples with low-ammonium concentrations. *Mar. Chem.* **60**, 235–243.

Horrigan, S. G., Montoya, J. P., Nevins, J. L., and Mccarthy, J. J. (1990a). Natural isotopic composition of dissolved inorganic nitrogen in the Chesapeake Bay. *Estuar. Coast. Shelf Sci.* **30**, 393–410.

Horrigan, S. G., et al. (1990b). Nitrogenous nutrient transformations in the spring and fall in the Chesapeake Bay. *Estuar. Coast. Shelf Sci.* **30**, 369–391.

IPCC (2007). Climate change 2007: Synthesis report. p. 104. Contribution of working groups I, II, and III to the Fourth Assessment Report of the Intergovernmental Panel on Climate Change.

Kator, H., Morrison, L. J., Wetzel, R. L., and Koepfler, E. T. (1992). A rapid chromatographic method for recovery of $^{15}NO_2^-$ and $^{15}NO_3^-$ produced by nitrification in aqueous samples. *Limnol. Oceanogr.* **37,** 900–907.

Kim, K.-R., and Craig, H. (1990). Two-isotope characterization of N_2O in the Pacific Ocean and constraints on its origin in deep water. *Nature* **347,** 58–61.

Klotz, M. G., et al. (2006). Complete genome sequence of the marine, chemolithoautotrophic, ammonia-oxidizing bacterium Nitrosococcus oceani ATCC 19707. *Appl. Env. Microbiol.* **72,** 6299–6315.

Klotz, M. G., Schmid, M. C., Strous, M., Op Den Camp, H. J. M., Jetten, M. S. M., and Hooper, A. B. (2008). Evolution of an octahaem cytochrome c protein family that is key to aerobic and anaerobic ammonia oxidation by bacteria. *Environ. Microbiol.* **10,** 3150–3163.

Kumar, S., Nicholas, D. J. D., and Williams, E. H. (1983). Definitive N-15 NMR evidence that water serves as a source of O during nitrite oxidation by *Nitrobacter agilis*. *FEBS Lett.* **152,** 71–74.

Laws, E. A., Bidigare, R. R., and Popp, B. N. (1997). Effect of growth rate and CO_2 concentration on carbon isotopic fractionation by the marine diatom *Phaeodactylum tricornutum*. *Limnol. Oceanogr.* **42,** 1552–1560.

Mariotti, A., et al. (1981). Experimental determination of nitrogen kinetic isotope fractionation: Some principles; Illustration for the denitrification and nitrification processes. *Plant Soil* **62,** 413–430.

Mariotti, A., Lancelot, C., and Billen, G. (1984). Natural isotopic composition of nitrogen as a tracer of origin for suspended organic-matter in the Scheldt Estuary. *Geochim. Cosmochim. Acta* **48,** 549–555.

Martens-Habbena, W., Berube, P. M., Urakawa, H., De La Torre, J. R., and Stahl, D. A. (2009). Ammonia oxidation kinetics determine niche separation of nitrifying Archaea and Bacteria. *Nature* **461,** 976, U234.

Mayer, B., Bollwerk, S. M., Mansfeldt, T., Hutter, B., and Veizer, J. (2001). The oxygen isotope composition of nitrate generated by nitrification in acid forest floors. *Geochim. Cosmochim. Acta* **65,** 2743–2756.

McIlvin, M. R., and Altabet, M. A. (2005). Chemical conversion of nitrate and nitrite to nitrous oxide for nitrogen and oxygen isotopic analysis in freshwater and seawater. *Anal. Chem.* **77,** 5589–5595.

McIlvin, M. R., and Casciotti, K. L. (2006). Method for the Analysis of $\delta^{18}O$ in Water. *Anal. Chem.* **78,** 2377–2381.

McIlvin, M. R., and Casciotti, K. L. (2010). Fully automated system for stable isotopic analyses of dissolved nitrous oxide at natural abundance levels. *Limnol. Oceanogr. Methods* **8,** 54–66.

Miyake, Y., and Wada, E. (1971). The isotope effect on the nitrogen in biochemical oxidation-reduction reactions. *Rec. Oceanogr. Works Japan* **11,** 1–6.

Needoba, J. A., Sigman, D. M., and Harrison, P. J. (2004). The mechanism of isotope fractionation during algal nitrate assimilation as illuminated by the N-15/N-14 of intracellular nitrate. *J. Phycol.* **40,** 517–522.

Northrop, D. B. (1981). The expression of isotope effects on enzyme-catalyzed reactions. *Annu. Rev. Biochem.* **50,** 103–131.

Norton, J. M., et al. (2008). Complete genome sequence of Nitrosospira multiformis, an ammonia-oxidizing bacterium from the soil enviroment. *Appl. Env. Microbiol.* **74,** 3559–3572.

O'leary, M. H. (1981). Carbon isotope fractionations in plants. *Phytochemistry* **20,** 553–567.

Olson, R. J. (1981). ^{15}N tracer studies of the primary nitrite maximum. *J. Mar. Res.* **39**, 203–226.

Ostrom, N. E., Russ, M. E., Popp, B., Rust, T. M., and Karl, D. M. (2000). Mechanisms of nitrous oxide production in the subtropical North Pacific based on determinations of the isotopic abundances of nitrous oxide and di-oxygen. *Chemosphere–Global Change. Science* **2**, 281–290.

Popp, B. N., *et al.* (2002). Nitrogen and oxygen isotopomeric constraints on the origins and sea-to-air flux of N$_2$O in the oligotrophic subtropical North Pacific gyre. *Global Biogeochem. Cycles* **16**, ARTN 1064.

Ravishankara, A. R., Daniel, J. S., and Portmann, R. W. (2009). Nitrous Oxide (N$_2$O): The dominant ozone-depleting substance emitted in the 21st Century. *Science* **326**, 123–125.

Revesz, K. M., and Böhlke, J. K. (2002). Comparison of δ^{18}O measurements in nitrate by different combustion techniques. *Anal. Chem.* **74**, 5410–5413.

Scott, K. M., Schewedock, J., Schrag, D. P., and Cavanaugh, C. M. (2004). Influence of form 1A RubisCO and environmental dissolved inorganic carbon on the δ^{13}C of the clam-chemoautotroph symbiosis *Solemya velum*. *Environ. Microbiol.* **6**, 1210–1219.

Sigman, D. M., Altabet, M. A., Michener, R. H., Mccorkle, D. C., Fry, B., and Holmes, R. M. (1997). Natural abundance-level measurement of the Nitrogen isotopic composition of oceanic nitrate: an adaptation of the ammonia diffusion method. *Mar. Chem.* **57**, 227–242.

Sigman, D. M., Altabet, M. A., Francois, R., Mccorkle, D. C., and Fischer, G. (1999). The δ^{15}N of nitrate in the Southern Ocean: Consumption of nitrate in surface waters. *Glob. Biogeochem. Cycles* **13**, 1149–1166.

Sigman, D. M., Casciotti, K. L., Andreani, M., Barford, C., Galanter, M., and Böhlke, J. K. (2001). A bacterial method for the nitrogen isotopic analysis of nitrate in seawater and freshwater. *Anal. Chem.* **73**, 4145–4153.

Sigman, D. M., *et al.* (2005). Coupled nitrogen and oxygen isotope measurements of nitrate along the eastern North Pacific margin. *Global Biogeochem. Cycles* **19**, GB4022.

Sigman, D. M., Difiore, P. J., Hain, M. P., Deutsch, C., and Karl, D. M. (2009a). Sinking organic matter spreads the nitrogen isotope signal of pelagic denitrification in the North Pacific. *Geophys. Res. Lett.* **36**, L08605.

Sigman, D. M., Karsh, K. L., and Casciotti, K. L. (2009b). Ocean process tracers: Nitrogen isotopes in the ocean. Encyclopedia of Ocean Sciences, pp. 4138–4153. Academic Press, New York.

Silva, S. R., Kendall, C., Wilkison, D. H., Ziegler, A. C., Chang, C. C. Y., and Avanzino, R. J. (2000). A new method for collection of nitrate from fresh water and analysis of nitrogen and oxygen isotope ratios. *J. Hydrol.* **228**, 22–36.

Solorzano, L. (1969). Determination of ammonia in natural waters by the phenolhypochlorite method. *Limnol. Oceanogr.* **14**, 799–801.

Soriano, S., and Walker, N. (1968). Isolation of ammonia-oxidizing autotrophic bacteria. *J. Appl. Bacteriol.* **31**, 493–497.

Stein, L. Y., and Yung, Y. L. (2003). Production, isotopic composition and atmospheric fate of biologically produced nitrous oxide. *Annu. Rev. Earth Planet. Sci.* **31**, 329–356.

Strickland, J. D. H., and Parsons, T. R. (1972). A practical handbook of seawater analysis, 2nd ed. *Bull. Fish. Res. Bd. Canada* **167**, 1–310.

Thornton, E. K., and Thornton, E. R. (1978). Scope and limitations of the concept of the transition state. In "Transition States of Biochemical Processes," (R. D. Gandour and R. L. Schowen, eds.), pp. 3–75. Plenum Press, New York.

Velinsky, D. J., Pennock, J. R., Sharp, J. H., Cifuentes, L. A., and Fogel, M. L. (1989). Determination of the isotopic composition of ammonium nitrogen at the natural abundance level from estuarine waters. *Mar. Chem.* **26**, 351–361.

Walker, C. B., et al. (2010). *Nitrosopumilus maritimus* genome reveals unique mechanisms for nitrification and autotrophy in globally distributed marine crenarchaea. *Proc. Natl. Acad. Sci. USA* **107**, 8818–8823.

Wankel, S. D., Kendall, C., Pennington, J. T., Chavez, F. P., and Paytan, A. (2007). Nitrification in the euphotic zone as evidenced by nitrate dual isotopic composition: Observations from Monterey Bay, California. *Global Biogeochem. Cycles* **21**, GB2009.

Ward, B. B., and O'Mullan, G. D. (2005). Community Level Analysis: Genetic and biogeochemical approaches to investigate community composition and function in aerobic ammonia oxidation. *Meth. Enzymol.* **397**, 395–413.

Watson, S. W. (1965). Characteristics of a marine nitrifying bacterium, *Nitrosocystis oceanus* sp. n. *Limnol. Oceanogr.* **10**(Suppl.), R274–R289.

Whittaker, M., Bergmann, D. J., Arciero, D. M., and Hooper, A. B. (2000). Electron transfer during the oxidation of ammonia by the chemolithotrophic bacterium *Nitrosomonas europaea*. *Biochim. Biophys. Acta* **1459**, 346–355.

Yamagishi, H., et al. (2007). Role of nitrification and denitrification on the nitrous oxide cycle in the eastern tropical North Pacific and Gulf of California. *J. Geophys. Res. Biogeosci.* **112** ARTN G02015.

Yoshida, N., Hattori, A., Saino, T., Matsuo, S., and Wada, E. (1984). $^{15}N/^{14}N$ ratio of dissolved N_2O in the eastern tropical Pacific Ocean. *Nature* **307**, 442–444.

Yoshinari, T. (1976). Nitrous oxide in the sea. *Mar. Chem.* **4**, 189–202.

Yoshinari, T., et al. (1997). Nitrogen and oxygen isotopic composition of N_2O from suboxic waters of the eastern tropical North Pacific and the Arabian Sea-measurement by continuous-flow isotope-ratio monitoring. *Mar. Chem.* **56**, 253–264.

Zhang, L., Altabet, M. A., Wu, T., and Hadas, O. (2007). Sensitive measurement of NH_4^+ $^{15}N/^{14}N$ ($\delta^{15}NH_4^+$) at natural abundance levels in fresh and saltwaters. *Anal. Chem.* **79**, 5297–5303.

CHAPTER TWELVE

Identification of Diazotrophic Microorganisms in Marine Sediment via Fluorescence *In Situ* Hybridization Coupled to Nanoscale Secondary Ion Mass Spectrometry (FISH-NanoSIMS)

Anne E. Dekas *and* Victoria J. Orphan

Contents

1. Introduction	282
1.1. Background	282
1.2. Methods to investigate nitrogen fixation: Why FISH-NanoSIMS?	283
1.3. A case study: Coupling function and phylogeny in a diazotrophic deep-sea symbiotic archaea using FISH-NanoSIMS	284
2. Methods	285
2.1. ^{15}N-labeling sediment incubations	285
2.2. Subsampling and preservation	287
2.3. Selecting and preparing microprobe slides for sample deposition	287
2.4. Concentrating cells using density gradients and deposition onto microprobe slides	288
2.5. FISH on microprobe slide-deposited cells	289
2.6. Mapping sample targets	291
2.7. Sectioning samples prior to analysis	294
2.8. NanoSIMS analysis	295
2.9. Interpreting NanoSIMS results	297
2.10. Complementary additional analyses	299
2.11. Measuring metal cofactors as indicators of nitrogenase	300
3. Concluding Remarks	300
Acknowledgments	301
References	301

Division of Geological and Planetary Sciences, California Institute of Technology, Pasadena, California, USA

Abstract

Growing appreciation for the biogeochemical significance of uncultured microorganisms is changing the focus of environmental microbiology. Techniques designed to investigate microbial metabolism *in situ* are increasingly popular, from mRNA-targeted fluorescence *in situ* hybridization (FISH) to the "-omics" revolution, including metagenomics, transcriptomics, and proteomics. Recently, the coupling of FISH with nanometer-scale secondary ion mass spectrometry (NanoSIMS) has taken this movement in a new direction, allowing single-cell metabolic analysis of uncultured microbial phylogenic groups. The main advantage of FISH-NanoSIMS over previous noncultivation-based techniques to probe metabolism is its ability to directly link 16S rRNA phylogenetic identity to metabolic function. In the following chapter, we describe the procedures necessary to identify nitrogen-fixing microbes within marine sediment via FISH-NanoSIMS, using our work on nitrogen fixation by uncultured deep-sea methane-consuming archaea as a case study.

1. INTRODUCTION

1.1. Background

Nitrogen is required in numerous biomolecules including amino and nucleic acids and is therefore an essential element for life. Although nitrogen is extremely abundant on the surface of the earth, most organisms are unable to assimilate the majority of this nitrogen, gaseous N_2, due to the stability of its triple bond. Nitrogen limitation in natural biological communities is therefore common, and the abundance of bioavailable nitrogen sources (such as nitrate and ammonia) can regulate overall productivity and chemical exchange. Importantly, a small subset of phylogenetically diverse prokaryotes known as diazotrophs are able to convert N_2 into NH_3, using a well-conserved enzyme called nitrogenase and a large input of ATP. Biological nitrogen fixation, or simply nitrogen fixation, essentially unlocks the reservoir of nitrogen in the atmosphere and provides a source of bioavailable nitrogen for the rest of the food chain. This conversion is additionally important in the context of global nitrogen cycling, because it compensates for the production of N_2 by energy-generating microbial processes (e.g., denitrification and anammox).

From the onset of marine nitrogen fixation research over 100 years ago, a variety of habitats have been recognized to host diazotrophs, including both the water column and sediments (Capone, 1988; Herbert, 1999; Howarth *et al.*, 1988; Waksman *et al.*, 1933). However, most marine nitrogen fixation research and global rate modeling have focused on *Tricodesmium*, a pelagic cyanobacteria, due to its abundance and important role in biogeochemical cycling (Capone *et al.*, 1997). In the past decade this focus has broadened, as an increasing number of studies have highlighted the great and occasionally unexpected phylogenetic and geographic diversity of marine diazotrophs (Capone, 2001; Karl et al., 2002; Mahaffey et al., 2005). Specifically, the development and wide-scale application of culture-independent techniques, including DNA and complementary DNA (cDNA) clone libraries (Mehta et al., 2003; Miyazaki et al., 2009), net N_2 flux measurements (Fulweiler et al., 2007), quantitative PCR (QPCR) (Moisander et al., 2010), quantitative reverse transcriptase PCR (Q-RT-PCR) (Foster et al., 2009; Short and Zehr, 2007; Veit et al., 2006), and single-cell level ^{15}N-tracer studies (Dekas et al., 2009; Finzi-Hart et al., 2009; Lechene et al., 2007; Popa et al., 2007), have revealed a previously unrecognized breadth of diazotrophs both within the cyanobacteria and throughout the prokaryotic domains.

These findings show that nitrogen fixation is more widespread than previously appreciated and suggest that some marine habitats may have been overlooked as net sources of bioavailable nitrogen. This possibility is significant in the context of the global marine nitrogen cycle, which for many years has been reported to have incongruous sources and sinks of fixed nitrogen (Codispoti, 1995, 2007; Codispoti and Christensen, 1985; Codispoti et al., 2001; Ganeshram et al., 1995; Gruber and Sarmiento, 1997; Mahaffey et al., 2005; McElroy, 1983). The recent developments in our understanding of diazotroph diversity support an artifact explanation for the incongruence: the perceived imbalance in the nitrogen cycle may be a result of an underestimation of marine nitrogen fixation rather than a real and significant departure from steady state (Brandes and Devol, 2002; Brandes et al., 2007; Codispoti, 2007; Fulweiler, 2009; Gruber and Sarmiento, 1997; Mahaffey et al., 2005). The further study of nitrogen fixation, both on an organismal and global scale, is therefore extremely timely.

1.2. Methods to investigate nitrogen fixation: Why FISH-NanoSIMS?

While informative and relatively easy to employ, a limitation of most molecular techniques and standard geochemical analyses is their inability to link specific phylogenetically identified microorganisms in environmental samples with their metabolic processes. Bulk analyses, such as the acetylene reduction assay, ^{15}N tracer experiments (measured with an elemental analyzer-isotope ratio mass spectrometer, EA-IRMS), and N_2 flux experiments (measured with the N_2/Ar technique; Fulweiler et al., 2007), can demonstrate active nitrogen fixation, along with quantitative rate data, but cannot identify the specific phylogenetic groups of microorganisms responsible for the conversion. And, although the acetylene reduction assay is typically credited with higher sensitivity than bulk $^{15}N_2$ tracer studies, it is subject to several sources of error (described in Karl et al., 2002), in addition to causing inhibition of some environmentally relevant microbes, including methanogens and anaerobic methanotrophs (Dekas et al., 2009; Oremland and Taylor, 1975; Sprott et al., 1982).

Targeted molecular approaches similarly have drawbacks. DNA investigations of relevant metabolic genes and the *nifH* genes, in particular (Mehta et al., 2003), can be considered an investigation of the metabolic potential of the community, but not the activity. Investigations of transcripts, and in particular, quantitative reverse transcriptase PCR (Q-RT-PCR) studies of *nifH* (Foster et al., 2009; Short and Zehr, 2007; Veit et al., 2006), are generally considered robust and even quantitative indicators of active nitrogen fixation (Short and Zehr, 2005). However, posttranscription and posttranslational regulation are known to occur, and activity may deviate from what the transcript level suggests (Kessler and Leigh, 1999; Liang et al., 1991; Zhang et al., 1993). Additionally, neither a cDNA clone library nor Q-RT-PCR targeting particular *nif* genes can reliably reveal the microbial source of the gene in terms of its 16S rRNA identity due to incongruence between *nif* and 16S phylogenies attributed to lateral gene transfer (Kechris et al., 2006).

An extremely promising approach to simultaneously measuring diazotrophic activity and phylogenetic identity is stable isotope probing with $^{15}N_2$ ($^{15}N_2$-DNA-SIP) (Buckley et al., 2007a). Although DNA-SIP with ^{13}C-labeled compounds is relatively common (Dumont and Murrell, 2005), the additional complications related to density separation of labeled and unlabeled DNA from ^{15}N-labeling experiments, which confer a smaller mass difference than ^{13}C-labeling experiments, have only been resolved recently (Buckley et al., 2007b). Although this method is superior to bulk ^{15}N incorporation experiments in that it can indicate which organisms incorporate ^{15}N, it is limited by grouping organisms into density fractions, equivalent to ranges in ^{15}N-incorporation, instead of directly measuring a $\delta^{15}N$ value for each member of the community. In fact, CHIP-SIP, a method where the labeled nucleic acids are hybridized to biopolymer microarrays and then analyzed individually via nanoscale secondary ion mass spectrometry (NanoSIMS) has been employed to overcome the limitations of traditional DNA-SIP density fractions (CHIP-SIP; Pett-Ridge et al., 2010). However, if spatial information is also desired, including chemical interactions between symbionts or variation in activity within clustered cells, whole cell visualization and analysis (e.g. FISH-NanoSIMS) rather than investigation of extracted nucleic acids must be pursued.

1.3. A case study: Coupling function and phylogeny in a diazotrophic deep-sea symbiotic archaea using FISH-NanoSIMS

In recent work, we demonstrated the ability of methane-oxidizing archaea (ANME-2c) to fix nitrogen when physically connected to a metabolically active sulfate-reducing bacterial symbiont (Dekas et al., 2009). These microorganisms are found in methane-rich marine sediment, such as methane seeps along continental margins, and have not yet been obtained in pure culture. It was hypothesized that at least one of these archaeal–bacterial symbionts was diazotrophic after the genes encoding nitrogenase, the *nif* genes, were detected in a metagenomic study of these microorganisms (Pernthaler *et al.*, 2008). To test this hypothesis, bulk methane seep sediment was anaerobically incubated with methane and ^{15}N-labeled nitrogen sources over a period of 6 months at 8 °C, a temperature approximating *in situ* conditions. Using fluorescence *in situ* hybridization (FISH) to phylogenetically identify individual microbial cells within the incubated sediment, and NanoSIMS to measure the isotopic composition of the positively hybridized microorganisms, we were able to track ^{15}N from ^{15}N$_2$ into the ANME-2c biomass. This demonstrated their ability to fix nitrogen. A less enriched ^{15}N/^{14}N ratio was also observed in the bacterial symbiont (affiliated with the deltaproteobacterial sulfate-reducing *Desulfosarcina/Desulfococcus* lineage; Orphan *et al.*, 2001a), which in combination with molecular data collected in parallel suggested a passage of ^{15}N-labeled fixed nitrogen from the archaea to the bacteria. The FISH-NanoSIMS approach builds upon previous techniques by coupling phylogenetic identity to metabolic function at single-cell resolution. It can effectively test the hypotheses set forth by traditional bulk or molecular methods.

The following section describes the method we employed to identify the ANME-2c archaea as diazotrophs without requiring pure cultures or even sediment-free enrichments. The nanometer resolution of the NanoSIMS is particularly well suited to study small, closely associated symbiotic microorganisms and has been used in other studies to demonstrate metabolic capabilities of microorganisms as well, including nitrogen fixation in cocultures, free-living planktonic microorganisms, and symbiotic metazoans (Finzi-Hart *et al.*, 2009; Halm *et al.*, 2009; Lechene *et al.*, 2006, 2007; Ploug *et al.*, 2010; Popa *et al.*, 2007) (see Boxer *et al.*, 2009; Herrmann *et al.*, 2007b; Orphan and House, 2009 for reviews). With minor modifications (noted where possible), these procedures could be applied to a wide range of biogeochemical studies where the metabolic capability of specific microorganisms can be directly assessed on a single cell level.

2. Methods

2.1. ^{15}N-labeling sediment incubations

Methane seep sediment samples are combined with anoxic media approximately 1:1 by volume within an anaerobic chamber or under a stream of argon. Recommended media are seawater filtered through a 0.2-µm filter, preferably collected at the sampling site, or sterilized artificial seawater modified to enrich for organisms of interest, and/or eliminate alternative nitrogen sources (such as the media described in Widdel *et al.*, 2006). Once combined, the sediment slurry is homogenized. The mixture is then equally aliquoted via volumetric pipette into sterilized serum bottles, leaving at least half of the bottle volume empty for the addition of a gas headspace. (Wheaton Science Products, Millville, NJ, #223748). Bottles are capped with butyl stoppers (Geo-Microbial Technologies, Inc., Ochelata, OK, #1313) and metal crimp seals (Wheaton, #224178-01). Using an entrance needle, which is connected to the gas tank and inserted first, and an exit needle, which is open to the air, the headspace within the bottle is flushed with methane, or the headspace gas of choice (needles: Becton Dickinson, Frankling Lakes, NJ, #305167). Previous incubation studies have shown that the rate of growth of ANME-2/*Desulfosarcina* aggregates increases with increasing methane partial pressure (Nauhaus *et al.*, 2002). By setting the methane tank regulator to the desired pressure, removing the exit needle from the incubation bottle, and only removing the entrance needle once the ripples on the surface of the sediment slurry within the bottle cease, this defined higher pressure can be achieved. A pressure of 2 bar can be safely maintained in 125 ml serum bottles and is what we routinely use for our enrichment studies. Appropriate safety precautions should be taken when over-pressurizing glass incubation bottles, as bottles can explode.

$^{15}N_2$ is commercially available (Sigma-Aldrich Isotec, #364584; Cambridge Isotopes Laboratories, NLM-363) and can be transferred several ways into the serum bottles. It is advisable to consult with a technical representative to determine which regulator would be appropriate for the particular cylinder selected. We recommend connecting a septum-fitted syringe attachment (Supelco, Bellefonte, PA, #609010) to the $^{15}N_2$ bottle (or attached regulator) and using a gas tight syringe (Becton Dickinson, #309602; Hamilton Company, GASTIGHT 1000 Series, Reno NV) to remove and transfer quantitative amounts of $^{15}N_2$. The syringe and needle should be flushed first with methane or other gas to remove air. The $^{15}N_2$ drawn from the bottle can then be added directly to the incubation bottles. Capone and Montoya (2001) report that a 0.5-ml bubble of $^{15}N_2$ in a liquid-filled 250-ml bottle is sufficient to yield bulk enrichment values of 12–13 atm% in liquid cultures, and we have observed similar enrichment values in single cells after adding 3 ml $^{15}N_2$ to the methane headspace of our 125-ml sediment incubations (approximately 1:4 liquid:headspace ratio). Adding undiluted ^{15}N-labeled dinitrogen did not have an effect on sulfate reduction in our incubations, measured by sulfide production, compared to unlabeled N_2. However, diluting $^{15}N_2$ in unlabeled N_2 may be desirable in order to provide environmentally relevant N_2 concentrations at lower cost. In this case, the isotopic composition of the dinitrogen gas mix should be determined before the experiment begins if quantitative bulk rate data are also desired. Bulk rates of nitrogen fixation can be determined following the equations in Capone and Montoya (2001) and Montoya et al. (1996). Bottles should be stored inverted to minimize gas exchange across the stopper and at the *in situ* temperature. In our study of deep sea sediments the *in situ* temperature was ~8 °C.

2.2. Subsampling and preservation

Subsamples from the stoppered serum bottles can be collected outside of an anaerobic chamber using an Ar-flushed disposable needle and syringe. Opening each bottle completely at each time point is not recommended because it will require replenishing the headspace and $^{15}N_2$ gas. Care should be taken to control the plunger on the syringe when removing sediment slurry from over-pressured bottles.

Sediment subsamples for FISH-NanoSIMS analysis should be fixed with freshly prepared, 0.2 μm filtered paraformaldehyde (Electron Microscopy Sciences, Hatfield, PA, #15713) that has been diluted to 4% in 1× phosphate buffered saline (PBS). Formaldehyde, glutaraldehyde, and ethanol are alternate fixatives and can be used depending on the microbial target and sample. Subsampled sediment should be immediately dispensed into eppendorf tubes containing an equal volume of fixative, to a final paraformaldehyde concentration of 2%. Fixed samples should be incubated at room temperature for 1 h or alternatively, overnight at 4 °C. After incubation, samples are then briefly centrifuged to remove traces of fixative, the supernatant removed and disposed of as hazardous waste, and the sediment pellets resuspended and washed twice, once with PBS and once with a 1:1 mixture of PBS and ethanol. The final sediment pellets are brought up in 1 ml 100% ethanol and can be stored at −80 °C for several years (some cell loss and degradation may be observed over this time, making timely analysis of the samples preferred). Similar protocols for sediment preparation for FISH can be found in the references cited below.

2.3. Selecting and preparing microprobe slides for sample deposition

NanoSIMS sample cartridges are designed to hold round samples of 1 in., 0.5 in., or 10 mm in diameter. Samples must be conductive for analysis, making round silicon wafers (Ted Pella, Inc.) a good choice for pure cultures, when specific cells do not need to be identified as targets prior to NanoSIMS analysis. However, for analysis of seep sediment microorganisms, samples must be mapped with both epifluorescence (to recognize hybridized cells) and transmitted light (to facilitate relocating the target within the NanoSIMS via a CCD camera) prior to NanoSIMS analysis (see Section 2.6). To allow mapping with transmitted light, the sample surface must be clear. Optically clear and conductive materials are therefore ideal for this purpose, including indium tin oxide (ITO), which can be purchased commercially and cut to size on site, or, if proper equipment is available, sputtered directly onto precut round glass microprobe slides (Lakeside Brand, Monee IL, #458; Pokaipisit et al., 2007). Alternatively, cells can be mapped directly on round glass microprobe slides and then gold sputtered (30 nm) just prior to NanoSIMS analysis. Depending on the isotopes analyzed (e.g., natural abundance $\delta^{13}C$), glass microprobe slides should be precleaned and combusted to remove any organic residue and handled with gloves throughout the sample preparation. The microprobe slide of the desired material should be completely cleaned and dry before adding the sample. Using a diamond scribe, the surface can be etched with reference marks. These score marks are important to the mapping process and greatly facilitate the relocation of sample target with the NanoSIMS. We recommend that these score marks be unique and nonsymmetric and toward the center of the slide. Deep cuts that traverse the entire surface could impact conductivity and should be avoided. The deposition of regular metal grids with letter and number identifiers can also be added to the sample surface to aid in sample relocation.

2.4. Concentrating cells using density gradients and deposition onto microprobe slides

In order to streamline the FISH–NanoSIMS analysis, separating cells from sediment using a density gradient prior to deposition on the sample round is recommended. Prior separation can be particularly helpful for the detection and measurement of low abundance community members and minority morphologies and associations. Although many types of density gradients exist and may be useful for some samples, we routinely use Percoll (Sigma-Aldrich, #P4937) gradients for the concentration of ANME/SRB aggregates (>3 μm). For the separation of single cells and small cell aggregates, including the ANME-1, sucrose and Nycodenz are recommended alternatives to Percoll.

The protocol for establishing Percoll gradients has been previously described in Orphan et al. (2002). Briefly, Percoll and PBS are mixed 1:1 (final volume approximately 10 ml) and centrifuged at 17,000 rpm for 30 min at 4 °C. Meanwhile, fixed sediment slurry is diluted in PBS and sonicated 3 × 10 s on ice. The exact dilution ratio of slurry to PBS will depend on density of material in the sample, but a 1:20 dilution is standard. One milliliter of the sonicated, diluted sediment slurry is floated onto the top of the Percoll gradient by slowly pipet dispensing the sample down the tilted edge of the tube. The gradient is then centrifuged at 4400 rpm for 15 min at 4 °C in a swinging bucket centrifuge. For sucrose gradients: 1 volume sonicated, diluted sampled is added to 2.5 volumes 1.95 M sucrose, then centrifuged at 1000 rcf for 20 min (Sigma-Aldrich, #S0389). For Nycodenz gradients: 1 volume sonicated, diluted sample is added to 3 volumes 70% by weight Nycodenz in sterile PBS and centrifuged at 12,000 rpm for 15 min (Sigma-Aldrich, #37890).

The entire fluid of the Percoll density gradient (excluding the sediment pellet) is transferred to a clean glass filter tower containing a 3-μm white polycarbonate filter (Millipore, #TSTP02500) backed with a glass microfiber filter (Millipore, GF/F, #1825-025). Percoll is difficult to filter through a pore size less than 2 μm, but water soluble Nycodenz or sucrose gradients can be used with 0.2 μm filters for single-cell investigations, as stated earlier. A low vacuum should be used to draw down the fluid, and the filter washed with several rinses of sterile PBS. Once the filter is nearly dry, it should be removed immediately from the filter tower and inverted onto the cleaned and marked microprobe slide containing a drop of deionized sterile water added to the region of the slide where the sample will be deposited. In some cases, placing the slide on a frozen block during filter inversion can increase the percentage of cells successfully transferred from the filter to the slide. The filter can be sectioned with a razor blade before inverting onto the glass round to allow the creation of several microprobe slides from a single filter (e.g., intended for different FISH probes or treatments) and to leave space for the addition of other samples and standards on the microprobe. The filter should be allowed to dry on the glass round until almost dry, but still stuck to the glass, and then gently peeled off. Depending on the sample, most material will transfer and stick to the glass.

Flow cytometry, and in particular, fluorescence-activated cell sorting (FACS), is another method to concentrate microbial cells from environmental samples before SIMS analysis. Although no published studies have combined these methods to date, FACS has been successfully coupled with bulk isotopic analysis of $\delta^{13}C$ of sorted cells (Eek et al., 2007) and our lab has had success imaging ^{13}C enrichment in individual FACS sorted methanotrophic microorganisms from lake sediment using the NanoSIMS 50L (M. Kalyuzhnaya, M. Lidstrom, and V. Orphan, in preparation). However, FACS is not compatible with all samples, and sediment samples, in particular, can be problematic. Therefore, as long as mapping of hybridized target cells is performed and the desired cells are generally abundant in the sample, density gradient separation is a fast and effective method to concentrate cells from marine sediment for NanoSIMS analysis.

2.5. FISH on microprobe slide-deposited cells

There are currently several methods of FISH coupled to NanoSIMS analysis, including monolabeled FISH (Dekas et al., 2009), catalyzed reporter deposition FISH (CARD-FISH) (Fike et al., 2008), and FISH or CARD-FISH with halogenated probes (i.e., El-FISH, HISH-SIMS, SIMSISH) (Behrens *et al.*, 2008; Halm *et al.*, 2009; Li *et al.*, 2008; Musat *et al.*, 2008). In El–FISH, SIMSISH, and HISH–SIMS, a halogen (i.e., ^{19}F, ^{81}Br, ^{127}I) that is typically rare in biomass but readily detectable via NanoSIMS is attached to a particular phylogenetic group of microorganisms during a modified protocol of either monolabeled or CARD-FISH. The NanoSIMS measurement can therefore be conducted without prior mapping (Musat *et al.*, 2008) and can indicate phylogenetic identity (via detection of the halogen) and isotopic composition of the biomass simultaneously. Although a promising direction for the future, the results of this technique can be difficult to interpret in complex environmental samples, such as sediments, due to high background signal relative to the often low ($\sim$1–30 counts/pixel) signal of the added element. For this chapter, we therefore focus on standard monolabeled FISH and CARD-FISH, and how to link the phylogenetic information they provide with NanoSIMS results.

CARD-FISH allows the deposition of many fluorophores for each hybridized gene target, thereby amplifying the fluorescent signal many times over that achieved with mono label FISH. CARD-FISH is therefore particularly helpful in cells with a low number of ribosomes and samples with high autofluorescence background, such as microbial mats (Fike *et al.*, 2008). Although CARD-FISH can be successfully employed in deep-sea sediments, it is not usually necessary for the detection of AOM aggregates. AOM aggregates have a distinctive morphology and a bright mono label FISH signal due to the many ribosomes present in a single aggregate of cells. It should be noted that CARD-FISH does introduce more exogenous C and N into the cell and can therefore have an effect on sensitive isotopic analyses. The reader is referred to the following references for details on the reactions: mono-FISH (Pernthaler *et al.*, 2001) and CARD-FISH (Pernthaler and Pernthaler, 2007; Pernthaler *et al.*, 2002). The procedure for FISH hybridization on a microprobe slide is the same as that on a traditional microscope slide or filter. The cells do not typically wash off during the hybridization. Oligonucleotide probe sequences for ANME and SRB can be found in Boetius *et al.* (2000) and can be custom ordered from Sigma-Aldrich. Especially for samples that require Z-stack image collection in epifluorescence light, Alexa Fluors [rather than cyanine (Cy3 or Cy5) or fluorescein isothiocyanate (FITC)] are recommended.

In some cases, such as samples with distinct cell morphology or defined cultures, general phylogenic identifications can be inferred without FISH. Even for well-defined samples, however, some type of imaging is recommended to accompany the NanoSIMS image. For example, TEM sectioning and imaging prior to NanoSIMS analysis (Finzi-Hart *et al.*, 2009; Lechene *et al.*, 2007) or collecting scanning electron images within the NanoSIMS in parallel to the analysis (Popa *et al.*, 2007) can provide sufficient context to allow identification without FISH for some well-defined samples. Complex environmental samples, or studies targeting morphologically standard cells, will almost always require FISH to identify the community present.

2.6. Mapping sample targets

Mapping sample targets is perhaps the most critical and time-consuming portion of this method. The goal of mapping is to identify good target cells for NanoSIMS analysis using epifluorescence microscopy and to create a map of the sample, indicating selected targets, in transmitted light at a range of magnifications (Fig. 12.1). Stitched panel images of transmitted light micrographs collected at low magnification (10× or 20× objective) will resemble the view of the sample produced by the CCD camera once the sample is loaded into the NanoSIMS. Depending on how the sample is loaded, however, the CCD image may be rotated relative to the map.

2.6.1. Selecting cells for analysis

The round microprobe slide with the hybridized cells is placed on top of a standard microscope slide and the edges fixed to the slide using standard lab tape (seen in Fig. 12.2). Securing the microprobe slide with tape is not only important for imaging using an inverted microscope, but also stabilizes the microprobe slide to prevent changes in orientation throughout the mapping procedure. If high magnification oil microscope objectives will be used, a drop of glycerol-based, water-soluble DAPI mounting medium is added to the hybridized sample, and a square coverslip placed on top, seated between the edges of the tape. To make DAPI mounting medium, combine 10 ml

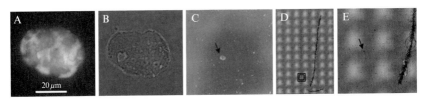

Figure 12.1 Targets for NanoSIMS analysis are identified at high magnification with epifluorescence light microscopy (A, an aggregate of cells hybridized with ANME-2 (red) and *Desulforsarcina/Desulfococcus* (green) targeted oligonucleotide probes) and then photographed in transmitted light at high (B) and lower (C) magnifications. A series of low magnification images are collected over the entire area of interest, and stitched together to create a sample map (D, double line box indicates the approximate size and location of panel C on the stitched image). The target is then located on the stitched image based on its position relative to reference-etched marks (the line visible in D and E) and distinctive neighboring particles (E). Black arrows point to the target identified in A and B. The etch marks will be visible in the NanoSIMS CCD camera, allowing the relocation of the identified target without epifluorescence microscopy. (For interpretation of the references to color in this figure legend, the reader is referred to the Web version of this chapter.)

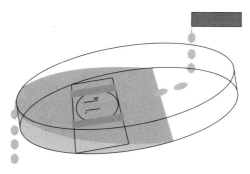

Figure 12.2 The round microprobe slide with etched reference marks is taped to a traditional rectangular microscope slide. After mapping with oil at high resolution, the cover slip can be slid off the side, and the DAPI mounting medium can be slowly rinsed with water indirectly in a tilted Petri dish.

citifluor mountant media (Ted Pella, #19470) with 1 ml 0.2 μm filtered PBS and 50 μl of 1 mg/ml DAPI (Sigma, #D8417). The sample should be stored in the dark for at least 15 min to allow the DAPI to fully permeate the cells prior to imaging with epifluorescence. If high resolution, high magnification epifluorescence images are not required, dry objectives can be used exclusively (Orphan et al., 2001b). Dry objectives have the advantage of not requiring the addition of mounting medium to the slide, eliminating the possibility of losing targets during the washing procedure or of incomplete removal of exogenous material.

We use an Applied Precision DeltaVision RT microscope equipped with the Softworx software package. In particular, the automated stage and ability to store and relocate X,Y locations as well as collect and stitch low magnification panel images in Softworx make the system ideal for sample mapping. However, it is possible to map samples with any microscope with a mounted camera (see our recommendations at the end of this section). If using a system like ours, the first step is to locate the etched reference lines on the microprobe slide (see previous section). Once these markings are located and the X,Y positions stored, the sample can be investigated for positively hybridized cells.

Preferred NanoSIMS targets are clean of contaminating materials such as undesired neighboring cells and sediment particles, but close to (~20–100 μm distance) reference marks or other distinctive particles. Proximity to reference marks allows for easier recognition with the NanoSIMS CCD camera. One advantage of NanoSIMS analysis is that cells do not have to be completely separated from other particles, since the measurement will not be averaged over the entire analysis space. During analysis with conventional ion microprobes, such as the CAMECA 1270 ims, with a 10–15 μm spot size for the primary beam, complete separation of the target cell is necessary to get an accurate measurement. The small beam diameter of the NanoSIMS and ion-imaging capabilities enables the selection of "regions of interest" from a complex field of view during postacquisition data analysis. Still, minimizing the material within the analysis frame simplifies identifying the target cell (see Fig. 12.1; a well-separated target). Good targets should be photographed at high magnification (we use a 60× oil objective for aggregates of cells, but a 100× objective may be preferred for single cells) (Fig. 12.1A), including acquisition of a z-stack (for 3D reconstruction), and their X,Y location should be saved. All targets should also be imaged in transmitted or reflected light at high magnification (Fig. 12.1B). At least five times the number of targets desired for analysis should be mapped due to sample loss in washing and difficulty refinding targets with the NanoSIMS.

2.6.2. Creating the map

Once all targets are identified, photographed in both epifluorescence and transmitted light, and their locations recorded, the microscope slide can be removed from the microscope and the taped-on microprobe slide gently cleaned of the DAPI mounting medium. Cleaning is accomplished by sliding the coverslip off, placing the attached slides in a slightly tilted Petri dish, and adding a slow stream of DI water to the raised end of the tilted dish (Fig. 12.2). The mounting medium will float in the water and wash out over the lower side of the dish with the water flow. Complete removal of the DAPI mounting medium requires about 5 min in the flowing water bath. Placing the slide directly in the stream of water or washing too rigorously could lead to sample loss or the detachment of cells and material from the slide. The microprobe slide (still taped to the microscope slide) is then dipped in ethanol (100%), air dried in the dark, and returned to the microscope in exactly the same position that it was previously.

All targets should then be revisited using a lower magnification, dry objective (e.g., 40× and/or 20× and 10×) and reimaged using transmitted or reflected light. These transmitted light micrographs more closely approximate the morphology of the target visible with the NanoSIMS CCD, assisting with confirmation of the target during analysis. Small adjustments in position may be necessary to ensure that the targets are in view when their stored X,Y positions are revisited with the software package. Finally, a 10× stitched image is created of the sample area containing both the reference marks and the sample targets (Fig. 12.1D). This can also be collected and stitched manually using image processing software such as Adobe Photoshop. It is useful to set the resolution of the stitched image such that individual features are still visible when zooming in on regions of interest. An image of the location of the saved targets and reference points as well as X,Y coordinates should also be saved. Then, starting with the highest magnification image, a target can be found by eye in the series of transmitted light images based on its morphology and orientation with respect to other nearby particles. Once located at high magnification, cell targets should be identified in the lower magnification photomicrographs and finally on the stitched image (Fig. 12.1E). Annotation of target locations on the map (consisting of the low magnification stitched image of the reference marks and targets) can be assembled in either Adobe Photoshop or Illustrator, or other image visualization software.

If using a system without point relocation capabilities, dry objectives are recommended. Each time a target is identified, the entire set of images ($60\times$, $40\times$, and $10\times$ magnifications) should be taken before moving to the next target. Additionally, a series of overlapping low magnification ($10\times$) images should be collected and manually aligned to create a single image between the nearest reference mark and the target. The entire process—the series of images at different magnifications and the overlapping path of images to the nearest reference mark—is repeated for each additional target.

2.7. Sectioning samples prior to analysis

Sectioning cells prior to NanoSIMS analysis is optional but can improve measurements. Perfectly flat samples are generally recommended for SIMS analysis to reduce artifacts due to variable sample topology (Orphan and House, 2009). Additionally, although the destructive nature of the NanoSIMS primary ion beam allows for analysis with depth, in some cases, higher resolution of internal cellular components can be achieved with prior sectioning. Focused ion beam (FIB) sectioning uses a narrow primary beam (such as Ga^+) to sputter away layers of material, exposing a flat and fresh interior surface (Fig. 12.3). This sputtering approach differs from that of the NanoSIMS because the beam can be applied parallel to the sample surface instead of normal to it. During perpendicular sputtering, the variable sputtering efficiency in a heterogeneous material such as a cell, or cell in sediment can result in uneven material removal and slightly irregular surface. With parallel sputtering, the beam can fully remove any material over a particular sample height (Fig. 12.3B). The FIB can be used to section samples in two ways: the first is by parallel sputtering, and the second is by creating a vertical thin section by removing the material on either side of the vertical section and then laying the section on its side (Weber et al., 2009). It is also possible to embed cells or tissue in epoxy resin and create thin sections using an ultramicrotome or other device (Herrmann et al., 2007a; Lechene et al., 2007; Slaveykova et al., 2009); however, in complex samples (i.e., natural sediments), locating the target cells of interest in the resin block is challenging and requires production and screening of many thin sections. For stable isotope tracer experiments not requiring high intracellular spatial resolution, for instance, when magnitude of incorporation is desired (i.e., evidence of nitrogen fixation), or gross trends in

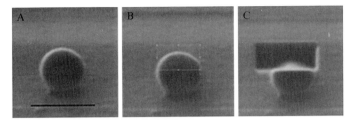

Figure 12.3 Cells can be sectioned prior to NanoSIMS analysis with a focused ion beam (FIB). Scale bar in A represents 1 μm. A bacterial spore is imaged via scanning electron microscopy (SEM) at a 60° angle from the surface (A). The area to be removed is identified (B), and a focused Ga^+ beam sputters away the material (C).

intracellular concentrations are sufficient, whole cell analysis with no prior sectioning is recommended (Ghosal et al., 2008).

2.8. NanoSIMS analysis

NanoSIMS instruments use a high-energy focused primary ion beam, typically Cs^+ or O^-, with a diameter ranging from 50 to 150 nm to bombard, embed within, and sputter secondary ions from the sample surface. These charged secondary ions are focused through a series of lenses, separated via differences in momentum due to mass and charge, and counted using electron multipliers or faraday cups. For the latest generation CAMECA NanoSIMS 50L instrument, up to seven ions can be simultaneously collected from the same sample volume. By rastoring the small primary ion beam across an area of the sample, an ion image is created, displaying the number of ions collected for each pixel of the image. As NanoSIMS analysis is a destructive process, repeated rastor cycles over the same area provide analysis with increasing depth into the sample. These individual images can be merged together to increase the number of ions per pixel, or alternatively, single layers can be compared to create a three-dimensional ion representation of the target (Fig. 4 in Finzi-Hart et al., 2009 and Fig. 3 in Dekas et al., 2009). The depth resolution of the NanoSIMS using a Cs^+ primary beam has been measured to be 14.3 nm, which is greater than the lateral resolution (Ghosal et al., 2008). This feature can be particularly useful when examining closely associated microorganisms in consortia, such as the ANME–SRB aggregates.

While interest and acquisition of NanoSIMS instruments is increasing at universities worldwide, there are currently only 27 instruments in operation at the time of this printing. However, many of the instruments at universities are available to visitors and the managers can be contacted with inquiries for use. Full-time support staff typically oversees the daily operation of the NanoSIMS instruments and may be available to assist with instrument operation and set up. We will not describe the setup, tuning, or use of the NanoSIMS here, but we refer the reader to the CAMECA website to learn more: www.cameca.com/instruments-for-research/NanoSIMS.aspx.

2.8.1. Sample preparation

The NanoSIMS operates under ultrahigh vacuum and biological samples that are not completely dehydrated or which are prepared with an inappropriate type of embedding resin can lead to analysis delays due to sample off gassing. Dehydration steps using ethanol, such as those which are routine after FISH, can assist with drying prior to SIMS analysis. If the microprobe slide is not conductive, or if the sample occupies a large area of the slide (such as a millimeter-scale microbial mat thin section; Fike et al., 2008), the sample must be sputter coated with a conductive metal (e.g., 30 nm gold) prior to analysis by SIMS, similar to standard preparations for SEM analysis. NanoSIMS facilities typically have sputter coaters on site for this purpose.

2.8.2. Sample analysis

Optimizing ion counts for single microbial cells can be a challenge because cells are small targets and contain elements with low sputtering efficiencies. For instance, the lack of useful secondary ion formation by the element nitrogen (mass 14, 15) has led to the routine analysis of the molecular cyanide ion (CN^-, mass 25, 26) as a proxy for nitrogen (McMahon et al., 2006; Peteranderl and Lechene, 2004). Additionally, some studies have collected the molecular C_2^- mass rather than C^- to increase the carbon ions collected (Orphan et al., 2001b, 2002). Furthermore, organic material, and particularly single cells, can sputter away quickly depending on the intensity of the primary beam. It is therefore recommended that tuning take place on a nearby nontarget cell rather than the cell of interest. The primary ion beam can also be adjusted to a lower intensity, more narrow focus, and shorter dwell time to allow the collection of the most highly resolved datasets in both the planar and Z dimensions. Analysis acquisition can be performed in two main modes: image acquisition and numerical isotope acquisition.

2.8.3. Standards

Element abundance and isotopic measurements acquired with the Nano-SIMS may be precise but not accurate. The inaccuracy can be due to several instrumental factors (Slodzian, 2004) as well as the matrix effects in the sample itself (Herrmann et al., 2007b). It is therefore necessary to measure standards not only of known elemental abundance and isotopic composition, but the standard should be composed of material as similar in physical structure and elemental composition as the sample itself. Information on selecting appropriate standards can be found in Davisson et al. (2008), Herrmann et al. (2007b), and Orphan and House (2009). We routinely use two standards. The first is a polished piece of graphite of known C isotopic composition. The second is a pure culture of cells (either *Escherichia coli* or *Clostridia* spores) prepared before each NanoSIMS session by dotting PBS or water-washed cells onto a clean area on the same microprobe slide as the sample. An aliquot of the same culture is then measured via elemental analysis isotope ratio mass spectrometry (EA-IRMS) for bulk elemental and isotope information. The variation within these pure cultures is typically within 10‰ in both $\delta^{13}C$ and $\delta^{15}N$ (Orphan and House, 2009). Greater than 10 cells in the pure cultures are measured with the same analysis parameters as the sample during each standard run, which is repeated before and after analysis sessions. These standard measurements enable the user to correct for matrix effects and instrumental mass fractionation during the analysis (Orphan and House, 2009). Precision on NanoSIMS measurements is usually controlled by counting statistics, which is described in Fitzsimons et al. (2000).

2.9. Interpreting NanoSIMS results

Several software programs are available to assist in analyzing NanoSIMS data, such as the program L'IMAGE PV-WAVE (written by Larry Nittler) and a free downloadable add-on to ImageJ (http://rsbweb.nih.gov/ij/, add-on OpenMIMS). Matlab scripts can also be created to analyze the raw output data directly, depending on the scientific question(s) and data processing required. Examples of strategies of data processing, including extracting data from individual regions of interest, and investigating isotopic composition with depth into the sample, can be found in the articles referenced in this chapter (for example, Dekas et al., 2009, Finzi-Hart et al., 2007, Lechene et al., 2007).

Correlating FISH and NanoSIMS images is critical for this method, as it provides the link between 16S identity and metabolic activity. For a well-mapped sample, this process only requires looking at the two corresponding images and adjusting for slight rotations. For roughly symmetrical targets, such as spherical ANME–SRB aggregates, it is often helpful to correlate the images of nearby asymmetrical targets analyzed on the same microprobe round first, to ensure that the correct rotations are applied.

Direct coregistration of FISH and NanoSIMS images can be challenging for some samples due to image aberrations and sample topography. A technique to aid in image correlation that is currently under development is the use of functionalized quantum dots (QD) for intracellular rRNA or protein detection. QD are nanometer-sized fluorescent particles with a heavy metal core, such as Se and Cd. Selenium (mass 78.9) is a moderate negative secondary ion producer and in theory can be used for ion imaging and spatial localization of QD-bound intracellular biomolecules or cell surface constituents in microbial cells. QD or gold particles, commonly applied in immunofluorescence staining, may also be used as cell-associated reference markers to assist in coregistering paired FISH and secondary ion images and identification of possible aberrations or shifts in the ion image during acquisition.

As mentioned previously, it is frequently recommended that flat and polished samples (e.g., thin sections or TEM preparations) be used for SIMS analysis for optimal spatial resolution and to minimize potential artifacts from uneven sample topology. While the benefit of prepolishing has been studied to some extent in geological materials and biological tissues, a comprehensive investigation of the specific effects of topography and uneven sputtering for microbial cell aggregates and heterogeneous environmental samples is still lacking. It is an important consideration when attempting to directly correlate a series of images of the same target generated with a NanoSIMS and generated with a microscope, or construct and compare 3-D representations. The NanoSIMS will generate secondary ions from the entire exposed surface of a target, which would be a curved, hemispherical plane in the case of a cell or aggregate of cells. Theoretically, the topology of the sample surface is then propagated throughout a depth analysis, yielding successive curved layers, and shrinking the lateral size of the target as the edges sputter to completion. (Although the edges may be subject to a lesser sputter rate than the center, due to the decreased angle of incidence of the primary beam, they will likely still sputter to completion before the center of the target because of the difference in thickness.) The resulting series of 2-D NanoSIMS images with depth may actually represent a series of more complex 3-D planes of analysis (Fig. 12.4, right). The NanoSIMS output with depth may therefore not be perfectly consistent with a paired Z-stack of images collected with a confocal microscope, or generated with deconvolution software, where each image represents a planar field of depth (Fig. 12.4).

We observed experimental support for this concept while analyzing seven roughly ellipsoidal aggregates with curved, dome-like surfaces. Although the diameters of these aggregates clearly increased with depth in the individual deconvolved Z-stack images collected by epifluorescence microscopy, the diameters of the of aggregates only decreased with increasing cycle number in the NanoSIMS image acquisition (Fig. 12.4). Further work is required to better understand the influence of geometry on the sputtering of biological targets, but for analyses requiring more than broad trends with depth, we encourage researchers to consider potential distortions.

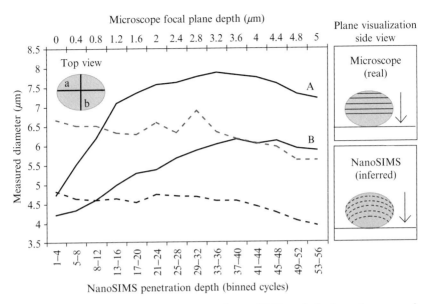

Figure 12.4 The geometry of the plane of NanoSIMS analysis depends on sample surface topology. Diameters of a single cellular aggregate measured on images collected with depth, created by deconvolved epifluorescence microscopy (solid line) and NanoSIMS (dashed line), show different trends. Diameters were collected in two directions perpendicular to each other for each type of analysis (A and B). A proposed explanation is that although the microscope images with depth represent parallel planes through the target, the NanoSIMS images may represent curved, peel-like layers as the curved surface topology is propagated throughout the sample.

2.10. Complementary additional analyses

Several analyses can be performed in addition to FISH and NanoSIMS imaging to support and enhance the data collected. The sample height can be measured with atomic force microscopy (AFM) or a profilometer before and after the SIMS analysis to quantify the depth of sample measured. The depth penetrated by the primary ion beam depends on instrument parameters and tuning as well as the nature of the sample material (as previously mentioned), making the sputter rate variable. The depth therefore must be determined on individual targets by measuring sample height before and after analysis if an exact value is desired. Additionally, sample visualization with either AFM or scanning electron microscopy (SEM) postanalysis can confirm the geometry of the rastor region. Although theoretically a rastor region is always a perfect square, nonoptimized instrument setup can result in a skewing of this region. The sputtered area can be visualized using these methods and any rastor area aberrations can be observed.

2.11. Measuring metal cofactors as indicators of nitrogenase

Traditional nitrogenase consists of an Fe protein as well as an MoFe protein; however, alternative nitrogenases do not require Mo and instead use V or additional Fe (Eady, 1996). Theoretically, these metals can be measured via NanoSIMS analysis in individual cells and used to provide independent information on the type of nitrogenase produced by specific microorganisms, and/or locate sites of active nitrogen fixation (e.g., Mo enrichment in heterocysts of cyanobacteria; Pett-Ridge et al., 2006). However, these analyses are challenging due to the low abundance of the metals of interest and possibility of background contamination. Additionally, the greatest ionization potential for most biologically relevant trace metals is typically as a positively charged secondary ion, which is created using a negative primary ion beam O^-, rather than Cs^+. Therefore, there is a trade-off between obtaining the maximal secondary ion yield from trace metals (using O^-) and the simultaneous analysis of other biologically important ions such as carbon, nitrogen and sulfur (by Cs^+).

In a pilot study performed on *Azotobacter*, a diazotroph with the genetic make-up to generate traditional nif, vnf, and anf forms of nitrogenase, we analyzed cells grown in different metal concentrations under nitrogen fixing conditions using a O^- primary ion beam with the CAMECA NanoSIMS 50L (James Howard, A. Dekas, and V. Orphan, unpublished data). The data collected in spot analysis mode indicated that the intracellular concentration of the three metals varied in three *Azotobacter* cultures, suggesting that the diazotrophic cells were synthesizing different forms of nitrogenase in each. While these controlled pure culture studies lack the complications of single-cell analysis in environmental samples, these and previously collected data (Pett-Ridge et al., 2006; Slaveykova et al., 2009) suggest that the NanoSIMS holds promise for quantifying the variation and spatial distribution of trace metals relevant to nitrogen fixation within and between microorganisms.

3. Concluding Remarks

FISH-NanoSIMS is a powerful tool in the investigation of metabolic activity in environmental microorganisms. The technique is particularly useful in single-cell observations of nitrogen fixation via ^{15}N incorporation from $^{15}N_2$ and has elucidated processes involved in nitrogen-based symbioses as well as identified new and unexpected diazotrophs. The strength of FISH-NanoSIMS is largely in its ability to bridge the gap between geochemical and molecular biological studies, linking cells of a particular species or group with a specific metabolism. However, the additional information accessible with this method, including patterns of substrate sharing between cells, variations in metabolic rates between members of the same community, and temporal and physical separation of particular intracellular activities, can also greatly improve our understanding of microbial ecosystems. With the installation of NanoSIMS at a growing number of institutes worldwide, we anticipate that the application of this technology to environmental microbiology will continue and expand for years to come.

ACKNOWLEDGMENTS

We thank Christopher House, Yunbin Guan, John Eiler, Kendra Turk, Annelie Pernthaler, Sameer Walavalkar, Kevin McKeegan, Raj Singh, and Mark Ellisman for helping to develop and optimize the method. We thank James Howard, Sameer Walavalkar, and Eric Matson for sample and data contributions to this chapter. We also thank two anonymous reviewers for helpful suggestions. This research was supported by grants from the Department of Energy (Career award, #DE-SC0003940) and Gordon and Betty Moore Foundation (to VJO) and the National Science Foundation (graduate fellowship to AED).

REFERENCES

Behrens, S., Losekann, T., Pett-Ridge, J., Weber, P. K., Ng, W.-O., Stevenson, B. S., Hutcheon, I. D., Relman, D. A., and Spormann, A. M. (2008). Linking microbial phylogeny to metabolic activity at the single-cell level by using enhanced element labeling-catalyzed reporter deposition fluorescence in situ hybridization (EL-FISH) and NanoSIMS. *Appl. Environ. Microbiol.* **74,** 3143–3150.

Boetius, A., Ravenschlag, K., Schubert, C. J., Rickert, D., Widdel, F., Gieseke, A., Amann, R., Jorgensen, B. B., Witte, U., and Pfannkuche, O. (2000). A marine microbial consortium apparently mediating anaerobic oxidation of methane. *Nature* **407,** 623–626.

Boxer, S. G., Kraft, M. L., and Weber, P. K. (2009). Advances in imaging secondary ion mass spectrometry for biological samples. *Annu. Rev. Biophys.* **38,** 53–74.

Brandes, J. A., and Devol, A. H. (2002). A global marine-fixed nitrogen isotopic budget: Implications for Holocene nitrogen cycling. *Global Biogeochem. Cycles* **16,** 1120–1134.

Brandes, J. A., Devol, A. H., and Deutsch, C. (2007). New developments in the marine nitrogen cycle. *Chem. Rev.* **107,** 577–589.

Buckley, D. H., Huangyutitham, V., Hsu, S. F., and Nelson, T. A. (2007a). Stable isotope probing with 15N2 reveals novel noncultivated diazotrophs in soil. *Appl. Environ. Microbiol.* **73,** 3196–3204.

Buckley, D. H., Huangyutitham, V., Hsu, S. F., and Nelson, T. A. (2007b). Stable isotope probing with 15N achieved by disentangling the effects of genome G+C content and isotope enrichment on DNA density. *Appl. Environ. Microbiol.* **73,** 3189–3195.

Capone, D. G., and Montoya, J. (2001). Nitrogen fixation and denitrification. *Methods Microbiol.* **30,** 501–513.

Capone, D. G. (1988). Benthic nitrogen fixation. *In* "Nitrogen Cycling in Coastal Marine Environments," (T. H. Blackburn and J. Sorensen, eds.), pp. 85–123. John Wiley & Sons, New York.

Capone, D. G. (2001). Marine nitrogen fixation: What's the fuss? *Curr. Opin. Microbiol.* **4,** 341–348.

Capone, D. G., Zehr, J. P., Paerl, H. W., Bergman, B., and Carpenter, E. J. (1997). Trichodesmium, a globally significant marine cyanobacterium. *Science* **276,** 1221–1229.

Codispoti, L. A. (1995). Is the ocean losing nitrate? *Nature* **376,** 724.
Codispoti, L. A. (2007). An oceanic fixed nitrogen sink exceeding 400 Tg Na-1 vs the concept of homeostasis in the fixed-nitrogen inventory. *Biogeosciences* **4,** 233–253.
Codispoti, L. A., Brandes, J. A., Christensen, J. P., Devol, A. H., Naqvi, S. W. A., Paerl, H. W., and Yoshinari, T. (2001). The oceanic fixed nitrogen and nitrous oxide budgets: Moving targets as we enter the anthropocene? *Sci. Mar.* **65,** 85–105.
Codispoti, L. A., and Christensen, J. P. (1985). Nitrification, denitrification, and nitrous oxide cycling in the eastern tropical South Pacific ocean. *Mar. Chem.* **16,** 277–300.
Davisson, M. L., Weber, P. K., Pett-Ridge, J., and Singer, S. (2008). Development of standards for NanoSIMS analyses of biological materials. pp. 1–39. L.L.N. Laboratory, Livermore, CA. DOI: 10.2172/945782.
Dekas, A. E., Poretsky, R. S., and Orphan, V. J. (2009). Deep-sea archaea fix and share nitrogen in methane-consuming microbial consortia. *Science* **326,** 422–426.
Dumont, M. G., and Murrell, J. C. (2005). Stable isotope probing—Linking microbial identity to function. *Nat. Rev. Microbiol.* **3,** 499–504.
Eady, R. R. (1996). Structureminus signfunction relationships of alternative nitrogenases. *Chem. Rev.* **96,** 3013–3030.
Eek, K. M., Sessions, A. L., and Lies, D. P. (2007). Carbon-isotopic analysis of microbial cells sorted by flow cytometry. *Geobiology* **5,** 85–95.
Fike, D. A., Gammon, C. L., Ziebis, W., and Orphan, V. J. (2008). Micron-scale mapping of sulfur cycling across the oxycline of a cyanobacterial mat: A paired NanoSIMS and CARD-FISH approach. *ISME J.* **2,** 749–759.
Finzi-Hart, J. A., Pett-Ridge, J., Weber, P. K., Popa, R., Fallon, S. J., Gunderson, T., Hutcheon, I. D., Nealson, K. H., and Capone, D. G. (2009). Fixation and fate of C and N in the cyanobacterium Trichodesmium using nanometer-scale secondary ion mass spectrometry. *Proc. Natl. Acad. Sci. USA* **106,** 6345–6350.
Fitzsimons, I. C. W., Harte, B., and Clark, R. M. (2000). SIMS stable isotope measurement: Counting statistics and analytical precision. *Mineralog. Mag.* **64,** 59–83.
Foster, R. A., Subramaniam, A., and Zehr, J. P. (2009). Distribution and activity of diazotrophs in the Eastern Equatorial Atlantic. *Environ. Microbiol.* **11,** 741–750.
Fulweiler, R. W. (2009). Microbiology. Fantastic fixers. *Science* **326,** 377–378.
Fulweiler, R. W., Nixon, S. W., Buckley, B. A., and Granger, S. L. (2007). Reversal of the net dinitrogen gas flux in coastal marine sediments. *Nature* **448,** 180–182.
Ganeshram, R. S., Pedersen, T. F., Calvert, S. E., and Murray, J. W. (1995). Large changes in oceanic nutrient inventories from glacial to interglacial periods. *Nature* **376,** 755–758.
Ghosal, S., Fallon, S. J., Leighton, T. J., Wheeler, K. E., Kristo, M. J., Hutcheon, I. D., and Weber, P. K. (2008). Imaging and 3D elemental characterization of intact bacterial spores by high-resolution secondary ion mass spectrometry. *Anal. Chem.* **80,** 5986–5992.
Gruber, N., and Sarmiento, J. L. (1997). Global patterns of marine nitrogen fixation and denitrification. *Glob. Biogeochem. Cycles* **11,** 235–266.
Halm, H., Musat, N., Lam, P., Langlois, R., Musat, F., Peduzzi, S., Lavik, G., Schubert, C. J., Sinha, B., LaRoche, J., and Kuypers, M. M. (2009). Co-occurrence of denitrification and nitrogen fixation in a meromictic lake, Lake Cadagno (Switzerland). *Environ. Microbiol.* **11,** 1945–1958.
Herbert, R. A. (1999). Nitrogen cycling in coastal marine ecosystems. *FEMS Microbiol. Rev.* **23,** 563–590.
Herrmann, A. M., Clode, P. L., Fletcher, I. R., Nunan, N., Stockdale, E. A., O'Donnell, A. G., and Murphy, D. V. (2007a). A novel method for the study of the biophysical interface in soils using nano-scale secondary ion mass spectrometry. *Rapid Commun. Mass Spectrom.* **21,** 29–34.

Herrmann, A. M., Ritz, K., Nunan, N., Clode, P. L., Pett-Ridge, J., Kilburn, M. R., Murphy, D. V., O'Donnell, A. G., and Stockdale, E. A. (2007b). Nano-scale secondary ion mass spectrometry—A new analytical tool in biogeochemistry and soil ecology: A review article. *Soil Biol. Biochem.* **39**, 1835–1850.

Howarth, R. W., Marino, R., and Lane, J. (1988). Nitrogen fixation in freshwater, estuarine, and marine ecosystems. 1. Rates and importance. *Limnol. Oceanogr.* **33**, 669–687.

Karl, D., Michaels, A., Bergman, B., Capone, D., Carpenter, E., Letelier, R., Lipschultz, F., Paerl, H., Sigman, D., and Stal, L. (2002). Dinitrogen Fixation in the world's oceans. *Biogeochemistry* **57/58**, 47–98.

Kechris, K. J., Lin, J. C., Bickel, P. J., and Glazer, A. N. (2006). Quantitative exploration of the occurrence of lateral gene transfer by using nitrogen fixation genes as a case study. *Proc. Natl. Acad. Sci. USA* **103**, 9584–9589.

Kessler, P. S., and Leigh, J. A. (1999). Genetics of nitrogen regulation in *Methanococcus maripaludis*. *Genetics* **152**, 1343–1351.

Lechene, C., Hillion, F., McMahon, G., Benson, D., Kleinfeld, A. M., Kampf, J. P., Distel, D., Luyten, Y., Bonventre, J., Hentschel, D., Park, K. M., Ito, S., *et al.* (2006). High-resolution quantitative imaging of mammalian and bacterial cells using stable isotope mass spectrometry. *J. Biol.* **5**, 20.

Lechene, C. P., Luyten, Y., McMahon, G., and Distel, D. L. (2007). Quantitative imaging of nitrogen fixation by individual bacteria within animal cells. *Science* **317**, 1563–1566.

Li, T., Wu, T. D., Mazeas, L., Toffin, L., Guerquin-Kern, J. L., Leblon, G., and Bouchez, T. (2008). Simultaneous analysis of microbial identity and function using NanoSIMS. *Environ. Microbiol.* **10**, 580–588.

Liang, J. H., Nielsen, G. M., Lies, D. P., Burris, R. H., Roberts, G. P., and Ludden, P. W. (1991). Mutations in the draT and draG genes of Rhodospirillum rubrum result in loss of regulation of nitrogenase by reversible ADP-ribosylation. *J. Bacteriol.* **173**, 6903–6909.

Mahaffey, C., Michaels, A. F., and Capone, D. G. (2005). The conundrum of marine N2 fixation. *Am. J. Sci.* **305**, 546–595.

McElroy, M. B. (1983). Marine biological controls on atmospheric CO_2 and climate. *Nature* **302**, 328–329.

McMahon, G., Saint-Cyr, H. F., Lechene, C., and Unkefer, C. J. (2006). CN- secondary ions form by recombination as demonstrated using multi-isotope mass spectrometry of 13C- and 15N-labeled polyglycine. *J. Am. Soc. Mass Spectrom.* **17**, 1181–1187.

Mehta, M. P., Butterfield, D. A., and Baross, J. A. (2003). Phylogenetic diversity of nitrogenase (nifH) genes in deep-sea and hydrothermal vent environments of the Juan de Fuca Ridge. *Appl. Environ. Microbiol.* **69**, 960–970.

Miyazaki, J., Higa, R., Toki, T., Ashi, J., Tsunogai, U., Nunoura, T., Imachi, H., and Takai, K. (2009). Molecular characterization of potential nitrogen fixation by anaerobic methane-oxidizing archaea in the methane seep sediments at the number 8 Kumano Knoll in the Kumano Basin, offshore of Japan. *Appl. Environ. Microbiol.* **75**, 7153–7162.

Moisander, P. H., Beinart, R. A., Hewson, I., White, A. E., Johnson, K. S., Carlson, C. A., Montoya, J. P., and Zehr, J. P. (2010). Unicellular cyanobacterial distributions broaden the oceanic N2 fixation domain. *Science* **327**, 1512–1514.

Montoya, J. P., Voss, M., Kahler, P., and Capone, D. G. (1996). A Simple, high-precision, high-sensitivity tracer assay for N(inf2) fixation. *Appl. Environ. Microbiol.* **62**, 986–993.

Musat, N., Halm, H., Winterholler, B. R., Hoppe, P., Peduzzi, S., Hillion, F., Horreard, F., Amann, R., Jørgensen, B. B., and Kuypers, M. M. M. (2008). A single-cell view on the ecophysiology of anaerobic phototrophic bacteria. *Proc. Natl. Acad. Sci. USA* **105**, 17861–17866.

Nauhaus, K., Boetius, A., Kruger, M., and Widdel, F. (2002). In vitro demonstration of anaerobic oxidation of methane coupled to sulphate reduction in sediment from a marine gas hydrate area. *Environ. Microbiol.* **4**, 296–305.

Oremland, R. S., and Taylor, B. F. (1975). Inhibition of methanogenesis in marine sediments by acetylene and ethylene: validity of the acetylene reduction assay for anaerobic microcosms. *Appl. Microbiol.* **30**, 707–709.

Orphan, V. J., Hinrichs, K. U., Ussler, W., 3rd, Paull, C. K., Taylor, L. T., Sylva, S. P., Hayes, J. M., and Delong, E. F. (2001a). Comparative analysis of methane-oxidizing archaea and sulfate-reducing bacteria in anoxic marine sediments. *Appl. Environ. Microbiol.* **67**, 1922–1934.

Orphan, V. J., and House, C. H. (2009). Geobiological investigations using secondary ion mass spectrometry: Microanalysis of extant and paleo-microbial processes. *Geobiology* **7**, 360–372.

Orphan, V. J., House, C. H., Hinrichs, K. U., McKeegan, K. D., and DeLong, E. F. (2001b). Methane-consuming archaea revealed by directly coupled isotopic and phylogenetic analysis. *Science* **293**, 484–487.

Orphan, V. J., House, C. H., Hinrichs, K. U., McKeegan, K. D., and DeLong, E. F. (2002). Multiple archaeal groups mediate methane oxidation in anoxic cold seep sediments. *Proc. Natl. Acad. Sci. USA* **99**, 7663–7668.

Pernthaler, A., Dekas, A. E., Brown, C. T., Geoffredi, S. K., Embaye, T., and Orphan, V. J. (2008). Diverse syntrophic partnerships from deep-sea methane vents revealed by direct cell capture and metagenomics. *Proc. Natl. Acad. Sci. USA* **105**, 7052–7057.

Pernthaler, A., and Pernthaler, J. (2007). Fluorescence in situ hybridization for the identification of environmental microbes. *Methods Mol. Biol.* **353**, 153–164.

Pernthaler, A., Pernthaler, J., and Amann, R. (2002). Fluorescence in situ hybridization and catalyzed reporter deposition for the identification of marine bacteria. *Appl. Environ. Microbiol.* **68**, 3094–3101.

Pernthaler, J., Glöckner, F.-O., Schönhuber, W., and Amann, R. (2001). Fluorescence in situ hybridization (FISH) with rRNA-targeted oligonucleotide probes. *Methods Microbiol.* **30**, 207–226.

Peteranderl, R., and Lechene, C. (2004). Measure of carbon and nitrogen stable isotope ratios in cultured cells. *Am. Soc. Mass Spectrom.* **15**, 478–485.

Pett-Ridge, J., Mayali, X., DeSantis, T., Karaoz, U., Andersen, G., Mabery, S., Hoeprich, P., Weber, P., and Brodie, E. (2010). CHIP-SIP: Linking Microbial Identity to Function in Marine and Terrestrial Systems. ISME-13 Stewards of a Changing Planet, Seattle, WA, Abstract CT11.001.

Pett-Ridge, J., Weber, P. K., Finzi, J., Hutcheon, I. D., and Capone, D. G. (2006). NanoSIMS Analyses of Mo Indicate Nitrogenase Activity and Help Solve a N and C Fixation Puzzle in a Marine Cyanobacterium. American Geophysical Union Annual Meeting, Abstract #B23D-1111. San Francisco, CA.

Ploug, H., Musat, N., Adam, B., Moraru, C. L., Lavik, G., Vagner, T., Bergman, B., and Kuypers, M. M. (2010). Carbon and nitrogen fluxes associated with the cyanobacterium *Aphanizomenon* sp. in the Baltic Sea. *ISME J.* **4**, 1215–1223.

Pokaipisit, A., Horprathum, M., and Limsuwan, P. (2007). Effect of films thickness on the properties of ITO thin films prepared by electron beam evaporation. *Kasetsart J.* **41**, 255–261.

Popa, R., Weber, P. K., Pett-Ridge, J., Finzi, J. A., Fallon, S. J., Hutcheon, I. D., Nealson, K. H., and Capone, D. G. (2007). Carbon and nitrogen fixation and metabolite exchange in and between individual cells of *Anabaena oscillarioides*. *ISME J.* **1**, 354–360.

Short, S. M., and Zehr, J. P. (2005). Quantitative analysis of nifH genes and transcripts from aquatic environments. *Methods Enzymol.* **397**, 380–394.

Short, S. M., and Zehr, J. P. (2007). Nitrogenase gene expression in the Chesapeake Bay Estuary. *Environ. Microbiol.* **9,** 1591–1596.

Slaveykova, V. I., Guignard, C., Eybe, T., Migeon, H. N., and Hoffmann, L. (2009). Dynamic NanoSIMS ion imaging of unicellular freshwater algae exposed to copper. *Anal. Bioanal. Chem.* **393,** 583–589.

Slodzian, G. (2004). Challenges in localized high precision isotope analysis by SIMS. *Appl. Surf. Sci.* **231–232,** 3–12.

Sprott, G. D., Jarrell, K. F., Shaw, K. M., and Knowles, R. (1982). Acetylene as an inhibitor of methanogenic bacteria. *J. Gen. Microbiol.* **128,** 2453–2462.

Veit, K., Ehlers, C., Ehrenreich, A., Salmon, K., Hovey, R., Gunsalus, R. P., Deppenmeier, U., and Schmitz, R. A. (2006). Global transcriptional analysis of Methanosarcina mazei strain Go1 under different nitrogen availabilities. *Mol. Genet. Genomics* **276,** 41–55.

Waksman, S. A., Hotchkiss, M., and Carey, C. L. (1933). Marine bacteria and their role in the cycle of life in the sea: II. Bacteria concerned in the cycle of nitrogen in the sea. *Biol. Bull.* **65,** 137–167.

Weber, P. K., Graham, G. A., Teslich, N. E., Chan, W. M., Ghostal, S., Leighton, T. J., and Wheeler, K. E. (2009). NanoSIMS imaging of Bacillus spores sectioned by focused ion beam. *J. Microsc.* **238,** 189–199.

Widdel, F., Boetius, A., and Rabus, R. (2006). Anaerobic biodegradation of hydrocarbons including methane. *Prokaryotes* **2,** 1028–1049.

Zhang, Y., Burris, R. H., Ludden, P. W., and Roberts, G. P. (1993). Posttranslational regulation of nitrogenase activity by anaerobiosis and ammonium in *Azospirillum brasilense. J. Bacteriol.* **175,** 6781–6788.

CHAPTER THIRTEEN

MEASUREMENT AND DISTRIBUTION OF NITRIFICATION RATES IN THE OCEANS

B. B. Ward

Contents

1. Introduction	308
1.1. Ammonia-oxidizing microorganisms	308
1.2. Nitrite-oxidizing bacteria	309
2. Nitrification Rate Measurements in Seawater (Table 13.1)	310
2.1. Dissolved inorganic nitrogen inventory approach	310
2.2. Specific inhibitors with the DIN inventory approach	312
2.3. Specific inhibitors with radioisotopes	313
2.4. N isotope methods	314
3. Nitrification Rate Measurements in Sediments	318
4. Distribution of Nitrification	319
References	320

Abstract

Nitrification is the process that converts ammonium to nitrate and thus links the regeneration of organic nitrogen to fixed nitrogen loss by denitrification. The first step, oxidation of ammonia to nitrite, is performed by a phylogenetically restricted group of proteobacteria (ammonia-oxidizing bacteria, AOB) and Crenarchaea (ammonia-oxidizing archaea, AOA). The second step is restricted to nitrite-oxidizing bacteria (NOB) as far as currently known. All three groups are assumed to be autotrophic and obligately aerobic, but the true extent of autotrophy and potential anaerobic pathways in these organisms is currently under investigation. Here, we describe methods for the measurement of nitrification rates in the marine environment, with a focus on seawater systems and stable isotopic tracer methods. The methods vary in analytical requirements but share the need for incubations, which must be optimized for different environments with different substrate concentrations. Recent advances in mass spectrometry now make it possible to minimize incubation artifacts and to achieve greatly improved sensitivity.

Department of Geosciences, Princeton University, Princeton, New Jersey, USA

1. Introduction

Nitrification is an essential step in the nitrogen (N) cycle in the ocean. While nitrification does not result in a direct change in the fixed N inventory, it is the critical link between organic N and its eventual loss from the system as N_2 via denitrification or anammox. Ammonia-oxidizing bacteria (AOB) and ammonia-oxidizing archaea (AOA) produce nitrite from ammonium, and nitrite-oxidizing bacteria (NOB) perform the final oxidation of nitrite to nitrate. No organism in culture is known to oxidize both ammonium and nitrite. The bacteria that oxidize nitrite or ammonia are not phylogenetically closely related to each other.

1.1. Ammonia-oxidizing microorganisms

The overall reaction for ammonia oxidation by AOB (Eq. 13.4) shows that the process consumes molecular oxygen and produces hydrogen ions, in addition to nitrite. A requirement for molecular oxygen occurs in the first step of the oxidation (Eq. 13.1), which is catalyzed by a monooxygenase (ammonia monooxygenase, AMO). Uncharged gaseous ammonia is the actual substrate for AMO, as demonstrated by the pH dependence of the reaction rate (Suzuki et al., 1974; Ward, 1987).

In AOB, the immediate product of AMO is hydroxylamine, which is further oxidized by hydroxylamine oxidoreductase (HAO) to nitrite (Eq. 13.2). Oxygen is also consumed by the terminal oxidase (Eq. 13.3), as a result of electron transport generating ATP for cellular metabolism.

$$NH_3 + O_2 + 2H^+ + 2\,e^- \rightarrow NH_2OH + H_2O \quad (13.1)$$

$$NH_2OH + H_2O \rightarrow NO_2^- + 5H^+ + 4e^- \quad (13.2)$$

$$2H^+ + 0.5\,O_2 + 2e^- \rightarrow H_2O \quad (13.3)$$

$$NH_3 + 1.5\,O_2 \rightarrow NO_2^- + H_2O + H^+ \quad (13.4)$$

The pathway by which AOA oxidize ammonia to nitrite is not entirely known. The first step is apparently catalyzed by protein homologous to the AMO of AOB, but no HAO has been found in AOA, suggesting that the product of AMO in AOA is not hydroxylamine. Based on the only complete AOA genome at this time, the pathway to nitrite is not clear (Walker et al., 2010). The stoichiometry of ammonium oxidation to nitrite (i.e., 1:1 ammonia consumed to nitrite formed) appears to be the same in AOA and AOB, however, which means that the two groups cannot be distinguished at this level. AOB require the recycling of reductant, generated at the hydroxylamine step, for the initial oxidation of ammonia. AOA do not appear to have this requirement, suggesting that their ammonia oxidation pathway is more efficient than that of the AOB. Once the AOA pathway is known, it may be possible to devise specific inhibitors to distinguish the contribution of each pathway to the overall process in the environment where both groups are present. In theory, antibiotics that are specific for bacteria versus archaea (e.g., in attacking the cell walls or preventing protein synthesis) should be useful in differentiating the contribution of the two groups during incubation measurements. They may not be effective in this application, however, as their mode of action, usually on the time scale of microbial generations, is too long to be effective in short-term incubations. Methods from molecular ecology which make it possible to enumerate and thus quantitatively distinguish AOB from AOA provide strong correlative evidence for dominance of AOA over AOB in measured nitrification rates (Beman et al., 2008) in the water column.

The AOB are found in the beta- and gamma-proteobacteria, and the beta genera are more common and widespread in the environment (Norton, in press). The AOA are found in the mesophilic Chrenarchaeota, in a proposed new lineage called Thaumarchaeota (Brochier-Armanet et al., 2008), in two broad phylogenetic groups distinguished as derived from aquatic and soil/sediment (Nicol et al., in press). In most environments so far investigated, the AOA outnumber the AOB by up to several orders of magnitude.

As far as is presently known, both AOA and AOB are primarily chemo-autotrophic, deriving their reducing power from the oxidation of ammonia and using it to fix CO_2. AOB use the Calvin cycle, while AOA use the 3-hydroxyproprionate/4-hydroxybutyrate pathway. Although the latter pathway is less efficient (less carbon fixed per reductant) than the Calvin cycle, both AOA and AOB require similar amounts of N oxidation for C fixation. For AOB, the ratio varies from about 8 to 42 (Billen, 1976; Carlucci and Strickland, 1968; Glover, 1985), that is, up to 42:1 mol NH_4^+ per mol C assimilated. For AOA, the value of 25 has been estimated (Konneke et al., 2005). Thus, although AOA and AOB use different pathways for both ammonia oxidation and CO_2 fixation, their stoichiometric impact on both C and N cycles is essentially the same.

1.2. Nitrite-oxidizing bacteria

The overall reaction for nitrite oxidation, best known in *Nitrobacter*, is one step in which one atom of oxygen is incorporated into the nitrite from water.

$$HNO_2 + H_2O \rightarrow HNO_3 + 2e^- + 2H^+ \qquad (13.5)$$

The obligately aerobic growth of most NOB is due to a dependence on aerobic respiration rather than the requirement of molecular oxygen in the nitrification reaction itself. Nitrite oxidation is directly linked to oxygen consumption through the consumption of protons by cytochrome oxidase. The nitrite oxidase enzyme, NXR, likely also functions in reverse to reduce nitrate to nitrite under anaerobic conditions, but the significance of this process in nature is unknown.

NOB are a more heterogeneous group than the AOB, and are found in the alpha-, beta-, delta-, and gamma-proteobacteria, as well as in a separate phylum, Nitrospirae (Daims *et al.*, in press). They are all considered to be primarily chemoautotrophic, but most possess limited capability to assimilate a small range of simple organic compounds. A recent report has demonstrated anaerobic anoxygenic photoautotrophy by an enrichment culture containing cells closely related to *Thiocapsa* (Griffin *et al.*, 2007). Phototrophic nitrite oxidation constitutes a new capability in the N cycle that is analogous to photoautotrophic sulfur oxidation. Like photoautotrophic sulfur oxidation, the new nitrite oxidation process is likely most important in relict environments, where light and nitrite, but not oxygen, are available.

The proteobacterial NOB use the Calvin cycle for CO_2 fixation, but the Nitrospirae appear to use the reverse TCA cycle (Daims *et al.*, in press), an ancient and less efficient pathway. For the proteobacterial chemoautotrophic NOB, the ratio of NO_2^- oxidized to CO_2 fixed is higher than for AOB or AOA: 32–71 measured in batch and chemostat cultures of *Nitrococcus mobilis* (Glover, 1985) and 48–81 in batch cultures of *Nitrobacter* (Laudelout *et al.*, 1968). This ratio of N oxidized to C fixed is important for application of some rate measurements, which rely on incorporation of $^{14}CO_2$ to estimate N oxidation (see below).

2. Nitrification Rate Measurements in Seawater (Table 13.1)

2.1. Dissolved inorganic nitrogen inventory approach

The most direct experimental design is to incubate samples and measure the concentrations of dissolved inorganic nitrogen (DIN) pools over time. This approach provided some of the earliest evidence for the occurrence of biologically mediated nitrification (Rakestraw, 1936; von Brand *et al.*, 1937). In such an experiment, decreasing ammonium and accumulation of nitrite or nitrate indicate net nitrification. It is possible to observe a decrease in the concentration if nitrite or nitrate over time even when nitrification is occurring, however, if consumption of nitrate or nitrite exceeds production in the incubation. This is often the case in bottom

Table 13.1 Methods for measurement of nitrification rates in water

	Analyte	Limitations	Advantages
DIN inventory with inhibitors	DIN	– Long incubations – Dark requirement to eliminate phytoplankton causes perturbations – Nonspecific or incomplete inhibition	– Simple measurement requirements
DIN inventory with $^{14}CO_2$ uptake	^{14}C-particulate material	– Radioisotope precautions – Need for conversion factor between CO_2 fixation and N oxidation	– Short incubations – High sensitivity
^{15}N tracers	$^{15}NO_2^-$ and $^{15}NO_3^-$	– Mass spectrometry required	– Direct measurement of specific N transformations – Short incubations – High sensitivity

water and sediment interface environments where nitrification and denitrification are tightly coupled, and in surface ocean waters, where the low concentration relative to high phytoplankton demand for fixed nitrogen assimilation means that large fluxes can be obscured by tight coupling between production and consumption terms. Similarly, simple measurements of ammonium concentration are usually not sufficient to determine nitrification rates because many other processes both consume and produce ammonium in the same bottle. Nevertheless, to demonstrate net nitrification, and to model the combined net rates of production and consumption processes, it is useful to measure DIN concentrations over time in long incubations, such as mesocosms. Explicit directions for the chemical methods for analyses of ammonium, nitrite, and nitrate can be found in standard handbooks (Grasshof et al., 1983; Parsons et al., 1984). High precision methods for measurement of DIN concentrations using chemiluminescence for nitrite and nitrate (Yoshizumi et al., 1985) or the OPA method for ammonium (Holmes et al., 1999) may improve the sensitivity of this approach.

2.2. Specific inhibitors with the DIN inventory approach

The addition of specific inhibitors is used as a modification of the nutrient inventory approach just described. In this approach, chemicals that specifically inhibit ammonia oxidation (e.g., acetylene, allylthiourea (ATU), methyl fluoride, N-serve) or nitrite oxidation (chlorate, azide) are added to replicate incubation vessels (Bianchi et al., 1997; Billen, 1976) and the difference in accumulation of DIN components is interpreted as rates in the presence and absence of inhibitors. The method assumes that the nitrite concentration is at steady state in the sample, and that nitrification is the only process that produces or consumes nitrite. Clearly, the vessels must be incubated in the dark to prevent DIN consumption by phytoplankton. One needs to measure only the concentration of nitrite over time in the bottles in which ammonium oxidation was inhibited to estimate the nitrite oxidation rate (equal to the rate of nitrite decrease). The rate of nitrite increase in the bottles to which nitrite-oxidation inhibitor was added represents the rate of ammonium oxidation.

There are some potential problems with this approach: (1) preventing photosynthesis by phytoplankton probably has cascading effects on the activities of other microbes in the bottle, such that the rate of ammonium mineralization is reduced, therefore changing the source term for the nitrification substrate. (2) Incubating in the dark may release the nitrifying organisms from light inhibition such that the measured rate exceeds the *in situ* rate. (3) Due to the necessity to detect small changes in DIN concentrations, incubations typically last 48 h, which is sufficient to overcome the lag induced by light inhibition, and is also long enough to create quite unnatural conditions. The search for precisely specific inhibitor compounds has been extensive and has resulted in a plethora of potentially useful compounds (Bedard and Knowles, 1989). Many are problematic for reasons not directly related to nitrification. For example, acetylene inhibits both nitrifiers and denitrifiers (Balderston et al., 1976; Berg et al., 1982). Thus, its use to measure one process will also inhibit the other, and when one depends on the other (as is the case when denitrification depends upon nitrification for nitrate at the sediment water interface), both rates are affected and the independent measurement of one is not possible. It is reported that the level or length of exposure to acetylene can be optimized to differentiate between its affects on nitrification and denitrification (Kester et al., 1996).

Many hydrocarbons act as alternative substrates for AMO and some of these are inhibitory to ammonia oxidation. The degree of both inhibition and inactivation of the enzyme by these compounds often depends on the concentration of ammonium (Arp and Stein, 2003), so it is difficult to optimize the treatment for diverse samples. ATU is a commonly used inhibitor for ammonia oxidation; it causes complete instantaneous inhibition at concentrations of 86 μM (10 mg L^{-1}) (Ginestet et al., 1998). Its mode of action is to bind copper, a trace metal involved in the active site of the AMO enzyme.

2.3. Specific inhibitors with radioisotopes

Specific nitrification inhibitors also serve as the basis for the sensitive $^{14}CO_2$ method for measurement of nitrification rates. Being mainly or totally chemolithoautotrophs, nitrifiers fix CO_2 while oxidizing nitrogen. The amount of CO_2 fixation due to nitrifiers can be computed by difference between incubations with and without addition of an inhibitor that specifically prevents CO_2 uptake by nitrifiers (Billen, 1976; Dore and Karl, 1996; Somville, 1978). Then, a conversion factor is used to translate the CO_2 fixation into ammonium and nitrite oxidation rates. This conversion factor has been shown to vary by a factor of five in pure cultures of AOB (Billen, 1976; Glover, 1985), and thus its use introduces some uncertainty, because the factor cannot be directly determined in field samples. The reason for the variation in conversion factor is not obvious, but the stoichiometry of N oxidation and C fixation in cultures (see above) suggests that it is probably not related to the relative contribution of AOA and AOB. N-serve, a commercial nitrification inhibitor (Goring, 1962), was used in the original application of this method (Billen, 1976), but its insolubility in water leads to nonspecific effects including stimulation of dark CO_2 uptake by non-nitrifiers (Owens, 1986; Priscu and Downes, 1985; Viner, 1990; Ward, 1986). ATU is now more often used as the inhibitor in the $^{14}CO_2$ approach (Rees et al., 2002). The advantage of the $^{14}CO_2$ method is that the radioisotope can be added at true tracer levels, allowing short incubations and thus minor perturbations.

2.3.1. Method

Oxidation of ammonium to nitrite and nitrate is estimated either from the change in DIN or by the incorporation of ^{14}C in the dark in the presence and absence of ATU. For the DIN method, fill six (or more) replicate 1-L polycarbonate bottles with sample. Add ATU to one-third of the bottles at a final concentration of 10 mg L^{-1} and 10 mM NaClO$_4$ to one-third of the bottles. Incubate in the dark for 12–48 h at *in situ* temperature. At 12 h intervals, remove 50 mL of water from each bottle and measure the DIN concentrations or store aliquots in the freezer for later analysis. If it is desired to preserve the *in situ* oxygen concentrations, the incubations should be performed in replicate small volume glass bottles with no headspace. In this case, sacrifice replicate bottles with and without inhibitors at each time point to remove water for DIN analysis.

For the radioisotope method (Rees *et al.*, 2002), fill six replicate 50-mL polycarbonate bottles with sample and add tracer quantities of NaH^{14}CO$_3$ to each bottle. Add ATU to three of the bottles at a final concentration of 10 mg L^{-1}. Incubate in the dark for 3–6 h at *in situ* temperature. Terminate the incubations by filtration onto 0.2 µm poresize membrane filters, expose the filters to fuming concentrated HCl for up to 10 min to remove any inorganic ^{14}C, and analyze by liquid scintillation counting. To convert carbon incorporation into ammonium oxidation, use a conversion factor to compute mol NH$_4^+$ oxidized per mol C fixed (Billen, 1976; Glover, 1985). Use of a low value (published conversion factors range 8.3–42 mol NH$_4^+$ per mol C) provides a conservative estimate of the nitrification rate.

2.4. N isotope methods

Both "tracer" and "dilution" approaches can be used to measure nitrification rates taking advantage of the sensitivity of stable isotope methods. In the tracer approach, a "trace" amount (an amount low enough to avoid perturbation of the ambient substrate concentration, generally taken as 10% of the ambient level) of labeled substrate (a radio or stable isotope) is added to a sample. After incubation, the amount of label in the product is used to compute the transformation rate. Unfortunately, a direct radioisotope tracer method is not very useful for measuring rates of nitrification in the environment. Capone *et al.* (1990) demonstrated the use of ^{13}N to quantify nitrification rates, but the isotope is so short-lived (half life = 10 min) that its use is usually impractical.

The main approach to measuring nitrification rates directly is to use the stable isotope, ^{15}N, as a tracer (Olson, 1981; Ward *et al.*, 1984). This approach has constraints that may limit its application, mainly due to the facts that ^{15}N has a significant natural abundance and that it must be measured using a mass spectrometer or emission spectrometer, both more expensive and difficult than using a scintillation counter for radioisotopes. Due to the great sensitivity of isotope ratio mass spectrometry, much shorter incubations (compared to the inhibitor and inventory methods) are possible (a few hours to 24 h are commonly used). The signal of transfer of the tracer from substrate to product pool (e.g., $^{15}NH_4^+$ to $^{15}NO_2^-$) can be detected regardless of what other processes are occurring in the incubation (so *in situ* light conditions can be used) and no assumptions of steady state need be made. The major drawback of this method is that it is not always possible to add the tracer at "trace" levels, due to the extremely low ambient concentrations of DIN. This problem has been largely overcome by the advent of more sensitive mass spectrometers and analysis methods, however, and estimates obtained under conditions approaching *in situ* are possible.

2.4.1. Method
2.4.1.1. *Measurement of ammonium and nitrite oxidation rates in seawater using ^{15}N*
Collect samples from the desired depth using Niskin bottles deployed on a rosette or with a peristaltic pump for shallower depths. Because nitrification, especially ammonia oxidation, is oxygen sensitive, take precautions to maintain the *in situ* oxygen conditions when working with low oxygen waters. This can be done by plumbing the Niskin bottle with CO_2 or N_2 while it is being emptied, and by overfilling the incubation bottles (preferably polycarbonate, to avoid potential trace metal contamination from glass, or ground glass stoppered bottles) and then sealing the caps without introducing bubbles into the bottle. Below the photic zone, incubations can be carried out in gas impermeable plastic bags (trilaminates produced by Pollution Measurement Corporation, Oak Park, IL, or in equivalent impermeable plastics). These are convenient and versatile incubation containers because they can be purchased in various sizes (we have found 500 mL to 10 L to be useful for various applications), filled directly from the Niskin sampler, and aseptically subsampled without perturbing the gas concentrations in the remaining sample.

Perform the incubations under simulated *in situ* conditions (using screening to simulate *in situ* light conditions) in controlled temperature incubators or in running seawater incubators (for surface samples). For single endpoint measurements, the length of the incubation should be kept as short as possible, usually 1–3 h. Even in short incubations, the atm.-% of the substrate pool can be diluted by ammonium regeneration occurring in the same bottle, and during long incubations, can make it impossible even to account for the dilution in the rate calculations (see below).

Measure the ambient substrate concentration prior to tracer addition so that you can add tracer at a level that increases the ambient pool by no more than 10%. This is not always possible, because the ambient pool size may be so low as to be impossible to add a large enough signal without overwhelming the *in situ* substrate levels. In that case, an addition of 50–100 nM is commonly used.

The sensitivity of the isotope tracer method is in the range of $nM\,d^{-1}$ and can be optimized by adjusting incubation volume, length of incubation, and level of tracer addition. Depending on the method used for final isotopic analysis, it may also be advisable to add about 250 nM natural abundance NO_2^- or NO_3^-. If the ambient product pool is quite small, any $^{15}NO_2^-$ or $^{15}NO_3^-$ that is produced is essentially lost immediately to oxidation to nitrate or uptake by phytoplankton. Addition of the carrier allows the recently produced product to be diluted into a larger pool, which can then be recovered at the end of the incubation. The amount of nitrite or nitrate added is too small to influence the ammonia oxidizers, although it may stimulate phytoplankton.

Add the tracer ($^{15}NH_4^+$ as $Na^{15}NH_4$ or $(^{15}NH_4)_2SO_4$ at 99 atm.-%; $^{15}NO_2^-$ as $Na^{15}NO_2$) by syringe through a silicone seal in the cap or with a pipettor just prior to sealing. In the latter case, it is a good idea to cool the tracer stock below the temperature of the incubation so that the added tracer solution sinks into the bottle and is not lost when the cap is secured. If using bags for incubation, add the tracer using a syringe while filling the bag from the sampler.

At the end of the incubation, filter the sample through precombusted glass fiber or silver filters to remove particulate material (which can be used for the determination of ammonium or nitrite uptake rates) and the filtrate can be frozen for storage until analysis.

2.4.1.2. $^{15}NO_2^-$ analysis by solute extraction Instructions for this method are provided in detail elsewhere (Ward and O'Mullan, 2005). Briefly, in order to assay its ^{15}N content, the dissolved nitrite or nitrate must be removed from solution and concentrated for introduction to the mass spectrometer. This is usually done by converting the nitrite to an azo dye that partitions into organic solvents (e.g., trichloroethylene, TCE) and then concentrating the dye by either solvent or solid-phase extraction (Kator *et al.*, 1992; Olson, 1981). If analyzing nitrate, this must be first reduced to nitrite, after removing the original nitrite using sulfamic acid (Ward *et al.*, 1984) or ascorbic acid (Granger *et al.*, 2006). The cleaned extract (the azo dye dissolved in TCE) can be stored at this step or evaporated and stored dry. To introduce the dye into the mass spectrometer, redissolve it into 100–200 μL TCE and transfer it onto a small fragment of precombusted glass fiber filter. Encapsulate the filter in foil manually or using a pellet press.

Because TCE dissolves many plastics, use glass or Teflon throughout this protocol for separatory funnels, graduated cylinders, and holding and storage vessels. All glass utensils and vessels should be combusted at 450 °C for 2 h prior to use. Forceps, foil, glass surfaces used for cutting filters, etc., should be cleaned with ethanol between samples. All of the extraction steps should be performed in a hood or very well-ventilated area. A variation on this method that uses different extraction chemistry has been applied in oligotrophic systems (Clark *et al.*, 2007).

2.4.1.3. $^{15}NO_2^-$ *analysis by conversion to* N_2O A method developed for natural abundance measurements of the isotopic content of nitrite (McIlvin and Altabet, 2005) can be adapted for use in tracer measurements (Bulow *et al.*, submitted). After incubation, the nitrite is chemically converted to nitrous oxide (N_2O), which can be analyzed with great sensitivity by mass spectrometry, thus increasing the sensitivity of the assay and reducing the volume required for analysis.

The method requires less than 40 nmol N, so measure the nitrite concentration in the sample prior to conversion and adjust the volume or add carrier nitrite if necessary. Place the water sample (50-mL) in a 60-mL vial and cap tightly with a Teflon-lined septum. Add 2 mL of azide/acetic acid buffer (prepared fresh by combining a 1:1 by volume mixture of mixture of 20% acetic acid and 2 M sodium azide) with a syringe and shake vigorously. After 15 min at room temperature, adjust the pH of the solution by adding 1.0 mL of 6 M NaOH and shake. Store the vials upside down until analysis of the headspace by mass spectrometry.

2.4.1.4. $^{15}NO_2^-$ + $^{15}NO_3^-$ *analysis by conversion to* N_2O To measure complete denitrification, the combined nitrate and nitrite pools can be analyzed with great sensitivity using the denitrifier method (Sigman *et al.*, 2001). Detailed instructions are provided by Sigman *et al.* (2001) and briefly summarized here. Prepare large volume cultures of *Pseudomonas chlororaphis* (ATCC# 43928) or *Pseudomonas aureofaciens* (ATCC# 13985) growing in liquid tryptic soy broth (Difco Laboratories). Maintain these cultures in constant growth if you will be doing these analyses frequently. On the day of sample preparation, wash the cells by centrifugation and resuspend them in spent medium at 10-fold cell concentration. Aliquot 2 mL of cell concentrate into 20-mL headspace vials and cap with Teflon-backed silicone septa and crimp seals. Purge the vials at 10–20 mL/min for 2 h or more with N_2 gas to remove residual N_2O.

Measure the NO_2^- + NO_3^- concentration of the sample and adjust the volume to contain 10–20 nmol N. Inject the sample of 4 mL or greater into the purged vial using a venting needle to prevent pressurization of the vial and subsequent loss of N_2O gas. Incubate the vials overnight to allow for complete conversion of sample NO_2^- + NO_3^- to N_2O. After the overnight incubation, inject 0.1–0.2 mL of 10 N NaOH. This raises the pH of the seawater sample to greater than 12, lyses the bacteria, and stops the reaction. Store the vials upside down until analysis by mass spectrometry as above.

2.4.1.5. Rate calculations   Using mass or emission spectrometry, determine the atm.-% ^{15}N of the particulate (dye extract) (Fiedler and Proksch, 1975) or gaseous (converted N_2O) (Sigman *et al.*, 2001) sample. It is very straightforward to compute the rate of $^{15}NO_2^-$ production if you can assume that the amount of ammonium oxidized represents an infinitesimal fraction of the ambient ammonium pool. This may be a reasonable assumption if very short incubations are used and true tracer additions were made. In seawater, where ammonium regeneration (ammonification of N-containing organic materials) often proceeds at a rate comparable to the rate of oxidation, the size and isotopic content of the substrate pool changes on the time scale of the incubation. In this case, the atm.-% of the substrate pool is continually diluted throughout the incubation and failure to account for this change will cause an underestimate of the ammonium oxidation rate. A comparison of the most commonly used equations and a comparison of the effect of accounting for or ignoring isotope dilution of the substrate pool can be found elsewhere (Ward *et al.*, 1989).

3. Nitrification Rate Measurements in Sediments

The ^{15}N approach is most useful in water samples because complete mixing of the tracer is possible. In sediments and soils, rate measurements are constrained by the heterogeneous nature of the sample and the dependence of rates on the structure of the environment. In this situation, fluxes between overlying water and sediment cores can be analyzed to obtain areal rates. Estimates of nitrification rates can be obtained from the isotopic dilution of NO_2^- or NO_3^- in the overlying water due to the production of NO_3^- or NO_2^- from processes in the sediments (Capone *et al.*, 1992). The isotope dilution approach is essentially the opposite of the tracer approach. The product pool is amended with isotopically labeled product. During the incubation, product with the natural abundance isotope signature is produced from naturally occurring substrate in the sample, effectively diluting the label in the product pool. The rate of dilution is used to compute the rate of product formation. The isotope pairing method for measurement of denitrification (Nielsen, 1992; Rysgaard *et al.*, 1993) is essentially a modification of an isotope dilution approach and provides information on the rates of both denitrification and nitrification, simultaneously.

Inhibitor approaches similar to those described above for water samples have been used in sediments (Henricksen *et al.*, 1981; Miller *et al.*, 1993). The methylfluoride method (Caffrey and Miller, 1995; Miller *et al.*, 1993) seems particularly promising because the gas can diffuse thoroughly into the core with minimal disturbance of microzones and gradients. This ammonium oxidation inhibitor is added to cores, and the accumulation of NH_4^+ over time is assumed to represent the net rate of nitrification. Other processes that consume ammonium would proceed without inhibition or competition for ammonium and therefore lead to an underestimate of the nitrification rate.

To overcome the biasing resulting from uneven dispersal of tracer or inhibitor, sediment rate measurements are often made in slurries that destroy the gradient structure of sediments, which is essential to the *in situ* fluxes. Slurries may provide useful information on potential rates, but not *in situ* rates. Even if rate measurements in sediments are made using whole core incubations, for example, when the inhibitor is a gas, it is still difficult to determine the depth distribution of the rate because an areal rate is usually obtained. A sophisticated measurement and model-based system that avoids direct rate measurements have been used to overcome this problem. Microelectrodes, which have very high vertical resolution, are used to measure the fine scale distribution of oxygen and nitrate in freshwater sediments. By assuming that the observed vertical gradients represent a steady-state condition, reaction-diffusion models can then be used to estimate the rates of nitrification, denitrification, and aerobic respiration and to compute the location of the rate processes in relation to the chemical profiles (e.g., Binnerup *et al.*, 1992; Jensen *et al.*, 1994). Recent advancements in biosensor design may overcome interferences that have been problematic for microelectrode measurements of dissolved nitrate in seawater.

4. Distribution of Nitrification

The magnitude and distribution of nitrification rates has been extensively reviewed elsewhere and the reader is referred to those publications for data on actual nitrification rates from many marine aquatic and sediment environments (Ward, 2008). In the ocean, nitrification is essentially the last step in the regeneration of inorganic nitrogen from organic matter decomposition, and it is closely coupled to organic matter flux in the water column. Thus in the open ocean, most nitrification occurs near the surface layer, and the rate decreases rapidly with depth as the organic matter flux decreases (Ward and Zafiriou, 1988). Although nitrifiers are widely reported to be inhibited by natural light levels, nitrification does occur within the euphotic zone. In a metadata analysis, nitrification was common in the euphotic zone (Yool *et al.*, 2007), but because the data did not extend below 250 m, the general depth trend mentioned above was not evident in this analysis.

In this chapter about methods for measuring nitrification rates, the distributions are important mainly for planning incubation experiments. Where ambient DIN levels are high, or where rapid rates are expected, shorter incubations are possible, and it is easier to meet the requirements of a true tracer experiment. In oligotrophic environments, even low level tracer experiments require long incubations (Clark *et al.*, 2008).

Nitrification has been detected as deep as 2500 m (Ward and Zafiriou, 1988), but these incubations have never been performed under *in situ* pressure conditions. New advances in pressurized incubation devices will be useful in determining whether published rates have been artifactually reduced by incubation at surface pressure and in determining the effect of pressure on microbial processes in general.

In sediment and coastal systems, the distribution of nitrification can be much more variable, but seasonal studies are rare (Ward, 2005). At least half of the denitrification occurring in the ocean is attributed to hemipelagic sediments (Devol, 2008), and this is largely supported by sedimentary nitrification, pointing to a very important role for nitrification in sediments. Nitrification can be a significant sink for oxygen in marine sediments (Murray and Grundmanis, 1980). In sediments, nitrification is restricted to the region of the oxic/anoxic interface and coupled to aerobic respiration above and to denitrification below the interface. This interface can occur across a gradient in vertically stratified sediments, or in conjunction with animal burrows, which introduce oxygen into deeper sediments and provide additional interface area for diagenesis (Pelegri *et al.*, 1994).

REFERENCES

Arp, D. J., and Stein, L. Y. (2003). Metabolism of inorganic N compounds by ammonia-oxidizing bacteria. *Crit. Rev. Biochem. Mol. Biol.* **38**, 471–495.

Balderston, W. L., Sherr, B., and Payne, W. J. (1976). Blockage by acetylene of nitrous oxide reduction in *Pseudomonas perfectomarinus*. *Appl. Environ. Microbiol.* **31**, 504–508.

Bedard, C., and Knowles, R. (1989). Physiology, biochemistry, and specific inhibitors of CH_4, NH_4^+ and CO oxidation by methanotrophs and nitrifiers. *Microbiology (UK)* **53**, 68–84.

Beman, M. J., Popp, B. N., and Francis, C. A. (2008). Molecular and biogeochemical evidence for ammonia oxidation by marine Chrenarchaeota in the Gulf of California. *ISME J.* **2**, 429–441.

Berg, P., Klemedtsson, L., and Roswall, T. (1982). Inhibitory effects of low partial pressure of acetylene on nitrification. *Soil Biol. Biochem.* **14**, 301–303.

Bianchi, M., Feliatra, M., Treguer, P., Vincendeau, M. A., and Morvan, J. (1997). Nitrification rates, ammonium and nitrate distribution in upper layers of the water column and in sediments of the Indian sector of the Southern Ocean. *Deep-Sea Res.* **44**, 1017–1032.

Billen, G. (1976). Evaluation of nitrifying activity in sediments by dark ^{14}C-bicarbonate incorporation. *Water Res.* **10**, 51–57.

Binnerup, S. J., Jensen, K., Revsbech, N. P., Jensen, M. H., and Sorensen, J. (1992). Denitrification, dissimilatory reduction of nitrate to ammonium, and nitrification in a bioturbated estuarine sediment as measured with N-15 and microsensor techniques. *Appl. Environ. Microbiol.* **58**, 303–313.

Brochier-Armanet, C., Boussau, B., Gribaldo, S., and Forterre, P. (2008). Mesophilic crenarchaeota: Proposal for a third archaeal phylum, the Thaumarchaeota. *Nat. Rev. Microbiol.* **6**, 245–252.

Bulow, S. E., Jayakumar, A., and Ward, B. B. (submitted). Ammonia oxidation rates and nitrification in the Arabian Sea.

Caffrey, J. M., and Miller, L. G. (1995). A comparison of two nitrification inhibitors used to measure nitrification rates in estuarine sediments. *FEMS Microbiol. Ecol.* **17,** 213–219.

Capone, D. G., Horrigan, S. G., Dunham, S. E., and Fowler, J. (1990). Direct determination of nitrification in marine waters by using the short-lived radioisotope of Nitrogen, N-13. *Appl. Environ. Microbiol.* **56,** 1182–1184.

Capone, D. G., Dunham, S. E., Horrigan, S. G., and Duguay, L. E. (1992). Microbial nitrogen transformations in unconsolidated coral-reef sediments. *Mar. Ecol. Prog. Ser. Ecol. Prog. Ser.* **80,** 75–88.

Carlucci, A. F., and Strickland, J. D. H. (1968). The isolation, purification and some kinetic studies of marine nitrifying bacteria. *J. Exp. Mar. Biol. Ecol.* **2,** 156–166.

Clark, D. R., Rees, A. P., and Joint, I. (2007). A method for the determination of nitrification rates in oligotrophic marine seawater by gas chromatography/mass spectrometry. *Mar. Chem.* **103,** 84–96.

Clark, D. R., Rees, A. P., and Joint, I. (2008). Ammonium regeneration and nitrification rates in the oligotrophic Atlantic Ocean: Implications for new production estimates. *Limnol. Oceanogr.* **53,** 52–62.

Daims, H., Lucker, S., Le Paslier, D., and Wagner, M. (in press). Diversity, environmental genomics and ecophysiology of nitrite-oxidizing bacteria. In "Nitrification," (B. B. Ward, M. G. Klotz, and D. J. Arp, eds.), American Society for Microbiology, Washington, DC.

Devol, A. H. (2008). Denitrification, including Anammox. In "Nitrogen in the Marine Environment," (D. G. Capone, D. A. Bronk, M. R. Mulholland, and E. J. Carpenter, eds.), pp. 263–301. Elsevier, Amsterdam.

Dore, J. E., and Karl, D. M. (1996). Nitrification in the euphotic zone as a source for nitrite, nitrate, and nitrous oxide at station ALOHA. *Limnol. Oceanogr.* **41,** 1619–1628.

Fiedler, R., and Proksch, G. (1975). Determination of nitrogen-15 by emission and mass-spectrometry in biochemical analysis—Review. *Anal. Chim. Acta* **78,** 1–62.

Ginestet, P., Audic, J. M., Urbain, V., and Block, J. C. (1998). Estimation of nitrifying bacterial activities by measuring oxygen uptake in the presence of the metabolic inhibitors allylthiourea and azide. *Appl. Environ. Microbiol.* **64,** 2266–2268.

Glover, H. E. (1985). The relationship between inorganic nitrogen oxidation and organic carbon production in batch and chemostat cultures of marine nitrifying bacteria. *Arch. Microbiol.* **74,** 295–300.

Goring, C. A. I. (1962). Control of nitrification by 2-chloro-6-(trichloromethyl)pyridine. *Soil Sci.* **93,** 211–218.

Granger, J., Sigman, D. M., Prokopenko, M. G., Lehmann, M. F., and Tortell, P. D. (2006). A method for nitrite removal in nitrate N and O isotope analyses. *Limnol. Oceanogr. Methods* **4,** 205–212.

Grasshof, K., Ehrhardt, M., and Kremling, K. (1983). Methods of Seawater Analysis. Verlag Chemie GmbH, Weinheim.

Griffin, B. M., Schott, J., and Schink, B. (2007). Nitrite, an electron donor for anoxygenic photosynthesis. *Science* **316,** 1870.

Henricksen, K., Hansen, J. I., and Blackburn, T. H. (1981). Rates of nitrification, distribution of nitrifying bacteria and nitrate fluxes in different types of sediment from Danish waters. *Mar. Biol.* **61,** 299–304.

Holmes, R. M., Aminor, A., Kerouel, R., Hooker, B. A., and Peterson, B. J. (1999). A simple and precise method for measuring ammonium in marine and freshwater ecosystems. *Can. J. Fish. Aquat. Sci.* **56,** 1801–1808.

Jensen, K., Sloth, N. P., Risgaard-Petersen, N., Rysgaard, S., and Revsbech, N. P. (1994). Estimation of nitrification and denitrification from microprofiles of oxygen and nitrate in model sediment systems. *Appl. Environ. Microbiol.* **60,** 2094–2100.

Kator, H., Morris, L. J., Wetzel, R. L., and Koepfler, E. T. (1992). A rapid chromatographic method for recovery of 15no2- and no3-produced by nitrification in aqueous samples. *Limnol. Oceanogr.* **37,** 900–907.

Kester, R. A., de Boer, L., and Laanbroek, H. J. (1996). Short exposure to acetylene to distinguish between nitrifier and denitrifier nitrous oxide production in soil and sediment samples. *FEMS Microbiol. Ecol.* **20,** 111–120.

Konneke, M., Berhnard, A. E., de la Torre, J. R., Walker, C. B., Waterbury, J. B., and Stahl, D. A. (2005). Isolation of an autotrophic ammonia-oxidizing marine archaeon. *Nature* **437,** 543–546.

Laudelout, H., Simonart, P. C., and Vandroog, R. (1968). Calorimetric measurement of free energy utilization by *Nitrosomonas* and *Nitrobacter*. *Arch. Mikrobiol.* **63,** 256.

McIlvin, M. R., and Altabet, M. A. (2005). Chemical conversion of nitrate and nitrite to nitrous oxide for nitrogen and oxygen isotopic analysis in freshwater and seawater. *Anal. Chem.* **77,** 5589–5595.

Miller, L. G., Coutlakis, M. D., Oremland, R. S., and Ward, B. B. (1993). Selective inhibition of nitrification (ammonium oxidation) by methylfluoride and dimethyl ether. *Appl. Environ. Microbiol.* **59,** 2457–2464.

Murray, J. W., and Grundmanis, V. (1980). Oxygen-consumption in pelagic marine-sediments. *Science* **209,** 1527–1530.

Nicol, G. W., Leininger, S., and Schleper, C. (in press). Distribution and activity of ammonia-oxidizing archaea in natural environments. *In* "Nitrification," (B. B. Ward, M. G. Klotz, and D. J. Arp, eds.), American Society for Microbiology, Washington, DC.

Nielsen, L. (1992). Denitrification in sediment determined from nitrogen isotope pairing. *FEMS Microbiol. Ecol.* **86,** 357–362.

Norton, J. M. (in press). The diversity and environmental distribution of ammonia-oxidizing bacteria. *In* "Nitrification," (B. B. Ward, M. G. Klotz, and D. J. Arp, eds.), American Society for Microbiology, Washington, DC.

Olson, R. J. (1981). ^{15}N tracer studies of the primary nitrite maximum. *J. Mar. Res.* **39,** 203–226.

Owens, N. J. P. (1986). Estuarine nitrification—A naturally-occurring fluidized-bed reaction. *Estuar. Coast. Shelf Sci.* **22,** 31–44.

Parsons, T. R., Maita, Y., and Lalli, C. M. (1984). A Manual of Chemical and Biological Methods for Seawater Analysis. Pergamon Press, Oxford.

Pelegri, S. P., Nielsen, L. P., and Blackburn, T. H. (1994). Denitrification in estuarine sediment stimulated by irrigation activity of the amphipod *Corophium volutator*. *Mar. Ecol. Prog. Ser.* **105,** 285–290.

Priscu, J. C., and Downes, M. T. (1985). Nitrogen uptake, ammonium oxidation and nitrous oxide (N_2O) levels in the coastal waters of Western Cook Strait, New Zealand. *Estuar. Coast. Shelf Sci.* **20,** 529–542.

Rakestraw, N. W. (1936). The occurrence and significance of nitrite in the sea. *Biol. Bull.* **71,** 133–167.

Rees, A. P., Malcolm, E., Woodward, S., Robinson, C., Cummings, D. G., Tarran, G. A., and Joint, I. (2002). Size-fractionated nitrogen uptake and carbon fixation during a developing coccolithophore bloom in the North Sea during June 1999. *Deep Sea Res. Part II Top. Stud. Oceanogr.* **49,** 2905–2927.

Rysgaard, S., Risgaardpetersen, N., Nielsen, L. P., and Revsbech, N. P. (1993). Nitrification and denitrification in lake and estuarine sediments measured by the N-15 dilution technique and isotope pairing. *Appl. Environ. Microbiol.* **59,** 2093–2098.

Sigman, D. M., Casciotti, K. L., Andreani, M., Barford, C., Galanter, M., and Bohlke, J. K. (2001). A bacterial method for the nitrogen isotopic analysis of nitrate in seawater and freshwater. *Anal. Chem.* **73,** 4145–4153.

Somville, M. (1978). A method for the measurement of nitrification rates in water. *Water Res.* **12,** 843–848.

Suzuki, I., Dular, U., and Kwok, S. (1974). Ammonia or ammonium ion as substrate for oxidation by *Nitrosomonas europaea* cells and extracts. *J. Bacteriol.* **120,** 556–558.

Viner, A. B. (1990). Dark ^{14}C uptake, and its relationships to nitrification and primary production estimates in a New Zealand upwelling region. *NZ J. Mar. Freshw. Res.* **24,** 221–228.

von Brand, T., Rakestraw, N., and Renn, C. (1937). The experimental decomposition and regeneration of nitrogenous organic matter in sea water. *Biol. Bull.* **72,** 165–175.

Walker, C. B., de la Torre, J. R., Klotz, M. G., Urakawa, H., Pinel, N., Arp, D. J., Brochier-Armanet, C., Chain, P. S. G., Chan, P. P., Gollabgir, A., Hemp, J., Hugler, M., *et al.* (2010). Nitrosopumilus maritimus genome reveals unique mechanisms for nitrification and autotrophy in globally distributed marine crenarchaea. *PNAS* **107,** 8818–8823.

Ward, B. B. (1986). Nitrification in marine environments. *In* "Nitrification," (J. I. Prosser, ed.), pp. 157–184. IRL Press, Oxford.

Ward, B. B. (1987). Kinetic studies on ammonia and methane oxidation by *Nitrosococcus oceanus*. *Arch. Microbiol.* **147,** 126–133.

Ward, B. B. (2005). Temporal variability in nitrification rates and related biogeochemical factors in Monterey Bay, California, USA. *Mar. Ecol. Prog. Ser.* **292,** 97–109.

Ward, B. B., and O'Mullan, G. (2005). Genetic and biogeochemical approaches to investigate community compostion and function in aerobic ammonia oxidation. *Methods Enzymol.* **397,** 395–413.

Ward, B. B., and Zafiriou, O. C. (1988). Nitrification and nitric-oxide in the oxygen minimum of the eastern tropical North Pacific. *Deep Sea Res. A* **35,** 1127–1142.

Ward, B. B., Talbot, M. C., and Perry, M. J. (1984). Contributions of phytoplankton and nitrifying bacteria to ammonium and nitrite dynamics in coastal water. *Cont. Shelf Res.* **3,** 383–398.

Ward, B. B., Kilpatrick, K. A., Renger, E., and Eppley, R. W. (1989). Biological nitrogen cycling in the nitracline. *Limnol. Oceanogr.* **34,** 493–513.

Ward, B. B. (2008). Nitrification in Marine Systems. *In* "Nitrogen in the Marine Environment," (D. G. Capone, D. A. Bronk, M. R. Mulholland, and E. J. Carpenter, eds.), 2 edn., pp. 199–262. Elsevier.

Yool, A., Martin, A. P., Fernandez, C., and Clark, D. R. (2007). The significance of nitrification for oceanic new production. *Nature* **447,** 999–1002.

Yoshizumi, K., Aoki, K., Matsuoka, T., and Asakura, S. (1985). Determination of nitrate by a flow system with a chemi-luminescent NOX analyzer. *Anal. Chem.* **57,** 737–740.

CHAPTER FOURTEEN

CONSTRUCTION OF STOX OXYGEN SENSORS AND THEIR APPLICATION FOR DETERMINATION OF O_2 CONCENTRATIONS IN OXYGEN MINIMUM ZONES

Niels Peter Revsbech,[*] Bo Thamdrup,[†] Tage Dalsgaard,[‡] *and* Donald Eugene Canfield[†]

Contents

1. Introduction	326
2. STOX Sensor Principle	327
3. Sensor Construction	328
4. Electronics for STOX Sensors	332
5. STOX Sensor Calibration and Performance	333
6. Calculation of Oxygen Concentrations from STOX Data	335
7. *In Situ* Deployment of STOX Sensors in OMZs	337
8. Future Fast Response STOX Sensors	339
9. Using STOX Sensors to Recalibrate Conventional Oxygen Sensors	339
Acknowledgments	340
References	340

Abstract

Until recently, it has not been possible to measure O_2 concentrations in oxygen minimum zones (OMZs) with sufficient detection limits and accuracy to determine whether OMZs are anoxic or contain 1–2 μM O_2. With the introduction of the STOX (switchable trace oxygen) sensor, the level for accurate quantification has been lowered by a factor of 1000. By analysis with STOX sensors, O_2 can be prevented from reaching the sensing cathode by another cathode (front guard cathode), and it is the amplitude in signal by polarization/depolarization of this front guard that is used as a measure of the O_2 concentration. The STOX sensors can be used *in situ*, most conveniently connected to a conventional CTD (conductivity, temperature, and depth analyzer) along with a conventional

[*] Department of Biological Sciences, Microbiology, Aarhus University, Aarhus, Denmark
[†] Nordic Center for Earth Evolution (NordCEE) and Institute of Biology, University of Southern Denmark, Odense, Denmark
[‡] National Environmental Research Institute, Aarhus University, Silkeborg, Denmark

oxygen sensor, and they can be used for monitoring O_2 dynamics during laboratory incubations of low-O_2 media such as OMZ water. The limiting factors for use of the STOX sensors are a relatively slow response, with measuring cycle of at least 30 s with the current design, and fragility. With improved procedures for construction, the time for a complete measuring cycle is expected to come down to about 10 s.

1. INTRODUCTION

Determination of O_2 concentration (or partial pressure) is one of the most frequently performed analyses in the biological sciences, probably only surpassed by pH measurements. The classical method is Winkler titration of the O_2 concentration, now improved by spectrophotometric determination of the titration endpoint (e.g., Broenkow and Cline, 1969; Labasque et al., 2004); but since Clark invented the membrane-covered O_2 sensor (Clark et al., 1953), most O_2 determinations have been performed by use of sensors. The electrochemical Clark sensors have recently been substituted by optical sensors (optodes) for many applications with the advent of fast-responding optical sensors based on O_2 quenching of fluorescence from a fluorophore (e.g., Wolfbeis et al., 1984).

In oceanographic applications, O_2 analysis has followed this general trend, with an increasing number of analyses being performed by optical sensors of which especially the Aanderaa optodes (Tengberg et al., 2006) have conquered a substantial part of the market. The optodes compare well with the Clark cells in terms of long-term stability, and for long-term deployments, they have the advantage of less power consumption (Martini et al., 2007), and they are also less affected by fouling as they do not consume O_2 and thereby are less influenced by changes in diffusional supply to the sensor surface. The O_2 consumption by a biofilm on the membrane has, however, the same effect on electrochemical sensors and optodes.

It is possible to perform a Winkler titration with an accuracy of 1‰, whereas the sensor-based methods have a somewhat lower accuracy that is rather at the 1% level in terms of absolute accuracy (Martini et al., 2007), although changes in relative concentration may be detected at a higher resolution. These levels of accuracy are satisfactory for most applications, but when it comes to determination of the very low O_2 concentrations associated with oxygen minimum zones (OMZs), all three methods have proven unable to quantify concentrations below 1–2 μM (Morrison et al., 1999). Paulmier and Ruiz-Pino (2009) thus define OMZs as oceanic water masses where the O_2 concentration reaches <20 μM and may get as low as 1 μM in the core, the 1 μM here representing the detection limit of the analytical methods applied. In theory, the Winkler titration should be able to resolve at least 0.2 μM O_2, but atmospheric contamination of sample and reagents becomes a critical factor in the analysis of waters with very low O_2 concentrations.

It could be argued that concentrations of < 1 μM are so low that they do not matter for in the overall transformations occurring in the sea. Many marine bacteria have, however, been shown to exhibit half-saturation values for O_2 that are about 1 μM (Devol, 1978), and the biogeochemical transformations based on O_2 respiration may thus continue almost unaffected in an OMZ where the O_2 concentration decreases to 1 μM O_2. It is therefore important to quantify these concentrations and subsequently be able to perform experiments with microbial communities or chemical transformations at the relevant low O_2 concentrations. The STOX (switchable trace oxygen) sensor was developed specifically for analysis of OMZ regions (Revsbech et al., 2009), and it can be used to detect and quantify O_2 concentrations < 10 nM. This chapter describes the construction and use of this sensor and associated electronics in detail, and how the STOX data compare with the data from a conventional O_2 sensor.

2. STOX Sensor Principle

The STOX sensor tip is shown schematically in Fig. 14.1. It is a Clark-type O_2 microsensor (Revsbech, 1989) positioned within an outer capillary inside which a porous gold cathode is positioned in front of the O_2

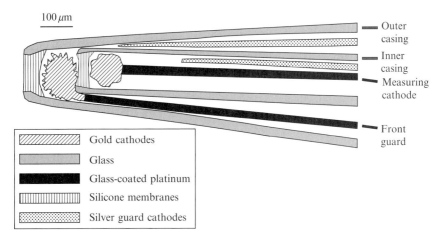

Figure 14.1 Design of fast-responding STOX sensor with a relatively short distance (about 130 μm) from tip to measuring cathode.

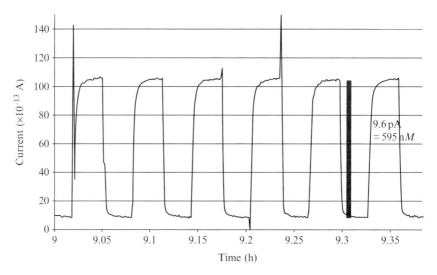

Figure 14.2 Signal from a STOX sensor operated on a research ship with engine vibrations and wave-induced movements. The sensor was inserted into a glass reactor containing low-oxygen seawater, and the amplitude of the signal (9.6 pA) corresponded to an O_2 concentration of 595 nM. The noise peaks were not caused by ship movements but by an imperfect switch-box for front guard polarization/depolarization.

microsensor. The diameter of the outer sensor tip may vary from 70 to 150 µm. When the front cathode (*front guard*) is polarized at -0.8 V relative to the inner Ag/AgCl reference, no or a little O_2 will enter the inner O_2 microsensor and a zero reading is obtained. Timer-controlled changes between polarized and unpolarized front guard result in data such as those shown in Fig. 14.2, and the amplitude in signal can be shown to be directly proportional to the oxygen partial pressure in the analyzed medium. Because the signal amplitude is independent of the zero current, which varies substantially in a real-life situation with fluctuating temperature and electronic offsets caused by various factors, oxygen can be measured at a much higher accuracy than with conventional sensors, and it is possible to detect and quantify O_2 concentrations below 10 nM by this technique (Revsbech *et al.*, 2009).

3. Sensor Construction

We have constructed the STOX sensors, which are commercially available from Unisense A/S (Aarhus, Denmark), with an outer shaft diameter of 8 mm, as this is the general sensor diameter applied by the deep-sea lander community (e.g., Glud and Reimers, 2000). It is thereby possible to use well-tested sensor holders and connectors. For the outer casing, we have used Schott AR glass with outer diameter (OD) of 8 mm and inner diameter (ID) of 7 mm. This glass is pulled in a flame and subsequently in an electrically heated nichrome loop (e.g., Revsbech and Jørgensen, 1986) to produce the desired conical tip shape. Finally, the glass tip is scored with a diamond knife (Amidia, Geneva, Switzerland) and broken at the desired tip diameter. A slight heating of the tip with a nichrome heating loop under the microscope (100× magnification) to close small cracks is recommended before the silicone membrane is applied by letting the tip touch uncured silicone rubber. Several types of clear silicone rubbers can be used, and we have had good results with Dow Corning 734 (Dow Corning Corp.) and Wacker Elastosil E43 (Wacker Chemie). A membrane thickness of 20–50 μm is suitable, the thinner membrane resulting in faster response but higher fragility. The curing of the membrane is faster at elevated temperatures, and 1 h at 100 °C suffices for complete curing.

The front guard gold-plated platinum cathode (Fig. 14.1) is made from a 0.5- to 0.8-mm thick AR glass capillary into which a tapered (Revsbech and Jørgensen, 1986) platinum wire is inserted. The terminal ~ 1 cm of the platinum wire is then coated with glass by melting the AR glass capillary around the platinum wire (Revsbech and Jørgensen, 1986). The tip of the glass-coated wire is then cut with sharp forceps where it is about 15 μm thick. The glass around the newly exposed tip is subsequently etched away for a length of 0.1–1 mm (depending on sensor design) by immersion into a droplet of 30% HF (extreme caution is needed during handling of HF) covered by mineral oil and contained in a polystyrene spectrophotometer cuvette. Etching gives flexibility to the tip, which is required during assembly of the sensor as described below. The etched tip is cleaned with acetone and subsequently coated with gold by rapid (~ 1 s) electroplating in 50% saturated $HAuCl_4$ at -0.7 V (relative to a platinum wire inserted into the solution). The tip is subsequently gold plated at first -0.8 V for a few minutes, until a solid gold lump corresponding to about half of the ID just behind the membrane of the outer casing is obtained. Finally, a high voltage of -2.3 V is applied for about 1 s to create a fluffy gold matrix with a diameter corresponding to the ID of the outer casing. This porous gold cathode is immersed into clean water for about 5 min and subsequently inserted into the 8 mm glass casing. The gold guard should touch the silicone membrane, and preferably even be pushed so far that the thin, etched part of the platinum wire bends a little. The glass capillary is fastened in place near the top of the outer casing with epoxy, or even better with a UV curing cement such as Loctite 190672 (Loctite Corporation).

The casing for the inner O_2 microsensor is made from 6 mm (OD), 5 mm (ID) Schott AR glass in a similar way as the outer casing, but here the terminal 2–4 mm are made almost parallel sided to facilitate insertion inside the tip of the outer, 8 mm glass casing. The OD of the inner sensor tip should be a little (~ 10 μm) smaller than the ID of the outer casing just behind the silicone membrane to allow for the platinum wire making contact with the porous gold cathode of the front guard. The tip of the front guard can be bent a little when the inner casing is pushed in place. The distance between the two membranes of outer and inner casing should be kept as short as possible to get a fast response from the finished STOX sensor.

The sensing cathode is made from highly insulating Schott 8533 glass. We applied 3.5 mm OD glass tubing that is pulled into a capillary and subsequently fused with a tapered platinum wire (Revsbech and Jørgensen, 1986). The procedure is basically as for the front guard, except that the exposed platinum should be kept short (can be done with forceps instead of the HF etching used for the front guard) and that the entire plating must be done at low voltage (-0.8 increasing to -1.0 V) to obtain a smooth gold surface. Faster plating at elevated voltages results in too high currents for zero O_2 measurements. The cathode should soak in water for about 5 min before insertion into the casing, where the gold cathode should touch or almost touch the silicone membrane. Fastening into position near the top of the casing is again best done with UV curing cement.

Into both inner and outer casings, it is necessary to insert back guards, made from 0.25 mm Teflon-coated silver wire (World Precision Instruments), from which the Teflon layer has been removed from the tip 1 cm. This exposed 1 cm piece of silver wire is subsequently etched to the desired tip diameter in saturated KCN (special safety procedures are required for work with KCN) while 5 V AC is applied (vs. a graphite rod). The tips of these back guards should be positioned as close to the front guard and sensing cathode as possible, without too much risk of short circuit. Where these wires are fastened with UV curing cement, it is necessary to have a knot on the wire, as basically nothing will glue onto Teflon. The back guards prevent diffusion of O_2 toward the sensor tip from the bulk electrolyte reservoirs within the casings. The internal anodes in each casing are also made from this Teflon-coated silver wire, but here the Teflon is removed from the tip 3 cm and no etching is performed. Instead of etching, the exposed silver is coated with AgCl by a short (1 s) immersion into 0.1 M HCl while $+1.5$ V is applied versus a silver wire. Electrical connection to the front guard is also made with the Teflon-coated wire. If the sensor is to be used in the laboratory, a thin low-noise quality (i.e., combined metal and graphite shielding) shielded cable is inserted into the 3.5-mm glass tube of the sensing cathode, and a silver wire soldered onto the inner connector of the cable makes contact with the platinum wire of the cathode. In this way, a total coaxial shielding—by either the shielding of the cable or the sensor electrolyte—of all signal conducting parts of the sensor is obtained. For marine *in situ* use, the sensor will be totally submersed into the water, and no shielding is needed. Here contact to the platinum wire of the sensing cathode is made by a Teflon-coated silver wire (from which the Teflon has been making contact with the platinum).

The most fragile part of the STOX sensor is the platinum connection to the gold front guard right at the tip of the inner capillary (Fig. 14.1). Vibrations can easily rip this tiny metal junction apart. We have tested various solutions to this problem, and the most efficient seems to be a combination of a relatively long (e.g., 0.5 mm) HF etched and thus exposed about 15-μm thick platinum wire behind the porous gold front guard that is bent into a slight Z shape (about 30°) by forceps under the microscope so that it can be stretched a little without rupture. However, due to the risk of short circuit, this design prevents a positioning of a silver back guard very close to the tip of the inner O_2 microsensor, and such a close positioning is necessary for a fast response. Otherwise, slow response results from O_2 diffusing through the gap between inner microsensor and outer casing when the front guard is depolarized, resulting in a gradual build-up of O_2 far behind the tip. Close proximity of the back guard, however, ensures that steady-state internal O_2 gradients are obtained relatively fast. The situation at the time of writing this chapter is thus that the choice is between fast-responding but very fragile sensors or sturdier but slower responding sensors. However, it should be possible to solve this problem so that it will be possible to make sturdy and still fast-responding sensors in the future.

Both inner and outer compartments are filled from behind with an aquatic electrolyte containing 0.3 M K_2CO_3, 0.2 M $KHCO_3$, and 0.5 M KCl. Filling all the way into the tips is done by placing the sensor in a vacuum flask and applying very strong vacuum where the electrolyte starts to boil. A more moderate vacuum will suffice if the tip of the sensor is immersed into hot (e.g., 80 °C) oil during the vacuum treatment. After filling with electrolyte, the sensor can be sealed completely at the top with UV curing cement if made for laboratory use. For *in situ* use, small glass capillaries sealed at the top are inserted into the inner and outer casings before the cement cures. These small capillaries serve for pressure compensation when opened by breaking off the top before deployment. For *in situ* sensors, it is advisable to fill the lower parts of the casings with electrolyte containing 1.5% agarose to avoid oil contamination of the interior of the tip during transport of sensors where previous deep-water use has resulted in injection of pressure-compensation oil from the sensor holder into the electrolyte compartments.

A cross section through a finished STOX sensor made ready for *in situ* deployment is illustrated in Fig. 14.3.

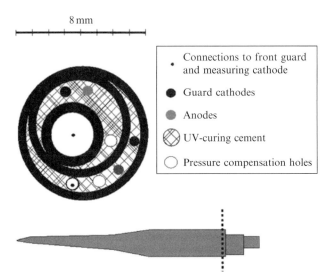

Figure 14.3 Cross section through the upper part of a STOX sensor showing the electrical connections (all made from 0.25 mm Teflon-coated silver wire) and pressure compensation holes.

4. Electronics for STOX Sensors

Standard high-sensitivity ammeters that resolve current changes down to 10^{-13} A can be used for laboratory use of the STOX sensors if a voltage source for polarization and a timer for polarization/depolarization of the front guard are added to the circuits. However, it is more convenient to use a picoammeter with built-in polarization voltage, such as those supplied from Unisense A/S. It is then a relatively simple task to add a timer for front guard polarization/depolarization, and this can also be supplied from Unisense.

For deep-water applications, all the electronics must be housed in a pressure-proof container (e.g., Glud and Reimers, 2000). Unisense markets a titanium cylinder containing the necessary electronics and with attached sensor holder (Fig. 14.4), which has been tested down to 6000 m. The unit can be attached directly to a standard Seabird CTD if not all add-on connectors on the Seabird are occupied. We have used Seabird 25 and 911 CTDs for STOX measurements. However, Seabird is at present applying only 12-bit AD converters in their CTDs, and their resolution is not satisfactory for use with the STOX sensor. Until now, this has been compensated for by logging only a fraction of the total signal from the STOX sensor, so that the AD converter is set to give the maximum of 4096 counts for say 30 µM. That would give a theoretical resolution of 7 nM. In a real-life situation, the resolution may be somewhat better if the signal is read

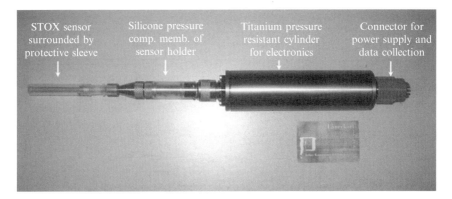

Figure 14.4 STOX sensor mounted on titanium cylinder that is pressure resistant up to 600 bar.

at high frequency, as a low level of electrical noise may lead to a higher definition average over many determinations. It is, however, highly unsatisfactory not to be able to resolve the signal over the full concentration range. One solution to this problem could be to use a separate data collection unit. One such unit is the controller unit of the Unisense, "eddy correlation" system, which is equipped with a 16-bit AD converter. By such an approach there will not be sampling of depth and O_2 information over the same AD converter, and the data from the CTD and the STOX sensor will have to be aligned using the internal clock information. Immersion into the sea could be an event that is sensed by both systems and thus usable for timer calibration. Seabird is, however, planning to release a Seabird 25 CTD with a 16 bit AD converter in 2011.

It can be recommended to suspend the STOX sensor and associated electronics in an elastic suspension (e.g., made from thin Nylon rope) to protect the STOX sensor from mechanical shocks. If the CTD is mounted on a rosette water sampler, the space in the center between the Niskin bottles is a suitable place for such an elastic suspension.

5. STOX Sensor Calibration and Performance

The STOX sensor can be calibrated down to about 1–2 nM concentration, and the calibration curve is linear from zero and up to atmospheric saturation (Revsbech *et al.*, 2009). The calibration data presented by Revsbech *et al.* (2009) were obtained in a 1-L bottle into which the STOX sensor was inserted through a butyl rubber stopper. Although the water within the bottle contained a little ascorbate to slowly consume any O_2 that might be present after the initial N_2 bubbling, it was not possible to get below 10 nM O$_2$. We now know that this was due to leakage of O$_2$ from the butyl rubber stopper, and that all polymers coming into contact with low-O$_2$ waters should be degassed before use—most efficiently by being placed in boiling water for 24 h followed by incubation in a jar kept anoxic by an envelope with an O$_2$ consuming substance (Anaerogen, Oxoid, England).

Given the linear response and the internal zero calibration from the switching front guard, a one-point calibration at a known oxygen concentration may be sufficient. However, we occasionally observe that the depolarized signal is slightly higher than the polarized signal under anoxic conditions. Such offsets, typically equivalent to oxygen concentrations of less than 10 nM, stem in some cases from the electronics, while in other cases, they are associated with the sensor and develop after extended use. Thus, a two-point calibration including an anoxic solution such as alkaline ascorbate is advisable. Although the zero current exhibits some drift (e.g., Fig. 14.6), the oxygen signal amplitude has good long-term stability as indicated by <2% drift over 12 weeks reported for two sensors (Revsbech *et al.*, 2009). For long continuous applications, we thus recommend calibration every day.

The most frequent cause of electrode failure is breakage of the connection to the front guard gold cathode (Fig. 14.1), which occurs particularly during transport. With occasional use and safe storage in between (dry and still), the sensors may function for a year or longer, while the typical lifetime during continuous use is a couple of months.

The response time of the sensors is dependent on both the distance between the sensor tip and the sensing cathode and on the proximities and efficiencies of the back guards as described above. Usually, the response time by the switch from front guard on to off is the slowest, whereas the decrease in signal when the front guard is polarized is very fast. At room temperature, a typical 90% rise time for a sensor by front guard off is about 20 s and the 90% decrease time for the guard on is about 5 s. Such a sensor could, for example, be used with a 60 s front guard off, 15 s on cycle, resulting in a total cycle time of 75 s. However, it is possible to make much faster sensors, and we have made sensors that could be used with a 10 s on, 5 s off cycle. The sensor output illustrated in Fig. 14.2 was obtained with a 110 s on, 110 s off cycle. The noise peaks on the signal were caused by electrical interference from the timer, and additional electrical noise may also occur, especially on a moving ship.

Important factors for the use of O$_2$ sensors are sensitivity to advection (stirring) in front of the sensor, effect of temperature changes, and effect of hydrostatic pressure (Glud *et al.*, 2000; Gundersen *et al.*, 1998). The STOX sensors have not yet been fully characterized for these factors, as experiments to find the optimal design are still in progress. Stirring effects result from the diffusive boundary layer around the sensor tip and are highly variable among sensors depending on design varying between 2% and 10%

(signal change from stagnant to stirred conditions). For oceanographic applications, sensitivity to stirring is, however, not a major problem as the "stirring" caused by movement of the ship in even gentle waves suffices to give full response. The effect of temperature has been measured for a few STOX sensors and is approximately linear with a slope of about 7–8% per degree (based on constant concentration—lower for constant partial pressure) relative to the extrapolated signal at 0 °C (Revsbech et al., 2009), that is, $U_t = U_0 + 0.08 \text{ degree}^{-1} \times U_0 \times t$, where $U_{t/0}$ is the signal at temperature t and 0 °C, respectively. The effect of hydrostatic pressure, attributed mainly to changes in the silicone membranes, has been investigated by Glud et al. (2000) for Clark-type microsensors, and a 3% decrease can be expected for a pressure increase of 50 bar, corresponding to 500 m water depth: $U_P = U_{\text{surface}} + (0.03/50) \text{ bar}^{-1} \times U_{\text{surface}} \times P$, where $U_{P/\text{surface}}$ is the signal at pressure P and at the surface, respectively. Until STOX sensors of a standardized design have been fully characterized for effect of temperature and pressure, it is recommended to use these sensors in combination with well-characterized standard oceanographic O_2 sensors and to intercalibrate the sensors in adjacent higher O_2 water layers to those being analyzed for low O_2 by a STOX sensor.

STOX sensors also offer themselves for applications in the laboratory such as analysis of oxygen respiration kinetics and for monitoring oxygen during studies of other microaerobic or anaerobic processes. Experimental systems should as extensively as possible be constructed in glass or other oxygen-impermeable materials, because substantial amounts of oxygen can be dissolved in plastic polymers in components such as Teflon-coated stir bars (Revsbech et al., 2009). The oxygen consumption of Clark-type microsensors is generally insignificant in closed incubations (Gundersen et al., 1998). In STOX sensors, oxygen consumption by the front guard is at the high end of the range of conventional microsensors. Thus, a sensor with dimensions as shown in Fig. 14.1 (60 μm diameter opening, 30 μm silicone membrane thickness, cathode located right behind the membrane) and 50% polarization time is predicted to consume 190 pmol O_2 h^{-1} in air-saturated water. Thus, even in a volume of 1 mL, the relative change in oxygen concentration due to consumption by the sensor is merely 0.00678 h^{-1} (calculated assuming 280 μM O_2 at air saturation).

6. Calculation of Oxygen Concentrations from STOX Data

When handling massive amounts of data from STOX sensors, it is necessary to have a standardized, computer-based method for calculating actual concentrations. We have chosen to show data from a laboratory

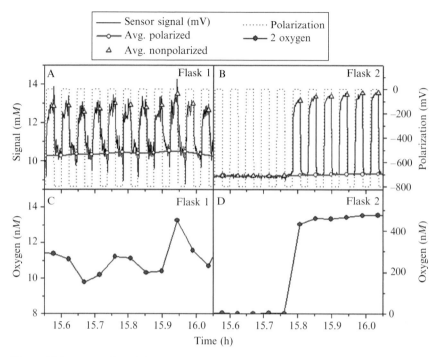

Figure 14.5 Examples of data calculation. In Flask 1, O_2 was kept at a constant low level, whereas in Flask 2, O_2 was added after 15.77 h. (A) and (B) show the polarization signal, the raw signal from the oxygen sensor, and the calculated signals at the end of each polarized (circles) or nonpolarized (triangles) period. (C) and (D) show the derived oxygen concentrations.

experiment with fixed O_2 concentrations to illustrate calculations and detection limit. The experiments shown in Fig. 14.5 were carried out in 2-L glass reactors with STOX sensors mounted in the lid and with a glass-coated stirring bar (20 × 5 mm) spinning at 200 rpm in the bottom. The reactors were filled with low-oxygen seawater from the OMZ ca. 50 km off Iquique, Chile, and kept at constant temperature (13 °C) in the dark. The signals from the STOX sensors were recorded using the AD converter ADC816 with the SensorTrace Basic software (Unisense). Data were logged every 5 s and the polarized and depolarized periods were 82 s each. The polarization voltage was recorded together with the electrode signals, so the exact time for the polarization shifts could be identified. Data processing was done in Microsoft Excel.

First step of the data processing was to identify the exact positions where the polarization voltage was shifted. The signal for each polarized and depolarized period was then calculated as the average of the last five points before the shift. The actual number of points to be averaged should in each case be adapted to the data acquisition frequency, the noise level of the electrode signal, and the length of the polarization cycles. The oxygen signal is the difference between the depolarized and polarized electrode signals and this can be calculated in different ways depending on the required response time and noise reduction. The signal for oxygen is calculated as the depolarized sensor signal minus the average of the polarized sensor signals before and after the depolarization (Fig. 14.5A and B).

The detection limit in the example in Fig. 14.5 was calculated, based on the standard deviation of the estimated oxygen concentrations close to the detection limit ($\sim$11 nM; Fig. 14.5C) multiplied with the t-value for the confidence interval required (11 degrees of freedom), yielding 1.61 and 2.44 nM for 95% and 99% confidence intervals, respectively. These detection limits were obtained for 30 min of continuous recording and will be higher if the period of measurement is shortened and will, furthermore, depend very much on the signal-to-noise ratio of each individual STOX sensor.

7. *In Situ* Deployment of STOX Sensors in OMZs

For *in situ* deployment, a pressure resistant electronics unit with attached sensor can be interfaced with normal oceanographic CTD systems such as the Seabird SBE911 for power supply and data acquisition as described above (Revsbech *et al.*, 2009). Figure 14.6 shows an example of data recorded with a STOX sensor during a cast to 500 m in the OMZ off Peru, together with simultaneous data from a conventional, laboratory-calibrated SBE43 oxygen sensor. During both down- and upcast, the SBE43 sensor exhibited a monotonously decreasing signal between $\sim$110 and 350 m depth, and the decrease continued independently of whether the CTD was profiling or poised at a given depth (Fig. 14.6A and insert), which suggests that the reading represented a drifting background signal rather than variations in the oxygen concentration in the OMZ water. This was confirmed by the data from the STOX sensor, which showed no difference between the signals obtained during front guard polarization and depolarization between 110 and 330 m depth, while it detected 30 nM oxygen at 350 m depth (Fig. 14.6B). The STOX sensor was operated with cycles of 20 s front guard polarization and 40 s depolarization, and from each phase the average signal (sampled at 25 Hz) from 6 to 1 s before switching was used for calculation of oxygen concentrations. The length of the cycle did not permit detailed profiling while the CTD was moving, typically at 0.5 m s^{-1}. Since the signal was also more noisy during movement, the most precise measurements were obtained when the CTD was stopped for three cycles at specific depths: 1.8 $\pm$ 4.6 and 1.8 $\pm$ 3.0 nM at 204 and 111 m, respectively, compared to $-$0.9 $\pm$ 7.8 and 0.2 $\pm$ 6.1 nM with the

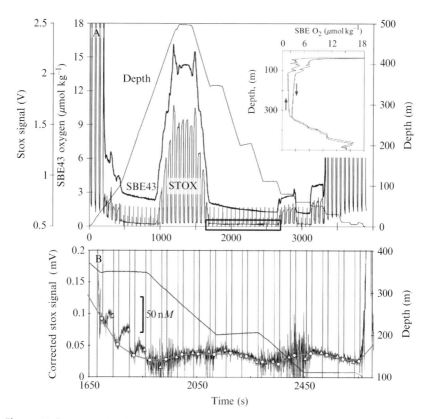

Figure 14.6 Example of *in situ* measurement with STOX sensor in the eastern tropical South Pacific OMZ. (A) Signals from STOX sensor, SBE43 O_2 sensor, and pressure transducer (depth) as a function of time (in seconds) after deployment. The instruments descended continuously down to 497 m during the downcast, but resting at depths of 497, 350, 200, 110, 60, 54, 40, and 15 m during the upcast. The insert shows the signal from the SBE43 during down- and upcast, as indicated by the arrows. The rectangle represents the part of the STOX signal shown in (B). (B) STOX signal after temperature and pressure correction, signals for front guard polarization/depolarization, and depth for the low-O_2 region between 110 and 350 m depth. ○: Calculated mean values for the final 5 s of the front guard polarized period. Δ: Calculated mean values for the final 5 s of the front guard depolarized period. The vertical lines in (B) are voltage spikes of 0.1–0.2 V in the STOX signal (also visible in (A)) arising from the shifts between polarization (20 s) and depolarization (40 s) of the front guard.

CTD rising through the intervals 350–204 and 204–111 m. The latter values do not represent means of the entire depth intervals but rather of the three or four 2.5 m intervals over which the STOX signal was sampled during front guard depolarization (5 s at 0.5 m s^{-1}).

Based on experience with profiles as those shown in Fig. 14.6 and in Revsbech *et al.* (2009), we recommend that *in situ* oxygen measurements in

OMZs with the present design and typical response time of STOX sensors are made with the sensors held at specific depths for several cycles. In order to minimize drift, this should be done during the upcast, where it might be combined with the stops made to release water samplers. Below, we will discuss the potential for combining STOX and conventional oxygen sensor data for continuous oxygen distributions.

8. Future Fast Response STOX Sensors

Preliminary experiments have shown that it is possible to construct STOX sensors with internally porous gold-plated silicone membranes. There are, however, problems in making a reproducibly porous gold layer and also in making a good and stable electrical contact with the gold-plated layer. However, when these problems have been solved, the diffusion distances within the STOX sensors can be dramatically shortened, and it should thereby be possible to make sensors with 98% response times of ~ 3 s for front guard polarization and ~ 7 s for front guard depolarization. If this potential is fulfilled, it will be possible to get a reading for every 10 s (98% response), corresponding to a reading for every 5 m at a CTD descent rate of 0.5 m s^{-1}. Whether this is sufficient will depend on the purpose of the investigation. It should be realized, however, that the time (and thereby also depth) resolution by a conventional sensor is even lower, although the continuous signal may seduce one to believe that all layers are analyzed at high spatial resolution. Tengberg et al. (2006) thus found a 90% response time of 10 s for Aanderaa optodes that had been optimized for fast response (optodes without optical isolation), and much slower response for standard electrochemical sensors.

9. Using STOX Sensors to Recalibrate Conventional Oxygen Sensors

In addition to the detailed factory calibration, conventional oxygen sensors such as the one used in Fig. 14.6A are normally checked and, if necessary, recalibrated against oxygen determinations by Winkler titration in the field. As discussed above, however, even the most sensitive adaptations of the Winkler method have a detection limit of ~ 1 μM, and STOX oxygen determinations thus provide a better reference for zeroing such sensors in oxygen-depleted waters. In our experience, such a correction includes a time-dependent offset. Thus, when first entering the OMZ core during the downcast, the SBE43 sensor in Fig. 14.6 gave a signal of almost 3 μM, and when the sensor finally left the core again during the upcast

2500 s later, the reading had drifted down close to 1 µM. We attribute the drift mainly to the release of oxygen from the polymers in the flow-through system in which the sensor is mounted, and the time dependence suggests that a stable reading could be reached if the instrument was allowed to equilibrate in the OMZ for an hour or longer. The signal should therefore depend on the extent to which the sensor system was exposed to oxygen before entering the OMZ. Nonetheless, we have observed a highly reproducible behavior of specific SBE43 sensors during profiling at several stations in an OMZ as long as casts were made to the same depth and the CTD had been on deck for some hours between deployments. The SBE43 oxygen reading in the oxygen-depleted core could be described as a combination of an offset and an exponential decay:

$$\text{Oxygen}_{SBE43} = a + b \exp(c \times \Delta t) \qquad (14.1)$$

where Δt represents the time elapsed since the sensor entered the OMZ core, and b is a sensor-specific constant. For the sensor used for Fig. 14.6, the best fit was obtained with an offset, a, of 1.1 µM and a decay constant, c, of -0.0015 s^{-1}, corresponding to a half-life of 470 s. Recalibration of the sensor with the above equation (oxygen = oxygen$_{SBE43}$ − a − b exp($c \times \Delta t$)) resulted in a detection limit of <0.1 µM and could be used to constrain oxygen levels in the gaps between depths with STOX data, and at stations where STOX sensors were not deployed (B. Thamdrup, unpublished results). The general applicability of this type of correction remains to be tested.

ACKNOWLEDGMENTS

The assistance in the laboratory and during cruises by Preben G. Sørensen is greatly acknowledged. The development and marine use of STOX sensors was supported by The Agouron Institute, The Gordon and Betty Moore Foundation, the Danish Natural Science Research Council, and the Danish National Research Foundation. The presented field data were obtained during the Galathea 3 expedition under the auspices of the Danish Expedition Foundation. This is Galathea 3 contribution no. P64.

REFERENCES

Broenkow, W. W., and Cline, J. D. (1969). Colorimetric determination of dissolved oxygen at low concentrations. *Limnol. Oceanogr.* **14,** 450–454.

Clark, L. C., Wulf, R., Granger, D., and Taylor, A. (1953). Continuous recording of blood oxygen tension by polarography. *J. Appl. Physiol.* **6,** 189–193.

Devol, A. H. (1978). Bacterial oxygen-uptake kinetics as related to biological processes in oxygen deficient zones of oceans. *Deep-Sea Res.* **25,** 137–146.

Glud, R. N., and Reimers, C. E. (2000). In situ chemical sensor measurements at the sediment–water interface. *In* "Chemical Sensors in Oceanography," (M. S. Varney, ed.), pp. 249–282. Gordon and Breach Science Publishers, New York.

Glud, R. N., Gundersen, J. K., and Ramsing, N. B. (2000). Electrochemical and optical oxygen microsensors for in situ measurements. *In* "In situ Monitoring of Aquatic Systems: Chemical Analysis and Speciation," (J. Buffle and G. Horvai, eds.), pp. 195–222. Wiley, New York.

Gundersen, J. K., Ramsing, N. B., and Glud, R. N. (1998). Predicting the signal of O_2 microsensors from physical dimensions, temperature, salinity, and O_2 concentration. *Limnol. Oceanogr.* **43,** 1932–1937.

Labasque, T., Chaumery, C., Aminot, A., and Kergoat, G. (2004). Spectrophotometric Winkler determination of dissolved oxygen: Re-examination of critical factors and reliability. *Mar. Chem.* **88,** 53–60.

Martini, M., Butman, B., and Michelson, M. J. (2007). Long-term performance of Aanderaa optodes and Seabird SBE-43 dissolved-oxygen sensors bottom mounted at 32 m in Massachusetts Bay. *J. Atmos. Ocean. Technol.* **24,** 1924–1935.

Morrison, J. M., Codispoti, L. A., Smith, S. H., Wishner, K., Flagg, C., Gardner, W. D., Gaurin, S., Naqvi, S. W. A., Manghnani, V., Prosperie, L., and Gundersen, J. S. (1999). The oxygen minimum zone in the Arabian Sea during 1995. *Deep Sea Res. II* **46,** 1903–1931.

Paulmier, A., and Ruiz-Pino, D. (2009). Oxygen Minimum Zones (OMZs) in the modern ocean. *Prog. Oceanogr.* **80,** 113–128.

Revsbech, N. P. (1989). An oxygen microelectrode with a guard cathode. *Limnol. Oceanogr.* **34,** 472–476.

Revsbech, N. P., and Jørgensen, B. B. (1986). Microelectrodes: Their use in microbial ecology. *Adv. Microb. Ecol.* **9,** 293–352.

Revsbech, N. P., Larsen, L. H., Gundersen, J., Dalsgaard, T., Ulloa, O., and Thamdrup, B. (2009). Determination of ultra-low oxygen concentrations in oxygen minimum zones by the STOX sensor. *Limnol. Oceanogr. Methods* **7,** 371–381.

Tengberg, A., Hovdenes, J., Andersson, J. H., Brocandel, O., Diaz, R., Hebert, D., Arnerich, T., Huber, C., Körtzinger, A., Khripounoff, A., Rey, F., Rönning, C., *et al.* (2006). Evaluation of a life time based optode to measure oxygen in aquatic systems. *Limnol. Oceanogr. Methods* **4,** 7–17.

Wolfbeis, O. S., Offenbacher, H., Kroneis, H., and Marsoner, H. (1984). A fast responding fluorescence sensor for oxygen. *Microchim. Acta* **1,** 153–158.

CHAPTER FIFTEEN

Regulation and Measurement of Nitrification in Terrestrial Systems

Jeanette M. Norton* and John M. Stark[†]

Contents

1. Introduction	344
2. Diversity of the Nitrification Process in Terrestrial Environments	345
3. Substrates and Products of Nitrification Reactions	346
4. Controls on Nitrification Rates in Soil Environments	347
4.1. Substrate availability	347
4.2. Ammonia sensitivity	348
4.3. Environmental controls: Oxygen, water potential, and temperature	348
4.4. Acidity and alkalinity of the soil environment	350
5. Measurements of Nitrification in Terrestrial Environments	350
6. Measurement of Nitrification Rates in Terrestrial Ecosystems	351
6.1. Use of ^{15}N to measure gross nitrification rates	352
6.2. Use of inhibitors to measure gross nitrification rates	355
7. Measurement of Nitrification Kinetics	357
8. Nitrifier Population Size	361
9. Modeling Approaches	361
10. Future Advancements in Nitrification Rate Research in Terrestrial Environments	362
References	362

Abstract

Understanding nitrification rates and their regulation continues as a key area of research for assessing human's increasing impact on the terrestrial N cycle. We review the organisms and processes responsible for nitrification in terrestrial systems. The control of nitrification by substrate availability is discussed with particular attention to the factors affecting ammonia/ammonium availability. The effects on nitrification rates of environmental controls including oxygen, water potential, temperature and pH are described. With this general understanding of the factors affecting nitrification rates as a basis, we present an in

* Department of Plants, Soils and Climate, Utah State University, Logan, Utah, USA
[†] Department of Biology and Ecology Center, Utah State University, Logan, Utah, USA

Methods in Enzymology, Volume 486   © 2011 Elsevier Inc.
ISSN 0076-6879, DOI: 10.1016/S0076-6879(11)86015-5   All rights reserved.

depth analysis of methods used to measure nitrification in terrestrial systems. Net, gross and potential nitrification rate measurements are explained including the use of isotopes and inhibitors to measure rates in soils. Methods for the estimation of nitrification kinetics and modeling are briefly described. Future challenges will require understanding the factors controlling nitrification across spatial scales from ecosystems to soil microsites if we are to sustainably manage reactive nitrogen in terrestrial environments.

1. INTRODUCTION

Nitrification is the biological conversion of reduced nitrogen (N) in the form of ammonia (NH_3) or ammonium (NH_4^+) or organic N to oxidized N in the form of nitrite (NO_2^-) or nitrate (NO_3^-). The conversion of the cation, NH_4^+, to an anion (NO_2^- or NO_3^-) determines the mobility of N through the generally negatively charged soil and therefore strongly influences the fate of N in terrestrial systems. Nitrate is more likely than NH_4^+ to move rapidly via mass flow to plant roots, leach out of the root zone, or be lost from the soil by denitrification (Fig. 15.1). Nitrification affects the

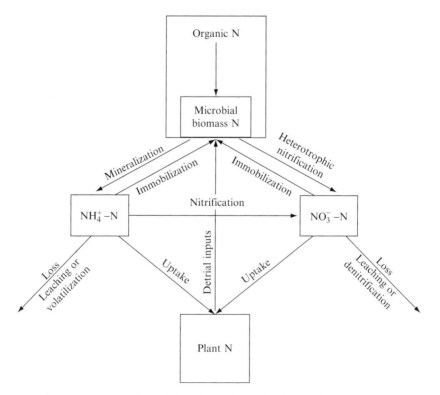

Figure 15.1 Overview of the role of nitrification in the plant–soil system.

coupled transport (e.g., diffusion) and reaction processes of the internal soil N cycle resulting in changes in N fertilizer use efficiency in agriculture and ecosystem N export. In the majority of agricultural soils, NH_4^+ is rapidly converted to NO_3^-, which may accumulate in the soil solution to high concentrations. For these reasons, it is often desirable to manage agricultural soils to reduce nitrification and increase N fertilizer use efficiency. The nitrification process is known to produce nitrogenous trace gases (NO and N_2O) important in greenhouse gas and ozone atmospheric chemistry (Godde and Conrad, 2000; Hutchinson and Davidson, 1993), both directly and indirectly through denitrification. It remains an important goal to understand the factors affecting N_2O production in order to devise appropriate mitigation strategies (Farquharson and Baldock, 2008; Stein and Yung, 2003). Understanding nitrification rates and their regulation continues as a key area of research for assessing human's increasing impact on the terrestrial N cycle (Schlesinger, 2009).

2. Diversity of the Nitrification Process in Terrestrial Environments

The reliable determination of nitrification rates depends upon our understanding of the nitrification process and benefits from at least some partial knowledge of the organisms mediating the individual steps. Classically, nitrification is thought of as a process mediated by chemolithoautotrophic bacteria that gain energy from the oxidation of N and fix inorganic C (CO_2 or bicarbonate). Recent evidence suggests that some archaea of the proposed phylum *Thaumarchaeota* (Brochier-Armanet *et al.*, 2008; Spang *et al.*, 2010) are also capable of ammonia oxidation, contain genes encoding the key enzymes of this process, and are widespread in soil environments (Leininger *et al.*, 2006; Treusch *et al.*, 2005). The isolation of a pure culture of an ammonia oxidizing autotrophic archaeon, *Nitrosopumilus maritimus* strain SCM1 (Konneke *et al.*, 2005), and its subsequent genome sequencing (Walker *et al.*, 2010) and physiological characterization (Martens-Habbena *et al.*, 2009) have confirmed its ability to oxidize ammonia to nitrite by a pathway distinct from that of the ammonia oxidizing bacteria. However, further characterization of ammonia oxidation pathway in soil ammonia oxidizing archaea awaits the isolation of a pure culture representative for this group.

Other nitrification processes have also been identified as important in specific environments (Hayatsu *et al.*, 2008). Heterotrophic nitrification may be broadly defined as the oxidation of reduced N compounds (including organic N) producing NO_2^- or NO_3^-. The oxidation of ammonium or organic N is not linked to cellular growth in this process (De Boer and Kowalchuk, 2001). Heterotrophic nitrification is catalyzed by a variety of microorganisms including fungi, actinomycetes, and bacteria, presumably using a wide variety of metabolic pathways (Prosser, 1989) and has been found to be significant in some forest and acidic pasture soils (Islam et al., 2007; Killham, 1990; Schimel et al., 1984; Trap et al., 2009). The anaerobic oxidation of ammonium in the presence of nitrite to dinitrogen gas termed the *anammox* process (Jetten et al., 1998, 2009; Schmidt et al., 2002) was originally identified from wastewater treatment systems but is now known to be widespread in marine and freshwater environments, although rarely identified in soil (Penton et al., 2006). The organisms responsible are nitrite-dependent anaerobic bacteria belonging to the phylum *Planctomycete* (Jetten et al., 2009). While the diversity of processes included under a broad definition of nitrification has expanded, the oxidation of ammonia and nitrite by chemolithotrophic bacteria and archaea remains the primary mechanism of nitrification in aerobic terrestrial systems (soil environments) examined to date.

3. SUBSTRATES AND PRODUCTS OF NITRIFICATION REACTIONS

In autotrophic nitrification by bacteria, the conversion of N takes place in two steps: the ammonia oxidizing bacteria such as *Nitrosomonas* or *Nitrosospira* convert NH_3 to NO_2^- according to Eq. (15.1):

$$NH_3 + O_2 + 2H^+ + 2e^- \rightarrow NH_2OH + H_2O \rightarrow NO_2^- + 5H^+ + 4e^-$$
(15.1)

while the nitrite oxidizing bacteria such as *Nitrobacter* or *Nitrospira* convert NO_2^- to NO_3^-:

$$NO_2^- + H_2O \rightarrow NO_3^- + 2H^+ + 2e^-$$
(15.2)

No bacteria have been found which can convert NH_3 to NO_3^- directly (Hooper et al., 1997). In general, NO_2^- does not accumulate in soils except under transient conditions that have decreased the population or inhibited the activity of nitrite oxidizers. The intensive application of ammonical fertilizers (i.e., urea or anhydrous NH_3) may result in NO_2^- accumulation due to the inhibition of nitrite oxidizers from the toxicity of high ammonia levels in the application zone (Schmidt, 1982) or from subsequent localized lowering of pH and production of nitrous acid (Venterea and Rolston, 2000a). The interaction of soil pH, buffering capacity, and ionization of ammonia and nitrite may be useful predictors of nitrite accumulation and production of N-oxides via nitrification (Venterea and Rolston, 2000b).

4. CONTROLS ON NITRIFICATION RATES IN SOIL ENVIRONMENTS

4.1. Substrate availability

The autotrophic ammonia oxidizing bacteria require NH_4^+/NH_3, CO_2, and O_2 to proliferate and grow. Since CO_2 is virtually always available in soil environments, and under well-drained soil conditions, O_2 concentrations are sufficient for nitrification to proceed, the availability of NH_4^+/NH_3 often limits both the rate of nitrification and the size of the resultant nitrifier populations in many soils (Fig. 15.2). While NH_3 rather than the cation, NH_4^+, is thought to be the actual substrate for the ammonia monooxygenase enzyme responsible for the initial step in ammonia oxidation (Arp et al., 2002), the ionic form is pH dependent ($pK_a = 9.25$) and conversions may occur close to or at the cell membrane. Ammonium production via mineralization, additions of ammonical fertilizers and animal wastes, and the atmospheric deposition of NH_4^+ increase substrate supply, while competing consumptive processes include microbial assimilation (immobilization), plant assimilation, and ammonia volatilization (Fig. 15.1). The ratio of nitrification to NH_4^+ immobilization has been used to indicate the potential

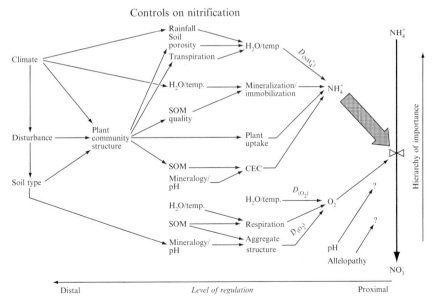

Figure 15.2 The major factors regulating nitrification in soil environments (Robertson, 1989; used with permission). Diffusion of NH_4^+ and of O_2 to the nitrifying organisms are indicated by $D_{(NH_4^+)}$ and $D_{(O_2)}$, respectively.

for N loss of both forest and agricultural ecosystems, ratios > 1 indicating higher potential for N loss (Stockdale et al., 2002). The proximal and distal factors controlling nitrification were summarized by Robertson (1989) as shown in Fig. 15.2. Rates of nitrification generally respond quickly to ammonium additions, while a significant lag occurs before ammonia oxidizer population increases are detectable (Mendum et al., 1999; Shi and Norton, 2000b). Kinetic parameters and models relating nitrification rates to NH_4^+/NH_3 supply are discussed below.

4.2. Ammonia sensitivity

Some, but not all, cultured ammonia oxidizing bacteria and ammonia oxidizing archaea are sensitive to high levels of their substrate, NH_4^+/NH_3. Substrate inhibition of nitrification has been observed in ammonia oxidizers isolated from diverse environments, but this trait may be more common than observed in cultures due to selection against this characteristic during standard isolation (Bollmann and Laanbroek, 2001; Tourna et al., 2010). Cultured *Nitrosomonas* strains sensitive to high ammonia concentrations are found in cluster 6A (Bollmann et al., 2002), and environmental strains of *Nitrosospira* enriched at low concentrations were also found to be sensitive to high levels of ammonium and have been designated as *Nitrosospira* cluster 3A (Tourna et al., 2010; Webster et al., 2005). Oligotrophic ammonia oxidation and ammonia sensitivity are also characteristics of the pure culture of the marine crenarchaeon ammonia oxidizer, *N. maritimus* strain SCM1 (Martens-Habbena et al., 2009). Enrichment cultures and soil slurries have also shown substrate inhibition starting within a range of approximately 1.0–5.0 mM NH_4^+/NH_3 concentrations in both wildland and agricultural soils (Koper et al., 2010; Shi and Norton, 2000a; Stark and Firestone, 1996). The ammonia tolerance of soil ammonia oxidizers from unfertilized systems is likely to be intermediate between those in heavily fertilized soils or sewage and the oligotrophic AOB of the open ocean (Martens-Habbena et al., 2009; Stark and Firestone, 1996). The product of ammonia oxidation, nitrite, is also known to cause product inhibition. Since both substrate and product inhibition of nitrification are widespread in the environment, concentrations of substrate and product accumulation are critical considerations during the design of experiments for determination of nitrification rates and kinetics.

4.3. Environmental controls: Oxygen, water potential, and temperature

Oxygen affects nitrification rates through its roles as a substrate for the ammonia monooxygenase enzyme and as the terminal electron acceptor from cytochrome c oxidases (Arp et al., 2002). Oxygen availability is controlled by the interaction of O_2 consumption and diffusion from the surface through the air-filled pores. Sufficient oxygen diffuses into most soils that are at field capacity or drier to maintain nitrification, although microsites lacking O_2 may frequently occur inside soil aggregates (Sexstone et al., 1985). In soils that remain wetter than field capacity for several days, nitrification rates generally decline. In three arable soils with increasing clay contents (108, 224, and 337 g clay kg^{-1} soil), maximum net nitrification was observed at -14, -17, and -43 kPa water potential, respectively (Schjønning et al., 2003), with rates decreasing at higher water potentials presumably due to oxygen depletion. Earlier studies in silt loam soils observed maximum rates for net nitrification at approximately -10 kPa (Saby, 1969). The observed decline in net nitrification rates may be due to either a decline in actual nitrification or an increase in denitrification (both of which should occur at lower oxygen concentrations). Some ammonia oxidizers have the ability to use NO_2^- as an alternative electron acceptor when oxygen is limiting and this is referred to as nitrifier-denitrification (Bock et al., 1995; Hayatsu et al., 2008; Shaw et al., 2006; Stein, 2010). Low oxygen availability may repress nitrite oxidizer activity before ammonia oxidation and result in the accumulation of nitrite (Laanbroek and Gerards, 1993).

In soils at field capacity and drier, water availability impacts nitrification rates through its direct effect on cell physiology and metabolic activity and through indirect effects on substrate availability (Stark and Firestone, 1995). In soil from a California oak woodland-annual grassland ecosystem, the relative importance of these two factors shifted over the range of water potentials; substrate diffusional limitation was the dominating factor at water potentials wetter than -0.6 MPa, while effects due to dehydration and its physiological impacts were more pronounced in soils drier than -0.6 MPa (Stark and Firestone, 1995).

The response of nitrification to temperature has been evaluated in a diverse range of soils, and the optimum temperature for nitrification has been found to be environment specific (Singh et al., 1993; Stark, 1996a; Stark and Firestone, 1996). Cultured AOB from soils generally have temperature optimum between 25 and 30 °C, but *Nitrosomonas cryotolerans* has a temperature optimum for growth at 22 °C and can grow at 0 °C (Koops et al., 1991). There is evidence for soil nitrifier activity under similarly cold temperatures typical of winter season soils (2–10 °C; Avrahami and Conrad, 2005; Avrahami et al., 2003; Cookson et al., 2002). Across a range of North American ecosystems, the community composition of ammonia oxidizing bacteria was correlated with temperature as indicated by mean annual temperature (Fierer et al., 2009). The temperature response of nitrification has been modeled using either the Arrhenius equation or a Poisson density function (Stark, 1996a).

4.4. Acidity and alkalinity of the soil environment

Both ammonia and nitrite oxidation are generally considered to be optimal at neutral to slightly alkaline soil pH values. For pure cultures of ammonia oxidizing bacteria, specific growth rates and activity are significantly reduced outside a relatively narrow pH range around the optimum for the organism (Prosser, 1989). However, contrary to previous conceptions that nitrification did not occur at extreme pH values, autotrophic nitrification has now been confirmed in soils with pH values from 3.0 (De Boer and Kowalchuk, 2001) up to 10.0 (Sorokin, 1998; Sorokin et al., 2001), and possible organisms have been isolated. The effect of acidity on ammonia oxidation is likely related to the exponential decrease in NH_3 availability with decreasing pH (pK_a of NH_3/NH_4^+ is 9.25). Microsites of higher pH such as those found near ammonifying bacteria or in cell aggregates may be responsible for continued ammonia oxidation in acid soils (De Boer and Kowalchuk, 2001). Ammonia oxidizers that can utilize urea to produce NH_3 intracellularly can maintain activity under more acid environments than when growing on NH_4^+ (Burton and Prosser, 2001; DeBoer and Laanbroek, 1989). Some NO_2^- produced under acid conditions equilibrates to HNO_2 ($pK_a = 3.15$) which increases its potential toxicity. Therefore, in acid soils, the coupling between ammonia oxidizers and nitrite oxidizers is even more crucial for continued nitrification. Isolated acid-tolerant ammonia oxidizers are generally urease positive *Nitrosospira* species (De Boer and Kowalchuk, 2001). Previously identified acid-tolerant nitrite oxidizers are *Nitrobacter* species (De Boer and Kowalchuk, 2001), but *Nitrospira* are likely to be important in acid soils as well. Contrary to these pure culture observations or to surveys within a limited area, when a global scale analysis of gross nitrification rates versus pH was performed based on data from a wide variety of terrestrial systems, no significant relationship between rates and pH was detected (Booth et al., 2005). It may be that shifts in nitrifier community composition or development of microsite structure in soils with acidic pH override the adverse effects of pH that are typically observed in laboratory experiments examining individual soils.

5. MEASUREMENTS OF NITRIFICATION IN TERRESTRIAL ENVIRONMENTS

The common uses of the term nitrification rate include at least three distinct experimentally defined meanings: net, gross, and potential nitrification rates. The net nitrification rate is the rate of NO_3^- accumulation and is equal to the rate of conversion of NH_4^+/NH_3 to NO_3^- minus any consumption. This is the most commonly measured rate in both laboratory and field conditions. The gross nitrification rate is the actual rate of conversion of NH_4^+/NH_3 to NO_2^-/NO_3^-, regardless of consumption. Determination of gross nitrification rates generally requires the use of isotopes (^{15}N) to give distinct estimates of simultaneous production and consumption of NO_3^-. The potential nitrification rate is the maximum rate of nitrification in a mixed system with nonlimiting substrate (NH_4^+/NH_3) supply. The details of the methods used for these rate determinations are available (Hart et al., 1994; Murphy et al., 2003; Schmidt and Belser, 1994; Stark, 2000), and pertinent details and applications are discussed in the following sections. Measurements of these different rates are often combined to give a clearer view of the nitrification process.

6. Measurement of Nitrification Rates in Terrestrial Ecosystems

In spite of decades of research examining nitrification, there is still substantial confusion regarding the appropriate method for measuring nitrification rates in soils. All approaches include measurement of changes in NO_3^- pool sizes over time; however, measurement of the change in NO_3^- pool alone (i.e., net nitrification) is not sufficient to quantify actual gross rates of nitrification. That is because net nitrification is the difference between the gross nitrification rate and the gross NO_3^- consumption rate (Fig. 15.3). Net nitrification underestimates gross nitrification if denitrification, NO_3^- immobilization by microbes, or other NO_3^--consuming processes occur. Also, if NO_3^- consumption rates change, net nitrification will change in spite of no change in gross nitrification rates. While denitrification may be inhibited in aerobic incubations, NO_3^- immobilization has been shown to be substantial in a wide range of soil types, frequently exceeding 50% of gross nitrification rates (Stark and Hart, 1997). As a result, net nitrification rates are poorly correlated with gross nitrification rates across a variety of ecosystems (Booth et al., 2005). Therefore, net nitrification alone should never be used to estimate gross nitrification. Instead, net nitrification measurements should be coupled with the use of either isotopes (^{15}N) or inhibitors.

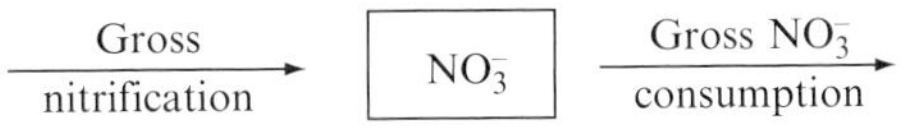

$$\xrightarrow{\text{Gross nitrification}} \boxed{NO_3^-} \xrightarrow{\text{Gross } NO_3^- \text{ consumption}}$$

$$\text{Net nitrification} = \frac{\Delta NO_3^-}{\Delta t} = \text{gross nitrification} - \text{gross } NO_3^- \text{ consumption}$$

Figure 15.3 Relationship between net nitrification rate and gross nitrification and NO_3^- consumption rates.

The most appropriate method depends on the objectives of the researcher. For example, if the goal is to estimate actual rates of nitrification in the field, ^{15}N isotope dilution measurements during field incubation of intact soil cores may be the best approach. If the researcher is more interested in comparing the effects on nitrification of a particular set of treatments, laboratory incubations of sieved soils may be appropriate. Care should be taken to carefully determine whether a particular approach will meet the objectives of the study, and whether the assumptions required for the approach can be satisfied under the experimental conditions.

6.1. Use of ^{15}N to measure gross nitrification rates

The stable isotope ^{15}N has been used in two approaches to estimate gross nitrification rates: tracer and isotope dilution techniques. In a tracer technique, ^{15}N is added to the substrate or source pool for nitrification (i.e., as ^{15}NH$_4^+$), and then the rate of appearance of ^{15}N in the product pool (i.e., ^{15}NO$_3^-$) is monitored. This approach has a number of shortcomings that severely limit its use. First, addition of ^{15}NH$_4^+$ may increase the substrate supply to nitrifiers and thus stimulate nitrification rates. Second, if nitrifiers are consuming urea or are receiving NH$_4^+$ directly from N-mineralizing microorganisms in isolated soil microsites, little of the added ^{15}NH$_4^+$ will be utilized. Third, if NO$_3^-$ immobilization or other NO$_3^-$-consuming processes occur, accumulation of ^{15}N in the NO$_3^-$ pool will underestimate flow of ^{15}N into the NO$_3^-$ pool. While the first problem results in an overestimate of actual nitrification rates, the second two problems will cause underestimation of rates. Unfortunately, which of these problems is occurring is not usually apparent from tracer measurements.

^{15}N isotope dilution: The most common approach for measuring gross nitrification rates is the isotope dilution technique (Kirkham and Bartholomew, 1954). In contrast to the tracer technique, ^{15}N is added to the pool that is the product of nitrification (i.e., NO$_3^-$) rather than to the source pool. As nitrification occurs, the added ^{15}NO$_3^-$ is "diluted out" by the ^{14}NO$_3^-$ produced during nitrification of the unlabeled NH$_4^+$. In addition, NO$_3^-$ consumption can be accounted for by measuring disappearance of ^{15}N from the NO$_3^-$ pool. Rates of dilution and disappearance of ^{15}N can then be used to calculate gross rates of nitrification and NO$_3^-$ consumption, even when both processes are occurring simultaneously.

During ^{15}N dilution measurements of nitrification, a solution containing K^{15}NO$_3^-$ (typically at high enrichment such as 99 atom % ^{15}N) is added to a soil sample. Sufficient liquid must be added to ensure reasonable distribution of the isotope throughout the soil, but not so much that increased soil moisture affects nitrification rates. Typically, 4–6 mL of solution is added per 100–120 g of moist soil (less solution for sandy soils, more solution for clay soils). Solutions must also contain sufficient ^{15}N to enrich the NO$_3^-$

pool to at least 25–50 atom %. Greater ^{15}N additions are advantageous because they ensure that sufficient N will be extractable at the end of the incubation period to allow mass spectrometric analyses. Higher $^{15}NO_3^-$ enrichments may also increase precision in heterogeneous soil samples. The ^{15}N solution may be added to sieved soil samples with a pipette, mixing the soil as the solution is added. When rates in intact soil cores are being measured, the solution is injected into the sample using a syringe with a long side-port needle. The needle is inserted fully into the soil core, and then the syringe plunger is depressed as the needle is withdrawn, distributing the solution throughout a column within the core. Multiple injections within a single core are usually required to promote uniform distribution of the solution within the core.

After the soil sample is injected with the ^{15}N solution, a subsample of soil is extracted with a salt solution, such as 2 M KCl, 0.5 M K_2SO_4, or 0.01 M CaCl (typically 10–20 g soil per 100 mL extractant). This initial extraction allows estimation of the "time-0" NO_3^- concentration and ^{15}N enrichment. A separate subsample is incubated for a period of time (typically 1–2 days) and then extracted to determine the "time-t" concentration and enrichment. The gross nitrification rate and the gross NO_3^- consumption rate are calculated from the following equations (Hart et al., 1994; Kirkham and Bartholomew, 1954):

$$\text{Gross nitrification rate} = \frac{(P_0 - P_t)}{t} \times \frac{\log(P_0/P_t)}{\log(I_0/I_t)} \quad (15.3)$$

$$\text{Gross } NO_3^- \text{ consumption rate} = \text{Gross nitrification rate} - \frac{(P_t - P_0)}{t} \quad (15.4)$$

where P is the soil NO_3^- concentration, I is the ^{15}N atom % excess (the ^{15}N enrichment above background, which in soil is typically 0.37 atom %), t is the length of the incubation, 0 and t indicate whether the measurement is made at time-0 or time-t.

When gross rate measurements are being made on intact soil cores, it is not possible to obtain a time-0 subsample from the same sample that is incubated. Instead, two soils cores must be injected with ^{15}N; one is harvested immediately and the second is incubated. Because the two cores will invariably differ in soil bulk density and possibly gravel content, the soil masses in the two cores will differ, and thus, the quantity of $^{15}NO_3^-$ added per unit mass of soil in the two cores will also differ. Instead of using the data from the first core to directly estimate time-0 ^{15}N enrichment in the incubated core, the data from the first core is used to estimate the $^{14}NO_3^-$ concentration and the ^{15}N extraction efficiency (the proportion of the added $^{15}NO_3^-$ that is extractable at time-0) in the incubated core. An estimate of extraction efficiency is necessary because not all of the ^{15}N added to soil can be recovered using soil extractants even when the sample is extracted within a few minutes of ^{15}N addition. This may be because extracting solutions are never 100% efficient, or because unexplained abiotic or biotic processes rapidly consume ^{15}N during the minutes immediately following addition (Davidson et al., 1991; Hart et al., 1994). For ^{15}NO$_3^-$, extraction efficiencies typically range from 90% to 95%; whereas for ^{15}NH$_4^+$, they are substantially lower (75–90%). Based on estimates of the initial ^{14}NO$_3^-$ concentration, the amount of ^{15}N injected, the ^{15}N extraction efficiency, and the mass of soil in the incubated core, the time-0 NO$_3^-$ concentration and enrichment can be calculated. Time-t measurements are then obtained directly by harvesting the incubated core, and these data are used with the calculated time-0 values and Eqs. (15.3) and (15.4) to estimate gross nitrification and NO$_3^-$ consumptions rates. Because soil heterogeneity is often great at relatively fine scales, many researchers prefer to collect additional core samples around the core to be incubated, so that a better estimate of the initial NO$_3^-$ concentration can be obtained. In some cases, a larger cylinder is driven into the soil around the core to be incubated, and the soil between the large cylinder and the inner core is harvested to provide a more accurate estimate of time-0 NO$_3^-$ concentrations. This "concentric core" approach along with the calculations for ^{15}N extraction efficiency is described in detail in Stark (2000). For examples of the application of this approach in combination with determination of mineralization rates in agricultural and grassland systems, see Habteselassie et al. (2006) and Hooker and Stark (2008), respectively.

A number of assumptions must be met for valid use of the isotope dilution technique. First, the added isotope must be uniformly distributed within the soil so that there is no preferential use of ^{14}NO$_3^-$ relative to ^{15}NO$_3^-$. Technically speaking, this assumption is never completely met because: (1) soil microsite heterogeneity makes it impossible to achieve completely uniform distribution of label and (2) discrimination between ^{14}N and ^{15}N by enzymes results in a slight amount of preferential use of ^{14}N. While errors due to nonuniform distribution of the label may be substantial if the soil sample is not well mixed, errors due to isotope discrimination are exceedingly small relative to other sources of error when high ^{15}N enrichments are used. Errors due to nonuniform distribution will also be small if the spatial distribution of isotope is random with regard to the spatial distribution of rates (Davidson et al., 1991). A second assumption is that rates of both nitrification and NO$_3^-$ consumption are constant throughout the incubation period. Keeping incubation times as short as possible will minimize changes in NH$_4^+$ and NO$_3^-$ concentrations which will help minimize changes in nitrification and NO$_3^-$ consumption resulting from reduced substrate availability. If concentrations vary widely, it may be desirable to use a more complex model than described above (Eqs. (15.3) and (15.4)). Simulation models coupled with numerical analysis (e.g.,

FLUAZ, ^{15}N; Mary et al., 1998; Murphy et al., 2003) are available which can be used to estimate multiple rates occurring simultaneously (nitrification, NO_3^- assimilation, N mineralization, NH_4^+ assimilation, denitrification, etc.), when either the NH_4^+ or NO_3^- pools are labeled. In addition, processes like nitrification can be described by zero-, first-, or mixed-order (e.g., Michaelis–Menten) kinetics, such that changes in substrate concentrations result in different rates. While these models provide additional flexibility and much more power in estimating rate constants, they do require additional sampling for parameter estimation and some basic modeling and programming skills.

For the isotope dilution method, NO_3^- concentrations in soil extracts are typically determined using colorimetric analyses, and $^{15}NO_3^-$ enrichments are determined by isotope-ratio mass spectrometry after concentrating the ^{15}N from the extracts on filter paper disks using a diffusion procedure (Stark and Hart, 1996).

6.2. Use of inhibitors to measure gross nitrification rates

Chemical inhibitors may be combined with net rate measurements to estimate nitrification rates. These inhibitors consist of either chemicals that block nitrification or chemicals that block NO_3^- consumption processes. Direct inhibitors of nitrification include acetylene (C_2H_2) and nitrapyrin (2-chloro-6-(trichloromethyl)pyridine). When nitrification rates are measured using nitrification inhibitors, two soil subsamples are used. One subsample is given enough inhibitor to completely block nitrification, and the second subsample is treated identically except no inhibitor is added. After obtaining time-0 estimates of NO_3^- concentrations, the two subsamples are incubated (1–2 days) and then extracted to obtain time-t estimates of NO_3^- concentrations. The net nitrification rates in the two samples (i.e., the change in NO_3^- concentration per unit time) are compared. The sample receiving the nitrification inhibitor will have lower net nitrification rates (negative rates) because only NO_3^- consumption occurred. The difference in net nitrification rates between the two samples equals the gross nitrification rate.

Two important assumptions must be met for this technique to work: the inhibitor must completely block nitrification, and rates of NO_3^- consumption must be identical in both subsamples. Acetylene and nitrapyrin are the two most commonly used nitrification inhibitors for ecological studies. Both of these substances act on ammonia oxidation at the level of the ammonia monooxygenase enzyme. Acetylene (C_2H_2) is a very potent nitrification inhibitor and concentrations in the headspace of incubation vessels as low as 1 Pa are effective in blocking nitrification if it penetrates sufficiently into the soil. Acetylene is highly water soluble, but effort must be made to ensure that it has time to reach all soil microsites and pores, especially when wet samples, large samples, or intact soil cores are being used. Generally, C_2H_2 is used at 10 Pa partial pressure in the headspace to inhibit ammonia oxidation in soil (compared to denitrification inhibition at 10 kPa; Bedard and Knowles, 1989; Berg et al., 1982). The C_2H_2 used for inhibition of ammonia oxidation should be acetone free, as there are potential side effects on N cycling from adding the rather large quantities of acetone found in commercial grade C_2H_2. Acetone-free purified C_2H_2 may be purchased or carefully made from small quantities of calcium carbide (CaC_2) by the addition of water (Weaver and Danso, 1994), although care with this flammable and potentially explosive gas is required. Acetylene is a suicide substrate for ammonia monooxygenase and the inactivation of the enzyme is irreversible (Hyman and Arp, 1992). This results in the requirement for the ammonia oxidizers to synthesize new enzyme before they can recover ammonia oxidizing activity and this may take more than 1 week (Kester et al., 1996). Typically, the C_2H_2 is left in the soil or in the headspace for the duration of the experiment, although in studies of nitrous oxide production from nitrifiers and denitrifiers, 24 h exposure to 100 Pa C_2H_2 resulted in 90% inhibition of nitrification over 6 days. Consumption of C_2H_2 in soil has been observed (DeBoer et al., 1993; Klemedtsson and Mosier, 1994) and may be responsible for decreased inhibition in longer experiments thus potentially requiring reapplication.

Nitrapyrin (2-chloro-6-(trichloromethyl)pyridine) is commercially available under the trade name of N-serve (Dow Chemical). Nitrapyrin is only sparingly soluble in water and therefore is typically mixed into soil as a powder or dissolved in an organic solvent before addition to soil. The carrier itself may have an inhibitory effect on nitrification or may affect the availability of NH_4^+ through increasing immobilization. Investigators should be careful to treat the control sample equivalently to the sample (equivalent mixing times, equivalent water or carrier addition, etc.). Inhibitory concentrations used vary from 0.5 to 20 mg nitrapyrin kg^{-1} soil (McCarty and Bremner, 1990) and is typically applied at 0.6–1.12 kg ha^{-1} rate (Wolt, 2000). Nitrapyrin is degraded in the soil with a half-life generally around 2 weeks at 20 °C although this varies with soil type, organic matter content, and especially temperature (McCarty and Bremner, 1990). Repeated use of nitrapyrin in agricultural systems has been associated with changes in the kinetic parameters for the nitrifier community (Shi and Norton, 2000a). Gross rates of N mineralization and immobilization are only slightly affected by nitrapyrin (Chalk et al., 1990). Sufficient NO_3^- concentrations must be present in the samples throughout the incubation so that NO_3^- consumption does not become limited by low NO_3^- availability in the samples where nitrification is blocked. This is likely to be a problem with longer incubation times, and thus incubations should be kept as short as possible.

Inhibitors that block NO_3^- consumption processes may also be used with net rate measurements to estimate gross nitrification rates. If NO_3^- consumption can be eliminated, then net nitrification equals gross nitrification. Examples of such inhibitors include O_2 for denitrification and NH_4^+ for NO_3^- assimilation. In a common assay for estimating the V_{max} for nitrification (see kinetic approaches below), the soil sample is incubated in a soil slurry. The slurry is shaken vigorously to ensure that O_2 is high enough to inhibit denitrification (and not inhibit nitrification) and NH_4^+ is provided in sufficient quantities to inhibit the activity of assimilatory NO_3^- reductase enzymes. Since these two processes are the only potential consumptive fates of NO_3^- in the assay, blocking them means that the rate of increase in NO_3^- concentration equals the gross nitrification rate.

7. Measurement of Nitrification Kinetics

While the kinetics of nitrification in soil samples have been measured using aerobic incubations of moist soil samples (Low et al., 1997), they are usually measured using shaken soil slurries. These assays are termed "nitrification potentials" and the high water content and shaking ensures complete aeration of the sample and promotes uniform distribution of substrates, thus minimizing diffusional limitations to substrate supply. Because the measurements are made on *in situ* microbial communities, consisting of whole cells imbedded in the soil matrix, the kinetic constants measured in these assays may not be comparable to kinetic constants determined during pure culture assays; however, the rates and kinetic constants determined during these whole soil assays have significant ecological value for comparing nitrifier communities from different soil types without suffering from the selectivity associated with the isolation of pure cultures. While the kinetic parameters describing the effect of NH_4^+ concentration on nitrification rates are not determined using cell-free enzyme assays, they are still described in the literature with the Michaelis–Menten constants, V_{max} and K_m, which define the rectangular hyperbolic function. We will also use those terms, recognizing that the values for the actual enzymes themselves may be quite different.

The nitrification potential assay is a shaken soil slurry assay that has been modified for use in determining the kinetic characteristics of nitrifier communities in soils. Nitrification potentials are generally measurements of NH_3/NH_4^+ oxidation, although NO_2^- oxidation potentials may also be determined. The original nitrification potential method included chlorate (10 mM) to inhibit NO_2^- oxidation so that the production of NO_2^- could be measured with increased sensitivity especially in the presence of high background NO_3^- (Belser and Mays, 1980; Schmidt and Belser, 1994); however, the following suggested method measures $NO_2^- + NO_3^-$ production in the absence of inhibitors to avoid problems with incomplete inhibition of NO_2^- oxidation by chlorate and possible inhibition of NH_3/NH_4^+ oxidation by chlorite (Hart et al., 1994; Schmidt and Belser, 1994). Soil samples are collected, sieved, and briefly stored at 4 °C until the assay can be performed. While ammonia oxidizer populations are thought to be relatively stable, storage should be minimized to prevent changes in potential rates. For soil samples that have very high ambient NO_3^- concentrations (e.g., >20 mg N kg^{-1} soil), it may be desirable to leach the soil with a dilute salt or phosphate buffer solution prior to the assay to reduce NO_3^- concentrations and increase precision of rate estimates. Typically 15 g of moist soil is suspended in a flask of 100 mL of a weak phosphate buffer (pH 7.2) containing 1 mM NH_4^+ (as $(NH_4)_2SO_4$). Since the buffering power of the suspended soil is often greater than that of the phosphate buffer, the pH for the assay should be considered, adjusted as needed, and documented. The samples are shaken at 200 rev. min^{-1} to maintain adequate aeration and uniform substrate distribution (Stark, 1996b). Subsamples are collected at a minimum of four time intervals during a 24-h incubation. The subsamples are collected while shaking the flask (to maintain a constant soil:solution ratio) and analyzed for NH_4^+ and ($NO_2^- + NO_3^-$) concentrations (see Table 15.1; Fig. 15.4). Because NO_3^- consumption processes are inhibited under these conditions, changes in $NO_2^- + NO_3^-$ concentrations can be assumed to equal the gross nitrification rate. Rates measured during this assay are assumed to approximate V_{max}, and have been used to estimate numbers of nitrifier cells in soils, based on per cell values for V_{max} from pure culture assays (Schmidt and Belser, 1994). In a similar fashion, NO_2^- oxidation potential can be determined by determining the consumption of NO_2^- in a shaken slurry to which NO_2^- has been added (8 µg NaNO mL^{-1}) and NH_3/NH_4^+ oxidation is inhibited by nitrapyrin (40 µg nitrapyrin mL^{-1}; Fuller and Scow, 1996; Schmidt and Belser, 1994). Typical nitrification potentials are in the range of 0.1–2 mmol N kg^{-1} day^{-1} (Fortuna et al., 2003; Hart and Stark, 1997; Koper et al., 2010; Stark and Firestone, 1996; Stienstra et al., 1994). The lower values are generally from unfertilized wildland ecosystems while higher values are found in heavily fertilized or manured agricultural soils.

In reality, the nitrification potential assay may substantially underestimate V_{max}. First, V_{max} is a theoretical rate, reached at infinitely high [NH_4^+], and thus rates at 1 mM may only be a fraction of this value. Second, many soils are populated by NH_4^+-sensitive nitrifiers, which display maximum nitrification rates at [NH_4^+] less than 1 mM. If the goal is to estimate V_{max}, then the assay should be repeated using a wide range of [NH_4^+], and V_{max} and K_m should be estimated by fitting the Michaelis–Menten equation to the data with nonlinear regression.

When the nitrification potential assay is modified to provide information on Michaelis–Menten constants, NH_4^+ concentrations ranging from <10 µM to >2 mM should be used because recent work has shown that K_m values in many natural environments are <50 µM and some are even

Table 15.1 Method for nitrification potential assay adapted from Hart et al. (1994)

Equipment and supplies
Orbital shaker with platform and clamps for 250-mL flasks
250-mL Flasks, wide mouth, thick walled preferred
Vented closures either double-hole stoppers with tube or use foil
15 mL Centrifuge tubes or filters (Whatman no. 40) and funnels
5 or 10 mL pipette with wide mouth tips (cut to 0.5 cm orifice)
Freezer tubes (10 mL) for samples

Reagents
0.2 M K_2HPO_4
0.2 M KH_2PO_4
0.05 M NH_4SO_4
Sieved (2 or 4 mm) field moist soil (determine gravimetric soil water content on separate subsample)

Procedure
1. Combine 1.5 mL 0.2 M KH_2PO_4, 3.5 mL 0.2 M K_2HPO_4, and 10 mL 0.05 M NH_4SO_4 in a 1 L volumetric, adjust to pH 7.2^a, bring to volume with deionized H_2O (nitrite/nitrate free). This solution is 1.0 mM NH_4^+ and 1 mM phosphate.
2. Place 15 g field moist soil into a 250-mL flask, add 100 mL of solution above. Place on shaker; cover with a vented cap (stopper with hole or other closure). Shake at 200 rpm for a total of 24 h.
3. Sample four times (suggest 2, 4, 22, 24 h). At least 1 h is necessary at the beginning of the incubation for equilibration. Shake the sample during sampling and use pipette with a wide mouth tip to sample an aliquot with the same soil/solution ratio as original. Centrifuge sample (8000 × g for 8 min) or filter through preleached Whatman no. 40. Cap and freeze tubes.
4. Analyze for $NO_2^- + NO_3^-$ using Griess–Illosvay method (Bundy and Meisinger, 1994). Calculate rate of $NO_2^- + NO_3^-$ production by linear regression of solution concentration over time, convert to a soil dry weight basis by correcting for the moisture content of the soil.

---

^{a} pH for assay may be adjusted to a value closer to soil pH as necessary by changing amounts of phosphate solutions.

< 10 μM NH_4^+ (Koper et al., 2010). If it is known with certainty that the response follows a Michaelis–Menten function, then it is most efficient to use a set of [NH_4^+] that are near K_m and another set of [NH_4^+] that are high enough to produce rates near V_{max}. This scheme will minimize the error around estimates of the two constants. Unfortunately, if the nitrifier community contains NH_4^+-sensitive nitrifiers, the response to NH_4^+ may instead follow a Haldane function that accounts for the suppressed nitrification rates at high [NH_4^+]. As a result, it is usually necessary to distribute the [NH_4^+] relatively evenly across a concentration range from below

Figure 15.4 Nitrification potential and kinetic assay setup. A wide mouth 250 mL flask with a two-hole stopper fitted with Tygon tubing (1 cm i.d. × 20 cm length) that may be attached to a 10-mL pipette is shown. The tubing allows collection of slurry samples without interrupting the shaking, while the second hole allowed air exchange.

10 μM to 1 or 10 mM [NH_4^+]. Because ambient soil NH_4^+ may be substantial, to lower the [NH_4^+], it may be necessary to wash the soil first with the phosphate buffer (followed by centrifugation). During the 24-h assay, NH_4^+ will be consumed by nitrification and microbial assimilation, and it will be produced by N-mineralizing microbes. As a result, NH_4^+ may either increase or decrease during the assay. If the concentration change is not too great, a time-weighted average may be used to calculate a single NH_4^+ solution concentration to which the nitrification rate can be related. Because linearizations of the Michaelis–Menten equation result in large biases in estimates of Michaelis–Menten constants (Robinson, 1985), the equation should be fit to NH_4^+ and nitrification rate data using nonlinear regression and numerical approximation. Fortunately, a number of statistical software packages are available that can perform this function (i.e., SigmaPlot® 11 with Enzyme Kinetics Module). Another important consideration when using the nitrification potential assay to determine Michaelis–Menten constants is that as NH_4^+ is lowered, inhibition of assimilatory NO_3^- reductase enzymes may be eliminated. As a result, at low NH_4^+ concentrations, net nitrification may underestimate gross nitrification and K_m will be overestimated. To prevent this artifact,

$^{15}NO_3^-$ isotope dilution (see above) should be used with the slurries to estimate gross nitrification rates.

8. Nitrifier Population Size

Earlier efforts at determination of the number of microbial cells accomplishing the process of nitrification have been estimated from a combination of most probable number determinations, immunofluorescent cell counts, and nitrification potentials (Belser and Mays, 1982; Prosser and Embley, 2002; Schmidt and Belser, 1994). Culture-based methods are still required as the basis for pure culture isolation and physiological investigations. However, recent advances in molecular methods targeting the genes encoding a key enzyme, ammonia monooxygenase, with quantitative real-time PCR are now the preferred method for quantifying both bacterial and archaeal ammonia oxidizers from DNA extracted from soil (Junier et al., 2010; Leininger et al., 2006; Prosser and Nicol, 2008). Similar methods are available for nitrite oxidizing bacteria (Bartosch et al., 2002; Poly et al., 2008; Spieck et al., 2006) but these are less developed and less practiced due to the concept that ammonia oxidation is the rate limiting step of nitrification. While technical questions remain concerning DNA extraction efficiency, primer design and selectivity, and inclusion of diverse ammonia oxidizers, molecular methods are generally considered a vast improvement over the culture-based approaches. Using the *amoA* gene as the target, both archaeal and bacterial gene copy numbers per unit soil may be estimated using the real-time PCR techniques calibrated with *amoA* genes of the respective groups. The ratio of archaeal to bacterial *amoA* may be determined. These ratios' values are extremely variable across agricultural and wildland ecosystems (Adair and Schwartz, 2008; Boyle-Yarwood et al., 2008; Hansel et al., 2008; Jia and Conrad, 2009; Leininger et al., 2006; Mertens et al., 2009; Nicol et al., 2008; Shen et al., 2008), and no definitive patterns have emerged unequivocally. The presence and abundance of *amoA* genes could be related to the activity of the different groups if selective inhibitors were available. Techniques to differentiate the flux of N through the different types of ammonia oxidizer prokaryotes in their environmental setting are in development in several laboratories.

9. Modeling Approaches

Recent assessments of the treatment of nitrification in the US, Canadian, and European nitrogen dynamic models are available (Grant, 2001; Ma and Shaffer, 2001; McGechan and Wu, 2001). The treatment of nitrification varies from extremely mechanistic, incorporating both ammonia and nitrite oxidizer growth and transformations (Grant, 1994; Maggi et al., 2008; Metivier et al., 2009), to cursory treatment that does not differentiate NH_4^+ from NO_3^- within the monthly time step (CENTURY; Parton et al., 1988). The desire to model soil N_2O fluxes (Del Grosso et al., 2005) has increased the interest in modeling nitrification. The most common treatment is to consider nitrification rates to be first order with respect to ammonium availability and to use laboratory values as starting points for the first-order rate constant parameterization. Rates are typically scaled by environmental factors of temperature, moisture (including low O_2 effects above field capacity), and pH (Grant, 2001; Ma and Shaffer, 2001; McGechan and Wu, 2001). Modeling of the soil N cycle is discussed in detail by Cabrera et al. (2008).

10. Future Advancements in Nitrification Rate Research in Terrestrial Environments

The measurement of nitrification rates and kinetics will be extended to both extremes of the spatial scale for terrestrial environments. The use of microprobes, isotopomers, and elegant fluorescent *in situ* hybridization (FISH) techniques will reveal activity at the single cell level within complex communities and biofilms. At the same time, we need to improve our ability to predict and manage nitrification across agricultural ecosystems that are a primary source of nitrogenous trace gases leading to global climate change. Understanding the role of functional redundancy in the nitrifying organisms and how this relates to the controlling factors on nitrification rates in the environment remains a challenging goal.

REFERENCES

Adair, K. L., and Schwartz, E. (2008). Evidence that ammonia-oxidizing archaea are more abundant than ammonia-oxidizing bacteria in semiarid soils of northern Arizona, USA. *Microb. Ecol.* **56,** 420–426.

Arp, D. J., Sayavedra-Soto, L. A., and Hommes, N. G. (2002). Molecular biology and biochemistry of ammonia oxidation by *Nitrosomonas europaea*. *Arch. Microbiol.* **178,** 250–255.

Avrahami, S., and Conrad, R. (2005). Cold-temperate climate: A factor for selection of ammonia oxidizers in upland soil? *Can. J. Microbiol.* **51,** 709–714.

Avrahami, S., Liesack, W., and Conrad, R. (2003). Effects of temperature and fertilizer on activity and community structure of soil ammonia oxidizers. *Environ. Microbiol.* **5,** 691–705.

Bartosch, S., Hartwig, C., Spieck, E., and Bock, E. (2002). Immunological detection of nitrospira-like bacteria in various soils. *Microb. Ecol.* **43,** 26–33.

Bedard, C., and Knowles, R. (1989). Physiology, biochemistry and specific inhibitors of CH4, NH4+, and CO, oxidation by methanotrophs and nitrifiers. *Microbiol. Rev.* **53**, 68–84.

Belser, L. W., and Mays, E. L. (1980). Specific Inhibition of nitrite oxidation by chlorate and its use in assessing nitrification in soils and sediments. *Appl. Environ. Microbiol.* **39**, 505–510.

Belser, L. W., and Mays, E. L. (1982). Use of nitrifier activity measurements to estimate the efficiency of viable nitrifier counts in soils and sediments. *Appl. Environ. Microbiol.* **43**, 945–948.

Berg, P., Klemedtsson, L., and Rosswall, T. (1982). Inhibitory effect of low partial pressure of acetylene on nitrification. *Soil Biol. Biochem.* **14**, 301–303.

Bock, E., Schmidt, I., Stuven, R., and Zart, D. (1995). Nitrogen loss caused by denitrifying *Nitrosomonas* cells using ammonium or hydrogen as electron donors and nitrite as electron acceptor. *Arch. Microbiol.* **163**, 16–20.

Bollmann, A., and Laanbroek, H. J. (2001). Continuous culture enrichments of ammonia-oxidizing bacteria at low ammonium concentrations. *FEMS Microbiol. Ecol.* **37**, 211–221.

Bollmann, A., Bar-Gilissen, M. J., and Laanbroek, H. J. (2002). Growth at low ammonium concentrations and starvation response as potential factors involved in niche differentiation among ammonia-oxidizing bacteria. *Appl. Environ. Microbiol.* **68**, 4751–4757.

Booth, M. S., Stark, J. M., and Rastetter, E. (2005). Controls on nitrogen cycling in terrestrial ecosystems: A synthetic analysis of literature data. *Ecol. Monogr.* **75**, 139–157.

Boyle-Yarwood, S. A., Bottomley, P. J., and Myrold, D. D. (2008). Community composition of ammonia-oxidizing bacteria and archaea in soils under stands of red alder and Douglas fir in Oregon. *Environ. Microbiol.* **10**, 2956–2965.

Brochier-Armanet, C., Boussau, B., Gribaldo, S., and Forterre, P. (2008). Mesophilic crenarchaeota: Proposal for a third archaeal phylum, the Thaumarchaeota. *Nat. Rev. Microbiol.* **6**, 245–252.

Bundy, L. G., and Meisinger, J. J. (1994). Nitrogen availability indices. *In* "Methods of Soil Analysis. Part 2. Microbiological and Biochemical Properties," (R. W. Weaver, S. Angle, P. Bottomly, D. Bezdicek, S. Smith, A. Tabatabai, and A. Wollum, eds.), pp. 951–984. Soil Science Society America, Madison, WI.

Burton, S. A. Q., and Prosser, J. I. (2001). Autotrophic ammonia oxidation at low pH through urea hydrolysis. *Appl. Environ. Microbiol.* **67**, 2952–2957.

Cabrera, M. L., Molina, J. A., and Vigil, M. (2008). Modeling the nitrogen cycle. *In* "Nitrogen in Agricultural Systems," (J. S. Schepers and W. R. Raun, eds.), pp. 695–730. American Society of Agronomy, Inc.; Crop Science Society of America, Inc.; Soil Science Society of America, Inc., Madison, WI.

Chalk, P. M., Victoria, R. L., Muraoka, T., and Piccolo, M. C. (1990). Effect of a nitrification inhibitor on immobilization and mineralization of soil and fertilizer nitrogen. *Soil Biol. Biochem.* **22**, 533–538.

Cookson, W. R., Cornforth, I. S., and Rowarth, J. S. (2002). Winter soil temperature (2-15 degrees C) effects on nitrogen transformations in clover green manure amended or unamended soils; a laboratory and field study. *Soil Biol. Biochem.* **34**, 1401–1415.

Davidson, E. A., Hart, S. C., Shanks, C. A., and Firestone, M. K. (1991). Measuring gross nitrogen mineralization, immobilization and nitrification by N-15 isotopic pool dilution in intact soil cores. *J. Soil Sci.* **42**, 335–349.

De Boer, W., and Kowalchuk, G. A. (2001). Nitrification in acid soils: Micro-organisms and mechanisms. *Soil Biol. Biochem.* **33**, 853–866.

DeBoer, W., and Laanbroek, H. J. (1989). Ureolytic nitrification at low pH by *Nitrosospira* spec. *Arch. Microbiol.* **152**, 178–181.

DeBoer, W., Gunneweik, P., Kester, R. A., Tietema, A., and Laanbroek, H. J. (1993). The effect of acetylene on N-transformations in an acid oak-beech soil. *Plant Soil* **149**, 292–296.

Del Grosso, S. J., Mosier, A. R., Parton, W. J., and Ojima, D. S. (2005). DAYCENT model analysis of past and contemporary soil N2O and net greenhouse gas flux for major crops in the USA. *Soil Tillage Res.* **83,** 9–24.

Farquharson, R., and Baldock, J. (2008). Concepts in modelling N2O emissions from land use. *Plant Soil* **309,** 147–167.

Fierer, N., Carney, K. M., Horner-Devine, M. C., and Megonigal, J. P. (2009). The biogeography of ammonia-oxidizing bacterial communities in soil. *Microb. Ecol.* **58,** 435–445.

Fortuna, A., Harwood, R. R., Robertson, G. P., Fisk, J. W., and Paul, E. A. (2003). Seasonal changes in nitrification potential associated with application of N fertilizer and compost in maize systems of southwest Michigan. *Agric. Ecosyst. Environ.* **97,** 285–293.

Fuller, M. E., and Scow, K. M. (1996). Effects of toluene on microbially-mediated processes involved in the soil nitrogen cycle. *Microb. Ecol.* **32,** 171–184.

Godde, M., and Conrad, R. (2000). Influence of soil properties on the turnover of nitric oxide and nitrous oxide by nitrification and denitrification at constant temperature and moisture. *Biol. Fertil. Soils* **32,** 120–128.

Grant, R. F. (1994). Simulation of ecological controls on nitrification. *Soil Biol. Biochem.* **26,** 305–315.

Grant, R. F. (2001). A review of the Canadian ecosystem model-ecosys. *In* "Modeling Carbon and Nitrogen Dynamics for Soil Management," (M. J. Shaffer, L. Ma, and S. K. Hansen, eds.), pp. 173–264. CRC Press LLC, Boca Raton, FL.

Habteselassie, M. Y., Stark, J. M., Miller, B. E., Thacker, S. G., and Norton, J. M. (2006). Gross nitrogen transformations in an agricultural soil after repeated dairy-waste application. *Soil Sci. Soc. Am. J.* **70,** 1338–1348.

Hansel, C. M., Fendorf, S., Jardine, P. M., and Francis, C. A. (2008). Changes in bacterial and archaeal community structure and functional diversity along a geochemically variable soil profile. *Appl. Environ. Microbiol.* **74,** 1620–1633.

Hart, S. C., and Stark, J. M. (1997). Nitrogen limitation of the microbial biomass in an old-growth forest soil. *Ecoscience* **4,** 91–98.

Hart, S. C., Stark, J. M., Davidson, E. A., and Firestone, M. K. (1994). Nitrogen mineralization, immobilization and nitrification. *In* "Methods of Soil Analysis. Part 2, Microbiological and Biochemical Properties," (R. W. Weaver, S. Angle, P. Bottomly, D. Bezdicek, S. Smith, A. Tabatabai, and A. Wollum, eds.), pp. 985–1018. Soil Science Society America, Madison, WI.

Hayatsu, M., Tago, K., and Saito, M. (2008). Various players in the nitrogen cycle: Diversity and functions of the microorganisms involved in nitrification and denitrification. *Soil Sci. Plant Nutr.* **54,** 33–45.

Hooker, T. D., and Stark, J. M. (2008). Soil C and N cycling in three semiarid vegetation types: Response to an in situ pulse of plant detritus. *Soil Biol. Biochem.* **40,** 2678–2685.

Hooper, A. B., Vannelli, T., Bergmann, D. J., and Arciero, D. M. (1997). Enzymology of the oxidation of ammonia to nitrite by bacteria. *Antonie Van Leeuwenhoek Int. J. Gen. Mol. Microbiol.* **71,** 59–67.

Hutchinson, G. L., and Davidson, E. A. (1993). Processes for production and consumption of gaseous nitrogen oxides in soil. *In* "Agricultural Ecosystem Effects on Trace Gases and Global Climate Change," (D. E. Rolston, ed.), pp. 79–93. ASA, CSSA, SSSA, Madison, WI.

Hyman, M. R., and Arp, D. J. (1992). $^{14}C_2H_2$-labeling and $^{14}CO_2$-labeling studies of the denovo synthesis of polypeptides by *Nitrosomonas europaea* during recovery from acetylene and light inactivation of ammoonia monooxygenase. *J. Biol. Chem.* **267,** 1534–1545.

Islam, A., Chen, D., and White, R. E. (2007). Heterotrophic and autotrophic nitrification in two acid pasture soils. *Soil Biol. Biochem.* **39,** 972–975.

Jetten, M. S. M., Strous, M., van de Pas-Schoonen, K. T., Schalk, J., van Dongen, U., van de Graaf, A. A., Logemann, S., Muyzer, G., van Loosdrecht, M. C. M., and Kuenen, J. G. (1998). The anaerobic oxidation of ammonium. *FEMS Microbiol. Rev.* **22,** 421–437.

Jetten, M. S. M., van Niftrik, L., Strous, M., Kartal, B., Keltjens, J. T., and Op den Camp, H. J. M. (2009). Biochemistry and molecular biology of anammox bacteria. *Crit. Rev. Biochem. Mol. Biol.* **44,** 65–84.

Jia, Z. J., and Conrad, R. (2009). Bacteria rather than Archaea dominate microbial ammonia oxidation in an agricultural soil. *Environ. Microbiol.* **11,** 1658–1671.

Junier, P., Molina, V., Dorador, C., Hadas, O., Kim, O. S., Junier, T., Witzel, K. P., and Imhoff, J. F. (2010). Phylogenetic and functional marker genes to study ammonia-oxidizing microorganisms (AOM) in the environment. *Appl. Microbiol. Biotechnol.* **85,** 425–440.

Kester, R. A., DeBoer, W., and Laanbroek, H. J. (1996). Short exposure to acetylene to distinguish between nitrifier and denitrifier nitrous oxide production in soil and sediment samples. *FEMS Microbiol. Ecol.* **20,** 111–120.

Killham, K. (1990). Nitrification in coniferous forest soils. *Plant Soil* **128,** 31–44.

Kirkham, D., and Bartholomew, W. V. (1954). Equations for following nutrient transformations in soil utilizing tracer data. *Soil Sci. Soc. Am. Proc.* **18,** 33–34.

Klemedtsson, L. K., and Mosier, A. R. (1994). Effect of long-term exposure of soil to acetylene on nitrification, denitrification and acetylene consumption. *Biol. Fertil. Soils* **18,** 42–48.

Konneke, M., Bernhard, A. E., de la Torre, J. R., Walker, C. B., Waterbury, J. B., and Stahl, D. A. (2005). Isolation of an autotrophic ammonia-oxidizing marine archaeon. *Nature* **437,** 543–546.

Koops, H. P., Bottcher, B., Moller, U. C., Pommerening-Roser, A., and Stehr, G. (1991). Classification of eight new species of ammonia-oxidizing bacteria: *Nitrosomonas communis* sp nov., *Nitrosomonas ureae* sp nov., *Nitrosomonas aestuarii* sp nov., *Nitrosomonas marina* sp nov., *Nitrosomonas nitrosa* sp nov., *Nitrosomonas oligotropha* sp nov., *Nitrosomonas halophila* sp nov. *J. Gen. Microbiol.* **137,** 1689–1699.

Koper, T. E., Habteselassie, M. Y., Stark, J. M., and Norton, J. M. (2010). Nitrification exhibits Haldane kinetics in an agricultural soil treated with ammonium sulfate or dairy waste compost. *FEMS Microbiol. Ecol.* **74,** 316–322.

Laanbroek, H. J., and Gerards, S. (1993). Competition for limiting amounts of oxygen between *Nitrosomonas europaea* and *Nitrobacter winogradsky* grown in mixed continuous cultures. *Arch. Microbiol.* **159,** 453–459.

Leininger, S., Urich, T., Schloter, M., Schwark, L., Qi, J., Nicol, G. W., Prosser, J. I., Schuster, S. C., and Schleper, C. (2006). Archaea predominate among ammonia-oxidizing prokaryotes in soils. *Nature* **442,** 806–809.

Low, A. P., Stark, J. M., and Dudley, L. M. (1997). Effects of soil osmotic potential on nitrification, ammonification, N-assimilation, and nitrous oxide production. *Soil Sci.* **162,** 16–27.

Ma, L., and Shaffer, M. J. (2001). A review of carbon and nitrogen cycling in nine U.S. soil nitrogen dynamic models. *In* "Modeling Carbon and Nitrogen Dynamics for Soil Management," (M. J. Shaffer, ed.)pp. 55–102. Lewis Publishers, Boca Raton, FL.

Maggi, F., Gu, C., Riley, W. J., Hornberger, G. M., Venterea, R. T., Xu, T., Spycher, N., Steefel, C., Miller, N. L., and Oldenburg, C. M. (2008). A mechanistic treatment of the dominant soil nitrogen cycling processes: Model development, testing, and application. *J. Geophys. Res. Biogeosci.* **113,** G02016. doi:10.1029/2007JG000578.

Martens-Habbena, W., Berube, P. M., Urakawa, H., de la Torre, J. R., and Stahl, D. A. (2009). Ammonia oxidation kinetics determine niche separation of nitrifying Archaea and Bacteria. *Nature* **461,** 976–979.

Mary, B., Recous, S., and Robin, D. (1998). A model for calculating nitrogen fluxes in soil using N-15 tracing. *Soil Biol. Biochem.* **30,** 1963–1979.

McCarty, G. W., and Bremner, J. M. (1990). Persistence of effects of nitrification inhibitors added to soils. *Commun. Soil Sci. Plant Anal.* **21,** 639–648.

McGechan, M. B., and Wu, L. (2001). A review of carbon and nitrogen processes in European soil nitrogen dynamic models. *In* "Modeling Carbon and Nitrogen Dynamics for Soil Management," (M. J. Shaffer, L. Ma, and S. K. Hansen, eds.), pp. 103–167. CRC Press LLC, Boca Raton, FL.

Mendum, T. A., Sockett, R. E., and Hirsch, P. R. (1999). Use of molecular and isotopic techniques to monitor the response of autotrophic ammonia-oxidizing populations of the beta subdivision of the class Proteobacteria in arable soils to nitrogen fertilizer. *Appl. Environ. Microbiol.* **65,** 4155–4162.

Mertens, J., Broos, K., Wakelin, S. A., Kowalchuk, G. A., Springael, D., and Smolders, E. (2009). Bacteria, not archaea, restore nitrification in a zinc-contaminated soil. *ISME J.* **3,** 916–923.

Metivier, K. A., Pattey, E., and Grant, R. F. (2009). Using the ecosys mathematical model to simulate temporal variability of nitrous oxide emissions from a fertilized agricultural soil. *Soil Biol. Biochem.* **41,** 2370–2386.

Murphy, D. V., Recous, S., Stockdale, E. A., Fillery, I. R. P , Jensen, L. S., Hatch, D. J., and Goulding, K. W. T. (2003). Gross nitrogen fluxes in soil: Theory, measurement and application of N-15 pool dilution techniques. *Adv. Agron.* **79,** 69–118.

Nicol, G. W., Leininger, S., Schleper, C., and Prosser, J. I. (2008). The influence of soil pH on the diversity, abundance and transcriptional activity of ammonia oxidizing archaea and bacteria. *Environ. Microbiol.* **10,** 2966–2978.

Parton, W. J., Stewart, J. W. B., and Cole, C. V. (1988). Dynamics of C, N, P and S in grassland soils—A model. *Biogeochemistry* **5,** 109–131.

Penton, C. R., Devol, A. H., and Tiedje, J. M. (2006). Molecular evidence for the broad distribution of anaerobic ammonium-oxidizing bacteria in freshwater and marine sediments. *Appl. Environ. Microbiol.* **72,** 6829–6832.

Poly, F., Wertz, S., Brothier, E., and Degrange, V. (2008). First exploration of Nitrobacter diversity in soils by a PCR cloning-sequencing approach targeting functional gene nxrA. *FEMS Microbiol. Ecol.* **63,** 132–140.

Prosser, J. I. (1989). Autotrophic nitrification in bacteria. *Adv. Microb. Physiol.* **30,** 125–181.

Prosser, J. I., and Embley, T. M. (2002). Cultivation-based and molecular approaches to characterisation of terrestrial and aquatic nitrifiers. *Antonie Van Leeuwenhoek Int. J. Gen. Mol. Microbiol.* **81,** 165–179.

Prosser, J. I., and Nicol, G. W. (2008). Relative contributions of archaea and bacteria to aerobic ammonia oxidation in the environment. *Environ. Microbiol.* **10,** 2931–2941.

Robertson, G. P. (1989). Nitrification and denitrification in humid tropical systems: Potential controls on nitrogen retention. *In* "Mineral nutrients in tropical forest and savanna ecosystems," (J. Proctor, ed.), pp. 55–69. Blackwell Scientific Publications, Oxford.

Robinson, J. A. (1985). Determining microbial kinetic parameters using nonlinear regression analysis. *In* "Advances in Microbial Ecology," (K. C. Marshall, ed.), pp. 61–114. Plenum Press, New York, NY.

Saby, B. R. (1969). Influence of moisture tension on nitrate accumulation in soils. *Soil Sci. Soc. Amer. Proc.* **33,** 263–266.

Schimel, J. P., Firestone, M. K., and Killham, K. S. (1984). Identification of heterotrophic nitrification in a Sierran forest soil. *Appl. Environ. Microbiol.* **48,** 802–806.

Schjønning, P., Thomsen, I. K., Moldrup, P., and Christensen, B. T. (2003). Linking soil microbial activity to water- and air-phase contents and diffusivities. *Soil Sci. Soc. Am. J.* **67**, 156–165.

Schlesinger, W. H. (2009). On the fate of anthropogenic nitrogen. *Proc. Natl. Acad. Sci. USA* **106**, 203–208.

Schmidt, E. L. (1982). Nitrification in soil. *In* "Nitrogen in Agricultural Soils," (F. J. Stevenson, ed.), pp. 253–288. American Society of Agronomy, Madison, WI.

Schmidt, E. L., and Belser, L. W. (1994). Autotrophic nitrifying bacteria. *In* "Methods of Soil Analysis. Part 2, Microbiological and Biochemical Properties," (R. Weaver, S. Angle, P. J. Bottomley, D. Bezdicek, S. J. Smith, A. Tabatabai, and A. Wollum, eds.), pp. 159–177. Soil Science Society America, Madison, WI.

Schmidt, I., Hermelink, C., de Pas-Schoonen, K., Strous, M., den Camp, H. J. O., Kuenen, J. G., and Jetten, M. S. M. (2002). Anaerobic ammonia oxidation in the presence of nitrogen oxides (NOx) by two different lithotrophs. *Appl. Environ. Microbiol.* **68**, 5351–5357.

Sexstone, A. J., Revsbech, N. P., Parkin, T. B., and Tiedje, J. M. (1985). Direct measurement of oxygen profiles and denitrification rates in soil aggregates. *Soil Sci. Soc. Am. J.* **49**, 645–651.

Shaw, L. J., Nicol, G. W., Smith, Z., Fear, J., Prosser, J. I., and Baggs, E. M. (2006). *Nitrosospira* spp. can produce nitrous oxide via a nitrifier denitrification pathway. *Environ. Microbiol.* **8**, 214–222.

Shen, J. P., Zhang, L. M., Zhu, Y. G., Zhang, J. B., and He, J. Z. (2008). Abundance and composition of ammonia-oxidizing bacteria and ammonia-oxidizing archaea communities of an alkaline sandy loam. *Environ. Microbiol.* **10**, 1601–1611.

Shi, W., and Norton, J. M. (2000a). Effect of long-term, biennial, fall-applied anhydrous ammonia and nitrapyrin on soil nitrification. *Soil Sci. Soc. Am. J.* **64**, 228–234.

Shi, W., and Norton, J. M. (2000b). Microbial control of nitrate concentrations in an agricultural soil treated with dairy waste compost or ammonium fertilizer. *Soil Biol. Biochem.* **32**, 1453–1457.

Singh, B., Singh, B., and Singh, Y. (1993). Potential and kinetics of nitrification in soils from semiarid regions of northwestern India. *Arid Soil Res. Rehabil.* **7**, 39–50.

Sorokin, D. Y. (1998). On the possibility of nitrification in extremely alkaline soda biotopes. *Microbiology* **67**, 335–339.

Sorokin, D., Tourova, T., Schmid, M. C., Wagner, M., Koops, H. P., Kuenen, J. G., and Jetten, M. (2001). Isolation and properties of obligately chemolithoautotrophic and extremely alkali-tolerant ammonia-oxidizing bacteria from Mongolian soda lakes. *Arch. Microbiol.* **176**, 170–177.

Spang, A., Hatzenpichler, R., Brochier-Armanet, C., Rattei, T., Tischler, P., Spieck, E., Streit, W., Stahl, D. A., Wagner, M., and Schleper, C. (2010). Distinct gene set in two different lineages of ammonia-oxidizing archaea supports the phylum Thaumarchaeota. *Trends Microbiol.* **18**, 331–340.

Spieck, E., Hartwig, C., McCormack, I., Maixner, F., Wagner, M., Lipski, A., and Daims, H. (2006). Selective enrichment and molecular characterization of a previously uncultured Nitrospira-like bacterium from activated sludge. *Environ. Microbiol.* **8**, 405–415.

Stark, J. M. (1996a). Modeling the temperature response of nitrification. *Biogeochemistry* **35**, 433–445.

Stark, J. M. (1996b). Shaker speeds for aerobic soil slurry incubations. *Commun. Soil Sci. Plant Anal.* **27**, 2625–2631.

Stark, J. M. (2000). Nutrient transformations. *In* "Methods in Ecosystem Science," (O. E. Sala, R. B. Jackson, H. A. Mooney, and R. W. Howarth, eds.), pp. 215–234. Springer-Verlag, New York.

Stark, J. M., and Firestone, M. K. (1995). Mechanisms for soil-moisture effects on activity of nitrifying bacteria. *Appl. Environ. Microbiol.* **61**, 218–221.

Stark, J. M., and Firestone, M. K. (1996). Kinetic characteristics of ammonium-oxidizer communities in a California oak woodland-annual grassland. *Soil Biol. Biochem.* **28**, 1307–1317.

Stark, J. M., and Hart, S. C. (1996). Diffusion technique for preparing salt solutions, Kjeldahl digests, and persulfate digests for nitrogen-15 analysis. *Soil Sci. Soc. Am. J.* **60**, 1846–1855.

Stark, J. M., and Hart, S. C. (1997). High rates of nitrification and nitrate turnover in undisturbed coniferous forests. *Nature* **385**, 61–64.

Stein, L. Y. (2010). Heterotrophic nitrification and nitrifier denitrification. *In* "Nitrification," (B. B. Ward, D. J. Arp, and M. G. Klotz, eds.), pp. 95–114. ASM Press, Washington.

Stein, L. Y., and Yung, Y. L. (2003). Production, isotopic composition, and atmospheric fate of biologically produced nitrous oxide. *Annu. Rev. Earth Planet. Sci.* **31**, 329–356.

Stienstra, A. W., Gunnewiek, P. K., and Laanbroek, H. J. (1994). Repression of nitrification in soils under a climax grassland vegetation. *FEMS Microbiol. Ecol.* **14**, 45–52.

Stockdale, E. A., Hatch, D. J., Murphy, D. V., Ledgard, S. F., and Watson, C. J. (2002). Verifying the nitrification to immobilisation ratio (N/I) as a key determinant of potential nitrate loss in grassland and arable soils. *Agronomie* **22**, 831–838.

Tourna, M., Freitag, T. E., and Prosser, J. I. (2010). Stable isotope probing analysis of interactions between ammonia oxidizers. *Appl. Environ. Microbiol.* **76**, 2468–2477.

Trap, J., Bureau, F., Vinceslas-Akpa, M., Chevalier, R., and Aubert, M. (2009). Changes in soil N mineralization and nitrification pathways along a mixed forest chronosequence. *For. Ecol. Manage.* **258**, 1284–1292.

Treusch, A. H., Leininger, S., Kletzin, A., Schuster, S. C., Klenk, H. P., and Schleper, C. (2005). Novel genes for nitrite reductase and Amo-related proteins indicate a role of uncultivated mesophilic crenarchaeota in nitrogen cycling. *Environ. Microbiol.* **7**, 1985–1995.

Venterea, R. T., and Rolston, D. E. (2000a). Mechanisms and kinetics of nitric and nitrous oxide production during nitrification in agricultural soil. *Glob. Change Biol.* **6**, 303–316.

Venterea, R. T., and Rolston, D. E. (2000b). Mechanistic modeling of nitrite accumulation and nitrogen oxide gas emissions during nitrification. *J. Environ. Qual.* **29**, 1741–1751.

Walker, C. B., de la Torre, J. R., Klotz, M. G., Urakawa, H., Pinel, N., Arp, D. J., Brochier-Armanet, C., Chain, P. S. G., Chan, P. P., Gollabgir, A., Hemp, J., Hugler, M., *et al.* (2010). *Nitrosopumilus maritimus* genome reveals unique mechanisms for nitrification and autotrophy in globally distributed marine crenarchaea. *Proc. Natl. Acad. Sci. USA* **107**, 8818–8823.

Weaver, R. W., and Danso, S. K. A. (1994). Dinitrogen fixation. *In* "Methods of Soil Analysis. Part 2, Biochemical and Microbiological Properties," (R. W. Weaver, S. Angle, P. Bottomly, D. Bezdicek, S. Smith, A. Tabatabai, and A. Wollum, eds.), pp. 1019–1045. Soil Science Society of America, Madison, WI.

Webster, G., Embley, T. M., Freitag, T. E., Smith, Z., and Prosser, J. I. (2005). Links between ammonia oxidizer species composition, functional diversity and nitrification kinetics in grassland soils. *Environ. Microbiol.* **7**, 676–684.

Wolt, J. D. (2000). Nitrapyrin behavior in soils and environmental considerations. *J. Environ. Qual.* **29**, 367–379.

CHAPTER SIXTEEN

Protocol for the Measurement of Nitrous Oxide Fluxes from Biological Wastewater Treatment Plants

Kartik Chandran

Contents

1. Introduction	370
2. Sampling Design for Full-Scale Monitoring	370
3. Sampling Procedures: Headspace Gas Measurement	373
4. Sampling Procedures: Measurement of Aqueous N_2O Concentrations	376
5. Sampling Procedures: Measurement of Advective Gas Flow Rate from Bioreactor Headspace	376
6. Principles of Real-Time N_2O Measurement	377
7. Data Analysis: Determination of Fluxes	377
8. Data Analysis: Determination of Emission Fractions	378
9. Data Analysis: Calculation of N_2O Emission Factors	378
10. Standardization of Protocol and Comparison with Established Emissions Flux Measurement Methods	379
11. N_2O Emission Fluxes from Activated Sludge Processes	381
12. Triggers for N_2O Emission from Wastewater Treatment Operations	382
13. Lab-Scale and Field-Scale Adaptation of Protocol N_2O Emission Measurements	383
14. Concluding Remarks	383
Acknowledgment	383
References	384

Abstract

The overarching goal herein was to develop a protocol that could be used to generate consistent information on the generation and emission of nitrous oxide (N_2O) from open-surface wastewater treatment bioreactors. The developed protocol was reviewed and endorsed by the United States Environmental Protection Agency (USEPA), whereupon it was used to determine N_2O emissions

Department of Earth and Environmental Engineering, Columbia University, New York, USA

Methods in Enzymology, Volume 486
ISSN 0076-6879, DOI: 10.1016/S0076-6879(11)86016-7

© 2011 Elsevier Inc.
All rights reserved.

from a wide array of wastewater treatment processes across the United States. Scaled-down variants of the protocol have also since been adopted for lab-scale measurements. The protocol consists of a combination of elements that entail real-time online measurement of headspace N_2O concentrations, supplemented by discrete measurements of liquid-phase N_2O and other routinely monitored wastewater and process parameters. Notably, the advective flow rate of headspace gas is also directly measured.

1. INTRODUCTION

Based on recent field-scale measurements, engineered biological nitrogen removal (BNR) plants, while effective to varying degrees in reducing *aqueous* nitrogen pollution could emit up to 7% of the influent nitrogen load as *gaseous* nitrous oxide (N_2O) and nitric oxide (NO) (Kampschreur *et al.*, 2008). Such emissions are deleterious to the environment. The greenhouse equivalence of N_2O is about 300 times that of carbon dioxide and both N_2O and NO contribute to depletion of the ozone layer (Ravishankara *et al.*, 2009). From a biological perspective, N_2O and NO are known intermediates in heterotrophic denitrification (Knowles, 1982; Zumft, 1997) and autotrophic nitrification and denitrification (Anderson and Levine, 1986; Anderson *et al.*, 1993; Kester *et al.*, 1997; Ritchie and Nicholas, 1972; Stuven *et al.*, 1992). However, the net contribution of processes such as denitrification wastewater treatment plants (WWTPs) to N_2O emissions has only recently been explicitly acknowledged (USEPA, 2009). Additionally, there is a real paucity of systematic protocols that enable collection of N_2O emission fluxes from open-surface activated sludge bioreactors using consistent methodology. The development and application of a detailed protocol to conduct plant-wide measurements of gaseous and aqueous N_2O concentrations is described herein. This protocol is intended to provide utilities and field sampling teams with a detailed description of the data collection methodology and analysis requirements to enable calculation of gaseous nitrogen fluxes from different zones of activated sludge trains in a wastewater treatment facility. The protocol was reviewed and endorsed by the United States Environmental Protection Agency (USEPA) during Fall 2008 and has since been implemented at different WWTPs in North America toward the quantification of N_2O emissions therein as described elsewhere (Ahn *et al.*, 2010a,b).

2. SAMPLING DESIGN FOR FULL-SCALE MONITORING

The N_2O emission fluxes of several BNR and non-BNR WWTPs were measured (Table 16.1). Testing was conducted at each plant during which gas-phase monitoring was performed in real-time continuous mode

Table 16.1 Summary of process schematics sampled

Plant configuration	Description
Separate-stage BNR	The low-rate separate-stage nitrification denitrification process at this WWTP was sampled. The process was configured as a sequence of five reactors in series. The influent to this process consisted of the clarified effluent from an upstream high-rate process, mainly engaged organic carbon removal. The influent was fed in a step-feed fashion to the first two aerobic zones. The last three zones of this process were nonaerated and the second nonaerated zone received methanol to promote denitrification. The effluent channel of this process was aerated prior to secondary clarification.
Four-stage Bardenpho	The four-stage Bardenpho process consisted of predenitrification (without external carbon addition) followed by a primary aerated zone. The effluent of the primary aerated zone was internally recycled to the anoxic zone. Following the primary aerated zone was a deoxygenation zone to scavenge dissolved oxygen, prior to methanol addition for enhanced denitrification. The final zone in this process was aerated primarily for stripping off the dinitrogen gas produced during denitrification, prior to secondary clarification.
Step-feed BNR 1	The four-pass step-feed BNR process sampled consisted of preanoxic zones comprising about 1/3 of the pass volume followed by aerated zones. The transition zone between each pass was nonaerated to facilitate deoxygenation. The approximate influent flow split was 10–40–30–20% to the four passes, respectively. The first pass also received presettled anaerobic digestion centrate, which constituted approximately 30% of the influent TKN load to the process. Return activated sludge was also fed to the first pass.
Step-feed non-BNR	The step-feed non-BNR process sampled was configured and operated in four-pass step-aeration mode. The process was completely covered primarily for odor control. The headspace off-gases were consolidated and fed to a biofilter. The approximate influent flow split was 10–40–30–20% to the four passes, respectively. Return activated sludge was fed to the first pass.
Separate centrate	The separate centrate treatment process was operated to process presettled anaerobic digestion centrate and partially convert the influent NH_4^+–N to NO_2^-–N. The separate centrate treatment process was operated in plug-flow mode. Effluent from the separate centrate tank was fed to the overall plant return activated sludge line for possible bioaugmentation with primarily ammonia oxidizing bacteria (AOB) and for nitrogen removal via the short-cut nitrite pathway similar to that described in (van Dongen *et al.*, 2001).
Plug-flow 1	The first plug-flow process sampled was designed and operated primarily for organic carbon removal and nitrification and did not have dedicated anoxic zones or external organic carbon addition. The process was configured in four-pass mode.

(continued)

Table 16.1 (continued)

Plant configuration	Description
Plug-flow 2	The second plug-flow process sampled was also designed and operated for organic carbon removal and nitrification and did not have dedicated anoxic zones or external organic carbon addition. The process was configured in two-pass mode.
MLE 1	The first modified Lutzack Ettinger (MLE) process sampled was originally designed for operation in enhanced biological phosphorous removal mode, but subsequently operated in MLE mode. The process consisted of predenitrification without external organic carbon addition.
MLE 2	The second modified Lutzack Ettinger (MLE) process sampled was also originally designed for operation in enhanced biological phosphorous removal mode, but subsequently operated in MLE mode. The process consisted of predenitrification without external organic carbon addition.
Step-feed BNR 2	The second step-feed process sampled was configured in four-pass mode. Each pass consisted of preanoxic zones comprising 1/3 of the pass volume followed by aerobic zones. The approximate influent flow split was 50–30–20–0% to the four passes, respectively. The anoxic zones were mixed via low intensity pulse aeration. The return activated sludge was fed to the first pass.
Oxidation ditch	The oxidation ditch process was operated to achieve simultaneous nitrification and denitrification by operation at uniformly low aeration intensities and dissolved oxygen concentrations. The influent flow to the process was fed to the inner loop and was mixed and circulated using surface mixers. No external organic carbon was added to enhance denitrification. Return activated sludge was fed to the inner loop of the process.
Step-feed BNR 3	The third four-pass step-feed BNR process sampled consisted of preanoxic zones comprising about 1/3 of the pass volume followed by aerated zones. The approximate influent flow split was 33.3–33.3–33.3–0% to the four passes, respectively. The first pass also received presettled anaerobic digestion centrate, which constituted approximately 40% of the influent TKN load to the process. Return activated sludge was also fed to the first pass. The reactors of this process were also covered and thus only composite measurements of the overall headspace could be performed.

and liquid-phase sampling was performed via a combination of plant online analyzers (where available) and discrete grab sampling conducted by plant operators and laboratory staff. The wastewater and process analytes sampled and the frequency and location of sampling at a typical WWTP are detailed in Table 16.2.

3. SAMPLING PROCEDURES: HEADSPACE GAS MEASUREMENT

The overall procedure for measuring N_2O, NO, and NO_2 fluxes from the headspace of activated sludge tanks involved a variant of the EPA/600/8-86/008 and the South Coast Air Quality Management District (SCAQMD) tracer methods. This variant was developed to measure those sources that have a relatively high surface flux rate when compared to diffusion (for instance, WWTPs). Commercially available replicas of the US EPA surface emission isolation flux chamber (SEIFC; Figs. 16.1 and 16.2) were used to measure gaseous N fluxes from activated sludge reactors. The SEIFC consisted of a floating enclosed space from which exhaust gas was collected in a real-time or discrete fashion. Since the surface area under the SEIFC could be measured, the specific flux of the gaseous compound of interest could be determined. The SEIFC "floated" on the activated sludge tank surface (Fig. 16.1, right panel) and several replicate measurements could be obtained at different locations in a single tank as well as from different tanks (nitrification, denitrification) along a treatment train. The SEIFC was also equipped with mixing via sweep gas circulation to ensure collection of representative gas-phase concentrations. The SEIFC is currently one of the few devices accepted by the USEPA for measuring gaseous fluxes (Tata et al., 2003).

Sampling was conducted at multiple locations of the activated sludge train in each wastewater treatment facility. These specific locations selected were the geometric center of each demarcated anoxic or aerobic zone in the WWTP, or alternately locations where nitrification could be inferred based on initial screening of NH_4^+-N and DO concentrations (as in the plug-flow processes). For discrete measurement at each of these locations, 30 replicate measurements of gaseous N_2O and one measurement of aqueous N_2O were obtained over a period of 30 min. During continuous measurement at each of these specific locations over a 24-h period, gaseous N_2O concentrations were still measured once per minute, while aqueous N_2O concentrations were measured about six times per day. Independent replication at each location (on different days) was not conducted owing to practical limitations associated with such an extensive campaign.

Table 16.2 Typical wastewater and process measurements conducted in parallel with gas-phase monitoring

Sample location	Analyte													
	TSS	VSS	Total cBOD$_5$	Soluble cBOD$_5$	Total COD	Sol. COD	ff COD	TKN	Sol. TKN	pH	Alk	NH$_3$–N	NO$_3$–N	NO$_2$–N
Primary effluent	8/d	2/d	8/d	8/d	8/d	8/d	8/d	8/d	8/d	8/d	8/d	8/d	8/d	8/d
Secondary effluent	8/d	–	8/d	8/d	8/d	8/d	8/d	8/d	8/d	8/d	8/d	8/d	8/d	8/d
RAS MLSS	8/d													
WAS MLSS	8/d													

Operating data	
Influent flow	Diurnal flow pattern at appropriate time intervals (15 min for periods of rapid diurnal increase, 1 h for stable periods)
RAS flow	Average daily RAS flow, Indicate location and type of flow measurement and variability of flow
WAS flow	Average daily WAS flow, Indicate location and type of flow measurement, times of WAS wasting if not continuous
Dissolved oxygen	1 h^{-1}, indicate location of DO measurement along basin length and time of measurement
Aeration rate	Daily average, indicate location of Air Flow Measurement and variability over the course of the day. SCADA output at short time intervals would be best

In-tank profiles								
TSS	VSS	pH	DO	ORP	Temp.	ff COD	Alk.	NH$_3$–N	NO$_3$–N	NO$_2$–N
8/d	2/d	8/d	8/d	8/d	8/d	8/d	8/d	8/d	8/d	8/d

TSS, total suspended solids; VSS, volatile suspended solids; cBOD$_5$, carbonaceous 5-day biological oxygen demand; COD, chemical oxygen demand; Sol COD, soluble chemical oxygen demand; ffCOD, filtered flocculated chemical oxygen demand (as described by Mamais et al., 1993); TKN, total Kjeldahl nitrogen; Alk, alkalinity; NH$_3$–N, ammoniacal nitrogen; NO$_3$–N, nitrate nitrogen; NO$_2$–N, nitrite nitrogen; RAS, return activated sludge; MLSS, mixed liquor suspended solids; WAS, waste activated sludge; DO, dissolved oxygen; ORP, oxidation–reduction potential.

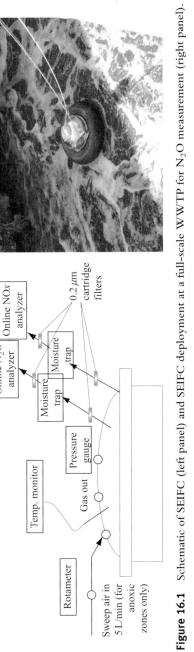

Figure 16.1 Schematic of SEIFC (left panel) and SEIFC deployment at a full-scale WWTP for N$_2$O measurement (right panel).

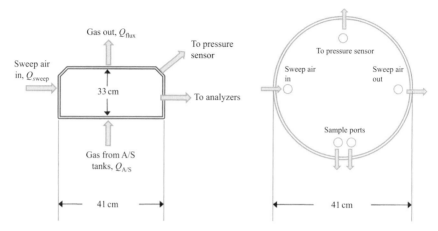

Figure 16.2 Schematic of gas flows in and out of the flux chamber in elevation (left panel) and plan (right panel) views.

4. SAMPLING PROCEDURES: MEASUREMENT OF AQUEOUS N_2O CONCENTRATIONS

Aqueous phase N_2O concentrations were measured using a miniaturized Clark-type sensor with an internal reference and a guard cathode (Unisense, Aarhus, Denmark). The sensor was equipped with an oxygen front guard, which prevented oxygen from interfering with the nitrous oxide measurements. The sensor was coupled to a high-sensitivity picoammeter to convert the current resulting from cathodic reduction of N_2O to an electric signal. Aqueous N_2O measurements were conducted right adjacent to the SEIFC location about six times per day, coincident with other liquid-phase wastewater measurements, which were indicative of the performance of the treatment plant.

5. SAMPLING PROCEDURES: MEASUREMENT OF ADVECTIVE GAS FLOW RATE FROM BIOREACTOR HEADSPACE

One of the most important developments included in this protocol is the explicit measurement of the advective flow of gases through the flux chamber. By measuring the gas flow rate, the actual "flux" of N_2O can be thus computed (as discussed below). Advective flow of gas through the flux chamber ($Q_{emission}$) *in aerated zones* was measured using a modification of American Society for Testing and Materials (ASTM) method D1946. Briefly, a tracer gas consisting of 100,000 ppmv ($C_{helium-tracer}$) He was introduced into the flux chamber at a known flow rate, Q_{tracer} (Eq. (16.1)). He concentrations in the off-gas from the flux chamber ($C_{helium-FC}$) were measured using a field gas-chromatograph equipped with a thermal conductivity detector (GC–TCD). $Q_{emission}$ was computed using Eq. (16.2).

$$Q_{tracer} \times C_{helium-tracer} = (Q_{tracer} + Q_{emission}) \times C_{helium-FC}. \quad (16.1)$$

$$Q_{emission} = \frac{Q_{tracer} \times (C_{helium-tracer} - C_{helium-FC})}{C_{helium-FC}}. \quad (16.2)$$

The only modification to the protocol to measure the emission flow rate from *nonaerated zones* was the introduction of sweep gas (air) or carrier gas through the headspace of the flux chamber at a known flow rate ($Q_{sweep} = 5$ L/min), in addition to the He tracer gas (Eq. (16.3)). The corresponding $Q_{emission}$ was computed using Eq. (16.4). Addition of sweep gas is needed to promote mixing of the SEIFC contents, owing to the low advective gas flow from the anoxic-zone headspace. Sweep-air N_2O concentrations were always measured and typically below the detection limits of the N_2O analyzer.

$$Q_{tracer} \times C_{helium-tracer} = (Q_{tracer} + Q_{sweep} + Q_{emission}) \times C_{helium-FC}. \quad (16.3)$$

$$Q_{emission} = \frac{Q_{tracer} \times (C_{helium-tracer} - C_{helium-FC})}{C_{helium-FC}} - Q_{sweep}. \quad (16.4)$$

During continuous N_2O measurements, $Q_{emission}$ was determined several times a day to match liquid-phase N_2O measurements.

6. Principles of Real-Time N_2O Measurement

Continuous N_2O measurements were performed via infra-red (IR) gas-filter correlation (Teledyne API Model 320E, San Diego, CA), which is based on the absorption of IR radiation by N_2O molecules at wavelengths near 4.5 μm.

7. Data Analysis: Determination of Fluxes

The net flux of gaseous N species (mg/min-m^2) was calculated based on the gas flow rate out of the flux chamber ($Q_{emission}$, L/min), headspace gas concentration (parts per million volume) and the cross-sectional area of the SEIFC (m^2) (Eq. (16.5)).

$$\text{Flux} = \frac{Q_{\text{emission}} \times C}{A}. \qquad (16.5)$$

8. DATA ANALYSIS: DETERMINATION OF EMISSION FRACTIONS

The surface flux calculated from Eq. (16.5) was translated into the flux of a given zone by multiplying with the area of the specific zone in the wastewater treatment reactor, where the measurements were conducted. The N_2O emission fractions (mass/mass) for each WWTP at any given time point were computed by normalizing the measured flux from each zone in the facility to the daily influent total Kjeldahl nitrogen (TKN) loading according to Eq. (16.6). Emission fractions were averaged over the course of the diurnal sampling period and reported as the average (avg.) ± standard deviation (sd.) for each individual process sampled.

During each campaign, wastewater nitrogen species concentrations including influent, bioreactor and effluent TKN, ammonium, nitrite, and nitrate were measured simultaneously at least six times per day according to Standard Methods (Eaton *et al.*, 2005) to supplement the gas-phase measurements. The discrete measurements were averaged to generate the emission fractions described in Eq. (16.6). Additionally, seven out of the twelve processes (Table 16.1) were sampled at minimum and maximum annual wastewater temperatures to examine seasonal temperature impacts on N_2O generation and emission.

$$\text{Emission fraction} = \frac{\sum_{i=1}^{n} \text{Flux}_i \times \text{Area}_i (\text{kg} N_2O - N)}{\text{Daily influent TKN load (kg} - N)}, \qquad (16.6)$$

where Flux_i, N_2O emission flux calculated from the ith zone (kg N_2O–N/m^2 d); Area_i, surface area of the ith zone (m^2); n, number of zones in a given facility from which N_2O fluxes are captured; Daily influent TKN load, average influent load (influent flow rate × influent TKN concentrations) over the course of 24 h.

9. DATA ANALYSIS: CALCULATION OF N_2O EMISSION FACTORS

N_2O emission factors were computed by normalizing the total reactor N_2O mass flux to the unit population equivalent flow rate (100 gal/PE/day for the United States; Tchobanoglous *et al.*, 2003) and were expressed in units consistent with the USEPA inventory report (g N_2O/PE/year) (USEPA, 2009). For aerobic zones, the helium-based advective gas-flow data were correlated to plant-recorded airflow rates for any given zone via linear regression and used to calculate diurnal N_2O emission factors. For anoxic (nonaerated) zones lacking associated plant airflow data, the average of the experimentally obtained helium-based gas flow rates was used to calculate diurnal N_2O emission factors.

10. Standardization of Protocol and Comparison with Established Emissions Flux Measurement Methods

The validity of the measurements using the protocol developed for this study was determined via a parallel sampling effort between two independent teams on September 9 and 10, 2008 at the wastewater treatment facility employing the step-feed BNR process 2 (Table 16.1). The Columbia University–Water Environment Research Foundation team (labeled WERF) used a flux chamber manufactured by St. Croix Sensory and measured N_2O off-gas concentrations via gas-filter correlation, described above. A second team (labeled CES) used an USEPA flux chamber and sampled the off-gas into opaque Tedlar® bags for subsequent Fourier-transform infra-red (FTIR) analysis (NIOSH 6660) by a commercial laboratory (Peak Analytical, Boulder, CO).

The possibility of "biasing" the measured N_2O concentrations by introduction of different sweep gas flow rates was also part of the validation testing. The successive-dilution method employed by the WERF team involved dilution of measured N_2O concentrations by virtue of introducing sweep gas at two different flow rates (4 and 8 L/min). To compare, the CES team employed ASTM method D1946, which involved introducing He tracer at 5 L/min. The equivalence of these two methods was determined by computing the headspace advective flow rate from the nonaerated zones.

Based on these parallel measurements conducted independently, similar results were obtained, with good correspondence in general in both the nitrous oxide fluxes (Fig. 16.3) and off-gas flow-rate (Fig. 16.4) in different zones of the selected activated sludge tank. The equivalence in the flow rates obtained using the two methods (successive dilution and He tracer) also helped to reject the possibility of "biasing" the measured N_2O concentrations due to changing hydrodynamic flow patterns in the headspace of the flux chamber by the introduction of sweep gas. Additionally, the following observations were made based on the results obtained and incorporated into subsequent full-scale measurement campaigns:

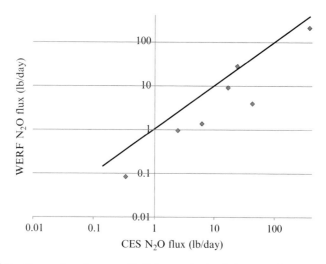

Figure 16.3 Comparison between N_2O fluxes obtained via two independent methods across eight zones of the four-pass step-feed BNR reactor 2.

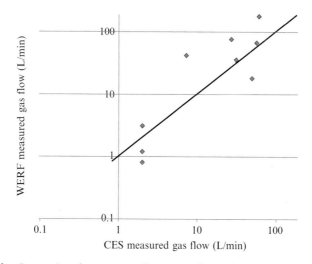

Figure 16.4 Comparison between gas flow rates obtained via the successive dilution (WERF) and tracer gas (CES) methods conducted at the step-feed BNR reactor 2.

a. The use of an inert gas tracer by the CES team was demonstrated to be an appropriate method to determine the advective off-gas flow rate. This was an operationally more facile and reliable method compared to the successive-dilution method developed by the Columbia University–WERF team based on successive dilution of the N_2O concentrations.

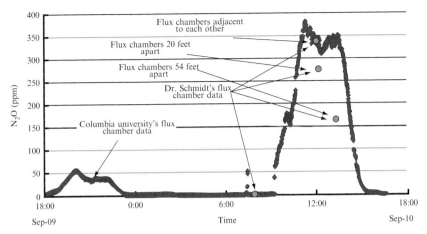

Figure 16.5 Illustration of spatial and temporal variability in N_2O concentrations in the headspace of an aerobic zone that necessitates real-time online monitoring. Columbia University–WERF's flux chamber data given by near continuous blue symbols, CES flux chamber data given by four discrete dots, as marked. Measurements were conducted at the step-feed BNR reactor 2 (courtesy, Dr. Charles E Schmidt).

Therefore, the successive-dilution method was discontinued following the parallel sampling study and replaced with He tracer-based method to determine advective flow rate.

b. Significant spatial and temporal variability in the measured concentrations of headspace N_2O was observed by the Columbia and the CES teams (Fig. 16.5). Therefore, for subsequent full-scale measurements, discrete measurements (initially proposed at a frequency of once a day) of N_2O at different locations in any given WWTP was discontinued. Instead, a significantly more involved sampling strategy that entailed 24 h "real-time online monitoring" of emissions at each location was initiated.

11. N_2O Emission Fluxes from Activated Sludge Processes

A wide range of N_2O emissions was measured across the 12 WWTPs operated at different temperatures, configurations, and influent characteristics. On average, N_2O emission fractions varied from 0.01% to 1.8% or 0.01% to 3.3%, when normalized to influent TKN load or influent TKN load processed, respectively (Ahn et al., 2010b). These emission fractions were on the lower end of the range reported by previous studies, which varied between 0% and 15% of influent TKN load (Czepiel et al., 1995; Kampschreur et al., 2008; Kimochi et al., 1998; Sommer et al., 1998; Sümer et al., 1995; Wicht and Beier, 1995).

Computed flow-normalized emission factors also varied in a wide range, over two orders of magnitude (Ahn et al., 2010b), and were mostly statistically higher (at the $\alpha = 0.05$ confidence level) than currently used values of 3.2 g N_2O/PE/yr (non-BNR processes; Czepiel et al., 1995) or 7.0 g N_2O/PE/yr (BNR processes; USEPA, 2009). A high degree of diurnal variability in emission factors was also observed and could be linked diurnal variations in influent N-loading as reported by Ahn et al. (2010a). Based on the observed variability either diurnally or across the range of WWTPs sampled, the use of a "single" universal emission factor to calculate N_2O emissions from all wastewater treatment processes is inadequate.

In general, N_2O emissions in aerated zones were higher than those in nonaerated zones (Ahn et al., 2010b). Therefore the currently held premise that N_2O emissions from WWTPs mostly occur in the anoxic zones (USEPA, 2009) is not accurate. Possible mechanisms for N_2O emissions via nitrification and denitrification have also been recently published (Lu and Chandran, 2010; Yu and Chandran, 2010; Yu et al., 2010). Both processes likely contributed to the measured N_2O emissions. Good correlation in general was also obtained between liquid-phase and gaseous-phase N_2O concentrations as discussed in Ahn et al. (2010a). However, the due to possible interference with dissolved oxygen and nitric oxide, a high level of confidence could not be placed in the aqueous N_2O concentrations even at lab-scale (Yu et al., 2010). Therefore, it is suggested that aqueous N_2O concentrations be alternately approximated based on estimated system-specific gas-liquid mass transfer coefficients, as described elsewhere (Yu et al., 2010).

12. Triggers for N_2O Emission from Wastewater Treatment Operations

The data obtained during the national survey were subjected to multivariate data mining to identify potential triggers for N_2O emission from wastewater treatment operations, described elsewhere (Ahn et al., 2010b). Based on this data mining approach, the triggers for N_2O emissions from aerobic zones were NH_4^+-N, NO_2^--N, and DO concentrations in isolation and NH_4^+-N and NO_2^--N concentrations in combination (Ahn et al., 2010b). However, high DO and NO_2^--N concentrations were positively correlated with N_2O emissions from anoxic zones (Ahn et al., 2010b). For more details on the plant specific data and data mining, please consult (Ahn et al., 2010b).

13. Lab-Scale and Field-Scale Adaptation of Protocol N$_2$O Emission Measurements

Although the original version of the protocol was developed for full-scale measurements of N$_2$O emissions from operational WWTPs, we have since successfully scaled down the protocol for lab-scale measurements. The scaled-down versions essentially employ smaller versions of the SEIFC to fit lab-scale bioreactors. Gas-phase and liquid-phase measurements are conducted in almost identical fashion to the full-scale version. The only difference is that the sampling frequency of liquid-phase variables, which is tailored depending on the experimental design. Using the scaled-down protocol, the magnitude and some novel mechanisms of N$_2$O emissions from both nitrifying and denitrifying lab-scale bioreactors have been recently reported (Lu and Chandran, 2010; Yu and Chandran, 2010; Yu et al., 2010).

The protocol has also been shared with research groups in Belgium, Portugal, and Spain to facilitate similar full-scale N$_2$O measurement campaigns. Efforts at training wastewater treatment operators in the implementation of the protocol at several additional plants in the United States are also underway.

14. Concluding Remarks

A protocol to measure N$_2$O emission fluxes from WWTPs was developed. This protocol represents the first systematic attempt to develop a consistent methodology for the measurement of such emissions. As the focus of the wastewater industry shifts increasingly toward environmentally sustainable treatment, the measurement of the overall greenhouse gas footprint of such treatment processes becomes more relevant. Broad application of this protocol will thus enable WWTPs to quantify their N$_2$O emissions and engineer approaches that are aimed at minimizing both aqueous and gaseous nitrogen pollution.

ACKNOWLEDGMENT

This protocol was developed as part of a larger study to develop a national database of N$_2$O emissions from WWTP, funded by the Water Environment Research Foundation (WERF) Climate Change Program.

REFERENCES

Ahn, J.-H., Kim, S., Pagilla, K., Katehis, D., and Chandran, K. (2010a). Spatial and Temporal Variability in Atmospheric Nitrous Oxide Generation and Emission from Full-Scale Biological Nitrogen Removal and Non-BNR Processes. *Water Environ. Res.* **82**(10), DOI: 10.2175/106143010X12681059116897.

Ahn, J. H., Kim, S., Park, H., Rahm, B., Pagilla, K., and Chandran, K. (2010b). N_2O Emissions from activated sludge processes, 2008–2009: Results of a national monitoring survey in the United States. *Environ. Sci. Technol.* **44**, 4505–4511.

Anderson, I. C., and Levine, J. S. (1986). Relative rates of nitric oxide and nitrous oxide production by nitrifiers, denitrifiers and nitrate respirers. *Appl. Environ. Microbiol.* **51**, 938–945.

Anderson, I. C., Poth, M., Homstead, J., and Burdige, D. (1993). A comparison of NO and N_2O production by the autotrophic nitrifier *Nitrosomonas europaea* and the heterotrophic nitrifier *Alcaligenes faecalis*. *Appl. Environ. Microbiol.* **59**, 3525–3533.

Czepiel, P., Crill, P., and Harriss, R. (1995). Nitrous oxide emissions from municipal wastewater treatment. *Environ. Sci. Technol.* **29**, 2352–2356.

Eaton, A. D., Clesceri, L. S., and Greenberg, A. E. (eds.), (2005). Standard Methods for the Examination of Water and Wastewater, APHA, AWWA, and WEF, Washington DC.

Kampschreur, M. J., van der Star, W. R. L., Wielders, H. A., Mulder, J. W., Jetten, M. S. M., and van Loosdrecht, M. C. M. (2008). Dynamics of nitric oxide and nitrous oxide emission during full-scale reject water treatment. *Water Res.* **42**, 812–826.

Kester, R. A., de Boer, W., and Laanbroek, H. J. (1997). Production of NO and N_2O by pure cultures of nitrifying and denitrifying bacteria during changes in aeration. *Appl. Environ. Microbiol.* **63**, 3872–3877.

Kimochi, Y., Inamori, Y., Mizuochi, M., Xu, K.-Q., and Matsumura, M. (1998). Nitrogen removal and N_2O emission in a full-scale domestic wastewater treatment plant with intermittent aeration. *J. Ferment. Bioeng.* **86**, 202–206.

Knowles, R. (1982). Denitrification. *Microbiol. Rev.* **46**, 43–70.

Lu, H., and Chandran, K. (2010). Factors promoting emissions of nitrous oxide and nitric oxide from denitrifying sequencing batch reactors operated with methanol and ethanol as electron donors. *Biotechnol. Bioeng.* **106**, 390–398.

Mamais, D., Jenkins, D., and Pitt, P. (1993). A rapid physical-chemical for the determination of readily biodegradable soluble COD in municipal wastewater. *Water Res.* **27**, 195.

Ravishankara, A. R., Daniel, J. S., and Portmann, R. W. (2009). Nitrous Oxide (N_2O): The dominant ozone-depleting substance emitted in the 21st century. *Science* **326**, 123–125.

Ritchie, G. A. F., and Nicholas, D. J. D. (1972). Identification of the sources of nitrous oxide produced by oxidative and reductive processes in *Nitrosomonas europaea*. *Biochem. J.* **126**, 1181–1191.

Sommer, J., Ciplak, A., Sumer, E., Benckiser, G., and Ottow, J. C. G. (1998). Quantification of emitted and retained N_2O in a municipal wastewater treatment plant with activated sludge and nitrification-denitrification units. *Agrobiol. Res.* **51**, 59–73.

Stuven, R., Vollmer, M., and Bock, E. (1992). The impact of organic matter on nitric oxide formation by *Nitrosomonas europaea*. *Arch. Microbiol.* **158**, 439–443.

Sümer, E., Weiske, A., Benckiser, G., and Ottow, J. C. G. (1995). Influence of environmental conditions on the amount of N_2O released from activated sludge in a domestic waste water treatment plant. *Cell. Mol. Life Sci.* **51**, 419–422.

Tata, P., Witherspoon, J., and Lue-Hing, C. (eds.), (2003). VOC Emissions from Wastewater Treatment Plants, Lewis Publishers, Boca Raton, FL.

Tchobanoglous, G., Burton, F. L., and Stensel, H. D. (2003). Metcalf and Eddy Wastewater Engineering: Treatment and Reuse McGraw Hill, New York, NY.

USEPA (2009). Inventory of U.S. Greenhouse Gas Emissions and Sinks: 1990-2006, EPA 430-R-08-005 (Washington DC).

van Dongen, U., Jetten, M. S. M., and van Loosdrecht, M. C. M. (2001). The SHARON-ANAMMOX process for treatment of ammonium rich wastewater. *Water Sci. Technol.* **44,** 153–160.

Wicht, H., and Beier, M. (1995). N_2O emission aus nitrifizierenden und denitrificierended Klaranlagen. *Korresp. Abwasser* **42**(404–406), 411–413.

Yu, R., and Chandran, K. (2010). Strategies of *Nitrosomonas europaea* 19718 to counter low dissolved oxygen and high nitrite concentrations. *BMC Microbiol.* **10,** 70.

Yu, R., Kampschreur, M. J., van Loosdrecht, M. C. M., and Chandran, K. (2010). Mechanisms and specific directionality of autotrophic nitrous oxide and nitric oxide generation during transient anoxia. *Environ. Sci. Technol.* **44,** 1313–1319.

Zumft, W. G. (1997). Cell biology and molecular basis of denitrification. *Microbiol. Mol. Biol. Rev.* **61,** 533–616.

SECTION FOUR

GENETICS, BIOCHEMISTRY AND BIOGEOCHEMISTRY

CHAPTER SEVENTEEN

GENETIC TRANSFORMATION OF AMMONIA-OXIDIZING BACTERIA

Luis A. Sayavedra-Soto* *and* Lisa Y. Stein[†]

Contents

1. Introduction	390
2. Transformation of AOB	391
2.1. Electroporation	391
2.2. Conjugation	392
2.3. Growth and selection	393
3. Gene Inactivation in AOB	393
3.1. Gene constructs for mutagenesis	393
3.2. Creating mutants with two genes inactivated	395
3.3. Screening and confirmation of recombination events in AOB	396
4. Use of Broad-Host Range Plasmids	398
4.1. Reporter gene constructs	398
4.2. Complementation of gene function	398
5. Strain Stability and Maintenance	399
6. Conclusions	399
References	400

Abstract

The study of traits of ammonia-oxidizing bacteria (AOB) by genetic transformation is an approach that is facilitated by the availability of AOB genome sequences. To transform an AOB, a vector construct is introduced into the cells by electroporation or conjugation to effect the inactivation, complementation, or expression of a selected gene. For inactivation studies, the vector construct should contain the gene of interest with an antibiotic resistance cassette and recombine into the cell's chromosome. For gene expression studies, a wide-host range vector with a transcriptional gene fusion can be used to test for gene roles. For gene complementation studies, a wide-host range vector expressing the inactivated gene can be used to recover the lost function in an AOB mutant strain. This chapter is a compilation of the methods that have been used to transform the AOB *Nitrosomonas europaea* and

---

* Department of Botany and Plant Pathology, Oregon State University, Corvallis, Oregon, USA
[†] Department of Biological Sciences, University of Alberta, Edmonton, Alberta, Canada

Nitrosospira multiformis and of the considerations and caveats to successfully produce, maintain, and store AOB transformants. The protocols may be applied to other AOB.

1. INTRODUCTION

Genetic transformation has been useful in studying the physiology of ammonia-oxidizing bacteria (AOB). This is most evident in the AOB *Nitrosomonas europaea* where the method was utilized in conjunction with the information from its genome sequence (Chain *et al.*, 2003). For example, gene inactivation studies were carried out to understand the roles of genes present in multiple copies (Hommes *et al.*, 1996, 1998, 2002; Stein *et al.*, 2000), to test metabolic capabilities that were suggested by the genome's composition (Hommes *et al.*, 2006), and in combination with expression microarrays, to determine regulatory gene functions (Cho *et al.*, 2006; Arp laboratory, unpublished). Genetic transformation has also been used to evaluate *N. europaea* bioreporter capabilities for common pollutants found in wastewater (Gvakharia *et al.*, 2009) and as an indicator of nitrification conditions (Iizumi *et al.*, 1998; Ludwig *et al.*, 1999).

Genetic modification of AOB present several challenges that need to be met: all known AOB have ≥8 h doubling times, achieve relatively low cell densities in batch cultures (≤0.1 $OD_{600\ nm}$), and none readily form colonies in solid medium. AOB are chemolithoautotrophs, a property that precludes the selection of a mutant by its inability to metabolize a substrate (as is often done for the selection of transformed heterotrophic bacteria). The current method to produce and isolate AOB transformants uses selection by acquired resistance to an antibiotic. However, the transformation efficiency of *N. europaea* is relatively low, and when exposed for extended periods of time to an antibiotic, *N. europaea* eventually acquires resistance to its effects. Although it is not known that other AOB develop antibiotic resistance, it is likely that they all have similar properties.

The pioneering work for the transformation of AOB was first done with *N. europaea* (Hommes *et al.*, 1996), and more recently with *Nitrosospira multiformis* (Stein laboratory, unpublished). In this chapter, we describe current methodology to successfully carry out genetic transformation of *N. europaea* and *N. multiformis*. The methods in this chapter aim to help researchers to overcome some of the most common hurdles encountered in the transformation of AOB and they may be directly applicable to transform other AOB whose genomes are known, for example, *Nitrosomonas eutropha* (Stein *et al.*, 2007) and *Nitrosococcus oceani* (Klotz *et al.*, 2006).

2. Transformation of AOB

All protocols developed to make the gene constructs that enable transformation of AOB are similar with small variations. The protocols to introduce the gene constructs into AOB vary, but all result in similar transformation efficiencies. These protocols are also used to introduce broad-host range plasmids, which can be maintained in AOB either for genetic complementation studies (Beaumont *et al.*, 2002) or for expression studies using a reporter gene (Gvakharia *et al.*, 2009; Hirota *et al.*, 2006). Because of the inherent physiological characteristics of AOB, the making of an AOB transformant usually takes 2–6 months depending on the metabolic trait that is being studied (Fig. 17.1).

2.1. Electroporation

The most common method to transform *N. europaea* is to introduce the vector construct into the cells by electroporation. For this purpose, cells of *N. europaea* (0.5 l) are grown in basic 25 mM (NH$_4$)$_2$SO$_4$ medium (Ensign *et al.*, 1993; Hommes *et al.*, 2003; Hyman and Arp, 1992) and harvested by centrifugation when they enter stationary phase (0.1 OD$_{600\ nm}$). The cells are then washed three times with sterile distilled H$_2$O. Extensive washing is

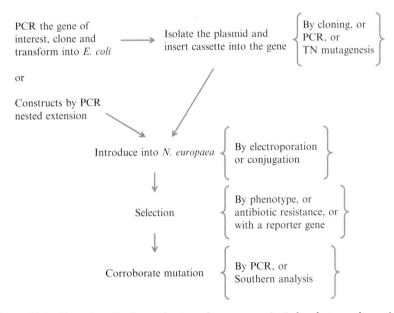

Figure 17.1 Flow chart for the production of a mutant strain. In brackets are alternatives for each step.

necessary to rid the cells of remnant salts from the growth medium that causes arcing in the electroporation cuvette as the voltage is applied (arcing kills the cells). For electroporation the cell pellet is suspended in sterile H_2O (~ 1.0 $OD_{600\ nm}$) and kept on ice until use. Approximately 100 μl of the cell suspension is mixed with 1 μg of DNA (vector construct at 1 mg/ml), placed in a 1-mm gap electroporation cuvette and pulsed (e.g., with an electroporation system from BTX Instrument Division, MA, USA). Typical electroporation parameters are about 1200 V, 25 mF and resistance larger than 100 Ohms (Hommes et al., 1996). The cells are transferred immediately to 0.5 l of fresh medium (25 mM $(NH_4)_2SO_4$ growth medium) and allowed to grow for 24 h under nonselective conditions at 30 °C with gentle agitation on a rotary shaker (e.g., ≤ 100 rpm). The antibiotic for selection is then added to a final concentration of 10–50 μg/ml. The antibiotic (e.g., kanamycin sulfate) is prepared as a 10-mg/ml stock which is sterilized by passing it through a sterile 0.2 μm syringe filter. Cell growth is monitored by the accumulation of NO_2^- or by optical density at 600 nm. The NO_2^- concentration is determined colorimetrically with the Griess reagent prepared from sulfanilamide and N-naphthylethylenediamine (Hageman and Hucklesby, 1971).

2.2. Conjugation

Plasmid constructs can also be transferred from Escherichia coli to N. europaea by conjugation (Ludwig et al., 1999). The donor strain E. coli S17-1 containing the plasmid construct is grown in Luria-Bertani medium to stationary phase, harvested by centrifugation and suspended in heterotrophic medium (KM1 medium composed of (g/l) sodium acetate, 1.2; L-asparagine, 0.4; yeast extract, 2.0; $MgCl_2 \cdot 6H_2O$, 0.2; $FeSO_4 \cdot 7H_2O$, 0.01; $CaCl_2 \cdot 2H_2O$, 0.02; and pH adjusted to 6.8; Usha et al., 2002) to a density of 10^6 cells/ml. The N. europaea recipient strain is grown in 25 mM $(NH_4)_2SO_4$ growth medium containing 30 μg nalidixic acid per ml to stationary phase, harvested by centrifugation and suspended in basal growth medium to an approximate density of 10^9 cells/ml. Equal volumes of donor and recipient strains are mixed (~ 150 μl total), spotted onto a cellulose nitrate or Nylon membrane (0.45 μM pore size, Nytran Schleicher & Schuell Bioscience, USA) that is placed on top of R2A agar or KM1 agar with 0.05% (w/v) glucose and incubated at 30 °C for 2 days. After the incubation, the bacteria are washed into 1 ml of growth medium containing 30 μg nalidixic acid per ml and transferred into fresh growth medium with the appropriate antibiotic for selection. The lithoautotrophic medium will select for growth of N. europaea instead of E. coli, and the antibiotic will select for transformed N. europaea cells.

To transform N. multiformis, the same protocols are followed as for N. europaea. However, each AOB has its own optimal conditions for substrate concentration, pH, temperature, and other factors. As such, *N. multiformis* tolerates no more than 20 m*M* ammonium and grows better in HEPES- than in phosphate-buffered medium. Furthermore, *N. multiformis* grows best at 25–28 °C and cannot tolerate higher temperatures. It should also be noted that to date of this chapter, transformation of *N. multiformis* by electroporation has not met with success by the authors. Conjugation is the recommended approach for introducing plasmid constructs into *N. multiformis* (Cantera and Stein, unpublished data).

2.3. Growth and selection

Growth after transformation of *N. europaea* in a liquid batch culture is observed after 1 or 2 weeks of incubation at 30 °C with gentle agitation on a rotary shaker (~100 rpm). Aliquots of actively growing cultures are plated onto a Petri dish with a Nylon membrane layered on top of solid-growth medium (e.g., 25 m*M* ammonium, 0.7% agarose and the corresponding antibiotic) to select for antibiotic-resistant cell colonies. Alternatively, cells from a conjugation transformation experiment are washed, diluted, and 100–250 µl are plated directly onto the membranes in Petri dishes. Colonies become evident in 2–4 weeks of incubation. The weekly transfer of the inoculated membranes onto fresh growth medium plates is recommended to prevent the accumulated nitrite and the acidification of the medium that tend to inhibit the growth of AOB (Stein and Arp, 1998). Colonies are picked up with a transfer loop and inoculated into liquid medium with the appropriate antibiotic from which clonal cell lines can be grown and harvested for physiological and genetic characterization.

3. Gene Inactivation in AOB

Vector constructs can be designed and used to inactivate a gene (suicide vectors) or to be maintained in the cell (broad-host range vectors). In an inactivation experiment, the target gene is cloned into a vector to interrupt the coding region with a marker, and transferred into the cell to recombine and inactivate the chromosomal copy (Fig. 17.2).

3.1. Gene constructs for mutagenesis

The most common approach to make a gene construct for AOB mutagenesis is to clone the gene of interest and interrupt (or replace) it with an antibiotic resistance cassette (Fig. 17.2). The plasmid with the gene-antibiotic resistance construct is then transformed into the AOB strain to allow for recombination into the chromosome. The recombinant strain is enriched in liquid

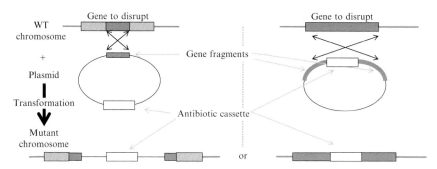

Figure 17.2 Diagram of the possible outcomes resulting from the recombination between a vector construct and the chromosome of a nitrifying bacterium.

medium in the presence of the selecting antibiotic and then transferred to solid medium to form individual cell colonies from which axenic cultures of a mutant are obtained.

The gene selected for inactivation is amplified by PCR and cloned into a high-copy plasmid (e.g., pSelect form Promega, Madison WI.). The PCR is carried out by standard protocols with primers flanking the open reading frame and with genomic DNA as template. Ideally the amplified fragment would include a restriction site near the center of the gene that facilitates the insertion of an antibiotic resistance cassette (e.g., Kan^r (*npt*) from pUC4 KSAC or from pKOK6.1 which encodes the neomycin/kanamycin-phosphotransferase) to cause the disruption of its function (Fig. 17.2). More often, a restriction site needs to be engineered in order to be compatible with restriction sites flanking an antibiotic resistance cassette. A restriction site can be created by site-directed mutagenesis with commercial kits, for example, by PCR with a high-fidelity DNA polymerase such as *Pfu* DNA polymerase, starting with a circular plasmid with the cloned gene as the template and the appropriate primers to introduce the restriction site as directed by the manufacturer (e.g., Invitrogen, Carlsbad, CA). This mutagenesis protocol was used, for instance, to introduce a *Bam*HI site into *amoA* in *N. europaea* (Hommes *et al.*, 1998). The kanamycin antibiotic resistance cassette was cloned into the new *Bam*HI site. The plasmid was then transformed into *E. coli* and selected by the resistance to the antibiotic. To confirm that the clones indeed contain the antibiotic resistance cassette, the plasmids are isolated by standard protocols (Sambrook *et al.*, 1989) and screened by restriction digest or by PCR. Preferably the antibiotic resistance cassette is inserted in the opposite orientation to the target gene to avoid read-through transcription. A variation to this protocol is to carry out the insertion of an antibiotic resistance cassette (e.g., Km^r) into a plasmid with the gene of interest using an *in vitro* transposon insertion kit such as EZ::TN <KAN2> from Epicenter (Madison, WI). The protocol results in multiple insertions in the plasmid. After transformation into *E. coli* viable colonies in the presence of kanamycin and ampicillin will contain successful transpositions; however, these need to be screened by restriction digests or nucleic acid sequencing to locate the point of insertion. This method was used to insert a transposon conferring kanamycin resistance (Km^r) into the promoter region of the gene encoding Fur (pFur-kanP; Arp laboratory, unpublished). The approach yields multiple points of insertion into the gene that can be used to test the effects of an insertion, whether at the promoter site, the 5′ or the 3′-end of the gene. Another variation is to clone an internal part of the gene of interest into a suicide vector such as pRVS3 (Beaumont *et al.*, 2002). The vector will recombine into the chromosome and interrupt the gene (Fig. 17.2).

The alternative method to design a gene construct for mutagenesis is to substitute a gene for an antibiotic resistance cassette and use it for allelic exchange. To achieve this, the antibiotic resistance cassette is flanked by sequences from the target gene (Berube *et al.*, 2007). The flanking segments are PCR amplified with primers that introduce a restriction site compatible with the antibiotic resistance cassette. The making of this type of gene construct requires sequential steps to flank the cassette. The resulting product is cloned, transferred into *E. coli*, and at each step the plasmids are isolated and checked for the correct insert. This method was applied to replace the gene for AmoC3 in *N. europaea* with *aacC1* encoding the gentamicin acetyltransferase 3-1 conferring resistance to gentamicin (Berube *et al.*, 2007). For electroporation into *N. europaea*, the plasmid in this experiment was linearized and introduced into *N. europaea* as described above.

A variation for generating DNA constructs for allelic exchange with a PCR approach (Baba *et al.*, 2006; Datsenko and Wanner, 2000) was adapted recently and used with *N. europaea* (Fig. 17.3). Primers (∼70 bp) homologous to the 5′- and 3′-ends of a gene of interest and overlapping an antibiotic resistance cassette are used. The resulting PCR product is reamplified with new *N. europaea* primers overlapping the sequences of the former primers, extending the flanking DNA up to 300 bp if performed two to three times with subsequent overlapping primers. The method shortens the design time of DNA constructs and assures efficient recombination in *N. europaea* after electroporation. The method was applied successfully to produce mutants of *N. europaea* to study sucrose synthesis in this AOB (Arp laboratory, unpublished). All gene constructs are introduced into *N. europaea* by one of the methods described above.

3.2. Creating mutants with two genes inactivated

On occasion, it is necessary to carry out the inactivation of two genes in an AOB. For example, through double mutagenesis the individual roles of the three gene copies encoding HAO in *N. europaea* were studied (Hommes *et al.*, 2002). The protocol to produce single-gene inactivation by insertion

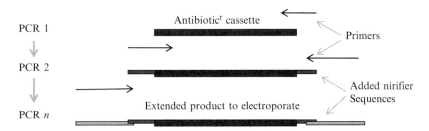

Figure 17.3 Accelerated production of DNA constructs for gene inactivation by PCR-nested extension. Each additional cycle adds 40–60 nucleotides up to the desired length that is determined by subsequent primers and PCR. The final product is ethanol precipitated, suspended in water and electroporated into *N. europaea*.

of an antibiotic resistance cassette also applies to inactivate multiple genes within a strain; however, the constructs should encode alternative selectable markers (e.g., Kan^r (*npt*) or Gen^r (*aacC1*) antibiotic resistance cassettes). The *aacC1* marker including the full coding region and promoter can be obtained from the pUCGM vector (Schweizer, 1993). The constructs are electroporated as described above into a *Nitrosomonas* strain containing a single mutation (e.g., in one of the three kanamycin-resistant single-*hao* mutant strains). In *N. europaea*, the single-*hao* mutant strains were transformed with specific copies of *hao* containing the Gen^r cassette so that all three double-*hao* mutant combinations resulted (Hommes *et al.*, 2002). Each double mutant required the expression of the third intact copy of *hao* for growth. In this experiment, following transformation and recovery (see above), liquid cultures of *N. europaea* were grown in the presence of 25 μg kanamycin sulfate per ml and 15 μg gentamicin sulfate per ml. Aliquots from these cultures were streaked on Nylon membranes laid onto solid *N. europaea* growth medium containing 10 μg kanamycin sulfate per ml and 6 μg gentamicin sulfate per ml to obtain axenic clonal mutant strains.

3.3. Screening and confirmation of recombination events in AOB

Recombinant clones selected from solid media are grown in liquid medium and harvested at stationary phase. Genomic DNA is prepared from these cells and checked by Southern hybridization or by PCR following standard protocols (Sambrook *et al.*, 1989). For Southern analysis, the DNA is digested with a restriction enzyme that does not cut the cassette and that was not used in the cloning scheme. Approximately 10 μg of digested DNA is electrophoretically separated on an agarose gel and blotted onto Nytran membranes (Sambrook *et al.*, 1989). DNA probes for the target gene (e.g., for kanamycin or gentamicin resistance genes) can be generated by PCR and labeled by random-priming with [^{32}P] dCTP (3000 Ci mmol^{-1};

DuPont NEN Products, Wilmington, Del.) and a kit (e.g., Promega's Prime-a-Gene) following the directions of the manufacturers. The hybridization protocols are standard (Sambrook *et al.*, 1989). Hybridization signals are analyzed by X-ray film or by a Phosphorimager and ImageQuant software (Molecular Dynamics, Sunnyvale, CA). For PCR corroboration, it suffices to carry out the amplification with genomic DNA from the putative mutants using primers flanking the cassette insertion site. The PCR fragment obtained with DNA from the mutant strain should be larger compared to that obtained with DNA from the wild type to reflect the size of the antibiotic resistance cassette. Alternatively, a primer for the antibiotic resistance cassette and a primer upstream or downstream of the region cloned for plasmid construction can be used. In this case, only the mutant strain with a proper recombination event will show a positive PCR result (Fig. 17.4).

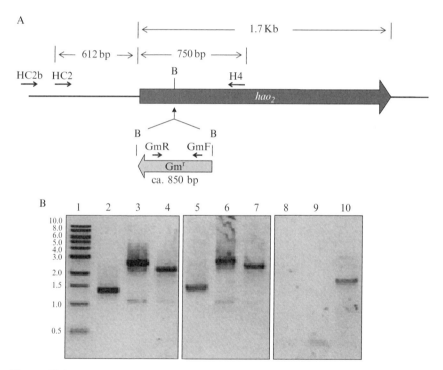

Figure 17.4 Example of a successful recombination event and its corroboration by PCR. Panel A: map depicting the insertion site of the antibiotic resistance cassette. (A) The *hao2* gene was PCR amplified from *N. europaea* genomic DNA using primers HC2 and H4. The *aacC1* was excised from pUCGM with *Bam*HI and inserted into the *Bam*HI site of cloned *hao2*. A second *hao2* mutant was constructed with pUC4 KSAC in place of *aacC1*. Panel B: corroboration the recombination event by PCR using primer combinations HC2 and H4 (lanes 2, 3, 4), HC2b and H4 (lanes 5, 6, 7), and HC2b and GmF (lanes 8–10). Lanes 2, 5, 8: wild type; lanes 3, 6, and 9 *hao2*::KSAC; lanes 4, 7, 10: *hao2*:: *aacC1*. Lane 1 shows DNA size markers in kilobases.

It is not uncommon when screening putative mutants to find that the antibiotic resistance cassette has been transposed elsewhere in the genome. In this case Southern hybridization and PCR results would show an intact copy of the targeted gene along with the presence of the antibiotic resistance cassette. Therefore, screening of several (10–20) putative mutant colonies may be required to locate one with a disrupted copy of the targeted gene. Transposition is especially problematic when targeting genes whose products are likely important for cell health but are not compensated for by other gene products (e.g., *nirK* in *N. multiformis*). In this event, study of gene function may be best accomplished through the use of reporter gene constructs (see below).

4. Use of Broad-Host Range Plasmids

Transcriptional or translational gene fusions in broad-host range plasmids can be used to study gene expression through a reporter gene (Gvakharia *et al.*, 2009) or to complement a *N. europaea* mutant strain (Beaumont *et al.*, 2002).

4.1. Reporter gene constructs

Vectors such as pPROBE allow transcriptional fusions with the green fluorescent protein encoded by *gfp* (Miller and Lindow, 1997; Miller *et al.*, 2000) and they have been used to test the promoters of two *N. europaea* genes that were highly upregulated after exposure to chloroform and chloromethane (Gvakharia *et al.*, 2009). To detect the florescence in *N. europaea* carrying *gfp*, the cells require 30–60 min following exposure to the inducing molecule to allow GFP expression and proper folding of the GFP protein (Franke *et al.*, 2007). Fluorescence is measured after excitation at 490 nm and detection of emission is at 510 nm. Fluorescence values are commonly reported in arbitrary units (raw fluorescent units or RFU) normalized by the cell density at 600 nm. Broad-host range plasmids with individual *hao*-LacZ fusions (in pKZ27) were used to characterize the expression of the three HAO copies in *Nitrosomonas* ENI-11 (Hirota *et al.*, 2006). The plasmids were transferred into the target cells by electroporation and the measurement of LacZ expression was as described by Miller (1972).

4.2. Complementation of gene function

For complementation studies, broad-host plasmids such as pEG400 are available to express a cloned gene. A *nirK*-deficient *N. europaea* mutant was complemented with this broad-host-vector containing *nirK* under the control of the promoter of *npt* (Beaumont *et al.*, 2002). In this experiment the kanamycin acetyltransferase gene promoter was apparently less active than the wild-type *nirK* promoter and partially restored the activity. This assumption was corroborated by the relative amounts of NirK in extracts from wild-type and complemented NirK-deficient cells as visualized by sodium dodecyl sulfate-polyacrylamide gel electrophoresis (SDS-PAGE; Garfin, 2009).

5. Strain Stability and Maintenance

During selection it is recommended to avoid incubations that last more than 30 days. *N. europaea* (and likely other AOB) can develop resistance to kanamycin, and some antibiotics, such as tetracycline or ampicillin, lose activity during the incubations. One needs to be wary of recombinant strains that have been recovered after 30 days of incubation. All AOB strains can be stored at $-80\,^{\circ}$C in 30% glycerol or 5% DMSO for extended periods (years). To recover a strain from a frozen stock, a fraction is transferred to fresh medium with the corresponding antibiotic; the strain will show appreciable growth after 2 weeks of incubation with gentle agitation at its appropriate incubation temperature (i.e., 30 $^{\circ}$C for *N. europaea*).

6. Conclusions

The production of an AOB mutant may require a 2–6-month investment; however, the possible outcomes can be highly valuable. AOB are important players in the nitrogen cycle initiating the oxidation of ammonia and in generating traces of the NO and N_2O gases. Through mutagenesis, it was possible to research the roles of the multiple copies of genes involved in the oxidation of NH_3 in *N. europaea*. Mutagenesis made possible the study of NirK in *N. europaea* and helped us to understand its roles in NO and N_2O generation (Beaumont *et al.*, 2002; Cantera and Stein, 2007). Genes in the metabolism of carbon were also studied by mutagenesis (Hommes *et al.*, 2006). In the metabolism of NH_3, *N. europaea* utilizes an array of Fe- and heme-containing proteins (Upadhyay *et al.*, 2003; Whittaker *et al.*, 2000). Mutagenesis made possible the study of the Fe uptake regulator, Fur (Arp laboratory, unpublished), and to characterize the specificity of the only annotated ABC-type siderophore transporter (Vajrala *et al.*, 2010) and the ferrioxamine siderophore transporter (Wei *et al.*, 2007).

The protocols described here should be equally applicable to other AOB isolates, although the length of mutant recovery times may be considerably longer for slower growing strains like *Nitrosomonas oligotropha*. One should keep in mind that growing conditions for other AOB isolates are frequently different from *N. europaea*, so substrate concentration, temperature, pH, and other limitations must be considered. Ultimately, to be useful, the inactivation of a gene ideally should produce a phenotype that is neither lethal nor hampers growth considerably.

Further development of AOB transformation methodology includes high-throughput screening approaches for selectable phenotypes, which would allow for random mutagenesis and pathway discovery. As described above, the largest hurdles to overcome include the slow growth rates and limited substrate use of AOB that precludes use of most techniques developed for heterotrophic bacteria. Recent developments in robotic liquid handling systems and single cell technologies may help overcome these limitations.

REFERENCES

Baba, T., Ara, T., Hasegawa, M., Takai, Y., Okumura, Y., Baba, M., *et al.* (2006). Construction of *Escherichia coli* K-12 in-frame, single-gene knockout mutants: The Keio collection. *Mol. Syst. Biol.* **2**, 0008.

Beaumont, H. J. E., Hommes, N. G., Sayavedra-Soto, L. A., Arp, D. J., Arciero, D. M., Hooper, A. B., *et al.* (2002). Nitrite reductase of *Nitrosomonas europaea* is not essential for production of gaseous nitrogen oxides and confers tolerance to nitrite. *J. Bacteriol.* **184**, 2557–2560.

Berube, P. M., Samudrala, R., and Stahl, D. A. (2007). Transcription of all *amoC* copies is associated with recovery of *Nitrosomonas europaea* from ammonia starvation. *J. Bacteriol.* **189**, 3935–3944.

Cantera, J. J., and Stein, L. Y. (2007). Role of nitrite reductase in the ammonia-oxidizing pathway of *Nitrosomonas europaea*. *Arch. Microbiol.* **188**, 349–354.

Chain, P., Lamerdin, J., Larimer, F., Regala, W., Lao, V., Land, M., *et al.* (2003). Complete genome sequence of the ammonia-oxidizing bacterium and obligate chemolithoautotroph *Nitrosomonas europaea*. *J. Bacteriol.* **185**, 2759–2773.

Cho, C. M., Yan, T., Liu, X., Wu, L., Zhou, J., and Stein, L. Y. (2006). Transcriptome of a *Nitrosomonas europaea* mutant with a disrupted nitrite reductase gene (*nirK*). *Appl. Environ. Microbiol.* **72**, 4450–4454.

Datsenko, K. A., and Wanner, B. L. (2000). One-step inactivation of chromosomal genes in *Escherichia coli* K-12 using PCR products. *Proc. Natl. Acad. Sci. USA* **97**, 6640–6645.

Ensign, S. A., Hyman, M. R., and Arp, D. J. (1993). *In vitro* activation of ammonia monooxygenase from *Nitrosomonas europaea* by copper. *J. Bacteriol.* **175**, 1971–1980.

Franke, G. C., Dobinsky, S., Mack, D., Wang, C. J., Sobottka, I., Christner, M., *et al.* (2007). Expression and functional characterization of gfpmut3.1 and its unstable variants in *Staphylococcus epidermidis*. *J. Microbiol. Methods* **71**, 123–132.

Garfin, D. E. (2009). One-dimensional gel electrophoresis. *In* "Methods in Enzymology," (R. R. Burges and M. P. Deutscher, eds.), pp. 497–513. Elsevier, San Francisco.

Gvakharia, B. O., Bottomley, P. J., Arp, D. J., and Sayavedra-Soto, L. A. (2009). Construction of recombinant *Nitrosomonas europaea* expressing green fluorescent protein in response to co-oxidation of chloroform. *Appl. Microbiol. Biotechnol.* **82**, 1179–1185.

Hageman, R. H., and Hucklesby, D. P. (1971). Nitrate reductase in higher plants. *Meth. Enzymol.* **23**, 491–503.

Hirota, R., Kuroda, A., Ikeda, T., Takiguchi, N., Ohtake, H., and Kato, J. (2006). Transcriptional analysis of the multicopy hao gene coding for hydroxylamine oxidoreductase in *Nitrosomonas* sp. strain ENI-11. *Biosci. Biotechnol. Biochem.* **70**, 1875–1881.

Hommes, N. G., Sayavedra-Soto, L. A., and Arp, D. J. (1996). Mutagenesis of hydroxylamine oxidoreductase in *Nitrosomonas europaea* by transformation and recombination. *J. Bacteriol.* **178**, 3710–3714.

Hommes, N. G., Sayavedra-Soto, L. A., and Arp, D. J. (1998). Mutagenesis and expression of *amo*, which codes for ammonia monooxygenase in *Nitrosomonas europaea*. *J. Bacteriol.* **180**, 3353–3359.

Hommes, N. G., Sayavedra-Soto, L. A., and Arp, D. J. (2002). The roles of the three gene copies encoding hydroxylamine oxidoreductase in *Nitrosomonas europaea*. *Arch. Microbiol.* **178**, 471–476.

Hommes, N. G., Sayavedra-Soto, L. A., and Arp, D. J. (2003). Chemolithoorganotrophic growth of *Nitrosomonas europaea* on fructose. *J. Bacteriol.* **185**, 6809–6814.

Hommes, N. G., Kurth, E. G., Sayavedra-Soto, L. A., and Arp, D. J. (2006). Disruption of *sucA*, which encodes the E1 subunit of alpha-ketoglutarate dehydrogenase, affects the survival of *Nitrosomonas europaea* in stationary phase. *J. Bacteriol.* **188**, 343–347.

Hyman, M. R., and Arp, D. J. (1992). $^{14}C_2H_2$- and $^{14}CO_2$-labeling studies of the *de novo* synthesis of polypeptides by *Nitrosomonas europaea* during recovery from acetylene and light inactivation of ammonia monooxygenase. *J. Biol. Chem.* **267**, 1534–1545.

Iizumi, T., Mizumoto, M., and Nakamura, K. (1998). A bioluminescence assay using *Nitrosomonas europaea* for rapid and sensitive detection of nitrification inhibitors. *Appl. Environ. Microbiol.* **64**, 3656–3662.

Klotz, M. G., Arp, D. J., Chain, P. S., El-Sheikh, A. F., Hauser, L. J., Hommes, N. G., *et al.* (2006). Complete genome sequence of the marine, chemolithoautotrophic, ammonia-oxidizing bacterium *Nitrosococcus oceani* ATCC 19707. *Appl. Environ. Microbiol.* **72**, 6299–6315.

Ludwig, C., Ecker, S., Schwindel, K., Rast, H. G., Stetter, K. O., and Eberz, G. (1999). Construction of a highly bioluminescent *Nitrosomonas* as a probe for nitrification conditions. *Arch. Microbiol.* **172**, 45–50.

Miller, J. H. (1972). Assay of β-galactosidase. *In*: "Experiments in Molecular Biology", pp. 352–355. Cold Spring Harbor Laboratory, Cold Spring Harbor, NY.

Miller, W. G., and Lindow, S. E. (1997). An improved GFP cloning cassette designed for prokaryotic transcriptional fusions. *Gene* **191**, 149–153.

Miller, W. G., Leveau, J. H., and Lindow, S. E. (2000). Improved *gfp* and *inaZ* broad-host-range promoter-probe vectors. *Mol. Plant Microbe. Interact.* **13**, 1243–1250.

Sambrook, J., Fritsch, E. F., and Maniatis, T. (1989). Molecular Cloning: A Laboratory Manual. Cold Springs Harbor Laboratory Press, Cold Springs Harbor, NY.

Schweizer, H. D. (1993). Small broad-host-range gentamycin resistance gene cassettes for site-specific insertion and deletion mutagenesis. *Biotechniques* **15**, 831–834.

Stein, L. Y., and Arp, D. J. (1998). Loss of ammonia monooxygenase activity in *Nitrosomonas europaea* upon exposure to nitrite. *Appl. Environ. Microbiol.* **64**, 4098–4102.

Stein, L. Y., Sayavedra-Soto, L. A., Hommes, N. G., and Arp, D. J. (2000). Differential regulation of *amoA* and *amoB* gene copies in *Nitrosomonas europaea*. *FEMS Microbiol. Lett.* **192**, 163–168.

Stein, L. Y., Arp, D. J., Berube, P. M., Chain, P. S., Hauser, L., Jetten, M. S., *et al.* (2007). Whole-genome analysis of the ammonia-oxidizing bacterium, *Nitrosomonas eutropha* C91: Implications for niche adaptation. *Environ. Microbiol.* **9**, 2993–3007.

Upadhyay, A. K., Petasis, D. T., Arciero, D. M., Hooper, A. B., and Hendrich, M. P. (2003). Spectroscopic characterization and assignment of reduction potentials in the tetraheme cytochrome c_{554} from *Nitrosomonas europaea*. *J. Am. Chem. Soc.* **125**, 1738–1747.

Usha, T., Sarada, L. R., and Ravishankar, G. A. (2002). Effect of culture conditions on growth of green alga—*Haematococcus pluvialis* and astaxanthin production. *Acta Physiol. Plant.* **24,** 323–329.

Vajrala, N., Sayavedra-Soto, L. A., Bottomley, P. J., and Arp, D. J. (2010). Role of *Nitrosomonas europaea* NitABC iron transporter in the uptake of Fe^{3+}-siderophore complexes. *Arch. Microbiol.* **192,** 899–908.

Wei, X., Sayavedra-Soto, L. A., and Arp, D. J. (2007). Characterization of the ferrioxamine uptake system of *Nitrosomonas europaea*. *Microbiology* **153,** 3963–3972.

Whittaker, M., Bergmann, D., Arciero, D., and Hooper, A. B. (2000). Electron transfer during the oxidation of ammonia by the chemolithotrophic bacterium *Nitrosomonas europaea*. *Biochim. Biophys. Acta* **1459,** 346–355.

CHAPTER EIGHTEEN

Dissecting Iron Uptake and Homeostasis in *Nitrosomonas europaea*

Luis A. Sayavedra-Soto, Neeraja Vajrala, *and* Daniel J. Arp

Contents

1. Introduction	404
2. Use of Bioinformatics to Plan Fe-Related Studies	406
3. Methods to Learn About the Physiological Responses to Fe Limitation	407
3.1. Preparation of Fe-free medium and glassware	407
3.2. Cell growth and analysis	408
3.3. Cellular Fe content and allocation	410
3.4. Heme *c* content determination	411
3.5. Protein identification by gel electrophoresis and mass spectrometry	412
3.6. Microscopy of cell structure	414
3.7. Microarrays to discover unexpected gene expression responses	415
3.8. Corroboration of gene expression by real-time qPCR	417
4. Methods to Study Siderophore Uptake	418
4.1. Growth experiments in the presence of siderophores	418
4.2. Uptake of fluorescently labeled siderophore analogs and localization in the cell	419
4.3. Labeled Fe feeding experiments	421
5. Genetic Complementation to Study *fur* Homologs	421
6. Methods for Gene Inactivation of Fe Uptake Systems	422
7. Lingering Questions that can be Answered by the Above Protocols	423
7.1. Siderophore-independent Fe uptake	423
7.2. Novel regulatory mechanisms for Fe homeostasis	424
8. Conclusion	425
References	425

Department of Botany and Plant Pathology, Oregon State University, Corvallis, Oregon, USA

Abstract

The chemolithoautotroph *Nitrosomonas europaea* oxidizes about 25 mol of NH_3 for each mole of CO_2 that is converted to biomass using an array of heme and nonheme Fe-containing proteins. Hence mechanisms of efficient iron (Fe) uptake and homeostasis are particularly important for this Betaproteobacterium. Among nitrifiers, *N. europaea* has been the most studied to date. Characteristics that make *N. europaea* a suitable model to study Fe uptake and homeostasis are as follows: (a) its sequenced genome, (b) its capability to grow relatively well in 0.2 μM Fe in the absence of heterologous siderophores, and (c) its amenability to mutagenesis. In this chapter, we describe the methodology we use in our laboratory to dissect Fe uptake and homeostasis in the ammonia oxidizer *N. europaea*.

1. Introduction

Iron is an important element required by almost all organisms but is of particular importance to *Nitrosomonas europaea*. To date, *N. europaea* is the best-studied nitrifier and is well suited as a model to research Fe uptake and metabolism among ammonia-oxidizing bacteria (AOB). As a chemolithoautotroph, *N. europaea* derives all of its energy and reductant for growth from the oxidation of ammonia (NH_3) to nitrite (NO_2^-; Fig. 18.1) and meets its requirements for carbon from the fixation of CO_2 (Arp *et al.*, 2002; Hooper *et al.*, 1990). For each mole of CO_2 that is converted to biomass, *N. europaea* oxidizes about 25 mol of NH_3. During this process, *N. europaea* utilizes an array of Fe-containing proteins such as hydroxylamine oxidoreductase, cytochromes $c554$, c_M552, and p460, as well as other enzymes involved in energy metabolism and electron transport such as heme/copper-type cytochrome oxidases and the heme bc complex (ubiquinol/cytochrome c oxidoreductase; Upadhyay *et al.*, 2003; Whittaker *et al.*, 2000). Many of these heme proteins are present at high concentrations in *N. europaea* leading to the reddish-brown color of the cells that is most evident at high cell densities (i.e., after harvesting by centrifugation). Accordingly, the Fe concentration in *N. europaea* is 19–80-fold higher than that in *Escherichia coli* (Wei *et al.*, 2006).

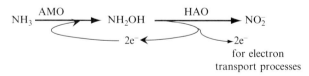

Figure 18.1 The pathway of ammonia oxidation in *N. europaea*. Ammonia monooxygenase (AMO) oxidizes NH_3 to hydroxylamine (NH_2OH) and hydroxylamine oxidoreductase (HAO) oxidizes NH_2OH to NO_2^-.

Several insights on *N. europaea* Fe uptake and metabolism (Fig. 18.2) were gained from its genome sequence and recent studies in our laboratory. The *N. europaea* genome encodes a relatively large number of proteins related to Fe homeostasis, including numerous Fe-siderophore transducers/receptors and Fe regulatory elements; however, its genome does not have any siderophore biosynthesis genes (Chain *et al.*, 2003). Cultures of *N. europaea* can grow relatively well at Fe concentrations as low as 0.2 μM even in the absence of heterologous siderophores (Wei *et al.*, 2006), suggesting an efficient siderophore-independent uptake system. In order to grow below 1 μM Fe concentrations, most other bacteria utilize siderophores, small molecular weight compounds that can efficiently chelate Fe^{3+} from the environment and make it available for uptake via specific outer membrane (OM) receptors (Andrews *et al.*, 2003; Hofte *et al.*, 1993; Neilands, 1993, 1995; Wiebe, 2002). Cells of *N. europaea* efficiently downregulate the expression of Fe-containing proteins while growing in Fe-limited conditions (Arp laboratory, unpublished).

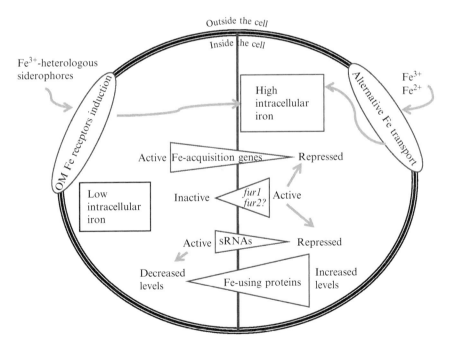

Figure 18.2 Possible mechanisms in *N. europaea* for Fe homeostasis. The diagram represents a cell in two scenarios: left at low intracellular Fe, and right, at high intracellular Fe. The triangles represent the elements at play at high levels of expression (wider side) and at low levels (narrow side). The ovals represent membrane-associated components active at low (left) and at high (right) Fe levels.

The ability of *N. europaea* to grow well under Fe-limited conditions gives it an advantage when competing for scarcely available Fe resources in its natural environment (Wei *et al.*, 2006). The mechanisms by which this Betaproteobacterium acquires sufficient Fe for its metabolism under Fe-limiting conditions in the absence of heterologous siderophores is indeed intriguing and highlights unique aspects of nitrifiers that are different from other model organisms (e.g., *E. coli*, *Bacillus subtilis*, or *Pseudomonas aeruginosa*).

2. Use of Bioinformatics to Plan Fe-Related Studies

The analysis of the *N. europaea* genome helps to generate hypotheses to understand how it is able to grow in a wide range of Fe concentrations. Bioinformatic analysis with software at NCBI (http://www.ncbi.nlm.nih.gov/) or at JGI (http://genome.jgi-psf.org/) indicated that the *N. europaea* genome has about 90 genes for Fe uptake (Chain *et al.*, 2003), a relatively large number compared to ~11 in *Nitrosococcus oceani*, ~12 in *Nitrosomonas eutropha*, ~12 in *Nitrosospira multiformis* (all AOBs), 11 in *Acidithiobacillus ferroxidans* (a chemolithotroph; Quatrini *et al.*, 2005), 14 in *P. aeruginosa*, 11 in *Pseudomonas putida* (pathogens; Andrews *et al.*, 2003) and six in *E. coli* (Andrews *et al.*, 2003).

Out of these 90 genes in *N. europaea*, 66 genes encode 22 sets of putative TonB-dependent transducer systems organized similarly to Fe regulatory *fecI/fecR* (σ-factor/anti-σ or sensor/regulator)-*fecA* (receptor/transducer) systems (Koebnik, 2005) with the TonB-dependent transducer genes located downstream of these σ and anti-σ factors or, in one case, upstream (Wei *et al.*, 2006). Three sets have unrelated genes between the *fecR* and *fecA* homologs, a feature also observed in other Betaproteobacteria and Gammaproteobacteria (Koebnik, 2005). This gene organization is similar to the *fecIR* and *fecABCDE* (Fe-dicitrate) systems in *E. coli* (Visca *et al.*, 2002), and to the *fpvIR* and *fpvA/pvd* (pyoverdin) system in *P. aeruginosa* (Ravel and Cornelis, 2003). In addition, there are three sets of Fe-ABC transporter genes in *E. coli* and four sets in *P. aeruginosa* (Andrews *et al.*, 2003). Unlike these bacteria, there are no Fe-ABC transporter genes associated with the *fec*-like genes in *N. europaea*.

In addition to the TonB-dependent transducer systems, the *N. europaea* genome also has 16 putative TonB-dependent receptor genes that are not directly associated with σ-/anti-σ factor genes, though some are in proximity to TonB-dependent transducers with σ- and anti-σ factor genes. Putative TonB boxes (Cadieux and Kadner, 1999; Gudmundsdottir *et al.*, 1989) were identified in 17 OM siderophore transducer/receptor gene sequences, but not in the remaining OM transducer/receptor gene sequences. Given the dissimilarities to *fecA*, alternatives to the consensus sequence of characterized TonB boxes may exist in *N. europaea* and may influence the rates of siderophore uptake. As with other Gram-negative bacteria (Andrews *et al.*, 2003), genes for the energy-transducing TonB–ExbBD protein complex are present in the *N. europaea* genome.

In contrast to siderophore-producing bacteria, the strategy of *N. europaea* is probably to conserve energy by not synthesizing its own siderophores, but instead to use its numerous OM transducer/receptors of diverse specificity to harvest heterologous Fe-loaded siderophores. Other AOB appear to be similar to *N. europaea* (e.g., *N. oceani*; Klotz *et al.*, 2006) as they also have multiple OM transducer/receptors and no complete pathways for siderophore production. Heterotrophic bacteria such as *Pseudomonas* sp. usually become Fe-deficient at 1–2 μM (Hofte *et al.*, 1993; Mossialos *et al.*, 2000), even though they produce siderophores. To prevent Fe deficiency, *E. coli* (Neilands, 1995) and *Fusarium venenatum* (Wiebe, 2002) turn on siderophore synthesis when Fe is ≤ 1 μM. In other species that require high bio-available Fe, siderophore synthesis is turned on when Fe is ≤ 5 μM (Calugay *et al.*, 2003; Tindale *et al.*, 2000).

3. METHODS TO LEARN ABOUT THE PHYSIOLOGICAL RESPONSES TO FE LIMITATION

Iron in the environment although ubiquitous is scarcely bio-available because of its insolubility at neutral pH (Andrews *et al.*, 2003) and, as explained above, *N. europaea* is particularly adept at scavenging Fe at concentrations as low as 0.2 μM. In order to study Fe uptake, special attention must be paid as to how *N. europaea* cells are grown and prepared. In this section, we describe the protocols used in our laboratory that are necessary to study Fe uptake and homeostasis in *N. europaea*.

3.1. Preparation of Fe-free medium and glassware

All media, buffers and other reagents are made with ultra-pure water. Ultra-pure water is prepared with commercially available organic-extraction and deionization cartridges (i.e., Millipore, MA, USA) that produce water with 18 MΩ cm resistivity at 25 °C. Glassware that is Fe-free is prepared by soaking them in 1.0% HNO_3 overnight and rinsing thoroughly with ultra-pure water. Medium for *N. europaea* made from reagent-grade chemicals still contains approximately 0.2 μM Fe and is adequate to grow Fe-limited cells. Media completely devoid from Fe is made by treatment with a metal chelator such as Chelex (i.e., from Sigma, St. Louis MO, USA).

The addition of 0.2 µM Fe to this Fe-free medium is necessary to permit the growth of $N.$ $europaea$. It is best to use dedicated reagents, plastic ware and glassware while studying Fe metabolism in $N.$ $europaea$.

3.2. Cell growth and analysis

Cell growth of $N.$ $europaea$ in media containing 25 mM $(NH_4)_2SO_4$, 3.9 mM Na_2CO_3 and trace minerals (Ensign et al., 1993) can usually be monitored by optical density (Fig. 18.3) at 600 nm(OD_{600}). $N.$ $europaea$ attains approximately 0.07 OD_{600} in normal medium and 0.03 OD_{600} in Fe-limited medium in approximately 72 h (Fig. 18.3), typical stationary cell densities that are in contrast to the very low final cell densities attained in cultures of other nitrifiers such as $N.$ $multiformis$ or $N.$ $oceani$. An alternative method to monitor growth of nitrifiers is to monitor the production of the NO_2^- that results from the oxidation of ammonia (Fig. 18.1). The accumulation of NO_2^- correlates well with cell density and is suitable to assess small differences in growth. In addition, the measurement of NO_2^- in the medium requires a sample of only a few microliters and is well suited to working with the small culture volumes that are necessary (i.e., ~10 ml) when high-cost siderophores are used for growth studies (see below). The NO_2^- concentration can be determined colorimetrically at nanomolar levels with the Griess reagent that is made with sulfanilamide and N-naphtylethylenediamine (Hageman and Hucklesby, 1971). Cell growth can also be estimated by protein quantification by the Biuret method (Gornall et al., 1949) or by the commercially available Micro BCA Protein Assay Kit (Pierce, Rockford, IL, USA) following the directions of the manufacturer.

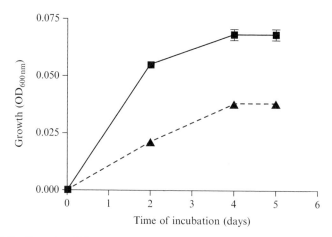

Figure 18.3 Growth of $N.$ $europaea$ (as measured by optical density) in media containing 10 µM (solid line) and 0.2 µM (dashed line) Fe.

The *N. europaea* cells for most analyses are harvested by centrifugation and suspended in phosphate-magnesium buffer (50 mM KH$_2$PO$_4$–2 mM MgSO$_4$, pH 7.8) to ~20 times of the original cell concentration. Phosphate-magnesium buffer is the buffer of choice to suspend cells of *N. europaea* to prepare them for most analyses.

The overall physiological status of *N. europaea* cultures are assessed by measuring the specific NH$_3$-dependent O$_2$ uptake activity (AMO activity), which requires both ammonia monooxygenase and hydroxylamine oxidoreductase activity in addition to the electron transport chain and NH$_2$OH-dependent O$_2$ uptake (HAO activity), which requires only hydroxylamine oxidoreductase activity coupled to the electron transport chain (Fig. 18.1). These activities reflect the cell's physiological status and are measured with a dissolved-oxygen meter equipped with a Clark-type O$_2$ probe (e.g., Yellow Springs Instruments Co, OH, USA). The AMO activity assay is usually carried out in no more than 2 ml of a washed ~1 OD cell suspension (i.e., equivalent to ~0.1 mg protein) in phosphate-magnesium buffer with 10 mM (NH$_4$)$_2$SO$_4$. After inhibiting AMO activity with 100 μM allylthiourea (Juliette *et al.*, 1993), the HAO activity is measured by adding 1 mM NH$_2$OH (the substrate for HAO) or 1 mM hydrazine hydrochloride (an alternative HAO substrate). The rate of O$_2$ consumption in these enzyme activity assays is expressed in nmol mg protein^{-1} min^{-1} assuming 230 μM O$_2$ in air-saturated buffer at 30 °C (Hommes *et al.*, 2006). Cells of *N. europaea* grown in 0.2 μM Fe exhibit lower O$_2$ uptake rates than cells grown in 10 μM Fe-medium (Table 18.1).

To prepare cell-free extracts, a concentrated cell suspension (2–20 ml) can be disrupted by sonication, for example, at 30% energy output for 10 s with a microprobe on an Ultrasonic Processor XL202 sonicator (Heat Systems-Ultrasonics, Inc., Farmingdale, NY), or alternatively, by passing through a French Pressure cell (e.g., Thermo Scientific, MA, USA) at 6000–7000 psi. It is recommended to keep the cell suspension cold during these procedures (e.g., by chilling the equipment and glassware prior to use) to avoid protein

Table 18.1 The NH$_3$-dependent and NH$_2$OH-dependent O$_2$ uptake activities of *N. europaea* cells grown in Fe-replete (10 μM) and Fe-limited (0.2 μM) media

Activity	Fe-replete	Fe-limited
Whole Cell		
NH$_3$-dependent O$_2$ uptake (AMO activity)[a]	89.17 ± 3.0	31.27 ± 2.4
NH$_3$-dependent O$_2$ uptake (AMO activity)[b]	1437 ± 34	676 ± 11
NH$_2$OH-dependent O$_2$ uptake (HAO activity)[a]	22 ± 0.8	8.64 ± 3.1
NH$_2$OH-dependent O$_2$ uptake (HAO activity)[b]	395.2 ± 4.1	169 ± 14

[a] nmol/min/OD.
[b] nmol/min/mg protein.

denaturation. Cell samples can be prepared simultaneously for different assays such as total protein content estimation in membrane and soluble fractions, quantification of soluble-heme content (see below), and peptide composition analysis (see below).

3.3. Cellular Fe content and allocation

Cells of *N. europaea* grown in Fe-limited medium acquire a pale-yellowish color in contrast to the reddish-brown colored cells grown in Fe-replete medium (Fig. 18.4). This easy visual inspection assures that the cells grown in Fe-limited medium are in fact iron deficient. The cells collected by centrifugation are washed at least twice in phosphate-magnesium buffer to remove nonspecifically bound Fe. The total cellular Fe content is determined by the ferrozine assay in a spectrophotometer at 550 nm. The ferrozine assay is linear between 1 and 30 nmol when $FeCl_3$ is used as standard (Riemer *et al.*, 2004). The cell lysis in 1 ml concentrated suspension is carried out with 200 µl of 50 mM NaOH for a few minutes until the suspension becomes transparent. To prevent Fe contamination, aliquots of the lysate are dispensed into polypropylene microcentrifuge tubes and used for total Fe and protein content determinations. To 100 µl of cell lysate, an equal volume of Fe-releasing reagent (1.4 M HCl and 4.5% (w/v) $KMnO_4$ in H_2O) is added and incubated 2 h at 60 °C in a fume hood. The addition of the Fe-releasing reagent is necessary, as the Fe^{2+}-ferrozine complex does not form in untreated samples due to unavailability of protein-bound Fe to

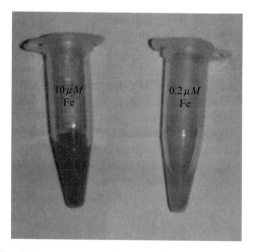

Figure 18.4 Cells of *N. europaea* grown in Fe-replete (10 µM) medium have a reddish darker color (in the picture dark gray) and those grown on Fe-limited (0.2 µM) medium have a pale-yellowish color (in the picture light gray).

ferrozine. After the sample cools, 30 μl of the ferrozine-reagent (6.5 M ferrozine, 6.5 mM neocuprine (Cu chelator), 2.5 M ammonium acetate, and 1 M ascorbic acid (the reducing agent) in water) are added and incubated for 30 min at room temperature and the absorbance is read at 550 nm. Alternatively, for total Fe determinations, the cells are digested in HNO_3 (1 M) until the cell suspension is completely transparent. In these digested samples, hydroxylamine•HCl (NH_2OH) is used as the reducing agent (e.g., 100 μl of digested cells, 100 μl of 0.5 mM NH_2OH, 10 μl of 50 mM ferrozine and H_2O to 1 ml). The samples are incubated 30 min as above and the absorbance is read at 550 nm. A consideration for acid digestion is that the cells suspension being digested must use almost all HNO_3 in the process. The digestion with HNO_3 has the advantage that samples with Fe concentrations below 10 μM can be submitted directly to analysis by inductively coupled-optical emission spectroscopy (ICP-OES). Facilities that use ICP-OES usually are available and are shared among multiple users at large institutions such as Oregon State University. Spent medium samples containing low Fe concentrations collected from Fe-limited growth experiments are acidified with HNO_3 similarly and sent to ICP-OES analysis. Using this methodology, it was possible to determine that the cells of N. europaea take up ~0.1 nmol/ml in Fe-limited cultures and ~1.1 nmol/ml in Fe-replete cultures (Table 18.2) and attests to its extraordinary adaptability to subsist in a wide range of Fe concentrations.

3.4. Heme c content determination

An important Fe-containing moiety in the utilization of NH_3 by N. europaea is heme c and was shown to vary in abundance depending on the availability of Fe in the cell (Wei et al., 2006). To measure heme c content after cell disruption, the preparation is subjected to ultracentrifugation at 150,000 RCF for 1 h at 4 °C, for example using a SW50.1 swinging bucket rotor

Table 18.2 Total Fe content and heme c content in the soluble membrane-free fraction of N. europaea cells grown in Fe-replete (10 μM) and Fe-limited (0.2 μM) media

Fe content and heme c	Fe-replete	Fe-limited
Whole Cell		
Fe (nmol/ml culture)	1.14 ± 0.19	0.11 ± 0.04
Fe (nmol/mg protein)	84.34 ± 4.71	21.16 ± 3.2
Cellular Fe concentration (mM)	9.12 ± 0.44	1.64 ± 0.09
Soluble Fraction		
Heme c (nmol/ml culture)	0.76 ± 0.06	0.31 ± 0.15
Heme c (nmol/mg protein)	8.17 ± 0.23	4.43 ± 0.51

in an L8-70 Ultracentrifuge, both from Beckman Instruments Inc (CA, USA). The supernatant containing soluble fractions is separated from the membrane pellet fraction and stored separately. The pellets containing crude membrane preparations are washed by thorough homogenization in TRIS buffer (0.1 M; pH 7.8) containing 1 M KCl (e.g., to a total volume no larger than the size of centrifuge tube), pooling and collecting the supernatant again after ultracentrifugation. The pellets can be suspended in TRIS buffer (50 mM; pH 7.8) if further analysis of the membranes is desired. Heme c content in the soluble fractions is quantified by dissociating the heme from the proteins using pyridine to form pyridine hemochromes as described by Berry and Trumpower (Berry and Trumpower, 1987). To assure complete oxidation of the heme, 0.5 ml of a solution containing 0.2 M NaOH, 40% (by volume) pyridine and 3 µl of ferricyanide (0.1 M $K_3Fe(CN)_6$) are added for heme concentrations between 1 and 2 µM. Since *N. europaea* has relatively high heme c concentrations, it is recommended to use 0.5 M NaOH, 20% pyridine and 6 µl of 0.1 M ferricyanide. The samples are thoroughly mixed and the oxidized spectrum is recorded within a minute. Solid sodium dithoionite ($Na_2S_2O_4$; 2–5 mg) from a fresh bottle is added to the cuvette, mixed, and several spectra of the reduced pyridine hemochromes are recorded until no differences are obtained. The absorption maxima of pyridine hemochrome c are recorded by scanning between 500 and 600 nm. To calculate the heme c concentration, the millimolar extinction coefficient (e) of 23.97 (reduced-oxidized at 550–540 nm) for pyridine hemochrome c in a 1-ml cuvette is used. The difference in absorbance (A) between the reduced form at 550 nm and oxidized form at 540, and the e value are used to calculate heme c content using below formula:

$$A \bullet e^{-1} = mM \text{ heme.}$$

It is to be noted that millimolar e-values for heme c vary between 20 and 24 in different reports (Berry and Trumpower, 1987), but the method described is adequate to document heme c concentration changes in *N. europaea*. Protein content in the original crude preparation is estimated as described above to normalize for cell mass. The heme c content (Table 18.2) of cells grown in Fe-limited medium decreased to ∼50% of that in cells grown in Fe-replete medium (Wei *et al.*, 2006).

3.5. Protein identification by gel electrophoresis and mass spectrometry

The cell's protein composition is dramatically affected by the Fe availability during growth (Fig. 18.5). Under low Fe environments, the cell modulates down the expression of Fe-containing proteins and increases the expression

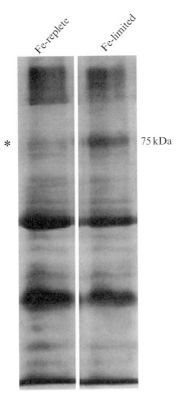

Figure 18.5 Analysis by SDS-PAGE of total membrane proteins of *N. europaea* cells grown in Fe-replete (10 μM) and Fe-limited (0.2 μM) media. The amount of total membrane protein loaded was ~6 μg per lane. Membrane proteins that were differentially expressed under different conditions are indicated in the figure by a star at 75 kDa.

of Fe-uptake proteins, events mostly controlled by Fur and small RNAs (Fig. 18.2 and below). Changes in cell-protein composition can be detected by SDS-polyacrylamide gel electrophoresis (SDS-PAGE) as described (Hyman and Arp, 1993), and selected proteins can be identified by LC tandem MS (LC/MS/MS) analysis. For a detailed method to perform one-dimensional gel electrophoresis, the reader is advised to consult *Methods of Enzymology*, Volume 463, Chapter 29 (Garfin, 2009).

For LC/MS/MS analysis, we recommend that protein bands in a polyacrylamide gel be visualized by staining with Coomassie Brilliant Blue R 250 and to follow the protocol used at the Mass Spectroscopy Facility at Oregon State University. The selected gel portions containing the proteins of interest are excised (e.g., to obtain a 5 mm by 1 mm by 0.5 mm slice), washed with 50 μl ultra-pure water two times and washed with 50% acetonitrile-25 mM ammonium bicarbonate two to four times until all the stain is removed. The gel slices are dehydrated with 100% acetonitrile until they turn opaque (about 15 min), decanted, and dried under vacuum. For protease digestion, a trypsin stock solution is prepared with sequencing grade-modified trypsin (Promega, WI, USA) in 50 mM acetic acid to 100 ng/μl final concentration. The working trypsin solution is prepared by diluting the stock solution 10 times with 25 mM ammonium bicarbonate, increasing the pH to 8 and is used immediately. To the gel slice, 25 μl of working solution is added and incubated 45 min on ice allowing infusion of the gel with trypsin solution. The excess trypsin solution is removed and 10–20 μl of 25 mM ammonium bicarbonate is added to keep it hydrated while digestion takes place at 37 °C for at least 4 h and up to 24 h. After digestion, the peptides in the solution are collected and the gel is extracted twice with 50% acetonitrile/25 mM ammonium carbonate for 15 min each to maximize the recovery of peptides. The extracts are pooled and concentrated to 5 μl or dried in a vacuum centrifuge for mass spectrometric analysis. The data in the MS/MS spectra are searched using a computer program such as Mascot (Matrix Science, London, UK) against a general protein database and the *N. europaea* sequence database (Chain *et al.*, 2003). This approach allowed us to identify the proteins of OM receptor proteins that were uniquely expressed in *N. europaea* grown in Fe-limited medium (Wei *et al.*, 2006).

3.6. Microscopy of cell structure

The growth conditions of *N. europaea* can affect the cell's structure, changes that can be visualized by electron microscopy (Fig. 18.6). The cells are harvested by centrifugation and washed two to three times in phosphate-magnesium buffer. The concentrated cells are fixed in glutaraldehyde and formaldehyde, postfixed in 1% OsO_4 at 20 °C (Schwalbach *et al.*, 1963), and then dehydrated in buffer-acetone mixtures (e.g., a gradual increase from 30% to 90% acetone content and finally in anhydrous acetone). The bacterial cells are infiltrated with Vestopal-W-embedding medium (Electron Microscopy Sciences PA, USA; for example, 1:3, 1:1; 3:1 to undiluted Vestopal, for 30 min at each concentration) and allowed to polymerize at 60 °C for 12–24 h. After polymerization, 1–2 μm sections are cut and dried onto slides; the slices are then floated in 1% uranyl acetate and Pb citrate as previously described (Murray and Watson, 1965). The thin sections of cells are examined usually in shared facilities equipped, for example, with a Philips CM-12 STEM transmission electron microscope. Differences in the cell membranes of *N. europaea* were apparent between cells grown in Fe-replete (Fig. 18.6A) and Fe-limited (Fig. 18.6B) media, mostly in the internalized membranes of the cells (Wei *et al.*, 2006). The membranes were irregular in shape in the Fe-limited cells, likely the result of lower contents of membranes with Fe-containing proteins (Fig. 18.6).

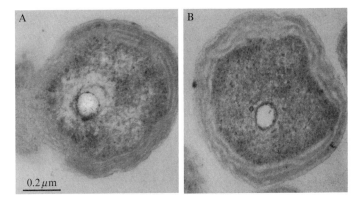

Figure 18.6 Transmission electron microscopic images of sections of *N. europaea* cells grown in (A) Fe-replete (10 μ*M*) medium and (B) Fe-limited (0.2 μ*M*) media. The images were taken at ×72,500 magnifications by the electron microscope facility at Oregon State University.

3.7. Microarrays to discover unexpected gene expression responses

With the available genome sequences of nitrifiers, and current technology, it is now possible to test global gene expression at a reasonable cost and in a relatively short time. For our Fe studies, we used NimbleChip 4-plex Made-to-Order microarrays for *N. europaea* (Roche NimbleGen Systems, Inc.), which are based on the published genome sequence (AL954747; Chain *et al.*, 2003). The genes in these microarrays are represented by probe sets with 14 pairs of 60-mer perfect match/mismatch oligo probes. We typically prepare *N. europaea* cells from three independent cultures grown at the same time and harvest at mid exponential phase by centrifugation (e.g., three iron-replete [300 ml culture at 0.04 OD] and three iron-limited cultures [500 ml culture at 0.025 OD]). Total cellular RNA is extracted in the presence of an RNAse inhibitor (e.g., Trysol; Ambion Inc., Austin, TX, USA). The extracted RNA is then purified using commercial kits (e.g., RNeasy Mini kit from Qiagen Inc. CA, USA) and treated with RNase-free DNase I to digest residual chromosomal DNA. It is important to use a known concentration (e.g., ≥ 20 μg) of good quality RNA for microarrays. The concentration of purified RNA is typically determined by spectrophotometry (e.g., with a Nanodrop spectrophotometer, Rockland, DE) and should have an A260/A280 ratio quality ≥ 1.85. The integrity of the RNA can be determined using a 6000-Nano LabChip kit on an Agilent Bioanalyzer 2100 (Agilent Technologies Inc., CA, USA) which is available at university laboratory core facilities. Intact, good quality RNA (>9.0 RIN score from the Bioanlyzer) is used for the experiments.

For convenience and for consistency, it is recommended to use dedicated facilities for cDNA synthesis, labeling, hybridization, scanning and data normalization of the microarrays. We used the services of Roche Nimblegen Core Facility, Iceland for processing of our RNA samples. To analyze normalized gene expression, data software such as DNASTAR ArrayStar v2.1 (Roche Nimblegen) was used. Differentially expressed genes are detected using a 2.0-fold change as a minimum for up- or downregulation. Student's t-Test with a cutoff p-value of 0.05 is used to compare the means of gene expression values for two individual replicates or for two groups of replicates for a given gene.

Microarray analysis revealed differential expression of at least 247 genes in *N. europaea* upon iron limitation (Arp laboratory, unpublished). Genes with higher transcript levels in Fe-limited cells included those with confirmed or assigned roles in iron acquisition, and genes with lower transcript levels included those encoding proteins containing iron or that use heme as a cofactor. The microarray studies enabled us to identify potential iron acquisition systems and proteins of unknown functions (Arp laboratory, unpublished and below). Specifically it was possible to identify the transcripts of six gene clusters with higher transcript levels under Fe-limiting conditions, some of which appear particularly interesting (Table 18.3). Gene cluster NE1540/1539/1538 was the most positively regulated under Fe-limiting conditions (Arp laboratory, unpublished). NE1540 encodes motifs typical of heme receptors in a TonB-dependent heme receptor homolog highly similar to HugA, the putative heme receptor of *Plesiomonas shigelloides* (Bracken et al., 1999). Following up with these results, we have demonstrated that *N. europaea* can use iron bound to hemoglobin as a source of Fe (Arp laboratory, unpublished). The protein encoded by NE1539 containing a chelatase domain may also play a role in acquiring iron from heme by distorting its ring structure, a possibility to be

Table 18.3 Genes highly expressed under Fe-limited conditions in *N. europaea*. Roles of these genes in *N. europaea* Fe metabolism are yet to be determined

Gene	Putative function	Fold expression
NE1540/1539/1538	Heme receptor/ Ferrochelatase/ Metalopeptidase	195/21/3
NE0726/0727	Liopocalin signature/ dihydrooratase	50/9
NE2038/2039	Myeloperoxidase/ hypothetical	8/3
NE2124/2125/2126	TonB-dependent PiuC	12/2/4
NE0508/0509/0510	SCO1/ hypotheticals	13/26/10
NE230/231	pdz domains/ laminin	10/9
NE0315	Cu oxidase	3
NE1543	Cu oxidase	8

investigated. NE1538 encodes a hypothetical protein with similarity to a membrane-bound metallopeptidase that may degrade peptide fragments of hemoglobin (Eggleson et al., 1999).

The transcripts NE0508/0509/0510-encoding hypothetical proteins were also present at higher levels under Fe-limiting conditions. Hypothetical proteins encoded by NE0508 and NE0510 appear to be conserved only in ammonia oxidizers such as *N. europaea*, *N. eutropha*, *N. multiformis*, and *N. oceani*, whereas a hypothetical protein encoded by NE0509 contains a Sco1-conserved domain and its homologs are present in other organisms including ammonia oxidizers. Sco1 proteins are copper chaperones that usually bind and insert copper ions into the active sites of cytochrome c oxidase subunits 1 and 2 (Palumaa et al., 2004). Copper chaperones were also shown to be necessary for incorporation of copper into multicopper ferroxidase Fet3, required for high affinity iron uptake in *S. cerevisiae* (Markossian and Kurganov, 2003; Taylor et al., 2005).

Iron limitation in *N. europaea* led to higher transcript levels of two multicopper oxidase genes (NE0315 and NE1543; Table 18.3). It will be interesting to study the possibility that NE0508/0509/0510 gene cluster might play a copper chaperone role in biogenesis of multicopper oxidases that are upregulated in Fe-limited conditions. Interestingly, the putative Fe-storage protein bacterioferritin (NE0863), in *N. europaea* was also at higher levels in Fe-limiting conditions (Arp laboratory, unpublished).

3.8. Corroboration of gene expression by real-time qPCR

It is customary to select a few key genes to validate the data that were obtained by microarrays. Often the comparisons between microarrays and real-time qPCR show differences in magnitude but should not show differences in the trend. For real-time qPCR, a commercial mix (e.g., SYBR-Green-I-detection system; Bio-Rad) and a real-time thermocycler (e.g., ICycler, Bio-Rad) were used following the directions of the manufacturers. The relative expression levels are calculated using dedicated software for data analysis.

The RNA for reverse-transcription PCR or real-time qPCR is isolated using a commercial kit, as above, or alternatively as described here. A quick method to isolate total RNA from *N. europaea* is to add 250 µl acidic phenol, 250 µl chloroform, SDS to 1%, and sodium acetate to 0.3 M–500 µl concentrated cells suspended in buffer in a microcentrifuge tube. After thoroughly mixing on a vortex for 1–2 min (Wei et al., 2004), the mix is centrifuged for 5 min at $16000 \times g$ to recover approximately 500 µl of the supernatant that is transferred to a clean microcentrifuge tube and the RNA is precipitated with 1 ml ethanol and recovered by centrifugation. The RNA pellet is dried almost completely, either by leaving the tube cap open on a bench, or by using a vacuum centrifuge (e.g., Speedvac, Savant, Thermo

Fisher USA). The pellets are suspended in about 50 μl of buffer or water. If RNAse contamination is a concern, the suspension liquid can be treated with 0.01% diethylpyrocarbonate overnight and autoclaved prior to use. The RNA samples are treated with DNase until no DNA product is detected by direct PCR. The cDNA is produced with M-MLV reverse transcriptase (Promega, Madison, WI, USA) and with the selected primer and a 50 °C extension temperature and PCR amplification of the cDNA with the second primer. The amount of transcript can be estimated by real-time qPCR as described above or by agarose gel electrophoresis. With this quick RNA extraction method, we have also shown that *N. europaea nitABC* genes have lower transcript levels in Fe-limited conditions compared to Fe-replete conditions. The *nitBC* genes in the *N. europaea* NitABC siderophore transporter mutant indeed are inactivated due to insertion of Km^r cassette and the NitABC siderophore transporter is not functional (Vajrala *et al.*, 2010).

4. Methods to Study Siderophore Uptake

The many siderophore transducers/receptors encoded in the *N. europaea* genome (Chain *et al.*, 2003) are indeed diverse (e.g., for ferrichrome, desferrioxamine, coprogen, pyoverdin, and for catechol/catecholate-type siderophores; Matzanke, 1991; Matzanke *et al.*, 1991). Transcript and protein expression experiments in our laboratory showed that at least seven siderophore transducer/receptors are induced in response to Fe limitation (Wei *et al.*, 2006). It is likely that the expression of other siderophore transducers/receptors is also enhanced by a dedicated siderophore while retaining specificity within a siderophore type. Concerted regulation of multiple specialized genes may be expected for a rapid coordination of expression. In this section, we describe the approach we use to study exogenous siderophore uptake in *N. europaea*.

4.1. Growth experiments in the presence of siderophores

A simple way to study uptake of a siderophore is to test whether its addition enhances the growth of Fe-limited cells. Often, this approach will give the first indications of correspondence between a siderophore and a transducer/receptor (see below). Cells grown in 0.2 μ*M* Fe-medium (to assure no carry over of excess Fe from the spent medium) are used to start a fresh culture in 0.2 μ*M* Fe-medium with a concentration of the siderophore equivalent to 10 μ*M*. An important consideration in these experiments is to identify a reliable source for siderophores (e.g., from EMC Microcollections, Germany) as on site siderophore production in some instances becomes cost prohibitive.

Exposure to a siderophore in Fe-limited medium may require an induction period that will be reflected in an extended lag phase during growth (Wei et al., 2006, 2007), so patience is required in siderophore incubations. Once growth is elicited, the detection of the corresponding siderophore transducer/receptor protein is performed by gel electrophoresis (in cell membranes) and identified by mass spectroscopy as explained above. Alternatively, quantitative reverse transcriptase PCR (real-time qPCR; see above) can be used to test for the expression of the candidate corresponding transducer/receptor gene of a particular siderophore type. We have characterized the Fe-DFX (desferal or ferrioxamine) uptake system (encoded by NE1097/1088) using this methodology. The induction of NE2433 (hydroxamate-type transducer/receptor associated with FecIR-like genes) was also detected specifically in response to ferrioxamine siderophore (Wei et al., 2007).

The definitive correspondence between a receptor and a siderophore is made by gene inactivation experiments. For example, three null mutants were constructed and characterized (Chapter 17). The null mutant in NE1097 and the double null mutant in NE1097/1088 failed to grow in Fe-limited medium containing 10 μM DFX. The null NE1088 mutant could only grow at Fe levels ≥ 0.4 μM when in the presence of 10 μM DFX. The gene inactivation experiments demonstrated that the efficient acquisition of Fe-loaded DFX required both TonB-dependent transducer/receptor genes NE1097 and NE1088.

4.2. Uptake of fluorescently labeled siderophore analogs and localization in the cell

There are siderophore analogs that can be used to test for the uptake of Fe through an OM siderophore receptor (Ardon et al., 1998; Meijler et al., 2002; Nudelman et al., 1998). The siderophore analogs are coupled to a fluorescently labeled probe, whose fluorescence is quenched when it chelates Fe. We have carried out experiments with a fluorescently labeled ferrichrome analog coupled to the fluorescent moiety naphthalic diimide (Fhu-NI) (Meijler et al., 2002) to test the specificity of the only annotated siderophore transporter encoded by *nitABC* in *N. europaea* (Vajrala et al., 2010). The analog Fhu-NI is not fluorescent in its iron-complex state and is fluorescent in the Fe-free form. Therefore, once iron is removed from the complex, an increase in fluorescence can be measured. The Fhu-NI analog is complexed with Fe^{3+} using $FeCl_3$ for 24 h and filter sterilized through a 0.2-μm cellulose acetate membrane. For Fhu-N1 uptake experiments, 2–3 day-old Fe-replete and Fe-limited cultures of *N. europaea* are harvested by centrifugation and suspended to 1.0 $OD_{600\ nm}$ in Fe-limited media. The ferrated fluorescently labeled ferrichrome analog is added to this cell suspension to a final 10 μM concentration and incubated with shaking for up to 24 h at 30 °C in the dark. During the incubation, 0.1 ml aliquots are harvested in a time course by centrifugation and suspended to the original volume in phosphate–magnesium buffer. The fluorescence emission of the cells is measured immediately by excitation at 400 nm and emission at 540 nm using a spectrofluorometer (e.g., Spectra- MAX GeminiXS; Molecular Devices, Sunnyvale, CA, USA). In growth experiments, we have demonstrated that the Fhu-NI analog promotes the growth of *N. europaea* in Fe-limited medium as efficiently as the natural ferrichrome, indicating that in *N. europaea*, the Fhu-NI analog is recognized and transported as efficiently as the native compound (Vajrala et al., 2010).

To demonstrate internalization of the Fhu-NI analog, spheroplasts are prepared as described (Suzuki and Kwok, 1969) during the time course. Cell aliquots are harvested, washed three times in 0.1 M potassium phosphate buffer, pH 7.5 at 4 °C and suspended in 0.25 M sucrose–0.1 M potassium phosphate, pH 7.5, 1 mM EDTA containing 10 µg of lysozyme to a final volume of 1 ml. The suspension is incubated for 2 h at 28 °C with gentle swirling. Subsequent centrifugation at 2000 × g for 20 min at 4 °C yields a pellet of spheroplasts, which is washed twice in 0.25 M sucrose–0.1 M potassium phosphate buffer pH 7.5 and suspended in the same buffer. The fluorescence emission in the spheroplasts is measured as described above. In order to compare spheroplasts with whole cells, an aliquot of cells not treated with lysozyme and EDTA are saved prior to preparation of the spheroplasts. Fluorescence of the Fhu-NI analog was detected in spheroplasts of *N. europaea* wild-type cells as early as 4 h after incubation with the ferrated Fhu-NI analog, and the intensity of the fluorescence increased with time indicating accumulation of the Fhu-NI analog in a deferrated state within the cytoplasm of *N. europaea* wild-type. The *N. europaea* NitABC siderophore transporter mutant (*nit::kan*) is incapable of transporting Fe-loaded hydroxamate-type siderophores (Table 18.4). When this mutant was grown in Fe-limited media and incubated with the ferrated Fhu-NI it

Table 18.4 Growth phenotypes of *N. europaea* wild-type, *nitB::kan* mutant in media containing different levels of Fe and siderophores

Treatment			Growth of *N. europaea* strains	
Fe	Siderophore	Siderophore type	Wild-type	*nitB::kan*
10.0 µM	0 µM	–	+	+
0.2 µM	0 µM	–	+	+
0.2 µM	10 µM Desferal	Hydroxamate	+	–
0.2 µM	10 µM Ferrichrome	Hydroxamate	+	–
0.2 µM	10 µM Aerobactin	Citrate-hydroxamate	+	+
0.2 µM	10 µM Pyoverdine	Mixed-chelating	+	–
0.2 µM	10 µM Enterobactin	Catecholate	+	+

showed no fluorescence increase incubated under the same conditions suggesting that the ferrated Fhu-NI analog enters the *N. europaea* cells via NitABC transporter. As expected, no fluorescence was detected from spheroplasts of *N. europaea* NitABC siderophore transporter mutant cells even after incubation for 48 h. The Fhu-NI ligand used in this study was found to be stable under the experimental conditions described (Vajrala *et al.*, 2010).

4.3. Labeled Fe feeding experiments

To test for the uptake of iron, isotopic iron (^{59}Fe) in ferric chloride form [specific activity >5 Ci g^{-1} (185 GBq g^{-1})] can be purchased from commercial suppliers such as PerkinElmer Inc. The ^{59}Fe isotope should be used in a designated lead-shielded area and handled with caution to prevent hazardous exposure, avoiding direct eye or skin exposure. The ^{59}Fe isotope is chelated to a siderophore (e.g., desferal, abbreviated DFX) in ultra-pure water pH 6.0. Cultures (OD$_{600}$ approx. 0.02) of the wild-type (and in the corresponding mutant if available) grown in Fe-limited medium are divided into treatment and control and tested for the uptake of ^{59}Fe-siderophore. To the control culture (to test for nonspecific binding of ^{59}Fe), 200 µM allylthiourea (an inhibitor of AMO activity) and 3 mM NaN$_3$ (inhibiting O$_2$ uptake by the cells) are added before the addition of ^{59}Fe-DFX (0.1 µM ^{59}Fe in 10 µM DFX). Fractions of the cultures are taken over a time course, and filtered through a 0.45-µm type HA filter (Millipore) to separate cells from unincorporated ^{59}Fe-DFX. The cells retained on the filters are washed with 5 ml 10 mM sodium citrate solution (pH 7), followed by 20 ml double-deionized water. The radioactivity in the cells attached to the filters is measured by liquid scintillation counting. Wild-type cells are able to uptake ^{59}Fe-DFX, but mutant cells deficient in the ferrioxamine transporter (encoded by NE1097/1088) are not. The mutant cells in the ferrioxamine transporter are still competent to uptake free ^{59}Fe (Wei *et al.*, 2007).

5. GENETIC COMPLEMENTATION TO STUDY *FUR* HOMOLOGS

Using genetic complementation studies, it is possible to demonstrate whether putative *fur* homologs encode functional Fur proteins. Using a Fur titration assay (FURTA) we have demonstrated that one *fur* homolog (NE0616) out of three in *N. europaea* encodes a functional Fur protein.

The gene to be tested with its promoter is cloned into a high copy plasmid and transformed into the test strain *E. coli* H1780 (Hantke, 1987). A negative control with the plasmid with no cloned gene is also transformed into the test strain. The *E. coli* H1780 strain was engineered to be *fur* deficient and to include the Fur-regulated gene *fiu* fused to a promoterless *lacZ* gene, which cannot be repressed, causing a Lac$^+$ phenotype (Hantke, 1987). A plasmid with a functional Fur transformed into this strain will produce a Lac$^-$ phenotype. The transformed strains are evaluated for the Lac phenotype on McConkey Lactose plates. Lac$^+$ strains (nonfunctional Fur) will produce a reddish color in the agar while Lac$^-$ strains will be colorless (functional Fur). The McConkey Lactose plates are supplemented with 30 μM Fe to ensure that a functional Fur is detected (Escolar *et al.*, 1999). In our studies, *E. coli* H1780 was transformed with four constructs to test three *fur* homologs and a negative control: H1780 (pFur616), H1780 (pFur616-kanC), H1780 (pFur730), and H1780 (pFur1722). The Lac$^+$ phenotype for *E. coli* H1780, whether grown in the presence or absence of added Fe supplement, should always be observed. Of the constructs, the only one that rescued the Fur defect in *E. coli* H1780 (a Lac$^-$ phenotype) was pFur616. Our results demonstrated that the *N. europaea* NE0616 *fur* homolog is expressed in *E. coli* in a functional form and is capable of regulating the Fur-dependent *fiu* promoter in H1780. The other *N. europaea fur* homologs (NE0730 and NE1722) were not capable of regulating the *fiu* promoter in H1780 (Arp laboratory, unpublished).

6. Methods for Gene Inactivation of Fe Uptake Systems

The Betaproteobacterium *N. europaea* is amenable to mutagenesis by electroporation and recombination allowing gene function validation of the many putative genes in Fe acquisition. The transformation parameters, and other transformation protocols and considerations are discussed in detail in Chapter 17 of this book. Basically, the gene of interest is cloned and inactivated by the insertion of an antibiotic cassette (kamamycin cassette). The construct is then transformed into *N. europaea* by electroporation. For this, *N. europaea* cells (0.5 l) from a culture in early-stationary phase ($A_{600}= 0.1$) are harvested by centrifugation and washed three times with sterile H_2O to rid them of residual buffer. Salts in electroporation procedures tend to cause arching in the cuvettes, killing the cells. The cells collected by centrifugation are suspended in 1.5 ml of sterile H_2O and kept on ice until use. Approximately 100 μl of the cell suspension with an approximate OD = 1 are mixed with 1 mg of DNA (1 mg/ml), placed in a 1-mm-gap electroporation cuvette and pulsed (e.g., with an electroporation system from BTX Instrument Division, MA, USA). Typical electroporation parameters are about 1200 V, 25 mF and resistance larger than 100 Ω. The cells are transferred immediately to 0.5 l of fresh medium and allowed to grow for 24 h under nonselective conditions at 30 °C with gentle agitation in a rotary shaker. The antibiotic for selection (e.g., kanamycin sulfate) is then added to a final concentration of 10 μg/ml. Cell growth is monitored by the accumulation of NO_2^- or by optical density at 600 nm.

A mutant constructed in our laboratory pertinent to Fe homeostasis is the *N. europaea fur* promoter knockout mutant strain (*fur:kanP*) that has enable us to test the expression of Fe-regulated genes and its physiological effects. The *fur:kanP* mutant strain accumulated 1.5-fold more Fe compared to the wild-type when grown in Fe-replete media. The *fur:kanP* mutant was also more sensitive to 500 μM Fe than the wild-type and was capable of utilizing Fe-bound ferrioxamine without any lag phase. Wild-type *N. europaea* requires an induction period of 2–3 days to utilize Fe-bound ferrioxamine (Arp laboratory, unpublished).

7. Lingering Questions that can be Answered by the Above Protocols

Catecholate-type siderophores such as enterobactin present less structural variation compared to hydroxamate-type siderophores (Matzanke, 1991; Matzanke *et al.*, 1991). In *N. europaea*, only two genes (NE0617 and NE1205) encode putative enterobactin receptors and are not associated with FecIR-like genes (Chain *et al.*, 2003). The siderophore receptor encoded by NE1205 was present at higher levels in Fe-limited treatments with no siderophore present (Wei *et al.*, 2006). In *N. europaea* there are at least 12 transducer/receptors for which bioinformatics cannot assign specific siderophore correspondence and which still need to be tested. Incubations with diverse siderophores might establish specificity for one of these unassigned genes provided they are induced by one of the siderophores (as confirmed by protein or transcript detection).

In the competition for biologically available Fe, the mechanisms for Fe acquisition, binding it tightly, and controlling its intracellular accumulation are all equally important to preserve Fe homeostasis, particularly in adverse conditions. Nitrifying bacteria often encounter environments with imbalances of a given essential substrate (i.e., Fe or O_2), thus proteins to prevent oxidative stress, or to store metals, are an advantage for growing under a wide range of Fe concentrations (0.2 to ∼500 μM).

7.1. Siderophore-independent Fe uptake

The *N. europaea* genome encodes putative high affinity siderophore-independent Fe uptake systems. Examples are the presence of seven multi-copper oxidase genes in its genome. Multicopper oxidases are involved in Fe acquisition in bacteria (Huston *et al.*, 2002), in yeast (Kosman, 2003;

Wang et al., 2003), and in *Chlamydomonas reinhardtii* (Herbik et al., 2002). In *N. europaea*, the transcripts of two putative multicopper oxidases were at higher levels in Fe-limited conditions and are likely involved in Fe uptake (Arp laboratory, unpublished).

Some pathogenic bacteria such as *N. meningitidis* utilize heme as an iron source (Linz et al., 2000). Interestingly, the transcript of a gene encoding a putative TonB-dependent heme receptor (NE1540) in *N. europaea* was expressed at high levels in Fe-limited conditions suggesting that it might use Fe bound to exogenous hemes (Arp laboratory, unpublished). Another possible mechanism to chelate Fe is through lipocalins, which are a functionally diverse family of proteins that generally bind small, hydrophobic ligands (Flower, 2000) and interact with cell-surface receptors (Flower et al., 2000). A protein called lipocalin 2 secreted by mammalian liver, spleen and macrophages can efficiently chelate Fe-enterobactin to keep bacterial infections in check (Flo et al., 2004). Lipocalin 2 was shown to have high affinity for enterochelin (10^{-10} M) hence it has been renamed siderocalin (Goetz et al., 2002; Yang et al., 2002). Lipocalin from human tears was also shown to bind bacterial and fungal siderophores (Fluckinger et al., 2004). In *N. europaea*, NE0726 encodes a hypothetical protein that shows motifs with a lipocalin signature domain and is highly expressed in Fe-limited conditions (Arp laboratory, unpublished). Whether *N. europaea* secretes a lipocalin-like peptide to sequester and obtain Fe remains to be tested.

The transcript from NE2323 with similarity to a Fe^{2+} transporter *feoB* was at lower level in Fe-replete than in Fe-deplete conditions. In an aerobic growth environment, Fe^{2+} would be scarce. The function of *feoB* and the level of contribution of Fe^{2+} to *N. europaea* Fe homeostasis may exist in environments with high competition for O_2. A putative gene encoding the Fe-storage protein bacterioferritin (NE0863) was at higher levels in Fe-replete conditions (Arp laboratory, unpublished).

7.2. Novel regulatory mechanisms for Fe homestasis

Through 36 *N. europaea* expression microarray experiments from our laboratory and OSU collaborators (available at Gene Expression Omnibus; GEO; NCBI) that include noncoding regions, the expression of 5 sRNAs was detected. Possible targets for these sRNAs included mRNAs for Fe-containing proteins such as the Fe-S containing protein succinate dehydrogenases (SDH) and other proteins of unknown function.

A higher level of the putative Fur-regulated sRNA (psRNA11) for SDH was detected in *N. europaea* grown in Fe-limited medium than in Fe-replete medium (Gvakharia et al., 2010). The cells grown in Fe-limited medium also showed low levels of the transcripts for Fe-containing proteins, including for SDH (Gvakharia et al., 2010). On the other hand, in the *N. europaea fur:kanP* strain the transcript for SDH was at a high level in Fe-replete and in

Fe-limited conditions suggesting that a SDH-specific sRNA may be regulated by Fur. The psRNA11 in the mutant had higher levels in Fe-replete medium than in Fe-limited medium (Gvakharia *et al.*, 2010). There is precedent in other microorganisms for posttranscriptional control of SDH expression by a Fur-regulated psRNA11 (Gaballa *et al.*, 2008). The mRNAs encoding enzymes involved in branched chain amino acid synthesis of Fe-S proteins such as 3-isopropylmalate dehydratase (NE0685), were also at lower levels in *N. europaea* under Fe limitation, but not in *N. europaea fur: kanP strain* (Arp laboratory, unpublished) suggesting regulation by Fur or by a Fur-regulated sRNA.

8. Conclusion

The methodology in this chapter offers the tools to expand Fe homeostasis studies to nitrifying bacteria other than *N. europaea* and to add to the understanding of their critical roles in global biochemical cycles (e.g., the nitrogen and carbon cycles). The continuation of genome sequencing programs of more nitrifying bacteria will most certainly help generate interesting hypotheses to study about Fe uptake and homeostasis in other ammonia-oxidizing microorganisms which could be tested with the methodology described here.

REFERENCES

Andrews, S. C., Robinson, A. K., and Rodriguez-Quinones, F. (2003). Bacterial iron homeostasis. *FEMS Microbiol. Rev.* **27**, 215–237.

Ardon, O., Nudelman, R., Caris, C., Libman, J., Shanzer, A., Chen, Y., and Hadar, Y. (1998). Iron uptake in Ustilago maydis: Tracking the iron path. *J. Bacteriol.* **180**, 2021–2026.

Arp, D. J., Sayavedra-Soto, L. A., and Hommes, N. G. (2002). Molecular biology of ammonia oxidation by *Nitrosomonas europaea*. *In* "Nitrogen Fixation: Global Perspectives. Proceedings of the 13th International Conference on Nitrogen Fixation," (T. M. Finan, M. R. O'Brian, D. B. Layzell, J. K. Vessey, and W. Newton, eds.), pp. 299–304. CABI Publishing, New York.

Berry, E. A., and Trumpower, B. L. (1987). Simultaneous determination of hemes *a*, *b*, and *c* from pyridine hemochrome spectra. *Anal. Biochem.* **161**, 1–15.

Bracken, C. S., Baer, M. T., Abdur-Rashid, A., Helms, W., and Stojiljkovic, I. (1999). Use of heme-protein complexes by the *Yersinia enterocolitica* HemR receptor: Histidine residues are essential for receptor function. *J. Bacteriol.* **181**, 6063–6072.

Cadieux, N., and Kadner, R. J. (1999). Site-directed disulfide bonding reveals an interaction site between energy-coupling protein TonB and BtuB, the outer membrane cobalamin transporter. *Proc. Natl. Acad. Sci. USA* **96**, 10673–10678.

Calugay, R. J., Miyashita, H., Okamura, Y., and Matsunaga, T. (2003). Siderophore production by the magnetic bacterium *Magnetospirillum magneticum* AMB-1. *FEMS Microbiol. Lett.* **218**, 371–375.

Chain, P., Lamerdin, J., Larimer, F., Regala, W., Lao, V., Land, M., et al. (2003). Complete genome sequence of the ammonia-oxidizing bacterium and obligate chemolithoautotroph Nitrosomonas europaea. J. Bacteriol. **185,** 2759–2773.

Eggleson, K. K., Duffin, K. L., and Goldberg, D. E. (1999). Identification and characterization of falcilysin, a metallopeptidase involved in hemoglobin catabolism within the malaria parasite Plasmodium falciparum. J. Biol. Chem. **274,** 32411–32417.

Ensign, S. A., Hyman, M. R., and Arp, D. J. (1993). In vitro activation of ammonia monooxygenase from Nitrosomonas europaea by copper. J. Bacteriol. **175,** 1971–1980.

Escolar, L., Perez-Martin, J., and de Lorenzo, V. (1999). Opening the iron box: Transcriptional metalloregulation by the Fur protein. J. Bacteriol. **181,** 6223–6229.

Flo, T. H., Smith, K. D., Sato, S., Rodriguez, D. J., Holmes, M. A., Strong, R. K., et al. (2004). Lipocalin 2 mediates an innate immune response to bacterial infection by sequestrating iron. Nature **432,** 917–921.

Flower, D. R. (2000). Beyond the superfamily: The lipocalin receptors. Biochim. Biophys. Acta **1482,** 327–336.

Flower, D. R., North, A. C., and Sansom, C. E. (2000). The lipocalin protein family: Structural and sequence overview. Biochim. Biophys. Acta **1482,** 9–24.

Fluckinger, M., Haas, H., Merschak, P., Glasgow, B. J., and Redl, B. (2004). Human tear lipocalin exhibits antimicrobial activity by scavenging microbial siderophores. Antimicrob. Agents Chemother. **48,** 3367–3372.

Gaballa, A., Antelmann, H., Aguilar, C., Khakh, S. K., Song, K. B., Smaldone, G. T., and Helmann, J. D. (2008). The Bacillus subtilis iron-sparing response is mediated by a Fur-regulated small RNA and three small, basic proteins. Proc. Natl. Acad. Sci. USA **105,** 11927–11932.

Garfin, D. E. (2009). One-dimensional gel electrophoresis. In "Methods in Enzymolohy," (R. R. Burges and M. P. Deutscher, eds.), pp. 497–513. Elsevier, San Francisco.

Goetz, D. H., Holmes, M. A., Borregaard, N., Bluhm, M. E., Raymond, K. N., and Strong, R. K. (2002). The neutrophil lipocalin NGAL is a bacteriostatic agent that interferes with siderophore-mediated iron acquisition. Mol. Cell **10,** 1033–1043.

Gornall, A. G., Bardawill, C. J., and David, M. M. (1949). Determination of serum proteins by means of the Biuret reaction. J. Biol. Chem. **177,** 751–766.

Gudmundsdottir, A., Bell, P. E., Lundrigan, M. D., Bradbeer, C., and Kadner, R. J. (1989). Point mutations in a conserved region (TonB box) of Escherichia coli outer membrane protein BtuB affect vitamin B12 transport. J. Bacteriol. **171,** 6526–6533.

Gvakharia, B. O., Tjaden, B., Vajrala, N., Sayavedra-Soto, L. A., and Arp, D. J. (2010). Computational prediction and transcriptional analysis of sRNAs in Nitrosomonas europaea. FEMS Microbiol. Lett. **312,** 46–54.

Hageman, R. H., and Hucklesby, D. P. (1971). Nitrate reductase in higher plants. Meth. Enzymol. **23,** 491–503.

Hantke, K. (1987). Selection procedure for deregulated iron transport mutants (fur) in Escherichia coli K 12: fur not only affects iron metabolism. Mol. Gen. Genet. **210,** 135–139.

Herbik, A., Bolling, C., and Buckhout, T. J. (2002). The involvement of a multicopper oxidase in iron uptake by the green algae Chlamydomonas reinhardtii. Plant Physiol. **130,** 2039–2048.

Hofte, M., Buysens, S., Koedam, N., and Cornelis, P. (1993). Zinc affects siderophore-mediated high affinity iron uptake systems in the rhizosphere Pseudomonas aeruginosa 7NSK2. Biometals **6,** 85–91.

Hommes, N. G., Kurth, E. G., Sayavedra-Soto, L. A., and Arp, D. J. (2006). Disruption of sucA, which encodes the E1 subunit of alpha-ketoglutarate dehydrogenase, affects the survival of Nitrosomonas europaea in stationary phase. J. Bacteriol. **188,** 343–347.

Hooper, A. B., Arciero, D. M., DiSpirito, A. A., Fuchs, J., Johnson, M., LaQuier, F., et al. (1990). Production of nitrite and N_2O by the ammonia-oxidizing nitrifiers. In "Nitrogen

Fixation: Achievements and Objectives," (P. M. Gresshoff, L. E. Roth, G. Stacey, and W. E. Newton, eds.), pp. 387–392. Chapman and Hall, New York.

Huston, W. M., Jennings, M. P., and McEwan, A. G. (2002). The multicopper oxidase of *Pseudomonas aeruginosa* is a ferroxidase with a central role in iron acquisition. *Mol. Microbiol.* **45,** 1741–1750.

Hyman, M. R., and Arp, D. J. (1993). An electrophoretic study of the thermal-dependent and reductant-dependent aggregation of the 27 kDa component of ammonia monooxygenase from *Nitrosomonas europaea*. *Electrophoresis* **14,** 619–627.

Juliette, L. Y., Hyman, M. R., and Arp, D. J. (1993). Mechanism-based inactivation of ammonia monooxygenase in *Nitrosomonas europaea* by allylsulfide. *Appl. Environ. Microbiol.* **59,** 3728–3735.

Klotz, M. G., Arp, D. J., Chain, P. S., El-Sheikh, A. F., Hauser, L. J., Hommes, N. G., *et al.* (2006). Complete genome sequence of the marine, chemolithoautotrophic, ammonia-oxidizing bacterium *Nitrosococcus oceani* ATCC 19707. *Appl. Environ. Microbiol.* **72,** 6299–6315.

Koebnik, R. (2005). TonB-dependent trans-envelope signalling: The exception or the rule? *Trends Microbiol.* **13,** 343–347.

Kosman, D. J. (2003). Molecular mechanisms of iron uptake in fungi. *Mol. Microbiol.* **47,** 1185–1197.

Linz, B., Schenker, M., Zhu, P., and Achtman, M. (2000). Frequent interspecific genetic exchange between commensal *Neisseriae* and *Neisseria meningitidis*. *Mol. Microbiol.* **36,** 1049–1058.

Markossian, K. A., and Kurganov, B. I. (2003). Copper chaperones, intracellular copper trafficking proteins. Function, structure, and mechanism of action. *Biochemistry Mosc.* **68,** 827–837.

Matzanke, B. F. (1991). Structures, coordination chemistry and functions of microbial iron chelates. *In* "CRC Hanbook of Microbial Iron Chelates," (G. Winkelmann, ed.), pp. 15–65. CRC Press, Boca Raton.

Matzanke, B. F., Berner, I., Bill, E., Trautwein, A. X., and Winkelmann, G. (1991). Transport and utilization of ferrioxamine-E-bound iron in *Erwinia herbicola* (*Pantoea agglomerans*). *Biol. Met.* **4,** 181–185.

Meijler, M. M., Arad-Yellin, R., Cabantchik, Z. I., and Shanzer, A. (2002). Synthesis and evaluation of iron chelators with masked hydrophilic moieties. *J. Am. Chem. Soc.* **124,** 12666–12667.

Mossialos, D., Meyer, J. M., Budzikiewicz, H., Wolff, U., Koedam, N., Baysse, C., *et al.* (2000). Quinolobactin, a new siderophore of *Pseudomonas fluorescens* ATCC 17400, the production of which is repressed by the cognate pyoverdine. *Appl. Environ. Microbiol.* **66,** 487–492.

Murray, R. G., and Watson, S. W. (1965). Structure of *Nitrosocystis Oceanus* and Comparison with *Nitrosomonas* and *Nitrobacter*. *J. Bacteriol.* **89,** 1594–1609.

Neilands, J. B. (1993). Siderophores. *Arch. Biochem. Biophys.* **302,** 1–3.

Neilands, J. B. (1995). Siderophores: Structure and function of microbial iron transport compounds. *J. Biol. Chem.* **270,** 26723–26726.

Nudelman, R., Ardon, O., Hadar, Y., Chen, Y., Libman, J., and Shanzer, A. (1998). Modular fluorescent-labeled siderophore analogues. *J. Med. Chem.* **41,** 1671–1678.

Palumaa, P., Kangur, L., Voronova, A., and Sillard, R. (2004). Metal-binding mechanism of Cox17, a copper chaperone for cytochrome *c* oxidase. *Biochem. J.* **382,** 307–314.

Quatrini, R., Jedlicki, E., and Holmes, D. S. (2005). Genomic insights into the iron uptake mechanisms of the biomining microorganism *Acidithiobacillus ferrooxidans*. *J. Ind. Microbiol. Biotechnol.* **32,** 606-614.

Ravel, J., and Cornelis, P. (2003). Genomics of pyoverdine-mediated iron uptake in pseudomonads. *Trends Microbiol.* **11,** 195–200.

Riemer, J., Hoepken, H. H., Czerwinska, H., Robinson, S. R., and Dringen, R. (2004). Colorimetric ferrozine-based assay for the quantitation of iron in cultured cells. *Anal. Biochem.* **331,** 370–375.

Schwalbach, G., Lickfeld, K. G., and Hoffmeister, H. (1963). Differentiated staining of osmium tetroxide-fixed, Vestopal-W-embedded tissue in thin sections. *Stain Technol.* **38,** 15–21.

Suzuki, I., and Kwok, S. C. (1969). Oxidation of ammonia by spheroplasts of *Nitrosomonas europaea*. *J. Bacteriol.* **99,** 897–898.

Taylor, A. B., Stoj, C. S., Ziegler, L., Kosman, D. J., and Hart, P. J. (2005). The copper-iron connection in biology: Structure of the metallo-oxidase Fet3p. *Proc. Natl. Acad. Sci. USA* **102,** 15459–15464.

Tindale, A. E., Mehrotra, M., Ottem, D., and Page, W. J. (2000). Dual regulation of catecholate siderophore biosynthesis in *Azotobacter vinelandii* by iron and oxidative stress. *Microbiology* **146**(Pt 7), 1617–1626.

Upadhyay, A. K., Petasis, D. T., Arciero, D. M., Hooper, A. B., and Hendrich, M. P. (2003). Spectroscopic characterization and assignment of reduction potentials in the tetraheme cytochrome c_{554} from *Nitrosomonas europaea*. *J. Am. Chem. Soc.* **125,** 1738–1747.

Vajrala, N., Sayavedra-Soto, L. A., Bottomley, P. J., and Arp, D. J. (2010). Characterization of *Nitrosomonas europaea nitABC* iron transporter. *Arch. Microbiol.* **192,** 899–908.

Visca, P., Leoni, L., Wilson, M. J., and Lamont, I. L. (2002). Iron transport and regulation, cell signalling and genomics: Lessons from *Escherichia coli* and *Pseudomonas*. *Mol. Microbiol.* **45,** 1177–1190.

Wang, T. P., Quintanar, L., Severance, S., Solomon, E. I., and Kosman, D. J. (2003). Targeted suppression of the ferroxidase and iron trafficking activities of the multicopper oxidase Fet3p from *Saccharomyces cerevisiae*. *J. Biol. Inorg. Chem.* **8,** 611–620.

Wei, X., Sayavedra-Soto, L., and Arp, D. (2004). The transcription of the *cbb* operon in *Nitrosomonas europaea*. *Microbiology* **150,** 1869–1879.

Wei, X. M., Vajrala, N., Hauser, L., Sayavedra-Soto, L. A., and Arp, D. J. (2006). Iron nutrition and physiological responses to iron stress in *Nitrosomonas europaea*. *Arch. Microbiol.* **186,** 107–118.

Wei, X., Sayavedra-Soto, L. A., and Arp, D. J. (2007). Characterization of the ferrioxamine uptake system of *Nitrosomonas europaea*. *Microbiology* **153,** 3963–3972.

Whittaker, M., Bergmann, D., Arciero, D., and Hooper, A. B. (2000). Electron transfer during the oxidation of ammonia by the chemolithotrophic bacterium *Nitrosomonas europaea*. *Biochim. Biophys. Acta* **1459,** 346–355.

Wiebe, M. G. (2002). Siderophore production by *Fusarium venenatum* A3/5. *Biochem. Soc. Trans.* **30,** 696–698.

Yang, J., Goetz, D., Li, J. Y., Wang, W., Mori, K., Setlik, D., *et al.* (2002). An iron delivery pathway mediated by a lipocalin. *Mol. Cell* **10,** 1045–1056.

CHAPTER NINETEEN

PRODUCTION OF RECOMBINANT MULTIHEME CYTOCHROMES C IN *WOLINELLA SUCCINOGENES*

Melanie Kern *and* Jörg Simon

Contents

1. Introduction	430
2. Bacterial Cytochrome c Biogenesis Systems	433
3. Recombinant Cytochrome c Production	435
4. Cytochromes c in the Epsilonproteobacterium *W. succinogenes*	435
5. Three Cytochrome c Heme Lyase Isoenzymes in *W. succinogenes*	439
6. Heterologous Production of Cytochromes c in *W. succinogenes*	439
7. Conclusions and Perspectives	442
Acknowledgments	443
References	443

Abstract

Respiratory nitrogen cycle processes like nitrification, nitrate reduction, denitrification, nitrite ammonification, or anammox involve a variety of dissimilatory enzymes and redox-active cofactors. In this context, an intriguing protein class are cytochromes c, that is, enzymes containing one or more covalently bound heme groups that are attached to heme c binding motifs (HBMs) of apocytochromes. The key enzyme of the corresponding maturation process is cytochrome c heme lyase (CCHL), an enzyme that catalyzes the formation of two thioether linkages between two vinyl side chains of a heme and two cysteine residues arranged in the HBM. In recent years, many multiheme cytochromes c involved in nitrogen cycle processes, such as hydroxylamine oxidoreductase and cytochrome c nitrite reductase, have attracted particular interest. Structurally, these enzymes exhibit conserved heme packing motifs despite displaying very different enzymic properties and largely unrelated primary structures. The functional and structural characterization of cytochromes c demands their purification in sufficient amounts as well as the feasibility to generate site-directed enzyme variants. For many interesting

Institute of Microbiology and Genetics, Department of Biology, Technische Universität Darmstadt, Darmstadt, Germany organisms, however, such systems are not available, mainly hampered by genetic inaccessibility, slow growth rates, insufficient cell yields, and/or a low capacity of cytochrome c formation. Efficient heterologous cytochrome c overproduction systems have been established using the unrelated proteobacterial species *Escherichia coli* and *Wolinella succinogenes*. In contrast to *E. coli*, *W. succinogenes* uses the cytochrome c biogenesis system II and contains a unique set of three specific CCHL isoenzymes that belong to the unusual CcsBA-type. Here, *W. succinogenes* is presented as host for cytochrome c overproduction focusing on a recently established gene expression system designed for large-scale production of multiheme cytochromes c.

1. INTRODUCTION

The conversion of compounds that are part of the biogeochemical nitrogen cycle is catalyzed by a diverse set of metalloenzymes that contain iron, copper, and molybdenum cofactors (Bothe *et al.*, 2007; Moir, 2011). The most widespread class of such enzymes are cytochromes c, that is, proteins that contain one or more covalently bound heme c groups. Such enzymes are crucial for the catalysis of several respiratory processes within the nitrogen cycle such as nitrification, periplasmic nitrate reduction, denitrification, nitrite ammonification, and anaerobic ammonium oxidation (Simon, 2011; see Table 19.1 for an overview). Presumably, the most important multiheme cytochromes c of the nitrogen cycle are cytochrome c nitrite reductase (NrfA), hydroxylamine oxidoreductase (Hao), and hydrazine oxidoreductase (Hzo) that are members of a family of evolutionary-related hemoproteins (Klotz *et al.*, 2008). These enzymes often interact with other electron-transferring multiheme cytochromes c and some of them are also involved in electron transfer between the membranous quinone/quinol pool and the site of nitrogen compound conversion (Simon and Kern, 2008; Simon *et al.*, 2008). Prominent examples are cytochrome c_{554} and cytochrome c_{m552} from nitrifying bacteria, members of the NrfH/NapC family as well as the NrfB and NapB multiheme cytochromes c (Table 19.1).

Cytochromes c are characterized by the covalent attachment of heme b (Fe-protoporphyrin IX) to an apo-cytochrome c via a so-called heme c binding motif (HBM), most commonly a Cys-X-X-Cys-His (CX_2CH) sequence. They are typically located in extracytoplasmic compartments like the periplasm or the anammoxosome, either as soluble proteins or attached to the corresponding membrane system. Classification of multiheme cytochromes c is mainly based on structural heme c group arrangements known as heme c packing motifs (Einsle *et al.*, 1999, 2000; Iverson *et al.*, 1998). Frequently occurring diheme c packing motifs involve parallel or perpendicular arrangements of the two heme porphyrin rings which are

Table 19.1 Compilation of selected cytochromes c involved in respiratory N-cycle processes

Respiratory pathway and cytochrome c designation	Properties and function
Nitrification	
Hao (hydroxylamine oxidoreductase)	Periplasmic octaheme cytochrome c catalyzing hydroxylamine oxidation to nitrite: $NH_2OH + H_2O \rightarrow NO_2^- + 4e^- + 5H^+$.
Cyt. c_{554}	Periplasmic redox mediator catalyzing electron transfer from Hao to Cyt. c_{m552}, 4 heme c groups.
Cyt. c_{m552}	Membrane-bound quinone reductase of the NapC/NrfH family; 4 heme c groups.
Nitrite ammonification	
NrfA (cytochrome c nitrite reductase)	Periplasmic pentaheme cytochrome c catalyzing nitrite reduction to ammonium: $NO_2^- + 6e^- + 8H^+ \rightarrow NH_4^+ + 2H_2O$.
NrfH	Membrane-bound quinol dehydrogenase of the NapC/NrfH family; 4 heme c groups; redox partner of NrfA.
NrfB	Periplasmic pentaheme cytochrome c; redox partner of NrfA (alternative to NrfH in, e.g., enterobacteria).
Periplasmic nitrate reduction	
NapB	Periplasmic diheme cytochrome c; redox partner of periplasmic nitrate reductase (NapA).
NapC	Membrane-bound quinol dehydrogenase of the NapC/NrfH family; 4 heme c groups; redox partner of NapB.
Denitrification	
NirS	Cytochrome cd_1 nitrite reductase catalyzing nitrite reduction to nitric oxide: $NO_2^- + 1e^- + 2H^+ \rightarrow NO + H_2O$.
dNirK	Copper-dependent nitrite reductase containing a C-terminal monoheme cytochrome c domain that probably acts as an electron entry point. Catalyzes the same reaction as NirS.
dNor	Membrane-bound complex containing a monoheme cytochrome c subunit (NorC) catalyzing nitric oxide reduction to nitrous oxide: $2NO + 2e^- + 2H^+ \rightarrow N_2O + H_2O$.

(*continued*)

Table 19.1 (continued)

Respiratory pathway and cytochrome c designation	Properties and function
cNosZ (cytochrome c nitrous oxide reductase)	Periplasmic nitrous oxide reductase containing a C-terminal monoheme cytochrome c domain that probably acts as an electron entry point; catalyzes nitrous oxide reduction to dinitrogen: $N_2O + 2e^- + 2H^+ \rightarrow N_2 + H_2O$.
Cyt. c_{550}	Periplasmic monoheme cytochrome c that mediates electron transfer from the cytochrome bc_1 complex to denitrifying reductases catalyzing reduction of nitrite, nitric oxide, or nitrous oxide.
Anaerobic ammonium oxidation (Anammox)	
Hzo (hydrazine oxidoreductase)	Hao-type octaheme cytochrome c that catalyzes hydrazine oxidation to dinitrogen: $N_2H_4 \rightarrow N_2 + 4e^- + 4H^+$.
Hydrazine hydrolase	Diheme cytochrome c thought to catalyze hydrazine production from nitric oxide and ammonium: $NO + NH_4^+ + 3e^- + 2H^+ \rightarrow N_2H_4 + H_2O$.

then combined to form extended redox chains that enable fast and efficient intramolecular electron transport. Three-dimensional structure models of cytochromes *c* are therefore clearly superior to primary structure alignments in order to predict distinct functional properties. Unfortunately however, for many organisms sufficient amounts of purified cytochromes *c* suitable for structural and functional characterization are often hard to obtain, hampered, for example, by slow growth rates of cells, insufficient cell yields and/or a low capacity of cytochrome *c* formation. For these reasons, recombinant production of cytochromes *c* using appropriate bacterial host cells is highly desirable. Furthermore, this approach offers the opportunity to easily produce cytochrome *c* variants containing site-directed modifications.

In Section 2 of this chapter, a brief overview is given on how cytochromes *c* are synthesized in different bacteria as this knowledge is crucial to identify bacterial hosts suitable for recombinant cytochrome *c* production. Section 3 gives a short overview on how this task is experimentally addressed in *Escherichia coli* whereas Sections 4–6 describe the Epsilonproteobacterium *Wolinella succinogenes* as a potential alternative to *E. coli*. Based on the metabolic properties of *W. succinogenes* and on our understanding of its cytochrome *c* biogenesis machinery, a genetic system is presented that allows efficient and reliable overproduction of multiheme cytochromes *c*.

2. BACTERIAL CYTOCHROME *c* BIOGENESIS SYSTEMS

Heme *c* attachment is a posttranslational modification that requires a complex enzymic maturation system. The most important enzyme in this process is cytochrome *c* heme lyase (CCHL) that catalyzes the formation of two thioether linkages derived from the two vinyl side chains of a heme molecule and two thiol groups of the cysteine residues arranged in the HBM. Two distinct maturation systems for bacterial cytochrome *c* synthesis have been described (see Ferguson *et al.*, 2008; Kranz *et al.*, 2009 for recent reviews). These system are known as system I (or the Ccm system, found, e.g., in Alpha- and Gammaproteobacteria) and system II (the Ccs system; present, e.g., in Epsilonproteobacteria and Gram-positives, see Fig. 19.1). Common to both is the export of the unfolded apo-cytochrome *c* via the membrane-integral Sec apparatus as well as a redox-active protein machinery that catalyzes transmembrane delivery of electrons in order to reduce the cysteine thiol groups within the HBM. The Ccm system of *E. coli* consists of eight proteins that are encoded by the *ccmABCDEFGH* gene cluster (Richard-Fogal *et al.*, 2009; Sanders *et al.*, 2010). In contrast, a maximum of four proteins (CcsB, CcsA, ResA, and CcdA) seem to be involved in system II (Fig. 19.1). Most likely, the CCHL in system I is CcmF whereas a complex of CcsB and CcsA seems to carry out this function in system II (Ahuja *et al.*, 2009). Notably,

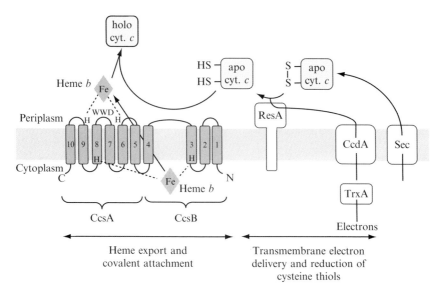

Figure 19.1 Overview of the cytochrome *c* biogenesis system II in Epsilonproteobacteria. Note that these bacteria contain one or more CcsBA-type CCHLs that most likely function in both heme export and periplasmic heme attachment to the reduced cysteine thiol groups of the heme *c* binding motif. The transmembrane segments of CcsBA are numbered 1–10 and the N- and C-termini are indicated. WWD and H respectively designate the WWD domain and essential histidine residues thought to be involved in axial heme ligation during export of the cofactor to the periplasmic space. See text for details.

epsilonproteobacterial genomes encode fusion proteins called CcsBA that form 10 transmembrane segments (Fig. 19.1) (Frawley and Kranz, 2009; Hartshorne *et al.*, 2006; Kern *et al.*, 2010a). The CcsBA proteins from *Helicobacter pylori* and *H. hepaticus* were both found to restore cytochrome *c* maturation in system I-deficient *E. coli* strains indicating that CcsBA-type enzymes function as CCHL (Feissner *et al.*, 2006; Goddard *et al.*, 2010; Richard-Fogal *et al.*, 2007). *H. hepaticus* CcsBA was purified from *E. coli* and shown to contain reduced heme *b* (Frawley and Kranz, 2009; Merchant, 2009). Thus, it was suggested that the CcsBA-type CCHL mediates heme export to the periplasmic space. According to this model, two pairs of conserved histidine residues are part of one cytoplasmic and one periplasmic heme binding pocket that are essential to facilitate heme export and might also help to maintain heme *b* in the reduced state. Variants of several CcsBA-type CCHLs from Epsilonproteobacteria in which any of the four histidines was replaced by alanine were unable to support cytochrome *c* biogenesis (Frawley and Kranz, 2009; Kern *et al.*, 2010b). After export it is assumed that heme is bound by an essential extracellular tryptophan-rich motif (known as the WWD domain, present in CcsBA as well as in CcmC and CcmF) that might be crucial to present the cofactor in a structural orientation suitable for CCHL-catalyzed attachment to the HBM

(Frawley and Kranz, 2009; Hamel et al., 2003; Kern et al., 2010b; Ren et al., 2002; Richard-Fogal and Kranz, 2010; Schulz et al., 2000). The molecular details of CCHL function are not understood in both bacterial maturation systems and no high-resolution structural model of a CCHL is available.

3. Recombinant Cytochrome c Production

Efficient heterologous production of multiheme cytochromes c requires host cells that can be easily grown to high cell densities and produce a cytochrome c biogenesis system exhibiting a high maturation capacity. This last point is of particular importance as overburdening of the biogenesis system would result in an inhomogeneous population of cytochromes c containing proteins which may lack one or more of its heme c groups. In addition, the ideal host is accessible to genetic engineering and known for stable heterologous protein production.

The organism that is most commonly used for heterologous cytochrome c production is the Gamma proteobacterium E. coli. These cells are easy to manipulate and suitable vector systems for gene overexpression are widely used. However, wild-type E. coli cells produce cytochromes c as well as the Ccm maturation system only under anaerobic growth conditions. Six different endogenous cytochromes c are encoded on the E. coli genome: components of the cytochrome c nitrite reductase system (NrfA and NrfB, Table 19.1), electron transport proteins (NapC, TorC, DmsC), and YhjA which is similar to bacterial diheme cytochrome c peroxidases. Since cell yields of E. coli are low under anaerobic growth conditions, a genetic system was developed by Linda Thöny-Meyer and colleagues that allowed cytochrome c production during aerobic cell cultivation (Arslan et al., 1998). In this experimental approach, E. coli cells contain plasmid pEC86 that promotes constitutive expression of the entire ccm gene cluster. This procedure turned out to be successful for the production of some (predominantly soluble) multiheme cytochromes c (see, e.g., Londer et al., 2005; Hoffmann et al., 2009) but failed for others. At present, the applicability of this method for a given cytochrome cannot be predicted which makes an alternative genetic system desirable, preferably using system II-dependent cells. In the following sections, key features of W. succinogenes are described that characterize this organism as such a candidate host.

4. Cytochromes c in the Epsilonproteobacterium W. succinogenes

The nonfermentative rumen bacterium W. succinogenes is a well-established and thoroughly investigated model organism of bacterial anaerobic respiration (Kern and Simon, 2009a; Kröger et al., 2002). The organism uses a diverse range of terminal electron acceptors like nitrate (which is reduced via nitrite to ammonium), nitrous oxide, fumarate, dimethyl sulfoxide, or polysulfide and can be grown in cheap minimal media to optical cell densities of about 1–1.5 (measured at 578 nm) (Kern and Simon, 2009a; Klimmek et al., 2004; Kröger et al., 2002; Simon, 2002; Simon et al., 2004). Prominent electron donors are formate or hydrogen gas. The genome sequence of W. succinogenes encodes 23 endogenous cytochromes c (Baar et al., 2003). Twelve of these are multiheme cytochromes c that harbor between two and eight HBMs (Table 19.2; see Fig. 1 in Kern et al., 2010a for a comprehensive list of the properties of each W. succinogenes cytochrome c). Judged from spectrophotometric analysis, cells grown by respiratory nitrate ammonification contained 3.1 ± 0.2 µmol heme c per gram of cell protein which is about three times more than under fumarate-respiring conditions (Simon et al., 2000). Heme staining of cell homogenates from W. succinogenes cells grown by anaerobic respiration with nitrate, fumarate, or polysulfide as terminal electron acceptor revealed different cytochrome c patterns (Simon et al., 2000, 2001). A representative gel shown in Fig. 19.2 demonstrates that the pentaheme cytochrome c nitrite reductase NrfA is by far the most prominent cytochrome c during nitrate respiration (Kern et al., 2010a; Simon et al., 2000). NrfA reduces nitrite to ammonium in the periplasmic space and is anchored to the membrane via the tetraheme cytochrome c NrfH, a member of the NapC/NrfH family (Einsle et al., 2000; Kern and Simon, 2009a; Kern et al., 2008; Simon and Kern, 2008; Simon et al., 2001). A metabolic function has been assigned to about half of the W. succinogenes multiheme cytochromes c (Table 19.2). W. succinogenes cells tolerate high Fe^{2+} concentration in the medium and are able to use hemin as sole iron source (our unpublished results).

Genetic manipulation of W. succinogenes is feasible and convenient despite the fact that plasmid replication has never been observed. Genomic deletion mutants lacking up to several kilo base pairs were initially described by Krafft et al. (1995). This study reported the deletion of polysulfide reductase-encoding psr genes which was achieved through double homologous recombination between a deletion plasmid and the W. succinogenes wild-type genome. Using this approach, an accurately defined part of the genome can be replaced by an antibiotic resistance gene cassette conferring resistance to either kanamycin (kan gene) or chloramphenicol (cat gene). Deletion mutants have been constructed for many of the genes encoding prominent electron transport enzymes in W. succinogenes (Gross et al., 1998a; Simon et al., 1998, 2000, 2003). To enable expression of genes that had been subjected to site-directed gene mutagenesis, deletion mutants of W. succinogenes were complemented by plasmid integration, that is, by single DNA recombination events between a suitable plasmid and the genome. In this way, entire operons encoding

Table 19.2 Multiheme cytochromes c encoded on the *W. succinogenes* genome [a]

Cytochrome c designation	Predicted molecular mass of mature holocytochrome c	Number of CX$_2$CH HBMs	Properties and function
MccA (Ws0379)	79,684	7 plus CX$_{15}$CH motif	Octaheme cytochrome c of unknown function
NrfA (Ws0969)	58,331	4 plus CX$_2$CK motif	Cytochrome c nitrite reductase (nitrite-reducing subunit) [b]
NrfH (Ws0970)	22,131	4	Menaquinol-oxidizing redox partner of NrfA [b]
FccC (Ws0121)	26,776	4	Member of the NapC/NrfH family
FccB (Ws0122)	16,073	4	Possible subunit of a periplasmic flavocytochrome c complex
Ws0710	13,186	4	Function not known
NapB (Ws1174)	18,886	2	Redox partner of periplasmic nitrate reductase NapA [b]
Ccp (Ws1491)	36,513	2	Cytochrome c peroxidase
CytA (Ws0009)	40,028	2	Function not known
PetC (Ws2154)	32,955	2	Diheme cytochrome c subunit of cytochrome bc_1 complex [c]
CcoP (Ws0181)	32,166	2	Subunit III of cytochrome cbb_3 oxidase [c]
Ws0705	45,434	2	Function not known

[a] Data taken from Kern *et al.* (2010a).
[b] See Table 19.1.
[c] Cytochrome c components of a microaerobic respiratory electron transport chain typical for Epsilonproteobacteria.

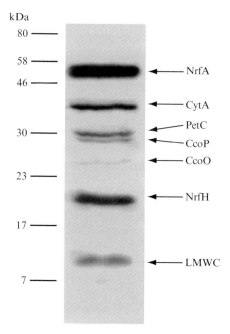

Figure 19.2 Cytochrome *c* profile of *W. succinogenes* cells grown by respiratory nitrate ammonification. Cells were grown anaerobically as described (Kern and Simon, 2009b) and an aliquot of the corresponding cell homogenate (100 μg of protein) was subjected to SDS-PAGE. Heme staining of the polyacrylamide gel was performed with 3,3-dimethoxybenzidine according to the method of Francis and Becker (1984). LMWC, low-molecular-weight cytochromes *c*.

several subunits of membrane-bound complexes were restored on the genome of *W. succinogenes* gene deletion mutants (Gross *et al.*, 1998a,b; Kern *et al.*, 2007; Pisa *et al.*, 2002; Simon *et al.*, 1998). Subsequently, variants of many enzymes were produced in *W. succinogenes* including modified subunits of the membrane-bound nitrite reducing NrfHA complex (Gross *et al.*, 2005; Kern *et al.*, 2008; Pisa *et al.*, 2002). Plasmid integration was also used to replace promoter regions on the *W. succinogenes* genome in order to synthesize gene products that are not detectable under standard laboratory growth conditions. For example, the multiheme cytochrome *c* gene *mccA* encoding an octaheme protein of unknown function was brought under the control of the fumarate reductase promoter which allowed MccA purification from cells grown by fumarate respiration (Hartshorne *et al.*, 2007). This study was the first example of genetically engineered cytochrome *c* overproduction in *W. succinogenes*.

5. Three Cytochrome c Heme Lyase Isoenzymes in W. succinogenes

W. succinogenes cells use the cytochrome c biogenesis system II and contain an inventory of three specific CCHLs called NrfI, CcsA1, and CcsA2 in addition to CcdA and ResA proteins (Fig. 19.1). The presence of multiple CCHL genes in a genome is unusual and points to the presence of unconventional HBMs in some of the encoded cytochromes c (Hartshorne et al., 2006). All three CCHLs of W. succinogenes belong to the CcsBA-type of fusion proteins described earlier and were shown to play specific roles in cytochrome c maturation (Kern et al., 2010a). NrfI was shown to be dedicated to the attachment of the unique active site heme c group in cytochrome c nitrite reductase NrfA that is bound via a CX_2CK HBM (Pisa et al., 2002). The lysine residue of this special HBM serves as an unusual proximal heme c iron ligand, a function which is normally accomplished by the histidine residue of the canonical CX_2CH HBM (Einsle et al., 1999, 2000). CcsA1 is apparently dedicated to mature the octaheme cytochrome c MccA which contains a conserved $CX_{15}CH$ sequence in addition to seven conventional CH_2CH HBMs and it was shown recently by mass spectroscopy that both cysteine residues of the $CX_{15}CH$ motif are involved in covalent heme binding (Hartshorne et al., 2007). In contrast to the *nrfI* and *ccsA1* genes, it proved impossible to delete *ccsA2* from the genome of anaerobically respiring W. succinogenes cells suggesting that cytochrome c biogenesis is essential for cell proliferation (Kern et al., 2010a). CcsA2 most likely recognizes the CX_2CH HBM. When each of the three CCHLs from W. succinogenes was produced in a *ccm*-deficient E. coli strain, only CcsA2 was found to be functional in heme attachment to the CH_2CH motifs of a diheme cytochrome c reporter protein (Kern et al., 2010a). In contrast, NrfI and CcsA1 were unable to attach heme to engineered CX_2CK or $CX_{15}CH$ motifs implying that the specialized CCHLs recognize specific structural features in their cognate apo-cytochromes in addition to the HBM (Kern et al., 2010a). All three CcsBA-type isoenyzmes of W. succinogenes contain the essential histidine residues and the WWD domain described above (Kern et al., 2010b).

6. Heterologous Production of Cytochromes c in W. succinogenes

The high capacity to produce cytochromes c during anaerobic respiration combined with its genetic amenability makes W. succinogenes a suitable candidate for heterologous cytochrome c overproduction. A recent study describes the heterologous production of the cytochrome c nitrite reductases (NrfA proteins) from *E. coli* and *Campylobacter jejuni* in *W. succinogenes* (Kern et al., 2010a). In the genomes of the corresponding *W. succinogenes* mutants, the endogenous *nrfA* gene was replaced by a 5′-truncated *nrfA* gene from either *E. coli* or *C. jejuni* in order to retain the signal peptide-encoding DNA stretch of *W. succinogenes nrfA* (Fig. 19.3). The constructed fusion genes encode two consecutive Strep-tag sequences at their 3′-ends to enable affinity chromatography of overproduced cytochromes c. Growth of *W. succinogenes* by nitrate respiration induced the formation of the heterologous NrfA proteins which were obtained as functional proteins active in nitrite reduction (Fig. 19.3). The NrfA proteins from *E. coli* and *C. jejuni* were both found in the soluble, that is, the periplasmic cell fraction,

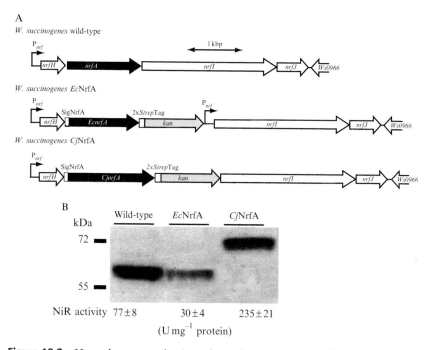

Figure 19.3 Heterologous production of cytochrome c nitrite reductases in *W. succinogenes*. (A) Partial physical maps of the genomes of *W. succinogenes* wild type, *W. succinogenes* EcNrfA and *W. succinogenes* CjNrfA (Kern et al., 2010a). (B) Heme stain analysis of cell homogenates from the indicated *W. succinogenes* strains grown by nitrate respiration (100 μg of protein per lane). NiR activity refers to specific nitrite reductase activities measured photometrically with reduced benzyl viologen as artificial electron donor (Kern et al., 2010a). One unit of enzyme activity is defined as the oxidation of 2 μmol benzyl viologen per minute. P_{nrf}, *W. succinogenes* cytochrome c nitrite reductase promoter element; SigNrfA, DNA stretch encoding the signal peptide of *W. succinogenes* NrfA; 2x*Strep*Tag, DNA region encoding two consecutive Strep-tags.

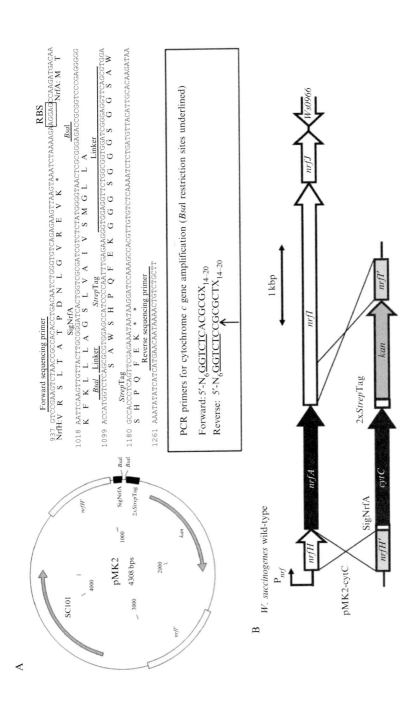

Figure 19.4 Strategy for heterologous expression of cytochrome *c* genes in *W. succinogenes*. (A) Map and key features of the expression vector pMK2 designed for cytochrome *c* gene cloning. Incorporating a cytochrome *c* gene of interest creates a fusion gene that encodes a pre-apocytochrome *c* containing the signal peptide of *W. succinogenes* NrfA and two consecutive C-terminal Strep-tags separated by a linker sequence. Prior to cloning, a cytochrome *c* gene of interest is amplified using a pair of primers that contain *Bsa*I restriction sites at their 5'-ends in addition

indicating that they did not form a complex with *W. succinogenes* NrfH. Production of active NrfA from *E. coli* required coexpression of the *W. succinogenes nrfI* gene indicating that this dedicated CCHL can also handle *E. coli* NrfA (Kern et al., 2010a). In contrast, *C. jejuni* NrfA does not contain a special CX_2CK motif but five CX_2CH motifs, an exceptional feature found in some proteobacteria. Accordingly, the maturation of active NrfA from *C. jejuni* in *W. succinogenes* was independent of NrfI (Kern et al., 2010a).

Based on the genetic strategy described in Fig. 19.3, a specifically designed low-copy expression vector (pMK2) was constructed to facilitate cloning of cytochrome *c* genes in *E. coli* (Fig. 19.4A). Two *Bsa*I cloning sites in pMK2 allow the seamless incorporation of any PCR-amplified cytochrome *c* gene. The PCR primer pair described in Fig. 19.4A ensures that the NrfA signal peptide and two Strep-tags are attached at either end of the encoded cytochrome *c*. After transformation of *W. succinogenes*, double homologous recombination between the genome and the pMK2 derivative results in the replacement of endogenous *nrfA* by the cytochrome *c* gene of interest and the kanamycin resistance gene (Fig. 19.4B; note that pMK2 contains partial *nrfH* and *nrfI* nucleotide sequences flanking the cytochrome *c* gene to promote recombination). To ensure the production of substantial amounts of cytochromes *c*, synthesis of a custom-made codon-optimized cytochrome *c* gene is recommended in order to accurately reflect the codon usage of *W. succinogenes* (which is rather different from that of *E. coli* or *Campylobacter* species). Further optimized expression vectors that, for instance, will utilize the fumarate reductase promoter for cytochrome *c* gene expression are currently constructed in our laboratory.

7. Conclusions and Perspectives

The results presented in this chapter demonstrate that *W. succinogenes* is a valuable host for cytochrome *c* production and a promising alternative to commonly used *E. coli* cells. The capacity to produce multiheme cytochromes *c* with special catalytic functions and/or heme attached to

---

to template-specific sequences of 14–20 bases (boxed). Note that *Bsa*I is an unusual restriction enzyme as it cleaves one nucleotide downstream of its recognition site (the cleavage site is indicated by a vertical arrow). Thus, the restricted PCR fragment carries one 5′-CGCG and one 5′-GCGC overhang suitable for directed ligation. Plasmid pMK2 contains a kanamycin resistance gene (*kan*) and two DNA regions (designated *nrfH′* and *nrfI′*) necessary for recombination with the *W. succinogenes* genome. (B) Double homologous recombination between the *W. succinogenes* wild-type genome and a pMK2 derivative (pMK2-cytC) results in a modified *nrf* gene cluster and the expression of the introduced cytochrome *c* fusion gene is induced under nitrate-respiring conditions. RBS, ribosome-binding site; see legend to Fig. 19.3 for further abbreviations.

nonstandard HBMs makes work with *W. succinogenes* even more exciting. Possible future cytochrome *c* targets are multiheme cytochromes *c* listed in Table 19.1, members of the MccA cytochrome *c* family, or the various multiheme cytochromes *c* involved in respiratory processes that are part of the biogeochemical sulfur and iron cycles. Furthermore, it is more than likely that future metagenomic studies will reveal a wealth of uncharacterized cytochrome *c* genes that might encode hemoproteins with novel catalytic functions. Such cytochromes can be conveniently identified by searching for HBM-encoding DNA sequences or for CCHL genes situated in the vicinity of cytochromes *c* genes. Finally, one might also consider to produce heterologous respiratory networks in *W. succinogenes* that comprise various cytochromes *c* (and possibly other metalloproteins) in a synthetic microbiology approach.

ACKNOWLEDGMENTS

The authors thank Juliane Scheithauer and Marlen Lepper for their contributions to the work presented in this chapter and the Deutsche Forschungsgemeinschaft for financial support (grants SI 848/1-1 and 2-1 to JS).

REFERENCES

Ahuja, U., Kjelgaard, P., Schulz, B. L., Thöny-Meyer, L., and Hederstedt, L. (2009). Haem-delivery proteins in cytochrome *c* maturation system II. *Mol. Microbiol.* **73,** 1058–1071.

Arslan, E., Schulz, H., Zufferey, R., Künzler, P., and Thöny-Meyer, L. (1998). Overproduction of the *Bradyrhizobium japonicum* c-type cytochrome subunits of the *cbb*$_3$ oxidase in *Escherichia coli*. *Biochem. Biophys. Res. Commun.* **251,** 83–87.

Baar, C., Eppinger, M., Raddatz, G., Simon, J., Lanz, C., Klimmek, O., Nandakumar, R., Gross, R., Rosinus, A., Keller, H., Jagtap, P., Linke, B., *et al.* (2003). Complete genome sequence and analysis of *Wolinella succinogenes*. *Proc. Natl. Acad. Sci. USA* **100,** 11690–11695.

Bothe, H., Ferguson, S. J., and Newton, W. E. (eds.), (2007). Biology of the Nitrogen Cycle, Elsevier, Amsterdam, The Netherlands.

Einsle, O., Messerschmidt, A., Stach, P., Bourenkov, G. P., Bartunik, H. D., Huber, R., and Kroneck, P. M. H. (1999). Structure of cytochrome *c* nitrite reductase. *Nature* **400,** 476–480.

Einsle, O., Stach, P., Messerschmidt, A., Simon, J., Kröger, A., Huber, R., and Kroneck, P. M. H. (2000). Cytochrome *c* nitrite reductase from *Wolinella succinogenes* - Structure at 1.6 Å resolution, inhibitor binding, and heme-packing motifs. *J. Biol. Chem.* **275,** 39608–39616.

Feissner, R. E., Richard-Fogal, C. L., Frawley, E. R., Loughman, J. A., Earley, K. W., and Kranz, R. G. (2006). Recombinant cytochromes *c* biogenesis systems I and II and analysis of haem delivery pathways in *Escherichia coli*. *Mol. Microbiol.* **60,** 563–577.

Ferguson, S. J., Stevens, J. M., Allen, J. W. A., and Robertson, I. B. (2008). Cytochrome *c* assembly: A tale of ever increasing variation and mystery? *Biochim. Biophys. Acta* **1777,** 980–984.

Francis, R. T., and Becker, R. R. (1984). Specific indication of hemoproteins in polyacrylamide gels using a double-staining process. *Anal. Biochem.* **136,** 509–514.

Frawley, E. R., and Kranz, R. G. (2009). CcsBA is a cytochrome c synthetase that also functions in heme transport. *Proc. Natl. Acad. Sci. USA* **106,** 10201–10206.

Goddard, A. D., Stevens, J. M., Rondelet, A., Nomerotskaia, E., Allen, J. W. A., and Ferguson, S. J. (2010). Comparing the substrate specificities of cytochrome *c* biogenesis systems I and II: Bioenergetics. *FEBS J.* **277,** 726–737.

Gross, R., Simon, J., Theis, F., and Kröger, A. (1998a). Two membrane anchors of *Wolinella succinogenes* hydrogenase and their function in fumarate and polysulfide respiration. *Arch. Microbiol.* **170,** 50–58.

Gross, R., Simon, J., Lancaster, C. R. D., and Kröger, A. (1998b). Identification of histidine residues in *Wolinella succinogenes* hydrogenase that are essential for menaquinone reduction by H_2. *Mol. Microbiol.* **30,** 639–646.

Gross, R., Eichler, R., and Simon, J. (2005). Site-directed modifications indicate differences in axial haem *c* iron ligation between the related NrfH and NapC families of multihaem *c*-type cytochromes. *Biochem. J.* **390,** 689–693.

Hamel, P. P., Dreyfuss, B. W., Xie, Z., Gabilly, S. T., and Merchant, S. (2003). Essential histidine and tryptophan residues in CcsA, a system II polytopic cytochrome *c* biogenesis protein. *J. Biol. Chem.* **278,** 2593–2603.

Hartshorne, R. S., Richardson, D. J., and Simon, J. (2006). Multiple haem lyase genes indicate substrate specificity in cytochrome *c* biogenesis. *Biochem. Soc. Trans.* **34,** 146–149.

Hartshorne, R. S., Kern, M., Meyer, B., Clarke, T. A., Karas, M., Richardson, D. J., and Simon, J. (2007). A dedicated haem lyase is required for the maturation of a novel bacterial cytochrome *c* with unconventional covalent haem binding. *Mol. Microbiol.* **64,** 1049–1060.

Hoffmann, M., Seidel, J., and Einsle, O. (2009). CcpA from *Geobacter sulfurreducens* is a basic di-heme cytochrome *c* peroxidase. *J. Mol. Biol.* **393,** 951–965.

Iverson, T. M., Arciero, D. M., Hsu, B. T., Logan, M. S. P., Hooper, A. B., and Rees, D. C. (1998). Heme packing motifs revealed by the crystal structure of the tetra-heme cytochrome c_{554} from *Nitrosomonas europaea*. *Nat. Struct. Biol.* **5,** 1005–1012.

Kern, M., and Simon, J. (2009a). Electron transport chains and bioenergetics of respiratory nitrogen metabolism in *Wolinella succinogenes* and other Epsilonproteobacteria. *Biochim. Biophys. Acta* **1787,** 646–656.

Kern, M., and Simon, J. (2009b). Periplasmic nitrate reduction in *Wolinella succinogenes*: cytoplasmic NapF facilitates NapA maturation and requires the menaquinol dehydrogenase NapH for membrane attachment. *Microbiology* **155,** 2784–2794.

Kern, M., Mager, A. M., and Simon, J. (2007). Role of individual *nap* gene cluster products in NapC-independent nitrate respiration of *Wolinella succinogenes*. *Microbiology* **153,** 3739–3747.

Kern, M., Einsle, O., and Simon, J. (2008). Variants of the tetrahaem cytochrome *c* quinol dehydrogenase NrfH characterize the menaquinol binding site, the haem *c* binding motifs and the transmembrane segment. *Biochem. J.* **414,** 73–79.

Kern, M., Eisel, F., Scheithauer, J., Kranz, R. G., and Simon, J. (2010a). Substrate specificity of three cytochrome *c* haem lyase isoenzymes from *Wolinella succinogenes*: Unconventional haem *c* binding motifs are not sufficient for haem *c* attachment by NrfI and CcsA1. *Mol. Microbiol.* **75,** 122–137.

Kern, M., Scheithauer, J., Kranz, R. G., and Simon, J. (2010b). Essential histidine pairs indicate conserved haem binding in epsilonproteobacterial cytochrome *c* haem lyases. *Microbiology* **156,** 3773–3781.

Klimmek, O., Dietrich, W., Dancea, F., Lin, Y.-L., Pfeiffer, S., Löhr, F., Rüterjans, H., Gross, R., Simon, J., and Kröger, A. (2004). Sulfur respiration. *In* "Respiration in

Archaea and Bacteria," (D. Zannoni, ed.) pp. 217–232. Springer, Dordrecht, The Netherlands.

Klotz, M. G., Schmid, M. C., Strous, M., op den Camp, H. J. M., Jetten, M. S. M., and Hooper, A. B. (2008). Evolution of an octahaem cytochrome *c* protein family that is key to aerobic and anaerobic ammonia oxidation by bacteria. *Env. Microbiol.* **10,** 3150–3158.

Krafft, T., Gross, R., and Kröger, A. (1995). The function of *Wolinella succinogenes psr* genes in electron transport with polysulphide as the terminal electron acceptor. *Eur. J. Biochem.* **230,** 601–606.

Kranz, R. G., Richard-Fogal, C., Taylor, J.-S., and Frawley, E. R. (2009). Cytochrome *c* biogenesis: Mechanisms for covalent modifications and trafficking of heme and for heme-iron redox control. *Microbiol. Mol. Biol. Rev.* **73,** 510–528.

Kröger, A., Biel, S., Simon, J., Gross, R., Unden, G., and Lancaster, C. R. D. (2002). Fumarate respiration of *Wolinella succinogenes*: Enzymology, energetics and coupling mechanism. *Biochim. Biophys. Acta* **1553,** 23–38.

Londer, Y. Y., Pokkuluri, P. R., Erickson, J., Orshonsky, V., and Schiffer, M. (2005). Heterologous expression of hexaheme fragments of a multidomain cytochrome from *Geobacter sulfurreducens* representing a novel class of cytochromes *c*. *Protein Expr. Purif.* **39,** 254–260.

Merchant, S. (2009). His protects heme as it crosses the membrane. *Proc. Natl. Acad. Sci. USA* **106,** 10069–10070.

Moir, J. W. B. (ed.) (2011). The Molecular Basis of the Nitrogen Cycle, Horizon Scientific Press, Norwich, UK.

Pisa, R., Stein, T., Eichler, R., Gross, R., and Simon, J. (2002). The *nrfI* gene is essential for the attachment of the active site haem group of *Wolinella succinogenes* cytochrome *c* nitrite reductase. *Mol. Microbiol.* **43,** 763–770.

Ren, Q., Ahuja, U., and Thöny-Meyer, L. (2002). A bacterial cytochrome *c* heme lyase. CcmF forms a complex with the heme chaperone CcmE and CcmH but not with apocytochrome c. *J. Biol. Chem.* **277,** 7657–7663.

Richard-Fogal, C., and Kranz, R. G. (2010). The CcmC:heme:CcmE complex in heme trafficking and cytochrome *c* biosynthesis. *J. Mol. Biol.* **401,** 350–362.

Richard-Fogal, C. L., Frawley, E. R., Feissner, R. E., and Kranz, R. G. (2007). Heme concentration dependence and metalloprotein inhibition of the system I and II cytochrome *c* assembly pathways. *J. Bacteriol.* **189,** 455–463.

Richard-Fogal, C. L., Bonner, E. R., Zhu, H., San Francisco, B., and Kranz, R. G. (2009). A conserved haem redox and trafficking pathway for cofactor attachment. *EMBO J.* **28,** 2349–2359.

Sanders, C., Turkarslan, S., Lee, D.-W., and Daldal, F. (2010). Cytochrome *c* biogenesis: The Ccm system. *Trends Microbiol.* **18,** 266–274.

Schulz, H., Pellicioli, E. C., and Thöny-Meyer, L. (2000). New insights into the role of CcmC, CcmD and CcmE in the haem delivery pathway during cytochrome *c* maturation by a complete mutational analysis of the conserved tryptophan-rich motif of CcmC. *Mol. Microbiol.* **37,** 1379–1388.

Simon, J. (2002). Enzymology and bioenergetics of respiratory nitrite ammonification. *FEMS Microbiol. Rev.* **26,** 285–309.

Simon, J. (2011). Organisation of respiratory electron transport chains in nitrate-reducing and nitrifying bacteria. *In* "The Molecular Basis of the Nitrogen Cycle," (J. W. B. Moir, ed.).Horizon Scientific Press, Norwich, UK, (in press).

Simon, J., and Kern, M. (2008). Quinone-reactive proteins devoid of haem *b* form widespread membrane-bound electron transport modules in bacterial anaerobic respiration. *Biochem. Soc. Trans.* **36,** 1011–1016.

Simon, J., Gross, R., Ringel, M., Schmidt, E., and Kröger, A. (1998). Deletion and site-directed mutagenesis of the *Wolinella succinogenes* fumarate reductase operon. *Eur. J. Biochem.* **251,** 418–426.

Simon, J., Gross, R., Einsle, O., Kroneck, P. M. H., Kröger, A., and Klimmek, O. (2000). A NapC/NirT-type cytochrome *c* (NrfH) is the mediator between the quinone pool and the cytochrome *c* nitrite reductase of *Wolinella succinogenes. Mol. Microbiol.* **35,** 686–696.

Simon, J., Pisa, R., Stein, T., Eichler, R., Klimmek, O., and Gross, R. (2001). The tetraheme cytochrome *c* NrfH is required to anchor the cytochrome *c* nitrite reductase (NrfA) in the membrane of *Wolinella succinogenes. Eur. J. Biochem.* **268,** 5776–5782.

Simon, J., Sänger, M., Schuster, S. C., and Gross, R. (2003). Electron transport to periplasmic nitrate reductase (NapA) of *Wolinella succinogenes* is independent of a NapC protein. *Mol. Microbiol.* **49,** 69–79.

Simon, J., Einsle, O., Kroneck, P. M. H., and Zumft, W. G. (2004). The unprecedented *nos* gene cluster of *Wolinella succinogenes* encodes a novel respiratory electron transfer pathway to cytochrome *c* nitrous oxide reductase. *FEBS Lett.* **569,** 7–12.

Simon, J., van Spanning, R. J. M., and Richardson, D. J. (2008). The organisation of proton motive and non-proton motive redox loops in prokaryotic respiratory systems. *Biochim. Biophys. Acta* **1777,** 1480–1490.

CHAPTER TWENTY

Techniques for Investigating Hydroxylamine Disproportionation by Hydroxylamine Oxidoreductases

A. Andrew Pacheco,* Jennifer McGarry,* Joshua Kostera,* and Angel Corona[†]

Contents

1. Introduction	448
2. General Considerations	450
2.1. The need for anaerobicity	450
2.2. Data analysis using complete spectral information	453
3. Ammonia Concentration Determination	454
3.1. Principle	454
3.2. Materials	454
3.3. Procedure	455
3.4. Comments	456
4. Nitric Oxide Concentration Determination	457
4.1. Principle	457
4.2. Materials	457
4.3. Procedure	457
4.4. Comments	458
5. Nitrite Concentration Determination	458
5.1. Principle	458
5.2. Materials	459
5.3. Procedure	459
5.4. Comments	460
6. Nitrous Oxide and Dinitrogen Concentration Determination	460
6.1. Principle	460
6.2. Materials and procedure	460
6.3. Comments	462
References	462

* Department of Chemistry and Biochemistry, University of Wisconsin-Milwaukee, Milwaukee, Wisconsin, USA
[†] Rufus King High School, Milwaukee, Wisconsin, USA

Abstract

Hydroxylamine, an important intermediate in ammonia oxidation by ammonia oxidizing bacteria (AOB), is inherently unstable with respect to disproportionation. The process is slow in neutral solutions, but could potentially be catalyzed by enzymes such as the hydroxylamine oxidoreductases, which normally catalyze the oxidation of ammonia to nitrite in the AOB. Disproportionation could be physiologically important to some AOB under microaerobic conditions, and could also confound *in vitro* analyses if it occurs and is not taken into consideration. This chapter presents methods for detecting ammonia, nitric oxide, nitrite, nitrous oxide, and isotopically labeled dinitrogen, which are the most thermodynamically favored products of hydroxylamine disproportionation.

1. Introduction

By definition, disproportionation is an oxidation–reduction reaction in which the same compound acts as both the oxidizing and reducing agent (Atkins *et al.*, 2010). Compounds susceptible to disproportionation will be inherently unstable, even in the absence of other redox-active compounds. Hydroxylamine is an example of a compound that is susceptible to disproportionation in aqueous solution, at every pH from 0 to 14. In ammonia oxidizing bacteria (AOB), the hydroxylamine oxidoreductases (HAOs) typically catalyze the four-electron oxidation of NH_2OH to NO_2^-, which is the second step in ammonia-dependent respiration. The possibility that HAO enzymes will catalyze NH_2OH disproportionation as well as its oxidation by a physiological electron acceptor is ever present, and should be tested for whenever a new HAO is isolated, for two reasons. First, disproportionation could potentially confound *in vitro* analyses if not accounted for. Second, disproportionation could prove physiologically relevant for some AOB living in microaerobic conditions. This chapter summarizes some of the techniques that can be used to test for HAO-catalyzed NH_2OH disproportionation.

There are numerous reaction pathways by which NH_2OH could disproportionate, of which Scheme 20.1 summarizes the four that are thermodynamically most favored. Figure 20.1 shows how the apparent standard cell potentials for these four processes vary with pH. All four ε_{cell}^0 values exhibit maxima at around pH 6, but remain high over the entire pH range. Disproportionation to produce N_2 or N_2O is considerably more favorable than the reactions that produce NO or NO_2^-; however, the former reactions require the formation of an N–N bond, which is likely to be kinetically slower at lower NH_2OH concentrations (Kostera *et al.*, 2008).

Figure 20.2 shows a schematic of HAO from *Nitrosomonas europaea*, the only HAO for which a structure is available to date (Igarashi *et al.*, 1997). This enzyme is a homotrimer that contains eight hemes/monomer. The

Reduction half reaction

$$NH_2OH + 3H^+ + 2e^- \longrightarrow NH_4^+ + H_2O$$

Oxidation half reactions

$$NO + 3H^+ + 3e^- \longrightarrow NH_2OH$$

$$NO_2^- + 5H^+ + 4e^- \longrightarrow NH_2OH + H_2O$$

$$N_2O + H_2O + 4H^+ + 4e^- \longrightarrow 2NH_2OH$$

$$N_2 + 2H_2O + 2H^+ + 2e^- \longrightarrow 2NH_2OH$$

Net disproportionation reactions

$$5NH_2OH + 3H^+ \longrightarrow 3NH_4^+ + 2NO + 3H_2O$$

$$3NH_2OH + H^+ \longrightarrow 2NH_4^+ + NO_2^- + H_2O$$

$$4NH_2OH + 2H^+ \longrightarrow N_2O + 2NH_4^+ + 3H_2O$$

$$3NH_2OH + H^+ \longrightarrow N_2 + NH_4^+ + 3H_2O$$

Scheme 20.1 The most thermodynamically favorable disproportionation reactions of NH_2OH. The corresponding oxidative and reductive half reactions are also given.

active sites are three unique hemes referred to as P_{460}s, which have a vacant site where NH_2OH molecules can bind and react. The remaining 21 hemes are typical 6-coordinate c-hemes that under physiological conditions serve to efficiently move electrons away from the active sites. As seen in Fig. 20.2 though, the circular arrangement of the c-hemes could in principle also allow facile electron transfer *between* the P_{460} active sites, in a way that would lead to catalytic disproportionation. Our experiments with *N. europaea* HAO have shown that in this particular form of HAO, disproportionation is suppressed, despite the apparently favorable enzyme architecture (Kostera *et al.*, 2010). The biochemical assays described herein will allow future researchers to test HAOs from other AOB for their ability to catalyze NH_2OH disproportionation. In addition, because these assays allow detection of many nitrogen cycle intermediates, they can also be readily adapted to monitor product distributions from other HAO-catalyzed reactions. We limit ourselves to describing assays for detection of the most likely products of NH_2OH disproportionation, NH_4^+, NO, NO_2^-, N_2O, and N_2. Hydroxylamine itself can be detected using an assay described by Frear and Burrell (1955). Other disproportionation products such as N_2H_4, NO_3^-, N_2O_4, and HNO are also possible, but these would be expected to disproportionate or comproportionate[1] in their turn to give the more stable species mentioned above.

---

[1] Comproportionation is the reverse of disproportionation: two species with the same element in different oxidation states react to form a product in which the element is in an intermediate oxidation state (Atkins *et al.*, 2010).

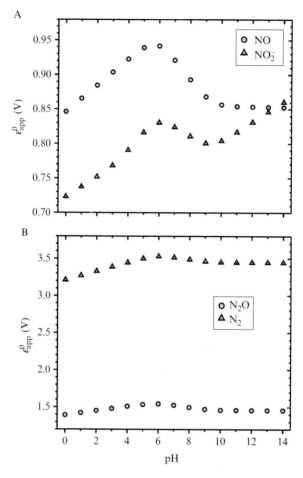

Figure 20.1 Dependence of the apparent standard cell potentials (ε_{app}^{0}) for various NH$_2$OH disproportionation reactions on pH. In all cases the NH$_2$OH reduction product is NH$_4^+$, while the oxidation products are NO, NO$_2^-$, N$_2$O, or N$_2$, as shown.

2. GENERAL CONSIDERATIONS

2.1. The need for anaerobicity

In any experiment designed to identify disproportionation, all oxidizing or reducing agents with the exception of NH$_2$OH must be rigorously excluded. This means that all such experiments should be done in an anaerobic environment, since oxygen is a potent oxidizer. A glovebox provides the most convenient way of carrying out anaerobic experiments. If a glovebox is not available, the reaction solution can be thoroughly

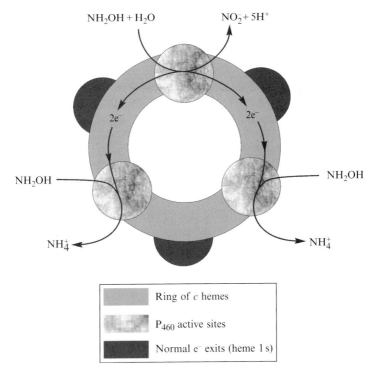

Figure 20.2 Schematic representation of the *N. europaea* HAO hemes, showing how the heme arrangement could in principle make the enzyme a good NH_2OH disproportionation catalyst. The figure shows production of the NH_2OH oxidation product NO_2^-, but similar schemes could lead to production of NO, N_2O, or N_2.

deoxygenated with H_2O-saturated, purified Ar, using an inert gas manifold, and if available, a vacuum pump (Shriver and Drezdzon, 1986). Argon is preferred over nitrogen because it is heavier than air, and more effectively blankets the reaction solution. A unique problem with protein solutions is that bubbling gas through them often leads to denaturation of the protein at the bubble interfaces. Figures 20.3 and 20.4 illustrate two strategies that can be used to overcome this problem. The figures show apparatuses developed for UV–vis spectroscopy, but such apparatuses can be easily adapted to other types of experiments with minor modifications.

If a vacuum pump is available, the most rigorous way to deoxygenate a solution is to use some variation of the apparatus shown in Fig. 20.3, which in older literature is referred to as a "Thunberg tube" or "Thunberg cuvette" (e.g., the work of Hooper and Terry, 1979). The solution to be deoxygenated is put in the bulb at the top of the apparatus, and the assembled apparatus is connected to a vacuum-inert atmosphere dual manifold (Shriver and Drezdzon, 1986). Vacuum is gently drawn until fine

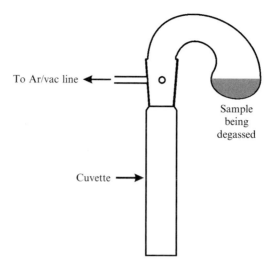

Figure 20.3 Apparatus for deoxygenating a solution via pump-purge cycles, in preparation for a UV–visible experiment (see text for details). The solution to be deoxygenated is first put in the bulb at the top of the apparatus, in order to increase the surface area exposed to the atmosphere being replaced.

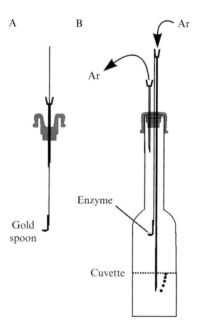

Figure 20.4 A simple method for deoxygenating a solution in preparation for a UV–visible experiment, when a vacuum source is unavailable. The protein is held on a gold spoon while the solution is sparged with argon, so that it is protected from the sparging bubbles (see text for details). (A) Method for threading the wire from the gold spoon assembly through the septum, using a syringe needle. (B) Cuvette set up for degassing. It is best to drive the inlet syringe needle down until it touches the bottom of the cuvette.

bubbles appear in the solution, but being careful to avoid foaming or vigorous bubbling. At this point argon is introduced into the apparatus, and the solution is gently rocked back and forth. This completes one "pump-purge" cycle. Repeating with multiple pump-purge cycles will ensure any desired degree of anaerobicity.

If one has access to inert gas but not to a vacuum pump, the simple apparatus shown in Fig. 20.4 allows the deoxygenation of all solution components in a UV–vis cuvette, without the need to blow bubbles directly through the protein solution. First, the cuvette is loaded with all of the reaction components except the protein(s). A small volume (≤ 5 µL) of concentrated protein solution is placed in the bowl of a home-built 14 kt gold spoon, attached to a length of 26 gauge stainless steel piano wire, which has previously been threaded through a rubber septum as shown in Fig. 20.4a. The septum with the gold spoon assembly is carefully fitted onto the cuvette as shown in Fig. 20.4b. The entire assembly can now be attached to an Ar line to sparge the bulk solution for about 50–60 min (Fig. 20.4b); in that time the protein is also deoxygenated because O_2 can rapidly diffuse out of the tiny drop. After deoxygenation is complete, apply a touch of vacuum grease to the septum in the area around the piano wire shaft, then push down on the shaft to lower the gold spoon into the buffer. Mix the reaction solution thoroughly, after which point it will be ready for the experiment. This apparatus is less reliable than the one shown in Fig. 20.3, because of the possibility of air leakage through the septum. However, it requires only minimal equipment, and can give good results if suitable care is taken.

2.2. Data analysis using complete spectral information

Many of the assays described below involve UV–vis spectroscopy. Such assays can be done by collecting single data points at one wavelength after a specified amount of time; however, in our laboratories we invariably prefer to collect spectra instead of single-wavelength data, and to monitor reaction mixtures as they change over time. This can be done using a diode array spectrophotometer or analogous equipment (we use the Cary 50 rapid scanning spectrophotometer for routine experiments). If one has the equipment available, it is usually no more time consuming to collect data as a function of wavelength and time than it is to collect a single data point. In most cases, the extra data will give no additional information; however, in some occasions they will reveal unexpected problems with the assay, which would have remained hidden if only a single data point had been collected. For example, collecting data at multiple wavelengths can reveal the presence of interfering colored species. Similarly, collecting data as a function of time can verify that a reaction is proceeding smoothly and that assumptions such as linearity in absorbance versus time traces are justified. All the assays presented below can be carried out at single wavelengths. For readers interested in analyzing data collected at multiple wavelengths and times, the authors have a suite of software tools available at http://alchemy.chem.uwm.edu/research/pacheco/.

3. AMMONIA CONCENTRATION DETERMINATION

3.1. Principle

To determine the amount of ammonia produced in NH_2OH disproportionation, we use the well-known method summarized in Scheme 20.2 (Bergmeyer and Beutler, 1985). Ammonia reacts with α-ketoglutarate (α-KG) and NADH in the presence of L-glutamate dehydrogenase (GDH) to form L-glutamate and NAD^+. NADH absorbs at 340 nm ($\varepsilon = 6300\ M^{-1} cm^{-1}$), whereas NAD^+ does not. Therefore, the change in absorbance at 340 nm is a direct measure of how much NADH has been oxidized during the reaction. The assay is always performed with NADH and α-KG in excess of NH_4^+, so that the absorbance change at 340 nm also shows how much ammonia is available to react.

3.2. Materials

The principal assay components can be obtained from Sigma: β-NADH disodium salt, 98% purity (Cat. No. N8129); α-ketoglutaric acid disodium salt, 98% purity (K3752); L-GDH from bovine liver, in phosphate Buffer pH 7.3, containing 50% glycerol (Cat. No. G2626). We prepared all our assay solutions in HEPES buffer (50 mM, pH 7.45), though comparable buffers should work. The only critical requirement is that all stock solutions be ammonia free. Either ammonium chloride or ammonium sulfate can be used to make ammonia standard solutions. Hydroxylamine hydrochloride solutions are not stable for prolonged periods, and should be made up fresh

$$\alpha\text{-KG} + NH_4^+ + 2e^- \longrightarrow Glu + H_2O$$

$$NADH \longrightarrow NAD^+ + H^+ + 2e^-$$

$$\alpha\text{-KG} + NH_4^+ + NADH \xrightarrow{\text{GDH}} Glu + H_2O + NAD^+ + H^+ \text{ net reaction}$$

Scheme 20.2 Ammonia assay reaction, and the associated half reactions. The assay follows the UV–vis absorbance change at 340 nm, due to NADH oxidation. See text for details.

every day. Unlike other stocks, which are made up in assay buffer, we recommend making NH_2OH stocks by dissolving solid hydroxylamine hydrochloride in nanopure water, as the stock is more stable at lower pHs.

Prepare a stock solution of 2.76 mM β-NADH by dissolving the contents of a 100 mg vial in 50 mL of HEPES buffer (50 mM, pH = 7.45). Divide the β-NADH stock solution into 50 1-mL aliquots, and keep them frozen below −25 °C until needed. β-NADH slowly decomposes in air, so keeping frozen stocks eliminates the need to make fresh solutions each time they are needed.

For use in a series of assays we recommend mixing 50 μL of the GDH commercial stock with 450 μL Hepes buffer to make a 10× diluted GDH stock. The diluted sample is easier to pipette accurately than the concentrated stock, which contains 50% glycerol.

Ammonia assay kits based on the GDH method are available from a variety of sources, including Sigma (Cat. No. AA0100). Most of these kits use NADPH as the reducing agent instead of the substantially cheaper NADH. In general, we found no difference for our applications. However, we did note that, when present in millimolar or greater concentrations, NH_2OH reacts slowly with NADPH but not with NADH. The reaction rate is independent of GDH and can be readily corrected for as described later, if a commercial assay kit containing NADPH is used.

3.3. Procedure

The assays are run at 25 °C. The final concentrations of the assay components used to test HAO from *N. europaea* are listed below (Kostera *et al.*, 2010). Other HAOs could be tested using comparable conditions.

	Blank	Standard curve	Sample
β-NADH	140 μM	140 μM	140 μM
α-KG	3.5 mM	3.5 mM	3.5 mM
GDH	~10^{-2} units	~10^{-2} units	~10^{-2} units
NH_4^+ standard	0	5–100 μM	0
HAO	0	0	10–100 nM
NH_2OH	5–100 μM	0	5–100 μM

Reactions are initiated by adding either the GDH or NH_2OH to a 1.00 cm semimicro cuvette already containing the other components. The typical assay volume is 1 mL. Place the cuvette in the spectrometer, add all components except the last one, and begin monitoring the absorbance at 340 nm as a function of time. After 60 s, add 100 μL of diluted GDH or NH_2OH stock, mix thoroughly with a bent glass rod, and continue monitoring the absorbance at 340 nm until it stops changing, typically within 5 min. The amount of ammonia present will be given by

$$[NH_4^+] = \frac{A_\infty - A_0 \times 0.9}{\varepsilon_{NADH} \times l} = \frac{A_\infty - A_0 \times 0.9}{6300 M^{-1}} \quad (20.1)$$

If the volume of GDH or NH_2OH added to initiate the reaction is other than 100 μL, the factor that multiplies A_0 in Eq. (1) should be adjusted accordingly.

3.4. Comments

While NH_2OH disproportionation can lead to many oxidation products, the only likely reduction product is ammonia (hydrazine production is also possible, but HAO is known to catalyze its subsequent oxidation to N_2; Hooper et al., 1984). Thus, a single ammonia assay can reveal that disproportionation has occurred, where one would have to test for a whole range of oxidation products. This is the biggest strength of the ammonia assay, and the primary reason that we recommend investigators use this assay as the initial screen for NH_2OH disproportionation.

HAO concentrations of 100 nm or greater can be used in the assay; however, at such high concentrations, HAO will contribute spectroscopically, possibly complicating data analysis. Global analysis of complete spectral evolution over time, as discussed in the previous section, will be of benefit in such situations.

GDH does not catalyze the oxidation of NADH or NADPH by NH_2OH and α-KG, even when NH_2OH is present in very high (millimolar) concentrations. However, NADPH does appear to react with NH_2OH without the need of the enzyme, though very slowly. Any nonenzymatic reaction can be corrected for by collecting absorbance versus time data for 1 min prior to adding the enzyme, and then using the slope of this data to calculate the nonenzymatic contribution of NADPH reduction. NADH does not appear to react with NH_2OH.

HAO does not utilize NADH as an electron donor, so it is not necessary to remove the enzyme before running the ammonia assay. Indeed, the putative disproportionation reaction and corresponding ammonia assay can be run concurrently in the same solution at pH 7.4. Though this allows continuous monitoring of ammonia production, under typical assay conditions all kinetic information about HAO is lost because the subsequent GDH reaction is slow. In principle, this limitation could be eliminated by using a greater GDH concentration, but to our knowledge the strategy has not been tested.

4. Nitric Oxide Concentration Determination

4.1. Principle

Numerous direct and indirect methods have been used in the past to detect NO in solution (see, e.g., Archer et al., 1995; Hevel and Marletta, 1994). A direct method that we have found particularly useful in our studies of HAO from *N. europaea* is to use catalase as a NO scavenger. Catalase binds NO rapidly ($k_{on} = (1.3 \pm 0.1) \times 10^7$ M^{-1}s^{-1}), and reasonably tightly (binding constant $K = (8.3 \pm 0.7) \times 10^6$), leading to a distinct spectral change (Purwar et al., 2010). Thanks to the high value of k_{on}, this assay can be used to directly monitor the rate at which NO is produced by another process, such as NH$_2$OH disproportionation.

4.2. Materials

Bovine liver catalase can be purchased from Sigma. Several forms are available, but we recommend using Cat. No. C3155, which is provided as a solution of 2× recrystallized enzyme, containing less than 0.2% thymol. In our experience, the forms provided as suspensions contain substantial amounts of insoluble solid. Some batches of C3155 also contain trace solid, but this is readily removed using a 0.2 μm filter. The concentration of the catalase can be determined from its absorbance at 403 nm ($\varepsilon_{403} = 1.2 \times 10^5$ M^{-1}cm^{-1}; Vlasits et al., 2007).

4.3. Procedure

The assays are run at 25 °C. The final concentrations of the assay components used to test HAO from *N. europaea* are listed below (Kostera et al., 2010). Other HAOs could be tested using comparable conditions. We used 50 mM phosphate, pH 7.4, but similar buffers should work equally well.

Catalase	8 μM
NH$_2$OH	5–100 μM
HAO	10–100 nM

Reactions are initiated by adding either the HAO or NH$_2$OH to a 1.00 cm semimicro cuvette already containing the other components, and mixing thoroughly using a bent glass stirring rod. The typical assay volume is 1 mL. If following the reaction at a single wavelength, the biggest changes in absorbance are detected at 403 and 433 nm ($\Delta\varepsilon_{403} = -4.28 \times 10^4$ M^{-1}cm^{-1}; $\Delta\varepsilon_{433} = 4.79 \times 10^4$ M^{-1}cm^{-1}).

4.4. Comments

The major strengths of this assay are its simplicity and the fact that it can be readily used to monitor appearance of an NH_2OH disproportionation product in real time. HAO concentrations of 100 nM or greater can be used in the assay; however, at such high concentrations, HAO will contribute spectroscopically, possibly complicating data analysis. Global analysis of complete spectral evolution over time, as discussed in Section 2, will be of benefit in such situations.

The major weakness of the assay is that catalase itself is an NH_2OH reductase, albeit an inefficient one (Kostera et al., 2010). Thus, catalase and HAO together might catalyze NH_2OH disproportionation, when HAO alone would not. Co-catalysis can be detected by comparing the results of the NO assay with those from an ammonia assay. Unexpectedly high rates of NO formation compared to the rates of NH_4^+ formation will suggest co-catalysis. Note that catalase by itself does not catalyze NH_2OH disproportionation.

Myoglobin binds NO as rapidly as catalase and much more tightly (Hoshino et al., 1993; Hoshino et al., 1996). However, it is not recommended for NH_2OH disproportionation tests because it is a very efficient NH_2OH reductase (300× more efficient than catalase; Kostera et al., 2010). Thus, myoglobin and HAO could well catalyze NH_2OH disproportionation efficiently enough to swamp the contribution of HAO alone.

5. Nitrite Concentration Determination

5.1. Principle

There are two well-established methods for detecting nitrite. The Griess colorimetric method is most common, and well suited to measuring NO_2^- concentrations greater than ~ 2 μM (Vogel et al., 1978). For studying disproportionation, we recommend a fluorometric method that was reported more recently and is best suited for detection of very low NO_2^- concentrations, in the range 0.020–10 μM (Misko et al., 1993; Nussler et al., 2006). In this method 2,3-diaminonaphthalene (2,3-DAN) reacts with NO_2^- under acidic conditions to generate the fluorescent product, 1(H)-naphtotriazole (Scheme 20.3). Nussler et al. (2006) provides a very detailed protocol for applying the method, so herein we outline only the aspects directly pertinent to studying NH_2OH disproportionation. Note that this method can be readily adapted for nitrate determination, as described by Nussler et al. (2006).

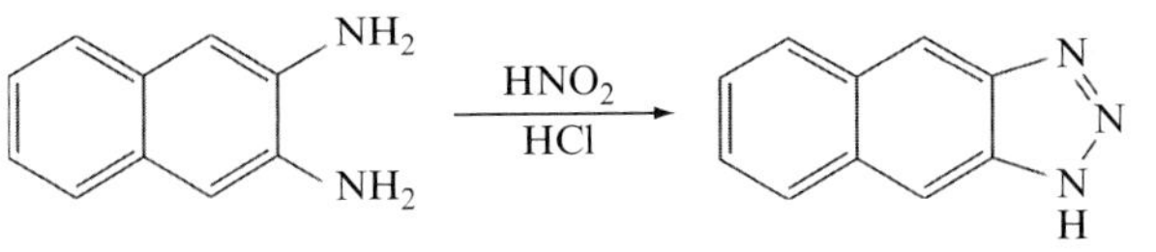

Scheme 20.3 Nitrite assay reaction. The assay monitors the fluorescence emission at 410 nm of the product 1(*H*)-naphtotriazole (excitation wavelength = 365 nm). See text for details.

5.2. Materials

The principal assay component, 2,3-DAN, is available from Sigma-Aldrich (Cat. no. 88461). In solution it has a limited shelf life, and hence we recommend preparing stock aliquots and keeping them frozen. Make a 158 µM solution of 2,3-DAN by dissolving the contents of a 25 mg vial in 1 L of 0.62 M HCl. Split the stock solution into 5-mL aliquots, and freeze below $-25\ °C$ until needed. One thawed aliquot will provide enough 2,3-DAN for 20 assays.

5.3. Procedure

This is a discontinuous assay, because the assay conditions do not match the conditions under which HAO is active. Prior to running the assays for NH$_2$OH disproportionation, a standard fluorescence versus [NO$_2^-$] curve should be obtained as described in Nussler *et al.* (2006).

Step 1: Prepare a 1-mL sample containing 10–100 nM HAO and 2.5–100 µM of NH$_2$OH, and then allow to stand for 15–60 min at 25 °C. We recommend that the optimal concentrations and times be determined beforehand using the ammonia assay described above.

Step 2: After 15–60 min, remove the HAO by ultrafiltration using a Centricon or equivalent filtration device, with 10 kDa molecular weight cutoff. Combine the following in a fluorescence cuvette:

2,3-DAN stock	250 µL (final conc. 39.5 µM)
Filtered reaction mixture	500 µL
1.5 M HCl stock	250 µL

Incubate the cuvette for 5 min in the dark at 30 °C; then add 30 µL of 3 M NaOH and mix thoroughly. *Immediately* measure the fluorescence using an excitation wavelength of 365 nm, and an emission wavelength of 410 nm. Calculate the nitrite concentrations from the standard curve. *Note*: proteins generate false positives in this assay, hence their removal prior to the assay is critical (Nussler *et al.*, 2006).

5.4. Comments

A major strength of this assay is its sensitivity, down to 10 nM NO_2^-. With a few extra steps, the assay can also be used for NO_3^- determination (Nussler *et al.*, 2006). The main disadvantages of the assay are that it is discontinuous, and that all protein must be removed prior to assaying. Given the added complications associated with the assay, it is recommended that nitrite analysis be undertaken only after NH_2OH disproportionation has been confirmed using the ammonia assay, and the optimal conditions for studying the reaction have been identified.

6. Nitrous Oxide and Dinitrogen Concentration Determination

6.1. Principle

The production of N_2O and N_2 by various *N. europaea* enzymes has been analyzed by gas chromatography and mass spectrometry (Hooper and Terry, 1979). A Poropak Q column run at 70 °C is used to separate the gaseous components. In order to distinguish N_2 generated by NH_2OH oxidation from atmospheric N_2, isotopically labeled $^{15}NH_2OH$ must be used (Hooper and Terry, 1979).

6.2. Materials and procedure

In their previous paper, Hooper *et al.* provide detailed instructions for detecting N_2O and N_2 under a variety of circumstances (Hooper and Terry, 1979). Herein most details are omitted, and we limit ourselves to one important and pertinent observation about N_2O. This gas has the unusually high solubility in water of 56.7 g/100 cm^3 at 25 °C, compared for example to N_2 ($\sim$2 g/100 cm^3), or NO ($\sim$5 g/100 cm^3) (Weast, 1982). The published procedure calls for sampling the gaseous phase, so special attention must be paid to extracting the N_2O from the aqueous phase, particularly if only small quantities of the gas are present.

Figure 20.5 suggests a small-scale apparatus and a strategy for extracting the gases from solution that is based on well-known gas-handling techniques (Shriver and Drezdzon, 1986). Prepare separate solutions containing twice the desired final concentrations of HAO and NH_2OH, in a phosphate

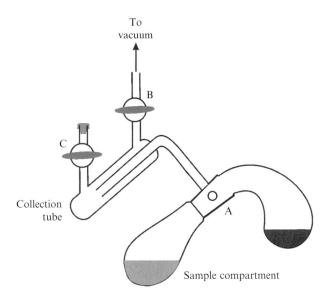

Figure 20.5 Apparatus for use in the detection of N_2O and N_2, generated by NH_2OH disproportionation. Initially the enzyme solution is in one bulb of the sample compartment, and the NH_2OH solution in the other. The two solutions are mixed once they have been thoroughly deoxygenated. See text for details.

or similar buffer pH ~7.4. Put 500 μL of the HAO solution in the top bulb of the apparatus and 500 μL of the NH_2OH solution in the bottom bulb (Fig. 20.5). Assemble the apparatus with stopcocks A and B open and stopcock C closed; then purge the entire assembly free of oxygen as described in Section 2. At this point, the reaction chamber is isolated from the collection chamber by closing stopcock A, and the enzyme and substrate are mixed to initiate disproportionation. Use the results of previous ammonia or NO assays to estimate the amount of time required for the reaction. At the end of the allotted time, the collection chamber is evacuated using a vacuum pump, and then cooled using a CO_2/butanol bath (Hooper and Terry, 1979), or if available a liquid nitrogen bath. Stopcock B is closed and stopcock A opened, in order to draw the gaseous products out of solution. N_2O has a freezing point of -90.86 °C (Cotton et al., 1999), and so will solidify in the collection chamber if liquid nitrogen is used to cool it. Even a CO_2/butanol bath, as used in the original reference (Hooper and Terry, 1979), should lead to extraction of most of the gaseous products from solution. Periodic agitation of the solution will help with the degassing process. After 10 min, stopcock A can be closed, and the collection chamber warmed to room temperature. Finally, the gaseous contents of the collection chamber can be sampled via the septum-covered stopcock C, using a gas-tight syringe.

6.3. Comments

In the original reference the authors were able to estimate the amounts of N_2 and N_2O generated in various reactions by using apparatuses of known volume, and taking into consideration the relative solubilities of the various gases (Hooper and Terry, 1979). This will probably only be possible if fairly large amounts of N_2O are being generated; for most cases the method should be considered a qualitative one.

REFERENCES

Archer, S. L., et al. (1995). Preparation of standards and measurement of nitric oxide, nitroxyl, and related oxidation products. *Methods* **7**, 21–34.

Atkins, P., et al. (2010). Shriver & Atkins' Inorganic Chemistry. W. H. Freeman, New York, NY, pp. 147–178.

Bergmeyer, H. U., and Beutler, H. O. (1985). *In* "Methods of Enzymatic Analysis," (H. U. Bergmeyer, ed.), Vol. 8, pp. 454–461. Academic Press, New York, NY.

Cotton, F. A., et al. (1999). Advanced Inorganic Chemistry. Wiley, New York, NY, pp. 309–379.

Frear, D. S., and Burrell, R. C. (1955). Spectroscopic method for determining hydroxylamine reductase activity in higher plants. *Anal. Chem.* **27**, 1664–1665.

Hevel, J. M., and Marletta, M. A. (1994). Meth. Enzymol. *In* "Oxygen Radicals in Biological Systems," (L. Packer, ed.), Vol. 233, pp. 250–258. Academic Press, San Diego, CA.

Hooper, A. B., and Terry, K. R. (1979). Hydroxylamine oxidoreductase of nitrosomonas: Production of nitric oxide from hydroxylamine. *Biochim. Biophys. Acta.* **571**, 12–20.

Hooper, A. B., et al. (1984). Kinetics of reduction by substrate or dithionite and heme-heme electron transfer in the multiheme hydroxylamine oxidoreductase. *Eur. J. Biochem.* **141**, 565–571.

Hoshino, M., et al. (1993). Photochemistry of nitric-oxide adducts of water-soluble iron(III) porphyrin and ferrihemoproteins studied by nanosecond laser photolysis. *J. Am. Chem. Soc.* **115**, 9568–9575.

Hoshino, M., et al. (1996). Studies on the reaction mechanism for reductive nitrosylation of ferrihemoproteins in buffer solutions. *J. Am. Chem. Soc.* **118**, 5702–5707.

Igarashi, N., et al. (1997). The 2.8 angstrom structure of hydroxylamine oxidoreductase from a nitrifying chemoautotrophic bacterium, Nitrosomonas europaea. *Nat. Struct. Biol.* **4**, 276–284.

Kostera, J., et al. (2008). Kinetic and product distribution analysis of NO reductase activity in *Nitrosomonas europaea* hydroxylamine oxidoreductase. *J. Biol. Inorg. Chem.* **13**, 1073–1083.

Kostera, J., et al. (2010). Enzymatic interconversion of ammonia and nitrite: The right tool for the job. *Biochemistry* **49**, 8546–8553.

Misko, T. P., et al. (1993). A fluorometric assay for the measurement of nitrite in biological samples. *Anal. Biochem.* **214**, 11–16.

Nussler, A. K., et al. (2006). Fluorometric measurement of nitrite/nitrate by 2, 3-diaminonaphthalene. *Nat. Protocols.* **1**, 2223–2226.

Purwar, N., et al. (2010). The Interaction of Nitric Oxide with Catalase: Structural and Kinetic Analysis. *J. Am. Chem. Soc.* (submitted for publication).

Shriver, D. F., and Drezdzon, M. A. (1986). The Manipulation of Air-Sensitive Compounds. Wiley, New York, NY.

Vlasits, J., et al. (2007). Hydrogen peroxide oxidation by catalase-peroxidase follows a non-scrambling mechanism. *FEBS Lett.* **581,** 320–324.

Vogel, A. I., et al. (1978). Vogel's Textbook of Quantitative Inorganic Analysis. Longman, New York, NY, pp. 755–756.

Weast, R. C. (1982). *In* "CRC Handbook of Chemistry and Physics," (R. C. Weast and M. J. Astle, eds.), p. B125. CRC Press, Boca Raton, FL.

CHAPTER TWENTY-ONE

Liquid Chromatography—Mass Spectrometry-Based Proteomics of *Nitrosomonas*

Hans J. C. T. Wessels,[*] Jolein Gloerich,[†] Erwin van der Biezen,[‡,1] Mike S. M. Jetten,[‡,§] and Boran Kartal[‡]

Contents

1. Introduction	466
2. LC–MS/MS Instrument Setup	468
3. Growth of *N. eutropha* C91 Pure Culture	469
4. Sample Preparation	470
5. C18 Reversed Phase LC–MS/MS Analysis	471
6. Database Searches and Validation of Results	472
7. Dataset Description and Protein Identification Example	476
8. Conclusions	479
References	479

Abstract

During the last century, the research on aerobic ammonium-oxidizing bacteria (AOB) lead to many exciting physiological and biochemical discoveries. Nevertheless the molecular biology of AOB is not well understood. The availability of the genome sequences of several *Nitrosomonas* species opened up new possiblities to use state of the art transcriptomic and proteomic tools to study AOB. With the currect technology, thousands of proteins can be analyzed in several hours of measurement and translated proteins can be detected at femtomole and attomole concentrations. Moreover, it is possible to use mass spectrometry-based proteomics approach to analyze the expression, subcellular localization, posttranslational modifications, and interactions of translated proteins. In this

[*] Nijmegen Centre for Mitochondrial Disorders, Department of Laboratory Medicine, Nijmegen Proteomics Facility, Radboud University Nijmegen Medical Centre, Geert Grooteplein-Zuid 10, Nijmegen, The Netherlands
[†] Department of Laboratory Medicine, Nijmegen Proteomics Facility, Radboud University Nijmegen Medical Centre, Geert Grooteplein-Zuid 10, Nijmegen, The Netherlands
[‡] Department of Microbiology, Institute of Water and Wetland Research, Radboud University Nijmegen, Heyendaalseweg 135, The Netherlands
[§] Department of Biotechnology, Delft University of Technology, Delft, The Netherlands
[1] Present address: Merck Sharp & Dohme BV, 5342 CC Oss, The Netherlands

Methods in Enzymology, Volume 486
ISSN 0076-6879, DOI: 10.1016/S0076-6879(11)86021-0

© 2011 Elsevier Inc.
All rights reserved.

chapter, we describe our LC–MS/MS methodology and quality control strategy to study the protein complement of *Nitrosomonas eutropha C91*.

1. INTRODUCTION

Ammonium is a major pollutant (toxic to animals and plants, causes eutrophication) in municipal and industrial wastewaters and its discharge to water bodies is regulated worldwide by strict legislation. In natural ecosystems and in conventional wastewater treatment plants, in the presence of O_2, ammonium is oxidized to nitrate via nitrite by two clades of microorganisms. The first step of this process is mediated by aerobic ammonium-oxidizing bacteria (AOB or ammonium-oxidizing archaea, AOA) and is followed by the oxidation of nitrite to nitrate by nitrite-oxidizing bacteria (NOB). AOB oxidize ammonium via hydroxylamine to nitrite and there are two major groups of bacteria that are thought to be important: beta- and gammaproteobacterial AOB.

In the betaproteobacterial linage, many strains affiliated to two species of *Nitrosomonadaceae* (*Nitrosomonas europaea* and *Nitrosomonas eutropha*) are among the AOB that have been studied over a 100 years since the first isolation of an ammonium-oxidizing microorganism by Winogradsky (1892). Physiological and biochemical studies have resulted in the identification and, in some cases, isolation of main catabolic enzymes of AOB (Arp *et al.*, 2002); still, the molecular biology of AOB is not well understood. For example, several strains of *N. eutropha* have been shown to oxidize ammonium with NO_2 under anoxic or microaerophilic conditions (Schmidt and Bock, 1997) and H_2-dependent growth was shown for *N. eutropha* N904 (Bock *et al.*, 1995; Kampschreur *et al.*, 2008; Schmidt *et al.*, 2001, 2004). However, it is currently unknown how these alternate pathways function or are regulated.

Recent publications of the genome sequences of several *Nitrosomonas* (Chain *et al.*, 2003; Stein *et al.*, 2007), *Nitrosococcus* (Klotz *et al.*, 2006), and *Nitrosospira* (Norton *et al.*, 2008) species have facilitated the application of state-of-the-art technologies for the in depth analysis of the molecular mechanism of AOB. The first publication of such an approach was on the transcriptomic analysis of *N. europaea* (Beyer *et al.*, 2009). This study revealed that mRNA concentrations of catabolic proteins such as amoA (ammonia monooxygenase subunit A) and hao (hydroxylamine oxidoreductase) could be correlated to activity and growth rates of bacteria under aerobic conditions. Under anaerobic conditions, the transcription levels of these genes were reduced and those of nirK (nitrite reductase), norB (nitric oxide reductase), and nsc (nitrosocyanin, red copper protein) were significantly increased.

The recent elucidation of the *N. eutropha C91* genome sequence enables the use of a mass spectrometry database search driven proteomics to analyze the protein complement of *Nitrosomonas* in great detail. Analyses of the expression (Lacerda and Reardon, 2009; Phillips and Bogyo, 2005), subcellular localization (Foster *et al.*, 2006), posttranslational modifications (Farley and Link, 2009; Rogers and Foster, 2009), and interactions (Andersen *et al.*, 2003; Monti *et al.*, 2009; Wessels *et al.*, 2009; Wittig and Schagger, 2009) of translated proteins by high-throughput mass spectrometry approaches provide a wealth of information to help unravel biochemical pathways and other cellular processes.

Although the analysis of the transcriptome of a given microorganism gives us valuable insights, the current technology allows us to look one step further: amount of translated proteins can now be detected at very low concentrations (i.e., *fmol*, *amol*). The sensitive and robust performance of LC–MS/MS-based proteomics makes it an attractive system to use for studying prokaryotic proteomes. Low femtomole to attomole protein sensitivity and high-throughput analysis of samples are crucial when biomass is limited or when multiple samples need to be analyzed within a short time span. Modern LC–MS/MS equipment (when handled correctly) is now capable of identifying and quantifying up to thousands of proteins across multiple samples within just hours of measurement time (Wisniewski *et al.*, 2009). Performance is not only instrument type dependent, but it is also determined by the quality of sample preparation, chromatographic separation, mass spectrometer operation, data processing, and fractionation approach. It is therefore important to optimize (and monitor) the performance of the entire procedure in advance to any large-scale proteomics experiments. Although sample fractionation at protein or peptide level can be used to increase the proteome coverage, it also significantly increases the amount of time and costs needed to perform the experiment (Fang *et al.*, 2010). The tradeoff between an increased proteome coverage and required throughput is thus unique for each research group and question, and is therefore an important aspect to consider when designing experiments. The approach described here was used to study the top 500–1000 most abundant proteins in *N. eutropha* cells of which many are involved in primary and secondary cellular processes, but may easily be applied to other microorganisms. It is important to realize that parts of the method described here might be equipment specific. Nevertheless we believe that the majority of our approach and quality control can be incorporated into existing protocols using different instrumentation. The method described in this communication is currently our optimal approach for the analysis of the unfractionated *N. eutropha C91* proteome.

2. LC–MS/MS Instrument Setup

The LC–MS/MS setup uses a splitless nanoflow liquid chromatograph (Easy nano LC, Proxeon) for C18 reversed phase peptide separations coupled online to a 7 T hybrid quadrupole linear ion trap Fourier-transform ion cyclotron resonance mass spectrometer (Syka et al., 2004; LTQ FT Ultra, Thermo scientific) via a modified nanoelectrospray source (Thermo scientific). Generic performance of the LC–MS/MS system is assessed on a daily basis by evaluating more than 46 performance metrics obtained from a standard *Escherichia coli* proteome LC–MS/MS measurement using the software of Rudnick et al. (2010). Peptide separations are performed using 15-cm long 100-μm internal diameter emitters (PicoTip emitter FS360-100-8-N-5-C15, New Objective) packed in-house with 3 μm particles (120 Å pore) C18-AQ ReproSil-Pur reversed phase material (Dr. Maisch GmbH) shown in Fig. 21.1. Column packing is achieved by pressure loading (40 bar) of the C18 particles into the emitter using a pressure bomb and high purity He gas (Ishihama et al., 2002). Upon packing, the emitter outlet forces the particles to form a self-assembled frit as long as the outlet size to particle diameter ratio remains below a 5:1 (Ishihama et al., 2002). The advantages of using these reproducible in-house packed emitters include versatility, very low costs, and, most importantly, high

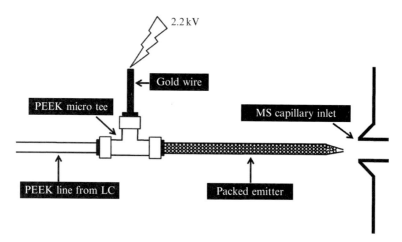

Figure 21.1 *Schematic representation of the nanoelectrospray setup.* The 50 μm internal diameter (ID) × 360 μm outer diameter (OD) flow line is connected via a micro tee to both the column and a gold wire. The column, which is a 15 cm long 100 μm ID × 360 μm OD emitter packed with 3 μm C18 reversed phase material, is placed just a few millimeters away from the mass spectrometer capillary inlet. The gold wire is used to apply a 2.2 kV electrospray voltage directly to the liquid flow in front of the column via a zero-dead volume micro tee fitting.

chromatographic resolution. Hence, there is virtually no possibility of post-column peak broadening as peptides are immediately ionized and analyzed when they leave the column outlet. Signal stability and intensity of the nanoelectrospray source depend (amongst other factors) on the quality of the emitter outlet which needs to be evaluated prior to any analytical LC–MS/MS analysis. We consider our ion source stable when the average total ion current (TIC) signal deviation for buffer ions is less than 12% for at least 150 survey spectra acquired at a constant flow of 300 nl min^{-1} 0.5% acetic acid. We recommend to document and monitor performance characteristics of the system on a regular basis. This may help as an early diagnosis for possible performance deterioration and ensures consistent quality throughout large-scale experiments.

3. Growth of *N. eutropha C91* Pure Culture

A 7-l glass vessel with a working volume of 4 l is used for growing *N. eutropha C91* (Fig. 21.2A). The reactor is fed continuously with a flow rate of 1.4 ml min^{-1}. The influent mineral medium is modified from Schmidt and Bock (1997) and contains 50 mM of NH_4Cl and following nutrients (per 1 l of medium): 585 mg NaCl, 147 mg $CaCl_2 \cdot 2H_2O$, 74 mg KCl, 54 mg KH_2PO_4, 49 mg $MgSO_4 \cdot 7H_2O$, 12 g HEPES, and 1 ml trace elements solution containing 0.02 M HCl, 973 mg $FeSO_4 \cdot 7H_2O$, 49 mg H_3BO_3, 43 mg $ZnSO_4 \cdot 7H_2O$, 37 mg $(NH_4)_6Mo_7O_{24} \cdot 4H_2O$, 34 mg $MnSO_4 \cdot H_2O$, 16 mg $CuSO_4$. The pH of the medium is then adjusted

Figure 21.2 (A) *N. eutropha C91* pure culture. (B) Fluorescence *in situ* hybridization micrograph of the *N. eutropha C91* pure culture. A triple hybridization with the probes NEU (Wagner *et al.*, 1995), Nso1225 (Mobarry *et al.*, 1996), and EUB338 (Daims *et al.*, 1999) is applied. Bar = 5 μm.

to 7. A high ammonium concentration (40–50 mM) is recommended for a rapid start-up of the bioreactor. For the homogeneous distribution of substrates, the vessel is stirred at 200 rpm with one six-bladed Rushton and one marine impeller (60 mm). Disappearance of ammonium and production of nitrite should be monitored daily. We use the rapid nitrite and ammonium determination protocols as described in Kartal *et al.* (2006). A heating blanket is used to keep temperature at 30 °C. A gas mixture of Air/N_2 (90/10%) with a flow of 500 ml l^{-1} is sparged through the reactor for O_2 supply. A pH controller unit is used to supply a solution of Na_2CO_3 (100 g l^{-1}) when necessary to keep the pH stable at 7.0. If desired, a dissolved O_2 probe may also be used to monitor the dissolved O_2 concentration.

4. Sample Preparation

The *N. eutropha* cells (5 l, OD_{600} ~ 0.2; Fig. 21.2B) are centrifuged at 5 °C and 4000 × g. After centrifugation, the pellet is resuspended in 1 volume 20 mM potassium phosphate buffer, pH 8. Cell suspensions are passed three times through a French pressure cell operated at 138 MPa. After centrifugation for 15 min at 1700 × g at 4 °C, the cell-free fraction is obtained as clarified supernatant. Proteins are diluted 1:1 (v/v) in 8 M urea 10 mM Tris–HCl pH 8 and placed at room temperature for 30 min to allow for protein denaturation. Reduction of cysteine disulfide bridges is performed by adding 2 μl 10 mM dithiothreitol solution and incubating for 30 min at room temperature. Reforming of disulfide bridges is prohibited via alkylation of reduced cysteines by adding 2 μl of 50 mM chloroacetamide and subsequent incubation for 30 min at room temperature in the dark. The use of chloroacetamide is preferred over iodoacetamide as Nielsen *et al.* (2008) reported that iodoacetamide induces artifacts that mimic ubiquitination. As trypsin digestion is incompatible with the 4 M urea concentration, a primary incubation of the sample with the protease "LysC" can be performed in a 1:50 (w/w) LysC:protein ratio for 3 h at room temperature. This step results in the cleavage of proteins into smaller peptides through cleavage at the C-terminal side of lysine residues under strong denaturing conditions. Next, the sample is diluted 1:3 (v/v) with 50 mM ammonium bicarbonate to allow trypsin digestion to take place in a less denaturing environment. Trypsin is added in a 1:50 (w/w) trypsin:protein ratio and the sample is placed at 37 °C for overnight digestion. Subsequently, the resulting peptide mixture can be concentrated and desalted using stop and go elution (STAGE) tips as described by Rappsilber *et al.* (2003). A popular alternative to this STAGE solid phase extraction step is the commercially available alternatives such as Millipore ZipTips. However, we prefer the use of in-house constructed STAGE tips because of the low costs and reduced handling steps. Briefly, acidified peptides bind to the activated C18 reversed phase material which is fixed in a standard pipettor tip allowing the peptides to be washed using 0.5% acetic acid and eventually eluted in 20 µl 80% acetonitrile 0.5% acetic acid. The acetonitrile buffer can be removed by centrifugation under vacuum after which samples are taken up in a total volume of 20 µl 0.5% acetic acid prior to LC–MS/MS analysis. We use acetic acid here to dissolve the sample as it is used later on as ion pair reagent in buffers used for C18 reversed phase LC–MS/MS analysis.

5. C18 Reversed Phase LC–MS/MS Analysis

For RP LC–MS/MS analysis, we load samples directly onto the analytical column at maximum flow rates determined by a set maximum pressure of 240 bar. Peptides are eluted from the analytical column directly toward the mass spectrometer capillary inlet via a nanoelectrospray source which is used to apply a distal electrospray voltage of 2.2 kV to the liquid flow using a standard micro tee setup (Fig. 21.1). Peptide elution is best performed using an increasing linear gradient of 10–40% acetonitrile in 90 min with a constant concentration of 0.5% acetic acid as ion pair reagent. To prevent carryover, the system is washed by increasing the acetonitrile concentration from 40% to 80% in 5 min followed by a 10 min isocratic flow of 80% acetonitrile at a flow rate of 600 nl·min^{-1}. Finally, the system is equilibrated with 3% acetonitrile at 600 nl min^{-1} for 10 min. The mass spectrometer settings described hereafter are specific for the type of instrument and manufacturer. In general, they should be optimized for the type of application and sample characteristics and may thus differ significantly between experiments and research groups. For the experimental design described here, we program the LTQ FT Ultra instrument to acquire four data-dependent MS/MS spectra of the four most abundant peptide ions (with charge $z = 2+$ or $3+$) from a survey scan by the ion cyclotron resonance cell. The acquisition of MS/MS spectra in the ion trap is performed in parallel to a long transient ICR cell survey scan. This parallel acquisition is achieved by selecting precursor ions for collision activated dissociation experiments in the linear ion trap from a first short part of the ICR cell measurement. This allows the ion trap to perform multiple MS/MS analyses while the ICR cell completes the accurate (long transient) measurement. Repetitive MS/MS analysis of identical ions is prohibited by the use of dynamic exclusion with settings optimized for the typical chromatographic performance of our system. Our instrument method uses a dynamic exclusion time of 180 s in combination with early expiration settings that remove m/z values from the dynamic exclusion list when 10 spectra are recorded with a signal-to-noise ratio below 2 for each individual ion. We highly recommend users to optimize the dynamic exclusion parameters as it may yield more than 100% performance increase for the number of identified unique peptides. Survey scans by the ICR cell are set to cover a range of 350–1600 m/z with a resolution power of 100,000 full width at half maximum (FWHM) at 400 m/z. The low mass cutoff of 350 m/z is used to prevent the detection of buffer ions (that reduce the effective dynamic range for peptide ions) and peptides shorter than six amino acids sequence length. In general, shorter peptides have low protein specificity and significantly contribute to the relative proportion of false positive identifications (Cox and Mann, 2008). Tryptic peptides are generally observed below m/z 1600 which is therefore used as high mass cutoff. Raw data files are evaluated manually to check for chromatographic (over)loading, resolution, and absence of interfering chemical contaminants like polyethylene-glycols. In our opinion, ion maps are ideally suited for this purpose as these (unlike total ion chromatograms) are capable of visualizing all critical characteristics for individual LC–MS/MS runs of very complex samples as shown in Fig. 21.3.

6. Database Searches and Validation of Results

Depending on the database search software and the type of mass spectrometer used, raw data needs to be converted to a format that can be used by the software of choice. Many different search engines are available nowadays (both commercial and open source; Dagda et al., 2010; Sadygov et al., 2004) that can be used for protein identification or quantification purposes. Some well-known search engines are Mascot (Perkins et al., 1999), X!Tandem (Craig and Beavis 2004), SEQUEST (Yates et al., 1995), OMSSA (Geer et al., 2004), Peaks studio (Bioinformatics Solutions: http://www.bioinformaticssolutions.com), ProsightPTM (LeDuc et al., 2004), and Phenyx (Colinge et al., 2003). We prefer to use Matrix Science Mascot for which we convert the Thermo raw files into generic mascot files via in-house developed Perl scripts (this can also be performed using freely available software packages; Cox and Mann, 2008; DTA supercharge: http://msquant.sourceforge.net; Pedrioli 2010). All database searches are performed using the Refseq *N. eutropha C91* sequences supplemented with sequences of known contaminant proteins. It is important to add these contaminant proteins to the database to ensure that corresponding spectra do not erroneously match *N. eutropha C91* sequences. Some of these contaminants are added on purpose during digestion (LysC and Trypsin) whereas other proteins (e.g., skin proteins such as keratins) contaminate samples via the air or reagents. We first perform a loose mass accuracy search of the data allowing a mass error of 20 ppm (which is a relatively large error

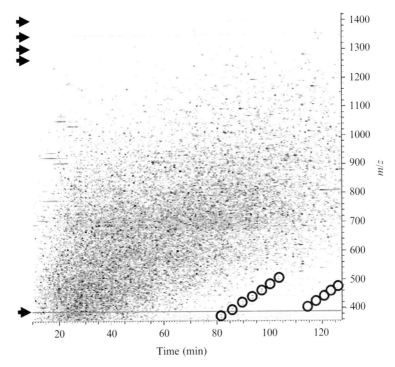

Figure 21.3 Ion map view of the analytical LC–MS/MS data acquired from a *N. eutropha C91* proteome sample. This map plots the *m/z* values versus elution time with corresponding intensity in gray scale values. Ions are visible as dots using this large time range. Zooming in on the data allows for manual evaluation of chromatographic performance and intensity-dependent mass accuracy. Constant background ions from fused silica or buffer components are visualized as horizontal lines that may span the complete time range (prominent ions are indicated using arrow heads). Sample contamination by polyethylene-glycols and other (similar) polymers is easily recognizable as a series of diagonal dots with a consistent increase of *m/z* value and retention time in the analytical part of a reversed phase separation. Polyethylene-glycol ions are encircled in this figure for ease of interpretation. In our experience, we always find low amounts of polyethylene-glycols to be present in reversed phase separations, presumably originating from laboratory plastics or reagents.

for our type of instrument). This database search result is then used to assess the mass accuracy of each measurement and can be used to recalibrate the data offline prior to conclusive database searches. The best suitable method for offline mass recalibration depends on the type of instrument and error (e.g., general transposition, *m/z* dependent mass error or signal intensity-dependent mass error). For this particular measurement, we calculated the average parent ion *m/z* errors for precursor *m/z* bins of 100 Th. This data was then used to obtain Eq. (1) which was used for *m/z*-dependent parent ion *m/z* recalibrations.

$$(m/z)_{\text{calibrated}} = (m/z)_{\text{measured}} - \left(-0.0000005 \times (m/z)_{\text{measured}} - 0.0009 \right). \tag{21.1}$$

This precursor ion recalibration improved mass accuracy from -2.0 (± 1.17) to -0.17 (± 1.13) ppm as visualized in Fig. 21.4.

For the LC–MS/MS dataset of *N. eutropha C91* samples, we specified a precursor mass tolerance of 10 ppm for ICR cell spectra of precursor ions and 0.8 Da mass tolerance for fragment ions analyzed by the linear ion trap. Database search parameters include specific trypsin cleavage with a tolerance of one missed cleavage. Furthermore, carboxamidomethylation of cysteines by chloroacetamide is set as fixed modification and variable modifications include N-terminal protein acetylation, deamidation of glutamine and asparagine, and oxidation of methionine. Result files are downloaded automatically from the server using a custom-made Perl script which is triggered by the Mascot Daemon search tool after each search completion.

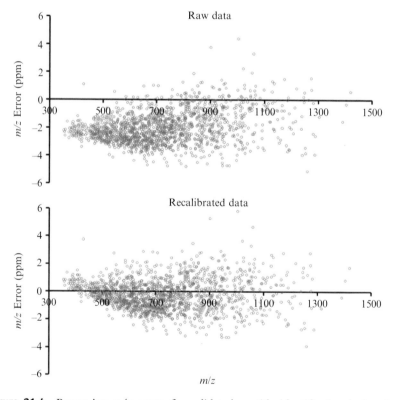

Figure 21.4 Parent ion m/z errors for validated peptide identification before (top graph) and after (bottom graph) offline recalibration. Recalibration improved the average precursor ion m/z accuracy from -2.00 (± 1.17) to -0.17 (± 1.13) ppm.

Subsequent validation of search results is of utmost importance considering the nature and amount of retrieved data. By now there are numerous software packages available for search data parsing and validation purposes such as MSQuant (DTA Supercharge: http://msquant.sourceforge.net), Trans-proteomic pipeline (Keller et al., 2002; Nesvizhskii et al., 2003; Pedrioli, 2010), PROVALT (Weatherly et al., 2005), MaxQuant (Cox and Mann, 2008), Scaffold (Searle, 2010), cot Distiller (Matrix Science: www.matrixscience.com). Many more excellent software solutions may be found in literature, but it is beyond the scope of this chapter to list and describe them all. Successful approaches for large dataset validation often use false discovery rate validation or even posterior error probabilities (local false discovery rates). Both approaches require an additional database search of the data against reversed or randomized sequences from the normal database (Bianco et al., 2009). These decoy search results are used to calculate the false positive rate or posterior error probabilities. In general, both methods require high numbers of matched MS/MS spectra to be accurate and are therefore not suited for every type of sample and/or purpose. We generally use false discovery rates at the protein level for datasets that encompass 10,000–100,000 MS/MS spectra and posterior error probabilities for datasets that include 100,000 or more MS/MS spectra. Validation of smaller datasets can be performed using the random match probabilities calculated by the search engine in combination with additional parameters to ensure confident protein identifications.

We currently cannot use FDR-based validation strategies for datasets generated using the described experimental design due to the low number of protein identifications found in decoy database searches (using reversed protein sequences). The decoy database search using the *N. eutropha C91* data yielded only 44 protein identifications (exclusively single peptide identifications) with Mascot identification scores in the range of 16–33 (we do not parse peptides with identification scores below 16). When we use this decoy database search result to calculate Mascot score thresholds to achieve a 1% FDR, we find an exceptionally low Mascot peptide identification score threshold of just 27. We therefore apply a low sample complexity validation strategy using the in-house developed PROTON software package (proteomics tools Nijmegen, H. J. C. T. Wessels, J. Gloerich et al., unpublished). The criteria used by PROTON for data validation were empirically determined from previously acquired (much larger) datasets from various organisms and cell types to achieve an overall protein false discovery rate of 1% or better. Criteria used for validation require a minimum Mascot score of 30 for peptides originating from proteins identified with multiple unique peptide sequences. Single peptide matches require a minimum Mascot score of 49 and a modified delta score of at least 10. The modified delta score is calculated as the difference in score between the first and second best unique sequence match for a given spectrum. This value differs from the original delta score which is calculated between the first and second match, regardless of the peptide sequences, and may thus differ only by the position of a variable modification site. We further allow a maximum of three variable modifications for individual peptide matches. When we apply the same validation criteria to the decoy database search result, we can calculate an infinitely small false discovery rate for our data, which is an underestimation of the true FDR due to the low number of decoy database hits. Our software is also used to calculate estimated protein abundance for all identified proteins using the exponentially modified protein abundance index (emPAI; Ishihama et al., 2005). This spectral count method uses the correlation between the number of identified unique peptide ions (normalized between proteins for the number of theoretically detectable peptides) and protein concentration (Ishihama et al., 2005; Rappsilber et al., 2002). The number of theoretically detectable peptides can be calculated *in-silico* for each protein using the physicochemical properties of predicted peptides in combination with the mass spectrometer settings. Ideally, the physicochemical cutoff values are calculated *a priori* using the properties of identified peptides from large datasets that are generated using an identical protocol. The exponential modification of the PAI value is needed to obtain a linear correlation between PAI and protein concentration (Ishihama et al., 2005). Calculated emPAI values may be used to compare abundances between different proteins in one or multiple samples. It is important to note that a comparison of protein abundances between samples requires the samples to be comparable in terms of complexity and data quality. A limitation of this technique can be its accuracy which depends on the number of theoretically detectable peptides and the number of identified unique ions. We therefore emphasize that care should be taken when using emPAI values of low abundant and/or small proteins.

7. DATASET DESCRIPTION AND PROTEIN IDENTIFICATION EXAMPLE

A single LC–MS/MS analysis of nonfractionated *N. eutropha C91* proteome using our system identified about 2000 unique peptides corresponding to ~600 unique proteins (~24% coverage of the predicted proteome; Kartal et al., unpublished). On average, each protein is identified using four unique peptides. Representative protein sequence coverages are shown as a relative frequency histogram in Fig. 21.5. These were calculated as the percentage of identified amino acid residues using unique valid peptide identification sequences. As an example, one of the most abundant proteins identified by our LC–MS/MS analysis is the alkyl hydroperoxide

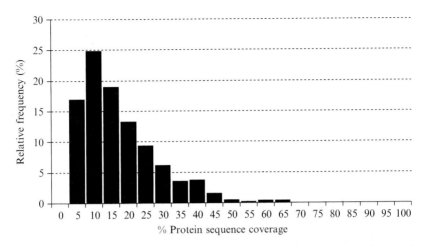

Figure 21.5 Protein sequence coverage histogram of *N. eutropha C91*. Values on the x-axis represent upper values for each bin whereas the y-axis is used to show the relative frequency for each respective bin. This figure shows, for example, that about 25% of all proteins in the dataset were identified with a sequence coverage between 5% and 10%.

reductase (Neut_2366, abundance based on emPAI). This protein is likely to function as a hydrogen peroxide scavenger to protect *N. eutropha C91* cells from oxidative damage. This protein is identified using nine unique peptides with combined protein sequence coverage of 60% (Table 21.1) and a Mascot protein identification score of 513. This protein identification score implies a chance of less than 1×10^{-51} for this protein to be a random event. The data in Table 21.1 shows all relevant characteristics generally used to describe protein identifications. Details for the protein identification at both the protein and peptide level are included in this table. From information at the protein level, it may be deduced that we predicted a total of 11 peptides to be theoretically detectable using our approach, of which nine peptides are identified using 13 unique MS/MS spectra. These data are used to calculate a protein abundance of 14.2 using emPAI. The peptide level information shows each best nonredundant peptide match data in detail. The first series of columns describe each peptide identification at the mass spectrometry level and include the mass over charge (m/z) value of each peptide ion, measured peptide mass as Mr(exp), theoretical mass of the peptide Mr(thr), peptide mass error in parts per million, and charge state of the ion. The next series of columns contain the database search information. First, the number of missed cleavage sites by trypsin is shown (miss) followed by the peptide match score and the expected value for the match to be a random event. The modified delta score is also shown (mDelta) which is the difference in Mascot score between the best and second best peptide sequence match. Please note that there is no other candidate peptide

Table 21.1 Identification data for the abundant alkyl hydroperoxide reductase (Neut_2366) of *N. eutropha* C91

Detectable peptides: 11				Detected peptides: 9		Matched ions: 13			Score: 513		emPAI: 14.2
m/z	Mr(exp)	Mr(thr)	ppm	Charge	Miss	Score	Expect	mDelta score	Residues	Peptide match	
826.879	1651.7434	1651.7478	−2.69	2	0	67	3.4E-07	67	13–27	K.ATAYHNGDFVEVSDK.T	
631.3103	1890.9091	1890.9112	−1.1	3	0	35	0.00035	34.5	64–80	K.AGIEVYSVSTDTHFTHK.A	
572.28	1142.5455	1142.5469	−1.19	2	0	48	0.000017	47.6	81–90	K.AWHDSSSVVR.K	
646.8183	1291.6221	1291.6231	−0.74	2	0	37	0.00022	36.5	92–102	K.VQYPMIGDPTR.Q + Oxidation (M)	
809.921	1617.8275	1617.825	1.57	2	0	90	1.5E-09	86.9	107–120	R.DFDVLIEEEGLALR.G	
651.8566	1301.6987	1301.6979	0.6	2	0	34	0.00064	33.9	121–132	R.GSFIINPEGOIK.A	
600.8325	1199.6505	1199.651	−0.44	2	0	74	3.6E-08	74.4	133–143	K.ALEVQDLSIGR.N	
393.7266	785.4387	785.4395	−1.07	2	0	53	5.6E-06	43.1	144–150	R.NAAELLR.K	
852.9034	1703.7922	1703.7937	−0.91	2	0	75	3.1E-08	75.1	154–169	K.AAQYVSTHSGEVCPAK.W	

Protein: YP_748545.1—alkyl hydroperoxide reductase (21 kDa).

match if the modified delta score equals the Mascot match score. Here, we also report the amino acid residue numbers from the protein sequence that correspond with each peptide match. Finally, the peptide match sequence is shown along with detected variable modifications (if any) and residues from the protein sequence that flank the peptide match.

8. Conclusions

Here, we describe the application of LC–MS/MS-based proteomics for the detection of nonfractionated proteome of *N. eutropha C91*. With the implementation of such state of the art technology, future studies could focus on the molecular mechanism of ammonium oxidation by *N. eutropha C91* under different growth conditions. Moreover, it is also possible to apply similar protocols to any other microorganism with an available genome sequence.

REFERENCES

Andersen, J. S., Wilkinson, C. J., Mayor, T., Mortensen, P., Nigg, E. A., and Mann, M. (2003). Proteomic characterization of the human centrosome by protein correlation profiling. *Nature* **426**, 570–574.

Arp, D. J., Sayavedra-Soto, L. A., and Hommes, N. G. (2002). Molecular biology and biochemistry of ammonia oxidation by Nitrosomonas europaea. *Arch. Microbiol.* **178**, 250–255.

Beyer, S., Gilch, S., Meyer, O., and Schmidt, I. (2009). Transcription of genes coding for metabolic key functions in Nitrosomonas europaea during aerobic and anaerobic growth. *J. Mol. Microbiol. Biotechnol.* **16**, 187–197.

Bianco, L., Mead, J. A., and Bessant, C. (2009). Comparison of novel decoy database designs for optimizing protein identification searches using ABRF sPRG2006 STANDARD MS/MS data sets. *J. Proteome Res.* **8**, 1782–1791.

Bock, E., Schmidt, I., Stuven, R., and Zart, D. (1995). Nitrogen loss caused by denitrifying Nitrosomonas cells using ammonium or hydrogen as electron donors nitrite as electron acceptor. *Arch. Microbiol.* **163**, 16–20.

Chain, P., Lamerdin, J., Larimer, F., Regala, W., Lao, V., Land, M., Hauser, L., Hooper, A., Klotz, M., Norton, J., Sayavedra-Soto, L., Arciero, D., *et al.* (2003). Complete genome sequence of the ammonia-oxidizing bacterium and obligate chemolithoautotroph *Nitrosomonas europaea*. *J. Bacteriol.* **185**, 2759–2773.

Colinge, J., Masselot, A., Giron, M., Dessingy, T., and Magnin, J. (2003). OLAV: Towards high-throughput tandem mass spectrometry data identification. *Proteomics* **3**, 1454–1463.

Cox, J., and Mann, M. (2008). MaxQuant enables high peptide identification rates, individualized p.p.b.-range mass accuracies and proteome-wide protein quantification. *Nat. Biotechnol.* **26**, 1367–1372.

Craig, R., and Beavis, R. C. (2004). TANDEM: Matching proteins with tandem mass spectra. *Bioinformatics* **9**, 1466–1467.

Dagda, R. K., Sultana, T., and Lyons-Weiler, J. (2010). Evaluation of the consensus of four peptide identification algorithms for tandem mass spectrometry based proteomics. *J. Proteomics Bioinform.* **3,** 39–47.

Daims, H., Bruehl, A., Amann, R., Schleifer, K. H., and Wagner, M. (1999). The domain-specific probe EUB338 is insufficient for the detection of all bacteria: Development and evaluation of a more comprehensive probe set. *Syst. Appl. Microbiol.* **22,** 434–444.

Fang, Y., Robinson, D. P., and Foster, L. J. (2010). Quantitative analysis of proteome coverage and recovery rates for upstream fractionation methods in proteomics. *J. Proteome Res.* **9,** 1902–1912.

Farley, A. R., and Link, A. J. (2009). Identification and Quantification of Protein Posttranslational Modifications. *Methods Enzymol.* **463,** 725–763.

Foster, L. J., de Hoog, C. L., Zhang, Y. L., Zhang, Y., Xie, X. H., Mootha, V. K., and Mann, M. (2006). A mammalian organelle map by protein correlation profiling. *Cell* **125,** 187–199.

Geer, L. Y., Markey, S. P., Kowalak, J. A., Wagner, L., Xu, M., Maynard, D. M., Yang, X., Shi, W., and Bryant, S. H. (2004). Open mass spectrometry search algorithm. *J. Proteome Res.* **3,** 958–964.

Ishihama, Y., Rappsilber, J., Andersen, J. S., and Mann, M. (2002). Microcolumns with self-assembled particle frits for proteomics. *J. Chromatogr. A* **979,** 233–239.

Ishihama, Y., Oda, Y., Tabata, T., Sato, T., Nagasu, T., Rappsilber, J., and Mann, M. (2005). Exponentially modified protein abundance index (emPAI) for estimation of absolute protein amount in proteomics by the number of sequenced peptides per protein. *Mol. Cell. Proteomics* **4,** 1265–1272.

Kampschreur, M. J., Tan, N. C. G., Kleerebezem, R., Picioreanu, C., Jetten, M. S. M., and Loosdrecht, M. C. M. (2008). Effect of dynamic process conditions on nitrogen oxides emission from a nitrifying culture. *Environ. Sci. Technol.* **42,** 429–435.

Kartal, B., Koleva, M., Arsov, R., van der Star, W., Jetten, M. S. M., and Strous, M. (2006). Adaptation of a freshwater anammox population to high salinity wastewater. *J. Biotechnol.* **126,** 546–553.

Keller, A., Nesvizhskii, A. I., Kolker, E., and Aebersold, R. (2002). Empirical statistical model to estimate the accuracy of peptide identifications made by MS/MS and database search. *Anal. Chem.* **74,** 5383–5392.

Klotz, M. G., Arp, D. J., Chain, P. S., El-Sheikh, A. F., Hauser, L. J., Hommes, N. G., Larimer, F. W., Malfatti, S. A., Norton, J. M., Poret-Peterson, A. T., Vergez, L. M., and Ward, B. B. (2006). Complete genome sequence of the marine, chemolithoautotrophic, ammonia-oxidizing bacterium *Nitrosococcus oceani* ATCC 19707. *Appl. Environ. Microbiol.* **72,** 6299–6315.

Lacerda, C. M., and Reardon, K. F. (2009). Environmental proteomics: Applications of proteome profiling in environmental microbiology and biotechnology. *Brief. Funct. Genomic. Proteomic.* **8,** 75–87.

LeDuc, R. D., Taylor, G. K., Kim, Y. B., Januszyk, T. E., Bynum, L. H., Sola, J. V., Garavelli, J. S., and Kelleher, N. L. (2004). ProSight PTM: An integrated environment for protein identification and characterization by top-down mass spectrometry. *Nucleic Acids Res.* **32**(Web Server issue), W340–W345.

Mobarry, B. K., Wagner, M., Urbain, V., Rittmann, B. E., and Stahl, D. A. (1996). Phylogenetic probes for analyzing abundance and spatial organization of nitrifying bacteria. *Appl. Environ. Microbiol.* **62,** 2156–2162.

Monti, M., Cozzolino, M., Cozzolino, F., Vitiello, G., Tedesco, R., Flagiello, A., and Pucci, P. (2009). Puzzle of protein complexes in vivo: A present and future challenge for functional proteomics. *Expert Rev. Proteomics* **6,** 159–169.

Nesvizhskii, A. I., Keller, A., Kolker, E., and Aebersold, R. (2003). A statistical model for identifying proteins by tandem mass spectrometry. *Anal. Chem.* **75,** 4646–4658.

Nielsen, M. L., Vermeulen, M., Bonaldi, T., Cox, J., Moroder, L., and Mann, M. (2008). Iodoacetamide-induced artifact mimics ubiquitination in mass spectrometry. *Nat. Methods* **5,** 459–460.

Norton, J. M., Klotz, M. G., Stein, L. Y., Arp, D. J., Bottomley, P. J., Chain, P. S., Hauser, L. J., Land, M. L., Larimer, F. W., Shin, M. W., and Starkenburg, S. R. (2008). Complete genome sequence of *Nitrosospira multiformis*, an ammonia-oxidizing bacterium from the soil environment. *Appl. Environ. Microbiol.* **74,** 3559–3572.

Pedrioli, P. G. (2010). Trans-proteomic pipeline: A pipeline for proteomic analysis. *Methods Mol. Biol.* **604,** 213–238.

Perkins, D. N., Pappin, D. J. C., Creasy, D. M., and Cottrell, J. S. (1999). Probability-based protein identification by searching sequence databases using mass spectrometry data. *Electrophoresis* **20,** 3551–3567.

Phillips, C. I., and Bogyo, M. (2005). Proteomics meets microbiology: Technical advances in the global mapping of protein expression and function. *Cell. Microbiol.* **7,** 1061–1076.

Rappsilber, J., Ryder, U., Lamond, A. I., and Mann, M. (2002). Large-scale proteomic analysis of the human splicesome. *Genome Res.* **12,** 1231–1245.

Rappsilber, J., Ishihama, Y., and Mann, M. (2003). Stop and go extraction tips for matrix-assisted laser desorption/ionization, nanoelectrospray, and LC/MS sample pretreatment in proteomics. *Anal. Chem.* **75,** 663–670.

Rogers, L. D., and Foster, L. J. (2009). Phosphoproteomics—finally fulfilling the promise? *Mol. Biosyst.* **5,** 1122–1129.

Rudnick, P. A., Clauser, K. R., Kilpatrick, L. E., Tchekhovskoi, D. V., Neta, P., Blonder, N., Billheimer, D. D., Blackman, R. K., Bunk, D. M., Cardasis, H. L., Ham, A. J. L., Jaffe, J. D., *et al.* (2010). Performance metrics for liquid chromatography-tandem mass spectrometry systems in proteomics analyses. *Mol. Cell. Proteomics* **9,** 225–241.

Sadygov, R. G., Cociorva, D., and Yates, J. R. (2004). Large-scate database searching using tandem mass spectra: Looking up the answer in the back of the book. *Nat. Methods* **1,** 195–202.

Schmidt, I., and Bock, E. (1997). Anaerobic ammonia oxidation with nitrogen dioxide by Nitrosomonas eutropha. *Arch. Microbiol.* **167,** 106–111.

Schmidt, I., Bock, E., and Jetten, M. S. M. (2001). Ammonia oxidation by Nitrosomonas eutropha with NO2 as oxidant is not inhibited by acetylene. *Microbiology UK* **147,** 2247–2253.

Schmidt, I., van Spanning, R. J. M., and Jetten, M. S. M. (2004). Denitrification and ammonia oxidation by Nitrosomonas europaea wild-type, and NirK- and NorB-deficient mutants. *Microbiology UK* **150,** 4107–4114.

Searle, B. C. (2010). Scaffold: A bioinformatic tool for validating MS/MS-based proteomic studies. *Proteomics* **10,** 1265–1269.

Stein, L. Y., Arp, D. J., Berube, P. M., Chain, P. S. G., Hauser, L., Jetten, M. S. M., Klotz, M. G., Larimer, F. W., Norton, J. M., den Camp, H., Shin, M., and Wei, X. M. (2007). Whole-genome analysis of the ammonia-oxidizing bacterium, *Nitrosomonas eutropha* C91: Implications for niche adaptation. *Environ. Microbiol.* **9,** 2993–3007.

Syka, J. E. P., Marto, J. A., Bai, D. L., Horning, S., Senko, M. W., Schwartz, J. C., Ueberheide, B., Garcia, B., Busby, S., Muratore, T., Shabanowitz, J., and Hunt, D. F. (2004). Novel linear quadrupole ion trap/FT mass spectrometer: Performance characterization and use in the comparative analysis of histone H3 post-translational modifications. *J. Proteome Res.* **3,** 621–626.

Wagner, M., Rath, G., Amann, R., Koops, H. P., and Schleifer, K. H. (1995). In-situ identification of ammonia-oxidizing bacteria. *Syst. Appl. Microbiol.* **18,** 251–264.

Weatherly, D. B., Atwood, J. A., Minning, T. A., Cavola, C., Tarleton, R. L., and Orlando, R. (2005). A heuristic method for assigning a false-discovery rate for protein identifications from mascot database search results. *Mol. Cell. Proteomics* **4,** 762–772.

Wessels, H., Vogel, R. O., van den Heuvel, L., Smeitink, J. A., Rodenburg, R. J., Nijtmans, L. G., and Farhoud, M. H. (2009). LC-MS/MS as an alternative for SDS-PAGE in blue native analysis of protein complexes. *Proteomics* **9,** 4221–4228.

Winogradsky, S. (1892). Contributions a la morphologie des organismes de la nitrification. *Arch. Sci. Biol.* **1,** 86–137.

Wisniewski, J. R., Zougman, A., Nagaraj, N., and Mann, M. (2009). Universal sample preparation method for proteome analysis. *Nat. Methods* **6,** 359–362.

Wittig, I., and Schagger, H. (2009). Native electrophoretic techniques to identify protein-protein interactions. *Proteomics* **9,** 5214–5223.

Yates, J. R., Eng, J. K., McCormack, A. L., and Schieltz, D. (1995). An approach to correlate tandem mass spectral data of peptides with amino acid sequences in a protein database. *J. Am. Soc. Mass Spectrom.* **5,** 976–989.

CHAPTER TWENTY-TWO

The Geochemical Record of the Ancient Nitrogen Cycle, Nitrogen Isotopes, and Metal Cofactors

Linda V. Godfrey* *and* Jennifer B. Glass[†]

Contents

1. Introduction	484
1.1. Evolution of the nitrogen cycle	484
1.2. Metal requirements for the nitrogen cycle	485
2. Reconstructing the Nitrogen Cycle and Associated Trace Metal Abundances through Time	489
2.1. Nitrogen	489
2.2. Trace metals	491
3. Determining Changes in the N-Cycle from the Geological Record of N Isotopes and Metal Availability	492
3.1. Site selection and origin of the $\delta^{15}N$ signature	492
3.2. Preservation of sediment $\delta^{15}N$ signals	493
3.3. Trace metals	495
4. Measurement	497
4.1. Nitrogen isotopes	497
4.2. Metals	498
5. Concluding Remarks	499
Acknowledgments	499
References	499

Abstract

The nitrogen (N) cycle is the only global biogeochemical cycle that is driven by biological functions involving the interaction of many microorganisms. The N cycle has evolved over geological time and its interaction with the oxygen cycle has had profound effects on the evolution and timing of Earth's atmosphere oxygenation (Falkowski and Godfrey, 2008). Almost every enzyme that microorganisms use to manipulate N contains redox-sensitive metals. Bioavailability of these metals has changed through time as a function of varying redox conditions,

---

* Institute of Marine and Coastal Sciences, Rutgers University, New Brunswick, New Jersey, USA
[†] School of Earth and Space Exploration, Arizona State University, Tempe, Arizona, USA

and likely influenced the biological underpinnings of the N cycle. It is possible to construct a record through geological time using N isotopes and metal concentrations in sediments to determine when the different stages of the N cycle evolved and the role metal availability played in the development of key enzymes. The same techniques are applicable to understanding the operation and changes in the N cycle through geological time. However, N and many of the redox-sensitive metals in some of their oxidation states are mobile and the isotopic composition or distribution can be altered by subsequent processes leading to erroneous conclusions. This chapter reviews the enzymology and metal cofactors of the N cycle and describes proper utilization of methods used to reconstruct evolution of the N cycle through time.

1. INTRODUCTION

Nitrogen is essential for life due to its presence in proteins and nucleic acids. It is the fourth most abundant element in biomass following hydrogen, carbon and oxygen. Unlike other macronutrient biogeochemical cycles like P and Si, the most important fluxes of the global N cycle are biologically controlled. The atmosphere is by far the largest surface reservoir of N on Earth, comprising 78% N_2. While atmospheric N_2 is readily accessible to living organisms, most are unable to directly use N_2. Nitrogenase, the enzyme that allows biological N_2 fixation to take place, is found only in prokaryotes (Raymond et al., 2004). Since eukaryotes cannot fix N_2, they require a source of fixed N. It has been postulated that the radiation of eukaryotes and evolution of complex life could only occur once there was an established source of fixed N (Anbar and Knoll, 2002; Knoll, 1992; Raymond et al., 2004). A subject of current interest is the response of the N cycle to changes in redox state and oxygenation of Earth's atmosphere. The dominant species of fixed N in seawater differs depending on its redox state. Ammonium (NH_4^+) is the major form of fixed N in anoxic systems and nitrate (NO_3^-) is the dominant form in oxic conditions. At redox boundaries, gaseous forms of N such as nitrous oxide (N_2O) or N_2 are produced during nitrification and denitrification, leaving seawater depleted in fixed N. Such conditions would have prevailed during the transition from anoxic to fully oxic ocean conditions when the partial pressure of O_2 (pO_2) was low or when the oceans were chemically stratified. The loss of fixed N from the ocean has important ramifications for both evolution and strength of the biological pump.

1.1. Evolution of the nitrogen cycle

The fundamental biological processes in the N cycle almost certainly have not changed since they evolved in the Archean (Falkowski and Godfrey, 2008) although O_2-requiring pathways within the cycle have become more critical since the Great Oxidation Event ~2.4 billion years ago (Falkowski, 1997). On the early anoxic Earth, N cycled between the major reservoirs of the atmosphere, ocean, and lithosphere as neutral N_2, NH_4^+ or organic bound-N (Godfrey and Falkowski, 2009; Papineau et al., 2005). While abiotic N_2 fixation occurred by oxidation of atmospheric N_2 to NO and then to NO_2^- during lightning discharge, bolide impacts, and volcanic eruptions (Mancinelli and McKay, 1988; Mather et al., 2004; Navarro-Gonzalez et al., 2001), the flux was probably small relative to the biological demand for N. Although the majority of the NO_2^- delivered to the oceans would have been reduced abiotically to NH_4^+ by Fe(II) (Summers, 1999; Summers and Chang, 1993), its presence in the surface ocean could have initiated new biological pathways such as anaerobic ammonium oxidation (or "anammox"; Klotz and Stein, 2008; Kuenen, 2008).

The greatest change in the N cycle occurred in the late Archean, following the onset of oxygenic photosynthesis by cyanobacteria (Anbar et al., 2007; Garvin et al., 2009; Godfrey and Falkowski, 2009; Waldbauer et al., 2009). Using newly available O_2, microbes were able to oxidize ammonia as an energy source (Hollocher et al., 1981). Oxygen was also used as a terminal electron acceptor, but unlike the oxidation of NH_4^+ in which O_2 is essential, nitrogen oxides (N_{ox}) could substitute in this reaction if the supply of O_2 became inadequate to support respiration. Metabolic pathways that may have arisen in response to the small flux of abiotically formed N_{ox} would have acquired widespread importance due to the much larger rates of N_{ox} formed using photosynthetically produced O_2. The respiratory use of N_{ox} (denitrification) and the catabolic anammox pathway bypassed the slower subduction-volcanic loop of the N cycle by enabling the immediate return of N to the atmosphere from the water column and surface sediments. Until rates of nitrification exceeded those of denitrification and anammox, the oceans would have remained N-limited because any NH_4^+ oxidized to NO_2^- or to NO_3^- could have been used in place of O_2 and denitrified (Fennel et al., 2005). If N_2-fixation rates were lower than those of coupled nitrification and denitrification, fixed N may have been stripped from seawater leading to widespread and pervasive N-limitation. During times when rates of nitrification exceeded denitrification, NO_3^- could have accumulated. Eventually NO_3^- became the dominant species of dissolved fixed N in the ocean and the N cycle acquired its modern characteristics.

1.2. Metal requirements for the nitrogen cycle

In reconstructing the evolution of the N cycle, it is important to consider changing trace metal bioavailability over geologic time. Almost every enzymatic pathway in the N cycle involves a metal cofactor: almost all contain iron (Fe) and many contain copper (Cu) or molybdenum (Mo); see Fig. 22.1 for a compilation of all metals involved in the N cycle.

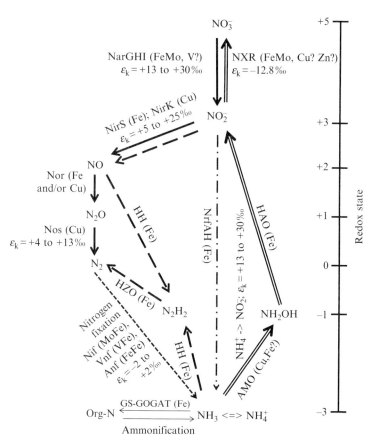

Figure 22.1 The nitrogen cycle showing the metalloenzymes involved at each step and, where data is available, the associated N isotope fractionation. Metals involved include Fe, Mo, Cu, and in rare cases V and Zn. If the N isotope enrichment factor ε_k is positive, the product will be isotopically depleted by that amount relative to the substrate. The vertical scale indicates the redox state of chemical species of N in bold. The double line arrows represent nitrification; the thick solid arrows, denitrification; and the large dashed arrows, anammox. Other transfers within the N cycle N$_2$-fixation (small dashed arrow), nitrite ammonification (dash-dot arrow), ammonium assimilation, and ammonification (thin solid arrows) are included to complete the cycle. NO_3^- and NH_4^+ are both assimilated to produce organic(Org)-N that is lower than the substrate by 5–10‰ and 14–27‰, respectively. Isotope enrichment factors from Casciotti (2009) and references therein. Figure modified from Buick (2007, Fig. 2). Enzyme name abbreviations: NarGHI, membrane-bound dissimilatory nitrate reductase; NXR, nitrite oxidoreductase; NirS, Fe-containing nitrite reductase; NirK, Cu-containing nitrite reductase; Nor, nitric oxide reductase; Nos, nitrous oxide reductase; Nif, Mo–Fe form of nitrogenase; Vnf, V–Fe form of nitrogenase; Anf, Fe–Fe form of nitrogenase; HH, hydrazine hydrolase; NrfAH, cytochrome c nitrite reductase; AMO, ammonia monooxygenase; HAO, hydroxylamine oxidoreductase; GS–GOGAT, glutamine synthetase–glutamate synthase (see text for more detailed descriptions of enzymes).

These metals are effective catalysts for the redox reactions that are the backbone of the N cycle. In this section, we review the current knowledge of metal requirements for nitrification, denitrification, and anammox. We then discuss methodology for using the geochemical record to track changes in Fe, Cu, and Mo abundances through time, with an emphasis on marine environments. Readers looking for a more in-depth review of N-cycle enzymatic structures are referred to Butler and Richardson (2005). For more about N acquisition in terms of metal requirements and assay techniques, we refer readers to two recent reviews: Berges and Mulholland (2008) and Glass *et al.* (2009).

1.2.1. Nitrification

The process of nitrification is composed of three oxidative steps: $NH_4^+ \rightarrow NH_2OH \rightarrow NO_2^- \rightarrow NO_3^-$. The first step is performed by ammonia monooxygenase (AMO), an enzyme with no crystal structure due to difficulties in purification and stabilization. However, AMO is similar to particulate methane monooxygenase (pMMO) in sequence, catalysis, and structure (Arp and Stein, 2003), suggesting that—like pMMO—AMO contains three Cu atoms per monomer (Rosenzweig, 2008 and references therein) and possibly additional Fe (Zahn *et al.*, 1996). The Cu requirement has been further substantiated by experiments in which addition of Cu activated AMO (Ensign *et al.*, 1993). The next step in nitrification, NH_2OH oxidation to NO_2^-, is catalyzed by the enzyme hydroxylamine oxidoreductase (HAO) which contains 24 Fe atoms in *c*-type cytochromes (Igarashi *et al.*, 1997). The final nitrification enzyme, nitrite oxidoreductase (NXR), has received much less attention, but its sequence suggests that it is a membrane-bound nitrate reductase protein (see Section 1.2.2) run in reverse (Kroneck and Abt, 2002). In addition to Mo and Fe, Cu and Zn have also been reported in NXR (Meincke *et al.*, 1992), although this has not been verified by crystallization. The high Fe requirements for nitrification are supported by experiments with the freshwater ammonia-oxidizing bacterium *Nitrosomonas europaea* in which the bacterium grown at 0.2 μM Fe displayed lower AMO and HAO activity than Fe-replete (10 μM) cultures (Wei *et al.*, 2006). While most attention has focused on Fe limitation of marine N_2-fixation (Carpenter and Capone, 2008; Krishnamurthy *et al.*, 2008, 2010), it is quite possible that nitrification in the ocean is also Fe limited, since dissolved marine Fe is typically in the picomolar range (Johnson *et al.*, 1997) and the sum of Fe atoms in the three nitrification proteins (AMO, HAO, and NXR) is ~ 50 (assuming the same Fe content in NXR as NarGHI; see Section 1.2.2), even greater than the 38 Fe atoms in nitrogenase enzyme.

1.2.2. Denitrification

Denitrification involves four reductive steps: $NO_3^- \rightarrow NO_2^- \rightarrow NO \rightarrow N_2O \rightarrow N_2$. The first two steps can be catalyzed by multiple enzymes, allowing some flexibility in metal cofactors, whereas the last two steps can only to be performed by one enzyme respectively and therefore have stricter metal requirements. Bacterial dissimilatory nitrate reduction usually involves a membrane-bound protein complex (NarGHI) containing two b-type cytochromes (each binding 1 Fe atom), a Fe_3S_4 cluster, four Fe_4S_4 clusters, and a molybdopterin (Mo) active site (Bertero et al., 2003; Jormakka et al., 2004). Alternative Mo-less nitrate reductases containing V and Fe in place of Mo have been discovered in vanadate and Fe(III)-reducing bacteria, respectively (Antipov et al., 1998; Naik et al., 1993). These discoveries suggest that ancient denitrification could have evolved even in the absence of Mo as long as metallo-alternatives like V and Fe were bioavailable.

The next step in denitrification, nitrite reduction, can produce either nitric oxide (NO) in the case of conventional denitrification to N_2 or NH_4^+ by dissimilatory nitrate reduction (DNRA) or respiratory nitrite ammonification (Simon, 2002). In the case of conventional denitrification, this process is catalyzed by either a cytochrome cd_1 (4 Fe)-containing nitrite reductase (cd_1NIR or NirS) (Fulop et al., 1995) or a Cu-containing nitrite reductase (CuNIR or NirK) that possesses either 6 or 24 Cu atoms depending on its subunit content (Godden et al., 1991; Nojiri et al., 2007). The proliferation of prokaryotic genome sequences in recent years has revealed even greater diversity of NiRs, but thus far all studies concur that NiRs contain either Fe or Cu, or possibly both (Ellis et al., 2007). In the case of DNRA, only Fe is required in the form of 10 cytochrome c heme cofactors for the enzyme NrfAH (Einsle et al., 1999). The following step in the respiratory pathway, NO reduction to N_2O by the enzyme nitric oxide reductase (NorBC), also requires only Fe, in the form of cytochrome bc hemes as well as nonheme Fe (Zumft, 2005 and references therein). Thus, all steps in the denitrification pathway through the production of N_2O can be catalyzed almost entirely by Fe along with a very small amount of Mo in NarG. The next step, N_2O reduction to N_2, catalyzed by the enzyme nitrous oxide reductase (Nos), is unique in the denitrification pathways because it contains no Fe, but binds 12 atoms of Cu per homodimer (Brown et al., 2000). This high Cu requirement has been shown to result in N_2O accumulation when denitrifiers are grown in the absence of Cu (Granger and Ward, 2003; Iwasaki and Terai, 1982). Coping with low Cu is likely an environmentally relevant handicap for denitrifiers living proximal to sulfidic environments where Cu is often low in concentration. These findings led Buick (2007) to surmise that denitrification to N_2 might be Cu-limited in sulfidic environments where Cu is removed by sulfide, including many modern marine sediments and basins such as the Black Sea, as well as the ancient ocean during the mid-Proterozoic

(~1.8–0.8 billion years ago), possibly resulting in the build-up of a N_2O "laughing gas" atmosphere at that time.

1.2.3. Anammox

Anammox is the most-recently discovered pathway in the N cycle, and is currently thought to involve three steps: $NO_2^- \rightarrow NO + NH_4^+ \rightarrow N_2H_4$ (hydrazine) $\rightarrow N_2$ (Kuenen, 2008). The first step involves the Fe-form of nitrite reductase (NirS) based on gene identification in the recently sequenced anammox bacterium *Kuenenia stuttgartiensis* (Kuenen, 2008). The next step involves hydrazine hydrolase (HH), composed of eight currently uncharacterized proteins which likely bind numerous Fe atoms as cytochromes (Strous *et al.*, 2006). The last step in anammox involves hydrazine oxidoreductase (HZO) which converts hydrazine to N_2 (Kuenen, 2008). Although it has not been crystallized, the protein sequence of HZO contains the same eight heme binding sites as HAO (Klotz *et al.*, 2008). The high Fe requirement for anammox is further demonstrated by the deep reddish/pink color of anammox bacteria (Kuenen, 2008). The lack of a Cu requirement for anammox led Klotz and Stein (2008) to speculate that anammox may predate conventional denitrification, and that Nos and NirK did not evolve until oxic conditions prevailed and Cu levels rose.

2. Reconstructing the Nitrogen Cycle and Associated Trace Metal Abundances through Time

2.1. Nitrogen

Nitrogen possesses two stable isotopes, ^{14}N and less abundant ^{15}N. Since the $^{15}N/^{14}N$ ratio of the atmosphere is 0.0036765, sample data are reported relative to the atmosphere in parts per thousand, $\delta^{15}N_{sample} = (^{15}N/^{14}N_{sample}/^{15}N/^{14}N_{atm} - 1) \times 1000$ in order to better highlight isotopic differences. In 1947, Urey published his work describing the physicochemical basis of isotope fractionation in which the dissociation energy of molecule, or zero-point energy minimum, is related to the vibrational energy of a molecular bond and to a lesser extent the translational and rotational frequencies. Bond strengths are thus a function of mass. As a consequence, bonds associated with lighter nuclei are weaker and these molecules react faster. Furthermore, since the stronger bonds associated with heavier nuclei persist longer than those of light nuclei, heavier isotopes are frequently partitioned into condensed phases. Biologically mediated reactions can be considered as irreversible, kinetic reactions and are frequently associated with large isotopic fractionation effects.

Each transfer of N within the N-cycle partitions the two isotopes of N. As with physicochemical transformations of N, biological reactions partition ^{14}N and ^{15}N with the rate-limiting step determining the isotope fractionation factor for the entire transfer. In nearly all N transformations, ^{15}N is retained in the substrate. An important exception is a large inverse isotope effect associated with nitrite oxidation (Casciotti, 2009). A compilation of isotope effects on nitrogen can be found in Casciotti (2009) and is included in Fig. 22.1. The imprint of a reaction on the isotope composition of the substrate or product is only preserved if the reaction does not go to completion. If the reaction does go to completion, mass balance requires the substrate and product to have the same isotopic composition. An example of this is denitrification. Microbial denitrification of NO_3^- to N_2 carries with it a ^{15}N-enrichment of 20–30‰. The rate-limiting step of water column denitrification is the activity of nitrate reductase. In modern upwelling systems, NO_3^- is advected through O_2 minimum zones (OMZ) and complete consumption of NO_3^- rarely ever occurs. Incomplete denitrification in these systems results in $\delta^{15}N-NO_3^-$ values of up to 16‰, and biological assimilation leads to preservation of the denitrification signal in underlying sediments. Respiration of organic matter (OM) in sediments also leads to denitrification, but the diffusion of seawater NO_3^- into sediments becomes the rate-limiting step due to low permeability and all NO_3^- is consumed during denitrification. Since all NO_3^- is converted to N_2, there is no change in the $\delta^{15}N$ value of seawater, and although the NO_3^- inventory is decreased. Average seawater NO_3^- $\delta^{15}N$ reflects the extent and intensity of water column denitrification (Altabet, 2007). Readers looking for a more in-depth review of the isotopic N-cycle in the modern ocean and on glacial–interglacial timescales are referred to a review by Altabet (2005).

Many of the functions of the N cycle are dependent on redox state of the environment. The strongest signal in $\delta^{15}N$ records comes from water column denitrification. In the modern ocean, denitrification occurs in areas of high productivity associated with upwelling or coastal eutrophication (Altabet *et al.*, 1999; Brandes *et al.*, 1998; Cline and Kaplan, 1975; Lui and Kaplan, 1989; Pride *et al.*, 1999; Sigman *et al.*, 2003; Voss *et al.*, 2000; Voss *et al.*, 2001). The extent and intensity of water column denitrification depends on the supply of O_2 to the water column beneath the chlorophyll maximum relative to respiration of OM. Oxygen supply and respiration rates can be affected by circulation, flux of OM from the surface ocean, or pO_2 and all of these factors can change through time, thereby changing $\delta^{15}N$ (e.g., Algeo *et al.*, 2008; Galbraith *et al.*, 2004; Papineau *et al.*, 2009; Robinson *et al.*, 2007). These parameters can be measured in modern environments, but in the geological record they need to be constrained by sedimentological investigations, tectonic-based simulation, geochemical or proxies (Berggren and Hollister, 1977; Goldberg and Arrhenius, 1958).

More recently, this approach has been reversed, and N isotopes have been used to constrain pO_2 (Garvin et al., 2009; Godfrey and Falkowski, 2009; Quan et al., 2008). However, if denitrification consumes all available NO_3^-, N_2-fixers gain a competitive edge and the OM that is exported to the sediments and preserved in the geological record has low $\delta^{15}N$ values (e.g., Cao et al., 2009; Kuypers et al., 2004; Paris et al., 2010).

2.2. Trace metals

Geochemical analyses of ancient rocks and modeling of metal abundances through earth history have thus far yielded reconstructions of two of the metals (Mo and Fe) involved in N-cycle (Anbar, 2008). Different types of sedimentary deposits are useful for studying Mo and Fe. Organic-rich shales deposited under anoxic conditions are used to reconstruct ancient marine Mo concentrations (Scott et al., 2008) because ratios of solid-phase Mo to total organic carbon increase linearly with deep water dissolved Mo in modern anoxic environments (Algeo and Lyons, 2006). Using this approach, Scott et al. (2008) calculated that Archean marine Mo concentrations were $\ll 2$–3 nM, Proterozoic concentrations were ~ 10–20 nM, and Mo rose to modern values of ~ 100 nM ~ 600 million years ago. Likewise, banded iron formations (BIFs) are used to constrain ancient Fe concentrations. In contrast to organic-rich shales, BIFs are not present in modern-day marine settings and in fact have been rare for the past 1.8 billion years. The commonly accepted theory for the origin of BIFs is upwelling of Fe-rich deep water combining with oxygenated surface water to precipitate the Fe oxide minerals in BIFs (Holland, 1973). Holland used a simple set of solubility equations to estimate the abundance of ferrous Fe (50–500 μM Fe^{2+}) that would have precipitated BIFs. Such high Archean Fe levels would have easily supported the high Fe demands of N cycle enzymes, whereas Fe has declined through time and now is <1 nM in seawater (Johnson et al., 1997). Similar reconstructions of Cu through time are lacking. Saito et al. (2003) used the thermodynamic software MINEQL$^+$ to model changes in a suite of trace metals through time and estimated that Cu concentrations in the Archean and Proterozoic would have been extremely low ($\ll 10^{-20}$ M) assuming that sulfide was present in the water column. Geochemical analyses are needed to independently test such models. One technique that has recently been shown to have potential for tracking metal concentrations through the first half of earth history is metal/Fe concentrations in BIFs followed up by experimental determinations of distribution coefficients between dissolved metal in solution and metal/Fe ratio in Fe oxide particles. This approach was used to reconstruct seawater Ni concentrations through time (Konhauser et al., 2009) and may serve as a useful paleoproxy for Cu concentrations as well.

3. DETERMINING CHANGES IN THE N-CYCLE FROM THE GEOLOGICAL RECORD OF N ISOTOPES AND METAL AVAILABILITY

The need to understand the processes that define and alter $\delta^{15}N$ of sedimentary OM (SOM) and the dominant reservoir of dissolved fixed N is particularly critical in deep time and for interpretation of the geologic record in context of the correct paleoenvironment. Records that work back in time from present day have the advantage of starting from an environment that can be directly studied. In order to interpret $\delta^{15}N$ records further back in time, it is important to first define the physical setting (e.g., open ocean vs. restricted basin, probable circulation, redox status) and constrain the system through other independent proxies. The second, and equally important, consideration is the possibility that one or other processes, early diagenesis, catagenesis, metamorphism, and fluid migration have altered the initial $\delta^{15}N$ signature of OM.

3.1. Site selection and origin of the $\delta^{15}N$ signature

The selection of sites and materials is dictated by the nature of the question one is asking. *In situ* marine sediments are limited to the Cenozoic and parts of the Mesozoic and there are multiple core repositories (Deep Sea Drilling Project, Ocean Drilling Project, and the International Ocean Drilling Project) as well as those specific to institutions, with coring capabilities. Opportunities may also arise to access industry cores. All of these materials are generally accompanied by site survey seismic data and geophysical logs. They often have associated preliminary data or have been studied with published results. For sediments older than the Mesozoic, one is invariably limited to the continents and uplifted/obducted sediment sequences and may not record open ocean conditions. Core material is preferred because one is less likely to encounter surface weathering and leaching of metals or oxidation of OM (e.g., Jaffe *et al.*, 2002; Sherman *et al.*, 2007) but cores are hard to come by, expensive to drill, and require a thorough understanding of the subsurface geology and strata relationships. However, useful information can often be obtained from local geological surveys, mines, or water well logs. Often it is prudent to identify an area where a core might have been taken in the past and redrill near it using desired techniques and practices. For outcrop samples, it is essential to chose rocks that show zero or minimal effects of weathering and then remove oxidation rinds in the laboratory (a practice that should also be applied to drill core material). Shale is an ideal material for N isotope studies because clays trap N during substitution of K^+ by NH_4^+ (Garrels and Christ, 1965). However, allochthonous sources of NH_4^+ in nearshore sediments must be considered (Sweeney and Kaplan, 1980) and shale is often poorly exposed in outcrop requiring trenches to be dug in order to reach bedrock before samples can be collected.

The residence time of NO_3^- in the modern ocean is slightly less than 2000 years, about 1.5–2× that of ocean mixing (Brandes and Devol, 2002). In general, the composition of OM $\delta^{15}N$ records subsurface $NO_3^- - \delta^{15}N$, which is either equal to average deep water $NO_3^- - \delta^{15}N$ or is (a) increased by N loss via denitrification or anammox in OMZs (Cline and Kaplan, 1975; Gruber and Sarmiento, 1997) or (b) decreased through surface inputs such as N_2 fixation. These processes are focused in the upper levels of the ocean or at redox boundaries, and changes in $\delta^{15}N$ can occur at rates much faster than suggested by nitrate's global oceanic residence time. If seeking a record of ocean average $\delta^{15}N$, it is important not to alias records by choosing sampling intervals that do not provide the full range in possible $\delta^{15}N$ values. It is also important to choose sites where $\delta^{15}N$ is unlikely to be strongly influenced by local variations in the N-cycle. This is often very difficult, but considerable insight into the workings of the N-cycle on a temporal basis can be gained by looking at multiple sites or at areas that might have been influenced by N laterally exported from other regions; for example, responses of N_2 fixation to changes in denitrification (Deutsch et al., 2004; Ren et al., 2009).

One also needs to consider whether the $\delta^{15}N$ signature transported to the sediments by OM reflects NO_3^- (or NH_4^+) that reaches the photic zone and not the fraction of available NO_3^- (or NH_4^+) that was consumed. The $\delta^{15}N$ of yearly averaged N entering the photic zone is recorded in surface sediments of the low and mid latitudes where nutrient consumption is complete. In high nutrient low chlorophyll (HNLC) regions, nutrients in surface water are not fully consumed and OM transports a $\delta^{15}N$ signal that is lower than that of fixed N (Altabet and Francois, 1994).

Marine sediments can also contain N which originated from terrestrial OM (Hedges et al., 1997). This is a particular problem in coastal environments due to input of terrestrial OM derived from vascular plants that usually have high C/N ratios and low $\delta^{15}N$ compared to typical marine OM (Peters et al., 1978). However, vascular plant material gains N during decay whereas marine plankton material preferentially loses N (Hedges and Oades, 1997). Therefore, marine OM can possess $\delta^{15}N$ compositions that encompass those of terrestrial OM.

3.2. Preservation of sediment $\delta^{15}N$ signals

It is critical to consider preservation of primary N isotope signals in studies of N in deep time because of the myriad diagenetic and metamorphic processes that can affect the distribution of N isotopes. In essence, the better the preservation of sedimentary organic matter (SOM), the more likely primary $\delta^{15}N$ signals have been preserved. There are numerous studies that have investigated preservation of sedimentary organic matter (SOM) in relation to the carbon cycle (e.g., Aller, 1994; Dauwe et al., 1999; Hedges and Keil, 1995; Mayer, 2004) and these approaches can be applied to reconstructing the ancient N cycle as well. In old sediments, the assumption that bulk material, kerogen, or layered silicates automatically yield reliable data cannot be made. In each study, the reliability of the $\delta^{15}N$ signal must be addressed. The reader can refer to reviews and methodologies to assess bulk $\delta^{15}N$ preservation in de Lange (1998), Freudenthal et al. (2001), Lehmann et al. (2002), Altabet (2005), and Junium and Arthur (2007). Other approaches for improving the fidelity of $\delta^{15}N$ records include the use of biogenic mineral-bound $\delta^{15}N$ found in diatoms (e.g., Robinson et al., 2005; Shemesh et al., 1993; Sigman et al., 1999) and foraminifera (Altabet and Curry, 1989; Ren et al., 2009; Uhle et al., 1997), or compound-specific (pigment/porphyrin) $\delta^{15}N$ extracted from bulk sediment (Kashiyama et al., 2007; Ohkouchi et al., 2005; Sachs and Repeta, 2000; Sachs et al., 1999). These methods will not be discussed further here.

3.2.1. Early diagenesis

Nitrogen is lost from complex molecules as ammonia (NH_3). Microbial deamination reactions occur regardless of redox state and should result in a decrease in bulk organic $\delta^{15}N$ (Macko, 1993; Macko et al., 1986). Since NH_4^+ is retained by clays, the $\delta^{15}N$ of bulk sediment may be unperturbed. As burial temperatures increase, organic molecules may degrade further to produce refractory OM (Durand, 1985; Ishiwatari et al., 1977). This can lead to additional release of N into fluids and subsequent capture by clays or condensation into more complex organic molecules. Studies have suggested that in general, there is little change in $\delta^{15}N$ of OM during early diagenesis under anoxic water columns, but substantial increases in $\delta^{15}N$ can occur under oxic conditions (e.g., Altabet, 2005).

3.2.2. Catagenesis and metamorphism

Marine algal biomass has a C:N ratio of ~ 7 and bacteria have a C:N ratio of ~ 4 (Hedges et al., 1986). Well preserved marine OM typically has a C:N ratio < 20 (Peters et al., 1978). Organic C is less mobile that N, and the C:N ratio of the bulk sediment can be used to assess the extent to which N may have been lost, a process that can also affect $\delta^{15}N$. Different sources of OM can be identified from kerogen (insoluble SOM) and also from the C, H, O, and N composition of OM. These measurements also yield information about the OM maturity and loss of H, O, and N. van Krevelen plots (van Krevelen, 1961) which use H/C, O/C, and N/C (or their converted H, O, and N index forms) are a common technique in the oil and gas industry but often underutilized in paleoceanographic studies. Kerogen is divided into four types (algal/bacterial, marine plankton, higher plants, and "oxidized"), and their migration toward the origin of van Krevelen plots indicates increasing maturity (van Gijzel, 1982). The reader can refer to the extensive petroleum literature to learn more about additional thermal alteration indices.

Fractionation with loss of ^{14}N from OM may occur during thermal maturation or catagenesis. The effect of N loss on the isotopic composition of OM $\delta^{15}N$ during catagenesis may amount to a couple of tenths of a per mil to ~ 2‰ depending on maturity and the amount of N released (Stiehl and Lehmann, 1980; Peters et al., 1981; Williams et al., 1995). Thermal processes alone may strongly affect $\delta^{15}N$: in a study of kerogen Boudou et al. (2008) suggested that thermal metamorphism removes N from OM but does not induce an isotopic effect on residual OM. However, devolatilization reactions can lead to substantial increases in bulk or kerogen $\delta^{15}N$ (e.g., Bebout and Fogel, 1992; Jia, 2006). Metamorphic loss of N can occur through $N_2-NH_4^+$ or $NH_3-NH_4^+$ exchanges which inversely correlate with temperature. The enrichment in ^{15}N during devolatilization and loss of N_2 is smaller (between 3‰ and 2‰ for temperatures between 350 and 630 °C) than for loss of NH_3 (between 12‰ and 6‰ for temperatures between 350 and 650 °C; Hanschmann, 1981; Haendel et al., 1986). Nitrogen loss through these exchange reactions can be considered as either following batch volatilization or Rayleigh distillation behavior, the difference being that during batch distillation, N exchanges in equilibrium between reactants and products whereas during Rayleigh distillation, product is continuously removed from the system. These models describe the change in the material losing N (mineral, kerogen) as $\delta_{final}-\delta_{initial} = 1000 \, (F^{(\alpha-1)}-1)$ for Rayleigh fractionation and $\delta_{final}-\delta_{initial} = (1-F)1000 \ln\alpha$ for batch distillation where F is the fraction of N remaining in the substrate (mineral or kerogen) and α is the fractionation factor. More often, it is found that the mode is some hybrid of these two mechanisms because devolatilization may be discontinuous on small time and spatial scales; see Fig. 22.2 and Bebout and Fogel (1992). If there is substantial loss of N, even relatively minor α values can lead to substantial increases in $\delta^{15}N$ (Dauphas and Marty, 2004; Jia and Kerrich, 2004; Kerrich and Jia, 2004). The opposite process, addition of N to kerogen, is equally concerning. Schimmelmann and Lis (2010) undertook 5 year incubations of ^{15}N-labeled NH_4^+ with different types of kerogen at temperatures between 100 and 200 °C. Their results indicate that NH_4^+ in migrating fluids can be abiotically sequestered into kerogen, thus introducing nonautochthonous $\delta^{15}N$ signals.

3.3. Trace metals

The trace metals of importance to the N cycle (Fe, Mo, and Cu) are redox sensitive and display different mobilities depending on redox state. Iron (III) forms an insoluble oxyhydroxide whereas Fe (II) is fluid mobile, but is

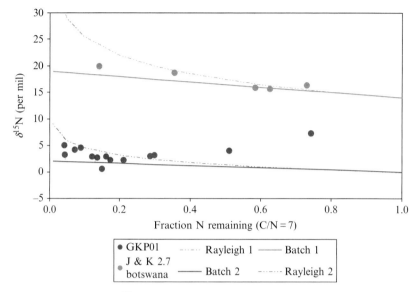

Figure 22.2 Kerogen $\delta^{15}N$ data from GKP01 and Jia and Kerrich (2004) plotted against the fraction of N remaining calculated from $\%C_{org}$. Data from Jia and Kerrich (2004) falls between Rayleigh 1 and batch 1 distillation model curves where $\varepsilon_p = 5$ and initial $\delta^{15}N = 14.0‰$. For GKP, most of the data fall between Rayleigh 2 and batch 2 distillation models where $\varepsilon_p = 2$ and initial $\delta^{15}N = 0.0‰$. The higher ε_p for the Jia and Kerrich data curves most likely reflects NH_3–NH_4^+ exchange, while for GKP data (Godfrey and Falkowski, 2009), it reflects N_2–NH_4^+ exchange. If postdeposition losses of ^{14}N are not considered and accounted for, measurement of high $\delta^{15}N$ could be misinterpreted as indicating the occurrence of biological or physical processes (e.g., Dauphas and Marty, 2004; Jia and Kerrich, 2004; Kerrich and Jia, 2004).

sequestered into sulfides. Molybdenum (VI) forms a soluble oxyanion (MoO_4^{2-}) in oxic water, but $MoO_{4-x}S_x^{2-}$ is formed in reduced settings and is particle reactive. Mo(IV) forms sulfides and the effect on Mo stable isotopes is included in a review by Anbar and Rouxel (2007). The distribution and redox chemistry of Cu is complex and is dominated by organic complexes. Copper's complexation with organic ligands affects its ocean distribution and is responsible for its benthic flux as OM decays (Boyle et al., 1977; Moffett and Zika, 1983).

The distribution of redox-sensitive metals can be used as a proxy for water column redox conditions (e.g., van Bentum et al., 2009). When these elements are normalized to Al, enrichments in sedimentary V and Mo occur if bottom water or sediments close to the seawater–sediment interface become anoxic (Calvert and Pedersen, 1993). At the same time, Mn/Al and Fe/Al ratios may decrease because Mn(II) and Fe(II) are solubilized. Enriched layers of Mn and Fe can form in sediments by precipitation of Mn

(II) and Fe(II) with sulfide produced by sulfate reduction (Calvert and Pedersen, 1993). The thickness and position of these layers depends on sulfate diffusion from overlying water and sulfate reduction rates. The relative positions of metal-rich layers combined with changes in δ^{15}N and simple observations of laminated or (for the Phanerozoic) bioturbated sediments can be very useful in recognizing changes in sediment redox state controlled by export productivity and changes in bottom water redox state which may be indicative of more general redox conditions. However, metals like Mn can migrate post-deposition out of suboxic layers as Mn(II) and then precipitate as MnO_2 in oxic conditions, likewise Mo and V can migrate from oxic environments toward suboxic fronts where they are enriched in solid phase. Classic examples of metal-rich layers are the rolling redox fronts associated with organic-rich turbidite emplacement in the Northeast Atlantic as well as climate-induced increases in sediment accumulation rates (e.g., Thomson et al., 1993; Wallace et al., 1988). In older rocks, groundwater may also change the redox conditions relative to those that existed during deposition. This condition is analogous to acid mine drainage, which an extreme version, but oxidation of sulfides (or leaching of Fe and Mn phases) may release chalcophile elements such as Mo and V, and are then mobile. Upon encountering reducing environments, groundwater redox state may decrease and these elements precipitate, producing postdepositional metal-rich layers that are not strictly associated with redox state or their availability at the time of sediment deposition.

4. Measurement

4.1. Nitrogen isotopes

Nitrogen isotopes are measured on gas source stable isotope ratio mass spectrometers (IRMS). The principle of the measurement is the same for all the IRMS instruments but sample introduction can differ. The most common method is continuous flow IRMS where a CNH analyzer is coupled to a mass spectrometer and the combustion products of the sample are transferred in a helium (He) stream. The sample is encapsulated in tin foil (but silver or aluminum can also be used) and dropped in to a furnace which is held around 1000–1050 °C. The He stream is briefly interrupted by a pulse of O_2, and as the sample ignites, the combustion flash reaches $\sim$1700 °C. The time that the sample remains at this temperature is short, and refractory samples may not fully combust. Addition of V_2O_5 flux may increase the efficiency of combustion, but needs to be checked for N blank (Bräuer and Hahne, 2005). The CF-IRMS method was originally developed for rapid analysis of organic compounds and its application to refractory OM and silicates introduces the possibilities of poor reproducibility arising from incomplete combustion.

Earlier off-line methods using sealed tube combustions and Cu/CuO combusted the sample at slightly lower temperatures (950-1050C), but for several hours. N_2 then needs to be cryogenically isolated from the other gases produced which is tedious and sample throughput is typically limited to < 10 samples a day rather than 50+ sample per day rate possible by CF-IRMS. Compared to CF-IRMS, the sensitivity was often lower although better precision could be obtained by the dual inlet (DI) IRMS used with this technique. Recent developments have increased sensitivity and substantially reduced sample size to ~10 nmol from ~1 μmol from traditional DI-IRMS or CF-IRMS (Bebout et al., 2007;Kashiyama et al., 2007; Polissar et al., 2009).

The techniques and instrumentation of Ar–Ar chronology laboratories have also been applied to N isotope measurement in silicates. This technique can analyze 10–300 nmol N but requires nonstandard static mass spectrometry (Cartigny et al., 2001 and references therein).

4.2. Metals

Inductively coupled plasma mass spectrometry (ICP-MS) is the most widely used method today for determination of metal concentrations in both biological and inorganic samples. The capability of ICP-MS to simultaneously measure the majority of elements in the periodic table has lead to its replacement of element-specific techniques such as atomic absorption or emission spectrometry. Other advantages of ICP-MS include its high sensitivity (down to concentrations subfemtogram metal per gram matrix solution) and large linear dynamic range (over several orders of magnitude). Several recent texts covering ICP-MS in great detail are available (Nelms, 2005; Taylor, 2001; Thomas, 2008). The simplest ICP-MS sample introduction method is as a solution in dilute nitric acid. The solution is pumped into the instrument and nebulized into a fine spray before hitting the plasma. This sample introduction method requires complete digestion of samples in concentrated acid (most commonly nitric, hydrochloric, and/or hydrofluoric) on hot plates. To avoid metal contamination, samples are digested in acid-washed Teflon containers with extremely low metal blanks in laminar flow hoods. Recently, other sample introduction methods have been optimized, allowing undigested rock samples and dried protein gels to be analyzed via laser ablation and liquid metalloprotein samples to be first separated by HPLC before ICP-MS analysis (Lobinski et al., 2006). Quantification of metal concentrations in samples is performed by correlation of metal isotope counts detected by the instrument to known concentrations in an ICP-MS calibration solution. Generally, a dilution factor is applied to convert the sample concentration introduced to the ICP-MS to the metal concentration in the original sample.

5. CONCLUDING REMARKS

The record of N isotope and authigenic metal concentration variations in ancient sediments can provide valuable insights into the evolution of the N cycle through geological time. Neither one of these proxies are without complexity which, if overlooked, can lead to erroneous interpretations and conclusions. The aim of this contribution was to both highlight the importance of these methodologies while simultaneously exposing their limitations.

ACKNOWLEDGMENTS

This work was funded by NSF EAR 0844252 and NASA NNX7AK14G-3/3 to LVG and support from the NASA Astrobiology Institute and NSF GRF (Geosciences-Geochemistry 2006038382) to JBG. J. D. Wright and C. K. Junium provided thoughtful comments and reviews which greatly improved this chapter. Thanks also to the organizers of the Nitrogen Agouron Meeting in October 2009 in Scottsdale, Arizona, which brought the authors together and set in motion the writing of this chapter.

REFERENCES

Algeo, T. J., and Lyons, T. W. (2006). Mo–total organic carbon covariation in modern anoxic marine environments: Implications for analysis of paleoredox and paleohydrographic conditions. *Paleoceanography* **21**, PA1016, doi: 10.1029/2004PA001112.

Algeo, T., Rowe, H., Hower, J. C., Schwark, L., Herrmann, A., and Heckel, P. (2008). Changes in ocean denitrification during Late Carboniferous glacial-interglacial cycles. *Nat. Geosci.* **1**, 709–714.

Aller, R. C. (1994). Bioturbation and remineralization of sedimentary organic matter: Effects of redox oscillation. *Chem. Geol.* **114**, 331–345.

Altabet, M. (2005). Isotopic tracers of the marine nitrogen cycle: Present and past. *In* "Marine Organic Matter: Biomarkers, Isotopes and DNA," (J. K. Volkman, ed.) pp. 251–293. Springer, Berlin.

Altabet, M. A. (2007). Constraints on oceanic N balance/imbalance from sedimentary ^{15}N records. *Biogeosciences* **4**, 75–86.

Altabet, M. A., and Curry, W. B. (1989). Testing models of past ocean chemistry using foraminifera ^{15}N/^{14}N. *Global Biogeochem. Cycles* **3**, 107–120.

Altabet, M. A., and Francois, R. (1994). Sedimentary nitrogen isotopic ratio as a recorder for surface nitrate utilization. *Global Biogeochem. Cycles* **8**, 103–116.

Altabet, M. A., Pilskaln, C., Thunell, R., Pride, C., Sigman, D., Chavez, F., and Francois, R. (1999). The nitrogen isotope biogeochemistry of sinking particles from the margin of the eastern North Pacific. *Deep Sea Res. Part I* **46**, 655–679.

Anbar, A. D., and Knoll, A. H. (2002). Proterozoic ocean chemistry and evolution: A bioinorganic bridge? *Science* **297**, 1137–1142.

Anbar, A. D. (2008). Elements and evolution. *Science* **322**, 1481–1483.

Anbar, A. D., and Rouxel, O. (2007). Metal stable isotopes in Paleoceanography. *Ann. Rev. Earth Planet. Sci.* **35**, 717–746.

Anbar, A. D., Duan, Y., Lyons, T. W., Arnold, G. L., Kendall, B., Creaser, R. A., Kaufman, A. J., Gordon, G. W., Scott, C., Garvin, J., and Buick, R. (2007). A whiff of oxygen before the great oxidation event? *Science* **317**, 1903–1906.

Antipov, A. N., Lyalikova, N. N., Khijniak, T. V., and L'Vov, N. P. (1998). Molybdenum-free nitrate reductases from vanadate-reducing bacteria. *FEBS Lett.* **441**, 257–260.

Arp, D. J., and Stein, L. Y. (2003). Metabolism of inorganic C compounds by ammonia-oxidizing bacteria. *Crit. Rev. Biochem. Mol. Biol.* **38**, 471–495.

Bebout, G. E., and Fogel, M. L. (1992). Nitrogen-isotope compositions of metasedimentary rocks in the Catalina Schist, California: Implications for devolatilization history. *Geochim. Cosmochim. Acta* **56**, 2839–2849.

Bebout, G. E., Idleman, B. D., Li, L., and Hilkert, A. (2007). Isotope-ratio-monitoring gas chromatography methods for high-precision isotopic analysis of nanomole quantities of silicate nitrogen. *Chem. Geol.* **240**, 1–10.

Berges, J. A., and Mulholland, M. (2008). Enzymes and nitrogen cycling. In "Nitrogen in the Marine Environment," (D. G. Capone, D. A. Bronk, M. Mulholland, and E. J. Carpenter, eds.), pp. 1361–1420. Elsevier, Amsterdam.

Berggren, W. A., and Hollister, C. D. (1977). Plate tectonics and paleocirculation - Commotion in the Ocean. *Tectonophysics* **38**, 11–48.

Bertero, M. G., Rothery, R. A., Palak, M., Hou, C., Lim, D., Blasco, F., Weiner, J. H., and Strynadka, N. C. J. (2003). Insights into the respiratory electron transfer pathway from the structure of nitrate reductase A. *Nat. Struct. Biol.* **10**, 681–687.

Boudou, J.-P., Schimmelmann, A., Ader, M., Masterlerz, M., Sebilo, M., and Gengembre, L. (2008). Organic nitrogen chemistry during low-grade metamorphism. *Geochim. Cosmochim. Acta* **72**, 1199–1221.

Boyle, E. A., Sclater, F. R., and Edmond, J. M. (1977). The distribution of dissolved copper in the Pacific. *Earth Planet. Sci. Lett.* **37**, 38–54.

Brandes, J. A., and Devol, A. H. (2002). A global marine-fixed nitrogen isotope budget: Implacations for Holocene nitrogen cycling. *Glob. Biogeochem. Cycles* **16**, doi: 10.1029/2001GB001856.

Brandes, J. A., Devol, A. H., Yoshinari, T., Jayakumar, D. A., and Naqvi, S. W. A. (1998). Isotopic composition of nitrate in the central Arabian Sea and eastern tropical North Pacific: A tracer for mixing and nitrogen cycles. *Limnol. Oceanogr.* **43**, 1680–1689.

Bräuer, K., and Hahne, K. (2005). Methodological aspects of the ^{15}N-analysis of Precambrian and Paleozoic sediments rich in organic matter. *Chem. Geol.* **218**, 361–368.

Brown, K., Tegoni, M., Prudencio, M., Pereira, A. S., Besson, S., Moura, J. J., Moura, I., and Cambillau, C. (2000). A novel type of catalytic copper cluster in nitrous oxide reductase. *Nat. Struct. Biol.* **7**, 191–195.

Buick, R. (2007). Did the Proterozoic Canfield Ocean cause a laughing gas greenhouse? *Geobiology* **5**, 97–100.

Butler, C. S., and Richardson, D. J. (2005). The emerging molecular structure of the nitrogen cycle: An introduction to the proceedings of the 10th annual N-cycle meeting. *Biochem. Soc. Trans.* **33**, 113–118.

Calvert, S. E., and Pedersen, T. F. (1993). Geochemistry of recent oxic and anoxic marine sediments: Implications for the geologic record. *Mar. Geol.* **113**, 67–88.

Cao, C., Love, G. D., Hays, L. E., Wang, W., Shen, S., and Summons, R. E. (2009). Biogeochemical evidence for euxinic oceans and ecological disturbance presaging the end-Permian mass extinction event. *Earth Planet. Sci. Lett.* **281**, 188–201.

Carpenter, E. J., and Capone, D. G. (2008). Nitrogen fixation in the marine environment. In "Nitrogen in the Marine Environment," (D. G. Capone, D. Bonk, M. Mulholland, and E. J. Carpenter, eds.), 2nd edition, pp. 141–198. Academic Press, Amsterdam.

Cartigny, P., De Corte, K., Shatsky, V. S., Ader, M., De Paepe, P., Sobolev, N. V., and Javoy, M. (2001). The origin and formation of metamorphic microdiamonds from the

Kokchetav massi, Kazakhstan: A nitrogen and carbon isotopic study. *Chem. Geol.* **176**, 265–281.
Casciotti, K. L. (2009). Inverse kinetic isotope fractionation during bacterial nitrite oxidation. *Geochim. Cosmochim. Acta* **73**, 2061–2076.
Cline, J. D., and Kaplan, I. R. (1975). Isotopic fractionation of dissolved nitrate during denitrification in the eastern tropical North Pacific. *Mar. Chem.* **3**, 271–299.
Dauphas, N., and Marty, B. (2004). "A large secular variation in the nitrogen isotopic composition of the atmosphere since the Archean?": Response to a comment on "The nitrogen record of crust-mantle interaction and mantle convection from Archean to Present" by R. Kerrich and Y. Jia, *Earth Planet. Sci. Lett.* **225**, 441–445.
Dauwe, B., Middelburg, J. J., Herman, P. H. J., and Heip, C. H. R. (1999). Linking diagenetic alteration of amino acids and bulk organic matter reactivity. *Limnol. Oceanogr.* **44**, 1809–1814.
de Lange, G. J. (1998). Oxic versus an oxic diagenetic alteration of turbiditic sediments in the Madeira Abyssal Plain, eastern North Atlantic. In "Proceedings of the Ocean Drilling Program, Scientific Results Volume 157," (P. P. E. Schmincke, H.-U. Firth, and W. Duffield, eds.), pp. 573–580. College Station, TX.
Deutsch, C., Sigman, D. M., Thunell, R. C., Meckler, A. N., and Haug, G. H. (2004). Isotopic constraints on glacial/interglacial changes in the oceanic nitrogen budget. *Glob. Biogeochem. Cycles* **18**, GB4012, doi: 10.1029/2003GB002189.
Durand, B. (1985). Diagenetic modification of kerogens. *Phil. Trans. R. Soc. A* **315**, 77–90.
Einsle, O., Messerschmidt, A., Stach, P., Bourenkov, G. P., Bartunik, H. D., Huber, R., and Kroneck, P. M. H. (1999). Structure of cytochrome *c* nitrite reductase. *Nature* **400**, 476–480.
Ellis, M. J., Grossman, J. G., Eady, R. R., and Hasnain, S. S. (2007). Genomic analysis reveals widespread occurrence of new classes of copper nitrite reductases. *J. Biol. Inorg. Chem.* **12**, 1119–1127.
Ensign, S. A., Hyman, M. R., and Arp, D. J. (1993). In vitro activation of ammonia monooxygenase from *Nitrosomonas europaea* by copper. *J. Bacteriol.* **175**, 1971–1980.
Falkowski, P. G. (1997). Evolution of the nitrogen cycle and its influence on the biological sequestration of CO_2 in the ocean. *Nature* **387**, 272–275.
Falkowski, P. G., and Godfrey, L. V. (2008). Electrons, life and the evolution of Earth's oxygen cycle. *Phil. Trans. R. Soc. B* **363**, 2705–2716, doi: 10.1098/rstb.2008.0054.
Fennel, K., Follows, M., and Falkowski, P. G. (2005). The co-evolution of the nitrogen, carbon and oxygen cycles in the Proterozoic ocean. *Amer. J. Sci.* **305**, 526–545.
Freudenthal, T., Wagner, T., Wenzhöfer, F., Zabel, M., and Wefer, G. (2001). early diagenesis or organic matter from sediments of the eastern subtropical Atlantic: Evidence from stable nitrogen and carbon isotopes. *Geochim. Cosmochim. Acta* **65**, 1795–1808.
Fulop, V., Moir, J. W. B., Ferguson, S. J., and Hajdu, J. (1995). The anatomy of a bifunctional enzyme: structural basis for reduction of oxygen to water and synthesis of nitric oxide by cytochrome cd1. *Cell* **81**, 369–377.
Galbraith, E. D., Kienast, M., Pedersen, T. F., and Calvert, S. E. (2004). Glacial-interglacial modulation of the marine nitrogen cycle by high-latitude O_2 supply to the global thermocline. *Paleoceanography* **19**, PA4007, doi: 10.1029/2003PA001000.
Garrels, R. M., and Christ, C. L. (1965). Solutions, Minerals and Equilibria. Freeman Cooper and Co. San Francisco, CA.
Garvin, J., Buick, R., Anbar, A. D., Arnold, G. L., and Kaufman, A. J. (2009). Isotopic evidence for an aerobic nitrogen cycle in the latest Archean. *Science* **323**, 1045–1048.
Glass, J. B., Wolfe-Simon, F., and Anbar, A. D. (2009). Coevolution of metal availability and nitrogen assimilation in cyanobacteria and algae. *Geobiology* **7**, 100–123.
Godden, J. W., Turley, S., Teller, D. C., Adman, E. T., Liu, M. Y., Payne, W. J., and LeGall, J. (1991). The 2.3 angstrom X-ray structure of nitrite reductase from *Achromobacter cycloclastes*. *Science* **253**, 438–442.

Godfrey, L. V., and Falkowski, P. G. (2009). The cycling and redox state of nitrogen in the Archean ocean. *Nat. Geosci.* **2**, 725–729.

Goldberg, E. D., and Arrhenius, G. O. S. (1958). Chemistry of pelagic sediments. *Geochim. Cosmochim. Acta* **13**, 153–212.

Granger, J., and Ward, B. B. (2003). Accumulation of nitrogen oxides in copper-limited cultures of denitrifying bacteria. *Limnol. Oceanogr.* **48**, 313–318.

Gruber, N., and Sarmiento, J. L. (1997). Global patterns of marine nitrogen fixation and denitrification. *Global Biogeochem. Cycles* **11**, 235–266.

Haendel, D., Mühle, K., Nitzsche, H. M., Stiehl, G., and Wand, U. (1986). Isotopic variations of the fixed nitrogen in metamorphic rocks. *Geochim. Cosmochim. Acta* **50**, 749–758.

Hanschmann, G. (1981). Berechung von Isotopieekkekten auf quantenschemischer Grundlage am Beispiel stick-stoffhaltiger Moleküle. *ZFI-Mitt.* **41**, 19–39.

Hedges, J. I., Clark, W. A., Quay, P. D., Richey, J. E., Devol, A. H., and Santos, U. M. (1986). Compositions and fluxes of particulate organic material in the Amazon river. *Limnol. Oceanogr.* **31**, 717–738.

Hedges, J. I., and Keil, R. G. (1995). Sedimentary organic matter preservation: An assessment and speculative synthesis. *Mar. Chem.* **49**, 81–115.

Hedges, J. I., and Oades, J. M. (1997). Comparative organic geochemistries of soils and marine sediments. *Org. Geochem.* **27**, 319–361.

Hedges, J. I., Keil, R. G., and Benner, R. (1997). What happens to terrestrial organic matter in the ocean? *Org. Geochem.* **27**, 195–212.

Holland, H. D. (1973). The oceans: A possible source of iron in iron-formations. *Econ. Geol.* **68**, 1169–1172.

Hollocher, T. C., Tate, M. E., and Nicholas, D. J. (1981). Oxidation of ammonia by *Nitrosomonas europea*. Definite ^{18}O-tracer eveidence that hydroxylamine formation involves a monooxygenase. *J. Biol. Chem.* **256**, 10834–10836.

Igarashi, N., Moriyama, H., Fujiwara, T., Fukumori, Y., and Tanaka, N. (1997). The 2.8 Å structure of hydroxylamine oxidoreductase from a nitrifying chemoautotrophic bacterium, *Nitrosomonas europaea*. *Nat. Struct. Biol.* **4**, 276–284.

Ishiwatari, R., Ishiwatari, M., Rohrback, B. G., and Kaplan, I. R. (1977). Thermal alteration experiments on organic matter from recent marine sediments in relation to petroleum genesis. *Geochim. Cosmochim. Acta* **41**, 815–828.

Iwasaki, H., and Terai, H. (1982). N_2 and N_2O produced during growth of denitrifying bacteria in copper-depleted and -supplemented media. *J. Gen. Appl. Microbiol.* **28**, 189–193.

Jaffe, L. A., Peucker-Ehrenbrink, B., and Petsch, S. T. (2002). Mobility of rhenium, platinum group metals and organic carbon during black shale weathering. *Earth Planet. Sci. Lett.* **198**, 339–353.

Jia, Y. (2006). Nitrogen isotope fractionations during progressive metamorphism: A case study from the Paleozoic Cooma metasedimentary complex, southeastern Australia. *Geochim. Cosmochim. Acta* **70**, 5201–5214.

Jia, Y., and Kerrich, R. (2004). Nitrogen 15-enriched Precambrian kerogen and hydrothermal systems. *Geochem. Geophys. Geosyst.* **5**, Q07005, doi: 10.1029/2004GC000716.

Johnson, K. S., Gordon, R. M., and Coale, K. H. (1997). What controls dissolved iron concentrations in the world ocean? *Mar. Chem.* **57**, 137–161.

Jormakka, M., Richardson, D., Byrne, B., and Iwata, S. (2004). Architecture of NarGH reveals a structural classification of Mo-*bis*MGD enzymes. *Structure* **12**, 95–104.

Junium, C. K., and Arthur, M. A. (2007). Nitrogen cycling during the Cretaceous, Cenomanian-Turonian Oceanic Anoxic Event II. *Geochem. Geophys. Geosyst.* **8**, 10.1029/2006GC001328.

Kashiyama, Y., Kitazato, H., and Ohkouchi, N. (2007). An improved method for isolation and purification of sedimentary porphyrins by high-performance liquid chromatography for compound specific isotopic analysis. *J. Chromatogr. A* **1138**, 73–93.

Kerrich, R., and Jia, Y. (2004). A comment on "The nitrogen record of crust-mantle ineraction and mantle convection from Archean to Present" by B. Marty and N. Dauphas [Earth Planet. Sci. Lett. 206 (2003) 397-410]. *Earth Planet. Sci. Lett.* **225**, 435–440.

Klotz, M. G., and Stein, L. Y. (2008). Nitrifier genomics and evolution of the nitrogen cycle. *FEMS Microbiol. Lett.* **278**, 146–156.

Klotz, M. G., Schmid, M. C., Strous, M., op den Camp, H. J. M., Jetten, M. S. M., and Hooper, A. B. (2008). Evolution of an octahaem cytochrome c protein family that is key to aerobic and anaerobic ammonia oxidation by bacteria. *Environ. Microbiol.* **10**, 3150–3163.

Knoll, A. H. (1992). The early evolution of eukaryotes: A geologic perspective. *Science* **256**, 622–627.

Konhauser, K. O., Pecoits, E., Lalonde, S. V., Papineau, D., Nisbet, E. G., Barley, M. E., Arndt, N. T., Zahnle, K., and Kamber, B. S. (2009). Oceanic nickel depletion and a methanogen famine before the great oxidation event. *Nature* **458**, 750–753.

Krishnamurthy, A., Moore, J. K., and Doney, S. C. (2008). The effects of dilution and mixed layer depth on deliberate ocean iron fertilization: 1-D simulations of the Southern Ocean iron experiment (SOFeX). *J. Mar. Syst.* **71**, 112–130.

Krishnamurthy, A., Moore, J. K., Mahowald, N., Luo, C., and Zender, C. S. (2010). Impacts of atmospheric nutrient inputs an marine biogeochemistry. *J. Geophys. Res.* **115**, 10.1029/2009JG001115.

Kroneck, P. M. H., and Abt, J. A. (2002). Molybdenum in nitrate reductase and nitrite oxidoreductase. *In* "Molybdenum and tungsten: Their roles in biological processes," (A. Sigel and H. Sigel, eds.), pp. 369–403. CRC Press.

Kuenen, J. G. (2008). Anammox bacteria: From discovery to application. *Nat. Rev. Microbiol.* **6**, 320–326.

Kuypers, M. M. M., van Breugel, Y., Schouten, S., Erba, E., and Sinninghe Damste, J. S. (2004). N_2-fixing cyanobacteria supplied nutrient N for Cretaceous oceanic anoxic events. *Geology* **32**, 853–856.

Lehmann, M. F., Bernasconi, S. M., Barbieri, A., and McKenzie, J. A. (2002). Preservation of organic matter and alteration of its carbon and nitrogen isotope composition during simulated and in situ early sedimentary diagenesis. *Geochim. Cosmochim. Acta* **66**, 3573–3584.

Lobinski, R., Schaumloffel, D., and Szpunar, J. (2006). Mass spectrometry in bioinorganic analytical chemistry. *Mass Spectrom. Rev.* **25**, 255–289.

Lui, K.-K., and Kaplan, I. R. (1989). The eastern tropical Pacific as a source of ^{15}N-enriched nitrate in seawater off southern California. *Limnol. Oceanogr.* **34**, 820–830.

Macko, S. A. (1993). Early diagenesis of organic matter in sediments: Assessment of mechanisms and preservation by the use of isotopic molecular approaches. *In* "Org. Geochem.," (M. H. Engel and S. A. Macko, eds.), pp. 211–236. Springer, New York.

Macko, S. A., Fogel Estep, M. L., Engel, M. H., and Hare, P. E. (1986). Kinetic fractionation of stable nitrogen isotopes during amino-acid transamination. *Geochim. Cosmochim. Acta* **50**, 2143–2146.

Mancinelli, R. L., and McKay, C. P. (1988). Evolution of nitrogen cycling. *Orig. Life* **18**, 311–325.

Mather, T. A., Pyle, D. M., and Allen, A. G. (2004). Volcanic source for fixed nitrogen in the early Earth's atmosphere. *Geology* **32**, 905–908.

Mayer, L. M. (2004). The inertness of being organic. *Mar. Chem.* **92**, 135–140.

Meincke, M., Bock, E., Kastrau, D., and Kroneck, P. M. H. (1992). Nitrite oxidoreductase from *Nitrobacter hamburgensis*: Redox centers and their catalytic role. *Arch. Microbiol.* **158**, 127–131.

Moffett, J. W., and Zika, R. G. (1983). Oxidation kinetics of Cu(I) in seawater: Implications for its existence in the marine environment. *Mar. Chem.* **13**, 239–251.

Naik, R. R., Murillo, F. M., and Stolz, J. F. (1993). Evidence for a novel nitrate reductase in the dissimilatory iron-reducing bacterium *Geobacter metallireducens*. *FEMS Microbiol. Lett.* **106**, 53–58.

Navarro-Gonzalez, R., McKay, C. P., and Mvondo, D. N. (2001). A possible nitrogen crisis for Archaean life due to reduced nitrogen fixation by lightning. *Nature* **412**, 61–64.

Nelms, S. M. (2005). *Inductively Coupled Plasma Mass Spectrometry Handbook*. Blackwell Publishing Ltd., Oxford.

Nojiri, M., Xie, Y., Inoue, T., Yamamoto, T., Matsumura, H., Kataoka, K., Deligeer, K., Yamaguchi, K., Kai, Y., and Suzuki, S. (2007). Structure and function of a hexameric copper-containing nitrite reductase. *Proc. Natl. Acad. Sci.* **104**, 4315–4320.

Ohkouchi, N., Nakajima, Y., Okada, H., Ogawa, N. O., Suga, H., Oguri, K., and Kitazato, H. (2005). Biogeochemical processes in a meromictic Lake Kaiike: Implications from carbon and nitrogen isotopic compositions of photosynthetic pigments. *Environ. Microbiol.* **7**, 1009–1016.

Papineau, D., Mojzsis, S. J., Karhu, J. A., and Marty, B. (2005). Nitrogen isotopic composition of ammoniated phyllosilicates: Case studies from Precambrian metamorphosed sedimentary rocks. *Chem. Geol.* **216**, 37–58.

Papineau, D., Purohit, R., Goldberg, T., Pi, D., Shields, G. A., Bhu, H., Steele, A., and Fogel, M. L. (2009). High primary productivity and nitrogen cycling after the Paleoproterozoic phosphogenic event in the Aravelli Supergroup, India. *Precambrian Res.* **171**, 37–56.

Paris, G., Beaumont, V., Bartolini, A., Clemence, M.-E., Gardin, S., and Page, K. (2010). Nitrogen isotope record of a perturbed paleo-ecosystem in the aftermath of the end-Triassic crisis, Doniford section (SW England). *Geochem. Geophys. Geosyst.* **11**, Q08021, doi: 10.1029/2010GC003161.

Peters, K. E., Sweeney, R. E., and Kaplan, I. R. (1978). Correlation of carbon and nitrogen stable isotope ratios in sedimentary organic matter. *Limnol. Oceanogr.* **23**, 598–604.

Peters, K. E., Rohrback, B. G., and Kaplan, I. R. (1981). Geochemistry of artificially heated humic and sapropelic sediments I: Protokerogen. *Bull. Am. Assoc. Petro. Geol.* **65**, 688–705.

Polissar, P. J., Fulton, J. M., Junium, C. K., Turich, C. C., and Freeman, K. H. (2009). Measurement of ^{13}C and ^{15}N isotopic composition on nanomolar quantities of C and N. *Anal. Chem.* **81**, 755–763.

Pride, R., Thunell, R., Sigman, D., Altabet, M., and Keigwin, L. (1999). Nitrogen isotopic variations in the Gulf of California since the latest deglaciation: Response to global climate change. *Paleoceanography* **14**, 397–409, doi:10.1029/1999PA900004.

Quan, T. M., van de Schootbrugge, B., Field, M. P., Rosenthal, Y., and Falkowski, P. G. (2008). Nitrogen isotope and trace metal analyses from the Mingolscheim core (Germany): Evidence for redox variations across the Triassic-Jurassic boundary. *Glob. Biogeochem. Cycles* **22**, GB2014, doi: 10.1029/2007GB002981.

Raymond, J., Siefert, J. L., Staples, C. R., and Blankenship, R. E. (2004). The natural history of nitrogen fixation. *Mol. Biol. Evol.* **21**, 541–554.

Ren, H., Sigman, D. M., Meckler, A. N., Plessen, B., Robinson, R. S., Rosenthal, Y., and Haug, G. H. (2009). Foraminiferal isotope evidence of reduced nitrogen fixation in the ice age Atlantic Ocean. *Science* **323**, 244–248.

Robinson, R. S., Sigman, D. M., DiFiore, P. J., Rohde, M. M., Mashiotta, T. A., and Lea, D. W. (2005). Diatom-bound ^{15}N/^{14}N: New support for enhanced nutrient consumption in the ice age subantarctic. *Paleoceanography* **20**, PA3003, doi: 10.1029/2004PA001114.

Robinson, R. S., Mix, A., and Martinez, P. (2007). Southern Ocean control on the extent of denitrification in the southeast Pacific over the last 70 ka. *Quat. Sci. Rev.* **26,** 201–212.

Rosenzweig, A. C. (2008). The metal centres of particulate methane mono-oxygenase. *Biochem. Soc. Trans.* **36,** 1134–1137.

Sachs, J. P., and Repeta, D. J. (2000). The purification of chlorins from marine particles and sediments for nitrogen and carbon isotopic analysis. *Org. Geochem.* **31,** 317–329.

Sachs, J. P., Repeta, D. J., and Goericke, R. (1999). Nitrogen and carbon isotopic ratios of chlorophyll from marine phytoplankton. *Geochim. Cosmochim. Acta* **63,** 1431–1441.

Saito, M. A., Sigman, D. M., and Morel, F. M. M. (2003). The bioinorganic chemistry of the ancient ocean: The co-evolution of cyanobacterial metal requirements and biogeochemical cycles at the Archean-Proterozoic boundary? *Inorg. Chim. Acta* **356,** 308–318.

Schimmelmann, A., and Lis, G. P. (2010). Nitrogen isotopic exchange during maturation of organic matter. *Org. Geochem.* **41,** 63–70.

Scott, C., Lyons, T. W., Bekker, A., Shen, Y., Poulton, S. W., Chu, X., and Anbar, A. D. (2008). Tracing the stepwise oxygenation of the proterozoic ocean. *Nature* **452,** 456–459.

Shemesh, A., Macko, S. A., Charles, C. D., and Rau, G. H. (1993). Isotopic evidence for reduced productivity in the glacial Southern Ocean. *Science* **262,** 407–410.

Sherman, L. S., Walbauer, J. R., and Summons, R. E. (2007). Improved methods for isolating and validating indigenous biomarkers in Precambrian rocks. *Org. Geochem.* **38,** 1987–2000.

Sigman, D. M., Altabet, M. A., Francois, R., McCorkle, D. C., and Gaillard, J. F. (1999). The isotopic composition of diatom-bound nitrogen in Southern Ocean sediments. *Paleoceanography* **14,** 118–134.

Sigman, D., Robinson, R., Knapp, A. N., van Geen, A., McCorkle, D. C., Brandes, J. A., and Thunell, R. C. (2003). Distinguishing between water column and sedimentary denirtification in the Santa Barbara Basin using the stable isotopes of nitrate. *Geochem. Geophys. Geosyst.* **4,** 1040, doi: 10.1029/2002GC000384.

Simon, J. (2002). Enzymology and bioenergetics or respiratory nitriet ammonification. *FEMS Microbiol. Rev.* **26,** 285–309.

Stiehl, G., and Lehmann, M. (1980). Isotopenvariationen des stickstoffs humoser und bituminöser natürlicher organischer substanze. *Geochim. Cosmochim. Acta* **44,** 1737–1746.

Strous, M., Pelletier, E., Mangenot, S., Rattei, T., Lehner, A., Taylor, M. W., Horn, M., Daims, H., Bartol-Mavel, D., Wincker, P., Barbe, V., and Fonknechten, N. (2006). Deciphering the evolution and metabolism of an anammox bacterium from a community genome. *Nature* **440,** 790–794.

Summers, D. P. (1999). Sources and sinks for ammonia and nitrite on the early Earth and the reaction of nitrite with ammonia. *Orig. Life Evol. Biosph.* **29,** 33–46.

Summers, D. P., and Chang, S. (1993). Prebiotic ammonia from reduction of nitrite by iron (II) on the early Earth. *Nature* **356,** 630–633.

Sweeney, R. E., and Kaplan, I. R. (1980). Natural abundances of ^{15}N as a source indicator for near-shore marine sedimentary and dissolved nitrogen. *Mar. Chem.* **9,** 81–94.

Taylor, H. E. (2001). *Inductively Coupled Plasma-Mass Spectrometry: Practices and Techniques* Academic Press.

Thomas, R. (2008). *Practical Guide to ICP-MS: A Tutorial for Beginners.* Second Edition CRC Press, Boca Raton, FL.

Thomson, J., Higgs, N. C., Croudace, I. W., Colley, S., and Hydes, D. J. (1993). Redox zonation of elements at an oxic/post-oxic boundary in deep-sea sediments. *Geochim. Cosmochim. Acta* **57,** 579–595.

Uhle, M. E., Macko, S. A., Spero, H. J., Engel, M. H., and Lea, D. W. (1997). Sources of carbon and nitrogen in modern planktonic foraminifera: the role of algal symbionts as determined by bulk and compound specific stable isotopic analyses. *Org. Geochem.* **27,** 103–113.

Urey, H. C. (1947). The thermodynamic properties of isotopic substances. *J. Chem. Soc.* **1947**, 562–581.

van Bentum, E. C., Hetzel, H.-J., Forster, A., Reichert, G.-J., and Sinninghe Damsté, J. S. (2009). Reconstruction of water column anoxia in the equatorial Atlantic during the Cenomanian-Turonian oceanic anoxic event using biomarker and trace metal proxies. *Palaeogeog. Palaeoclimatol. Palaeoecol.* **280**, 489–498.

van Gijzel, P. (1982). Characteization and identification of kerogen and bitumen and determination of thermal maturation by means of qualitative and quantitative microscopical techniques. *SEPM, Short Course Notes* **7**, 159–216.

van Krevelen, D. W. (1961). *Coal: Typology, Chemistry, Physics*. Constitution, Elsevier, Amsterdam.

Voss, M., Larsen, B., Leivuori, M., and Vallius, H. (2000). Stable isotope signals of eutrophication in Baltic Sea sediments. *J. Mar. Syst.* **25**, 287–298.

Voss, M., Dippner, J. W., and Montoya, J. P. (2001). Nitrogen isotope patterns in the oxygen deficient waters of the Eastern Tropical North Pacific Ocean. *Deep Sea Res. Part I* **48**, 1905–1921.

Waldbauer, J. R., Sherman, L. S., Sumner, D. Y., and Summons, R. E. (2009). Late Archean molecular fossils from the Transvaal Supergroup record the antiquity of microbial diversity and aerobiosis. *Precambrian. Res.* **169**, 28–47.

Wallace, H. E., Thomson, J., Wilson, T. R. S., Weaver, P. P. E., Higgs, N. C., and Hydes, D. J. (1988). Active diagenetic formation of metal-rich layers in N E. Atlantic sediments. *Geochim. Cosmochim. Acta* **52**, 1557–1569.

Wei, X., Vajrala, N., Hauser, L., Sayavedra-Soto, L. A., and Arp, D. J. (2006). Iron nutrition and physiological responses to iron stress in *Nitrosomonas europaea*. *Arch. Microbiol.* **186**, 107–118.

Williams, L. B., Ferrell, R. E., Hutcheon, I., Bakel, A. J., Walsh, M. M., and Krouse, H. R. (1995). Nitrogen isotpoe geochemistry of organic matter and minerals during diagenesis and hydrocarbon migration. *Geochim. Cosmochim. Acta* **59**, 765–779.

Zahn, J. A., Arciero, D. M., Hooper, A. B., and DiSpirito, A. A. (1996). Evidence for an iron center in the ammonia monooxygenase from *Nitrosomonas europaea*. *FEBS Lett.* **397**, 35–38.

Zumft, W. G. (2005). Nitric oxide reductases of prokaryotes with emphasis on the respiratory, heme–copper oxidase type. *J. Inorg. Biochem.* **99**, 194–215.

Author Index

A

Aakra, A., 68, 212
Aamand, J., 112
Abbas, B., 207–208
Abdur-Rashid, A., 416
Abeliovich, A., 262
Abell, G. C. J., 206–207
Abruna, H. D., 136, 140
Abt, J. A., 487
Achtman, M., 424
Acuna-Gonzalez, J., 91
Adachi, Y., 140
Adair, K. L., 155–156, 158, 160, 165, 361
Adam, B., 285
Adams, H., 207
Ader, M., 495, 498
Adman, E. T., 488
Aebersold, R., 475
Afshar, S., 177
Agogué, H., 208
Aguilar, C., 425
Ahn, J. H., 370, 381–382
Ahuja, U., 433, 435
Alawi, M., 111–112, 122, 124–125, 173, 175–176
Alcaman, M. E., 4, 20
Alexander, M., 70
Algeo, T. J., 490–491
Allen, A. G., 485
Allen, D. E., 91
Allen, J. I., 12
Allen, J. W. A., 433–434
Aller, R. C., 494
Alm, E. J., 136, 142
Altabet, M. A., 184, 255, 257, 262–263, 317, 490, 493–494
Altendorf, K., 123–126
Alzerreca, J. J., 212
Amann, R., 34, 43, 90–92, 99, 101–102, 121, 290, 469
Amend, J. P., 188
Aminor, A., 312
Aminot, A., 82, 326
Anbar, A. D., 484–485, 491, 496
Andersen, G., 284
Andersen, J. S., 467–468
Andersen, K., 212
Anderson, I. C., 370
Anderson, T. R., 13, 15, 18

Andersson, J. H., 326, 339
Andersson, K. K., 258
Andreani, M., 180, 184–185, 263–264, 317
André, J.-M., 20
Andrews, S. C., 405–407
Angelis, D. L., 17
Angly, F., 34, 49
Antelmann, H., 425
Antipov, A. N., 177, 488
Antonov, J., 5
Aoki, K., 312
Arad-Yellin, R., 419
Ara, T., 395
Archer, S. L., 457
Arciero, D. M., 38, 57, 147, 273, 346, 391, 395, 398–399, 404, 430, 466, 487
Ardon, O., 419
Arkin, A. P., 136, 142
Armstrong, R. A., 5, 13, 15
Arndt, N. T., 491
Arnerich, T., 326, 339
Arnold, G. L., 485, 491
Arp, D. J., 34, 57, 112, 116, 132–133, 136, 139, 141, 147, 189, 208, 212, 308, 313, 345, 347–348, 356, 390–396, 398–399, 403–409, 411–414, 417–419, 421–425, 466, 487
Arrhenius, G. O. S., 490
Arslan, E., 435
Arsov, R., 470
Arthur, M. A., 494
Asakura, S., 312
Ashi, J., 283
Atkins, P., 448–449
Attard, E., 120
Atwood, J. A., 475
Aubert, M., 346
Audic, J. M., 313
Auguet, J.-C., 212
Augustin, J., 190
Avanzino, R. J., 184, 263
Avrahami, S., 349

B

Baar, C., 436
Baba, M., 395
Baba, T., 395
Bae, J. W., 102

507

Baer, M. T., 416
Baggs, E. M., 133, 349
Bai, D. L., 468
Baird, M. E., 13
Bakel, A. J., 495
Bakken, L. R., 56, 68
Balderston, W. L., 187, 313
Baldock, J., 345
Bañeras, L., 212
Banfield, J. F., 34
Bano, N., 206, 208
Banta, A., 177
Barber, R. T., 19–20
Barbe, V., 92, 102, 489
Barbieri, A., 494
Bardawill, C. J., 408
Barford, C., 180, 184–185, 263–264, 317
Bär-Gilissen, M.-J., 25, 60, 67, 348
Barkay, T., 167
Barley, M. E., 491
Baross, J. A., 172–173, 283–284
Barraclough, D., 180, 186
Bartholomew, W. V., 352–353
Bartolini, A., 491
Bartol-Mavel, D., 92, 102, 489
Bartosch, S., 111, 122, 361
Bartossek, R., 212
Bartunik, H. D., 139, 430, 439, 488
Batchelor, S. E., 70
Bates, N. R., 20
Baysse, C., 407
Beardall, J., 6
Beaumont, H. J. E., 147, 391, 395, 398–399
Beaumont, V., 491
Beavis, R. C., 472
Bebout, G. E., 495, 498
Becker, R. R., 438
Bedard, C., 232, 313, 356
Beese, F., 190
Behrenfeld, M. J., 5–6
Behrens, D., 112, 176
Behrens, S., 290
Beier, M., 382
Beinart, R. A., 283
Béjà, O., 35
Bekker, A., 491
Bell, P. E., 406
Belser, L. W., 73, 115, 118–119, 351, 357–358, 361
Beman, J. M., 77–78, 80, 157, 160, 175, 195, 197, 207–211, 214–216
Beman, M. J., 309
Benckiser, G., 382
Beninca, E., 44
Benner, R., 493
Benson, D., 285
Berenguer, J., 177
Berges, J. A., 487

Berggren, W. A., 490
Bergman, B., 282–283, 285
Bergmann, D. J., 136, 139–140, 273, 346, 399, 404
Bergmeyer, H. U., 454
Berg, P., 313, 356
Berhnard, A. E., 309
Bernasconi, S. M., 494
Berner, I., 418, 423
Bernhard, A. E., 34, 57, 64, 67–68, 72–73, 76, 133, 172, 206–207, 211–212, 345
Berounsky, V. M., 207
Berry, E. A., 412
Bertero, M. G., 488
Berube, P. M., 25, 57, 74–75, 77, 136, 139, 141, 166, 175, 207, 275, 345, 348, 390, 395, 466
Berzins, N., 194–195
Bessant, C., 475
Besson, S., 488
Beutler, H. O., 454
Beveridge, T. J., 177
Beyer, S., 466
Bhandari, B., 267
Bhu, H., 490
Bianchi, M., 9–10, 312
Bianco, L., 475
Bickel, P. J., 284
Bidigare, R. R., 272
Biel, S., 435–436
Bigeleisen, J., 256, 272
Bill, E., 418, 423
Billen, G., 270, 309, 312–314
Billheimer, D. D., 468
Binnerup, S. J., 75, 319
Birch, J. M., 215
Birrien, J. L., 91, 174
Bissett, A. P., 206–207
Bissett, W. P., 14, 18
Blackall, L. L., 120
Blackburne, R., 121
Blackburn, T. H., 318, 320
Blackman, R. K., 468
Blainey, P. C., 208, 212
Blankenship, R. E., 484
Blasco, F., 488
Blevins, A., 206, 211
Blöchl, E., 174, 177
Block, J. C., 313
Bloem, J., 69, 73, 75
Blonder, N., 468
Bloor, K., 44
Bluhm, M. E., 424
Bock, E., 110–114, 116–117, 121–122, 173, 175–176, 266, 349, 361, 370, 466, 469, 487
Bodelier, P. L. E., 60, 74–76
Boehm, A. B., 64, 206–207, 209
Boetius, A., 44, 286, 290
Boettcher, B., 56, 68

Bogyo, M., 467
Bohland, M., 44
Böhlke, J. K., 180, 184–185, 258–261, 263–266, 274, 317
Bolling, C., 424
Bollmann, A., 25, 55, 60, 64–68, 74–76, 78, 81, 348
Bollwerk, S. M., 258
Bonaldi, T., 470
Bonner, E. R., 433
Bonventre, J., 285
Boon, N., 207
Booth, M. S., 350–351
Bordalo, A. A., 206, 211
Borregaard, N., 424
Borrego, C. M., 212
Botero, L. M., 177
Bothe, H., 430
Both, G. J., 111
Bottcher, B., 349
Bottomley, P. J., 57, 112, 116, 118, 156, 361, 390–391, 395, 398–399, 418–423, 425, 466
Bouchez, T., 290
Boudjellal, B., 10
Boudou, J.-P., 495
Bourenkov, G. P., 139, 430, 439, 488
Boussau, B., 309, 345
Boxer, S. G., 285
Boyce, D. G., 6
Boyd, E. S., 173
Boyd, P. W., 5, 8, 13
Boyer, P. D., 157
Boyer, T., 5
Boyle, E. A., 496
Boyle-Yarwood, S. A., 156, 361
Bracken, C. S., 416
Bradbeer, C., 406
Bradford, M. M., 76
Bradley, P. B., 8
Brady, N. C., 166
Braff, J., 208
Braman, R. S., 240
Branco, P., 44
Brandes, J. A., 255, 283, 490, 493
Braster, M., 44
Bräuer, K., 497
Bremner, J. M., 184, 356
Brettske, I., 104
Breuer, U., 45–46
Breznak, J. A., 133, 135, 139
Bricker, S. B., 206
Brinkhoff, T., 117
Brink, M., 208
Brocandel, O., 326, 339
Brochier-Armanet, C., 57, 133, 308–309, 345
Broda, E., 90
Brodie, E., 284
Broenkow, W. W., 326

Brondino, C. D., 138
Bronk, D. A., 8
Brooks, P. D., 184–185
Broos, K., 361
Brothier, E., 361
Brown, C. T., 285
Brown, K., 488
Brown, M., 215
Bruchert, V., 256
Bruehl, A., 99, 469
Brunori, M., 136, 140
Bryant, S. H., 472
Brzezinski, M., 20
Buchner, A., 104
Buchwald, C., 253, 256–261, 265, 269
Buckhout, T. J., 424
Buckley, B. A., 283
Buckley, D. H., 284
Budzikiewicz, H., 407
Buesseler, K. O., 5, 8
Buick, R., 485–486, 488, 491
Bulow, S. E., 317
Bundy, L. G., 350, 359
Bunk, D. M., 468
Burdige, D., 370
Bureau, F., 346
Burggraf, S., 174, 177
Burkill, P. H., 14
Burrell, R. C., 449
Burris, R. H., 284
Burr, M. D., 177
Burton, F. L., 378
Burton, S. A. Q., 350
Busby, S., 468
Bush, A., 206, 211
Butcher, J. C., 23
Butler, C. S., 487
Butler, J. H., 255
Butler, M. K., 34, 133
Butman, B., 326
Butterfield, D. A., 283–284
Buysens, S., 405, 407
Bynum, L. H., 472
Byrne, B., 488
Byrne, N., 91, 174

C

Cabantchik, Z. I., 419
Cabello, P., 177
Cabrera, M. L., 362
Caffrey, J. M., 206, 318
Caldeira, K., 6
Calugay, R. J., 407
Calvert, S. E., 283, 490, 496–497
Cambillau, C., 488
Campbell, D. H., 184
Campos, J. L., 94

Canfield, D. E., 90–91, 325
Cantera, J. J. L., 133, 136, 138, 141, 144, 399
Cantrell, L., 179
Cao, C., 491
Capone, D. G., 282–283, 285–287, 290, 295, 300, 314, 318, 487
Cardasis, H. L., 468
Carder, K. L., 5, 14, 18
Carey, C. L., 282
Caris, C., 419
Carlson, C. A., 283
Carlson, C. G., 20
Carlucci, A. F., 262, 309
Carney, K. M., 349
Carpenter, E. J., 282–283, 487
Cartigny, P., 498
Casamayor, E. O., 212
Casciotti, K. L., 4, 7, 135, 180, 184–186, 207–210, 212, 214–215, 253, 255–261, 263–266, 269–271, 274–275, 317, 486, 490
Castresana, J., 136, 140–141
Cava, F., 177
Cavanaugh, C. M., 256
Cavola, C., 475
Centola, F., 136, 140
Chai, F., 14, 18–20
Chain, P. S. G., 34, 38, 57, 112, 132–133, 136, 139, 141, 273, 308, 345, 390, 405–407, 414–415, 418, 423, 466
Chalk, P. M., 356
Chanapan, S., 173, 176
Chandran, K., 133, 369–370, 381–383
Chaney, S. G., 157
Chang, C. C. Y., 184, 263
Chang, L., 207
Chang, S., 485
Chang, Y. J., 80, 210–211
Chan, P. P., 57, 133, 308, 345
Chan, W. M., 294
Chapin, F. S., 132
Charles, C. D., 494
Chaumery, C., 326
Chavez, F. P., 207, 255, 270–271, 490
Chemie, W., 329
Chen, D., 346
Chen, J. W., 90, 173, 175, 193, 195
Chen, R. R., 184, 207
Chen, X. G., 90
Chen, Y., 419
Chernyh, N. A., 177
Chevalier, R., 346
Chhabra, S. R., 70
Chisholm, S. W., 13, 16
Cho, C. M.-H., 136, 139, 141, 390
Choi, P. S., 136, 140
Christ, C. L., 492
Christensen, B. T., 349
Christensen, J. P., 283

Christner, M., 398
Church, M. J., 207–208, 211, 214
Chu, X., 491
Cifuentes, L. A., 262, 270
Ciplak, A., 382
Cirpus, I. E. Y., 91, 93, 98
Clark, D. R., 9–10, 16, 20–22, 206–207, 316, 319–320
Clarke, K. R., 12
Clarke, T. A., 438–439
Clark, L. C., 326
Clark, R. M., 297
Clark, W. A., 494
Clauser, K. R., 468
Cleland, W. W., 255–256, 272
Clemence, M.-E., 491
Clement, C., 206
Clesceri, L. S., 138, 378
Cline, D. E., 215
Cline, J. D., 326, 490, 493
Clode, P. L., 285, 294, 296–297
Coale, K. H., 487, 491
Coatanoan, C., 10
Cociorva, D., 472
Cockell, C. S., 194
Codispoti, L. A., 224, 283, 326
Cohen, Y., 255
Cohn, M., 157
Cole, C. V., 362
Colinge, J., 472
Colley, S., 497
Commeaux, C., 120
Conan, P., 10
Conrad, R., 345, 349, 361
Cookson, W. R., 349
Coolen, M. J. L., 207–208
Cooper, M., 70
Coplen, T. B., 264
Cornelis, P., 405–407
Cornforth, I. S., 349
Corona, A., 447
Correia, C., 138
Costa, K. C., 176, 194–196, 198
Coste, B., 10
Cotton, F. A., 461
Cottrell, J. S., 472
Cottrell, M. T., 208
Coutlakis, M. D., 318
Coutts, G., 83–84
Cowan, D. A., 194
Coward, A. C., 15
Cox, J., 470, 472, 475
Cox, R. D., 262
Cozzolino, F., 467
Cozzolino, M., 467
Craig, H., 255
Craig, R., 472
Creaser, R. A., 485

Creasy, D. M., 472
Crepeau, V., 91, 174
Criddle, C. S., 175
Crill, P., 382
Cron, B. R., 194
Crossey, L. J., 194
Cross, R., 136, 140
Croudace, I. W., 497
Culbertson, C. W., 188
Cummings, D. G., 313–314
Curry, W. B., 494
Cutruzzola, F., 136, 140
Czepiel, P., 382
Czerwinska, H., 410

D

da Costa, M. S., 177
Dagda, R. K., 472
Dai, M. H., 194–195
Daims, H., 72, 76, 92, 99, 102, 112, 120, 122–126, 172–173, 175–176, 193–194, 310, 361, 469, 489
Daldal, F., 433
Dale, O. R., 91, 104
Dalsgaard, T., 90–91, 226–227, 231–232, 325, 327, 333–334, 338
Damsté, J. S. S., 90–92, 100
Dancea, F., 436
Dang, H., 207
Daniel, J. S., 254, 370
Danso, S. K. A., 356
Dapena, A., 91–92, 98
Datsenko, K. A., 395
Dauphas, N., 495–496
Dauwe, B., 494
David, M. M., 408
Davidson, E. A., 132, 345, 351, 353–354, 358–359
Davidson, S. K., 180
Davies, D., 4
Davisson, M. L., 297
de Beer, D., 34, 121, 133
de Boer, L., 313
DeBoer, W., 56–57, 77, 79–81, 345, 350, 356, 370
Debruijn, P., 90, 97
De Corte, D., 208
De Corte, K., 498
de Cuevas, B., 15
Degrange, V., 361
de Hoog, C. L., 467
Dekas, A. E., 281, 283–285, 289, 295, 297
de Lajudie, P., 124
de Lange, G. J., 494
de la Torre, J. R., 25, 34, 57, 64, 67–68, 72–77, 133, 166, 171–173, 175, 193–197, 206–207, 211–212, 215, 275, 308–309, 345, 348

Del Grosso, S. J., 362
Deligeer, K., 488
DeLong, E. F., 35, 77, 79–80, 207–208, 211–212, 214, 285, 288, 292, 296
de Lorenzo, V., 422
den Camp, H. J. O., 93, 102, 139, 346, 466
Deng, S. C., 194–195
Denison, R. F., 158, 211
Denman, K. L., 14, 19–20
De Paepe, P., 498
de Pas-Schoonen, K. V., 91, 98, 346
Deppenmeier, U., 283–284
De Rijk, P., 81
DeSantis, T., 284
de Sieyes, N. R., 64, 206–207, 209
Dessingy, T., 472
Detmers, J., 256
Deutsch, C., 255, 258, 283, 493
Devol, A. H., 90, 101, 255, 283, 320, 327, 346, 490, 493–494
de Vossenberg, J. V., 90–92, 98, 100
de Waal, E. C., 79, 81
de Wit, C. A., 132
Diaz, F., 10
Diaz, R., 326, 339
Dieterle, D. A., 14, 18
Dietrich, W., 436
DiFiore, P. J., 258, 494
Di, H., 175
Diller, S., 177
Dimitri, G., 91–92, 102
Dinasquet, J., 208
Dinsdale, E. A., 34, 49
Dipietro, K., 124
Dippner, J. W., 490
Dirmeier, R., 177
DiSpirito, A. A., 269, 404, 487
Distel, D. L., 283, 285, 290, 294, 297
Dittberner, P., 56, 68
Dittel, A. I., 208
Dittert, K., 184
Dixon, R., 142
Dobinsky, S., 398
Dodsworth, J. A., 171, 173–174, 176–177, 182, 188, 194
Dollhopf, S., 207
Doney, S. C., 16, 487
Dong, H. L., 173, 175, 193–195
Dong, L. F., 34
Donn, T., 207
Dorador, C., 79, 361
Dore, J. E., 9–10, 207, 255, 313
Dortch, Q., 16
Doucette, G. J., 215
Downes, M. T., 313
Drange, H., 15
Dreyfuss, B. W., 435
Drezdzon, M. A., 451, 460

Dringen, R., 410
Drobner, E., 174, 177
Duan, Y., 485
Dua, R. D., 267
Du, B., 91, 103
Dubchak, I. L., 136, 142
Ducklow, H. W., 12–13, 16, 22
Dudley, J., 104
Dudley, L. M., 357
Duffin, K. L., 417
Duffy, J. J., 157
Dugdale, R. C., 8, 12, 19–20
Duguay, L. E., 318
Dular, U., 308
Dumont, M. G., 157, 284
Dunham, S. E., 314, 318
Durand, B., 494
Durisch-Kaiser, E., 91
Dutilh, B. E., 34, 133
Dutkiewicz, S., 13
Duyts, H., 56
Dytczak, M. A., 121

E

Eady, R. R., 300, 488
Earley, K. W., 434
Eaton, A. D., 138, 378
Eberz, G., 390, 392
Ecker, S., 390, 392
Edmond, J. M., 496
Edwards, R. A., 34, 49
Eek, K. M., 289
Eenhoorn, H., 81
Eggleson, K. K., 417
Eggleston, B., 175
Egli, T., 72–73, 76
Ehlers, C., 283–284
Ehrenreich, A., 283–284
Ehrhardt, M., 312
Ehrich, S., 112, 176
Eichler, R., 436, 438–439
Eickhout, B., 6
Einsle, O., 139, 430, 435–436, 438–439, 488
Eisel, F., 434, 436, 439–440, 442
Eisen, J. A., 57, 132
Elderfield, H., 6
Elifantz, H., 208
Elkins, J. W., 255
Ellis, M. J., 488
Ellner, S. P., 44
Elmore, B. O., 136, 139–140
El-Sheikh, A. F., 57, 390, 407, 466
Embaye, T., 285
Embley, T. M., 57, 77, 79–81, 348, 361
Engel, M. H., 494
Eng, J. K., 472
Ensign, S. A., 391, 408, 487
Eppinger, M., 436

Eppley, R. W., 8, 318
Epstein, S., 264
Erba, E., 491
Erguder, T. H., 207
Erickson, J., 435
Escolar, L., 422
Ettwig, K. F., 34, 133
Eurwilaichitr, L., 173, 176
Evans, G. T., 12, 22
Eybe, T., 294, 300

F

Falkowski, P. G., 4–6, 12, 18, 224, 484–485, 491, 496
Fallon, S. J., 283, 285, 290, 295
Fang, F., 90, 101–102
Fang, Y., 467
Farhoud, M. H., 467
Farias, L., 4, 20
Farley, A. R., 467
Farquharson, R., 345
Farrell, R. E., 140
Farrow, D., 206
Fasham, M. J. R., 12–17, 22
Fath, B., 35
Fear, J., 133, 349
Federova, R. I., 187
Feissner, R. E., 434
Feldman, J., 215
Feliatra, F., 9–10
Feliatra, M., 312
Fendorf, S., 361
Fennel, K., 12, 18, 485
Fennel, W., 15
Ferguson, S. J., 430, 433–434, 488
Fernández, C., 4, 9–10, 16, 20–22, 206–207, 319
Fernandez-Herrero, L. A., 177
Fernández, I. C., 9–10
Ferrell, R. E., 495
Fiedler, R., 317
Field, M. P., 491
Fiencke, C., 111, 173, 175–176
Fierer, N., 349
Fijiwara, T., 139
Fike, D. A., 289–290, 296
Fillery, I. R. P., 351, 355
Finstein, M. S., 119
Finzi-Hart, J. A., 283, 285, 290, 295, 297
Finzi, J. A., 283, 285, 290, 300
Firestone, M. K., 180, 346, 348–349, 351, 353–354, 358–359
Fischer, G., 257
Fischer, T. P., 188
Fisk, J. W., 358
Fitzsimons, I. C. W., 297
Flagg, C., 326
Flagiello, A., 467

Flemming, C. A., 80, 210–211
Flessa, H., 190
Fletcher, I. R., 294
Flo, T. H., 424
Flower, D. R., 424
Fluckinger, M., 424
Flynn, K. J., 13
Fogel Estep, M. L., 494
Fogel, M. L., 262, 270, 490, 495
Folke, C., 132
Follows, M. J., 12–13, 18, 485
Fonknechten, N., 92, 102, 489
Forster, A., 496
Forster, W., 104
Forterre, P., 309, 345
Fortuna, A., 358
Foster, L. J., 467
Foster, R. A., 283–284
Fouts, D. E., 57, 132
Fowler, J., 314
Frame, C. H., 253, 256, 274
Francis, C. A., 64, 77–78, 80, 157, 160, 175, 195, 197, 205–212, 214–216, 227, 309, 361
Francis, R. T., 438
Francois, R., 257, 490, 493–494
Franke, G. C., 398
Frawley, E. R., 433–435
Frear, D. S., 449
Freeman, K. H., 173, 498
Freitag, A., 116
Freitag, T. E., 157, 207, 348
French, E., 55, 64
Freudenthal, T., 494
Friedman, S. H., 269
Friedrich, J., 178, 187
Friedrich, M. W., 157
Fritsch, E. F., 394, 396–397
Fry, B., 263, 272
Fuchs, B. M., 90
Fuchs, J., 404
Fuerst, J. A., 34, 90, 92, 97, 99–101, 172, 174
Fu, F., 25
Fuhrman, J. A., 207–208
Fu, H. X., 91, 103
Fujii, M., 14, 18
Fujii, T., 102
Fujimoto, A., 91
Fujisaki, K., 91
Fujiwara, T., 487
Fukui, A., 140
Fukumori, Y., 139, 487
Fuller, M. E., 358
Fulop, V., 488
Fulton, J. M., 498
Fulweiler, R. W., 283
Furukawa, K., 91, 102
Fushinobu, S., 140

G

Gaballa, A., 425
Gabilly, S. T., 435
Gachter, R., 178, 187
Gagliardi, G., 264
Gaillard, J. F., 494
Galanter Hastings, M., 185, 258, 263–264
Galanter, M., 180, 184–185, 263–264, 317
Galbraith, E. D., 490
Galperin, M. Y., 139
Gammon, C. L., 289–290, 296
Gandhi, H., 133, 135, 139
Ganeshram, R. S., 283
Garavelli, J. S., 472
Garcia, B., 468
García-Martínez, J., 214
Garcia, N., 10
Gardin, S., 491
Gardner, A. M., 137, 140–141
Gardner, P. R., 137, 140–141
Gardner, W. D., 326
Garfin, D. E., 399, 413
Garrels, R. M., 492
Garrido, J. M., 94
Garside, C., 262
Garver, S. A., 5
Garvin, J., 485, 491
Gaurin, S., 326
Gebreyesus, B., 158, 211
Geer, L. Y., 472
Geerts, W., 89, 91, 93–94
Gelfand, M. S., 136, 142
Gemeinhardt, S., 136, 140–141
Gengembre, L., 495
Gent, P. R., 22
Gentry, M. E., 112
Geoffredi, S. K., 285
Gerards, S., 60, 74–76, 349
Gerber, H., 81–82
Gessner, C. R., 137, 140–141
Ghostal, S., 294–295
Gianfrani, L., 264
Giblin, A. E., 206–207, 211
Gibson, T. J., 142
Gieseke, A., 290
Gihring, T. M., 91
Gilch, S., 466
Ginestet, P., 313
Giovannoni, S. J., 156
Giron, M., 472
Glasgow, B. J., 424
Glass, J. B., 483, 487
Glazer, A. N., 284
Glöckner, F.-O., 290
Gloerich, J., 34, 133, 465
Glover, D. M., 4
Glover, H. E., 309–310, 313–314

Glover, L. A., 70
Glud, A., 90
Glud, R. N., 90–91, 329, 332, 334–335
Goddard, A. D., 434
Godde, M., 345
Godden, J. W., 488
Godfrey, L. V., 483–485, 491, 496
Godfroy, A., 91
Goericke, R., 272, 494
Goering, J. J., 8
Goetz, D. H., 424
Golabgir, A., 133
Goldberg, D. E., 417
Goldberg, E. D., 490
Goldberg, T., 490
Goldewijk, K. K., 6
Goldrath, T., 81
Gollabgir, A., 57, 308, 345
Goncalves, V. L., 137
Gons, H. J., 81
González, P. J., 138
Gordon, G. W., 485
Gordon, L. I., 255
Gordon, R. M., 487, 491
Goring, C. A. I., 313
Gornall, A. G., 408
Gotz, D., 177
Gouy, M., 142
Graco, M., 91
Graham, C. E., 44
Graham, D. W., 44
Graham, G. A., 294
Graham, J. E., 136, 139
Grandy, A. S., 135
Granger, D., 326
Granger, J., 263, 271, 316, 488
Granger, S. L., 283
Grant, R. F., 361–362
Grant, S., 13, 194
Grant, W. D., 194
Grasshoff, K., 236–237, 248, 312
Greenberg, A. E., 138, 378
Gribaldo, S., 309, 345
Griffin, B. M., 34, 310
Grigoryants, V. M., 136, 140
Groffman, P. M., 178, 187–189
Grossman, J. G., 488
Gross, R., 435–436, 438–439
Gruber, N., 4, 6–7, 15, 224–225, 283, 493
Grundmanis, V., 320
Gu, C., 362
Guckert, J. B., 127
Gudmundsdottir, A., 406
Guerquin-Kern, J. L., 290
Guerrero, M. A., 212
Guignard, C., 294, 300
Gu, J. D., 102–103
Gulledge, J., 136, 139

Gundersen, J. K., 327–328, 333–335, 338
Gundersen, J. S., 326
Gunderson, T., 283, 285, 290, 295
Gunnewiek, P. K., 356, 358
Gunsalus, R. P., 283–284
Guo, L., 207
Gutiérrez, D., 34, 43, 91
Güven, D., 92
Gvakharia, B. O., 390–391, 398, 424–425

H

Haas, H., 424
Habicht, K. S., 256
Habteselassie, M. Y., 348, 354, 358–359
Hadar, Y., 419
Hadas, O., 79–80, 212, 262, 361
Haendel, D., 495
Hafenbradl, D., 174, 177
Hageman, R. H., 138, 392, 408
Hahn, C. M., 57
Hahne, K., 497
Hain, M. P., 258
Hajdu, J., 488
Hallam, S. J., 208, 212, 214
Hall, D., 34, 49
Hall, G. H., 232
Hall, J. R., 194
Hall, S. J., 120
Halm, H., 285, 290
Halpern, A. L., 57, 132
Halverson, 168
Ham, A. J. L., 468
Hamel, P. P., 435
Hamersley, M. R., 91, 225, 233–235, 241, 247–248
Hamilton, T. L., 173
Hankinson, T. R., 114
Hannig, M., 241
Hannon, J. E., 259–261, 263–266, 274
Hanschmann, G., 495
Hansel, C. M., 206, 361
Hansell, D. A., 20
Hansen, J. I., 318
Hantke, K., 421–422
Hare, P. E., 494
Harhangi, H. R., 139, 172, 174
Harms, H., 45–46, 56, 68, 111–112, 122, 266
Harris, J., 99, 101
Harrison, P. J., 271–272
Harrison, S. P., 16
Harriss, R., 382
Harte, B., 297
Hart, P. J., 417
Hart, S. C., 180, 184–185, 351, 353–355, 358–359
Hartshorne, R. S., 434, 438–439
Hartwig, C., 120, 124–126, 361
Harwood, R. R., 358

Author Index

Hasegawa, M., 395
Hasnain, S. S., 488
Hatch, D. J., 348, 351, 355
Hattori, A., 10, 255
Hatzenpichler, R., 72, 76, 172–173, 175, 275, 345
Haug, G. H., 255, 493–494
Hauser, L. J., 38, 57, 136, 139, 141, 390, 404–407, 411–412, 414, 418–419, 423, 466, 487
Hayatsu, M., 345, 349
Hayes, J. M., 285
Hays, L. E., 491
Heaphy, S., 194
Hebert, D., 326, 339
Heckel, P., 490
Hederstedt, L., 433
Hedges, J. I., 5, 493–494
Hedlund, B. P., 171, 176, 182, 188, 194–196, 198
Heerkloss, R., 44
Heidelberg, J. F., 57, 132
Heijmans, K., 139
Heijnen, J. J., 44, 90, 93
Heip, C. H. R., 19–20, 494
He, J. Z., 175, 361
Helmann, J. D., 425
Helms, W., 416
Hemp, J., 57, 133, 308, 345
Henderson, E. W., 12
Hendrich, M. P., 136, 139, 399, 404
Hendriks, J., 136, 140–141
Hendrix, S. A., 240
Henricksen, K., 318
Hentschel, D., 285
Herbert, R. A., 282
Herbik, A., 424
Herfort, L., 207–208
Herman, P. H. J., 494
Herman, P. M. J., 15, 19–20
Hermans, J. H. M., 93
Hermelink, C., 134, 346
Hermes, J. D., 272
Herndl, G. J., 208
Herrera, A., 194
Herrmann, A. M., 285, 294, 297, 490
Herrmann, M., 73
Hettich, R. L., 34
Hetzel, H.-J., 496
Hevel, J. M., 457
Hewson, I., 283
Higa, R., 283
Higgins, D. G., 142
Higgs, N. C., 497
Hilkert, A., 185, 258, 263–264, 498
Hillion, F., 285, 290
Hill, T., 99, 101
Hinrichs, K. U., 285, 288, 292, 296
Hirayama, H., 173, 175–176

Hirota, R., 391, 398
Hirsch, P. R., 158, 348
Hoepken, H. H., 410
Hoeprich, P., 284
Hoffmann, L., 294, 300
Hoffmann, M., 435
Hoffmeister, H., 414
Hofte, M., 405, 407
Hoidoshi, K., 173, 175–176
Holland, H. D., 491
Holland, M., 188
Holland, R. J., 14
Hollibaugh, J. T., 206–208
Hollister, C. D., 490
Hollocher, T. C., 267, 269, 485
Holloway, J. M., 188
Holmes, D. S., 406
Holmes, M. A., 424
Holmes, R. M., 82, 236, 248, 262–263, 312
Holstein, J. M., 24
Holtappels, M., 223
Hommes, N. G., 57, 136, 147, 347–348, 390–392, 394–396, 398–399, 404, 407, 409, 466
Homstead, J., 370
Hong, Y. G., 102–103
Honjo, S., 5
Hood, R. R., 13, 15
Hooker, B. A., 82, 312
Hooker, T. D., 354
Hooper, A. B., 38, 57, 66, 102–103, 111–112, 122, 136, 139–140, 147, 258, 269, 273, 275, 346, 391, 395, 398–399, 404, 430, 451, 456, 460–462, 466, 487, 489
Hopmans, E. C., 91, 172, 174
Hoppe, P., 290
Hornberger, G. M., 362
Horn, C., 177
Horner-Devine, M. C., 349
Horning, S., 468
Horn, M., 92, 102, 489
Horprathum, M., 288
Horreard, F., 290
Horrigan, S. G., 20, 207, 270, 314, 318
Hoshino, M., 458
Hotchkiss, M., 282
Hou, C., 488
Hou, S. B., 139
House, C. H., 285, 288, 292, 294, 297
Hovdenes, J., 326, 339
Hovey, R., 283–284
Howarth, R. W., 282
Hower, J. C., 490
Hristova, K. R., 158, 211
Hsu, B. T., 430
Hsu, S. F., 284
Hu, A., 157
Huang, Q., 173, 175, 193–195

Huangyutitham, V., 284
Huang, Z., 173, 175, 193, 195
Hubbell, P., 44
Huber, C., 326, 339
Huber, H., 177
Huber, R., 139, 174, 177, 430, 436, 439, 488
Hu, B. L., 90, 101–102
Hucklesby, D. P., 138, 392, 408
Huegler, M., 57
Hugenholtz, P., 120
Hughes, M. N., 137, 141
Hügler, M., 133, 308, 345
Huisman, J., 6, 44
Huismans, R., 81
Huizinga, K. M., 135
Hulth, S., 90–91
Hungate, B., 171
Hungate, R. E., 190
Hunt, D. F., 468
Huston, W. M., 423
Hutcheon, I. D., 283, 285, 290, 295, 300, 495
Hutchings, M. I., 142
Hutchins, D. A., 25
Hutchinson, G. L., 345
Hutter, B., 258
Hydes, D. J., 497
Hyman, M. R., 189, 356, 391, 408–409, 413, 487
Hyun, J., 207

I

Idleman, B. D., 498
Igarashi, N., 139, 448, 487
Iizumi, T., 390
Ikeda, T., 391, 398
Ilyuchina, N. I., 187
Imachi, H., 283
Imamura, Y., 56, 64, 68
Imhoff, J. F., 79–80, 212, 361
Inagaki, F., 173, 175–176
Inamori, Y., 382
Ineson, P., 157
Ingalls, A. E., 64, 72, 172–173, 175, 193–197, 215, 275
Ingvardsen, S., 172
Inoue, T., 488
Inskeep, W. P., 173, 175, 177
Ishihama, Y., 468, 470, 476
Ishiwatari, M., 494
Ishiwatari, R., 494
Islam, A., 346
Itoh, K., 140
Ito, S., 285
Iverson, T. M., 430
Iwasaki, H., 488
Iwata, K., 177
Iwata, S., 488

J

Jackson, B. J., 17
Jackson, L. E., 158, 211
Jackson-Weaver, O., 194
Jaeschke, A., 91, 172, 174
Jaffe, J. D., 468
Jaffe, L. A., 492
Jagtap, P., 436
Jannasch, H. W., 174, 177
Januszyk, T. E., 472
Jardine, P. M., 361
Jarrell, K. F., 284
Jarvis, B. D. W., 124
Javoy, M., 498
Jayakumar, A., 317
Jayakumar, D. A., 490
Jeanmougin, F., 142
Jeanthon, C., 177
Jedlicki, E., 406
Jenkins, D., 374
Jennings, M. P., 423
Jensen, K., 319
Jensen, L. S., 351, 355
Jensen, M. H., 319
Jensen, M. M., 34, 43, 91–92, 102, 211, 223, 229, 232, 235, 247
Jensen, S. D., 215
Jetten, M. S. M., 34, 43–44, 57, 89–94, 99–103, 117, 132, 134–136, 139, 141, 172, 174, 275, 346, 350, 370–371, 382, 390, 430, 465–466, 470, 489
Jiang, H. C., 171, 173, 175, 193–195
Jiang, M.-S., 19–20
Jiang, Q. Q., 56
Jiao, N. Z., 90
Jia, Y., 495–496
Jia, Z. J., 361
Jobb, G., 104
Joergensen, B. B., 36
John, K. D., 44
Johnson, K. S., 283, 487, 491
Johnson, M., 404
Joint, I., 9–10, 313–314, 316, 320
Jones, B. E., 194
Jones, R. D., 56, 68, 212
Jørgensen, B. B., 44, 90–91, 100–101, 290, 329
Jormakka, M., 488
Joye, S., 207
Jozsa, P. G., 112, 122, 173, 176
Juliette, L. Y., 409
Junier, P., 79–80, 212, 361
Junier, T., 79–80, 212, 361
Junium, C. K., 494, 498
Juretschko, S., 56–57, 99, 101, 212
Justino, M. C., 137

K

Kadner, R. J., 406
Kahler, P., 287
Kahru, M., 5
Kai, Y., 488
Kalanetra, K. M., 206, 208
Kamber, B. S., 491
Kampf, J. P., 285
Kampschreur, M. J., 370, 382–383, 466
Kandeler, E., 81–82
Kangur, L., 417
Kanokratana, P., 173, 176
Kaplan, I. R., 490, 492–495
Karaoz, U., 284
Karas, M., 438–439
Karhu, J. A., 485
Karl, D. M., 9–10, 207–208, 211, 214, 255, 258, 283, 313
Karner, M. B., 208
Karsh, K. L., 256
Kartal, B., 34, 44, 89–92, 94, 98, 100–102, 132, 174, 226, 231, 346, 465, 470
Kashiyama, Y., 494, 498
Kastrau, D., 121, 487
Kataoka, K., 136, 140, 488
Katehis, D., 370, 382
Kato, J., 391, 398
Kator, H., 263, 316
Kaufman, A. J., 485
Kawagoshi, Y., 91
Kawahara, Y., 102
Kawamiya, M., 20
Kawashima, H., 91
Kaya, M., 111
Kechris, K. J., 284
Keeney, D. R., 83, 184
Keigwin, L., 490
Keil, R. G., 493–494
Kelleher, N. L., 472
Keller, A., 475
Keller, H., 436
Keller, J., 120–121
Keller, M., 177
Keltjens, J. T., 34, 90, 132, 346
Kendall, B., 485
Kendall, C., 184, 207, 255, 263, 270–271
Kenedy, V. C., 184
Kergoat, G., 326
Kern, M., 429–430, 434–436, 438–440, 442
Kerouel, R., 82, 312
Kerrich, R., 495–496
Kerstel, E. R., 264
Kessler, P. S., 284
Kester, R. A., 313, 356, 370
Khakh, S. K., 425
Khanna, M., 166–167
Khijniak, T. V., 488

Khripounoff, A., 326, 339
Kieft, 168
Kienast, M., 490
Kikuchi, T., 140
Kilburn, M. R., 285, 296
Killham, K. S., 346
Kilpatrick, K. A., 318
Kilpatrick, L. E., 468
Kim, C., 177
Kim, K.-R., 255
Kimochi, Y., 382
Kim, O. S., 79–80, 212, 361
Kim, S. W., 140, 370, 381–382
Kimura, T., 91
Kim, Y. B., 472
Kindaichi, T., 91, 101–102
Kirchman, D. L., 208
Kirkham, D., 352–353
Kishi, M. J., 14, 18, 20
Kitazato, H., 90, 494, 498
Kjaer, T., 75
Kjelgaard, P., 433
Kleerebezem, R., 35, 466
Kleinfeld, A. M., 285
Klein, M., 99, 101
Klemedtsson, L. K., 178, 187–188, 190–191, 313, 356
Klenk, H. P., 57, 132, 212, 345
Kletzin, A., 57, 132, 212, 345
Klimiuk, A., 172, 174
Klimmek, O., 436
Klotz, M. G., 34, 38, 57, 102–103, 112, 132–133, 136, 139–140, 207–208, 212, 273, 275, 308, 345, 390, 407, 430, 466, 485, 489
Knapp, A. N., 490
Knapp, C. W., 44
Knoll, A. H., 484
Knowles, R., 139, 187, 232, 284, 313, 356, 370
Kobayashi, M., 140
Koch, H., 112
Koebnik, R., 406
Koedam, N., 405, 407
Koenneke, M., 64, 72
Koepfler, E. T., 263, 316
Koleva, N., 470
Kolker, E., 475
Kondeputi, D., 35
Konhauser, K. O., 491
Könneke, M., 34, 57, 64, 67–68, 72–73, 76, 133, 172–173, 175, 193–197, 212, 215, 275, 309, 345
Konstantinidis, K. T., 208
Koops, H. P., 56–57, 61, 63, 68, 71, 74, 79, 101, 111–112, 122, 212, 266, 349–350, 469
Kooser, A. S., 194
Koper, T. E., 348, 358–359
Körtzinger, A., 326, 339
Kosman, D. J., 417, 423–424

Kostera, J., 447–449, 455, 457–458
Kostka, J. E., 90, 207
Kostrikina, N. A., 177
Kovarova-Kovar, K., 72–73, 76
Kowalak, J. A., 472
Kowalchuk, G. A., 57, 77, 79–81, 156, 210–211, 345, 350, 361
Krafft, T., 436
Kraft, M. L., 285
Kramer, E. H. M., 34, 90, 97, 172, 174
Kranz, R. G., 433–436, 439–440, 442
Kremer, J. N., 15
Kremling, K., 312
Krishnamurthy, A., 487
Kristo, M. J., 295
Kröger, A., 430, 435–436, 438–439
Kroneck, P. M. H., 121, 139, 430, 436, 439, 487–488
Kroneis, H., 326
Krouse, H. R., 495
Kruger, M., 286
Krüger, S., 91
Kruse, M., 126
Kuenen, G. J., 90, 134
Kuenen, J. G., 44, 90–91, 93, 97–98, 100–101, 117, 133, 346, 350, 485, 489
Kuever, J., 256
Kuhn, D. N., 212
Kühn, W., 19
Kumar, S., 104, 258
Künzler, P., 435
Kurganov, B. I., 417
Kuroda, A., 391, 398
Kurth, E. G., 390, 399, 409
Kuwazaki, S., 140
Kuypers, M. M. M., 34, 43, 90–92, 100–103, 133, 223, 225–227, 233, 247, 285, 290, 491
Kwok, S. C., 308, 420

L

Laanbroek, H. J., 25, 55–56, 60–68, 74–76, 78, 81, 111, 313, 348–350, 356, 358, 370
Labasque, T., 326
Lacerda, C. M., 467
Lai, T., 104
Lalli, C. M., 312
Lalonde, S. V., 491
Lambin, E. F., 132
Lamborg, C. H., 5, 8
Lamerdin, J., 38, 57, 390, 405–406, 414–415, 418, 423, 466
Lamond, A. I., 476
Lamont, I. L., 406
Lam, P., 34, 43, 90–92, 102–103, 211, 225–226, 229, 232–233, 235, 248, 285, 290
Lancaster, C. R. D., 435–436, 438
Lancelot, C., 270

Land, M. L., 38, 57, 390, 405–406, 414–415, 418, 423, 466
Landry, Z. C., 206, 211
Lane, D. J., 77, 79–80
Lane, J., 282
Lane, T. B., 44
Langezaal, A. M., 172
Langlois, R., 285, 290
Langston, J., 184
Lanz, C., 436
Lanzen, A., 212
Lao, V., 38, 57, 390, 405–406, 414–415, 418, 423, 466
LaQuier, F., 404
Larimer, F. W., 38, 57, 112, 390, 405–406, 414–415, 418, 423, 466
LaRoche, J., 285, 290
Larsen, B., 490
Larsen, L. H., 75, 327–328, 333–334, 338
Larsen, P., 115
La Torre, J. R. D., 208, 212
Laudelout, H., 310
Laurent, F., 120
Laverman, A., 44
Lavik, G., 34, 43, 90–92, 100–102, 211, 223, 232, 241, 285, 290
Laws, E. A., 13, 15, 272
Lea, D. W., 494
Lebaron, P., 208
Lebauer, D., 158
Lebedeva, E. V., 72, 76, 111–112, 122, 126, 172–173, 175–176
Leblon, G., 290
Le Brun, N. E., 142
Lechene, C. P., 283, 285, 290, 294, 297–298
Leclerc, M., 35
Ledgard, S. F., 348
LeDuc, R. D., 472
Lee, C., 5
Lee, D.-W., 433
Leemans, R., 6
Lee, S. T., 102
LeGall, J., 488
Legendre, L., 14, 19
Lehmann, M. F., 316, 494–495
Lehner, A., 92, 102, 489
Leigh, J. A., 284
Leighton, T. J., 294–295
Leininger, S., 57, 132, 156, 165–166, 175, 195, 212–213, 309, 345, 361
Leivuori, M., 490
Lema, J., 35
Lens, S. I., 147
Lenton, T. M., 132–133
Leoni, L., 406
Le Paslier, D., 34, 112, 123, 133, 310
Le Quéré, C., 16
Le Roux, X., 120

Lesongeur, F., 91
Letelier, R., 283
Leung, K. T., 80, 210–211
Leutenegger, C. M., 158, 211
Leveau, J. H., 398
Levine, J. S., 370
Levitus, S., 5
Lévy, M., 20
Levy, S., 57, 132
Lewis, B. J., 57
Lewis, M. R., 6
L'Haridon, S., 177
Liang, J. H., 284
Libman, J., 419
Lickfeld, K. G., 414
Liesack, W., 57, 77–78, 80, 158, 160, 210–211, 349
Lies, D. P., 284, 289
Li, F., 135, 139
Li, J. Y., 207, 424
Li, L., 498
Li, M., 102–103
Limburg, P., 79–80, 212
Lim, D., 488
Limsuwan, P., 288
Lindow, S. E., 398
Lindsay, K., 16
Lindström, K., 124
Lin, J. C., 284
Link, A. J., 467
Linke, B., 436
Lin, Y.-L., 436
Linz, B., 424
Lipschultz, F., 20, 283
Lipski, A., 109, 111–112, 120, 122–126, 361
Lis, G. P., 495
Li, T., 290
Liu, M. Y., 488
Liu, X. W., 90, 101–102, 136, 139, 141, 390
Li, W., 173, 175, 193–195
Li, X. R., 91, 103
Li, Y. L., 173, 175
Lloyd, D., 136, 140
Lobinski, R., 498
Loftfield, N., 190
Logan, M. S. P., 430
Logemann, S., 34, 90, 97, 172, 174, 346
Löhr, F., 436
Lomas, M. W., 8
Londer, Y. Y., 435
Londry, K. L., 121
Loosdrecht, M. C. M., 466
Losekann, T., 290
Loughman, J. A., 434
Love, G. D., 491
Low, A. P., 357
Lucas, M., 8–9
Lücker, S., 112, 123, 126, 310

Ludden, P. W., 284
Ludwig, C., 390, 392
Ludwig, W., 57, 104, 112, 176, 210
Lueders, T., 157
Lue-Hing, C., 373
Lu, H., 382–383
Lui, K.-K., 490
Lundrigan, M. D., 406
Luo, C., 487
Luyten, Y., 283, 285, 290, 294, 297
Lv, G., 194–195
L'Vov, N. P., 488
Lyalikova, N. N., 488
Ly, B. V., 139
Lyons, T. W., 485, 491
Lyons-Weiler, J., 472

M

Maas, B., 91–92, 98
Mabery, S., 284
Machado, A., 206, 211
Mack, D., 398
Macke, T. J., 57
Mack, L., 190
Macko, S. A., 494
Macnaughton, S. J., 80, 210–211
Madigan, M. T., 156
Magalhaes, C. M., 206, 211
Mager, A. M., 438
Maggi, F., 362
Magnin, J., 472
Mahaffey, C., 283
Mahmood, Q., 90
Mahowald, N., 487
Maita, Y., 312
Maixner, F., 112, 120, 122, 124–126, 173, 176, 361
Makarova, K. S., 139
Makolla, A., 123–126
Ma, L., 361–362
Malcolm, E., 313–314
Maldonado, M. T., 271
Malfatti, S. A., 57, 466
Malmstrom, R. R., 208
Mamais, D., 374
Mancinelli, R. L., 485
Mandelbaum, R. T., 81
Manefield, M., 157
Mangenot, S., 34, 38, 92, 102, 133, 489
Manghnani, V., 326
Maniatis, T., 394, 396–397
Mann, M., 467–468, 470, 472, 475–476
Mansfeldt, T., 258
Marino, R., 282
Marin, R., 215
Mariotti, A., 255, 259, 262, 270
Maritorena, S., 5

Markey, S. P., 472
Markossian, K. A., 417
Marletta, M. A., 457
Marra, M. A., 147
Marsoner, H., 326
Martens-Habbena, W., 25, 74–75, 77, 166, 175, 207, 275, 345, 348
Martin, A. P., 9–10, 14, 16, 20–22, 206–207, 319
Martinez, P., 490
Martini, M., 326
Martinko, J. M., 156
Marto, J. A., 468
Marty, B., 485, 495–496
Mary, B., 355
Marzorati, M., 207
Mashiotta, T. A., 494
Maskow, T., 45–46
Massana, R., 214
Massefski, W., 269
Masselot, A., 472
Massion, E. I., 215
Masterlerz, M., 495
Mather, T. A., 485
Matsui, T., 177
Matsumura, H., 488
Matsumura, M., 382
Matsunaga, T., 407
Matsuoka, T., 312
Matsuo, S., 255
Matulewich, V. A., 119
Matzanke, B. F., 418, 423
Ma, W. K., 140
Ma, Y., 194
Mayali, X., 284
Mayeda, T., 264
Mayer, B., 258
Mayer, L. M., 494
Maynard, D. M., 472
Mayor, T., 467
Mays, E. L., 357, 361
Mazeas, L., 290
McCaig, A. E., 80
Mccarthy, J. J., 270
McCarty, G. W., 356
McClain, C. R., 5, 19
Mcclelland, J., 262
McCorkle, D. C., 257, 263, 490, 494
McCormack, A. L., 472
McCormack, I., 120, 124–126, 361
McDermott, T. R., 177
McElroy, M. B., 283
McEwan, A. G., 423
McGarry, J., 447
McGechan, M. B., 361–362
McGinnis, D. F., 91, 102
McIlvin, M. R., 184, 255–261, 263–266, 270–271, 274, 317
McInteer, B. B., 184–185

McKay, C. P., 485
McKeegan, K. D., 288, 292, 296
McKelvie, S. M., 12–13, 16, 22
McKenzie, J. A., 494
McMahon, G., 283, 285, 290, 294, 296
McNally, P. M., 262
McWilliams, J. C., 22
Mead, J. A., 475
Meckler, A. N., 255, 493–494
Megonigal, J. P., 349
Mehrotra, M., 407
Mehta, M. P., 172–173, 283–284
Meier, H., 104, 210
Meijler, M. M., 419
Meincke, M., 121, 487
Meisinger, J. J., 350, 359
Meixner, T., 165
Melack, J. M., 165
Melekhina, E. I., 187
Mellinger, E., 215
Mémery, L., 20
Mendez, R., 92, 94
Mendum, T. A., 158, 348
Merchant, S., 434–435
Merschak, P., 424
Mertens, J., 361
Messerschmidt, A., 139, 430, 436, 439, 488
Metivier, K. A., 362
Metzger, J. W., 99, 101
Meyer, B., 438–439
Meyer-Dombard, D. R., 188
Meyer, J. M., 407
Meyer, O., 466
Meyer, R. L., 91–92
Meyer, W., 112
Michaels, A. F., 283
Michelson, M. J., 326
Michener, R. H., 263
Miclea, A. I., 90, 92–93, 96–97
Middelburg, J. J., 19–20, 224, 494
Migeon, H. N., 294, 300
Mikha, M. M., 165
Miller, A. E., 165
Miller, B. E., 354
Miller, J. H., 398
Miller, L. G., 188, 318
Miller, N. L., 362
Miller, W. G., 398
Milliken, G. A., 165
Mills, R., 175
Mincer, T. J., 207–208, 211–212, 214
Minning, T. A., 475
Minz, D., 81
Miroshnichenko, M. L., 177
Mishima, M., 177
Misko, T. P., 458
Mitchell, B. G., 5
Mitchell, K. R., 194

Mix, A., 490
Miyake, Y., 270
Miyashita, H., 407
Miyazaki, J., 283
Miyazaki, T., 10
Mizumoto, M., 390
Mizuochi, M., 382
Mobarry, B. K., 469
Moen, B., 212
Moffett, B., 99, 101
Moffett, J. W., 496
Moin, N. S., 206, 211
Moir, J. W. B., 136, 140, 430, 488
Moisander, P. H., 283
Mojzsis, S. J., 485
Moldrup, P., 349
Molina, J. A., 362
Molina, V., 79–80, 212, 361
Moller, U. C., 349
Monbouquette, H. G., 177
Mongin, M., 20
Monod, J., 58
Monti, M., 467
Montoya, J. P., 270, 272, 283, 286–287, 490
Moore, C. M., 8–9
Moore, J. K., 16, 487
Moore, L. R., 16
Mootha, V. K., 467
Moraru, C. L., 285
Morel, F. M. M., 491
Moreno-Vivian, C., 177
Mori, K., 424
Morita, R. Y., 56, 68
Moriyama, H., 139, 487
Moroder, L., 470
Morozova, O., 147
Morris, L. J., 316
Morrison, J. M., 326
Morrison, L. J., 263
Mortensen, P., 467
Morvan, J., 9–10, 312
Mosier, A. C., 205–209, 212
Mosier, A. R., 178, 187–188, 190–191, 356, 362
Mossialos, D., 407
Moura, I., 138, 488
Moura, J. J. G., 138, 488
Moutin, T., 10
Mroczkowski, S. J., 259–261, 263, 265–266, 274
Mühle, K., 495
Mulder, A., 90, 225
Mulder, J. W., 370, 382
Mulholland, M. R., 25, 487
Muller, B., 91, 102
Müller, S., 33
Mulvaney, R. L., 183
Muraoka, T., 356
Muratore, T., 468
Murillo, F. M., 488
Murphy, D. V., 285, 294, 296, 348, 351, 355
Murray, J. W., 283, 320
Murray, R. G., 414
Murrell, J. C., 157, 284
Murtugudde, R., 19
Musat, F., 285, 290
Musat, N., 285, 290
Muyzer, G., 79, 81, 90, 92–93, 96–97, 117, 172, 174, 346
Mvondo, D. N., 485
Myrold, D. D., 156, 361

N

Nagaraj, N., 467
Nagasu, T., 476
Naik, R. R., 488
Nakajima, J., 91
Nakajima, Y., 494
Nakamura, H., 140
Nakamura, K., 390
Nakamura, Y., 91
Namsaraev, B., 111, 173, 175–176
Nandakumar, R., 436
Naqvi, S. W. A., 90, 283, 326, 490
Nara, K., 177
Nauhaus, K., 286
Navarro-Gonzalez, R., 485
Navarro, J. B., 176, 194–196, 198
Nealson, K. H., 173, 175–176, 283, 285, 290, 295
Nedwell, D. B., 34
Needoba, J. A., 271–272
Neef, A., 99, 101
Neilands, J. B., 405, 407
Nei, M., 104
Nelms, S. M., 498
Nelson, D. M., 20
Nelson, D. W., 83
Nelson, K. A., 206, 211
Nelson, K. E., 57, 132
Nelson, T. A., 284
Nelson, W., 57, 132
Nercessian, O., 177
Nes, I. F., 56, 68, 212
Nesvizhskii, A. I., 475
Neta, P., 468
Neufeld, J. D., 157
Neumann, T., 15
Nevins, J. L., 270
Newton, W. E., 430
Ng, W.-O., 290
Ni, B. J., 90, 101–102
Nicholas, D. J. D., 258, 267, 370, 485
Nichols, P. D., 127
Nick, G., 124
Nicolaisen, M. H., 74–76, 81
Nicol, G. W., 81, 133, 156, 165–166, 175, 195, 212–213, 309, 345, 349, 361

Nielsen, G. M., 284
Nielsen, J. L., 115
Nielsen, L. P., 318, 320
Nielsen, M. L., 470
Nielsen, P. H., 115
Niftrik, L. V., 34
Nigg, E. A., 467
Nijtmans, L. G., 467
Nisbet, E. G., 491
Nishiyama, T., 102
Nitzsche, H. M., 495
Nixon, D. J., 191
Nixon, S. W., 15, 207, 283
Nojiri, M., 488
Nojiri, Y., 14, 18
Nomaki, H., 90
Nomerotskaia, E., 434
Noone, K., 132
Nordstrom, D. K., 188
North, A. C., 424
Northrop, D. B., 255–256
Norton, J. M., 38, 57, 212, 273, 309, 343, 348, 354, 356, 358–359, 466
Novick, A., 43
Nudelman, R., 419
Nunan, N., 285, 294, 297
Nunoura, T., 283
Nurser, G. A., 15
Nussler, A. K., 458–460
Nyerges, G., 139
Nykvist, B., 132

O

Oades, J. M., 493
Oakley, B. B., 77–78, 80, 157, 160, 175, 195, 197, 209–211, 214, 216
O'Brien, S., 207
Oda, Y., 476
O'Donnell, A. G., 285, 294, 297
Oenema, O., 138
Offenbacher, H., 326
Offre, P., 81
Off, S., 111–112, 126
Ogawa, N. O., 494
Ogram, A., 167
Oguri, K., 90, 494
Ohkouchi, N., 494, 498
Ohtake, H., 391, 398
Ojima, D. S., 362
Okabe, S., 91, 101–102
Okada, H., 494
Okamura, Y., 407
Okano, Y., 158, 211
Okumura, Y., 395
Oldenburg, C. M., 362
O'Leary, M. H., 273
Oleszkiewicz, J. A., 121

Olson, R. J., 10, 263, 314, 316
Omelchenko, M. V., 139
Omori, T., 177
O'Mullan, G. D., 178, 207, 263, 316
O'Neill, R. V., 17
Op den Camp, H. J. M., 34, 57, 90, 92, 100, 102–103, 132, 134, 139, 172, 174, 275, 346, 430, 489
O'Reilly, J. E., 5
Oremland, R. S., 188, 284, 318
Orlando, R., 475
Orlando, S., 206
Orphan, V. J., 281, 283–285, 288–290, 292, 294–298
Orshonsky, V., 435
Osborn, A. M., 34
Osburn, G. R., 188
Ostrom, N. E., 9–10, 133, 135, 139, 255
Ostrom, P. H., 9–10, 133, 135, 139
Ottem, D., 407
Ottengraf, S., 121
Ottow, J. C. G., 382
Oubrie, A., 136, 140–141
Ou, J., 156, 165
Ouverney, C. C., 208
Oved, T., 81
Owens, N. J. P., 313

P

Pacheco, A. A., 447
Paerl, H. W., 282–283
Page, K., 491
Page, W. J., 407
Pagilla, K., 370, 381–382
Palak, M., 488
Palmer, J. R., 23
Palumaa, P., 417
Papaspyrou, S., 34
Papineau, D., 485, 490–491
Pappin, D. J. C., 472
Pargett, D., 215
Paris, G., 491
Parkeh, N. R., 157
Parker, J., 156
Park, H. D., 175, 370, 381–382
Parkin, T. B., 349
Park, J. R., 102
Park, K. M., 285
Park, Y. H., 102
Parslow, J. S., 12, 15, 22
Parsons, T. R., 312
Parton, W. J., 362
Paster, B. J., 57
Pastor, J. J., 17
Pätsch, J., 19
Pattey, E., 362
Paul, E. A., 358

Paull, C. K., 285
Paulmier, A., 326
Paulsen, I., 57, 132
Paulsen, S., 188
Payne, W. J., 187, 313, 488
Paytan, A., 206–207, 255, 270–271
Pearson, A., 173, 175
Pecoits, E., 491
Pedersen, T. F., 283, 490, 496–497
Pedrioli, P. G., 472, 475
Pedrós-Alió, C., 214
Peduzzi, S., 285, 290
Pelegri, S. P., 320
Pelletier, E., 34, 38, 92, 102, 112, 133, 489
Pellicioli, E. C., 435
Peng, T.-H., 19–20
Pennington, J. T., 207, 255, 270–271
Pennock, J. R., 262, 270
Penton, C. R., 90, 101, 346
Pereira, A. S., 488
Perez-Martin, J., 422
Perkins, D. N., 472
Pernthaler, A., 208, 285, 290
Pernthaler, J., 208, 290
Perry, M. J., 314
Persson, A., 132
Petasis, D. T., 399, 404
Peteranderl, R., 296
Petersen, J., 90–91
Peters, J. W., 173
Peters, K. E., 493–495
Peterson, B. J., 8, 82, 312
Petsch, S. T., 492
Pett-Ridge, J., 283–285, 290, 295, 297, 300
Peucker-Ehrenbrink, B., 492
Pfannkuche, O., 290
Pfeiffer, E. M., 111–112, 122, 124–125
Pfeiffer, S., 436
Phanikumar, M. S., 9–10
Phillips, C. I., 467
Piccolo, M. C., 356
Picioreanu, C., 90, 92–93, 96–97, 466
Pi, D., 490
Pilskaln, C., 490
Pimenov, N. V., 177
Pinel, N., 57, 133, 208, 212, 308, 345
Pirhalla, D., 206
Pisa, R., 436, 438–439
Pitt, A. J., 135, 139
Pitt, P., 374
Pi, Y., 173, 175
Plessen, B., 493–494
Ploug, H., 285
Pokaipisit, A., 288
Pokkuluri, P. R., 435
Pol, A., 139
Polissar, P. J., 498
Poly, F., 120, 361

Pommerening-Röser, A., 56–57, 61, 63, 68, 71, 74, 79, 101–103, 212, 349
Pondaven, P., 13–14, 16, 18, 20
Poole, R. K., 136–137, 140–141
Pootanakit, K., 173, 176
Popa, R., 283, 285, 290, 295
Popova, E. E., 15
Popp, B. N., 207–208, 214–216, 255, 272, 309
Poret-Peterson, A. T., 57, 112, 136, 139, 466
Poretsky, R. S., 283–284, 289, 295, 297
Portmann, R. W., 254, 370
Post, A. F., 16
Post, W. M., 17
Potanina, A., 208, 212
Poth, M., 370
Poulton, S. W., 491
Powell, T. M., 15
Pratt, E. J., 183
Prentice, C., 16
Preston, C. M., 207–208, 211–212, 214–215
Preston, T., 184–185
Pride, C., 490
Pride, R., 490
Prieur, D., 91
Prigogine, I., 35
Priscu, J. C., 313
Prokopenko, M. G., 316
Proksch, G., 317
Prosperie, L., 326
Prosser, J. I., 57, 70, 77, 79–81, 118, 120, 133, 156–157, 165–166, 175, 207, 345–346, 348–350, 361
Prudencio, M., 488
Pucci, P., 467
Pujo-Pay, M., 10
Purdy, K. J., 194
Purkhold, U., 56–57, 61, 63, 68, 71, 74, 79, 212
Purohit, P., 490
Purwar, N., 457
Pyle, D. M., 485

Q

Qi, J., 175, 195, 213, 345, 361
Quake, S. R., 208, 212
Quan, T. M., 491
Quan, Z. X., 91, 102–103
Quatrini, R., 406
Quay, P. D., 494
Quintanar, L., 424

R

Rabus, R., 286
Rachel, R., 174, 177
Radajewski, S., 157
Raddatz, G., 436
Rahm, B., 370, 381–382
Raimbault, P., 10

Rakestraw, N. W., 310
Ramirez-Arcos, S., 177
Ramsing, N. B., 81, 334–335
Rappsilber, J., 468, 470, 476
Rastetter, E., 350–351
Rast, H. G., 390, 392
Rath, G., 56, 68, 469
Rattei, T., 38, 92, 102, 112, 345, 489
Rattray, J. E., 90–92, 94, 100
Rau, G. H., 494
Ravel, J., 406
Raven, J. A., 4, 6
Ravenschlag, K., 290
Ravishankara, A. R., 254, 370
Ravishankar, G. A., 392
Raymond, J., 484
Raymond, K. N., 424
Reardon, K. F., 467
Recous, S., 120, 351, 355
Redfield, A. C., 7
Redl, B., 424
Rees, A. P., 9–10, 313–314, 316, 320
Rees, D. C., 430
Regala, W., 38, 57, 390, 405–406, 414–415, 418, 423, 466
Reichert, G.-J., 496
Reigstad, L. J., 173, 175–176, 193–194, 198
Reimers, C. E., 329
Reinthaler, T., 208
Relman, D. A., 290
Remington, K., 57, 132
Renger, E., 318
Ren, H., 493–494
Renn, C., 310
Ren, Q., 435
Ren, Y., 139
Repeta, D. J., 494
Révész, K. M., 185–186, 263
Revill, A. T., 206–207
Revsbech, N. P., 75, 90–92, 233, 318–319, 326–330, 333–335, 337–338, 349
Rey, F., 326, 339
Reysenbach, A. L., 177
Rhee, S. K., 102
Rhine, E. D., 183
Rice, C. W., 165
Richard-Fogal, C. L., 433–435
Richards, O. C., 157
Richardson, D. J., 37, 430, 434, 438–439, 487–488
Richardson, P. M., 208, 212, 214
Richey, J. E., 494
Rich, J. J., 91
Richter, A., 72, 76, 172–173, 175–176, 193–194
Richter, L., 104, 210
Rickert, D., 290
Ridd, P. V., 7
Riemer, J., 410

Riggs, M., 184
Rijpstra, W. I. C., 99, 101
Riley, G. A., 12
Riley, W. J., 362
Rinaldo, S., 136, 140
Ringel, M., 436, 438
Risgaardpetersen, N., 318
Risgaard-Petersen, N., 90–92, 172, 319
Ritchie, G. A. F., 370
Rittmann, B. E., 469
Ritz, K., 285, 296
Roberts, G. P., 284
Roberts, K. J., 77–78, 80, 157, 160, 175, 195, 197, 208–212, 214, 216
Robertson, G. P., 135, 347–348, 358
Robertson, I. B., 433
Robertson, L. A., 90, 97
Robert, S. S., 206–207
Robin, D., 355
Robinson, A. K., 405–407
Robinson, C., 313–314
Robinson, D. P., 467
Robinson, J. A., 360
Robinson, R. S., 490, 493–494
Robinson, S. R., 410
Rocap, G., 16
Röckstrom, J., 132
Rodenburg, R. J., 467
Rodionov, D. A., 136, 142
Rodriguez, D. J., 424
Rodriguez, J., 35
Rodriguez-Quinones, F., 405–407
Rodríguez-Valera, F., 214
Roest, K., 81
Rogers, L. D., 467
Rohde, M. M., 271, 494
Rohrback, B. G., 494–495
Roldan, M. D., 177
Rolston, D. E., 346
Roman, B., 215
Romanek, C. S., 173
Rondelet, A., 434
Rönning, C., 326, 339
Rosenthal, Y., 491, 493–494
Rosenzweig, A. C., 487
Rosinus, A., 436
Rossnagel, P., 177
Rosswall, T., 356
Roswall, T., 313
Rothery, R. A., 488
Rotthauwe, J. H., 57, 77–78, 80, 158, 160, 210–211
Rouxel, O., 496
Rowarth, J. S., 349
Rowe, H., 490
Rozen, S., 146
Rudert, M., 116
Rudnick, P. A., 468

Ruiz-Pino, D., 326
Rusch, D., 57, 132
Rushdi, A. I., 177
Russell, S. A., 136
Russ, M. E., 255
Rust, T. M., 255
Rüterjans, H., 436
Ryden, J. C., 191
Ryder, U., 476
Rysgaard, S., 90–91, 318–319

S

Saby, B. R., 349
Sachdeva, R., 207
Sachs, J. P., 494
Sadygov, R. G., 472
Saino, T., 255
Saint-Cyr, H. F., 296
Saito, M. A., 345, 349, 491
Sakka, K., 91
Sakka, M., 91
Salmon, K., 283–284
Sambrook, J., 394, 396–397
Samudrala, R., 395
Sanders, C., 433
Sanders, R., 8–9
Sanders, T., 111–112, 119, 122, 124–125
San Francisco, B., 433
Sänger, M., 436
Sansom, C. E., 424
Sansone, F. J., 207, 255
Santoro, A. E., 64, 77–78, 80, 157, 160, 175, 195, 197, 206–211, 214–216, 253
Santos, U. M., 494
Sarada, L. R., 392
Saraiva, L. M., 137
Saraste, M., 136, 140–141
Sarmiento, J. L., 4, 6–7, 15, 224–225, 283, 493
Sasser, M., 126–127
Sato, S., 424
Sato, T., 476
Saunders, A. M., 73–76
Saw, J. H. W., 139
Sayavedra-Soto, L. A., 38, 57, 112, 136, 147, 347–348, 389–392, 394–396, 398–399, 403–406, 409, 411–412, 414, 417–425, 466, 487
Sayler, G. S., 167
Schagger, H., 467
Schalk, J., 92, 100, 346
Schaumloffel, D., 498
Scheffer, M., 44, 132
Scheithauer, J., 434–436, 439–440, 442
Schellnhuber, H. J., 132
Schenker, M., 424
Schewedock, J., 256
Schieltz, D., 472

Schiffer, M., 435
Schimel, J. P., 165, 346
Schimmelmann, A., 495
Schink, B., 34, 310
Schjønning, P., 349
Schleifer, K. H., 99, 101, 469
Schleper, C., 57, 132, 156, 165–166, 173, 175–176, 193–194, 208, 212, 214, 309, 345, 361
Schlesinger, W. H., 345
Schlesner, H., 99, 101
Schlosser, U., 112, 122
Schloter, M., 156, 165, 175, 195, 213, 345, 361
Schmid, M. C., 34, 43, 56–57, 90–92, 98–103, 139, 172, 174, 211–212, 275, 350, 430, 489
Schmidt, E. L., 73, 114–115, 118–119, 346, 351, 357–358, 361, 436, 438
Schmidt, I., 74–76, 91–92, 134–135, 346, 349, 466, 469
Schmitz, R. A., 283–284
Scholes, C. P., 136, 140
Scholin, C. A., 215
Schönhuber, W., 290
Schott, J., 34, 310
Schouten, S., 90–92, 99–101, 491
Schrag, D. P., 256
Schramm, A., 73, 76, 121
Schreiber, F., 34, 133
Schreier, H. J., 104
Schroder, I. I., 177
Schubert, C. J., 91, 102, 285, 290
Schubert, T., 45–46
Schulz, B. L., 433
Schulz, H. N., 36, 435
Schuster, S. C., 57, 132, 156, 165, 212, 345, 361, 436
Schwalbach, G., 414
Schwark, L., 156, 165, 173, 175–176, 193–195, 213, 345, 361, 490
Schwartz, E., 155–158, 160, 165, 361
Schwartz, J. C., 468
Schweizer, H. D., 396
Schwindel, K., 390, 392
Sclater, F. R., 496
Scott, C., 485, 491
Scott, K. M., 256
Scow, K. M., 158, 358
Searle, B. C., 475
Searle, P. L., 183
Sebilo, M., 495
Seeyave, S., 8–9
Seidel, J., 435
Seitzinger, S. P., 206
Sejr, M. K., 90–91
Senin, P., 139
Senko, M. W., 468
Sessions, A. L., 289
Setlik, D., 424

Severance, S., 424
Sexstone, A. J., 349
Shabanowitz, J., 468
Shaffer, M. J., 361–362
Shagzhina, A., 126
Shand, C., 83–84
Shanks, C. A., 180, 354
Shanzer, A., 419
Shapleigh, J. P., 136, 140
Sharp, J. H., 262, 270
Shatsky, V. S., 498
Shaviv, A., 81
Shaw, K. M., 284
Shaw, L. J., 133, 349
Shemesh, A., 494
Sheng, G. P., 90, 101–102
Shen, J. P., 175, 361
Shen, S., 491
Shen, Y., 491
Sherman, L. S., 485, 492
Sherr, B., 187, 313
Shields, G. A., 490
Shigeno, T., 177
Shigetomo, H., 102
Shi, J. H., 91, 103
Shimamura, M., 102
Shin, M. W., 57, 466
Shiro, Y., 140
Shi, W., 348, 356, 472
Shock, E. L., 176, 188, 194–196, 198
Short, S. M., 283–284
Shoun, H., 140
Shriver, D. F., 451, 460
Shu, Q. L., 90
Siciliano, S. D., 140
Sickman, J. O., 165
Siefert, J. L., 484
Siegal, D. A., 5
Sigman, D. M., 135, 180, 184–185, 255–259, 262–264, 270–272, 275, 283, 316–317, 490–491, 493–494
Signorini, S., 19
Sillard, R., 417
Silva, S. R., 184, 263
Simkins, S., 178, 187–189
Simonart, P. C., 310
Simoneit, B. R., 177
Simon, J., 37, 429–430, 434–436, 438–440, 442, 488
Sims, G. K., 183
Singer, S., 297
Singh, B., 349
Singh, Y., 349
Sinha, B., 285, 290
Sinigalliano, C. D., 212
Sinninghe Damsté, J. S., 90–92, 94, 98–101, 112, 491, 496
Sintes, E., 208

Siyambalapitiya, N., 120
Skaletsky, H. J., 146
Skinner, J. H., 191
Slaveykova, V. I., 294, 300
Slawyk, G., 10
Sliekers, A. O., 90–91, 100–101
Slodzian, G., 296
Sloth, N. P., 319
Smaldone, G. T., 425
Smeitink, J. A., 467
Smets, B. F., 120
Smith, A., 207
Smith, C. J., 34, 206–207
Smith, C. L., 19–20
Smith, K. D., 424
Smith, R. L., 188, 263
Smith, S. H., 326
Smith, Z., 133, 348–349
Smolders, E., 361
Sobolev, N. V., 498
Sobottka, I., 398
Sockett, R. E., 158, 348
Soetaert, K., 15, 19–20
Sola, J. V., 472
Solomon, E. I., 424
Solorzano, L., 262
Sommer, J., 382
Somville, M., 313
Song, B. K., 91, 104
Song, K. B., 425
Song, Z., 173, 175, 193, 195
Sorensen, J., 188, 319
Soriano, S., 259
Sorokin, D. Y., 117, 350
Soukup, D., 176, 194–196, 198
Sozen, S., 92
Spang, A., 345
Spear, J. R., 178
Speksnijder, A. G. C. L., 81
Spero, H. J., 494
Spieck, E., 72, 76, 109–112, 118, 120, 122–126, 172–173, 175–176, 345, 361
Spiro, S., 142
Spormann, A. M., 290
Springael, D., 361
Springer, A. L., 20, 207
Sprott, G. D., 284
Spudich, E. N., 35
Spudich, J. L., 35
Spurr, A. R., 123
Spycher, N., 362
Stach, P., 139, 430, 436, 439, 488
Stackebrandt, E., 57, 112–113, 117
Stahl, D. A., 25, 57, 64, 67–68, 72–77, 133, 166, 172–173, 175, 193–197, 206–207, 211–212, 215, 275, 309, 345, 348, 395, 469
Stahl, H., 90
Stal, L., 283

Staples, C. R., 484
Starkenburg, S. R., 57, 112, 116, 118, 466
Stark, J. M., 184–185, 343, 348–351, 353–355, 357–359
Steefel, C., 362
Steele, A., 490
Steele, J. H., 12
Steffen, W., 132
Stehfest, E., 6
Stehr, G., 56, 68, 349
Steingruber, S. M., 178, 187
Stein, L. Y., 57, 112, 131–134, 136, 138–139, 141, 274, 313, 345, 349, 389–390, 393, 399, 466, 485, 487, 489
Steinmüller, W., 114
Stein, T., 436, 438–439
Stensel, H. D., 378
Stephen, J. R., 57, 77, 79–81, 156, 210–211
Steppi, S., 104
Stetter, K. O., 174, 177, 390, 392
Stevens, J. M., 433–434
Stevenson, B. S., 290
Stewart, G. S. A. B., 70
Stewart, J. W. B., 362
Stiehl, G., 495
Stienstra, A. W., 358
Stinchcombe, M., 8–9
Stockdale, E. A., 285, 294, 297, 348, 351, 355
Stocker, T. F., 6–7
Stoecker, K., 72, 76, 172–173, 175
Stoj, C. S., 417
Stojiljkovic, I., 416
Stolz, J. F., 488
Stott, A., 34
Stotzky, G., 166–167
Strady, E., 235
Stramma, L., 226
Streit, W., 345
Strickland, J. D. H., 309
Strohm, T. O., 34
Strom, P. F., 119
Strong, R. K., 424
Strous, M., 33–34, 38, 44, 90–94, 97–102, 132, 134, 139, 172, 174, 275, 346, 430, 470, 489
Strunk, O., 104, 210
Strynadka, N. C. J., 488
Stuven, R., 349, 370, 466
Subramaniam, A., 283–284
Suga, H., 494
Suginohara, N., 20
Sultana, T., 472
Sümer, E., 382
Summers, D. P., 485
Summons, R. E., 485, 491–492
Sumner, D. Y., 485
Sundermeyer-Klinger, H., 112–113, 117
Sutherland, S. C., 4

Suthers, I. M., 13
Sutka, R. L., 9–10, 133, 135, 139
Suwa, Y., 56, 64, 68, 212
Suzuki, I., 308, 420
Suzuki, M., 173, 175–176
Suzuki, S., 136, 140, 488
Suzuki, T., 56, 64, 68
Svendsen, T. C., 115
Sweeney, R. E., 492–494
Syka, J. E. P., 468
Sylva, S. P., 285
Szilard, L., 43
Szpunar, J., 498

T

Tabata, T., 476
Tago, K., 345, 349
Takacs-Vesbach, C. D., 194
Takahashi, T., 4
Takai, K., 173, 175–176, 283
Takai, Y., 395
Takaya, N., 140
Takiguchi, N., 391, 398
Talbot, M. C., 314
Tal, Y., 104
Tamura, K., 104
Tanaka, N., 139, 487
Tang, C. J., 90
Tan, N. C. G., 466
Tappe, W., 44
Tarleton, R. L., 475
Tarran, G. A., 313–314
Tashiro, T., 56, 64, 68
Tata, P., 373
Tate, M. E., 267, 485
Taylor, A. B., 326, 417
Taylor, A. H., 15
Taylor, B. F., 284
Taylor, G. K., 472
Taylor, H. E., 498
Taylor, J.-S., 433
Taylor, L. T., 207–208, 211, 214, 285
Taylor, M. W., 92, 102, 489
Tchekhovskoi, D. V., 468
Tchobanoglous, G., 378
Tedesco, D., 139
Tedesco, R., 467
Tegoni, M., 488
Teira, E., 208
Teixeira, M., 137
Teller, D. C., 488
Tengberg, A., 326, 339
Terada, A., 120
Terai, H., 488
Terry, K. R., 66, 451, 460–462
Teslich, N. E., 294
Tett, P., 19–20

Thacker, S. G., 354
Thamdrup, B., 90–91, 102, 225, 227, 231, 325, 327–328, 333–334, 338
Thauer, R. K., 177
Theis, F., 436, 438
Thelen, M. P., 34
Thomas, R., 498
Thomas, S., 7
Thompson, J. D., 142
Thompson, T. M., 255
Thomsen, I. K., 349
Thomson, J., 497
Thöny-Meyer, L., 433, 435
Thornton, E. K., 256
Thornton, E. R., 256
Thunell, R. C., 255, 490, 493
Tian, R. C., 14, 19
Tiedje, J. M., 90, 101, 178, 187–189, 191, 346, 349
Tietema, A., 356
Tighe, S. W., 124
Timmermann, G., 56–57, 61, 63, 68, 71, 74, 79, 101
Timmers, P., 207–208
Tindale, A. E., 407
Tindall, B. J., 177
Tischler, P., 345
Tjaden, B., 424–425
Tobias, C. R., 91, 104
Toffin, L., 290
Toki, T., 283
Topp, E., 139
Tortell, P. D., 271, 316
Totterdell, I. J., 15, 23
Tourna, M., 157, 348
Tourova, T. P., 177, 350
Toyomoto, T., 102
Trap, J., 346
Trautwein, A. X., 418, 423
Tréguer, P., 9–10, 20, 312
Treusch, A. H., 57, 132, 212, 345
Trigo, C., 94
Trimmer, M., 227
Trincone, A., 174, 177
Trull, T. W., 4
Trumpower, B. L., 412
Tsunogai, U., 283
Tsushima, I., 91, 101–102
Tucker, N. P., 142
Turich, C. C., 498
Turkarslan, S., 433
Turley, S., 488
Twachtmann, U., 99, 101
Tyrrell, T., 7, 12

U

Ueberheide, B., 468
Uhle, M. E., 494
Uitterlinden, A. G., 79, 81

Ulloa, O., 327–328, 333–334, 338
Umberger, C., 188
Unden, G., 435–436
Unkefer, C. J., 296
Upadhyay, A. K., 136, 139, 399, 404
Urakawa, H., 25, 57, 74–75, 77, 133, 166, 175, 207–208, 212, 275, 308, 345, 348
Urbach, E., 156
Urbain, V., 313, 469
Urbani, A., 136, 140–141
Urey, H. C., 489
Urich, T., 156, 165, 173, 175–176, 193–195, 213, 345, 361
Urushigawa, Y., 56, 64, 68
Usha, T., 392
Ussler, W. III., 285
Utåker, J. B., 56, 68, 212

V

Vacherie, B., 112
Vadivelu, V. M., 121
Vagner, T., 285
Vajrala, N., 390, 395, 399, 403–406, 411–412, 414, 418–425, 487
Valentine, D. L., 50
Vallius, H., 490
Valois, F. W., 112, 122
van Agterveld, M. P., 81
van Aken, H. M., 208
van Bentum, E. C., 496
van Beusichem, M. L., 138
van Bleijswijk, J., 207–208
van Breugel, Y., 491
van Burgel, M., 264
Van de Graaf, A. A., 90, 97, 346
van den Heuvel, J. C., 121
van den Heuvel, L., 467
van de Pas-Schoonen, K. T., 90, 92, 97–99, 101, 134, 172, 174, 346
Van de Peer, Y., 81
van der Biezen, E., 465
Van der Star, W. R. L., 90, 92–93, 96–97, 370, 382, 470
Van der Waal, D. B., 6
van de Schootbrugge, B., 491
van de Vossenberg, J., 34, 43, 90–92, 94, 100, 102, 211
van Dongen, U., 90, 92–93, 96–97, 346, 371
van Donk, E., 6
van Donselaar, E. G., 91, 94
van Drecht, G., 6
Vandroog, R., 310
van Geen, A., 490
VanGerven, E., 93
van Gijzel, P., 495
van Hannen, E. J., 81
van Krevelen, D. W., 494
van Loosdrecht, M. C. M., 90, 92–93, 96–97, 346, 370–371, 382–383

Van Minnen, J. G., 6
Vannelli, T., 346
Van Nes, E. H., 44
van Niftrik, L., 90–92, 94, 98, 100, 132, 346
van Schooten, B., 147
van Spanning, R. J. M., 37, 134–135, 147, 430, 466
Van Vleck, E. S., 44
Varela, M. M., 208
Veit, K., 283–284
Veizer, J., 258
Veldkamp, H., 58–59
Velinsky, D. J., 262
Velthof, G. L., 138
Venterea, R. T., 346, 362
Venter, J. C., 57, 132
Ventosa, A., 194
Verberkmoes, N. C., 34
Verbruggen, M. J., 99, 101
Vergez, L. M., 57, 466
Vergin, K. L., 156
Verhagen, F. J. M., 60–63, 67, 74–76
Vermeulen, M., 470
Verschoor, A. M., 6
Verspagen, J. M. H., 6
Verstraete, W., 207
Veth, C., 208
Vézina, A. F., 14, 19
Vicente, J. B., 137
Vick, T. J., 194
Victoria, R. L., 356
Vigil, M., 362
Vincedeau, M.-A., 9–10, 312
Vinceslas-Akpa, M., 346
Viner, A. B., 313
Visca, P., 406
Vitiello, G., 467
Vlasits, J., 457
Vogel, A. I., 458
Vogel, R. O., 467
Vohra, J., 157
Völkl, P., 174, 177
Volkman, J. K., 206–207
Vollmer, M., 370
von Brand, T., 310
von Fischer, J. C., 135
Vorholt, J. A., 177
Voronova, A., 417
Vos, A., 69, 73, 75
Voss, M., 287, 490
Voytek, M. A., 80

W

Wada, E., 10, 255, 270
Wagner, L., 472
Wagner, M., 56–57, 61, 63, 68, 71–72, 74, 76, 79,
    91, 98–99, 101, 112, 120, 124–126,
    172–173, 175, 212, 310, 345, 350, 361, 469

Wagner, T., 494
Wai, B., 208, 211
Wakagi, T., 140
Wakeham, S. B., 5
Wakelin, S. A., 361
Waksman, S. A., 282
Waldbauer, J. R., 485, 492
Walker, C. B., 34, 57, 64, 67–68, 72–73, 76, 133,
    172–173, 175, 193–197, 208, 212, 215, 273,
    275, 308–309, 345
Walker, N. J., 73, 259
Wallace, H. E., 497
Walsh, J. J., 14, 18
Walsh, K., 99, 101
Walsh, M. M., 495
Wand, U., 495
Wang, C. H., 90
Wang, C. J., 398
Wang, F., 173, 175, 193–195
Wang, J. M., 139
Wang, L., 207
Wang, P., 173, 175, 193–195
Wang, R. F., 91, 103
Wang, T. P., 424
Wang, W., 424, 491
Wang, Y., 91, 103
Wankel, S. D., 206–207, 255, 270–271
Wanner, B. L., 395
Wanninkhof, R., 4
Ward, B. B., 25, 57, 80, 90–91, 112, 135, 178,
    207, 212, 226, 233, 256, 259, 263, 270, 275,
    307–308, 313–314, 316–320, 466, 488
Warninghoff, B., 112
Waterbury, J. B., 57, 64, 67–68, 72–73, 76,
    111–113, 122, 133, 172, 212, 309, 345
Watson, C. J., 348
Watson, S. W., 56, 68, 111–113, 122, 259, 414
Watts, J. E. M., 104
Weast, R. C., 460
Weatherly, D. B., 475
Weaver, P. P. E., 497
Weaver, R. W., 356
Webb, R. I., 90, 92, 97, 99–101
Weber, P. K., 283–285, 290, 294–295, 297, 300
Webster, G., 348
Wefer, G., 494
Wehrli, B., 91, 178, 187
Wehrmann, H., 56, 68
Weiner, J. H., 488
Weisburg, W. G., 57
Weiske, A., 382
Weiss, P. M., 272
Wei, X. M., 57, 399, 404–406, 411–412, 414,
    417–419, 421, 423, 466, 487
Wells, G. F., 175
Wenzhöfer, F., 90, 494
Wertz, S., 361
Wessels, H. J. C. T., 34, 133, 465, 467

Westerhoff, H. V., 147
Westram, R., 104, 210
Wetzel, R. L., 263, 316
Wheeler, K. E., 294–295
White, A. E., 283
White, D. C., 80, 127, 210–211
White, R. E., 346
Whittaker, M., 273, 399, 404
Whittlesey, L. H., 179
Wicht, H., 382
Widdel, F., 286, 290
Wiebe, M. G., 405, 407
Wiegel, J., 173, 175
Wielders, H. A., 370, 382
Wild-Allen, K., 19–20
Wilderer, P. A., 116
Wilkerson, F. P., 19–20
Wilkinson, C. J., 467
Wilkison, D. H., 184, 263
Williams, B., 83–84
Williams, E. H., 258
Williams, L. B., 495
Williams, M. A., 165
Williamson, R., 8–9
Williams, P., 70
Wilson, M. J., 406
Wilson, T. R. S., 497
Wincker, P., 92, 102, 489
Winkelmann, G., 418, 423
Winogradsky, S., 56, 466
Winterholler, B. R., 290
Wirtz, K. W., 24
Wishner, K., 326
Wisniewski, J. R., 467
Witherspoon, J., 373
Wittebolle, L., 207
Witte, U., 290
Wittig, I., 467
Witzel, K. P., 57, 77–80, 158, 160, 210–212, 361
Woebken, D., 34, 43, 90–92, 102–103, 211, 233
Woese, C. R., 57
Woldendorp, J. W., 57, 77, 79–81
Woldendorp, V. W., 111
Wolfbeis, O. S., 326
Wolfe-Simon, F., 487
Wolff, U., 407
Wolfsberg, M., 256
Wolf, Y. I., 139
Wolgast, I., 111, 122
Wolt, J. D., 356
Woodward, S., 313–314
Worm, B., 6
Wrage, N., 138
Wright, D. K., 215
Wroblewski, J., 16–17
Wuchter, C., 207–208
Wu, D. Y., 57, 132
Wuebbles, D. J., 132

Wu, L., 136, 139, 141, 361–362, 390
Wulf, R., 326
Wu, T. D., 262, 290

X

Xia, K., 165
Xie, W. M., 90, 101–102
Xie, X. H., 467
Xie, Y., 488
Xie, Z., 435
Xu, K.-Q., 382
Xu, M. G., 175, 472
Xu, T., 362

Y

Yadhukumar, H., 104, 210
Yamagishi, H., 255
Yamaguchi, K., 136, 140, 142, 488
Yamamoto, T., 488
Yamamoto, Y., 173, 175–176
Yamanaka, Y., 14, 18, 20
Yang, J., 424
Yang, X., 472
Yang, Y., 102
Yan, T., 136, 139, 141, 390
Yates, J. R., 472
Ye, Q., 173, 175, 193, 195
Yeung, C. H., 175
Yokokawa, T., 208
Yool, A., 3, 7, 9–10, 16, 20–22, 206–207, 319
Yoshida, N., 255
Yoshinari, T., 133, 187, 255, 283, 490
Yoshizumi, K., 312
Young, M. J., 177, 179
Yuan, Z., 121
Yu, B. S., 194–195
Yu, H. Q., 90, 101–102
Yung, Y. L., 132, 274, 345
Yu, R., 133–134, 382–383
Yutin, N., 139
Yu, Y., 90

Z

Zabel, M., 494
Zablein, L. B., 57
Zafiriou, O. C., 319
Zafra, O., 177
Zahn, J. A., 487
Zahnle, K., 491
Zart, D., 349, 466
Zedelius, J., 34, 133
Zehr, J. P., 90, 282–284
Zender, C. S., 487
Zhang, C. L., 173, 175–176, 193–196, 198
Zhang, J. B., 175, 361
Zhang, L. M., 90, 175, 262, 361

Author Index

Zhang, Y. L., 284, 467
Zhang, Z., 207
Zhao, W., 173, 175
Zheng, P., 90, 101–102
Zheng, Y. M., 175
Zhou, J., 136, 139, 141, 390
Zhou, S. M., 140
Zhou, Z. M., 139
Zhu, H., 433
Zhu, P., 424
Zhu, Y. G., 175, 361
Ziebis, W., 289–290, 296

Ziegler, A. C., 184, 263
Ziegler, L., 417
Zika, R. G., 496
Zougman, A., 467
Zubkov, M. V., 14
Zufferey, R., 435
Zumbrägel, S., 126
Zumft, W. G., 34, 132, 134, 136, 140–141, 225–226, 370, 436, 488
Zuo, J. E., 102
Zwart, G., 81

Subject Index

A

Acetylene block method
  analysis and rate calculations
    factory-ready protocol, 191
    GC-ECD system, 190
  application protocols, 188
  description and limitations, 187
  equipment, 189–190
  incubation, 190
  measurement, 187–188
  N_2O, 188
  preparation
    equipment, 189–190
    synthetic mineral salts medium, 189
    ultrafiltration, 188–189
  standard factory-ready protocol, 191
Aerobic respiration, denitrification, 43
Affinity
  nitrite oxidizers, 36
  Nitrosopumilus maritimus SCM1, 207
  phytoplankton, 17
American Society for Testing and Materials (ASTM) method, 376–377
Ammonia monooxygenase (AMO)
  crenarchaeal, 208
  isotope effects, 266
  NH_3 oxidation, 273
  N isotopic fractionation, 258
  nitrification, 487
Ammonia monooxygenase subunit A *(amoA)*
  amplification, 78–79
  AOA abundance, 206–207
  AOB abundance, 206
  archaeal and bacterial, quantification
    qPCR, 210–211
    sample nucleic acids, 211
    SYBR assays, 211–212
  "*Candidatus* Nitrosocaldus yellowstonii," 194–196
  crenarchaea, 208
  DNA, archaeal and bacterial, 163
  gene, 175
  group A and B sequences, 214–215
  mortality index, 166
  N-cycle processes, 173–174
  quantitative real-time PCR analysis, 160
  RNA extracts, 209–210
  shallow and deep archaeal clades, 217
  sheer number, sequences, 212
Ammonia oxidation
  AOA kinetics, 77
  environmental factors, 76–77
  microbe groups, 206
  N and O isotopic fractionation and exchange
    cells, harvest, 259
    $\delta^{18}OH_2O$, 259
    time course experiments, 260
  N isotope effect, 270
  N isotopic fractionation
    nitrite oxidation, 265
    Rayleigh model, 264
  nitrous oxide production, 254
  O isotope fractionation and exchange
    $\delta^{18}OH_2O$, 266–267
    $\delta^{18}O_{NO2}$ values, 265
    H_2O incorporation, 265–266
    NH_2OH, 267
  pathway, AOA and AOB, 275
Ammonia oxidizers
  activity kinetics
    AOA, O_2 sensor, 77
    environmental factors, 76–77
    K_m and V_{max}, determination, 76
    NO_2^- plus NO_3^- biosensor, 75–76
    O_2 sensor, AOB, 76
  enrichment
    batch culture, 63–65
    continuous culture, 65–67
    medium preparation, 67–68
  growth kinetics
    calculation, 73
    environmental factors, 74
    K_s and μ_{max}, batch culture, 73–74
    medium, 72
    sampling pattern, 73
  identification, 16S rRNA and functional genes
    amplification, 78–79
    community composition, 81
    DNA extraction, 77–78
    enrichment cultures, 79
    pure cultures, 79–81
  isolation
    dilution, liquid media, 69–70
    long-term storage, 71
    pour plates, 68–69
    serial dilution, liquid media, 70–72
    stock culturing, 71
Ammonia-oxidizing archaea (AOA). *See also*
  Marine and coastal AOA and AOB,

distribution determination; Stable isotope probing; Thermophilic AOA and *Thermus thermophilus* denitrification
activity kinetics, O_2 sensor, 77
ammonia oxidation, determination, 74–75
discovery, 57
enrichment, 64–65
growth kinetics, calculation, 73
indices
  growth, 160–161, 163
  mortality, 161, 164
isolation, serial dilution, 72
mineral salt growth medium, 61–62
molecular analysis, 65
primers and PCR conditions, identification, 79–80
role, 172, 175
substrate availability, 207
Ammonia-oxidizing bacteria (AOB). See also Marine and coastal AOA and AOB, distribution determination; Stable isotope probing
activity kinetics
  NO_2^- plus NO_3^- biosensor, 75–76
  O_2 sensor, 76
batch culture, enrichment
  repeated transfers, 64
  steps, 63–64
betaproteobacterial linage, 466
broad-host range plasmids
  gene function complementation, 398–399
  reporter gene constructs, 398
conjugation
  optimal conditions, 392–393
  plasmid constructs, 392
continuous culture, enrichment
  batch culture, 67
  chemostats, set-up, 65–66
  steady state, 66–67
  volume change, 66
cultivation methods, 57
discovery, 56
electroporation, *N. europaea*
  cell pellet, 392
  cell washing, 391–392
  mutant strain production, 391
  vector construct introduction, 391
gene inactivation
  construct, mutagenesis, 393–395
  mutant creation, 395–396
  recombination events, 396–398
growth and selection, 393
growth kinetics, calculation, 73
high-throughput screening approaches, 400
identification, 79–81
index
  growth, 163, 166
  mortality, 164

isolation
  dilution, liquid media, 69–70
  pour plates, 68–69
  serial dilution, liquid media, 70–71
  stock culturing, 71
K_s and μ_{max}, batch culture, 73–74
long-term storage, 71
mineral salt growth medium, 61–62
modification, 390
molecular analysis, 65
mutagenesis, 399
mutant recovery times, 399–400
N_2O production, 274
nutrient-limited cultivation, 59
ocean
  AOA, 308–309
  molecular oxygen, 308
$^{18}O_2$ enrichment, 267
stock culturing, 71
strain stability and maintenance, 399
Ammonia-oxidizing microorganisms (AOM)
batch and continuous culture cultivation
  chemostat, nutrient-limited cultivation, 59
  growth-limiting substrate concentration, 58, 60
  microbes, 57
  set-up, 60–61
determination
  NH_4^+, 81–82
  NO_2, 83
  NO_3^-, 84
  NO_2^- plus NO_3^-, 83–84
enrichment, oxidizers
  batch culture, 63–65
  continuous culture, 65–67
  medium preparation, 67–68
isolation, oxidizers
  dilution, liquid media, 69–70
  long-term storage, 71
  pour plates, 68–69
  serial dilution, liquid media, 70–72
  stock culturing, 71
media
  contamination test media, 63
  mineral salt medium, 61–62
  trace elements solution, preparation, 62–63
prokaryotes, ammonia oxidizers
  activity kinetics, 74–77
  growth kinetics, 72–74
  growth *vs.* activity kinetics, 72
AMO activity, *N. europaea* cultures, 409
Anaerobic ammonium-oxidizing bacteria. See Anammox
Anaerobic oxidation of methane (ANME)
  ANME-2c (see Diazotrophic microorganisms identification, FISH-NanoSIMS)
  and SRB, 290, 295, 297

Subject Index

Anammox. *See also* ^{15}N-labeling experiments, heterotrophic denitrification and anammox
"*Candidatus* Scalindua," 90–91
denitrification activity, 91–92
description, 225
direct tracer-based measurements, 227
enrichment, planktonic cell suspensions
mechanism, 96–97
membrane maintenance, 96
physical separation, cells, 97
reactor operation, 94–96
SBR, 93
sludge, 93–94
substrates and macro and micronutrients, 94
FISH, 91
molecular detection
approaches, 97
core catalytic proteins, 102
functional genes amplification, 103
melting curve analysis, 104
primers, planctomycete and anaerobic ammonium oxidizer, 101
probes, *in situ* hybridization, 98–100
qPCR primers, 102, 104
N-cycle processes, 174
OMZs, 493
Planctomycetes phylum, 90
redox reactions, 92
steps, 489
ANME. *see* Anaerobic oxidation of methane
AOA. *See* Ammonia-oxidizing archaea
AOB. *see* Ammonia-oxidizing bacteria
AOM. *See* Ammonia-oxidizing microorganisms
Atomic force microscopy (AFM), 299

B

Bacteria. *See also* Ammonia-oxidizing bacteria; Nitrite-oxidizing Bacteria
ammonia oxidation, 175
ammonia-oxidizing (*see* Ammonia-oxidizing bacteria)
ammonium-oxidizing, anaerobic (*see* Anammox)
amoA genes, 57
and archaeal, PCR amplification, 195
elimination, enrichment culture, 72
nitrite-oxidizing (*see* Nitrite-oxidizing bacteria)
N$_2$O-production, survey (*see* Nitrous oxide (N$_2$O)-producing pathways, bacteria)
pathogenic, 424
thermophilic, 177
Bacterial cytochrome *c* biogenesis system
CCHL, 433
system II, 433–435

Bacterial cytochrome *c* biogenesis system I (Ccm)
E. coli cells
deficient strain, 439
gene cluster, 435
Bacterial cytochrome *c* biogenesis system II (Ccs)
A1 and *A2* genes, 439
CcsBA, 433–434
Banded iron formations (BIFs), 491
Basic local alignment search tool (BLAST), 135
Batch culture
ammonia oxidizers enrichment, 63–64
centrifuge cells, 117–118
cultivation
chemostat, 58–59
and continuous culture, 57–58
growth-limiting substrate, 58, 60
microbes, 57
issues, 143–144
kinetic fractionation factors, 264
K_s and μ_{max} determination, 73–74
N and O isotope effects, 260–261
set-up, 60
BIFs. *see* Banded iron formations
Bioinformatics
N. europaea, FE-related studies
NCBI, 406
siderophores, 407
TonB-dependent receptor genes, 406–407
Bioinformatics, N$_2$O-production
aerobic hydroxylamine oxidation pathway, 138–140
cytochrome P460, 139–140
HURM, 139
monooxygenase, 138
BLAST, 135
characterized inventory, 136–137
detoxification inventory, 141
dissimilatory reduction, nitrite
classes, 140
NORs, 140–141
qNOR and sNOR, 141
true denitrifying bacteria, 138, 140
gene function, 135, 138
isotopic techniques, 135
motifs and regulatory sequence identification algorithms, 142
promoters, 142–143
Biological pump
anthropogenic CO$_2$, 6
description, 5
DIC, 4–5
nutrient supply, 7
BLAST. *See* Basic local alignment search tool
Broad-host range plasmids, *N. europaea*
gene function complementation, 398–399
reporter gene constructs, 398

C

Calorimetry
  description, 45
  isothermal, 45–46
  nitrogen cycle, 50
  sensitivity and resolution, 46
Calvin cycle, 309
Candidatus Kuenenia, 90–91, 99, 101
Candidatus Scalindua
  detection, 90
  enrichment, 91, 95
Catalyzed reporter deposition–fluorescence *in situ* hybridization (CARD–FISH), 208
CCHL. *see* Cyctochrome *c* heme lysase
Chelex, 407
Chemolithoautotroph, 92, 225–226
  N. europaea, 404
Chemostat
  ammonia oxidizers, 58
  AOB enrichment, 65–66
  cultivation, 58
  and natural ecosystems, 44
  nutrient-limited cultivation, AOB, 59
  O_2 partial pressure, 61
  stirrer systems, 60–61
Complementation
  gene function, 398–399
Concentration determination, NH_2OH disproportionation
  ammonia
    assay reaction, 454
    GDH, 456
    HAO concentrations, 456
    materials, 454–455
    principle, 454
    procedure, 455–456
  nitric oxide
    HAO concentrations, 458
    myoglobin, 458
    principle and materials, 457
    procedure, 457
    weakness, 458
  nitrite
    assay reaction, 459
    materials, 459
    principle, 458–459
    procedure, 459–460
    sensitivity, 460
  nitrous oxide and dinitrogen
    apparatus, detection, 461
    gases solubilities, 462
    materials and procedure, 460–461
    principle, 460
Conjugation, AOB transformation, 392–393
Continuous cultivation and thermodynamic aspects, nitrogen cycle
  experimental approach
    bioenergetic efficiency, 45
    bioinformatic pipeline, 49
    calorimetric sensitivity and resolution, 46
    chemostat, 43
    data, 46–47
    dilution rate, 44
    isothermal calorimetry, 45–46
    metagenomic sequencing, 49
    peristaltic pumps, 48
    pressure sensing, 49
    pseudo steady state, 44
    "selection" and "evolution," 45
    steady state, 43–44
    theoretical heat effect, gas mixing, 47
    vessels design, 47–48
  hypotheses generation, 35
  mathematical modeling, 50
  microbial processes, 33–34
  minimal model, microbial behavior
    cellular respiration, 37
    concentration matrixes, 41–42
    denitrification, 43
    metabolic pathways, 40
    Monod kinetics, 39, 41
    oxygen influx, 42
    periodic high substrate concentrations, 36–37
    respiratory chains, nitrogen cycle processes, 38
    steady state, 35–36
    substrate concentration, 35
    substrate influxes, 36
Continuous culture
  AOB enrichment, 65–67
  and batch culture, 43
  chemostat, 58
  microbes, 57
  set up, 60–61
  volumes, 46
Cultivation. *See also* Continuous cultivation and thermodynamic aspects, nitrogen cycle
  NOB
    dissimilatory nitrate reduction, growth, 116
    incubation, 114
    reagents, Griess–Ilosvay spot test, 115
    storage, 115–116
Cyctochrome *c* heme lysase (CCHL)
  CcsBA-type enzymes, 434
  thioether linkage, 433
  W. succinogenes cells, 439
Cytochromes *c*
  biogenesis systems
    cyctochrome *c* heme lysase (CCHL), 433–435
    epsilonproteobacteria, 434
  c'-beta, 140
  characterization, 430, 433

Subject Index

complication, respiratory N-cycle processes, 431–432
recombinant production, 435
Wolinella succinogenes
 electron acceptors, 435–436
 genetic manipulation, 436–438
 heme lyase isoenzymes, 439
 heterologous production, 439–442
 respiratory nitrate ammonification, 438

D

Data processing
 N_2 peak areas to concentrations
  $^{29}A_N$, 244–245
  isotopic composition, 244
  $^{29}N_2$ and $^{30}N_2$, 246
  systematic error, 245
 peak areas, 243
 rate calculations and interpretations
  ammonium profile, lower depth, 248
  anammox, 247–248
  Exetainers, 246–247
  incubation, 246
  $^{29}N_2$ and $^{30}N_2$ concentrations, 247
Deep-sea, diazotrophic symbiotic archae, 284–285
Denitrification
 and anammox, heterotrophic (*see* ^{15}N-labeling experiments, heterotrophic denitrification and anammox)
 canonical, 224
 chemolithotrophic, 232
 classical, 133
 method (*see* Denitrification measurement, methods)
 N_2 isotopes, 230
 and nitrification, 485
 "nitrifier–denitrification," 274
 nitrogen cycle, 488–489
 nonammonia-oxidizing organisms, 134
Denitrification measurement, methods
 acetylene block (*see* Acetylene block method)
 $^{15}NO_3^-$ tracer approach
  anaerobic sediment slurry amendment, 191–192
  analysis and calculations, 193
  incubation, 192
  preparation, 192
Diazotrophic microorganisms identification, FISH-NanoSIMS
 AFM/SEM, 299
 ANME-2c
  identification, 285
  sulfate-reducing bacterial symbiont, 284–285
 cells, density gradients and deposition
  FACS, 289

Percoll, 288–289
FISH, microprobe slide-deposited cells
 CARD-FISH, 289–290
 image samples, 290–291
mapping sample targets
 cell selection, 291–293
 description, 291
 map creation, 293–294
metal cofactors, nitrogenase, 300
microprobe slides, sample deposition
 conductive materials, 287–288
 epifluorescence and transmitted light, 287
NanoSIMS (*see* NanoSIMS)
NanoSIMS instrument analysis, 295–297
nitrogen, 282
nitrogen fixation (*see* Nitrogen fixation)
^{15}N-labeling incubations
 methane seep and anoxic media, 285–286
 $^{15}N_2$, 286–287
sample
 analysis, 296
 preparation, 296
 sectioning, 294–295
 standards, 296
 subsampling and preservation, 287
DIN. *See* Dissolved inorganic nitrogen
Disproportionation. *see* Hydroxylamine disproportionation, HAOs
Dissimilatory nitrate reduction to ammonia (DNRA)
 anammox rates, 231
 and denitrification, 43
 description, 226
Dissolved inorganic nitrogen (DIN) inventory approach
 inhibitors
  ammonia/nitrite oxidation, 312
  hydrocarbons, 313
  problems, 312–313
 mesocosms, 312
 nitrification, 310, 312
 occurrence, 8
DNRA. *See* Dissimilatory nitrate reduction to ammonia

E

Ecosystem models
 marine, 15
 non-nitrogen currency, 16
 physical model, 22
 plankton
  biogeochemical cycles, global scale, 12
  diagrammatic representation, 26
  diversity, 28
  1D models, 14
  identical configuration, 14–15
  mixed layer, 13

Ecosystem models (cont.)
  NPZ systems, 12–13
  "pools," 13
  temperature-and oxygen-dependent functionality, 21
Electron microscopy
  culture, monitoring, 123–124
  ICM, presence and arrangement, 122
  N. europaea cells, 415
Environmental sample processor (ESP), 215
Epsilonproteobacteria, CcsBA, 433–434
ESP. *See* Environmental sample processor
Export production
  DIN, 8
  geographical distribution, 10
  nitrate and ammonium, 8–9
  organic material, 7–8
  parameterized ecosystem models, 12
  pulse-chase experiment, 9
  "Redfield ratio," 7
  specific rate, 10–11
  surface nitrate, 11

F

Fatty acid methyl esters (FAMEs)
  extraction, 126
  identification, 127
Fatty acid profiles
  Alphaproteobacteria, 124
  cultivated strains and enrichment cultures, *Nitrospira*, 126
  FAME extracts, 127
  nitrite-oxidizing genera, 125
  Nitrospira, 125
  Nitrotoga, 124
  pellets, 126
Fe-ABC transporter genes, 406
Fe-homeostasis
  banded iron formations (BIFs), 491
  N. europaea, 405–406, 424
FISH. *See* Fluorescence in situ hybridization; Fluorescence *in situ* hybridization
Fluorescence-activated cell sorting (FACS), 289
Fluorescence in situ hybridization (FISH)
  anammox bacteria
    quantification, 91
    16S rRNA, 98
  AOA and AOB, quantification, 214
  CARD–FISH, 208
Fluorescence *in situ* hybridization (FISH). *See also* Diazotrophic microorganisms
  identification, FISH-NanoSIMS; Nitrogen fixation
  microprobe slide-deposited cells
    CARD-FISH, 289–290
    image samples, 290–291
  and NanoSIMS images correlation, 297
Fur titration assay (FURTA), 421

G

Gene inactivation, AOB
  construct, mutagenesis
    allelic exchange, 395
    antibiotic resistance cassette, 394–395
    DNA, PCR-nested extension, 396
    PCR, 394
    plasmid, 393
    recombinant strain, 393–394
  mutants creation, two genes
    double mutagenesis, 395
    hao strains, 396
    single-gene inactivation, 395–396
  recombination events, screening and confirmation
    antibiotic resistance cassette, 398
    genomic DNA, 396–397
    PCR corroboration, 397
Genomic DNA preparation, 396–397
Griess–Ilosvay spot test, 115
Growth
  vs. activity kinetics, 72
  AOB, 393
  experiment, siderophores
    gene inactivation, 419
    incubation, 419
    0.2 μM Fe-medium, 418
  index, AOA and AOB, 160–161, 163–164
  kinetics, ammonia oxidizers
    calculation, 73
    environmental factors, 74
    K_s and μ_{max}, 73–74
    medium preparation, 72
  Liebig's law, 17
  N. europaea cell and analysis, Fe limitation
    ammonia oxidation, 408–409
    AMO activity, 409
    sonication, 409–410
  N. eutropha C91 pure culture, 469–470
  nitrifier, 27
  NOB
    dissimilatory nitrate reduction, 116
    mineral salts medium, lithoautotrophic, 113
    parameters, 120
  phytoplankton, 23
Growth medium
  mineral salt
    AOA and AOB, 62
  N. europaea, $(NH_4)_2 SO_4$, 392

H

HAO. *see* Hydroxylamine oxidoreductase
Heme c binding motifs (HBMs)
  CX_2CH, 439
  description, 430
  W. succinogenes, 437
Heme *c* packing motif, 430

Subject Index 539

Hydroxylamine disproportionation, HAOs
  anaerobicity
    Argon, 450–451
    bubbling gas problem, 451–452
    glovebox, 450
    inert gas, 452–453
    "Thunberg tube"/"Thunberg cuvette," 451–453
  concentration determination
    ammonia, 454–456
    nitric oxide, 457–458
    nitrite, 458–460
    nitrous oxide and dinitrogen, 460–462
  data analysis, spectral information, 453–454
  definition, 448
  HAO, 448–449
  NH_2OH, 448, 450
Hydroxylamine oxidation
  aerobic pathway, 138–140
  N_2O, 144
Hydroxylamine oxidoreductase (HAO)
  AOB, 308
  disproportionation
    ammonia, 454–456
    anaerobicity, 450–453
    data analysis, spectral information, 453–454
    definition, 448
    NH_2OH, 448, 450
    nitric oxide, 457–458
    nitrite, 458–460
    nitrous oxide and dinitrogen, 460–462
  N. europaea cultures, 409, 487
  pentaheme nitrite reductase and, 139
Hydroxylamine-ubiquinone redox module (HURM), 139

I

ICP-MS. see Inductively coupled plasma mass spectrometer
Incubation, ^{15}N
  $^{15}NH_4^+$ addition, 229
  nitrification detection, 232–233
  $^{15}NO_2^-$ addition
    anammox, 229
    denitrification, 229–230
    DNRA, 231
  $^{15}NO_3^-$ addition, 232
  $^{15}NO_2^-$ and $^{14}NH_4^+$ addition
    anammox rates, 232
    $^{30}N_2$ production, 231
Inductively coupled plasma mass spectrometer (ICP-MS), 498
Inhibition of ammonia oxidation, C_2H_2, 356
IRMS. See Isotope ratio mass spectrometry
Iron (Fe) limitation. see also N. europaea physiological responses, Fe limitation physiological responses
  cell growth and analysis, 408–410
  cellular content and allocation, 410–411
  free medium and glassware preparation, 407–408
  gene expression, microarrays, 415–417
  heme c content, 411–412
  microscopy, cell structure, 414–415
  protein identification, 412–414
  real-time qPCR, gene expression, 417–418
Iron uptake and homeostasis, Nitrosomonas europaea
  advantage, 406
  bioinformatics use
    genes, 406
    OM transducer/receptors, 407
    TonB-dependent receptor genes, 406–407
    TonB-dependent transducer systems, 406
  fur homologs, genetic complementation
    E. coli H1780 strain, 421–422
    FURTA, 421
  gene inactivation methods
    electroporation, 422–423
    fur:kanP mutant strain, 423
    transformation protocols, 422
  limitation, physiological responses
    cell growth and analysis, 408–410
    cellular content and allocation, 410–411
    free medium and glassware preparation, 407–408
    gene expression, microarrays, 415–417
    heme c content, 411–412
    microscopy, cell structure, 414–415
    protein identification, 412–414
    real-time qPCR, gene expression, 417–418
  mechanisms, 405
  NE1205, 423
  regulatory mechanism
    putative Fur-regulated sRNA (psRNA11), 424–425
    SDH, 424
  siderophore-independent
    multicopper oxidases, 423–424
    NE2323, 424
    TonB-dependent heme receptor (NE1540), 424
  siderophore uptake
    fluorescent analog and cell localization, 419–421
    growth experiment, 418–419
    labeled Fe feeding experiments, 421
Isolation
  AOA, serial dilution, 72
  AOB
    dilution, liquid media, 69–70
    pour plates, 68–69
    serial dilution, 70–71
  NOB
    density gradient centrifugation, 117–118

Isolation (cont.)
  dilution series, 117
  enrichment, 116
  plating technique, 117
Isotope ratio mass spectrometry (IRMS)
  denitrifier method, 185
  N composition, N$_2$O, 193
  ^{15}N-NO$_x$ determination, 180, 184
Isotopes
  N$_2$, 230
  N and O, nitrite oxidation, 265
  nitrogen cycle (see Isotopes and metal, nitrogen cycle)
  O isotope systematics, 268
  stable, marine N cycle, 255
Isotopes and metal, nitrogen cycle
  δ^{15}N signature, site selection and origin
    Cenozoic and Mesozoic, 492
    NO$_3^-$ residence time, 493
    shale, 492–493
    terrestrial OM, 493
  preservation, δ^{15}N signals
    catagenesis and metamorphism, 494–495
    diagenesis, 494
    sedimentary organic matter (SOM), 493–494
  trace metals, 495–497
Isotopic fractionation
  analyses
    concentration, 262
    δ^{15}N$_{NH3+NH4}$, 262
    δ^{15}N$_{NO2}$ and δ^{18}O$_{NO2}$, 263
    δ^{15}N$_{NO3}$ and δ^{18}O$_{NO3}$, 263–264
    δ^{18}O$_{H2O}$, 264
  archaea, isotope effects, 275
  bond-breaking and bond-forming reactions, 256
  data analysis
    ammonia and nitrite oxidation, 264–265
    exchange, ammonia oxidation, 265–267
    mixed cultures and field populations, 269
    nitrite oxidation, 268–269
    O$_2$ enrichment, 267–268
    tracking, nitrification, 270–272
  expression, closed system, 257
  H$_2$O and O$_2$ incorporation
    archaeal ammonia oxidation, 274
    first-order prediction, 273
  kinetic fractionation factors, 255–256
  N and O
    ammonia oxidation, 259–260
    nitrite oxidation, 260–261
  nitrification and δ^{18}O$_{NO3}$
    bacterial, 257
    and exchange, 258
    O atom donors, 258
  nitrification and marine N cycle
    ammonia oxidation, 254–255

  N$_2$O source, 255
  steps, 254
  N$_2$O production, 274
  O exchange, mixed culture and field populations, 261–262
  Rayleigh model
    evolution, isotope ratio, 256–257
    kinetic isotope effect, 272
    NH$_2$OH, 273
    transport balance and enzyme-level effects, 272–273
  stable isotopes, marine N cycle, 255

K

Kinetics
  ammonia oxidizers
    activity, 74–77
    growth, 72–74
    growth vs. activity, 72
  Monod, 39 42
  nitrification
    cell-free enzyme assays, 357
    Michaelis–Menten constants, 358–361
    potential assays, 357–358

L

Labeled Fe feeding experiment, 421
Labeling experiment
  ^{13}C, 284
  ^{15}N, 284 (see ^{15}N-labeling experiments, heterotrophic denitrification and anammox)
  ^{15}NO$_3^-$, 232
LC-MS/MS
  C18 reversed phase, nitrosomonas proteomics
    acetonitrile, 471
    ICR cell survey scan, 471–472
    peptide elution, 471
    tryptic peptides, 472
  instrument setup, nitrosomonas proteomics
    advantages, 468–469
    nanoelectrospray, 468–469
    peptide separation, 468
Lipid marker
  fatty acid 16:1 cis-11, 125
  nitrite-oxidizing genera, 124
  Nitrospira species, 125

M

Marine and coastal AOA and AOB, distribution determination
  abundance measurement methods
    amoA genes and transcripts, quantification, 210–212
    DNA extraction, 209
    PCR screening and gene sequencing, 210

Subject Index

RNA extraction and cDNA synthesis, 209–210
sample collection, 209
target genes, ammonia oxidizers, 212
ammonia oxidizers, 206
amoA genes, 206–207
amoA qPCR primers, 208
crenarchaea, 207–208
ecotypes, target
  group A and B, 214–215
  Gulf of California, water column, 215
  primers and probes, *amoA* assays, 216
  shallow and deep clades, 217
methodological considerations
  bead-beating times, 213
  FISH, 214
  qPCR assay, 213–214
nitrification, 206
substrate availability, 207
Membrane bioreactor (MBR)
  nutrition and growth conditions, anammox, 94
  physical separation, anammox cells, 97
Metagenomics
  community members, 34
  "experimental metagenomics," 50
  sequencing, 49–50
  targeted molecular approaches, 45
Metals. *see also* Nitrogen (N) cycle
  cofactor, nitrogenase, 300
  dissimilatory nitrite reductase, 140
Metamorphosis and catagenesis, 494–495
Microarrays
  gene expression, 415–417
  hybridization, 147
Microscopy
  cell structure, 414–415
  electron, 123–124
  phase contrast, 98, 122
Modeling. *See also* Nitrification, open ocean productivity
  N and O isotope deviations, 271
  nitrification
    ammonium uptake, 27
    elemental fluxes, 25
    "nitrifying microbes" state variable, 26
    ocean ecosystem models, 24–25
Mutagenesis, 134
  AOB, gene construct, 393–395

N

^{15}N. *see also* N isotope methods
  depleted nitrification, 270
  gross nitrification rate, 352–355
  isotope dilution technique
    colorimetric analysis, 355
    description, 352
    gross rate measurement, 353–354
  nitrification measurement, 352–353
  soil sample, 353
  use validation, 354–355
  labeling experiments (*see* ^{15}N-labeling experiments, heterotrophic denitrification and anammox)
  labeling incubations
    methane seep and anoxic media, 285–286
    ^{15}N$_2$, 286–287
  nitrate assimilation, 271
NanoSIMS
  analysis
    advantage, 292
    plane geometry, 296
    sample preparation, 296
    standards, 296
    targets, 291
  results, interpreting
    flat and polished samples, 298
    Matlab scripts, 297
    quantum dots (QD), 297
    software programs, 297
    Z-stack images, 298
N. europaea physiological responses, Fe limitation
  cell growth and analysis
    ammonia oxidation, 404, 408
    NH$_3$-dependent O$_2$ uptake activity, 409
    NO$_2^-$, 408–409
    optical density, 408
    sonication, 409–410
  cell structure, 414–415
  cellular content and allocation
    Fe-replete medium, 410
    ferrozine assay, 410–411
  free medium and glassware preparation
    reagent-grade chemicals, 407–408
    ultra-pure water, 407
  gene expression, microarrays
    DNASTAR, 416
    NE1540/1539/1538, 416–417
    NE0508, NE0509 and NE0510, 417
    NimbleChip 4-plex Made-to-Order microarrays, 415
  heme *c* content
    pellets, 412
    ultracentrifugation, 411–412
  protein identification
    Fur and small RNAs, 412–413
    LC/MS/MS analysis, 413–414
    SDS-PAGE analysis, 413
  real-time qPCR, gene expression
    vs. microarrays, 417
    RNA, 417–418
^{15}N isotope dilution technique
  colorimetric analysis, 355
  description, 352
  gross rate measurement, 353–354
  nitrification measurement, 352–353

^{15}N isotope dilution technique (*cont.*)
  soil sample, 353
  use validation, 354–355
N isotope methods
  description, 314–315
  $^{15}NO_2^-$ analysis
    N_2O, 317
    $^{15}NO_3^-$, N_2O, 317
    solute extraction, 316
  ^{15}N, seawater ammonia and nitrite oxidation
    ambient substrate concentration, 315
    incubations, 315
    isotope tracer method, 316
    rate calculation, 317–318
    tracer/dilution approach, 314
Nitric oxide (NO)
  NH_2OH disproportionation
    HAO concentrations, 458
    myoglobin, 458
    principle and materials, 457
    procedure, 457
    weakness, 458
  reductase gene, 212
Nitrification
  description, 344–345
  distribution, ocean
    incubation experiments, 319–320
    nitrifiers, 319
    sediment and coastal systems, 320
  diversity, terrestrial environment, 345
    ammonium anaerobic oxidation, 346
    chemolithoautotrophic bacteria, 345
    heterotrophic, 345–346
  kinetics, measurement
    cell-free enzyme assays, 357
    potential assay (*see* Nitrification potential assay)
  method (*see* Nitrification measurement, methods)
  modeling approaches, 361–362
  nitrifier population size, 361
  nitrogenous trace gases, 345
  plant–soil system, 344
  rate (*see* Nitrification rate)
  reaction, substrates and products, 346
Nitrification inhibitors
  gross rate
    acetylene (C_2H_2), 355–356
    chemical, 355
    nitrapyrin, 356
    NO_3^- consumption, 356–357
Nitrification measurement, methods
  atom% ^{15}N-NO_x determination, 184
  calculation, gross nitrification rate, 186–187
  denitrifier method, ^{15}N-NO_x analysis
    description and culture, *Pseudomonas aureofaciens*, 185
    procedure, 185–186
  diffusion technique, ^{15}N-NO_x analysis
    procedure steps, 184
    PTFE tape, 184–185
  $^{15}NO_3^-$ pool dilution approach (*see* $^{15}NO_3^-$ pool dilution)
  NO_x and $NH4^+$ quantification
    Berthelot reaction, 183
    sample preserves, 183–184
  pool dilution technique, 179–180
Nitrification, open ocean productivity and nitrogen cycle
  ammonia-oxidizing organisms, 23
  ammonium uptake, 27
  carbon cycle
    biogeochemical cycles, 7
    CO_2 emissions, 4
    geographical distribution, primary production, 6
    nutrient availability, 5–6
    "pumps," 4–5
  elemental fluxes, 25
  export production and *f*-ratio
    DIN, 8
    geographical distribution, 10
    nitrate and ammonium, 8–9
    organic material, 7–8
    parameterized ecosystem models, 12
    pulse-chase experiment, 9
    "Redfield ratio," 7
    specific rate, 10–11
    surface nitrate, 11
  modeling
    ambient temperature, 20
    ammonium concentration, 19–20
    euphotic zone nitrification, 20
    functions, 18–19
    generic NPZD model, 15
    Liebig's law, 17
    partial differential equations, 16
    processes and parameters, 16–17
    simulated distribution, 21
  "nitrifying microbes" state variable, 26
  ocean ecosystem models, 24–25
  operational considerations
    checking, 24
    numerical method, biological equations, 23
    physical framework, 22
    temporal evolution, 23
  plankton ecosystem models
    biogeochemical cycles, global scale, 12
    1D models, 14
    identical configuration, 14–15
    mixed layer, 13
    NPZ systems, 12–13
    "pools," 13
Nitrification potential assay
  description, 357–358
  method for, 359

Subject Index

Michaelis–Menten constants
   function, 359
   Haldane function, 359–360
   NH_4^+ concentrations, 358–361
   $^{15}NO_3^-$ isotope dilution, 360–361
Nitrification rate
   ammonia addition, 158
   control, soil environment
      acidity and alkalinity, 350
      ammonia sensitivity, 348
      oxygen, water potential and temperature, 348–349
      substrate availability, 347–348
   ecosystems, terrestrial
      inhibitors, gross, 355–357
      ^{15}N, gross, 352–355
      NO_3^-, 351
   gross, calculation, 186–187
   measurement, sediments
      inhibitor approach, 318–319
      ^{15}N approach, water sample, 318
      slurries, 319
   N_2O accumulation, 255
   ocean
      AOB, 308–309
      DIN inventory approach (see Dissolved inorganic nitrogen (DIN) inventory approach)
      inhibitors, radioisotopes, 313–314
      N isotope methods, 314–318
      NOB, 309–310
   open ocean measurements, 11
   terrestrial environment
      advancements, 362
      gross and potential, 351
      net, 350–351
   variability, 25
Nitrite (NO_2^-)
   aerobic oxidizers, 36
   ammonia conversion, 206
   and ammonia oxidation, sea water, 315–316
   ammonification, 226
   anaerobic oxidation, ammonium, 90
   anammox activity test, 96
   concentration profiles, 248
   dissimilatory reduction pathway, 140–141
   NH_2OH disproportionation
      assay reaction, 459
      materials, 459
      principle, 458–459
      procedure, 459–460
      sensitivity, 460
   NOB (see Nitrite-oxidizing bacteria)
   oxidation, 175–176
      and ammonia, seawater, 315–316
      $\delta^{18}O_{NO3}$, 267
      N isotopic fractionation, 260, 264–265
      Nitrospira, 115
      O isotopic fractionation, 260, 268–269
      temperature dependence, 121
   oxidizing bacteria (see Nitrite-oxidizing bacteria)
   photometric method, 236
   phototrophic oxidizers, 34
Nitrite-oxidizing activity
   measurement, 118
   potential, 119
Nitrite-oxidizing bacteria (NOB), 175–176
   aerobic growth, 310
   and AOB, 310
   chemical analyses, 121
   cultivation procedure
      growth, dissimilatory nitrate reduction, 116
      incubation, 114
      reagents, Griess–Ilosvay spot test, 115
      storage, 115–116
   culture, recipes
      acidic medium, 114
      heterotrophic medium, 113–114
      marine medium, 113
      media preparation, 112
      mineral salts medium, lithoautotrophic growth, 113
      mixotrophic medium, 113
      purity tests, 114
   fatty acid analyses
      Alphaproteobacteria, 124
      cultivated strains and enrichment cultures, Nitrospira, 126
      FAME extracts, 127
      nitrite-oxidizing genera, 125
      Nitrospira, 125
      Nitrotoga, 124
      pellets, 126
   investigations
      cultivation conditions, 111–112
      isolates, 110
      nitrite concentration, 111
   isolation
      density gradient centrifugation, 117–118
      dilution series, 117
      enrichment and end-point-dilution, 116
      plating technique, 117
   monitoring, cultures
      acid biomass hydrolysis, 123
      cell shape and ultrastructure, 122
      electron microscopy, 123–124
      genera, 121
      Nitrospira, morphology change, 123
   nitrobacter, 309
   physiological investigations
      growth parameters, 120–121
      marine biofilter, colonized plastic biocarriers, 119
      permafrost soil, nitrifying activity, 119–120

Nitrite-oxidizing bacteria (NOB) (*cont.*)
   potential nitrite-oxidizing activity, 119
   short-term nitrifying activity, 118
  proteobacterial, 310
Nitrogen
  cycle (*see* Nitrogen cycle)
  DIN (*see* Dissolved inorganic nitrogen)
  fixation (*see* Nitrogen fixation)
  inorganic inventory approach, 310–312
  isotopes (*see* Nitrogen isotopes, N-cycle)
  liquid
   AOB storage, 71
   freezing, 116
   trap, 243
Nitrogen (N) cycle *See also* Continuous cultivation and thermodynamic aspects, nitrogen cycle; Nitrification, open ocean productivity and nitrogen cycle
  activity measurement, terrestrial geothermal habitats
   description, 177–178
   heterogeneity, 178
   hot spring research, 179
   sample and incubation site, 178
  changes, N isotopes and metal
   δ^{15} N signature, site selection and origin, 492–493
   preservation, δ^{15} N signals, 493–495
   trace metals, 495–497
  evolution
   biological processes, 484–485
   O_2, 485
  marine
   nitrification, 254–255
   stable isotopes, 255
  measurement
   metals, 498
   nitrogen isotopes, 497–498
  metal requirement
   anammox, 489
   cofactor, 485
   denitrification, 488–489
   metalloenzymes, 486
   nitrification, 487
   redox reactions, 487
  microbial, 133
  processes, 172–174
  reconstruction and trace metals
   denitrification, 492
   isotopes, 489
   Mo and Fe, 490–491
   physicochemical transformations, 490
Nitrogen fixation
  FISH-NanoSIMS
   acetylene reduction assay, 283–284
   isotope probing, $^{15}N_2$-DNA-SIP, 284
   molecular approaches, 284
   global marine nitrogen cycle, 283

   Tricodesmium, 282–283
Nitrogen isotopes, N-cycle
  changes and metal
   δ^{15} N signature, site selection and origin, 492–493
   preservation, δ^{15} N signals, 493–495
   trace metals, 495–497
  measurement
   CF-IRMS, 497–498
   dual inlet (DI) IRMS, 498
   in silicates, 498
Nitrogen removal. *See* ^{15}N-labeling experiments, heterotrophic denitrification and anammox
Nitrosocaldus, 72
  "*Candidatus* N. yellowstonii"
   protocol, primers targeting putative *amoA*, 195–197
Nitrosomonas europaea. see also *N. europaea*
  physiological responses, Fe limitation
   ammonia oxidation, 404
   gene encoding NorS, 139
   N. eutropha C91 (*see Nitrosomonas eutropha C91*)
   NH_2OH, 267
Nitrosomonas eutropha, 57
Nitrosomonas eutropha C91
  alkyl hydroperoxide reductase, 478
  genome sequence, 466–467
  growth, 469–470
  ion map view, 473
  LC-MS/MS, 474
  sequences, 472
Nitrosomonas proteomics
  ammonium, 466
  C18 reversed phase LC–MS/MS analysis
   acetonitrile, 471
   ICR cell survey scan, 471–472
   peptide elution, 471
   tryptic peptides, 472
  dataset description and protein identification
   characteristics, 477–478
   N. eutropha C91, 476–477
   peptide, 477, 479
  LC–MS/MS instrument setup
   advantages, 468–469
   nanoelectrospray, 468–469
   peptide separation, 468
  N. eutropha C91
   genome sequence, 466–467
   pure culture growth, 469–470
  result, database searches and validation
   decoy, 475
   engines, 472
   LC-MS/MS dataset, 474
   Mascot and delta score, 475–476
   N. eutropha C91 sequences, 472–473
   parent ion *m/z* errors, 473–474
   PROTON, 475

Subject Index

software packages, 475
spectral count method, 476
sample preparation, 470
"LysC," 470
N. eutropha cell centrifugation, 470
STAGE, 470–471
Nitrous oxide. *see also* N isotope methods
and dinitrogen concentration determination
  apparatus, detection, 461
  gases solubilities, 462
  materials and procedure, 460–461
  principle, 460
fluxes, WWTPs (*see* Nitrous oxide (N_2O) fluxes, WWTPs)
Nitrous oxide (N_2O)-producing pathways, bacteria
aerobic production, 133–134
bacterial isolates, *in silico* identification, 134
bioinformatics
  aerobic hydroxylamine oxidation pathway, 138–140
  BLAST, 135
  characterized inventory, 136–137
  detoxification inventory, 141
  dissimilatory reduction, nitrite, 140–141
  gene function, 135, 138
  isotopic techniques, 135
  motifs and regulatory sequence identification, 142–143
industrialization, 132
microbial nitrogen cycle, process, 133
Nitrosopumilus maritimus, 133
pathogenic microbiologists, 134–135
surveying function
  activity and gene expression, 146–147
  batch culture studies, 143–144
  following N, 144–146
  isotopic tracers, 143
Nitrous oxide (N_2O) fluxes, WWTPs
activated sludge processes, emission
  aerated and nonaerated zones, 382
  computed flow-normalized factors, 382
  range, 381–382
advective gas flow rate, 376–377
aqueous concentration, 376
determination, data analysis, 377–378
emission factor calculation
  aerobic and anoxic zones, 379
  USEPA inventory report, 378–379
emission fractions, 378
full-scale monitoring
  BNR and non-BNR, 370–373
  gas-phase, 374
headspace gas measurement
  EPA/600/8–86/008 and SCAQMD, 373
  gas flows, 376
  sampling, 373

SEIFC, 373, 375
lab-scale and field-scale adaptation, emission, 383
protocol standardization and comparison
  "biasing," 379
  gas flow rates, 379
  inert gas tracer, 380–381
  spatial and temporal variability, 381
real-time measurement principle, 377
triggers, emission, 382
^{15}N-labeling experiments, heterotrophic denitrification and anammox
data processing and interpretation
  N_2, peak areas to concentrations, 244–246
  peak areas, 243
  rate calculations and interpretations, 246–248
incubation
  degassing, 239
  Exetainers, 239–240
  helium, 237, 239
  microbial activity, Exetainers, 240
  $^{15}NH_4^+$ addition, 229
  nitrification detection, 232–233
  N-loss pathways, 228
  $^{15}NO_2^-$ addition, 229–231
  $^{15}NO_3^-$ addition, 232
  $^{15}NO_2^-$ and $^{14}NH_4^+$ addition, 231–232
  N-substrates, 227
  pump-CTD, 237
  sample transfer, 237
  tracer-based measurements, 227
mass spectrometry measurement
  inlet systems, 241–242
  mass spectrometers, 241
  N_2 isotope composition, 240
  preparation line, 242–243
  sample, handling, 243
NH_4^+, NO_2^- and NO_3^-, sampling
  CTD-Rosette, 235
  fluorometric method, 236
  photometric method, 236–237
OMZ, boundary detection
  density and oxygen profiles, 233–234
  nutrient profiles, 235
  oxygen sensing, 233
pitfalls
  anaerobic incubations, 248
  headspace setting, 249
  microbial processes, 248–249
$^{15}NO_3^-$ pool dilution
incubation, 182–183
preparation, 181
protocol, 180
NOB. *See* Nitrite-oxidizing bacteria
N_2O measurement protocol, 377

O

Ocean
 Candidatus Scalindua, 90–91
 carbon cycle, 4–7
 crenarchaea, 207–208
 ecosystem models, 24
 "export production," 7
 nitrification measurements, 10
 nitrification rates
  AOB, 308–309
  DIN inventory approach (see Dissolved inorganic nitrogen (DIN) inventory approach)
  inhibitors, radioisotopes, 313–314
  N isotope methods, 314–318
  NOB, 309–310
  sediments, 318–319
 nitrogen cycle, 9
OMZ. See Oxygen minimum zone
^{18}O-water
 ammonia pools, 165
 Escherichia coli, 157
 growth index, AOA/AOB, 160–161
 SIP, ammonia availability, 158
 soil incubation, 159
Oxygen determination, Winkler titration, 339–340
Oxygen isotopes, C-113a, 274
Oxygen minimum zone (OMZ)
 anammox, 225
 nitrogen removal, ocean (see ^{15}N-labeling experiments, heterotrophic denitrification and anammox)
 occurrence, 224
 oxygen-free, 226
 SBE43 sensor, 339–340
 STOX oxygen sensor in situ deployment
  CTD systems, 337–338
  oxygen measurement, 338–339

P

Percoll density gradient centrifugation, 98
Plankton ecosystem models
 biogeochemical cycles, global scale, 12
 diagrammatic representation, 26
 diversity, 28
 1D models, 14
 identical configuration, 14–15
 mixed layer, 13
 NPZ systems, 12–13
 "pools," 13
Planktonic cell suspensions, anammox bacteria enrichment
 mechanism
  aggregated cells, 96
  calcium and magnesium, concentrations, 96–97
  yeast extract and low shear stress, 97
  membrane maintenance, 96
  physical separation, cells, 97
 reactor operation
  activity test, ammonium and nitrite, 96
  freshwater species, 95
  MBR, 94
 SBR, 93
 sludge, 93–94
 substrates and macro and micronutrients, 94
Protein
 anammox metabolism, 102
 BCA method, determination, 120
 flavorubredoxin, 141
 identification
  and dataset, 476–479
  gel electrophoresis and MS, 412–414
  by MS, 413–414
  by SDS-PAGE, 413
Proteomics
 community members, selection, 34
 metagenomic sequencing, 49
 Nitrosomonas (see Nitrosomonas proteomics)
PROTON, 475

Q

Quantitative PCR (qPCR)
 amoA qPCR primers, 208
 archaeal and bacterial amoA gene, 210
 Clade-specific analysis, 215
 efficiency, 210–211
 experiment set up, 213
 and PCR
  "Candidatus Nitrosocaldus yellowstonii," 195–197
  primers targeting putative amoA, 195–197
  protocol, 196
  T. thermophilus, 196–197
 PCR mixture and reaction protocol, 102–103
 premixes, 213
 protocol, 211–212
 quantification, anammox bacteria, 104

R

Real-time qPCR, gene expression, 417–418
Reporter gene constructs, 398
Reversed phase LC–MS/MS analysis, C18, 471–472

S

Scanning electron microscopy (SEM), 299
Siderophore uptake
 fluorescent labeled analog and cell Fhu-NI, 419–421

Subject Index

growth phenotype, 420
OM receptor, 419
growth experiment
  cell, 0.2 μM Fe-medium, 418
  gene inactivation, 419
  incubations, 419
labeled Fe feeding experiments, 421
transducer/receptors, 419
SIP. *See* Stable isotope probing
South Coast air quality management district (SCAQMD), 373
Stable isotope
^{15}N abundance, 227
probing (*see* Stable isotope probing)
ratio measurements, 271
Stable isotope probing (SIP)
AOA and AOB, 165–166
bands, 157–158
CHIP-SIP, 284
DNA extraction, 159
DNA-SIP, 284
experimental design, 158
growth and mortality indices, calculation
  AOA/AOB, 160–161
  control samples, 166
  delta, 161
incubation, ^{18}O water, 159
isopycnic centrifugation, DNA, 165
labeled and nonlabeled DNA, separation, 159–160
labeled DNA, 157
nitrate and ammonia pools, soil, 158–159
oxygen atoms, 157
quantitative real-time PCR analysis, 160
results
  ammonia pools, size, 161
  ammonium and nitrate, concentration, 162
  amoA gene, 163
  Δ growth and mortality index, 164
  growth indices, 163
  ultracentrifuge tubes, 162
soil moisture, 167
STOX oxygen sensor
calibration and performance
  data, 333–334
  electrode failure and response time, 334
  oxygen respiration kinetics, 335
  use, factors, 334–335
construction
  casing, Teflon-coated silver wire, 330–331
  cross section, 332
  front guard gold-plated platinum cathode, 329
  O_2 microsensor casing, 329–330
  platinum connection and gold front guard, 331
  Schott AR glass, 329
  sensing cathode, Schott 8533 glass, 330

Unisense A/S, 328–329
conventional oxygen sensor recalibration
  SBE43, 340
  Winkler titration, 339–340
electronics
  deep-water application, 332–333
  "eddy correlation" system, 333
fast-responding, 327
fast response, 339
in situ deployment, OMZs
  CTD systems, 337–338
  measurement, 338–339
O_2 concentration determination, 326
oxygen concentrations calculation
  data, 335–336
  detection limit, 337
  polarization voltage, 336–337
principle
  polarized and unpolarized front guard, 328
  tip description, 327–328
on titanium cylinder, 333
Winkler titration, 326–327
Surface emission isolation flux chamber (SEIFC), 373

T

Terrestrial ecosystems, nitrification rate
gross
  acetylene (C_2H_2), 355–356
  chemical, 355
  nitrapyrin, 356
  NO_3^- consumption, 356–357
inhibitors, gross, 355–357
^{15}N, gross
  isotope dilution (*see* ^{15}N isotope dilution technique)
  tracer technique, 352
NO_3-consumption, 351
Terrestrial geothermal ecosystems
AOA and *Thermus thermophilus*, 193–197
AOB and AOA role, 172, 175
denitrification
  description, 176–177
  measurement, methods, 185–193
N-cycle activity measurement and processes (*see* Nitrogen (N) cycle)
nitrification measurement, methods (*see* Nitrification measurement, methods)
nitrification step, 175
Nitrospira, 176
NOB, 175–176
NO_2^- oxidation, 175–176
thermophile
  AOA and denitrifying *Thermus thermophilus*, 193–197
  and hyperthermophile, 176–177

Thermodynamics. *See* Continuous cultivation and thermodynamic aspects, nitrogen cycle
Thermophilic AOA and *Thermus thermophilus* denitrification
  "*Candidatus* Nitrosocaldus yellowstonii," 194
  functional genes, 193
  PCR and qPCR
    amoA genes/transcripts, hot springs, 193–194
    "*Candidatus* Nitrosocaldus yellowstonii," 195–197
    primers targeting putative amoA, 195–197
    16S rRNA gene and putative amoA, 196
    T. thermophilus, 196–197
  sample collection and nucleic acid extraction
    crude RNA samples preparation, 195
    freezing, 194
    protocol and kit, 194–195
  Thermus thermophilus denitrification, 193–194
Thunberg tube, 451
TonB
  dependent heme receptor, 416, 424
  dependent receptor gene, 406–407
  dependent transducer gene, 406
Total Kjeldahl nitrogen (TKN), 378
Trace metals, nitrogen cycle
  Mo and Fe, 491
  redox-sensitive metals, 495–496
  water column redox conditions, 496–497

W

Wastewater treatment
  anoxic plant, 90
  cold-adapted NOB, 111
  exogenous input, 207
  sludge, 93
Wastewater treatment plants (WWTPs)
  nitrous oxide (N_2O) fluxes
    activated sludge processes, emission, 381–382
    advective gas flow rate, 376–377
    aqueous concentration, 376
    determination, data analysis, 377–378
    emission factor calculation, 378–379
    emission fractions, 378
    full-scale monitoring, 370–374
    headspace gas measurement, 373, 375–376
    lab-scale and field-scale adaptation, emission, 383
    protocol standardization and comparison, 379–381
    real-time measurement principle, 377
    triggers, emission, 382
Winkler titration, 326–327, 339
Wolinella succinogenes
  CCHLs isoenzymes, 439
  electron acceptors, 435–436
  genetic manipulation, 436–438
  heterologous production
    description, 439–440
    expression, 441
    NrfA protein, 440, 442
    pMK2, 442
    Strep-tag sequences, 440
  multiheme encoded, 437
  respiratory nitrate ammonification, 438
WWTPs. *see* Wastewater treatment plants

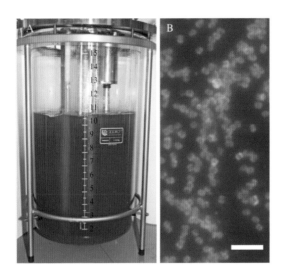

Boran Kartal et al., Figure 4.1 (A) The anammox membrane bioreactor. (B) Fluorescence *in situ* hybridization micrograph depicting highly enriched "*Candidatus* Scalindua sp." in pink (combination of BS820 probe counterstained with DNA stain DAPI); scale bar = 5 μm.

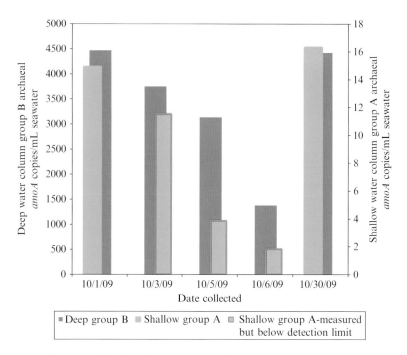

Annika C. Mosier and Christopher A. Francis, Figure 9.1 Abundance of the shallow (group A) and deep (group B) archaeal *amoA* clades within seawater collected at a depth of 891 m in Monterey Bay, CA. Note difference in scales (*y*-axis) for group A versus B qPCR data.